Springer Collected Works in Mathematics

Bombay, 1973

Harish-Chandra

Collected Papers IV
1970 – 1983

Editor

Veeravalli Seshadri Varadarajan

Reprint of the 1984 Edition

 Springer

Author
Harish-Chandra (1923 Kanpur,
    India – 1983 Princeton, USA)

Editor
Veeravalli Seshadri Varadarajan
University of California
Los Angeles, USA

ISSN 2194-9875
ISBN 978-3-662-45435-0   (Softcover)
      978-0-387-90782-6   (Hardcover)
DOI 10.1007/978-3-662-45436-7
Springer Heidelberg New York Dordrecht London

Library of Congress Control Number: 2012954381

Printed on acid-free paper

Springer is part of Springer Science+Business Media (www.springer.com)

Harish-Chandra
Collected Papers

Volume IV
(1970–1983)

Edited by V. S. Varadarajan

Springer-Verlag Berlin Heidelberg GmbH

Harish-Chandra
1923–1983

Editor

V. S. Varadarajan
Department of Mathematics
University of California
Los Angeles, CA 90024
U.S.A.

AMS Subject Classifications: 01A75, 22-XX

Library of Congress Cataloging in Publication Data
Harish-Chandra.
  Harish-Chandra collected papers.
  Contents: v. 1. 1944–1954—v. 2. 1955–1958—
[etc.]—v. 4. 1970–present.
  1. Mathematics—Collected works. I. Varadarajan,
V. S.   II. Title.
QA3.H294   1983   510   83-4828

9 8 7 6 5 4 3 2 1

ISBN 978-1-4899-7409-9   ISBN 978-1-4899-7407-5 (eBook)
DOI 10.1007/978-1-4899-7407-5

Table of Contents
Volume IV

Bibliography of Harish-Chandra

[1944a]   (with Bhabha, H. J.) On the theory of point-particles. *Proc. Royal Soc. A.* **183**, 134–141.

[1944b]   On the removal of the infinite self-energies of point-particles. *Proc. Royal Soc. A.* **183**, 142–167.

[1945a]   On the scattering of scalar mesons. *Proc. Indian Acad. Sci. Sect. A.* **21**, 135–146.

[1945b]   Algebra of the Dirac-matrices. *Proc. Indian Acad. Sci. Sect. A.* **22**, 30–41.

[1946a]   (with Bhabha, H. J.) On the fields and equations of motion of point particles. *Proc. Royal Soc. A.* **185**, 250–268.

[1946b]   On the equations of motion of point particles. *Proc. Royal Soc. A.* **185**, 269–287.

[1946c]   A note on the σ-symbols. *Proc. Indian Acad. Sci. Sect. A.* **23**, 152–163.

[1946d]   The correspondence between the particle and the wave aspects of the meson and the photon. *Proc. Royal Soc. A.* **186**, 502–525.

[1947a]   On the algebra of the meson matrices. *Proc. Camb. Phil. Soc.* **43**, 414–421.

[1947b]   On relativistic wave equations. *Phys. Rev.* **71**, 793–805.

[1947c]   Equations for particles of higher spin. Report of an International Conference on Fundamental particles and Low temperatures held at the Cavendish Laboratory, Cambridge (1946). Vol. I, Fundamental Particles, 185–188. The Physical Society, London.

[1947d]   Infinite irreducible representations of the Lorentz group. *Proc. Royal Soc. A.* **189**, 372–401.

[1948a]   Relativistic equations for elementary particles. *Proc. Royal Soc. A.* **192**, 195–218.

[1948b]   Motion of an electron in the field of a magnetic pole. *Phys. Rev.* **74**, 883–887.

[1949a]   Faithful representations of Lie algebras. *Ann. of Math.* **50**, 68–76.

[1949b]   On representations of Lie algebras. *Ann. of Math.* **50**, 900–915.

[1950a]   On the radical of a Lie algebra. *Proc. Amer. Math. Soc.* **1**, 14–17.

[1950b]   On faithful representations of Lie groups. *Proc. Amer. Math. Soc.* **1**, 205–210.

[1950c]   Lie algebras and the Tannaka duality theorem. *Ann. of Math.* **51**, 299–330.

[1951a]   On some applications of the universal enveloping algebra of a semisimple Lie algebra. *Trans. Amer. Math. Soc.* **70**, 28–96.

[1951b]   Representations of semisimple Lie groups on a Banach space. *Proc. Nat. Acad. Sci. U.S.A.* **37**, 170–173.

BIBLIOGRAPHY OF HARISH-CHANDRA

[1951c]   Representations of semisimple Lie groups. II. *Proc. Nat. Acad. Sci. U.S.A.* **37**, 362–365.

[1951d]   Representations of semisimple Lie groups. III. Characters. *Proc. Nat. Acad. Sci. U.S.A.* **37**, 366–369.

[1951e]   Representations of semisimple Lie groups. IV. *Proc. Nat. Acad. Sci. U.S.A.* **37**, 691–694.

[1951f]   Plancherel formula for complex semisimple Lie groups. *Proc. Nat. Acad. Sci. U.S.A.* **37**, 813–818.

[1952]   Plancherel formula for the 2×2 real unimodular group. *Proc. Nat. Acad. Sci. U.S.A.* **38**, 337–342.

[1953]   Representations of a semisimple Lie group on a Banach space. I. *Trans. Amer. Math. Soc.* **75**, 185–243.

[1954a]   Representations of semisimple Lie groups. II. *Trans. Amer. Math. Soc.* **76**, 26–65.

[1954b]   Representations of semisimple Lie groups. III. *Trans. Amer. Math. Soc.* **76**, 234–253.

[1954c]   The Plancherel formula for complex semisimple Lie groups. *Trans. Amer. Math. Soc.* **76**, 485–528.

[1954d]   On the Plancherel formula for the right K-invariant functions on a semisimple Lie group. *Proc. Nat. Acad. Sci. U.S.A.* **40**, 200–204.

[1954e]   Representations of semisimple Lie groups. V. *Proc. Nat. Acad. Sci. U.S.A.* **40**, 1076–1077.

[1954f]   Representations of semisimple Lie groups. VI. *Proc. Nat. Acad. Sci. U.S.A.* **40**, 1078–1080.

[1955a]   Integrable and square-integrable representations of a semi-simple Lie group. *Proc. Nat. Acad. Sci. U.S.A.* **41**, 314–317.

[1955b]   On the characters of a semisimple Lie group. *Bull. Amer. Math. Soc.* **61**, 389–396.

[1955c]   Representations of semisimple Lie groups. IV. *Amer. J. of Math.* **77**, 743–777.

[1956a]   Representations of semisimple Lie groups. V. *Amer. J. of Math.* **78**, 1–41.

[1956b]   Representations of semisimple Lie groups. VI. Integrable and square-integrable representations. *Amer. J. of Math.* **78**, 564–628.

[1956c]   The characters of semisimple Lie groups. *Trans. Amer. Math. Soc.* **83**, 98–163.

[1956d]   On a lemma of F. Bruhat. *J. Math. Pures. Appl.* (9) **35**, 203–210.

[1956e]   Invariant differential operators on a semisimple Lie algebra. *Proc. Nat. Acad. Sci. U.S.A.* **42**, 252–253.

[1956f]   A formula for semisimple Lie groups. *Proc. Nat. Acad. Sci. U.S.A.* **42**, 538–540.

[1957a]   Representations of semisimple Lie groups. *Proceedings of the International Congress of Mathematicians*, Amsterdam (1954). Vol. 1, 299–304. Erven P. Noordhoff N. V., Groningen, North Holland Publishing Company.

[1957b]   Differential operators on a semisimple Lie algebra. *Amer. J. of Math.* **79**, 87–120.

[1957c]   Fourier transforms on a semisimple Lie algebra. I. *Amer. J. of Math.* **79**, 193–257.

[1957d]   Fourier transforms on a semisimple Lie algebra. II. *Amer. J. of Math.* **79**, 653–686.

[1957e]   A formula for semisimple Lie groups. *Amer. J. of Math.* **79**, 733–760.

[1957f]   Spherical functions on a semisimple Lie group. *Proc. Nat. Acad. Sci. U.S.A.* **43**, 408–409.

[1958a]   Spherical functions on a semisimple Lie group. I. *Amer. J. of Math.* **80**, 241–310.

[1958b]   Spherical functions on a semisimple Lie group. II. *Amer. J. of Math.* **80**, 553–613.

[1959a]   Automorphic forms on a semisimple Lie group. *Proc. Nat. Acad. Sci. U.S.A.* **45**, 570–573.

[1959b]   Some results on differential equations and their applications. *Proc. Nat. Acad. Sci. U.S.A.* **45**, 1763–1764.

BIBLIOGRAPHY OF HARISH-CHANDRA

[1960a]   Some results on differential equations.

[1960a']   Supplement to "Some results on differential equations"

[1960b]   Differential equations and semisimple Lie groups.

[1961]   (with Borel, A.) Arithmetic subgroups of algebraic groups. *Bull. Amer. Math. Soc.* **67**, 579–583.

[1962]   (with Borel, A.) Arithmetic subgroups of algebraic groups. *Ann. of Math.* **75**, 485–535.

[1963]   Invariant eigendistributions on semisimple Lie groups. *Bull. Amer. Math. Soc.* **69**, 117–123.

[1964a]   Invariant distributions on Lie algebras. *Amer. J. of Math.* **86**, 271–309.

[1964b]   Invariant differential operators and distributions on a semi-simple Lie algebra. *Amer. J. of Math.* **86**, 534–564.

[1964c]   Some results on an invariant integral on a semisimple Lie algebra. *Ann. of Math.* **80**, 551–593.

[1965a]   Invariant eigendistributions on a semisimple Lie algebra. *Publ. Math. IHES* No. **27**, 5–54.

[1965b]   Invariant eigendistributions on a semisimple Lie group. *Trans. Amer. Math. Soc.* **119**, 457–508.

[1965c]   Discrete series for semisimple Lie groups. I. Construction of invariant eigendistributions. *Acta Math.* **113**, 241–318.

[1966a]   Two theorems on semisimple Lie groups. *Ann. of Math.* **83**, 74–128.

[1966b]   Discrete series for semisimple Lie groups. II. Explicit determination of the characters. *Acta. Math.* **116**, 1–111.

[1966c]   Harmonic analysis on semisimple Lie groups. Some recent advances in the basic sciences. Vol. 1 (1962, 1963, 1964), 35–40. *Belfer Graduate School of Science Annual Science Conference Proceedings.* Edited by A. Gelbart. Belfer Graduate School of Science, Yeshiva University, New York, New York.

[1967]   Characters of semisimple Lie groups. *Symposia on Theoretical Physics*, 4, 137–142. (Lectures presented at the 1965 Third Anniversary Symposium of the Institute of Mathematical Sciences, Madras, India. Edited by A. Ramakrishnan.) Plenum Press, New York.

[1968a]   Harmonic analysis on semisimple Lie groups. *Proceedings of the International Congress of Mathematicians*, Moscow (1966), 89–94. "Mir", Moscow.

[1968b]   Automorphic forms on semisimple Lie groups. Notes by J. G. M. Mars. *Lecture Notes in Mathematics*, No. 62, Springer-Verlag, Berlin-Heidelberg-New York.

[1970a]   Some applications of the Schwartz space of a semisimple Lie group. *Lectures in Modern Analysis and Applications. II*, 1–7. Edited by C. T. Tamm. *Lecture Notes in Mathematics.* No. **140**, Springer-Verlag, Berlin-Heidelberg-New York.

[1970b]   Eisenstein series over finite fields. Functional analysis and related fields, 76–88. *Proceedings of a Conference in honor of Professor Marshall Stone held at the University of Chicago, May 1968.* Edited by F. E. Browder. Springer-Verlag, Berlin-Heidelberg-New York.

[1970c]   Harmonic analysis on semisimple Lie groups. *Bull. Amer. Math. Soc.* **76**, 529–551.

[1970d]   Harmonic analysis on reductive p-adic groups. Notes by G. van Dijk. *Lecture Notes in Mathematics*, No. **162**, Springer-Verlag, Berlin-Heidelberg-New York.

[1972]   On the theory of the Eisenstein integral. Conference on Harmonic Analysis, College Park, Maryland (1971), 123–149. Edited by D. Gulick and R. L. Lipsman. *Lecture Notes in Mathematics*, No. **266**, Springer-Verlag, Berlin-Heidelberg-New York.

[1973]   Harmonic analysis on reductive p-adic groups. Harmonic analysis on homogeneous spaces, 167–192. Edited by C. C. Moore. *Proceedings of Symposia in Pure Mathe-*

matics, Vol. **XXVI**, Amer. Math. Soc., Providence, R. I., U.S.A.

[1975] Harmonic analysis on real reductive groups. I. The theory of the constant term. *J. of Functional Analysis* **19**, 104–204.

[1976a] Harmonic analysis on real reductive groups. II. Wave packets in the Schwartz space. *Inventiones Math.* **36**, 1–55.

[1976b] Harmonic analysis on real reductive groups. III. The Maass–Selberg relations and the Plancherel formula. *Ann. of Math.* **104**, 117–201.

[1977a] The characters of reductive *p*-adic groups. *Contributions to Algebra*, 175–182. A collection of papers dedicated to Ellis Kolchin. Edited by H. Bass, P. J. Cassidy, and J. Kovacic. Academic Press, New York–San Francisco–London.

[1977b] The Plancherel formula for reductive *p*-adic groups.

[1977b'] Corrections to "The Plancherel formula for reductive *p*-adic groups"

[1978] Admissible invariant distributions on reductive *p*-adic groups. Lie theories and their applications, 281–347. *Proceedings of the 1977 annual seminar of the Canadian Mathematical Congress.* Edited by W. Rossmann. Queen's papers in Pure and Applied Math. No. **48**. (Editors: A. J. Coleman and P. Ribenboim). Kingston, Ontario.

[1980] A submersion principle and its applications. Geometry and Analysis, 95–102. Papers dedicated to the memory of V. K. Patodi, Indian Academy of Sciences, Bangalore, and the Tate Institute of Fundamental Research, Bombay.

[1983] Supertempered distributions on real reductive groups. *Studies in Applied Mathematics, Advances in Mathematics, Supplementary Studies Series*, Vol. 8, Academic Press Inc., pp. 139–153, edited by Victor Guillemin.

Abbreviations Used in the Bibliography

Proc. Royal Soc. A. ...	Proceedings of the Royal Society, A...
Proc. Indian Acad. Sci. Sect. A	Proceedings of the Indian Academy of Sciences, Section A
Proc. Camb. Phil. Soc.	Proceedings of the Cambridge Philosophical Society
Phys. Rev.	The Physical Review
Ann. of Math.	Annals of Mathematics
Proc. Amer. Math. Soc.	Proceedings of the American Mathematical Society
Trans. Amer. Math. Soc.	Transactions of the American Mathematical Society
Proc. Nat. Acad. Sci. U.S.A.	Proceedings of the National Academy of Sciences of the United States of America
Bull. Amer. Math. Soc.	Bulletin of the American Mathematical Society
Amer. J. of Math.	American Journal of Mathematics
J. Math. Pures Appl.	Journal de Mathématiques Pures et Appliquées
Publ. Math. IHES	Publications Mathématiques, Institut des Hautes Études Scientifiques
Acta Math.	Acta Mathematica
J. Functional Analysis	Journal of Functional Analysis
Inventiones Math.	Inventiones Mathematicae

Preface

These volumes of the Collected Papers of Harish-Chandra are being brought out in response to a widespread feeling in the mathematical community that they would immensely benefit scholars and researchworkers, especially those in analysis, representation theory, arithmetic, mathematical physics, and other related areas. It is hoped that in addition to making his contributions more accessible by collecting them in one place, these volumes would help focus renewed attention on his ideas and methods as well as lend additional perspective to them.

The papers are arranged chronologically. Harish-Chandra is still an extraordinarily productive mathematician, and so I feel that there is no need to impose any structure on his output (past and current) other than that coming from the evolution of his own thinking. It was also decided to divide the papers into four volumes so that each volume would have a reasonable size. However, I have attempted to make the lines of division in such a way that each volume has a certain unity within itself.

It is impossible to mention all the individuals and organisations that helped me with this project. Nevertheless, I do want to record my special thanks to Phyllis Parris, for typing the two handwritten manuscripts on differential equations; to Walter Kaufmann-Bühler, for his constant enthusiasm and encouragement; and to Elise Oranges for her tireless work in producing the final copy. In addition I am extremely grateful to Roger Howe and to Nolan Wallach for agreeing to write commentaries on selected parts of Harish-Chandra's work, and doing it within the time frame that I gave them. Finally, I want to thank Mrs. Lily Harish-Chandra for providing me with the photographs of Harish-Chandra that appear in these volumes.

Pacific Palisades, California
May, 1983

V. S. VARADARAJAN

Biographical Note

Harish-Chandra was born on October 11, 1923 in Kanpur, India. He received the M.Sc degree from Allahabad University in 1943 and spent the years 1943–1945 at the Indian Institute of Science, Bangalore, India. He went to Cambridge University in England as a graduate student of Dirac in 1945 and received his Ph.D from there in 1947. He was at Columbia University, New York, from 1950 to 1963. In 1963 he became a permanent member of the Institute for Advanced Study at Princeton, New Jersey, where he has been the IBM von Neumann Professor from 1970 onwards.

He was a Guggenheim Fellow from 1957 to 1958 and a Sloan Fellow from 1961 to 1963. He was elected a Fellow of the Royal Society in 1973 and a member of the National Academy of Sciences of the U.S.A. in 1981. He was awarded honorary doctorates by Delhi University in 1973 and by Yale University in 1981.

He and his wife live in Princeton. They have two daughters. Although he has not returned to India except for brief visits, he has always been profoundly concerned with India and Indian science. He remains a figure of great inspiration in his native land.

Postscript

Harish-Chandra is no more. He suffered a fatal heart attack while out for a walk on the evening of Sunday, October 16, 1983. He survives in his work, which is a faithful reflection of his personality—lofty, intense, uncompromising.

Princeton
October 19, 1983

V. S. VARADARAJAN

Introduction

V. S. Varadarajan

> *Bottom*: Masters, I am to discourse wonders:
> but ask me not what; for if I tell you, I am
> no true Athenian. I will tell you every-
> thing, right as it fell out.
>
> —Shakespeare, *A Midsummer Night's
> Dream*, Act Four, Scene II.

§0. The first scientific papers of Harish-Chandra were in theoretical physics. However, his interests shifted to mathematics rather early in his career. His prolonged and intense preoccupation with group representations began in the late 1940's when, at the suggestion of Dirac, he started investigating the infinite dimensional unitary representations of the Lorentz group [1947d]. Since then he has erected, almost singlehandedly, a monumental theory of harmonic analysis on reductive groups and their homogeneous spaces. His work is a profound synthesis of algebra, geometry, and analysis. The great force and continued resonance of his ideas have inspired a generation of mathematicians. The singlemindedness and courage with which he has pursued his goals, the beauty of his results, the power and originality of his methods, and the ultimate simplicity of his philosophy, as well as the conviction that sustains it, compel our admiration. There can be no doubt that his achievement is one of the greatest in mathematics in our time.

This introduction attempts to give an overview of this work. I have sketched, in an impressionistic way, the main lines of it in §2; the remaining sections contain additional detail on selected parts of his work so that one of his great achievements —the harmonic analysis of $L^2(G)$ for a real reductive group G—may come into clearer focus. I hope that this account, although incomplete and imperfect, may still be of value as a rough guide to the prospective reader.

§1. The early sources of motivation for representation theory and harmonic analysis were the classical theory of Fourier series and integrals and the theory of linear representations of finite groups. After the rise of topology and functional analysis in the early decades of this century, far-reaching generalizations of these classical themes began to emerge.

For a locally compact abelian group G let $\hat{G}$ be the group of characters of G. $\hat{G}$ is also locally compact abelian and the Pontryagin–van Kampen theory pairs G and $\hat{G}$ in a canonical duality. Given any function $f \in L^1(G)$ its Fourier transform is the function $\hat{f}$ on $\hat{G}$ given by

$$\hat{f}(\hat{x}) = \int_G f(x)\langle x, \hat{x} \rangle \, dx \qquad (\hat{x} \in \hat{G}, \, dx \text{ a Haar measure})$$

and $f \mapsto \hat{f}$ is a homomorphism of the convolution algebra $L^1(G)$ into the algebra of continuous functions on $\hat{G}$ that vanish at infinity. The Fourier transform can also be introduced on $L^2(G)$; it is then a unitary isomorphism of $L^2(G)$ with $L^2(\hat{G})$ where the Haar measure $d\hat{x}$ on $\hat{G}$ is uniquely determined by dx (the measure dual to dx). In particular we have the *Plancherel formula*

$$\int_G |f(x)|^2 \, dx = \int_G |\hat{f}(\hat{x})|^2 \, d\hat{x} \qquad (f \in L^2(G))$$

and the *inversion formula*

$$f(x) = \int_{\hat{G}} \hat{f}(\hat{x})\langle x, \hat{x} \rangle^{\mathrm{conj}} \, d\hat{x} \qquad \left(f \in L^1(G), \hat{f} \in L^1(\hat{G}) \right)$$

which generalize the corresponding formulae from classical Fourier analysis on $\mathbb{R}^n$ ([W 1] [Lo] [Bl]).

The classical Poisson Summation formula can also be generalized to this setting. Let $\Gamma \subset G$ be a lattice in G, i.e., a discrete subgroup of G such that G/Γ is compact. Then $\Gamma^\vee$, the annihilator of Γ, is also a lattice in $\hat{G}$. If Fourier transforms are defined using the Haar measure that gives the measure 1 to G/Γ, then the Poisson formula takes the form

$$\sum_{\gamma \in \Gamma} f(\gamma) = \sum_{\gamma^\vee \in \Gamma^\vee} \hat{f}(\gamma^\vee)$$

where f and $\hat{f}$ are suitably restricted ([W 2]).

The Poisson formula has always been a powerful tool in number theory. The importance of Fourier analysis on general locally compact abelian groups became very clear with the work of Tate who, at the suggestion of Artin, used the general Poisson formula in the setting provided by the adele group of a number field to derive the analytic theory of the Hecke L-series attached to a Grössencharakter ([T] [W 2]).

Suppose now that G is compact. Let dx be the normalized Haar measure on G so that $\int_G dx = 1$. Let $\hat{G}$ be the set of equivalence classes of irreducible (finite dimensional) unitary representations of G. The general facts concerning linear representations of finite groups extend essentially unchanged to G. Thus if $\pi \in \omega \in \hat{G}$ and ϕ, ψ are unit vectors in the representation space of π,

$$\int_G |(\pi(x)\phi, \psi)|^2 \, dx = \frac{1}{d(\omega)}$$

where $d(\omega)$ is the degree of ω. Moreover, for $\omega \in \hat{G}$ let Θ_ω be the corresponding character; then

$$\int_G \Theta_\omega \Theta_{\omega'}^{\mathrm{conj}} \, dx = \delta_{\omega\omega'} \qquad (\omega, \omega' \in \hat{G}).$$

The fundamental result is the Peter–Weyl theorem. It asserts the completeness of the irreducible representations and can be stated as the completeness formula

$$\int_G |f|^2\, dx = \sum_{\omega \in \hat{G}} d(\omega)\|\pi_\omega(f)\|_2^2$$

where $f \in L^2(G)$, $\pi_\omega \in \omega$, and $\|\cdot\|_2$ is the Hilbert–Schmidt norm. The measure on $\hat{G}$, which assigns the mass $d(\omega)$ to ω, may thus be viewed as the *Plancherel measure* of G.

Let G be a compact Lie group. For $f \in L^2(G)$ let $\hat{f}$ be its Fourier transform, defined as the function on $\hat{G}$ given by

$$\hat{f}(\omega) = \int_G f\Theta_\omega^{\mathrm{conj}}\, dx \qquad (\omega \in \hat{G}).$$

The completeness formula can then be restated as

$$f(e) = \sum_{\omega \in \hat{G}} d(\omega)\hat{f}(\omega) \qquad (f \in C^\infty(G)).$$

In other words, we have the relation

$$\delta = \sum_{\omega \in \hat{G}} d(\omega)\Theta_\omega$$

in the space of *distributions* on the manifold G, δ being the delta function at the identity element of G.

It is natural to ask for an explicit determination of the irreducible representations of special groups such as the simple Lie groups which had been classified in the late 1890's by Killing and Élie Cartan. The work of Cartan and Hermann Weyl in the 1920's gave a complete solution of this problem. For Cartan it was a question of determining the irreducible finite dimensional representations of the (complex) simple *Lie algebras*; he obtained them as highest weight modules. Weyl had a more global view and worked with the *compact simple Lie groups*. He discovered beautiful formulae for the characters and dimensions of the irreducible representations. Moreover his discovery of the existence of a compact form of a complex simple Lie algebra ("unitarian trick") showed that the representation theory of a complex simple Lie algebra is essentially the same as that of the (simply connected) compact form associated to it, thus unifying his theory with that of Cartan ([We 2]; see also [V 1]).

The notion of distributions, which came into widespread use with the work of Laurent Schwartz [S], led to a deeper understanding of the notion of Fourier transform and to a great extension of the domain of its applicability. In the language of distributions the Plancherel formula for $\mathbb{R}^n$ became the "expansion" of the delta function at the origin in terms of the characters:

$$\delta = (2\pi)^{-n/2}\int_{\mathbb{R}^n} e(\lambda)\, d\lambda \qquad \left(e(\lambda)(x) = e^{i\lambda \cdot x}\right).$$

Suppose now that G is a locally compact group, unimodular and second countable, which is neither compact nor abelian. Let $\hat{G}$ be the set of equivalence classes of its irreducible unitary representations (finite or infinite dimensional). There was no systematic theory of representations of such groups in the 1930's. But in Quantum Physics unitary representations of the various symmetry groups had already begun to play a prominent role. This is because the covariance properties of a system with a (connected) group of symmetries are expressed by means of a unitary representation of the group in question or an extension of it. In relativistic quantum physics this meant the inhomogeneous Lorentz group; its physically significant irreducible unitary representations (both single and double valued in the physicists' terminology) were classified by Wigner in a famous memoir in 1939 [Wi].

The situation changed completely in the 1940's. For, by then, the epoch-making papers of von Neumann (by himself and with Murray) on operator algebras in Hilbert spaces had appeared [MvN] [vN]. This theory provided the framework for the study of infinite dimensional unitary representations; in addition it revealed that new phenomena such as type II and type III algebras, direct integral decompositions, and so on, would have to be taken into account.

The systematic beginnings of unitary representation theory of general locally compact groups go back to 1943 when Gel'fand and Raikov proved that every such group had enough irreducible unitary representations to separate the points of the group [GR]. The subject developed rapidly in the late 1940's and early 1950's mainly at the hands of Godement, Mackey, Mautner, and Segal. Their point of view was functional analytic, and their main tool was the theory of operator algebras. Their work led, among other things, to a general decomposition theory of unitary representations, abstract Plancherel theorems, and to the recognition that a reasonable harmonic analysis can be expected only for type I groups, namely groups whose unitary representations were all of type I [Ma 1] [Ma 2] [Ma 3]. It was also in this period that the papers of Mackey on systems of imprimitivity and induced representations appeared, in which he extended the classical Frobenius theory to all separable locally compact groups. Mackey's work, which went far beyond what is suggested in these remarks, has proved to be very influential in the representation theory of nilpotent and solvable groups (cf. [AM] [Mo]).

These developments led to the following formulation of the central problem of harmonic analysis: given a separable locally compact group G acting transitively on a space X with an invariant measure μ, and an irreducible unitary representation τ of the stabilizer in G of a given point of X, decompose into irreducible constituents the representation of G induced by τ. If τ is the trivial representation, the induced representation is just the natural representation of G in $L^2(X, \mu)$; if $G = X$ and the action is by left translations, the problem is the decomposition of the left regular representation of G. The decomposition may not be entirely discrete so that it has to be understood as a direct integral.

In this generality this problem is hopeless because the category of all separable locally compact groups is simply too vast for classical harmonic analysis to be anything more than a very uncertain guide. New examples were clearly needed, and it was natural to look to the simple Lie groups once again for this purpose. The papers of Gel'fand and Neumark [GN 1] and Bargmann [B], dealing with SL(2, C) and SL(2, R), and published independently of each other in 1947, mark the beginning of the subject of infinite dimensional representations of real semisimple Lie groups.

In [GN 1] Gel'fand and Neumark determined all the irreducible unitary representations of SL(2, $\mathbb{C}$) upto equivalence, discovered an explicit Plancherel formula, and proved that any unitary representation of SL(2, $\mathbb{C}$) could be decomposed as a direct integral of irreducible representations in a natural way. They discovered the surprising fact that this group had *additional* irreducible unitary representations which played no role in the harmonic analysis of $L^2(G)$. They called these the representations of the *supplementary series*. They followed this up with a remarkable and seminal series of papers in which they extended their theory to SL(n, $\mathbb{C}$) and, to a lesser extent, to the other *complex classical groups*. This work, which was presented in detail in their beautiful monograph [GN 2], completely changed the perspective of the theory.

The group SL(n, $\mathbb{C}$) acts transitively on a number of projective varieties such as the Grassmannians and the various flag spaces. The spaces of *holomorphic* sections of the complex analytic line bundles on these homogeneous spaces carry the finite dimensional complex analytic representations of SL(n, $\mathbb{C}$). With remarkable insight Gel'fand and Neumark realized that the *smooth* sections of the *real analytic* line bundles on the same homogeneous spaces would transform (roughly speaking) according to the *infinite dimensional* irreducible representations of the group. These representations would be unitary under appropriate conditions. In this way, one had a single scheme for *all* irreducible representations, which contained the finite dimensional as well as the unitary ones as special cases. For example, $G = \mathrm{SL}(n, \mathbb{C})$ acts transitively on the space of all flags (e_m) $(1 \leqslant m \leqslant n, e_1 \subset e_2 \subset \cdots \subset e_n, e_m$ is an m-dimensional linear subspace of $\mathbb{C}^n$). Then $X \approx G/B$ where B is the subgroup of the upper triangular matrices. Gel'fand and Neumark proved that the unitary representations of G induced by the unitary one dimensional representations of B, which they called the *principal series* of representations, are all irreducible, and that the regular representation of the group is a direct integral of them. They extended to SL(n, $\mathbb{C}$) (as well as to the other complex classical groups) the construction of the supplementary series.

Although these representations are all infinite dimensional, Gel'fand and Neumark succeeded in associating *characters* to them. If ξ is any one dimensional representation of B and π_ξ the representation of G induced by ξ, the operator

$$\pi_\xi(x) = \int_G x(g)\pi_\xi(g)\,dg$$

is of trace class if x is suitably restricted, and there is a (unique) class function θ_ξ on G such that

$$\mathrm{tr}\big(\pi_\xi(x)\big) = \int_G x(g)\theta_\xi(g)\,dg.$$

It is natural to call θ_ξ the *character* of π_ξ. They evaluated θ_ξ on the subgroup D of diagonal matrices and found for it a formula of the same general structure as Weyl's character formula. In terms of these characters they obtained the *Plancherel formula* for $SL(n, \mathbb{C})$ in the form

$$x(e) = \int_{\hat{D}} \hat{x}(\xi) \omega(\xi)\, d\xi \qquad (x \in C_c^\infty(G))$$

where

$$\hat{x}(\xi) = \text{trace}\big(\pi_\xi(x)\big) \qquad \text{(Fourier transform!)}$$

and $\omega(\xi)$ is an explicitly determined function $\geqslant 0$. If we rewrite this in the form

$$\delta = \int \omega(\xi) \theta_\xi\, d\xi,$$

its analogy with the Plancherel formula for $\mathbb{R}^n$ and compact Lie groups becomes clear; $\omega(\xi)\, d\xi$ is the Plancherel measure.

Let me now turn to the work of Bargmann which treated both $SL(2, \mathbb{C})$ and $SL(2, \mathbb{R})$. So far as $SL(2, \mathbb{C})$ was concerned the results were essentially the same as those in [GN 1]; however $SL(2, \mathbb{R})$ presented new phenomena. In addition to the principal and supplementary series, defined essentially as for the complex unimodular group, there now appeared a discretely parametrized series of irreducible unitary representations characterized by the remarkable fact that their matrix elements were already square integrable on the group manifold. This property meant that these representations would occur as *discrete direct summands* of the regular representation. The property of having square integrable matrix coefficients makes sense for irreducible unitary representations of arbitrary separable locally compact unimodular groups and is equivalent to their occurring as direct summands of the regular representation; they form the so-called *discrete series* of the group, and the set of the corresponding equivalence classes is denoted by $\hat{G}_d$. Their appearance in the case of $SL(2, \mathbb{R})$ meant that the decomposition of its regular representation had both discrete and continuous parts.

A highlight of the paper of Bargmann is the emergence of another new theme in our subject. He viewed the matrix elements of the irreducible representations as eigenfunctions of the Casimir differential operator defined on the group manifold (originally introduced by the physicist Casimir for $SU(2)$). In this way the problem of harmonic analysis in $L^2(G)$ became one of eigenfunction expansions for the Casimir operator. The matrix elements have simple transformation properties relative to the action (from both left and right) of the rotation subgroup of $SL(2, \mathbb{R})$; and so they are essentially determined by their values at the elements $\begin{pmatrix} e^t & 0 \\ 0 & e^{-t} \end{pmatrix} (t > 0)$ of the group. Thus they may be viewed as functions on the half-line of positive reals; moreover the action of the Casimir operator coincides with that of a second order differential operator on this half-line. The spectral theory of such differential operators goes back to Hermann Weyl [We 1]. Bargmann used Weyl's theory to prove the completeness theorem, i.e., the fact that the linear space, spanned by the matrix elements of the discrete series and the "wave packets" of the matrix elements of the principal series, is dense in $L^2(G)$.

For the matrix elements of the discrete series Bargmann discovered remarkable generalizations of the Schur orthogonality relations. For the principal series the corresponding problem is the explicit determination of the Plancherel measure.

Bargmann found a beautiful expression for it in terms of the leading coefficients of the asymptotic expansions of the matrix elements.

The papers of Gel'fand and Neumark and Bargmann form a great watershed in our subject. Their work showed clearly, at least in retrospect, that to do harmonic analysis on real semisimple Lie groups it is absolutely necessary to operate at a level that allowed for a true interaction among the algebraic, geometric, and analytic properties of these groups.

This was the state of affairs in the early 1950's when Harish-Chandra entered the field. His methods combined in a deep and original manner the infinitesimal and global aspects of these problems, and, through an astonishing sequence of papers spanning almost three decades, led to a profound understanding of harmonic analysis of real reductive groups. Sometime in the 1960's he realized that his work contained deep analogies with the spectral theory of automorphic forms that was then being built through the efforts of Selberg, Gel'fand (and his collaborators), and Langlands. His exploration of these analogies has led in recent years to significant results in the harmonic analysis of reductive groups over local, global, and even finite fields, and has revealed the essential unity of the structure of Fourier analysis in these diverse contexts.

§2. Before starting with a description of Harish-Chandra's work we introduce a few definitions and some notation. G is a real semisimple Lie group, connected and with finite center; K is a maximal compact subgroup of G; and θ, the involution that fixes K pointwise; $\mathfrak{g} = \mathrm{Lie}(G)$, $\mathfrak{k} = \mathrm{Lie}(K)$. For any real Lie algebra $\mathfrak{m}$, $U(\mathfrak{m})$ is the universal enveloping algebra of the complexification $\mathfrak{m}_c$ of $\mathfrak{m}$.

Manifolds are smooth and second countable. They need not be connected but all the connected components will have the same dimension. For any manifold M, $C^\infty(M)$ (resp. $C_c^\infty(M)$) is the space of complex valued functions on M that are smooth (resp. smooth with compact support). If ω is a volume element (= exterior differential form of maximal degree which is nowhere zero) on M, then given any smooth differential operator D on M its transpose D^t is the differential operator such that $\int_M Df \cdot g\omega = \int_M f \cdot D^t g\omega$ for all $f, g \in C_c^\infty(M)$. If T is a distribution on M the distribution DT is defined by $\langle DT, f \rangle = \langle T, D^t f \rangle$ ($f \in C_c^\infty(M)$). Any function F on M that is locally integrable with respect to ω will be usually identified with the distribution $f \mapsto \int_M Ff\omega$. The actions of differential operators on smooth functions and distributions are compatible under this identification. If M is a Lie group we shall always choose ω to be a left invariant volume element. In this case $U(\mathfrak{m}) = U(\mathfrak{m})^t$ is canonically isomorphic to the algebra of all left invariant differential operators on M; $X^t = -X$ for $X \in \mathfrak{m}$.

We write $\mathfrak{Z}$ for the center of $U(\mathfrak{g})$. Then $\mathfrak{Z}$ is canonically isomorphic to the algebra of all differential operators on G that are both left and right invariant. $\mathfrak{Z}$ is a polynomial algebra in l indeterminates where l is the rank of G.

The centralizer in G of a Cartan subalgebra of $\mathfrak{g}$ is called a Cartan subgroup. Cartan subgroups are parameter spaces for conjugacy classes, and so are absolutely fundamental objects in character theory which deals with class functions and distributions. They are abelian if G is linear and are maximal tori if G is compact. A compact Cartan subgroup is always connected and is in fact a torus. An element $x \in G$ is regular if its centralizer in $\mathfrak{g}$ is a Cartan subalgebra. If $l = rk(G)$, t is an indeterminate and $D(x)$ is the coefficient of t^l in $\det(\mathrm{Ad}(x) - 1 + t)$, x is regular if and only if $D(x) \neq 0$. The set of regular points is denoted by G'; it is a dense open subset of G. D is often called the discriminant of G.

Parabolic subgroups (psgrps) are normalizers in G of parabolic subalgebras of $\mathfrak{g}$; a subalgebra $\mathfrak{p}$ of $\mathfrak{g}$ is parabolic if $\mathfrak{p}_c$ contains a Borel subalgebra of $\mathfrak{g}_c$. Any psgrp P has a canonical decomposition $P = MAN \simeq M \times A \times N$. Here N is the unipotent radical of P, namely the maximal normal subgroup of P that consists entirely of unipotent elements; $M_1 = MA$ is reductive and is defined as $P \cap \theta(P)$, so that $P = M_1 N$ is essentially the Levi decomposition of P, while A is the maximal $\mathbb{R}$-split* subgroup lying in the center of M_1. The subgroup M is the intersection of the kernels of all the continuous homomorphisms of M_1 into the positive reals. The decomposition $P = MAN$ is called the Langlands decomposition of P. The parabolic subgroups are needed for defining the various series of irreducible unitary representations of G; moreover the formulation of many of the asymptotic questions in harmonic analysis depend in a fundamental way on the geometry of the set of all parabolic subgroups.

Harish-Chandra's early work was largely devoted to the development of algebraic methods for studying infinite dimensional representations of real semisimple Lie groups. These methods led him to discover the fundamental facts about unitary representations of G. Among these are the results that assert that G is a type I group and that in every irreducible unitary representation of G the irreducible representations of K occur only with finite multiplicities.

Let us define a $(\mathfrak{g}, K)$-module to be a module V for $\mathfrak{g}$ which has a compatible structure as a K-module such that V is the algebraic sum of finite dimensional K-modules. V is unitary if there is a positive definite scalar product $(\ ,\ )$ for V such that $(Xu, v) = -(u, Xv)$ for all $u, v \in V$, $X \in \mathfrak{g}$. One of Harish-Chandra's major discoveries was that the space $V(\pi)$ of K-finite vectors in an irreducible unitary representation π is an irreducible $(\mathfrak{g}, K)$-module and that the assignment $\pi \mapsto V(\pi)$ induces a bijection of $\hat{G}$ with the set of isomorphism classes of irreducible $(\mathfrak{g}, K)$-modules that are unitary. He also proved that any irreducible $(\mathfrak{g}, K)$-module is of the form $V(\pi)$ for some irreducible representation π acting on a Hilbert space and possessing an infinitesimal character, thus showing that in the notion of a $(\mathfrak{g}, K)$-module which is irreducible we have the correct formulation of the vague idea of an "arbitrary" irreducible representation of G.

In spite of the promise and power of the algebraic method he never pursued it beyond his initial investigations and was content to use his early work as the point of departure for the analytical theory.

The basic step in the analytical theory is the introduction of characters. The algebraic treatment had already shown that the multiplicities of the irreducible representations of K that occur in an irreducible unitary representation of G do not exceed a constant multiple of the degrees of the representations of K. It follows from this that for any function $f \in C_c^\infty(G)$ the operator

$$\pi(f) = \int_G f(x)\pi(x)\,dx$$

---

*This means that $\mathrm{Ad}(a)$ is diagonalizable over $\mathbb{R}$ for all $a \in A$.

is of trace class, and the linear functional

$$\Theta_\pi : f \mapsto \mathrm{tr}(\pi(f))$$

is a distribution on the group G. Θ_π depends only on the equivalence class ω of π; it is denoted by Θ_ω and is called the character of π (or ω). It is an invariant distribution, i.e., a distribution that is invariant under all inner automorphisms of G. It determines ω completely. If $(\phi_i)_{i \geqslant 1}$ is an orthonormal basis of K-finite vectors for π and f_i is the matrix element defined by ϕ_i, i.e., $f_i(x) = (\pi(x)\phi_i, \phi_i)$ $(x \in G)$,

$$\Theta_\omega = \sum_{i \geqslant 1} f_i,$$

this equation being understood in the sense of distributions, i.e., for all $u \in C_c^\infty(G)$,

$$\Theta_\omega(u) = \sum_{i \geqslant 1} \int_G f_i u \, dx.$$

Since any matrix element satisfies the differential equation

$$zf = \chi_\omega(z)f \qquad (z \in \mathfrak{Z})$$

where χ_ω is the infinitesimal character of ω, it follows that

$$z\Theta_\omega = \chi_\omega(z)\Theta_\omega \qquad (z \in \mathfrak{Z}).$$

Here we recall that the infinitesimal character of ω is the homomorphism according to which $\mathfrak{Z}$ acts (by Schur's lemma) on the space $V(\pi)$. In other words, Θ_ω is an *invariant eigendistribution* on G.

Using the notion of characters one may formulate the problem of harmonic analysis on G as follows (cf. the introductory remarks to [1957e] [1970c]): given any invariant distribution T on G, express T as a "linear combination" of the irreducible characters. For $f \in C_c^\infty(G)$ define its *Fourier transform* $\hat{f}$ to be the function on $\hat{G}$ given by

$$\hat{f}(\omega) = \Theta_\omega(f) \qquad (\omega \in \hat{G}).$$

The problem is then to define a "distribution" $\hat{T}$ on $\hat{G}$ such that

$$T(f) = \hat{T}(\hat{f}) \qquad (f \in C_c^\infty(G)).$$

$\hat{T}$ will be the "Fourier transform" of T. If $T = \delta$, the delta function at the identity element of G, the determination of $\hat{\delta}$ is the problem of finding the explicit Plancherel formula for G; $\hat{\delta}$ will have to be a nonnegative measure, the Plancherel measure. The derivation by Gel'fand and Neumark of the Plancherel formula for $\mathrm{SL}(n, \mathbf{C})$, Harish-Chandra's generalization of this to all complex semisimple Lie groups [1951f] [1954c], and his treatment of the case of $\mathrm{SL}(2, \mathbf{R})$ [1952], and the work of Gel'fand and Graev on $\mathrm{SL}(n, \mathbf{R})$ [GG 1] [GG 2], may be regarded as initial evidence for this view.

A second approach to harmonic analysis is through *eigenfunction expansions*; this goes back to Bargmann as we saw in §1. Let $H \subset G$ be a closed unimodular subgroup, $X = G/H$, and let $\mathcal{H} = L^2(X)$ be the Hilbert space of functions square integrable with respect to the invariant measure on X. Let λ be the natural representation of G in $\mathcal{H}$. If $\mathcal{H}^\infty$ is the space of differentiable vectors in $\mathcal{H}$ for λ, we have a representation $a \mapsto \lambda(a)$ of $U(\mathfrak{g})$ in $\mathcal{H}^\infty$. If $z \in \mathfrak{Z}$ is Hermitian, i.e., $(z^{\text{conj}})^t = z$, the operator $\lambda(z)$ is essentially self adjoint on $\mathcal{H}^\infty$ and is invariant under G. So, in the first approximation, one may view the problem of decomposition of λ as an eigenfunction expansion problem for the commuting algebra of differential operators $\lambda(z)$, $z \in \mathfrak{Z}$ (see the introductory remarks to [1968b]). The eigenspaces are parametrized by the points of the spectrum of the commutative algebra $\mathfrak{Z}$, i.e., by the homomorphisms $\chi: \mathfrak{Z} \to \mathbf{C}$. For a given χ, the corresponding eigenspace would be a G-module which, typically, would be infinite dimensional. In view of the relationship between the eigenvalue problem and the representation theoretic question, we should expect that only those eigenfunctions will contribute to the spectral expansion in $L^2(X)$ whose χ are the infinitesimal characters of irreducible *unitary* representations of G (not necessarily all of them either).

The basic example in this approach is the case $X = G$, viewed as a homogeneous space for G (left action) or $G \times G$ (two-sided action). A variant of this in which the eigenspaces are finite dimensional is obtained as follows. Let U be a finite dimensional Hilbert space which is a bimodule for K, i.e., which carries a unitary representation τ of $K \times K^\vee$, $K^\vee$ being the group opposite to K. The Hilbert space is now the space $L^2(G, \tau)$ of functions $f: G \to U$ which are square integrable and satisfy

$$f(k_1 x k_2) = k_1 f(x) k_2$$

for all $k_1, k_2 \in K$ and almost all $x \in G$. Such functions are called τ-*spherical*; if $U = \mathbf{C}$ and $K \times K^\vee$ acts trivially, one speaks of *spherical* or *strictly spherical* functions. The terminology reminds us that we have a generalization of the classical theory of spherical harmonics. If χ is a homomorphism of $\mathfrak{Z}$ into $\mathbf{C}$ and $f \in L^2(G, \tau)$ is such that $zf = \chi(z)f$ for all $z \in \mathfrak{Z}$, f is C^∞ (even analytic), and the space of all such f is of finite dimension (cf. [1960b]). One of Harish-Chandra's basic results is that the eigenfunctions needed for the L^2-expansion are precisely those that satisfy the so-called weak inequality ([1970c], p. 539).

The theory of automorphic forms gives another example of this point of view. The classical automorphic forms on the Poincaré half-plane may be interpreted as eigenfunctions for the Laplace–Beltrami operator. They can then be generalized to the context of nonanalytic forms and higher dimensional symmetric spaces, as was done by Maass [M] and Selberg [Se 1]. Following the work of Borel and Harish-Chandra [1962] on the structure of arithmetic subgroups of reductive groups and the associated homogeneous spaces, Gel'fand [G], Selberg [Se 1] [Se 2], and Langlands [L 1] constructed a general theory of automorphic forms. They approached the fundamental problem as the spectral theory of the invariant differential operators in $L^2(G/\Gamma)$, G being the group of real points of a reductive group defined over $\mathbf{Q}$, and Γ is the associated arithmetic subgroup of G. The classical Eisenstein series and its generalizations are examples of eigenfunctions of these differential operators. It is interesting to note that the homomorphisms χ in the spectrum of $\mathfrak{Z}$ that correspond to these eigenfunctions do not come from unitary representations, and so one must make an analytic continuation in χ in order to get the eigenfunctions needed for the spectral theory (cf. [1968b]).

Whichever point of view is adopted, it is not possible to go far without a good supply of irreducible unitary representations. The known results for complex groups and $SL(n,\mathbb{R})$ seemed to suggest that each conjugacy class of Cartan subgroups of G gives rise to a "series" of irreducible unitary representations, and that the Plancherel measure is a *disjoint sum* of the measures contributed by each of these series (see [1957a] [1954e]). As a partial justification of this principle Harish-Chandra sketched in [1954e] a method of constructing a series of irreducible unitary representations associated to any Cartan subgroup L. Given L, one can select a parabolic subgroup $P = MAN$ such that A is the maximal $\mathbb{R}$-split* component of L. If η is an irreducible unitary representation of M and $\nu \in \hat{A}$, the unitary representation $\pi_{\eta,\nu}$ of G is defined as the representation induced by the representation $m \cdot a \cdot n \mapsto \nu(a)\eta(m)$ of P. The $\pi_{\eta,\nu}$ depend only on the conjugacy class of L and are discretely decomposable into irreducible subrepresentations. If L is chosen so that $\dim(A)$ is maximum, P is a minimal parabolic subgroup, M is compact, and what we have is called the *principal series*; for complex G it *is* the principal series of Gel'fand and Neumark. Bruhat's work [Br 1] had shown that for minimal P the $\pi_{\eta,\nu}$ are irreducible for ν in "general position." So one could hope that this would be so in the case of arbitrary L also.

If P is not minimal, M is not compact and so we are still faced with the problem of determining $\hat{M}$. Harish-Chandra's remarkable idea, which is implicit but very clear in [1957a] and [1954e], was that if we restrict η to vary always over the *discrete series* of M, the various series arising from a complete system of mutually nonconjugate Cartan subgroups would determine essentially disjoint parts of $\hat{G}$; and these representations would suffice to obtain the Plancherel formula for G. The discrete series is thus associated to the (unique) conjugacy class of compact Cartan subgroups, and the general discussion makes it clear that these may have to be constructed a priori and by quite different methods.

Discrete series: beginnings. Already in 1947 Godement had proved that the matrix elements of the discrete series satisfy orthogonality relations [Go]. In particular, one can associate to each $\omega \in \hat{G}_d$ a positive number $d(\omega)$, the *formal degree of* ω, such that

$$\int_G |(\pi(x)\phi, \psi)|^2 \, dx = \frac{1}{d(\omega)}$$

where $\pi \in \omega$ and ϕ, ψ are arbitrary unit vectors in the representation space of π. Moreover if $^0L^2(G)$ is the closed linear span of the irreducible subspaces of $L^2(G)$ (= the "discrete part" of $L^2(G)$), and E is the orthogonal projection $L^2(G) \to {}^0L^2(G)$, then, for all $f \in L^1(G) \cap L^2(G)$,

$$\|Ef\|^2 = \sum_{\omega \in \hat{G}_d} d(\omega)\|\pi_\omega(f)\|_2^2$$

where $\pi_\omega \in \omega$ and $\|\cdot\|_2$ is the Hilbert-Schmidt norm; this shows that the Plancherel

---

*This means that A is the maximal subgroup lying in the connected component of the identity of L such that $\mathrm{Ad}(a)$ is diagonalizable over $\mathbb{R}$ for all $a \in A$.

measure assigns the mass $d(\omega)$ to the class ω [1956b]. These results are actually true for all second countable locally compact groups ([Ma 2], p. 640).

In a succession of beautiful papers Harish-Chandra added new evidence to the picture of harmonic analysis of $L^2(G)$ suggested above [1954f], [1955a], [1955c], [1956a], [1956b]. Assuming not only that $rk(G) = rk(K)$ but further that G/K is Hermitian symmetric (this means that G/K has a G-invariant complex structure) he constructed a whole family of discrete series representations. They were realized on certain Hilbert spaces of holomorphic functions on which G acts naturally; they are the *holomorphic discrete series*. When G/K is not Hermitian symmetric this method would give nothing; even in the Hermitian symmetric case it will not give the entire discrete series. Nevertheless the introductory remarks to [1955c] reveal his conviction that in these cases different methods would have to be invented.

These papers revealed profound analogies between the (holomorphic) discrete series and the finite dimensional representations. These representations were *highest weight modules* with respect to positive systems coming from the complex structure of G/K. Their formal degrees were given *exactly by Weyl's dimension formula* applied to their highest weights. Finally, their *characters made sense as point functions* on the set of regular points of a compact Cartan subgroup and *could be described there by precise analogues of the Weyl character formulae*.

These results were the first serious indication that the discrete series would be parametrized by the characters of the compact Cartan subgroup, and hence that the representations of the series associated to an arbitrary Cartan subgroup would be parametrized by the characters of that Cartan subgroup. One would thus have a natural map

$$\coprod_{1 \leqslant i \leqslant r} \hat{A}_i \to \hat{G}$$

where $A_1, A_2, \ldots, A_r$, is a complete system of mutually nonconjugate Cartan subgroups of G (cf. the introductory remarks to [1957e]).

It is thus clear that by 1956 Harish-Chandra had essentially visualized the lines along which the Fourier analysis of $L^2(G)$ would have to be constructed. In retrospect it is really remarkable that he saw so clearly and so far; the clues available to him were certainly very meager.

Discrete series: construction. The determination of the discrete series turned out to be a prodigious undertaking ([1957b]–[1957e], [1964a]–[1966b]). His method, which was entirely novel, was roughly in three parts.

The first part is a direct and explicit construction of the invariant eigendistributions Θ_ξ, parametrized by the (regular) characters ξ of a compact Cartan subgroup $B \subset K$; these would ultimately be the characters of the representations of the discrete series. In the second part the question is that of showing that the Fourier coefficients (with respect to K) of the Θ_ξ, which are clearly eigenfunctions for $\mathfrak{Z}$ that are K-finite, are in $L^2(G)$. The third part is concerned with the completeness of the Θ_ξ; and the main result here is that these eigenfunctions and their translates span a dense subspace of ${}^0L^2(G)$.

Before the invariant eigendistributions could be constructed it is absolutely essential to understand the nature of their *singularities*. An initial exploration [1956c] had already shown that any invariant eigendistribution Θ on G is an analytic function F_Θ on G', the set of regular points of G. This function is locally integrable around the singular points also and so defines an invariant distribution $f \mapsto \int F_\Theta f \, dx$ ($f \in C_c^\infty(G)$) on G, also denoted by F_Θ. Experience suggested that $\Theta = F_\Theta$ (cf. [1955b], although the statement there is somewhat guarded), and this actually turned out to be true in complete generality. This result, known as the *Regularity theorem*, lies at a great depth, and is the corner stone for the entire structure of harmonic analysis.

The function F_Θ will in general blow up in the neighborhood of a singular point of the group. This behavior however cannot be completely arbitrary because the differential equations satisfied by the characters remain valid in the neighborhood of the singular points also. Harish-Chandra's work revealed the true nature of the limitations on the behaviour of F_Θ, namely that F_Θ should satisfy conditions of the following form: for suitable (explicitly determined) invariant differential operators ∇ on G', ∇F_Θ should extend continuously to all of G.

The starting point of the construction of the Θ_ξ is now quite clear; one defines Θ_ξ at the regular points on the compact Cartan subgroup by the analogue of the Weyl character formula. But the problem of defining Θ_ξ on the other Cartan subgroups is nontrivial because an invariant eigendistribution is not uniquely determined by its values on $B' = B \cap G'$. Nevertheless Harish-Chandra discovered that there is uniqueness *if the distribution is suitably restricted at infinity on the group*. The condition is that the distributions should be tempered; this however is to anticipate later ideas. At this stage the condition had to be formulated in a more direct manner, and he did it by requiring that $|D|^{-1/2}$ be a *global majorant* for Θ_ξ, D being the discriminant of G. The existence problem, i.e., the construction of the Θ_ξ, is however much deeper and depends in a fundamental way on *Fourier transforms on the Lie algebra of G*. Harish-Chandra proved that the Θ_ξ can be obtained by exponentiating from the Lie algebra the Fourier transforms of the invariant measures on appropriately chosen regular semisimple orbits [1965c].

The regularity theorem is the key that opens all the doors. If we look at the smooth, almost magical manner in which the theory flows once the regularity theorem is established, and also remember that there was very little in the known history of the subject to guide him in his prolonged struggle or to reassure him during moments of doubt, we have no choice except to place this among his greatest creative achievements.

The idea now is to use the method of integration over the conjugacy classes ($=$ orbits) to reduce the harmonic analysis of the discrete series to that on the compact Cartan subgroup. This is of course exactly the method used by Weyl in his famous completeness proof for the compact case. However the present situation is much more complicated, for one must take into account the contributions of the other Cartan subgroups. Before explaining Harish-Chandra's method of doing this I shall discuss in some detail the method of orbital integrals.

It was Gel'fand and Neumark who first realized that the theory of orbital integrals can be made the basis for a proof of the Plancherel formula [GN 2]. They treated $SL(n, \mathbb{C})$ by this method. In [1951f], [1954c] Harish-Chandra pushed this method through for all *complex* semisimple Lie groups. To motivate this let us first assume that G is compact and let $T \subset G$ be a maximal torus. Using Weyl's formulae we can rewrite the Plancherel formula as a sum over $\hat{T}$:

$$f(e) = \sum_{\xi \in \hat{T}} \Omega(\xi) \int_G \Theta_\xi f\, dx \qquad (f \in C^\infty(G)).$$

Here $\Omega(\xi)$ is a polynomial function of ξ, namely a constant multiple of the product of the roots of a positive system; Θ_ξ is the Weyl character, given on T by

$$\Delta(t)\Theta_\xi(t) = \sum_{s \in W} \varepsilon(s)(s\xi)(t) \qquad (t \in T)$$

where W is the Weyl group and Δ is the Weyl denominator function. Integrating over the orbits and setting

$$F_f(t) = \Delta(t) \int_G f(xtx^{-1})\, dx \qquad (t \in T)$$

we have

$$\int_G \Theta_\xi f\, dx = (-1)^p \hat{F}_f(\xi)$$

where p is the number of positive roots and $\hat{F}_f$ is the Fourier transform of F_f (F_f is obviously in $C^\infty(T)$). So the Plancherel formula is equivalent to the assertion

$$f(e) = (-1)^p \sum_{\xi \in \hat{T}} \Omega(\xi)\hat{F}_f(\xi).$$

But, by Fourier transform theory on T, $(-1)^p \Omega(\xi)\hat{F}_f(\xi)$ is $\widehat{\Omega F_f}(\xi)$ where Ω *is now viewed as a constant coefficient differential operator on T*. So the Plancherel formula is equivalent to the relation

(L.F.) $$f(e) = (\Omega F_f)(e) \qquad (f \in C^\infty(G)).$$

We shall refer to (L.F.) as the *limit formula for the orbital integral*.

This derivation suggests a method for proving the Plancherel formula: first prove (L.F.) and then apply Fourier transform to it. It is remarkable that this method continues to work when G is complex. The maximal torus is now replaced by a Cartan subgroup L. For any $f \in C_c^\infty(G)$, $F_{f,L}(h)$ will denote the integral of f over the orbit of h:

$$F_{f,L}(h) = \Delta(h) \int_{G/L} f(xhx^{-1})\, d\dot{x}$$

where $d\dot{x}$ is the invariant measure on G/L. Initially this makes sense only for h regular, and in the regular subset of L, $F_{f,L}$ will be a smooth function. It can then be proved that $F_{f,L}$ extends to all of L as an element of $C_c^\infty(L)$. The character formulae for the principal series now show that, with appropriate normalization,

$$\langle T_\xi, f \rangle = \hat{F}_{f,L}(\xi) \qquad (\xi \in \hat{L})$$

where T_ξ is the distribution character of the principal series representation corresponding to the character ξ of L. But now the limit formula

(L.F.) $$f(e) = (\Omega F_{f,L})(e) \qquad (f \in C_c^\infty(G))$$

holds, with same Ω as before, and the Fourier transformation of this relation gives the Plancherel formula

$$f(e) = \int_L \hat{\Omega}(\xi)\langle T_\xi, f\rangle \, d\xi \qquad (f \in C_c^\infty(G))$$

where $\hat{\Omega}$ is the polynomial function which is the Fourier transform of the differential operator Ω.

The set $\{e\}$ is a very complicated singular point in the space of conjugacy classes. In particular it is not the limit of orbits in general position, and so $f(e)$ is not the limit of $F_{f,L}(h)$ as $h \to e$. It is therefore striking that one can still recover $f(e)$ as the limit as $h \to e$ of *suitable derivatives* of $F_{f,L}(h)$.

It is natural to try to extend this method to all cases by establishing the limit formula and relating the Fourier transforms of the orbital integrals to the irreducible characters. However for arbitrary G the orbital integrals and their derivatives have jumps at the singular points. Moreover there is more than one orbital integral to be considered since there is in general more than one conjugacy class of Cartan subgroups. Nevertheless calculations in special cases [1952] seemed to suggest that the Fourier transform of the orbital integral is very closely related to the irreducible characters, and so a limit formula should still represent a primitive version of the Plancherel formula (see the remarks at the beginning of [1957e]). In a series of papers in the 1950's ([1957b]–[1957e] and [1964c]) Harish-Chandra established the limit formula for the orbital integral associated to a *fundamental* Cartan subgroup, namely a Cartan subgroup whose maximal compact subgroup has the largest possible dimension (these form a single conjugacy class, and when $rk(G) = rk(K)$, fundamental means compact). In the limit formula

$$\text{(L.F.)} \qquad\qquad f(e) = (\Omega F_{f,L})(e) \qquad (L \text{ fundamental})$$

$F_{f,L}$ may still have jumps at singular points of L; *nevertheless the differential operator Ω, which is still the product of roots of a positive system, kills all the jumps, i.e., $\Omega F_{f,L}$ extends continuously to all of L, so that the right side makes sense.*

The jumps of $F_{f,L}$ make the Fourier analysis of the relation (L.F.) a very delicate matter (see [1952]). These jumps are in fact the orbital integrals associated to Cartan subgroups which are "less compact" than L, suggesting that the jumps are caused by the contributions to f from the representations associated to these less compact Cartan subgroups. Thus when L is compact, one must work with matrix elements of the discrete series representations in order to get rid of the jumps. This can only be done by working with a space of rapidly decreasing functions instead of $C_c^\infty(G)$.

For this purpose Harish-Chandra introduced in [1966b] the Schwartz space $\mathcal{C}(G)$ of G. Overcoming many technical difficulties he proved in [1966b] that the theory of orbital integrals can be extended to the Schwartz space, that the limit formula (L.F.) is true for all f in this space, and that the distributions Θ_ξ constructed earlier are *tempered*, with

$$\langle \Theta_\xi, f\rangle = \int_G \Theta_\xi f \, dx,$$

the integrals converging absolutely. Moreover he proved that *if f is an eigenfunction* *for $\mathfrak{z}$ in the Schwartz space, the integrals of f on the regular non-elliptic orbits are all zero*; here we recall that an orbit is said to be *elliptic* if it meets a compact Cartan subgroup. If G has no compact Cartan subgroup this means that all the orbital integrals of f are zero; the limit formula would then imply that $f = 0$. We shall see presently that eigenfunctions of $\mathfrak{z}$ in $L^2(G)$ are automatically in Schwartz space; so this argument already shows that the condition $rk(G) = rk(K)$ is necessary for the existence of the discrete series. If $rk(G) = rk(K)$, the same argument would still allow us to conclude that the orbital integral $F_{f,B}$ ($B \subset K$) has no jumps, i.e., it lies in $C^\infty(B)$. The Fourier transformation of the limit formula coupled with the formula for Θ_ξ on B would show that the harmonic analysis of f is entirely controlled by the Θ_ξ. This is substantially Harish-Chandra's argument for the completeness of the Θ_ξ.

To round out this picture it is necessary to know a method of determining when a given eigenfunction of $\mathfrak{z}$ belongs to the Schwartz space. In [1966b] Harish-Chandra developed a powerful theory to describe the asymptotic behaviour of τ-spherical eigenfunctions for $\mathfrak{z}$ that satisfy the weak inequality. If f is any τ-spherical eigenfunction satisfying the weak inequality, Harish-Chandra was able to associate to f, at each point at infinity of the group, a so called *constant term*, that was the dominant term in the asymptotic development of f there. These dominant terms are parametrized by the parabolic subgroups of G. If $P = MAN$ is a parabolic subgroup and f_P is the associated constant term, f_P is essentially a function of the same nature as f, but lives on the group MA; the difference between f and f_P has very rapid decay at infinity "in the direction of P." An important consequence of this theory is the result that f lies in $L^2(G, \tau)$ *if and only if all its constant terms are zero; and that in this case, f is in the Schwartz space.*

In order to apply this theory successfully two things have to be done. First, it must be shown that the Fourier coefficients of the Θ_ξ satisfy the weak inequality. This is technically rather subtle and is done in [1966a]. The second point is to show that all the constant terms of these Fourier coefficients are zero. Harish-Chandra's proof of this made use of a beautiful property of these constant terms, namely that they are *transitive* in a natural sense with respect to the partial ordering among the parabolic subgroups. If f is a τ-spherical eigenfunction built out of the Fourier coefficients of the Θ_ξ, and $P \neq G$ is chosen so that $f_P \neq 0$ and $\dim(M)$ is minimal, the transitivity mentioned above would show that the restrictions of f_P to M are eigenfunctions in the Schwartz space of M (actually linear combinations of such); so M has to have a compact Cartan subgroup, implying that P is associated to a Cartan subgroup of G itself. This Cartan subgroup cannot be compact, and hence as the eigenhomomorphism to which f belongs is regular and is associated to the compact Cartan subgroup, we have a contradiction.

These brief remarks cannot give an adequate idea of the pathbreaking nature of the papers [1965c] [1966b]. The audacity of their conceptions as well as the delicacy and craftsmanship that were needed in executing them, compel one to place them among the greatest works in Fourier analysis.

The continuous spectrum: the theory of the Eisenstein integral and the explicit Plancherel formula. With the discrete series completely determined Harish-Chandra turned to the problem of explicitly decomposing the regular representation of G. He went much farther than a mere L^2-theory and obtained a Fourier analysis of the

Schwartz space of G. The results were worked out in three long papers [1975], [1976a], and [1976b]; but the lectures [1970a], [1970c], and [1972] containing the announcements of these and many other results are interesting in themselves, for they give a bird's eye view of his entire approach.

Let L be a Cartan subgroup written (canonically) as $T \cdot A$ where T is compact and A is connected and split over $\mathbb{R}$. Let $\mathscr{P}(A)$ be the finite set of parabolic subgroups $P = MAN$. Since $rk(M) = \dim(T)$, $\hat{M}_d \neq \phi$. For $\eta \in \omega \in \hat{M}_d$ and $\nu \in \mathfrak{a}^*$ ($\mathfrak{a} = \mathrm{Lie}(A)$), $\pi_{\eta, \nu}$ is the unitary representation of G induced by the representation $m \cdot a \cdot n \mapsto e^{i\nu(\log a)} \cdot \eta(m)$ of P. The character $\Theta_{\omega, \nu}$ of $\pi_{\eta, \nu}$ is independent of P (M and A are the same for all $P \in \mathscr{P}(A)$). Then (see [1976b])

(a) $\pi_{\eta, \nu}$ is finitely decomposable; it is irreducible for regular ν.
(b) Let $W(A)$ be the Weyl group of A, i.e., the normalizer of A modulo the centralizer of A; then $W(A)$ acts on $\hat{M}_d \times \mathfrak{a}^*$ and

$$\Theta_{s\omega, s\nu} = \Theta_{\omega, \nu} \qquad (s \in W(A)).$$

Actually results much stronger than (a) were proved by Harish-Chandra. For example, if L is fundamental, *all* the $\pi_{\eta, \nu}$ are irreducible ([1976b]); special cases of this were known earlier, going all the way back to the irreducibility of the principal series of $\mathrm{SL}(n, \mathbb{C})$ due to Gel'fand and Neumark.

One can now associate to L the subspace $\mathcal{C}_L$ of $\mathcal{C}$ consisting of all functions that are orthogonal to the matrix elements of the representations $\pi_{\eta, \nu}$ attached to the Cartan subgroups that are not conjugate to L. $\mathcal{C}_L$ is closed in $\mathcal{C}$ and is invariant under translations. Central to Harish-Chandra's theory is the decomposition (Theorem 12, [1970c])

$$\mathcal{C} = \bigoplus_L \mathcal{C}_L \qquad (*)$$

where the sum is over a complete set of mutually nonconjugate Cartan subgroups; the sum is orthogonal and smooth (i.e., the projections are continuous in $\mathcal{C}$). It will turn out that $\mathcal{C}_L$ is exactly the space of "wave packets" of the matrix elements of the $\pi_{\eta, \nu}$ associated to L (this can be formulated more precisely). Thus $(*)$ is a beautiful and definitive formulation of the heuristic principle that "each Cartan subgroup makes a separate contribution to the Fourier analysis of G."

The splitting $(*)$ allows one to work on each $\mathcal{C}_L$ separately. Let $\Omega_L = \hat{M}_d \times \mathfrak{a}^*$ and let us define, for $f \in \mathcal{C}$, its Fourier transform $\hat{f} = (\hat{f}_L)$ by

$$\hat{f}_L(\omega, \nu) = \int_G \Theta_{\omega, \nu}^{\mathrm{conj}} f \, dx \qquad ((\omega, \nu) \in \Omega_L).$$

Then $\hat{f}_L$ is $W(A)$-invariant, and we have an initial formulation of the Plancherel formula ([1972], Theorem 11): there is a unique $W(A)$-invariant continuous function μ_L on Ω_L ($\hat{M}_d$ is given the discrete topology) such that for all $f \in \mathcal{C}_L$,

$$f(e) = \sum_{\omega \in \hat{M}_d} d(\omega) \int_{\mathfrak{a}^*} \hat{f}_L(\omega : \nu) \mu_L(\omega : \nu) \, d\nu$$

the series converging absolutely. The function μ_L is ≥ 0 and has other nice properties. For fixed ω, it extends to a meromorphic function on $\mathfrak{a}_c^*$ that is holomorphic on $\mathfrak{a}^*$ and > 0 at the regular points of $\mathfrak{a}^*$; moreover it is of at most polynomial growth on $\mathfrak{a}^*$.

The basic question now is of course the explicit calculation of μ_L. Harish-Chandra's method for doing this was based on a remarkable *product formula* for μ_L. To describe this, fix a parabolic subgroup $P \in \mathscr{P}(A)$; then one can associate in a canonical manner a set of pairs (M_i, Q_i) $(1 \leqslant i \leqslant m)$ where M_i is a reductive subgroup of G with no split component and $Q_i = MA_iN_i$ is a parabolic subgroup of M_i with $A_i \subset A$ and $\dim(A_i) = 1$. If $\mathfrak{a}_i = \mathrm{Lie}(A_i)$ and $\nu_i = \nu|_{\mathfrak{a}_i}$, then

$$\mu_L = c \prod_{1 \leqslant i \leqslant m} \mu_{L,i}, \qquad \mu_{L,i}(\omega : \nu) = \mu_{L,i}(\omega : \nu_i)$$

where $\mu_{L,i}$ is the μ-function associated to (M_i, Q_i) (see [1972], pp. 144–145 for a description of the (M_i, Q_i)); c is a constant > 0 independent of ω and ν and can be explicitly evaluated. This reduces the problem of explicit determination of μ_L to the case (G, P) where G has no split component, $P = MAN$, and $\dim(A) = 1$. If $rk(G) > rk(K)$, then L is fundamental and the limit formula, by Fourier transformation on L, leads to the determination of μ_L as a *polynomial function* on $\hat{L}$; here we must remember that we are operating not on the whole Schwartz space but only on $\mathcal{C}_L$ so that the contributions of the other Cartan subgroups of G do not enter the limit formula. Thus the analysis proceeds as if G has a single conjugacy class of Cartan subgroups (see §24, [1976b]). If $rk(G) = rk(K)$, the treatment follows the model where $rk(G/K) = 1$ (cf. [1966a] and §36, [1976b]).

It is impossible to describe except in bare outline the architecture of this beautiful theory. I shall however try to compensate by discussing two important sources of motivation behind Harish-Chandra's treatment. The first is the harmonic analysis of *spherical functions* [1958a] [1958b]. Many of the crucial features that are typical of the continuous spectrum were discovered first in this context. The second source is the spectral theory of $L^2(G/\Gamma)$, due to Selberg [Se 2], Gel'fand and Piatetsky-Shapiro [G] [GP], and Langlands [L 1], Γ being an arithmetic subgroup of G. Harish-Chandra's treatment of this in [1968b] is especially useful for this purpose.

The theory of spherical functions is essentially the harmonic analysis of $L^2(G/K)$. The fundamental objects are the elementary spherical functions $\varphi(\nu : x) = \varphi_\nu(x)$ and the transform that is defined by them. In [1958a] and [1958b] Harish-Chandra made the remarkable discovery that the φ_ν, regarded as functions on A (of the Iwasawa decomposition), look like a sum of plane waves at infinity on A; and that all of these have the same amplitude, in terms of which the Plancherel measure can be determined rather simply. More precisely,

$$d(a)\varphi(\nu : a) \sim \sum_{s \in W(A)} c_s(\nu) e^{is\nu(\log a)},$$

d being a certain homomorphism of A into the positive reals; the amplitudes satisfy

$$|c_s(\nu)| = |c(\nu)| \qquad (c = c_1)$$

and

$$|c(\nu)|^{-2} d\nu$$

is the Plancherel measure. Actually we have the relations

$$c_s(\nu) = c(s\nu)$$

which mirror the *functional equations*

$$\varphi_{s\nu} = \varphi_\nu$$

satisfied by the elementary spherical functions.

The computation of the spherical Plancherel measure thus reduces to determining the function c, which is a special case of the famous c-function of Harish-Chandra. It turned out that this function is meromorphic on $\mathfrak{a}_c^*$ and that it can be expressed as a product of the analogous c-functions associated to the real rank one subgroups of G defined by the root system of (G, A) (the Gindikin–Karpelevič product formula). Since the c-function can be explicitly evaluated for groups of real rank one, the story is complete.

For the harmonic analysis of arbitrary (not necessarily spherical) functions in the Schwartz space Harish-Chandra introduced the eigenfunctions that generalized the elementary spherical functions. Let U be a finite dimensional Hilbert space carrying a unitary double representation τ of K. Let $\tau_M = \tau|_M$ and let $\mathcal{L}$ be the space of τ_M-spherical maps of M into U whose components (with respect to some basis of U) are in $\mathcal{C}(M) \cap {}^0 L^2(M)$. Then $\dim(\mathcal{L})$ is finite and for each $\psi \in \mathcal{L}$, $\nu \in \mathfrak{a}_c^*$, and a given choice of P, a parabolic subgroup in $\mathscr{P}(A)$, Harish-Chandra defined an eigenfunction for $\mathfrak{Z}$ in $C^\infty(G, \tau)$, denoted by $E_{\psi, \nu}$ or $E_{P, \psi, \nu}$, called the *Eisenstein integral*. If P is minimal, $U = \mathbf{C}$, τ is the trivial representation, and ψ is 1, this is just the elementary spherical function (see [1972], p. 133).

The Eisenstein integrals satisfy the weak inequality, and so one can determine their behaviour at infinity on the group by applying the asymptotic theory of tempered eigenfunctions developed in [1966b] [1975]. If Q is a parabolic subgroup in $\mathscr{P}(A)$ which may be different from P, then one finds that the constant term of the Eisenstein integral $E_{P, \psi, \nu}$ associated to Q is of the form

$$\sum_{s \in W(A)} (\psi_{s, \nu})(m) e^{is\nu(\log a)}$$

where ν is a regular element of $\mathfrak{a}^*$ and $\psi_{s, \nu} \in \mathcal{L}$. The maps

$$c_{Q|P}(s:\nu): \psi \mapsto \psi_{s, \nu}$$

are uniquely determined endomorphisms of $\mathcal{L}$ and the relation

$$E_{P, \psi, \nu} \underset{Q}{\sim} \sum_{s \in W(A)} c_{Q|P}(s:\nu) \psi e^{is\nu}$$

expresses the fact that $E_{P, \psi, \nu}$ looks like a sum of plane waves at infinity in the direction of Q. This is clearly analogous to what happens in the spherical case.

Let us now fix ω and choose τ so that the subspace $\mathcal{L}(\omega)$ of $\mathcal{L}$ that "transforms according to ω" is nonzero; this is always possible. It can then be shown that for regular $\nu c_{Q|P}(1:\nu)$ is a bijection of $\mathcal{L}(\omega)$ onto itself. Let $c_{Q|P, \omega}(1:\nu)$ be the restriction of $c_{Q|P}(1:\nu)$ to $\mathcal{L}(\omega)$. Then the fundamental result expressing the relationship between the Plancherel formula and the asymptotics of the Eisenstein integrals can be stated as follows (see [1976b], §13):

$$\mu_L(\omega:\nu)\,c_{Q|P,\omega}(1:\nu)^\dagger\,c_{Q|P,\omega}(1:\nu) = c\cdot id_\omega$$

where id_ω is the identity endomorphism of $\mathcal{L}(\omega)$, $c>0$ is a constant independent of ω and ν which can be explicitly calculated. One must also note that as $\mathcal{L}\subset{}^0L(M,\tau_M)$, $\mathcal{L}$ has a natural structure as a Hilbert space, and $\dagger$ is the adjoint operation in this structure.

The c-functions are closely related to the intertwining operators between the induced representations. The latter have natural product formulae which then lead to product formulae for the c-functions and thence to the μ-functions. This is the origin of the product formula for the Plancherel measure.

A central role in Harish-Chandra's treatment of these questions is played by the functional equations of the Eisenstein integrals. The analogy with the theory of automorphic forms is most significant here, and so it may be useful to make a few remarks comparing the Eisenstein integral with the Eisenstein series. The starting point of this comparison is the following characterization of the "discrete part" of the Schwartz space of G, obtained by Harish-Chandra: if ${}^0\mathcal{C} = \mathcal{C}(G)\cap{}^0L^2(G)$, then an element f in $\mathcal{C}(G)$ is in ${}^0\mathcal{C}$ if and only if

$$\int_N f(xn)\,dn = 0 \qquad (x\in G)$$

for all parabolic subgroups $P = MAN \neq G$ (Theorems 14 and 15, [1970c]). This is obviously analogous to the condition that defines a cusp form on G/Γ, so that it is natural to speak of the elements of ${}^0\mathcal{C}$ as *cusp forms* on G. Thus $\mathcal{L}$ may be viewed as the space of τ_M-spherical cusp forms on M; and the definition of the Eisenstein integral imitates that of the Eisenstein series with integration over K replacing the summation over Γ (cf. [1968b], p. 29). Both the characterization of ${}^0\mathcal{C}$ and the finite dimensionality of $\mathcal{L}$ are deep lying consequences of the theory of the discrete spectrum.

The functional equations of the Eisenstein integrals involve the comparison of $E_{P,\psi,\nu}$ and $E_{Q,\psi',s\nu}$, and Harish-Chandra obtained them as a consequence of the remarkable principle asserting that *if just one summand is common to the constant terms of two such functions, they must be identical.* This principle follows from what he called the *Maass–Selberg relations* which may be described as follows: if f is a τ-spherical eigenfunction satisfying the weak inequality and suitable additional conditions (which are satisfied by the Eisenstein integrals), the intensities ($=$ norms in ${}^0L^2(M,\tau_M)$) of the plane waves that occur in its asymptotic forms along the parabolic subgroups in $\mathcal{P}(A)$ are all equal (cf. Theorem 14.1 and the results of §17, especially Lemmas 17.2 and 17.3 of [1976b]). The Maass–Selberg relations originally arose in the theory of automorphic forms ([1968b], Chap. IV, §2); the discovery and proof of their analogues in the present context is a highlight of [1976b].

Representation theory of $\mathfrak{p}$-adic groups. The Lefschetz principle and the philosophy of cusp forms. The beginnings of the representation theory of reductive $\mathfrak{p}$-adic groups go back to the late 1950's when Mautner's paper [Mau 1] on the representations of $GL(2,\Omega)$ and $PGL(2,\Omega)$ appeared, Ω being a local field (also called a $\mathfrak{p}$-adic field), i.e., a locally compact non-discrete field. It developed rapidly in the early 1960's at the hands of Bruhat, Gel'fand–Graev, Satake and others. Harish-Chandra became actively interested in this area in the mid 1960's; his ideas and results since then have proved to be of fundamental importance for the subject. I shall first try to motivate these developments.

Let k be an algebraic number field. Then one can attach to k the local fields k_v which are its completions at the various places v, and the adele ring $\mathbf{A}$ which is a "restricted" direct product of the $k_v \cdot \mathbf{A}$ is an abelian ring which is locally compact in a natural topology; k is diagonally imbedded in it as a discrete subring and $\mathbf{A}/k$ is compact. By the 1950's it had become a well understood principle in number theory that many arithmetic questions on k could be studied over each k_v, and then by working over $\mathbf{A}$, the local data coming from the various k_v could often be put together to reach global results. This is so in classfield theory, in the theory of simple algebras and Brauer groups, and in the arithmetic theory of quadratic forms [W 2].

After the development of the theory of linear algebraic groups in the 1950's it was realized that the formalism of adeles could be applied to any algebraic group defined over k. Let $\mathbf{G}$ be such a group, and for any k-algebra R, let $\mathbf{G}(R)$ be the group of R-points of $\mathbf{G}$. We then have the groups $G_v = \mathbf{G}(k_v)$ for each v and the adele group $G_{\mathbf{A}} = \mathbf{G}(\mathbf{A})$. G_v and $G_{\mathbf{A}}$ are locally compact, separable, and $G_k = \mathbf{G}(k)$ is (diagonally) imbedded naturally as a discrete subgroup of $G_{\mathbf{A}}$. With the availability of the Chevalley theory of reductive algebraic groups and the work of Borel and Harish-Chandra [1962] (see also [Bo 1]) on the structure of $G_{\mathbf{A}}/G_k$ it was natural to hope that one could begin studying the arithmetic aspects of the theory of reductive groups. A succession of closely related ideas and papers, especially those of Weil [W 3] [W 4] [W 5] and Langlands [L 1] [JL] (with Jacquet) [L 2] [L 3] led to an explosive development of this point of view in the 1960's. As a consequence, the supreme importance of the harmonic analysis of $L^2(G_{\mathbf{A}}/G_k)$ for arithmetic was recognized. For example, if $\mathbf{G} = GL(1)$, $G_{\mathbf{A}} = \mathbf{I}$ the idele group and $G_{\mathbf{A}}/G_k = \mathbf{I}/k^\times$ is the idele class group; a character of $\mathbf{I}/k^\times$ is essentially a Grössencharakter of Hecke. Tate's work (cf. §1) in 1947 had derived the Hecke theory of the associated L-series from the Poisson formula for $\mathbf{A}/k$. Also one knows from classfield theory that the characters of $\mathbf{I}/k^\times$ of finite order define (through Artin's reciprocity law) cyclic extensions of k. If $\mathbf{G} = GL(2)$, the modular forms on the Poincaré half plane which are eigenfunctions for the Hecke operators define (under suitable spectral conditions) irreducible representations of $G_{\mathbf{A}}$ that occur in $L^2(G_{\mathbf{A}}/G_k)$, showing that in the notion of an irreducible representation occurring in $L^2(G_{\mathbf{A}}/G_k)$ we have a far reaching generalization of the classical automorphic forms and Eisenstein series. The work of [L 3] suggested that for any integer $n \geqslant 1$ and any n-dimensional representation of the Galois group $\mathrm{Gal}(\bar{k}/k)$ ($\bar{k}$ = an algebraic closure of k) there would be associated an irreducible unitary representation occurring in $L^2(G_{\mathbf{A}}/G_k)$.

The irreducible unitary representations of $G_{\mathbf{A}}$ are tensor products of irreducible unitary representations of the local groups G_v. It is therefore natural to try to understand the representation theory of the groups G_v as a first step. Of course these representations would reflect the arithmetic of the local fields k_v. For example, following [L 3], one would expect any n-dimensional representations of $\mathrm{Gal}(\bar{k}_v/k_v)$ to determine an irreducible unitary representation of $GL(n, k_v)$. This is a very rough sketch of some of the ideas that led to the great interest in the 1960's for constructing a theory of representations of groups $G = \mathbf{G}(\Omega)$ where Ω is a p-adic field and $\mathbf{G}$ is a reductive group defined over Ω.

From the very beginning Harish-Chandra emphasized the striking similarities between the theory for real groups and the newly developing theory for the p-adic groups [1968a]. In [1970d] he formulated this as a broad heuristic principle, which he called the *Lefschetz principle*, asserting that whatever is true for real groups is also true for the p-adic ones. Its thrust was to use the rich theory of Fourier analysis on real reductive groups as a guide in the search for discovering the basic facts to be proved in the representation theory of the p-adic groups. In [1970d], as a first illustration of this principle, Harish-Chandra explored the theory of characters and orbital integrals on $G = \mathbf{G}(\Omega)$ as above. These investigations were completed in [1978] where he showed, assuming char $\Omega = 0$, that the theory of characters of G resembles the theory for real groups to a remarkably close extent. Thus, characters of irreducible admissible representations are locally summable functions on G which are locally constant on G'; and they can be obtained by exponentiating the Fourier transforms of the invariant measures on the (nilpotent) orbits in the Lie algebra. One should recall that in the real case such theorems came out of the profound study of the differential equations satisfied by the characters. In the present case there are no differential equations, and so it is quite surprising that such results are true. Harish-Chandra's work uses a beautiful theorem of Howe on invariant distributions on the Lie algebra of G as a substitute for the differential equations.

The philosophy of cusp forms, which he had already introduced in [1970b] as the unifying principle for Fourier analysis on reductive groups is an even more striking illustration of the Lefschetz principle. In [1970b] he had illustrated it for reductive groups over finite fields, and it was the driving force in his work on the Fourier analysis of the Schwartz space of a real reductive group. He showed that the harmonic analysis of these groups can be worked out quite explicitly, based on the concept of cusp forms and parabolic induction. In [1973] and [1977a] he succeeded in pushing this method through for reductive p-adic groups. The cusp forms are once again (essentially) the matrix elements of the discrete series and this method led him to the theory of the Eisenstein integral, in particular, to the Maass–Selberg relations and Plancherel formula, in virtually complete analogy with the real case.

For groups over p-adic fields the discrete series is still not completely determined. Nevertheless, by taking Fourier analysis of reductive groups to this stage, Harish-Chandra has shown that its inner structure can be understood in terms of a small number of simple and yet general principles. This is a truly extraordinary accomplishment.

§3. *Orbital Integrals and Limit Formula.* I now wish to supplement this general survey with a little more detailed discussion of selected parts of Harish-Chandra's work. I shall begin with the theory of orbital integrals and the limit formula. The bulk of this work is contained in [1957b]–[1957e] and [1964c]; the extension of the theory to the Schwartz space of G is treated in [1965b], [1966a], and [1966b].

We take G to be the real form of a complex simply connected semi simple group. Let $L \subset G$ be a Cartan subgroup with $\mathfrak{h} = \text{Lie}(L)$; P, a positive system of roots of $(\mathfrak{g}_c, \mathfrak{h}_c)$; and Δ, the Weyl denominator on L. Define ϖ and π by $\varpi = \prod_{\alpha \in P} H_\alpha$, $\pi = \prod_{\alpha \in P} \alpha$. The orbital integral associated to L is then defined for $f \in C_c^\infty(G)$, $h \in L' = L \cap G'$ by

$$F_{f,L}(h) = \varepsilon_R(h)\Delta(h)\int_{G/L = G^*} f(xhx^{-1})\, dx^* \tag{1}$$

where $\varepsilon_R(h)$ is a locally constant sign factor (§22, [1965b]). The analogous definition on $\mathfrak{g}$ is (see §5, [1964c])

$$\psi_{f,\mathfrak{h}}(H) = \varepsilon_R(H)\pi(H)\int_{G^*} f(H^x)\, dx^* \qquad (f \in C_c^\infty(\mathfrak{g})). \tag{2}$$

These are well defined because the regular orbits are closed, and it is easy to show that $F_{f,L}$ and $\psi_{f,\mathfrak{h}}$ are C^∞ on L' and $\mathfrak{h}'$, respectively.

The key to the entire theory is the system of differential equations satisfied by F_f and ψ_f:

$$F_{zf,L} = \mu_{\mathfrak{g}/\mathfrak{h}}(z)F_{f,L} \qquad (z \in \mathfrak{Z}) \tag{3G}$$

$$\psi_{\partial(p)f,\mathfrak{h}} = \partial(p_\mathfrak{h})\psi_{f,\mathfrak{h}} \qquad (p \in I(\mathfrak{g})). \tag{3g}$$

Here, $I(\mathfrak{g})$ is the algebra of G-invariant elements in $S(\mathfrak{g}_c)$ and $p_\mathfrak{h}$ is the "restriction" of p to $\mathfrak{h}$ while $\mu_{\mathfrak{g}/\mathfrak{h}}$ is as in §12 of [1965b]. To illustrate this let us examine how they control the growth of the orbital integrals near a singular point. Consider for instance $\psi_{f,\mathfrak{h}}$. The equations (3g) and the formula for integration on $\mathfrak{g}$ give the estimate

$$\int_{\mathfrak{h}'} |\pi||\partial(p_\mathfrak{h})\psi_{f,\mathfrak{h}}|\, d\mathfrak{h} \leqslant \int_{(\mathfrak{h}')^G} |\partial(p)f|\, d\mathfrak{g}. \tag{4}$$

From this one can conclude that all derivatives of $\psi_{f,\mathfrak{h}}$ are locally bounded, and in fact even more. The question is the often encountered one of obtaining point estimates in terms of L^1-norms of a function and its derivatives; but there is a technical complication because of the weight function $|\pi|$ which vanishes on $\mathfrak{h} \setminus \mathfrak{h}'$. In [1965b] (cf. Theorem 3) Harish-Chandra obtained a global estimate of the Sobolev type that can be used in the present situation. This can be formulated as follows (see [V 3], Proposition 7, p. 157). Let V be a real finite dimensional vector space; $w = \prod_{1 \leqslant j \leqslant q} |\lambda_j|^{a_j}$ where $a_j > 0$, $\lambda_j \in V_c^*$, and $V' =$ the set where $w \neq 0$; let S_0 be a subalgebra of $S = S(V_c)$ such that S is a finite module over S_0. Write $\mathfrak{U}^1(S_0: w)$ (resp. $\mathfrak{U}^\infty$) for the Frechet space of all $g \in C^\infty(V')$ such that $\int |\partial(u)g|w\, dV < \infty$ (resp. $\sup|\partial(u)g| < \infty$) for all $u \in S_0$ (resp. $u \in S$). Then

$$\mathfrak{U}^1(S_0: w) \subset \mathfrak{U}^\infty$$

and the inclusion is continuous. Since $\psi_{f,\mathfrak{h}} \in \mathfrak{U}^1(I(\mathfrak{h}):|\pi|)$ by (4), we must have $\psi_{f,\mathfrak{h}} \in \mathfrak{U}^\infty$. Actually, replacing f by qf where q is an invariant polynomial on $\mathfrak{g}$, we find the much stronger result that $\psi_{f,\mathfrak{h}}$ lies in the Schwartz space $\mathcal{C}(\mathfrak{h}')$ of $\mathfrak{h}'$ and that the map $f \mapsto \psi_{f,\mathfrak{h}}$ is continuous with respect to the collection of seminorms of the form

$$f \mapsto \int_{(\mathfrak{h}')^G} |\partial(p)(qf)| \, d\mathfrak{g} \qquad (f \in C_c^\infty(\mathfrak{g})).$$

Since *these are all continuous on* $\mathcal{C}(\mathfrak{g})$, the orbital integral extends to a continuous map $\mathcal{C}(\mathfrak{g}) \to \mathcal{C}(\mathfrak{h}')$; in particular, the invariant measures on the orbits are all tempered. The original proof in [1957c] was essentially a variant of this type of argument.

The same argument works (up to a point) on the group also. The crucial case is when $L = B$ is compact and let us make this assumption. Let $\mathcal{U}^\infty$ now be the space of all $g \in C^\infty(B')$ such that $\sup|ug| < \infty$ for all $u \in S(\mathfrak{b}_c)(\mathfrak{b} = \mathrm{Lie}(B))$, regarded as a Frechet space in the usual way. Then the above argument will show that $F_{f,B} \in \mathcal{U}^\infty$ and that $f \mapsto F_{f,B}$ will be continuous with respect to the topology induced on $C_c^\infty(G)$ by the seminorms

$$\nu_z : f \mapsto \int_{G_e} |zf| \, dx \qquad (z \in \mathfrak{Z})$$

where $G_e = (B')^G$ is the elliptic set. If one can show that

$$V(f) = \int_{G_e} |f| \, dx < \infty$$

for any $f \in \mathcal{C}(G)$ and that V is a continuous seminorm on $\mathcal{C}(G)$, the continuity of the ν_z will follow, leading to the extension of the theory of orbital integrals to $\mathcal{C}(G)$. Let φ_B be the characteristic function of the set G_e and let

$$b(x) = \int_K \varphi_B(xk) \, dk \qquad (x \in G).$$

Since

$$\int_{G_e} |f| \, dx \leqslant \int_G |f| b \, dx$$

it is enough to prove that b satisfies the weak inequality.

Harish-Chandra does this in [1966a] (Theorem 5). It is obvious that $b(x)$ is the measure of the set of all k in K for which xk is elliptic. One can use the boundedness of the finite dimensional characters on G_e to conclude that $b(x) \to 0$ when x goes to infinity on G. For the sharper result needed here, one needs class functions with good growth properties on *all* Cartan subgroups. Harish-Chandra's proof depends on the existence of a locally summable class function Θ such that

(i) Θ is a nonzero constant on G_e.

(ii) If $L = T \cdot A$ is any Cartan subgroup, and $\mathfrak{z}$ is the centralizer of A in $\mathfrak{g}$,

$$|\Theta(x)| \leqslant \mathrm{const.} |\det(1 - \mathrm{Ad}(x^{-1}))_{\mathfrak{g}/\mathfrak{z}}|^{-1/2} \qquad (x \in L').$$

(iii) $\int_K \Theta(xk) \, dk = 0 \qquad (x \in G).$

Using his theory of the distributions $\Theta_\xi(\xi \in \hat{B})$ he shows that $\Theta = \sum_{s \in W} \varepsilon(s) \Theta_{s\rho}$ ($W = $ the Weyl group of $(\mathfrak{g}_c, \mathfrak{b}_c)$ and ρ has the usual meaning), will have the required properties. Θ is essentially the character of the sum of the discrete series representations which have the same infinitesimal character as the trivial representation. It is interesting to observe that the same argument works in the $\mathfrak{p}$-adic case also; Θ may then be taken as the Steinberg character ([1973]).

The analysis of the jumps of the orbital integrals at the singular points comes down, through general arguments, to the case when the singular point is semi regular, and hence, to a calculation in $\mathfrak{sl}(2,\mathbb{R})$. Here there are two Cartan subalgebras $\mathfrak{h} = \mathbb{R} \cdot H$, $\mathfrak{b} = \mathbb{R} \cdot (X - Y)$ where

$$H = \begin{pmatrix} 1 & 0 \\ 0 & -1 \end{pmatrix}, \qquad X = \begin{pmatrix} 0 & 1 \\ 0 & 0 \end{pmatrix}, \qquad Y = \begin{pmatrix} 0 & 0 \\ 1 & 0 \end{pmatrix}.$$

The orbit E_θ of $\theta(X - Y)$ is a two-sheeted elliptic hyperboloid, and the orbit H_t of tH is a single sheeted hyperbolic hyperboloid; these are separated by C, the cone of nilpotents (light cone!), composed of two halves $C_{\pm}$ and the vertex 0. $\psi_{f,\mathfrak{h}}$ is C^∞ while $\psi_{f,\mathfrak{b}}$ is not. To see why, let $\theta \to 0+$, $t \to 0$. Then the $E_{\pm\theta}$ tend to $C_{\pm}$ while H_t tends to C (draw a figure). Thus

$$\lim_{\theta \to \pm 0} \psi_{f,\mathfrak{b}}(\pm \theta(X - Y)) = \pm \int_{C_{\pm}} f \, dC$$

$$\lim_{t \to 0+} \psi_{f,\mathfrak{b}}(tH) = \int_C f \, dC$$

giving the "jump formula"

$$\psi_{f,\mathfrak{b}}(0+) - \psi_{f,\mathfrak{b}}(0-) = \psi_{f,\mathfrak{h}}(0). \tag{5}$$

Generalized to arbitrary $\mathfrak{g}$, this is the basic relation that links the orbital integrals associated to different Cartan subalgebras ([1964c] Theorem 2); there is of course a corresponding version on the group. It shows that jumps arise only when *crossing root hyperplanes corresponding to singular imaginary roots*, and that for derivatives that are skew with respect to these roots, the jumps are 0 (Theorem 1, [1964c]). The continuity of $\partial(\varpi)\psi_{f,\mathfrak{b}}$ and $\varpi F_{f,L}$ follow from this, as well as the vanishing of their values at 0 and e respectively if there are real roots, i.e., if L is not fundamental. The formula (5) or rather its generalization, namely Theorem 2 of [1964c], implies that if $F_{f,\bar{L}} = 0$ for Cartan subgroups $\bar{L}$ "more split" than L, then $F_{f,L}$ is of class C^∞ on L. We have already seen in §2 how this principle was decisive in the determination of the discrete spectrum. In particular, if G is complex, more generally, if L is "Iwasawa", $F_{f,L}$ is of class C^∞; of course one can prove this by a direct evaluation of $F_{f,L}(h)$:

$$F_{f,L}(h) = d_P(h) \iint_{K \times N} f(khnk^{-1}) \, dk \, dn$$

$$\left(d_P(h) = |\det(\mathrm{Ad}(h)_{\mathfrak{n}})|^{1/2} \right).$$

Let us now turn to the limit formula. It is enough to prove it on the Lie algebra. Write

$$T(f) = \left(\partial(\varpi)\psi_{f,\mathfrak{h}} \right)(0) \qquad (f \in \mathcal{C}(\mathfrak{g})). \tag{6}$$

T is a *tempered* invariant distribution on $\mathfrak{g}$ and the limit formula is the assertion that $T = c\delta$ where $c = c(\mathfrak{h})$ is a real constant which, for fundamental $\mathfrak{h}$, is $(-1)^q$ times a positive number, q being the integer $1/2\{\dim(G/K) - rk(G) + rk(K)\}$.

If $\mathfrak{g}$ is complex or if it has a single conjugacy class of Cartan subalgebras, one can show that (cf. [1975], §§35–36) that

$$\hat{\psi}_{f,\mathfrak{h}} = c\hat{\psi}_{f,\mathfrak{h}} \qquad (f \in \mathcal{C}(\mathfrak{g})) \tag{7}$$

where c is a constant $\neq 0$ (recall that $\psi_{f,\mathfrak{h}}$ is smooth and so lies in $\mathcal{C}(\mathfrak{h})$). The integration formula

$$\int_{\mathfrak{g}} u \, d\mathfrak{g} = \text{const.} \int_{\mathfrak{h}} \pi\psi_{u,\mathfrak{h}} \, d\mathfrak{h} \qquad (u \in \mathcal{C}(\mathfrak{g})),$$

on Fourier transformation, gives the limit formula ([1954c]). For arbitrary $\mathfrak{g}$ the relation (7) has to be suitably modified, and this requires a deeper study of the Fourier transforms $\hat{\mu}_H$ of the orbital measures $\mu_H \colon f \mapsto \int_{G^*} f(H^x) \, dx^*$. Since the invariant polynomials are constant on the orbits, the $\hat{\mu}_H$ are invariant eigendistributions:

$$\partial(p)\hat{\mu}_H = \tilde{p}(iH)\hat{\mu}_H \qquad (p \in I(\mathfrak{g})) \tag{8}$$

where $\tilde{p}$ is the polynomial corresponding to p. From the theory of invariant eigendistributions (see §4 below) it follows that $\hat{\mu}_H$ is an analytic function on $\mathfrak{g}'$, say $\hat{\mu}(H \colon \cdot)$, which can be studied in detail. If $\bar{\mathfrak{h}}$ is a Cartan subalgebra, $\bar{\mathfrak{h}}^+$ a connected component of $\bar{\mathfrak{h}}'$, $\overline{W}$ the Weyl group of $(\mathfrak{g}_c, \bar{\mathfrak{h}}_c)$, and y is an element of the complex adjoint group taking $\mathfrak{h}_c$ to $\bar{\mathfrak{h}}_c$,

$$\pi(H)\bar{\pi}(\overline{H})\hat{\mu}(H \colon \overline{H}) = \sum_{s \in \overline{W}} c_s(H) e^{i\langle sH^y, \overline{H}\rangle} \tag{9}$$

for $H \in \mathfrak{h}'$, $\overline{H} \in \bar{\mathfrak{h}}^+$; the equations (3g) now imply that *the c_s are locally constant on* $\mathfrak{h}'$ (Lemma 24, [1957c]). With this proviso, (9) is a generalization of (7) and leads directly to the conclusion that $\hat{T}$ is locally constant on $\mathfrak{g}'$. But $\text{supp}(T) \subset$ the set of nilpotents of $\mathfrak{g}$ so that $\hat{T}$ is $I(\mathfrak{g})$-finite. Since an invariant $I(\mathfrak{g})$-finite distribution on $\mathfrak{g}$, which is locally constant on $\mathfrak{g}'$, is constant on $\mathfrak{g}$, we have $\hat{T} = $ a constant. This argument (see §10, [1965a]) makes use of the regularity theorem on the Lie algebra; the earlier proof in [1957d] was more direct but was also more involved.

One should think of the $\hat{\mu}_H(H \in \mathfrak{h}')$ as the analogues of the irreducible characters of G associated to L by the inducing process. For example one can prove that if $\bar{\mathfrak{h}}$ is a Cartan subalgebra, $\hat{\mu}_H$ is 0 on $(\bar{\mathfrak{h}}')^G$ if $\bar{\mathfrak{h}}$ is not conjugate to a Cartan subalgebra of the centralizer of the split component $\mathfrak{h}_\mathbf{R}$ of $\mathfrak{h}$; in particular this will be the case if $\dim(\bar{\mathfrak{h}}_\mathbf{R}) < \dim(\mathfrak{h}_\mathbf{R})$, i.e., $\bar{\mathfrak{h}}$ is "more compact" than $\mathfrak{h}$.

It remains to look at the case when $rk(G) = rk(K)$ and $\mathfrak{h} \subset \mathfrak{k}$. Harish-Chandra uses a fundamental solution Ξ, due to de Rham, of $\partial(\Omega)^{[n/2]}$ ($\Omega = $ the Casimir element, $n = \dim(\mathfrak{g})$), whose restriction to $\mathfrak{h}$ is a fundamental solution to $(-1)^q \partial(\Omega_\mathfrak{h})^{l/2}$, $l = \dim(\mathfrak{h})$. He now chooses $f \in \mathcal{C}(\mathfrak{g})$ such that

$$f(0) = 1, \qquad \psi_{f,\bar{\mathfrak{h}}} = 0 \quad \text{if } \bar{\mathfrak{h}} \text{ is not conjugate to } \mathfrak{h}. \tag{10}$$

Then in view of earlier remarks on the jumps of ψ_f, $\psi_{f,\mathfrak{h}}$ will be smooth and so will lie in $\mathcal{C}(\mathfrak{h})$; and the relation

$$1 = \int_{\mathfrak{g}} \Xi \cdot \left(\partial(\Omega)^{[n/2]} f \right) d\mathfrak{g}$$

will reduce to an integral on $\mathfrak{h}$ that gives the limit formula ([1964c]).

Thus everything comes down to the choice of f. Harish-Chandra takes $f = \hat{g}$ where $g \in C_c^{\infty}((\mathfrak{h}')^G)$ and $\int_{\mathfrak{g}} g \, d\mathfrak{g} = 1$; by our earlier remarks on $\hat{\mu}_H$, $\psi_{f,\bar{\mathfrak{h}}} = 0$ for $\bar{\mathfrak{h}}$ not conjugate to $\mathfrak{h}$.

The functions such as f in the above proof are truly remarkable because their Fourier analysis is controlled completely by $\mathfrak{h}$. Their analogues on the group are clearly cusp forms. Since we do not have access to these until a substantial amount of Fourier analysis on the group is done, it is now clear why the proof of the limit formula takes place on the Lie algebra. Nevertheless the reader should note how closely this proof resembles the proof, on the group, of the separation of the discrete spectrum. I think the essential originality of Harish-Chandra's treatment of the limit formula was to have conceived of the transition to $\mathfrak{g}$, and to have realized that the theory of Fourier transforms on the Lie algebra could provide him with the "additive" analogues of the cusp forms that held the key to the solution of the problem.

§4. *The Regularity Theorem.* Announced in [1963], this is the central result in the long series of papers [1964a]–[1965b] on the structure of invariant eigendistributions. Harish-Chandra's approach was to prove it first on $\mathfrak{g}$ and then carry the result over to G. For mostly technical reasons he worked on the group with invariant distributions that were $\mathfrak{Z}$-finite (resp. $I(\mathfrak{g})$-finite on $\mathfrak{g}$). The essential problem is to investigate the structure of invariant distributions around semi simple points.

His main technique for doing this was the *method of descent*. Let M be a manifold on which a Lie group L acts, $m_0 \in M$, and let E be a submanifold of M which contains m_0 and is transversal to the orbit of m_0 at m_0. Then the descent mechanism is a method of reducing questions involving invariant distributions and differential operators defined on M around m_0, to similar questions on E around m_0. The classical example is the reduction of rotationally symmetric problems to problems on a half-line.

To set up the descent machinery (see [1964a]) we need volume elements dE, dM on E and M respectively, with dM invariant under L. Assume π: $L \times E \to M$ is submersive and let $M' = \pi(L \times E)$. Integrating on the fibers of π with respect to the volume element $dL \, dE/dM$ will then give a "pull-back" $T \mapsto \beta_T$, taking distributions T on M' to distributions β_T on $L \times E$. If T is invariant, β_T will be of the form $1 \otimes \sigma_T$, and σ_T will be the *transfer* of T; the map $T \mapsto \sigma_T$ is injective. If F is any L-invariant function and $F_E = F|_E$, F is locally integrable on M' if and only if F_E is so on E; and $\sigma_F = F_E$. If L is unimodular and L_0 is a closed subgroup of L such that E is L_0-stable and dE is L_0-invariant, σ_T is L_0-invariant. Suppose now that D is an analytic invariant differential operator on M'. If E is sufficiently small, we can choose an analytic differential operator D' on $L \times E$, invariant under the action x, $(x', m) \mapsto (xx', m)$ of L, such that D' and D are π-related. The analytic differential operator

$\Delta(D)$ on E, obtained by retaining only the terms in D' not involving differentiations with respect to L, is called a *radial component of D*. It is in general not unique because D' is not unique; but it always satisfies

$$\sigma_{DT} = \Delta(D)\sigma_T. \tag{1}$$

If, however, E is a local section at m_0 for the action of L, then $\Delta(D)$ is uniquely determined by the requirement

$$(DF)_E = \Delta(D)F_E \qquad \left(F \in (C^\infty(M'))^L\right) \tag{2}$$

and $D \mapsto \Delta(D)$ will be a homomorphism.

In the present context it is necessary to compute explicitly the radial components of $\mathfrak{Z}$ and $\partial(I(\mathfrak{g}))$ for the actions of G on itself and on $\mathfrak{g}$, around semisimple points. Let a (resp. X) be a semisimple element of G (resp. $\mathfrak{g}$); then its centralizer Z(resp. $\mathfrak{Z}$) in G (resp. $\mathfrak{g}$) is transversal to the orbit of a (resp. X). The radial components and distributions on Z (resp. $\mathfrak{Z}$) are then defined around a (resp. X) and invariant under $\Xi = Z^0$. If a or X is not central, $Z(X)$ will have lower dimension than $G(\mathfrak{g})$ and so induction on dimension is possible.

Harish-Chandra's main calculations may be summarized as follows.

(i) Here G acts on $\mathfrak{g}$, X is regular, and $\mathfrak{Z} = \mathfrak{h}$ is a Cartan subalgebra; $M = \mathfrak{g}'$. We take $E = \mathfrak{h}'$ and find ([1957b] Lemma 8)

$$\Delta(\partial(p)) = \pi^{-1} \circ \partial(p_\mathfrak{h}) \circ \pi \qquad (p \in I(\mathfrak{g})). \tag{3}$$

Let $\mathcal{D}(\mathfrak{g})$ (resp. $\mathcal{D}(\mathfrak{h})$) be the algebra of differential operators with polynomial coefficients on $\mathfrak{g}$ (resp. $\mathfrak{h}$). In [1964b] Harish-Chandra proved, as a partial extension of (3), that for $D \in \mathcal{D}(\mathfrak{g})$, $\delta(D) = \pi \circ \Delta(D) \circ \pi^{-1}$ is in $\mathcal{D}(\mathfrak{h})$, invariant under the Weyl group, and

$$\delta(\hat{D}) = \widehat{\delta(D)} \quad (\wedge = \text{Fourier transform}). \tag{4}$$

(ii) G acts on itself by conjugacy, $E = L'$ where L is a Cartan subgroup, and $M = G'$. Then the radial components are unique, and ([1956c] Theorem 2)

$$\Delta(z) = \Delta^{-1} \circ \mu_{\mathfrak{g}/\mathfrak{h}}(z) \circ \Delta \qquad (z \in \mathfrak{Z}) \tag{5}$$

where $\Delta = \Delta_L$ is the Weyl denominator and $\mathfrak{h} = \mathrm{Lie}(L)$.

(iii) G acts on $\mathfrak{g}$; $X \in \mathfrak{g}$ semisimple. Let $\zeta(Z) = \det(\mathrm{ad}\, Z)_{\mathfrak{g}/\mathfrak{z}}$; $\mathfrak{z}'$ is the subset of $\mathfrak{z}$ where ζ is not zero. The radial components are not unique; nevertheless, if Ω is an invariant open neighbourhood of X in $\mathfrak{g}$ and T an invariant distribution on Ω,

$$\Delta(\partial(p))\sigma_T = |\zeta|^{-1/2} \circ \partial(p_\mathfrak{z}) \circ |\zeta|^{1/2}\sigma_T \qquad (p \in I(\mathfrak{g})) \tag{6}$$

thus generalizing (3) ([1964b] Theorem 2). In particular, T is $I(\mathfrak{g})$-finite on $(\Omega \cap \mathfrak{z}')^G$ if and only if $|\zeta|^{1/2}\sigma_T$ is $I(\mathfrak{z})$-finite on $\Omega \cap \mathfrak{z}'$.

(iv) G acts on itself; $a \in G$ is semisimple. Let $\nu(y) = \det(\mathrm{Ad}((ay))^{-1} - 1)_{\mathfrak{g}/\mathfrak{z}})$ where $\mathfrak{z} = \mathrm{Lie}\,(Z)$. Let Ξ' be the subset of Ξ where ν is nonzero. If Ω is a completely invariant (cf. [1963]) open neighbourhood of a in G,

$$\Delta(z)\theta = \left(|\nu|^{-1/2} \circ \mu_{\mathfrak{g}/\mathfrak{z}}(z) \circ |\nu|^{1/2}\right)\theta \tag{7}$$

for all Ξ-invariant distributions θ on $\Omega_\Xi = \Xi' \cap (a^{-1}\Omega)$. In particular, an invariant distribution T on $(a\Omega_\Xi)^G$ is $\mathfrak{Z}$-finite if and only if $|\nu|^{1/2}\sigma_T$ is $\mathfrak{Z}_a$-finite on Ω_Ξ; $\mathfrak{Z}_a$ = center of $U(\mathfrak{z})$ ([1965b], Lemma 22).

(v) Transfer from $\mathfrak{g}$ to G via the exponential map. Let $z \mapsto p_z$ be the isomorphism (§12, [1965b]) of $\mathfrak{Z}$ with $I(\mathfrak{g})$ such that $\mu_{\mathfrak{g}/\mathfrak{h}}(z) = (p_z)_\mathfrak{h}$ for all Cartan subalgebras $\mathfrak{h} \subset \mathfrak{g}$. Define the invariant function $\xi = \xi_\mathfrak{g}$ on $\mathfrak{g}$ by

$$\xi(X) = |\det\{(e^{\operatorname{ad} X/2} - e^{-\operatorname{ad} X/2})/\operatorname{ad} X\}|^{1/2} \qquad (X \in \mathfrak{g}).$$

Then $dx = \xi^2 dX$ ($x = \exp X$) and $|\pi|\xi = |\Delta \circ \exp|$ on any Cartan subalgebra $\mathfrak{h}$. Let U be a completely invariant open neighbourhood of 0 in $\mathfrak{g}$ such that $U_G = \exp U$ is completely invariant open in G and $\exp: U \to U_G$ is a diffeomorphism. We then have a G-equivariant pull back isomorphism $T \mapsto \tilde{T}$ of distributions from U_G to U such that at the level of locally summable functions

$$\tilde{F} = (F \circ \exp)\xi. \tag{8}$$

The main formula here is

$$(zT)^{\tilde{}} = \partial(p_z)\tilde{T} \tag{9}$$

for all invariant distributions T (§14, [1965b]). Formula (9) implies that T is $\mathfrak{Z}$-finite if and only if $\tilde{T}$ is $I(\mathfrak{g})$-finite.

The formulae (3) and (6) on the Lie algebra are proved essentially by direct calculation. But such a method cannot handle (7) or (9). It is easy in both cases to determine how the transferred differential operators act on smooth invariant functions; but their actions on invariant distributions are quite difficult to determine directly. Harish-Chandra reduced everything to calculations with invariant *smooth* functions by first proving the following remarkably general and beautiful theorem (Theorem 1, [1965b]): if D is an invariant analytic differential operator on Ω such that $Df = 0$ for all invariant $f \in C^\infty(\Omega)$, then $DT = 0$ for all invariant distributions T on Ω. Its proof comes down, via descent, to an analogous result on $\mathfrak{g}$; the proof of this however uses the regularity theorem on $\mathfrak{g}$ ([1965a] Theorems 4 and 5; for an illuminating sketch of the argument see [1963]).

Fix a completely invariant open set Ω of $\mathfrak{g}$ and an invariant $I(\mathfrak{g})$-finite distribution T on Ω. Using (3) one finds that T is an analytic function F on $\Omega' = \Omega \cap \mathfrak{g}'$; F as well as $\partial(p)F(p \in I(\mathfrak{g}))$ are locally summable on Ω. The first step in proving the regularity theorem is to show that T_F, the distribution defined by F on Ω, is $I(\mathfrak{g})$-finite. If this is done, $\mathfrak{U} = T - T_F$ is invariant, $I(\mathfrak{g})$-finite, and supported on the singular set; the second step is to show that 0 is the only such distribution.

For the first step it is a question of proving that $\partial(p)T_F = T_{\partial(p)F}$ for all $p \in I(\mathfrak{g})$. A formal argument (cf. [1957b], p. 99) reduces this to $p = \omega$, the Casimir operator. Let $J_0 = \partial(\omega)T_F - T_{\partial(\omega)F}$. J_0 can be expressed rather explicitly in terms of boundary values of the orbital integrals ([1965a] p. 15). Theorem 4 of [1964c] guarantees that such a distribution is zero on Ω if it is zero around the semi regular points of the noncompact type in Ω. For J_0, this condition is equivalent, by descent, to the regularity theorem on $\mathfrak{sl}(2,\mathbb{R})$, which can be verified by direct calculation.

The second step, by (6) and induction on $\dim(\mathfrak{g})$, reduces to showing that 0 is the only invariant $I(\mathfrak{g})$-finite distribution on $\mathfrak{g}$ supported by the set $\mathfrak{N}$ of nilpotents. The key to Harish-Chandra's proof of this (Theorems 4 and 5, [1964a]) is to view the space $\mathfrak{J}$ of invariant distributions supported by $\mathfrak{N}$ as a module for a Lie subalgebra

$\mathcal{D} \subset \mathcal{D}(\mathfrak{g})$ which is isomorphic to $\mathfrak{sl}(2,\mathbf{C})$ with a standard basis $\{H', X', Y'\}$ where $2Y' = \partial(\omega)$, $2X' = $ multiplication by $\tilde{\omega}$, and $H' = D + \frac{1}{2}\dim(\mathfrak{g})$ where D is the Euler vector field. Using descent theory and the finiteness of $G \setminus \mathfrak{N}$ it is proved that $\mathfrak{I}$ is a weight module for $\mathcal{D}$ none of whose weights (= eigenvalues of H') is an integer $\geqslant 0$. This implies at once the much stronger result that 0 is the only $\partial(\omega)$-finite element of $\mathfrak{I}$. This completes the sketch of the proof of the regularity theorem on $\mathfrak{g}$; the formulae (7) and (9) then lead to the theorem on the group.

The behaviour around the singular points of the analytic functions defined by the invariant $\mathfrak{Z}$-finite (or $I(\mathfrak{g})$-finite) distributions can be studied in detail. Theorems 2, 3 and Lemma 25 of [1965a] do this on the Lie algebra while Lemmas 31, 34–37 of [1965b] treat the group situation. Let L be a Cartan subgroup and F the invariant locally summable function on Ω that defines a $\mathfrak{Z}$-finite distribution. If $\Phi_L = \Delta_L \cdot (F|_{\Omega \cap L'})$, then Φ_L extends analytically to $L'(R)$, the subset of L where no real (global) root takes the value 1. Moreover, $\varpi_L \Phi_L = \Psi_L$ extends continuously on all of $\Omega \cap L$ and the system of functions (Ψ_L) are compatible in the sense that

$$\Psi_{L_1} = \Psi_{L_2} \quad \text{on } L_1 \cap L_2 \cap \Omega. \tag{10}$$

These are generally known as the Harish-Chandra *matching conditions*; They are equivalent to the statement that $\nabla_G F$ extends to a continuous function on Ω (for the definition of ∇_G see Lemma 36 of [1965b]). Entirely analogous results are true on $\mathfrak{g}$.

§5. *Construction of the Invariant Eigendistributions* Θ_ξ. In [1965c] the regularity theorem and the theory of Fourier transforms are used to construct the distributions Θ_ξ that will be the characters of the discrete series. Let G be as in §3 with $rk(G) = rk(K)$; let $B \subset K$ be a Cartan subgroup, with $\mathfrak{b} = \text{Lie}(B)$. We select a compact form U of G_c such that $K = U \cap G$ and $B \subset U$. W and W_c denote the respective Weyl groups of (K, B) and (U, B). W_c operates on B and the $\xi \in \hat{B}$ that have trivial stabilizer in W_c are called regular. Theorem 3 of [1965c] then asserts that for any regular $\xi \in \hat{B}$ there is a unique invariant eigendistribution Θ_ξ on G such that

(a) $\quad |\Theta_\xi| \leqslant \text{const.}|D|^{-1/2} \quad$ on G'

(b) $\quad \Theta_\xi = \Delta^{-1} \Sigma_{s \in W} \varepsilon(s)(s\xi) \quad$ on $B' = B \cap G'$. $\qquad$ (1)

The Θ_ξ are obtained by exponentiating suitable invariant eigendistributions on the Lie algebra. The condition (a) is essentially equivalent to the *temperedness* of these distributions on the Lie algebra; it thus allows Fourier transforms to be used for constructing them. In fact, if η is the analogue of D on $\mathfrak{g}$ ($\eta = \pi^2$ on any Cartan subalgebra), the formula (8) of §4 shows that the conditions $|F| \leqslant \text{const.}|D|^{-1/2}$ and $|\tilde{F}| \leqslant \text{const.}|\eta|^{-1/2}$ are equivalent; the theory of orbital integrals can be used to show that the latter condition, for invariant eigendistributions with regular eigenvalues, is equivalent to temperedness. The fundamental result concerning the distributions on the Lie algebra is Theorem 2 of [1965c]: let $\lambda \in i\mathfrak{b}^*$ be regular; then there is a unique invariant distribution $T_\lambda = T_{\mathfrak{g},\lambda}$ on $\mathfrak{g}$ such that

(a) $\quad T_\lambda$ is tempered, $\partial(p)T_\lambda = p_\mathfrak{b}(\lambda)T_\lambda \quad (p \in I(\mathfrak{g}))$

(b) $\quad T_\lambda = \pi^{-1} \Sigma_{s \in W} \varepsilon(s)e^{s\lambda} \quad$ on $\mathfrak{b}' = \mathfrak{b} \cap \mathfrak{g}'$.

INTRODUCTION

Let $\mathfrak{T}_\lambda$ be the space of all tempered invariant distributions T on $\mathfrak{g}$ such that

$$\partial(p)T = p_\mathfrak{h}(\lambda)T \qquad (p \in I(\mathfrak{g})).$$

Let A_λ be the space of all functions on $\mathfrak{h}$ of the form $\sum_{s \in W_c} c_s e^{s\lambda}(c_s \in \mathbb{C})$ which are W-skew. If $T \in \mathfrak{T}_\lambda$, $\pi \cdot (T|_{\mathfrak{h}'})$ belongs to A_λ (Theorem 2, [1965a]), and the uniqueness part of the theorem above is the statement that the restriction map $R_\mathfrak{h}: T \mapsto \pi \cdot (T|_{\mathfrak{h}'})$ is injective. On the other hand, $-iH_\lambda$ is in $\mathfrak{h}$ and we have

$$(-iH_\lambda)^{G_c} \cap \mathfrak{g} = \coprod_{s \in W \setminus W_c} (-isH_\lambda)^G$$

showing that the Fourier transforms $\hat{\mu}_s(s \in W \setminus W_c)$ of the invariant measures on the G-orbits $(-isH_\lambda)^G$ are linearly independent elements of $\mathfrak{T}_\lambda$ (cf. (8) of §3). Since $\dim(A_\lambda) = [W_c: W]$, the injectivity of $R_\mathfrak{h}$ will imply that $\mathfrak{T}_\lambda$ is *spanned* by the $\hat{\mu}_s$ and that $R_\mathfrak{h}$ is an isomorphism of $\mathfrak{T}_\lambda$ with A_λ. This of course will prove Theorem 2 of [1965c].

The key result is thus the injectivity of $R_\mathfrak{h}$. To get a feeling for it consider $\mathfrak{g} = \mathfrak{sl}(2,\mathbb{R})$ with $\lambda \neq 0$ in $i\mathbb{R}$, $T \in \mathfrak{T}_\lambda$, $\partial(\omega)T = \lambda^2 T$ where $\omega = H^2 + 4YX$. Let $i\lambda > 0$. Then

$$i\theta T'(\theta(X - Y)) = c_\mathfrak{h}^+ e^{\lambda\theta} - c_\mathfrak{h}^- e^{-\lambda\theta} \qquad (0 \neq \theta \in \mathbb{R})$$

$$tT(tH) = a_\mathfrak{h}^+ e^{-i\lambda t} - a_\mathfrak{h}^- e^{i\lambda t} \qquad (t > 0).$$

The matching conditions give

$$c_\mathfrak{h}^+ + c_\mathfrak{h}^- = a_\mathfrak{h}^+ + a_\mathfrak{h}^- \tag{2}$$

while the temperedness of T implies that only bounded exponentials can appear in $tT(tH)$, so that $a_\mathfrak{h}^- = 0$. But then $a_\mathfrak{h}^+ = c_\mathfrak{h}^+ + c_\mathfrak{h}^-$, showing that T is completely determined by $T|_{\mathfrak{h}'}$.

Such an argument works in general. In Lemma 28 of [1965c] Harish-Chandra proves that if $\mathfrak{h}$ is a Cartan subalgebra and $\mathfrak{h}^+$ is a component of $\mathfrak{h}'(R)$, one can choose Cartan subalgebras $\tilde{\mathfrak{h}}$ "adjacent" to $\mathfrak{h}$ (i.e., $\dim(\tilde{\mathfrak{h}}_\mathbf{R}) = \dim(\mathfrak{h}_\mathbf{R}) - 1)$ and components $\tilde{\mathfrak{h}}^+$ of $\tilde{\mathfrak{h}}'(R)$ such that (a) $Cl(\tilde{\mathfrak{h}}^+)$ and $Cl(\mathfrak{h}^+)$ interface along $\tilde{\mathfrak{h}} \cap \mathfrak{h}$ (which is a hyperplane in both $\mathfrak{h}$ and $\tilde{\mathfrak{h}}$); (b) the coefficients of the exponentials occurring in the formula for T on $\mathfrak{h}^+$ and those coming from $\tilde{\mathfrak{h}}^+$ are related in a manner similar to (2); (c) only bounded exponentials occur. The proof that $T = 0$ on $\mathfrak{h}' \Rightarrow T = 0$ then goes by induction on $\dim(\mathfrak{h}_\mathbf{R})$. The regularity of λ is decisive here.

For technical reasons it is necessary to prove the uniqueness even when T is not everywhere defined. The domains Ω however cannot be arbitrary; roughly speaking $\Omega \cap \mathfrak{h}$ must be "unbounded in real directions." To make this precise note first that any semisimple $X \in \mathfrak{g}$ can be uniquely written as $X = X_e + X_h$ where (i) $X_h \in [\mathfrak{g}, \mathfrak{g}]$ (to allow for cases where $\mathfrak{g}$ is only reductive), X_e, X_h are semisimple and $[X_e, X_h] = 0$; (ii) ad X_e(resp. ad X_h) has only pure imaginary (resp. real!) eigenvalues; X_e is called the *elliptic component of* X. The uniqueness theorem is then true for the class $\mathcal{E}(\mathfrak{g})$ of open sets Ω with the following properties:

(a) Ω is completely invariant; if $X \in \Omega$, then $tX \in \Omega$ for $0 \leqslant t \leqslant 1$;
(b)   if $X \in \mathfrak{g}$ is semisimple, then Ω contains all semisimple elements X' of $\mathfrak{g}$ such that $X'_e = X_e$ (in particular $X_e \in \Omega$).

Typical examples of sets in $\mathcal{E}(\mathfrak{g})$ are $\mathfrak{g}(\varepsilon)$ ($\varepsilon > 0$), the subset of all X in $\mathfrak{g}$ such that $|\mathrm{Im}\,\lambda| < \varepsilon$ for any eigenvalue λ of $\mathrm{ad}\,X$. If $\mathfrak{z} \subset \mathfrak{g}$ is a reductive subalgebra containing $\mathfrak{b}$, then $\mathfrak{z} \cap \mathfrak{g}(\varepsilon) \in \mathcal{E}(\mathfrak{z})$. These are the domains that occur in [1965c].

In lifting results to G one can use the exponentiation only around *elliptic* elements of G; for, only these have centralizers in $\mathfrak{g}$ that admit Cartan subalgebras of compact type. For any $b \in B$ let $\mathfrak{z}_b$ be the centralizer of b in $\mathfrak{g}$; and for any $\varepsilon > 0$, let

$$G_\varepsilon^{(b)} = \left(b\exp(\mathfrak{z}_b \cap \mathfrak{g}(\varepsilon))\right)^G. \tag{4}$$

The sets $G_\varepsilon^{(b)}$, which are used by Harish-Chandra in the lifting process, have remarkable properties. They are of course completely invariant, open if $\varepsilon > 0$ is sufficiently small. Moreover, if $\varepsilon_b > 0$ are arbitrary numbers, $(b \in B)$,

$$G = \bigcup_{b \in B} G_{\varepsilon_b}^{(b)} \tag{5}$$

while for $b', b'' \in B$ and $\varepsilon', \varepsilon'' > 0$ (small enough),

$$G_{\varepsilon'}^{(b')} \cap G_{\varepsilon''}^{(b'')}$$

is the union of sets of the form $G_\varepsilon^{(b)}$. The uniqueness principle on the Lie algebras $\mathfrak{z}_b$ (in their localized version) will then give, in view of these properties, the uniqueness, not only on the whole group, but also for invariant $\mathfrak{z}$-finite distributions satisfying (a) and (b) of (1), but defined only on $G_\varepsilon^{(b)}$. For the existence, a suitable linear combination of $T_{\mathfrak{z}_b, s\lambda}$ ($s \in W$) will lift to a distribution $\Theta_\xi^{(b)}$ on $G_{\varepsilon_b}^{(b)}$ satisfying (a) and (b) of (1) on it if $\varepsilon_b > 0$ is sufficiently small; the compatibility of these on the overlaps of the $G_{\varepsilon_b}^{(b)}$ is a consequence of the strengthened form of the uniqueness mentioned above, and so, in view of (5), we would have constructed Θ_ξ on all of G (cf. [V 3], Part II, §§2, 5).

The technique of using the matching conditions on the interfaces of adjacent Cartan subalgebras, which was used in the proof of uniqueness, has other applications. As an illustration, let us consider the distribution

$$\Theta_\xi^* = \sum_{s \in W_c} \varepsilon(s)\Theta_{s\xi}. \tag{6}$$

On B, Θ_ξ^* coincides, up to a constant factor, with the character of an irreducible finite dimensional representation of U. In particular, it is bounded and W_c-invariant. If L is any other Cartan subgroup, and $W_{L,I}$ is the subgroup of the Weyl group of (G_c, L_c) generated by the reflexions associated to the imaginary roots, the technique mentioned above can be used to prove by induction on $\dim(A)$ (where A is the $\mathbb{R}$-split component of L), that $\Theta_\xi^*|_{L'}$ is $W_{L,I}$-invariant.[†] This will imply that

$$\sup_{L'} \left(|D_L|^{1/2}|\Theta_\xi^*|\right) < \infty \tag{7}$$

where

$$D_L(x) = \det\left(\left(Ad(x^{-1})-1\right)_{\mathfrak{g}/\mathfrak{z}}\right)$$

[†] It can be shown that $W_{L,I}$ operates on L and stabilizes each component of $L'(R)$; see p. 307, [1965c].

($\mathfrak{z}$ = centralizer of A in $\mathfrak{g}$) (§24, [1965c]). We have already seen in §3 the decisive role of the distributions Θ_{ξ}^{*} and the estimate (7) in the problem of extending the theory of orbital integrals to $\mathcal{C}(G)$.

§6. *Eigenfunction Asymptotics, Weak Inequality, and Analysis in the Schwartz Space.* I shall conclude this introduction with a few comments on the Schwartz space and Harish-Chandra's treatment of the asymptotic behaviour of the matrix elements of the irreducible unitary representations and their wave packets [1958a] [1958b] [1966b] [1975] [1976a]. Let $G = KA_0N_0$ be an Iwasawa decomposition with associated (minimal) parabolic subgroup $P_0 = M_0A_0N_0$; A_0^+, the positive chamber of A_0; $\mathfrak{a}_0 = \mathrm{Lie}(A_0)$ and $\rho_0 \in \mathfrak{a}_0^*$ defined as usual by $\rho_0(H) = \frac{1}{2}\mathrm{tr}(\mathrm{ad}\,H)_{\mathfrak{n}_0}$ where $\mathfrak{n}_0 = \mathrm{Lie}(N_0)$. We write σ for the spherical function on G such that $\sigma(a) = \|\log a\|(a \in A_0)$; $\sigma(x)$ is then the distance from K to xK in the Riemannian space G/K.

The concern in [1958a] [1958b] was with the elementary spherical functions φ_ν. The basic estimate for them is (Theorem 3, [1958a]) the following

$$|\varphi_\nu(a)| \leqslant Ce^{-\rho_0(\log a)}(1 + \sigma(a))^m \qquad (a \in A_0^+, \nu \in \mathfrak{a}_0^*)$$

where $C > 0$, $m \geqslant 0$ are constants. In [1958b] it was proved that wave packets f of the φ_ν satisfy, for each $m \geqslant 0$,

$$|f(a)| \leqslant C_m e^{-\rho_0(\log a)}(1 + \sigma(a))^{-m} \qquad (a \in A_0^+)$$

$C_m > 0$ being a constant. Since the Jacobian for the polar decomposition $G = K\,Cl(A_0^+)K$ behaves roughly as $e^{2\rho_0}$ on A_0^+, these estimates show that the φ_ν are in $L^{2+\varepsilon}(G)$ for every $\varepsilon > 0$ and that the wave packets are actually in $L^2(G)$. They motivate to some extent the following definition ([1970c], p. 539): a continuous function f on G is said to satisfy the *weak inequality* if there are constants $C > 0$, $m \geqslant 0$, such that

$$|f(k_1ak_2)| \leqslant Ce^{-\rho_0(\log a)}(1 + \sigma(a))^m \qquad (k_1, k_2 \in K,\ a \in A_0^+). \tag{1}$$

The Schwartz space $\mathcal{C}(G)$ is then defined as the space of all $f \in C^\infty(G)$ such that for $u, v \in U(\mathfrak{g})$ and any $m \geqslant 0$, the two-sided derivative ufv of f satisfies

$$|(ufv)(k_1ak_2)| \leqslant Ce^{-\rho_0(\log a)}(1 + \sigma(a))^{-m} \qquad (k_1, k_2 \in K,\ a \in A_0^+) \tag{2}$$

for some constant $C > 0$.

Actually, if we write $\varphi_0 = \Xi$, then

$$e^{-\rho_0(\log a)} \leqslant \Xi(a) \leqslant 1 \qquad (a \in A_0^+) \tag{3a}$$

so that in (1) and (2) we can replace $e^{-\rho_0(\log a)}$ by $\Xi(a)$. Since Ξ has good properties from the point of view of harmonic analysis, this change improves the formal aspects of the theory also (cf. the proof that $\mathcal{C}(G)$ is a topological algebra under convolution, in §14, [1975]). In particular

$$\int_G \Xi^2(1 + \sigma)^{-m}\,dx < \infty \tag{3b}$$

if $m \gg 0$.

The space $\mathcal{C}(G)$ gives rise to a natural notion of tempered distributions on G. The fundamental nature of the weak inequality is then revealed by Theorem 14.1 of [1975]: for a K-finite $\mathfrak{Z}$-finite function on G, weak inequality is equivalent to temperedness. We also mention the corresponding results for class functions. For invariant $\mathfrak{Z}$-finite distributions Θ temperedness is equivalent to the estimate

$$|\Theta| \leqslant \text{const.}|D|^{-1/2}(1+\sigma)^m \qquad \text{(pointwise on } G') \tag{4}$$

for some $m \geqslant 0$, and then

$$\Theta(f) = \int_G \Theta f \, dx \qquad (f \in \mathcal{C}(G)) \tag{5}$$

the integral converging absolutely; the counterpart to (3b) is

$$\int_G |D|^{-1/2} \Xi (1+\sigma)^{-m} dx < \infty \tag{6}$$

for $m \gg 0$ (cf. [1975], §§10–14; see also the introductory remarks to [1970a]). In (4) the factor $(1+\sigma)^m$ can be dropped if Θ is an eigenfunction for a *regular* infinitesimal character, thereby elucidating the meaning of the condition $\sup|D|^{1/2}|\Theta_\xi| < \infty$ in the construction of the distributions Θ_ξ. Temperedness (of the characters as well as the matrix elements) is the characteristic property of the series of representations $(\pi_{\eta,\nu})$ associated to the various Cartan subgroups of G.

In [1958b] Harish-Chandra discovered that the differential equations satisfied by the φ_ν become (roughly speaking), at infinity on A_0, differential equations with *constant coefficients*. This fact led to the result that φ_ν is asymptotically just a sum of plane waves. In [1966b] Harish-Chandra extended this method to handle all tempered K-finite $\mathfrak{Z}$-finite functions. Let τ be a finite dimensional double unitary representation of K and let $\mathcal{C}(G, \tau)$ be the space of all τ-spherical tempered $\mathfrak{Z}$-finite functions on G. Let $f \in \mathcal{C}(G, \tau)$. Fix a parabolic subgroup $P = MAN$. Then $G = K(MA)K$, and one can use the τ-sphericity of f to reduce the differential equations satisfied by f on G to a system of differential equations on MA; in the limit on MA, when $a \in A$ and $a \to {}_P\infty$ (this means that a goes to infinity in such a way that for some $\varepsilon > 0$, $\alpha(\log a) \geqslant \varepsilon\sigma(a)$ for all roots α of (P, A)), the latter differential equations become the ones that define elements of $\mathcal{C}(MA, \tau_M)$. This suggests that f may be approximated by an element of $\mathcal{C}(MA, \tau_M)$ at infinity on MA. That this is true is the content of Theorem 21.1 of [1975]; $f_P \in \mathcal{C}(MA, \tau_M)$ defined there is the constant term of f along P. The fact that f_P is a good approximation to f in a suitable regime is contained in the following estimate (Lemma 23.4 [1975]):

$$|e^{\rho_0(\log a)} f(a) - e^{\rho_{00}(\log a)} f_P(a)| \leqslant C(1+\sigma(a))^m e^{-\kappa\rho_0(\log a)}; \tag{7}$$

here ρ_{00} is the "ρ_0" of MA, $\kappa > 0$, $C > 0$, $m \geqslant 0$ are constants, and a is to be restricted to regions of the following form ($\varepsilon > 0$ can be arbitrary)

$$A_0^+(P:\varepsilon) = \{a \in Cl(A_0^+)|\alpha(\log a) \geqslant \varepsilon\rho_0(\log a) \quad \text{for all roots } \alpha \text{ of } (P, A)\}. \tag{8}$$

We have already seen in §2 how decisive the theory of the constant term is in the treatment of the discrete spectrum. In particular the estimates (7) on sets $A_0^+(P:\varepsilon)$

show that for an $f \in \mathcal{C}(G, \tau)$,

$$f \in L^2(G, \tau) \Leftrightarrow f \in \mathcal{C}(G, \tau) \Leftrightarrow f_P = 0 \quad \text{for all } P.$$

Let me mention now another illustration of the method of differential equations. Let g be in $\mathcal{C}(G) \cap {}^0L^2(G)$. We can write $g = \sum_j f_j$ where the f_j are K-finite eigenfunctions, the sum being convergent in $L^2(G)$. The f_j of course have vanishing constant terms and so decay exponentially in view of (7); by being careful in the analysis that led to (7) one can keep track of the dependence of (7) on the parameters of f_j (the class $\omega \in \hat{G}_d$ to which f_j belongs, and the K-types according to which f_j transforms). The result is that the series $\sum_j f_j$ is convergent *in the topology of the Schwartz space*. On the other hand, the f_j, being $\mathfrak{Z}$-finite functions in $\mathcal{C}(G)$, are cusp forms (Theorem 18.1, [1975]) so that g is a cusp form, showing that $\mathcal{C}(G) \cap {}^0L^2(G)$ is precisely the space of cusp forms ([1970c], §8; [V 3], Part II, §16).

In the theory of the continuous spectrum one considers families of eigenfunctions, for instance the Eisenstein integrals. Harish-Chandra introduces the concept of (τ-spherical) eigenfunctions (ϕ_ν) of type $\text{II}(\lambda)$ associated to a Cartan subgroup $L = T \cdot A$ (§8, [1976a]) here $\lambda \in it^*$ ($t = \text{Lie}(T)$) is regular, ν varies in $\mathfrak{a}^*$ and ϕ_ν is smooth in ν, satisfies the weak inequality in a uniform way, and satisfies the differential equation

$$z\phi_\nu = \mu_{\mathfrak{g}/\mathfrak{h}}(z)(\lambda + i\nu)\phi\nu \qquad (z \in \mathfrak{Z}).$$

The fundamental idea now is to form wave packets

$$\phi_\alpha = \int_{\mathfrak{a}^*} \alpha(\nu)\phi_\nu \, d\nu \qquad (\alpha \in \mathcal{C}(\mathfrak{a}^*)) \tag{9}$$

and to try to prove that $\phi_\alpha \in \mathcal{C}(G, \tau)$. This may not always be true even if $\alpha \in C_c^\infty(\mathfrak{a}^*)$. The point is that the constant terms $(\phi_{P,\nu})(P \in \mathcal{P}(A))$ are not eigenfunctions on MA but only linear combinations of such (this is because the asymptotic form on MA of the differential operator $z \in \mathfrak{Z}$ is $\mu_{\mathfrak{g}/\mathfrak{h}}(z)$, and $\mu_{\mathfrak{g}/\mathfrak{h}}(\mathfrak{Z}) \neq \mathfrak{Z}_{MA}$, the center of $U(\mathfrak{m} \oplus \mathfrak{a})$). For regular $\nu \in \mathfrak{a}^*$ one can write $\phi_{P,\nu}$ as $\sum_{s \in W(A)}\phi_{P,s,\nu}$; the $\phi_{P,s,\nu}$ will now be eigenfunctions (these statements have to be formulated with some care; see §§8–9 of [1976a]) but they may become singular in ν if ν approaches root hyperplanes in $\mathfrak{a}^*$ (for instance, in the spherical case, they are $c(s\nu)e^{is\nu}$). To overcome this, Harish-Chandra introduces the concept of eigenfunctions (ϕ_ν) of type $\text{II}'(\lambda)$ in §9, loc.cit, and proves that if (ϕ_ν) is of type $\text{II}'(\lambda)$, $\phi_\alpha \in \mathcal{C}(G, \tau)$ for any $\alpha \in \mathcal{C}(\mathfrak{a}^*)$, and $\alpha \mapsto \phi_\alpha$ is a continuous map (Theorem 13.1, loc.cit). The usefulness of this concept becomes clear from Theorem 9.1 of loc.cit asserting that if (ϕ_ν) is of type $\text{II}(\lambda)$, $(\varpi(\lambda + i\nu)\phi_\nu)$ is of type $\text{II}'(\lambda)$. Theorem 11.1 of loc.cit actually gives a necessary and sufficient condition that (ϕ_ν) which is of type $\text{II}(\lambda)$, is actually of type $\text{II}'(\lambda)$.

The wave packet theorem is proved by induction on $\dim(G)$. The basic point of course is to check that if (ϕ_ν) is of type $\text{II}'(\lambda)$ on G, the $(\phi_{P,s,\nu})$ lead to eigenfunctions of type $\text{II}'(\lambda)$ on MA (Lemma 9.1, [1976a]). If one had simply restricted oneself to Eisenstein integrals, the inductive step would have become much more complicated.

The wave packets of $E_{P,\psi,\nu}$ define *essentially* the inverse of the Fourier transform. Essentially, because we are using only $d\nu$ as the weight and not the Plancherel measure; or (equivalently), because we have not yet determined the true normalization of the $E_{P,\psi,\nu}$. To correct this one has to compute the Fourier transform of the wave packet ϕ_α and relate it to α. Theorem 13.2 of [1976a] does this. It is actually a generalization of the corresponding theorem 4 of [1958b] for spherical functions, and is based on the same circle of ideas. What it says is that if $\bar{P}$ is the parabolic subgroup opposite to P, the integral

$$\phi_\alpha^{\bar{P}}(m) = d_P(m) \int_{\bar{N}} \phi_\alpha(\bar{n}m)\, d\bar{n} \qquad (m \in MA)$$

can be evaluated by substituting for ϕ_α the "truncated wave packets"

$$\phi_{P,\alpha}(m) = \int_{\mathfrak{a}^*} \alpha(\nu)\phi_{P,\nu}(m)\, d\nu \qquad (m \in MA)$$

(which are obtained by replacing ϕ_ν by its constant term in the formation of the wave packet). If (ϕ_ν) is the Eisenstein integral associated to a $Q \in \mathscr{P}(A)$, $\phi_{P,\nu}$ is expressible as a sum of plane waves $c_{P|Q}(s: \nu)e^{is\nu}$, and the integral of the truncated wave packet has an explicit formula in which the c-functions enter directly (Theorem 19.2, [1976b]). The theory of the c-functions and the functional equations of the Eisenstein integrals are now combined to derive the explicit Plancherel formula. I should mention however one technical difficulty in forming wave packets with the properly normalized weight function, namely, the control of the growth of the Plancherel measure at infinity on $\mathfrak{a}^*$; to overcome this one already needs the product formula and the explicit calculations in the case when $\dim(A)$ is equal to 1. I refer the reader to [1976b] for details.

References

[AM]   Auslander, L., and Moore, C. C., Unitary representations of solvable Lie groups. *Memoirs of the Amer. Math. Soc.* **62** (1966), 1–199.

[B]   Bargmann, V., Irreducible unitary representations of the Lorentz group, *Ann. of Math.* **48** (1947), 568–640.

[Bl]   Blattner, R. J., *General Background. Harmonic Analysis and Representations of Semisimple Lie Groups.* Edited by J. A. Wolf, M. Cahen, and M. De wilde. D. Reidel Publishing Company, Holland, 1980, pp. 1–67.

[Bo 1]   Borel, A., *Introduction aux groupes arithmétiques.* Hermann, Paris, 1969.

[Bo 2]   —, *Formes automorphes et séries de Dirichlet.* Springer Lecture Notes **514** (1976), pp. 183–222.

[Br 1]   Bruhat, F., Sur les représentations induites des groupes de Lie. *Bull. Soc. Math. France,* **84** (1956), 97–205.

[Br 2]   —, Sur les représentations des groupes classiques p-adiques, I, II. *Amer. J. of Math.* **83** (1961), 321–338, 343–368.

[G]   Gel'fand, I. M., Automorphic functions and the theory of representations. *Proc. Int. Cong. of Mathematicians, Stockholm, 1962,* pp. 74–85.

INTRODUCTION

[GG 1]   Gel'fand, I. M., and Graev, M. I., On a general method of decomposition of the regular representation of a Lie group into irreducible representations. *Dokl. Akad. Nauk. SSSR*, **92** (1953), 221–224.

[GG 2]   —, The analogue of Plancherel's theorem for real unimodular groups. *Dokl. Akad. Nauk, SSSR* **92** (1953), 461–464.

[GN 1]   Gel'fand, I. M., and Neumark, M. A., Unitary representations of the Lorentz group. *Izvestiya. Akad. Nauk. SSSR* **11** (1947), 411–504.

[GN 2]   —, Unitary representations of the classical groups. *Trudy Mat. Inst. Steklova*, **36** (1950), 1–288.

[GP]   Gel'fand, I. M., and Piatetsky-Shapiro, I. I., Unitary representations in the G/Γ space, where G is the group of real n^{th} order matrices and Γ is the subgroup of integral matrices. *Dokl. Akad. Nauk. SSSR* **147** (1962), 275–278.

[GR]   Gel'fand, I. M., and Raikov, D. A., Irreducible unitary representations of arbitrary locally bicompact groups. *Mat. Sbornik* (N.S) **13** (55) (1943), 301–316.

[Go]   Godement, R., Sur les relations d'orthogonalité de V. Bargmann. I.II. I: Résultats préliminaires, *C.R. Acad. Sci. Paris.* **225** (1947), 521–523; II: Démonstration génerale, *C.R. Acad. Sci. Paris*, **225** (1947), 657–659.

[JL]   Jacquet, H., and Langlands, R. P., *Automorphic Forms on* GL (2), Springer Lecture Notes **114** (1970).

[L 1]   Langlands, R. P., *On the Functional Equations Satisfied by Eisenstein Series.* Springer Lecture Notes **544** (1976).

[L 2]   —, *Euler Products.* Yale University Press, 1967.

[L 3]   —, *Problems in the Theory of Automorphic Forms. Lectures in Modern Analysis and Applications*, Springer Lecture Notes, **170** (1970), 18–86.

[Lo]   Loomis, L. H., *An Introduction to Abstract Harmonic Analysis.* Van Nostrand, New York, 1953.

[M]   Maass, H., Über eine neue Art von nichtanalytischen automorphen Funktionen. *Math. Ann.* **121** (1949), 141–183.

[Mau 1] Mautner, F. I., Spherical functions over p-adic fields, I. *Amer. J. of Math.*, **80** (1958), 441–457.

[Mau 2] —, Spherical functions over p-adic fields, II. *Amer. J. of Math.* **86** (1964), 171–200.

[Ma 1]   Mackey, G. W., *The Theory of Unitary Group Representations. Chicago Lectures in Mathematics.* The Chicago University Press, Chicago and London, 1976.

[Ma 2]   —, Infinite dimensional group representations. *Bull. Amer. Math. Soc.* **69** (1963), 628–686.

[Ma 3]   —, *Unitary Group Representations in Physics, Probability, and Number Theory.* Benjamin, 1978.

[Mo]   Moore, C. C., Representations of Solvable and nilpotent groups and harmonic analysis on nil and solvmanifolds. Harmonic analysis on homogeneous spaces. *Proceedings of Symposia in Pure Mathematics*, **XXVI**. Edited by C. C. Moore. *Amer. Math. Soc.* 1973, pp. 3–44.

[MvN]   Murray, F. J., and von Neumann, J., On rings of operators. I, II, IV. I: *Ann. of Math.* **37** (1936), 116–229; II: *Trans. Amer. Math. Soc.* **41** (1937), 208–248; IV: *Ann. of Math.* **44** (1943), 716–808.

[Se 1]   Selberg, A., Harmonic analysis and discontinuous groups in weakly symmetric Riemannian spaces with applications to Dirichlet series. *J. Indian Math. Soc.* **20** (1956), 47–87.

[Se 2]   —, Discontinuous groups and harmonic analysis. *Proc. Int. Cong. of Mathematicians, Stockholm* (1962), 177–189.

[S]   Schwartz, L., *Théorie des Distributions*, Hermann, Paris, 1973 (Nouvelle édition).

[T]   Tate, J., Fourier analysis in number fields and Hecke's Zeta functions. Thesis (Princeton), 1950. Reproduced in *Algebraic Number Theory*, Edited by J. W. S. Cassels and A. Frölich, Thompson Book Company Inc., Washington D.C., 1967.

[vN]   von Neumann, J., (a) On rings of operators III. *Ann. of Math.* **41** (1940), 94–161. (b) On rings of operators. Reduction theory. *Ann. of Math.* **50** (1949), 401–485.

[V 1]   Varadarajan, V. S., *Lie Groups, Lie Algebras, and Their Representations*. Prentice Hall, Englewood Cliffs, N.J., 1974.

[V 2]   —, The theory of characters and the discrete series for semisimple groups. Harmonic analysis on homogeneous spaces, *Proceedings of Symposia in Pure Mathematics*, **XXVI**. Edited by C. C. Moore. *Amer. Math. Soc.* 1973, pp. 45–99.

[V 3]   —, *Harmonic Analysis on Real Reductive Groups*. Springer Lecture Notes **576** (1976).

[W 1]   Weil, A., *L'integration dans les groupes topologiques et ses applications*. Hermann, Paris, 1940.

[W 2]   —, *Basic Number Theory*. Springer-Verlag, New York, 1967.

[W 3]   —, (a) Sur certains groupes d'opérateurs unitaires *Acta Math.* **111** (1964), 143–211. (b) Sur la formule de Siegel dans la théorie des groupes classiques. *Acta Math.* **113** (1965), 1–87.

[W 4]   —, Über die bestimmung Dirichletscher Reihen durch Funktionalgleichungen. *Math. Ann.* **168** (1967), 149–156.

[W 5]   —, *Automorphic Forms and Dirichlet Series*. Springer Lecture Notes **189** (1971).

[We 1]   Weyl, H., Über gewöhnliche Differentialgleichungen mit singularitäten und die zugehorigen Entwicklungen willkürlicher Funktionen. *Math. Ann.* **68** (1910), 220–269.

[We 2]   —, Theorie der Darstellung kontinuierlicher halbeinfacher Gruppen durch lineare Transformationen. I, II, III, und Nachtrag. I: *Math. Zeist.* **23** (1925), 271–309; II: *Math. Zeist.* **24** (1926), 328–376; III: *Math. Zeist.* **24** (1926), 377–395; Nachtrag: *Math. Zeist.* **24** (1926), 789–791.

[Wi]   Wigner, E. P., On unitary representations of the inhomogeneous Lorentz group. *Ann. Math.* **40** (1939), 149–204.

Some Additional Aspects of Harish-Chandra's Work on Real Reductive Groups

Nolan R. Wallach

The main emphasis in the introduction to these volumes is the work of Harish-Chandra which led to the Plancherel theorem for reductive groups over local fields. This work did not historically follow a straight line and it included important results that are not directly related to the main theme. We include here a partial guide to some of these theorems and ideas. We label each paper to be discussed using the scheme of the Bibliography, e.g. [1951a].

[1951a] In this paper Harish-Chandra shows how to simultaneously construct a semisimple Lie algebra and all of its irreducible finite dimensional representations from its Cartan matrix. Recall that a Cartan matrix is an integral $l \times l$ matrix

$$A = (a_{ij})$$

with

(1) $a_{ii} = 2$, $a_{ij} \leqslant 0$, $i \neq j$;
(2) if $a_{ij} = 0$, then $a_{ji} = 0$;
(3) $\det(A) \neq 0$;
(4) the group generated by the transformations

$$s_i \colon x_j \mapsto x_j - a_{ji} x_i$$

     is finite.

Starting with such a matrix A he studies the algebra (either associative or Lie) with generators X_i, Y_i, H_i, $1 \leqslant i \leqslant l$, and the relations

(R)
$$[H_i, H_j] = 0, \qquad [X_i, Y_j] = \delta_{ij} H_i$$
$$[H_i, X_j] = a_{ji} X_j, \qquad [H_i, Y_j] = -a_{ji} Y_j.$$

"

At first Harish-Chandra studies the associative algebra subject to (R). He later looks at the Lie algebra subject to (R) (the Lie subalgebra generated by the X_i, Y_i, H_i in the associative algebra). Given a linear functional Λ on the span of the H_i with $\Lambda(H_i) = n_i$, $n_i \in \mathbb{Z}$, $n_i \geq 0$, he constructs a module for the algebra by taking a suitable quotient of what has come to be called a Verma module. He then takes its unique irreducible quotient (which we denote), $L(\Lambda)$. If $\mathfrak{g}'$ is the Lie algebra determined by the relations (R) and if $I = \{ X \in \mathfrak{g}' | X$ acts by zero on every $L(\Lambda)\}$, then $\mathfrak{g} = \mathfrak{g}'/I$ defines the semisimple Lie algebra. The $L(\Lambda)$ are the finite dimensional irreducible representations of $\mathfrak{g}$. Parts of this argument are attributed by Harish-Chandra to Chevalley.

It is of some interest that the conditions (3), (4) are used only to guarantee that $\dim \mathfrak{g} < \infty$ and $\dim L(\Lambda) < \infty$. If the conditions (3) and (4) are dropped, the construction of Harish-Chandra gives the Kac-Moody Lie algebras and their standard modules.

[1953] In the first part of this paper Harish-Chandra initiates the algebraic theory of $(\mathfrak{g}, \mathfrak{k})$-modules (or Harish-Chandra modules). Here $\mathfrak{g}$ is a semisimple Lie algebra over the real numbers, θ is a Cartan involution of $\mathfrak{g}$, and $\mathfrak{k}$ is the fixed point algebra of θ. A $(\mathfrak{g}, \mathfrak{k})$-module is a $\mathfrak{g}$-module, M, over $\mathbf{C}$ that splits into a (not necessarily finite) direct sum of irreducible $\mathfrak{k}$-modules. Harish-Chandra calls such modules quasi-semisimple for $\mathfrak{k}$. The basic theorem in the first part is Theorem 1 which may be stated as follows. Let $U(\mathfrak{g}_{\mathbf{C}}) \supset U(\mathfrak{k}_{\mathbf{C}})$ be the universal enveloping algebras of $\mathfrak{g}_{\mathbf{C}}$ and $\mathfrak{k}_{\mathbf{C}}$ respectively. Let $Z(\mathfrak{g}_{\mathbf{C}})$ be the center of $U(\mathfrak{g}_{\mathbf{C}})$. The theorem then says: if W_1, W_2 are finite dimensional semisimple $\mathfrak{k}_{\mathbf{C}}$-modules then

$$\mathrm{Hom}_{\mathfrak{k}}\left(W_1, U(\mathfrak{g}_{\mathbf{C}}) \underset{U(\mathfrak{k}_{\mathbf{C}})}{\otimes} W_2 \right)$$

is finitely generated as a $Z(\mathfrak{g}_{\mathbf{C}})$-module (the action of $Z(\mathfrak{g}_{\mathbf{C}})$ is $(z \cdot T)(v) = z \cdot T(v)$).

This theorem in particular implies that a $(\mathfrak{g}, \mathfrak{k})$-module M that is finitely generated and is $Z(\mathfrak{g}_{\mathbf{C}})$-finite ($\dim Z(\mathfrak{g}_{\mathbf{C}}) \cdot m < \infty$, $m \in M$) has finite $\mathfrak{k}$-multiplicities, i.e., is admissible. This implication is the starting point for the theory of admissible $(\mathfrak{g}, \mathfrak{k})$-modules. For the special case when $\mathfrak{g}$ is a complex Lie algebra looked upon as a real Lie algebra this result (as well as the basic idea of its proof) is in [1951a]. In [1951a] Harish-Chandra indicates that the result in this special case was suggested to him by Mautner.

In the second part of this paper Harish-Chandra initiates the theory of analytic vectors for Banach space representations of Lie groups. These vectors are called well-behaved in this paper. Let (π, H) be a Banach representation of a connected Lie group G. Let H^ω be the space of analytic vectors. His main results are that H^ω is dense in H and is $\mathfrak{g}$-invariant, and that the closure of a $\mathfrak{g}$-invariant subspace of H^ω is G-invariant.

Let G be connected semisimple with finite center and let $K \subset G$ be the analytic subgroup of G corresponding to $\mathfrak{k}$. Let (π, H) be a Banach representation of G such that if $v \in H^\omega \dim Z(\mathfrak{g}_{\mathbf{C}})v < \infty$, and suppose that there exist $v_i \in H^\omega \cap H_K$, $1 \leq i \leq N$ such that H is the closed linear span of $\pi(x)v_i$, $x \in G$, $1 \leq i \leq N$; here $H_K = \{v \in H | \pi(K) \cdot v$ spans a finite dimensional space$\}$. Then Theorem 1 and the theory of analytic vectors combine to prove

(1) $H_K \subset H^\omega$; all the K-multiplicities in H are finite.

(2) $V \mapsto \mathrm{Cl}(V)$ is a bijection between the lattice of $\mathfrak{g}$-stable subspaces of H_K and the lattice of G-stable closed subspaces of H.

These results are then used by him to prove that a connected semisimple Lie group is of type I in the sense of Murray and von Neumann.

[1954a] This paper contains the proof of his celebrated subquotient theorem. Let G be linear, connected and semisimple. The theorem asserts that any irreducible $(\mathfrak{g}, \mathfrak{k})$-module that integrates to a K-module is isomorphic to a subquotient of the space of K-finite vectors of a representation (in general nonunitarily) induced from a one dimensional representation of an Iwasawa subgroup of G. He was later able to drop the linearity condition by using his results on differential equations [1960a].

[1954c] The main result of this paper is the Plancherel theorem for semisimple Lie groups defined over the field of complex numbers. Theorems 1 and 2 give the formula for the characters of the principal series representations for arbitrary connected semisimple Lie groups over $\mathbb{R}$, with finite center. The Plancherel theorem is based on this character computation and an integro-differential formula (Lemma 14) which generalizes earlier work of Gel'fand and Naimark. He generalized Lemma 14 to arbitrary semisimple Lie algebras in [1957c, d]; this provided one of the main steps in his proof of the Plancherel theorem for real reductive groups (see Varadarajan's introduction). Lemma 10 and its Corollary are important ingredients in the proofs of Lemma 14 and Theorem 2. He later used this result in his proof [1956d] of the Bruhat lemma.

[1956b] Section 7 of this paper is titled "A Digression on a Theorem of Cartan". Harish-Chandra considers the case $\mathfrak{g}_\mathbb{C}$ simple, and $\mathfrak{g}_\mathbb{C} = \mathfrak{k}_\mathbb{C} \oplus \mathfrak{p}_\mathbb{C}$ (complexified Cartan decomposition) with $\mathfrak{p}_\mathbb{C} = \mathfrak{p}^+ \oplus \mathfrak{p}^-$ a direct sum of two Ad K-invariant nonzero subspaces. He had earlier shown [1955c] that there is $J \in \mathfrak{k}$ with ad $J|_{\mathfrak{p}^+} = iI$, ad $J|_{\mathfrak{p}^-} = -iI$, ad $J|_{\mathfrak{k}} = 0$. Thus $[\mathfrak{p}^+, \mathfrak{p}^+] = [\mathfrak{p}^-, \mathfrak{p}^-] = 0$. Let G be linear; then G is contained in $G_\mathbb{C}$, a connected Lie group with Lie algebra $\mathfrak{g}_\mathbb{C}$. By going to a covering one may assume that $G_\mathbb{C}$ is simply connected. In $G_\mathbb{C}$ there is the connected parabolic subgroup

$$K_\mathbb{C} \exp \mathfrak{p}^+.$$

In [1956a] it is shown that

$$G K_\mathbb{C} \exp \mathfrak{p}^+ \subset (\exp \mathfrak{p}^-) K_\mathbb{C} (\exp \mathfrak{p}^+)$$

is an open subset, and that

$$G \cap (K_\mathbb{C} \exp \mathfrak{p}^+) = K.$$

One therefore has

$$G K_\mathbb{C} \exp \mathfrak{p}^+ = \exp(\Omega) K_\mathbb{C} \exp \mathfrak{p}^+$$

with $\Omega \subset \mathfrak{p}^-$ an open subset.

This gives rise to a diffeomorphism

$$\psi\colon G/K \to \Omega$$

given by

$$(\exp\psi(g))K_{\mathbb{C}}\exp\mathfrak{p}^+ = gK_{\mathbb{C}}\exp\mathfrak{p}^+$$

Using ad J (defined above) one shows that G/K has a G-invariant complex structure and it is easy to see that $\psi\colon G/K \to \Omega$ is complex analytic. Using his theory of strongly orthogonal roots and Lemma 20 Harish-Chandra shows that Ω is a bounded, symmetric domain in $\mathfrak{p}^-$. This gives a proof without going through classification of Cartan's basic theorem that every Hermitian symmetric domain is biholomorphic with a bounded symmetric domain. The map ψ is now called the Harish-Chandra imbedding and is the starting point in the study of Hermitian symmetric domains.

In this paper there is also a very interesting argument which transfers the calculation of an integral on G to the calculation of a corresponding integral on a compact form of $G_{\mathbb{C}}$ (see Lemma 28).

[1960a] This previously unpublished paper extends work in [1958a, b]. In this paper Harish-Chandra generalizes the classical theory of regular singular points and Frobenius's method to several variables. Most notable is his handling of what is now called the asymptotic expansion along the walls. These results were basic to the Langlands classification of irreducible $(\mathfrak{g},\mathfrak{k})$-modules.

[1962] In this paper with A. Borel a general reduction theory of arithmetic subgroups of semisimple Lie groups over $\mathbb{R}$ is developed. Theorem 9.4 says that the volume of a fundamental domain for such a discrete subgroup is finite. Theorem 11.8 (which gives a positive answer to Godement's conjecture) is now called the Borel–Harish-Chandra criterion for compactness of a fundamental domain. It says that if $\Gamma \subset G$ is arithmetic then G/Γ is compact if Γ has no unipotent elements. This criterion is now the main method of proving compactness for arithmetic quotients.

[1976b] This paper is the culmination of Harish-Chandra's work on the Plancherel formula for real reductive groups. In addition to the completion of the proof of the Plancherel theorem this paper contains the completeness theorem for intertwining operators for cuspidal principal series representations (Theorem 37.1). This theorem is the basic ingredient in the analysis of the irreducibility of the cuspidal principal series representations.

[1983] The purpose of this paper is to study the basic ingredients that will be necessary for the spectral analysis of tempered, invariant distributions (for example the orbital integrals) on a real reductive group G. As in the case of the Plancherel theorem (which is concerned with the spectral analysis of the Dirac delta function at the identity of G) it is necessary to have the correct notion of the constant term. The theory of the constant term of a tempered, $Z(\mathfrak{g})$-finite, central distribution on G is one of the main themes of this article.

Let Θ be a tempered, invariant, $Z(\mathfrak{g})$-finite distribution on G. Let $K \subset G$ be a maximal compact subgroup of G. Let P be a proper parabolic subgroup of G with standard Levi decomposition $P = MN$. The constant term of Θ along P, Θ_P, should be a $Z(\mathfrak{m})$-finite tempered, central distribution on M which is the limit in a suitable sense of Θ along the positive chamber relative to P in A (= the standard split component of M). If (π, H) is an irreducible tempered representation of G and if Θ is the character of π then there is a natural notion of Θ_P which we now describe. If H_K is the space of K-finite vectors of H then

$$H_K / \mathfrak{n} \cdot H_K$$

is an admissible, finitely generated, $(\mathfrak{m}, K \cap M)$-module. As a module for $\mathfrak{a}$, it splits into a direct sum of generalized weight spaces for $\mathfrak{a}$:

$$\left(H_K / \mathfrak{n} \cdot H_K \right)_\lambda.$$

Let ρ be as usual the differential of the square root of the modular function, δ, of P. Let $\bar{P} = \theta(P)$. Then $\Theta_{\bar{P}}$ is $\delta^{1/2}$ times the sum of the characters of the $(\mathfrak{m}, K \cap M)$-modules

$$\left(H_K / \mathfrak{n} \cdot H_K \right)_\lambda$$

with

$$\mathrm{Re}(\lambda - \rho, \alpha) = 0$$

for α a root of (P, A). Harish-Chandra's definition in the general case is consistent with this definition for characters.

A tempered, invariant, $Z(\mathfrak{g})$-finite distribution Θ is called *super tempered* if $\Theta_P = 0$ for all proper parabolic subgroups P of G. Thus the super tempered distributions are the analogues of the cusp forms of G. Using the results in [1960a] it is not difficult to see that if Θ is the character of an irreducible representation π of G which is tempered and unitary, then Θ is super tempered if and only if π is square integrable. However, there are super tempered, invariant, eigendistributions that are not linear combinations of characters of discrete series representations. For example, if (π, H) is the reducible unitary principle series representation of SL $(2, \mathbb{R})$, if π^+ and π^- are the irreducible constituents of π, and if $\Theta^\pm$ are the characters of $\pi^\pm$, then $\Theta^+ - \Theta^-$ is super tempered, although neither Θ^+ nor Θ^- is super tempered. One of the main results of this paper is that all super tempered eigendistributions that are not linear combinations of discrete series characters are given in terms of distributions generalizing $\Theta^+ - \Theta^-$ for SL $(2, \mathbb{R})$.

The Work of Harish-Chandra on Reductive p-adic Groups

Roger Howe

Harish-Chandra's work on p-adic groups is broadly based on his experience with Lie groups. (In this survey, the term *Lie group* means an analytic group over **R**.) He himself made the connection quite explicit, formulating what he called the Lefschetz Principle and the Philosophy of Cusp Forms. His Lefschetz Principle [1970d] is a paraphrase for harmonic analysis of the rule of thumb of the same name from algebraic geometry. It asserts that the phenomena of representation theory for groups over p-adic fields, finite fields or adele rings are, suitably interpreted, essentially the same as for groups over **R**, i.e. Lie groups. In applying this principle the "suitable interpretation" is sometimes less than obvious; there are many respects in which p-adic harmonic analysis might seem rather different from real harmonic analysis. Nevertheless, it is a valuable principle, and there have been many examples where one theory will enlighten the other, or an attempt to reconcile apparent differences will lead to insight in both theories. For example, the cleanness of the theory of the constant term in the p-adic case led to a reworking of it for real groups [CM], [Wa]. It is a tribute to the robustness of Harish-Chandra's methods and to his tenacity that, despite serious obstacles including a lack of detailed understanding of the discrete series, characters and orbital integrals, he was able to reach a main goal, the Plancherel Formula [1977b].

The Philosophy of Cusp Forms is much more specific to harmonic analysis than is the Lefschetz Principle; it might be regarded as the primary concrete embodiment of that Principle in representation theory of reductive groups. It is most explicitly stated in [1970b] where it is traced back as far as a talk of Gel'fand at the Stockholm International Congress. It has become a basic feature of reductive harmonic analysis. In particular it is a unifying feature of Harish-Chandra's work.

The gist of the Philosophy of Cusp Forms is that the collection of all representations of a reductive group G should be partitioned into disjoint classes, with each class being attached to a certain parabolic subgroup. (Or more precisely, to a class of associated parabolic subgroups. Recall that two parabolic subgroups are *associated* if their Levi components (the reductive components in their Levi decomposions) are conjugate.) The representations attached to G itself are called *cuspidal* representations and the matrix coefficients of cuspidal representations are *cusp forms*. The representations associated to a proper parabolic subgroup P are further associated to Weyl group orbits of cuspidal representations of M, the Levi component of P, and this further association is finite-to-one. Moreover, the representations of G associated to a given cuspidal representation σ of M may be recovered as components of the induced representation

$$\mathrm{ind}_P^G\left(\sigma \otimes \delta_P^{1/2}\right)$$

where σ is extended from M to P by letting it be trivial on the unipotent radical of P, and δ_P is the modular function of P. Finally, the method of passing from a representation of G to its associated P and representation σ of M is explicit; it is based on what Harish-Chandra calls the "theory of the constant term".

Thus the Philosophy of Cusp Forms provides an inductive strategy for understanding the representations of G. The total problem is divided into two parts: to determine the cusp forms and to analyze induced representations. Both problems are quite difficult and neither is solved in all cases, but the division has provided a fruitful method of attack and substantial progress has been made on each partial problem.

It should perhaps be pointed out that what one means by "cusp form" varies slightly according to context. For purposes of the Plancherel formula for real or p-adic groups, "cusp form" has an analytic meaning: a representation is cuspidal if and only if it is in the discrete series (modulo the center). For purposes of the theory of admissible representations of p-adic groups only supercuspidal representations are cuspidal, while in the Langlands classification [BW], [L2], [Vo] all representations which are tempered on the commutator subgroup play the part of cuspidal representations. The discrepancy between the cusp forms and the discrete spectrum is perhaps most vexed in the case of automorphic forms. All cusp forms in $L^2(G_\mathbf{A}/G_k)$ (for notation see [Bo]) belong to the discrete spectrum, but the non-cuspidal or "residual discrete spectrum" is very rich and almost as poorly understood as the cusp forms. The variability of the notion of cusp form of course complicates the theory, but pays off amply in increased flexibility and applicability.

The determination of the discrete series of a semisimple Lie group was one of Harish-Chandra's great achievements. For semisimple p-adic groups this problem is still unsolved, except in some cases, and in fact it still seems quite far from solution. (See [MS] for some discussion of its current status.) An explicit description even of the representations of the compact group of norm units in a p-adic division algebra has not yet been given.

There are several reasons for the greater mystery of the discrete series, in the p-adic case. Recall that if G is a connected semisimple Lie group, then [V]:

1)   Up to conjugacy G contains at most one compact Cartan subgroup T.
2)   The discrete series of G are parametrized in a natural way by the regular characters of T.

Some groups, including complex groups and $SL_n(\mathbf{R})$, $n \geqslant 3$, have no compact Cartan subgroup and therefore no discrete series. By contrast, all p-adic groups have compact Cartan subgroups and discrete series in abundance. Further, while it is roughly true that characters of the compact Cartan subgroups parametrize discrete series representations, there can be no such clean correspondence as for real groups. Some characters (even "regular" characters, although the definition of that term is itself problematic) will not correspond to representations. Further, some representations would seem to correspond to characters of several tori. And finally, there are some discrete series, analogous to the unipotent cuspidal representations constructed by Lusztig [Lu1] for groups over finite fields, which seem not to correspond to a character of any torus. Langlands [Bo], [L1] has conjectured that the discrete series should be parametrized by certain homomorphisms of the Weil group (of the local field over which the group is defined) into an appropriate group, called the L-group. This has been verified in some cases [K], [M], [He], [KM], [T]. But even the conjectures on how to describe the discrete series of p-adic groups are still undergoing refinement [Ar], [Lu2].

Because of the relative intractability of the determination of cuspidal representations for p-adic groups, Harish-Chandra along with most other workers in this area concentrated mainly on the second part of the cusp form program, namely, the connection between representations of a reductive group G and its parabolic subgroups P. He also established what qualitative results he could about discrete series, and these were sufficient to yield a version of the Plancherel formula [1977b] quite analogous to, though somewhat less precise than, the real case. We will describe below his progress toward this goal in more detail.

Harish-Chandra's papers on p-adic groups are [1970d], [1973], [1977a], [1977b], [1978], [1980]. However, it should be noted that [1973] is only a summary of results, essentially an extended research announcement. The full proofs of the results in [1973] have not been published by Harish-Chandra. Instead they appear in Allan Silberger's book [Si], which is largely based on lectures of Harish-Chandra at the Institute for Advanced Study in the academic years 1971–1972 and 1972–1973. For understanding Harish-Chandra's work on p-adic groups, this book is an essential supplement to the papers appearing under his own name. Neither have the full proofs of some results stated in [1977b], [1978], or [1980] yet appeared.

The main topics treated in these papers are ones familiar from Harish-Chandra's work on real groups: the theory of characters, orbital integrals, the constant term, discrete series and the Schwartz space. All of these topics are mutually related via the Philosophy of Cusp Forms. In the real case all were subservient to the Plancherel Formula, but in the p-adic case neither character theory nor orbital integrals has been made as precise as in the real case, and eventually were bypassed by Harish-Chandra in his proof of the p-adic Plancherel Formula. We will discuss these topics in turn.

As a prelude to the more detailed discussion, a remark on technique is in order. Throughout his work on real groups, Harish-Chandra relies on the differential equations supplied by the universal enveloping algebra, especially its center. This fundamental avenue of approach is of course not available in the study of p-adic groups and replacing it is a major technical problem. For some purposes, the place

of the differential equations is taken by certain finiteness theorems. Thus a result of
Jacquet [J] forms the basis for the theory of the constant term, a result of Howe [Ho]
is useful in character theory, and a finiteness result in [1977b] provides the final step
in the Plancherel Theorem. We will discuss these in more detail below. But in fact,
the lack of the differential equations has not been completely made good, and as was
already noted, complete information about discrete series, character formulas, and
orbital integrals is not yet available.

Character theory is the primary topic of [1977a], [1978], and [1980], and occupies
the bulk of [1970d]. The paper [1977a] is an announcement of the results in [1978],
which contains the most precise results on characters, and largely supersedes [1970d].
However, the paper [1978] is restricted to groups over fields of characteristic zero; in
[1980] some of the results are extended to groups over fields of positive characteristic.

Before one can describe characters one must know in what sense they exist. For
this, and much of the rest of representation theory for p-adic groups, a basic notion
is that of *admissible representation*, first formalized in [JL]. Let G be a locally
compact topological group which is totally disconnected as topological space—a $t.d.$
group in Harish-Chandra's terminology [1973]. This is the same as to say G has a
basis of neighborhoods of the identity consisting of open compact subgroups. All
p-adic algebraic groups are t.d. groups. Let ρ be a representation of G on a vector
space V. A vector $v \in V$ is called a *smooth vector* for ρ if there is an open subgroup
$K \subseteq G$ such that v is invariant under $\rho(K)$, i.e., $\rho(k)v = v$ for $k \in K$. The set of
smooth vectors for ρ form a subspace $V^\infty \subseteq V$; this subspace V^∞ is invariant by G. If
V is a complete locally convex topological vector space, and ρ is a strongly
continuous representation of G, then V^∞ is a dense subspace of V.

For a given compact open group $K \subseteq G$, let V^K denote the subspace of vectors in
V fixed by $\rho(K)$. Then $V^K \subseteq V^\infty$, and V^∞ is a union of the V^K as K varies over all
compact open subgroups of G. If $V = V^\infty$, then the representation ρ is called *smooth*.
If in addition all the spaces V^K are finite dimensional, then ρ is called *admissible*. A
continuous representation of ρ on a locally convex space whose associated smooth
subrepresentation is admissible is also called admissible.

An admissible representation has a character in the following sense. Let $C_c^\infty(G)$
denote the space of locally constant, compactly supported, complex-valued functions
on G. Then $C_c^\infty(G)$ is a convolution algebra under the usual definition of convolution on G. If ρ is a smooth representation of G on V, then we can define in an
obvious and standard manner a representation, also denoted ρ, of $C_c^\infty(G)$ on V. If ρ
is an admissible representation, then $\rho(f)$ will be a finite rank operator. (If f is
invariant under left translation by the open subgroup $K \subseteq G$, then $\rho(f)(V) \subseteq V^K$.)
Hence the trace of $\rho(f)$ will be well defined, and will depend linearly on f. Thus

$$\theta_\rho(f) = \operatorname{trace} \rho(f)$$

defines a linear functional on $C_c^\infty(G)$. By definition a linear functional on $C_c^\infty(G)$ is
called a *distribution* on G. Thus we may say that an admissible representation has a
character defined by a distribution on G—its "distribution character".

Thus the first question to ask concerning character theory of representations of G
is, are all reasonable (say unitary, though one can be more general) irreducible

representations of G admissible? The analogous question for Lie groups was one of the first issues resolved by Harish-Chandra [1953]. In the case of a reductive p-adic group the answer was much longer in coming. That the answer was "yes" was first established by Bernstein [Be]. A stronger result was proved for groups over fields of characteristic zero by Harish-Chandra in [1978], generalizing and extending [Ho]. Both results were based on [1970d] and in particular were resolutions of conjectures made in [1970d]. We will explain these conjectures.

The conjectures concern the supercuspidal representations, which are the most characteristic and the most mysterious phenomenon of representation theory on reductive p-adic groups.

Let G be a locally compact group. Let f be a function on G. Define the left and right translate $\lambda_g f$ and $\rho_g f$ of f by g in the usual way:

$$\lambda_g(f)(x) = f(g^{-1}x) \qquad \rho_g(f)(x) = f(xg) \qquad x, g \in G. \tag{1}$$

Let $H \subseteq G$ be a subgroup. We say f is left (right) H-*invariant* if $\lambda_h(f) = f$ ($\rho_h(f) = f$) for $h \in H$. If f is complex- or vector-valued we say f is left (right) H-*finite* if the left translates $\lambda_h(f)$ (right translates $\rho_h(f)$) span a finite dimensional space. We say f is left (right) compactly supported mod H if there is a compact set $C \subseteq G$ such that $f = 0$ outside HC (outside CH). If H is central in G, we may omit the adjectives "left" and "right" in the definitions above.

Let G be a reductive p-adic group—the rational points of a connected reductive algebraic group $\mathbf{G}$ defined over a p-adic field Ω. Let Z be the center of G. Let P be a parabolic subgroup of G and let N be the unipotent radical of P. Let δ_P be the modular function of P. Let f be a smooth complex- or vector-valued function on G, compactly supported mod Z. Then in analogy with the real case we can define a function $f^{(P)}$ on P/N by the formula

$$f^{(P)}(m) = \delta_P(m)^{1/2} \int_N f(mn)\, dn \qquad m \in P.$$

Here dn is Haar measure on N.

We say a function f on G is a *supercusp form* if

i) f is smooth;
ii) f is compactly supported mod Z;
iii) f is Z-finite; and
iv) $(\lambda_g f)^{(P)} = 0$ for all proper parabolics $P \subseteq G$ and $g \in G$.

A *supercuspidal representation* is a representation whose matrix coefficients are supercusp forms. Let π be a smooth representation of G on the space V. Consider $T \in \text{End } V$. We say T is a *smooth operator* if there is an open subgroup $K \subseteq G$ such that

$$\pi(k)T = T = T_\pi(k) \qquad k \in K.$$

Let $\text{End}^\circ(V)$ denote the space of smooth operators which also have finite rank. (Observe that if π is admissible, then a smooth operator automatically has finite rank.) Then $\text{End}^\circ(V)$ is a subalgebra of V, and is invariant by right and left multiplication by $\pi(g)$, for $g \in G$. Set

$$f_T(g) = \operatorname{trace} T\pi(g).$$

The function f_T is called a *matrix coefficient* of ρ. It is easy to see that $f_T \in C^\infty(G)$, the space of locally constant functions on G. The space spanned by all matrix coefficients of all admissible representations is denoted $\mathscr{C}(G)$. If π is admissible and all matrix coefficients of π are supercusp forms, then π is called supercuspidal. It turns out ([1973] §6) that for π to be supercuspidal, it is sufficient that its matrix coefficients be compactly supported mod Z.

Supercuspidal representations are obviously discrete series representations (mod Z). A general argument ([1970d], p. 6) using the Schur orthogonality relations for discrete series shows that all discrete series representations are admissible. In fact if π is an irreducible square integrable representation (mod Z) of G on a space V, and V^K is the space of $\pi(K)$-invariant vectors in V for an open subgroup $K \subseteq G$, one has

$$\dim V^K \leqslant \left(d(\pi)\mu(K) \right)^{-1}$$

where $d(\pi)$ is the formal degree of π and $\mu(K)$ is the Haar measure of K. If 1_K denotes the trivial representation of K, then $\dim V^K$ is the multiplicity of the trivial representation of K in π. It will be denoted $[\pi:1_K]$.

Harish-Chandra shows in [1970d], Part II, that the question of admissibility hinges on the possibility of bounding $[\pi:1_K]$ independently of π. He considers three possible facts that would do this. Observe that if π is an irreducible admissible representation of G, then $\pi|Z$ is a multiple of some character $\chi_\pi = \chi$, called the *central character* of π. Consider the following statements.

i) For each compact open subgroup K there is a number $\delta(K)$ such that $[\pi:1_K] \leqslant \delta(K)$ for all supercuspidal irreducible representations π of G.

ii) For each compact open subgroup K there is a number $m(K)$ such that, for any character χ of Z,

$$\sum_\pi [\pi:1_K] \leqslant m(K)$$

where π ranges over all (equivalence classes of) irreducible supercuspidal representations of G with central character χ.

iii) There is $\varepsilon > 0$ such that $d(\pi) > \varepsilon$ for all irreducible supercuspidal representations π of G.

Obviously statement ii) implies i). The Schur orthogonality relations in fact imply that iii) implies ii). In [1970d], Part II, Harish-Chandra shows that statement i) implies that all irreducible unitary representations of G are admissible. In [Be] Bernstein shows by a simple argument based on the structure of $C^\infty(G//K)$, the K bi-invariant functions of compact support, that statement i) is true. In [1978] Harish-Chandra, as a consequence of a study of characters of supercuspidal representations, shows that if the ground field Ω has characteristic 0, then statement iii) is true. In fact, he shows that if Haar measure on G is appropriately normalized, then the formal degrees $d(\pi)$ of supercuspidal representations are integers.

Once the issue of existence of characters is settled, one would like to know more precisely what they look like. For supercuspidal characters and when the ground field is of characteristic 0, Harish-Chandra already establishes an important qualitative result in [1970d]. Let dg denote Haar measure on G, and let φ be a complex-valued function on G which is locally L^1, i.e. the integral

$$\int_C |\varphi|(g)\, dg$$

where $|\varphi|$ indicates the absolute value of φ and C is any open compact subset of G, is finite. Then φ defines a distribution D_φ on G by the obvious formula

$$D_\varphi(f) = \int_G \varphi(g) f(g)\, dg \qquad f \in C_c^\infty(G).$$

If a distribution D is equal to D_φ for some φ as above, we say D is *locally L^1*. If the function φ may be taken to be locally constant on some open set $S \subseteq G$, we say D is *locally constant* on S. In [1970d], Part V, Harish-Chandra shows that the character of an irreducible supercuspidal representation is a locally constant function on the set G' of regular elements of G. In Parts VII and VIII of [1970d] he goes on to show that if the ground field Ω has characteristic zero, then in fact the character of an irreducible supercuspidal representation is locally L^1 on G. These results are based on an integral formula ([1970d], Theorem 9) for the character of a supercuspidal representation and on a detailed analysis of the geometry of conjugacy classes, analogous to his work on real groups.

In [1978] Harish-Chandra extends these basic facts to all irreducible admissible representations. Furthermore he gives more precise information about the behavior of characters near singular points. These results make use of the exponential map from the Lie algebra $\mathfrak{g}$ of G to G, and so are valid only when the ground field Ω has characteristic zero. They depend on a finiteness result, proved first for GL_n in [Ho], and extended and generalized by Harish-Chandra in [1978]. This may be stated as follows. Let $L \subseteq \mathfrak{g}$ be a lattice, and let $C_c^\infty(\mathfrak{g}/L)$ be the functions on $\mathfrak{g}$ which are compactly supported and constant on cosets of L. Let $\mathrm{Ad}\, G$ denote the adjoint action of G on $\mathfrak{g}$. Let $\omega \subseteq \mathfrak{g}$ be a compact set, let $\mathrm{Ad}\, G(\omega)$ denote the subset of $\mathfrak{g}$ swept out by the action of $\mathrm{Ad}\, G$ on ω, and let $J(\omega)$ denote the space of distributions which are supported on $\mathrm{Ad}\, G(\omega)$ and which are invariant under $\mathrm{Ad}\, G$. Let $j_L J(\omega)$ denote the space of linear functionals on $C_c^\infty(\mathfrak{g}/L)$ obtained by restricting elements of $J(\omega)$ to $C_c^\infty(\mathfrak{g}/L)$. Then $j_L J(\omega)$ has finite dimension.

This finiteness result has two main consequences for characters. First, it gives a description of the possible singularities of characters near any point. It is in terms of Fourier transforms of invariant measures on nilpotent conjugacy classes. Let $x \in \mathfrak{g}$ be a nilpotent element, and let $\mathcal{O} = \mathcal{O}_x = \mathrm{Ad}\, G(x)$ denote the G-conjugacy class of x. There are only finitely many nilpotent conjugacy classes. On $\mathcal{O}$ there is supported a G-invariant measure ν_x, unique up to multiples. Furthermore, a result of Deligne and Rao [Ra] says that the measure ν_x assigns finite mass to bounded subsets of $\mathcal{O}$; consequently ν_x defines a distribution on $\mathfrak{g}$.

We want to consider the Fourier transforms of the measures ν_x. These may be defined in the usual way. If $f \in C_c^\infty(\mathfrak{g})$, its Fourier transform $\hat{f}$ is defined by

$$\hat{f}(x) = \int \hat{f}(x')\chi(B(x',x))\,dx'$$

where $B(\ ,\ )$ is an $\mathrm{Ad}\,G$-invariant, non-degenerate symmetric bilinear form on $\mathfrak{g}$ (the Killing form if $\mathfrak{g}$ is semisimple), χ is a unitary character of the (additive group of the) base field Ω, and dx' is a Haar measure on $\mathfrak{g}$, normalized to make the map $f \to \hat{f}$ unitary. We can then extend $\hat{\ }$ to distributions by the recipe

$$\hat{D}(f) = D(\hat{f}) \qquad f \in C_c^\infty(\mathfrak{g}),\ D \in C_c^\infty(\mathfrak{g})^*. \tag{2}$$

Recall that $J(\omega)$ is the space of $\mathrm{Ad}\,G$-invariant distributions supported on $\mathrm{Ad}\,G(\omega)$ where $\omega \subseteq \mathfrak{g}$ is some compact set. Harish-Chandra in [1978] has described the Fourier transforms of elements of $J(\omega)$. To state his result we need the functions $\eta_\mathfrak{g}$ and D_G, familiar from the real theory, which measure how regular elements of $\mathfrak{g}$ or G are.

Consider the polynomial $\det(t - \mathrm{ad}\,x)$ on $\mathfrak{g} \times \Omega$. Regard this as a polynomial in t with coefficients depending on x. Then if l is the rank of $\mathfrak{g}$ (i.e., the dimension of any Cartan subalgebra), the coefficient of t^b is identically zero if $b < l$. The function $\eta_\mathfrak{g}$ is the coefficient of t^l. Thus

$$\det(t - \mathrm{ad}\,x) = t^l\eta_\mathfrak{g}(x) + t^{l+1}R(x,t). \tag{3}$$

The function $\eta_\mathfrak{g}$ may be described in another way, as follows. Let $x \in \mathfrak{g}$ be regular, let $\mathfrak{t} \subseteq \mathfrak{g}$ be the kernel of $\mathrm{ad}\,x$ and let $\mathfrak{t}^\perp$ be the orthogonal complement of $\mathfrak{t}$ with respect to the bilinear form B. Then

$$\eta_\mathfrak{g}(x) = \det\left(\mathrm{ad}\,x | \mathfrak{t}^\perp\right).$$

The function D_G is the analogue on G of $\eta_\mathfrak{g}$. The polynomial $\det(1 + t - \mathrm{Ad}\,g)$ also begins with the term t^l and one writes

$$\det(1 + t - \mathrm{Ad}\,g) = t^l D_G(g) + t^{l+1}R'(g,t).$$

The functions $\eta_\mathfrak{g}$ and D_G take values in Ω, the ground field. Recall [W] that on Ω there is defined a natural absolute value

$$|\ \ |: \Omega \to \mathbf{R}.$$

We will need the real-valued functions gotten by taking absolute values of $\eta_\mathfrak{g}(x)$ or $D_G(x)$. Thus we write

$$|\eta_\mathfrak{g}|(x) = |\eta_\mathfrak{g}(x)| \qquad |D_G|(x) = |D_G(x)|.$$

Consider a distribution $D \in J(\omega)$. Harish-Chandra ([1978], Theorem 3) shows that $\hat{D}$ is locally L^1, locally constant on $\mathfrak{g}'$, the set of regular elements in $\mathfrak{g}$, and that $|\eta_\mathfrak{g}|^{1/2}\hat{D}$ is locally bounded on $\mathfrak{g}$. These results apply in particular to the Fourier transforms $\hat{\nu}_x$ of the invariant measures on nilpotent orbits.

Finally we can describe characters. Let θ_π be the character of an irreducible admissible representation π of G. Then θ_π is a locally L^1 function, locally constant on G', and the function $|D_G|^{1/2}\theta_\pi$ is locally bounded. Furthermore, if $\gamma \in G$ is a semisimple element, then the behavior of θ_π near γ is as follows. Let M be the centralizer of γ in G, and $\mathfrak{m}$ the Lie algebra of M. Let C be a small neighborhood of 0 in $\mathfrak{m}$, on which the exponential map exp is defined. Then $\gamma \exp C$ is a neighborhood of γ in M, and $\mathrm{Ad}\, G(\gamma \exp C)$ is an $\mathrm{Ad}\, G$-invariant neighborhood of γ in G. If C is chosen sufficiently small, then we may write

$$\theta_\pi(\gamma \exp Y) = \sum_\xi c_\xi \hat{\nu}_\xi(Y) \qquad Y \in C$$

where ξ runs over the nilpotent conjugacy classes *in* $\mathfrak{m}$, and the c_ξ are complex numbers depending on π and γ. Roughly, we may say that locally θ_π is a linear combination of Fourier transforms of nilpotent orbits.

This result is analogous to the knowledge that an invariant eigendistribution on a real reductive group is a locally L^1 function, and is given locally on each Cartan subgroup by $D_G^{-1/2}$ times a linear combination of certain exponentials. What is still missing, except in special cases, are results as precise as Harish-Chandra's character formula for discrete series of real groups.

As was already stated, the above results apply when the ground field Ω has characteristic zero. When Ω is of positive characteristic, much less is known. However, in [1980] it is shown that in all characteristics characters are locally constant on the regular set. The techniques of this paper are different from those of [1973] or [1978]. The basic fact used is the submersiveness of a certain map ([1980], Theorem 1). Also, the following interesting fact is established. Let K be a "good" maximal compact subgroup of G in the sense of Bruhat and Tits [BT]. For an admissible representation π of G on a vector space V and a regular element x in G, set

$$T_x = \int_K \pi(kxk^{-1})\, dk.$$

Then $T_x \in \mathrm{End}^\circ(V)$, and in particular T_x has finite rank. Then evidently $\theta_\pi(x) = \mathrm{tr}\, T_x$.

The study of the map F_f or, in common current parlance, orbital integrals, is entwined with and in some sense dual to the study of characters, and currently is in roughly the same state. One has substantial knowledge of the qualitative behavior of F_f, but precise control analogous to the "jump formula" for real groups is still lacking. This would be provided by an explicit description of the "Shalika germs" which will be defined and discussed below.

First recall the definition of F_f. Fix a Cartan subgroup $A \subseteq G$, and fix an invariant measure dg^* on G/A. Then for $f \in C_c^\infty(G)$ and $a \in A'$, where $A' = G' \cap A$ one sets

$$F_f(a) = |D_G(a)|^{1/2} \int_{G/A} f(gag^{-1})\, dg^*. \tag{4}$$

There is a close analogue of F_f for the Lie algebra. If $\mathfrak{a}$ is the Lie algebra of A and $x \in \mathfrak{a}$, put

$$\phi_f(x) = |\eta_\mathfrak{a}(x)|^{1/2} \int_{G/A} f(\mathrm{Ad}\, g(x))\, dg^* \qquad f \in C_c^\infty(\mathfrak{g}). \tag{5}$$

In formulas (4) and (5) the functions D_G and $\eta_\mathfrak{a}$ are the ones defined above in (2) and (3).

Formulas (4) and (5) are very near parallels of the definitions of F_f and ϕ_f for real groups. However, we should remark on one difference that has not been important up to now, but which needs clarification before these maps can be thoroughly understood. For real groups the function D_G is a smooth function. The function $|D_G|^{1/2}$ is not smooth. However, on any Cartan subgroup A, there is a function $D_G^{A\ 1/2}$ such that $(D_G^{A\ 1/2})^2 = D_G$ on A and which yields an optimal theory of orbital integrals. In other words, if one multiplies $|D_G|^{1/2}$ by appropriate phase factors, one obtains a better theory, in which the functions F_f are as smooth as possible. It is the function $D_G^{A\ 1/2}$ rather than the positive $|D_G|^{1/2}$ which is used in Harish-Chandra's definition of F_f for real groups. It would seem that some modification of $|D_G|^{1/2}$ by appropriate phases is also preferable in the p-adic case; but precisely how to do this has not yet been made clear.

The parts of Harish-Chandra's work devoted to F_f on p-adic groups are [1970d], Part VI; [1973], §16; [1978] §§3, 4, 8, 9; and [1980], §5. In [1970d] it is already established that when the ground field Ω has characteristic 0, the function F_f is bounded for any $f \in C_c^\infty(G)$. However, it is not shown that F_f is locally constant. This is done in [1973], §16 (for the proofs see [Si]) by the same technique that establishes local constancy of characters. In fact, local constancy of F_f is even established for f belonging to Harish-Chandra's Schwartz space $\mathcal{C}(G)$ (see below for the definition of $\mathcal{C}(G)$); however local boundedness of F_f for $f \in \mathcal{C}(G)$ is not proven. Local constancy of F_f for Ω of positive characteristic, and $f \in \mathcal{C}(G)$ is proved in [1980], but boundedness not.

For real groups the behavior of $F_f(x)$ as x approaches singular points is of much interest. This is also true for p-adic groups. The standard approach to the study of this limiting behavior is provided by an observation of Shalika [Sh]. Harish-Chandra's version of this result is Theorem 14 of [1978], which applies to the Lie algebra. As above, we let ν_x denote the invariant measures on the nilpotent G-conjugacy classes $\mathcal{O}_x$ in $\mathfrak{g}$. Fix a Cartan subalgebra $\mathfrak{a} \subseteq \mathfrak{g}$. Then there are functions $\Gamma_{\mathcal{O}}^{\mathfrak{a}} = \Gamma_{\mathcal{O}}$, for each nilpotent conjugacy class $\mathcal{O} = \mathcal{O}_x$ such that

$$\phi_f(a) - \sum \nu_x(f)\Gamma_{\mathcal{O}}(a)$$

vanishes in some neighborhood of the origin in $\mathfrak{a}$. The functions $\Gamma_{\mathcal{O}}$ are strictly speaking, only germs of functions, an equivalence class of functions equal in some neighborhood of 0; they are generally called "Shalika germs". However, they may be defined uniquely on all of $\mathfrak{a}$ by requiring them to satisfy an appropriate condition of homogeneity, namely

$$\Gamma_{\mathcal{O}}(t^2 a) = |t|^m \Gamma_{\mathcal{O}}(a) \qquad t \in \Omega^x,\ a \in \mathfrak{a}$$

where m is an integer depending on $\mathcal{O}$. There are analogous germ expansions around non-zero singular points of $\mathfrak{a}$, and for F_f.

Clearly the functions $\Gamma_{\mathcal{O}}$ control the singularities of ϕ_f. Considerable effort consequently has been devoted to determining them [Vi], [Ko], [Re1], [Re2], [Ro2]. However, they remain mysterious. Harish-Chandra shows in [1978], §9 that the $\Gamma_{\mathcal{O}}^{\mathfrak{a}}$ separate the nilpotent orbits $\mathcal{O}$ in the sense that the only linear combination $\Sigma c_{\mathcal{O}}\Gamma_{\mathcal{O}}^{\mathfrak{a}}$ which vanishes for all Cartan subalgebras $\mathfrak{a}$ is the trivial combination with all $c_{\mathcal{O}} = 0$. This is implied by his result ([1978], Theorem 10) that all invariant distributions annihilate any function $f \in C_c^\infty(\mathfrak{g})$ such that $\phi_f = 0$ for all Cartan subalgebras $\mathfrak{a}$. He also shows that $\Gamma_0^{\mathfrak{a}}$, the germ associated to the origin, is zero if $\mathfrak{a}$ is not an elliptic Cartan subalgebra, and on an elliptic Cartan is a constant times $|\eta_{\mathfrak{a}}|^{1/2}$. The value of the constant was conjectured by Harish-Chandra and determined by Rogawski [Ro1].

The third major topic in Harish-Chandra's work is the Plancherel Formula, and the associated analysis on G: the constant term, induced series of representations, wave packets, Eisenstein integrals, c-functions. For real groups, character theory and orbital integrals also contribute to the Plancherel Formula, but as explained above, for p-adic groups, they have not reached the state necessary for use in the Plancherel Formula. However certain technical aspects of the p-adic situation allow Harish-Chandra to complete his program without knowledge of the discrete series. The resulting Plancherel Formula is not as explicit as the one for real groups. But except for the lack of explicit knowledge of the discrete series, the Plancherel Formula for p-adic groups is quite parallel to that for real groups. Hence our discussion of it will be relatively brief, and will focus on the aspects which are particular to p-adic groups.

Since the theory of the constant term for real groups depends heavily on the differential equations supplied by the center of the enveloping algebra, its transference to p-adic groups requires new techniques. Harish-Chandra made a preliminary essay in [1970d], establishing the existence of the constant term contingent on a conjecture. However a firmer basis was provided shortly after by a result of Jacquet [J].

Let π be an admissible representation of G on the vector space V. Let $P \subseteq G$ be a parabolic subgroup with unipotent radical N. Let $V(N)$ be the subspace of V spanned by vectors of the form $\pi(n)v - v$. Since N is normalized by P, one can easily see that $V(N)$ is stable under P. Hence there is defined a representation of P on $V_N = V/V(N)$. From the definition of $V(N)$ it is clear that N acts trivially on V_N. Hence the action of P on V_N factors to an action π_M of $P/N \simeq M$ on V_N. Jacquet showed (for $G = GL_n(\Omega)$, but with a general proof) that V_N is also an admissible module for M. It turns out that π is supercuspidal if and only if $V_N = \{0\}$.

Recall that $\mathcal{Q}(G)$ is the space of all matrix coefficients of all admissible representations of G. Using Jacquet's theorem, Harish-Chandra establishes the existence of the constant term in a very strong form. Let M now denote a Levi component of P, so that $M \simeq P/N$. Let A denote the center of M. Let $\{\alpha_i\}$ be the simple roots of A acting on N by conjugation. Each α_i is a rational homomorphism

$$\alpha_i : A \to \Omega^\times.$$

For a given $t > 0$, set

$$A^+(t) = \{a \in A : |\alpha_i(a)| > t \text{ for each } \alpha_i\}.$$

As above let δ_P denote the modular function of P. Then Harish-Chandra shows ([1973], §6; see [Si], Chapter 2 for proofs) that given $f \in \mathcal{Q}(G)$, there is a unique $f_P \in \mathcal{Q}(M)$ such that, for any compact set $\omega \subseteq M$, one has

$$f(ma) = \delta_P(ma)^{1/2} f_P(ma) \qquad m \in \omega, \ a \in A^+(t)$$

for t sufficiently large. Thus in the p-adic case one eventually has actual equality between f and f_P, its *constant term along P*, and not merely an asymptotic relation.

Since f_P is in $\mathcal{C}(M)$, it is in particular A-finite. The finite dimensional space spanned by the A-translates of f_P will be spanned by generalized eigenspaces for A. If Y is such an eigenspace, then there is a quasicharacter ψ of A such that $\lambda_a - \psi(a)$ is nilpotent on Y. (Here λ_a is the left action of A, as in (1).) Harish-Chandra calls these characters ψ the *exponents* of f relative to the pair (P, A). The A-finiteness of f_P takes the place in many situations of the differential equations governing the constant term in the real case, and the exponents take the place of infinitesimal characters. To place limits on the elements of $\mathcal{C}(M)$ which can be constant terms of matrix coefficients of a fixed irreducible representation of G, Harish-Chandra develops a sharp form of the theory of intertwining operators for induced representations, pioneered by Bruhat for real groups [Br1].

With the constant term, Harish-Chandra can proceed toward the Plancherel Formula along very much the same road used in the real case. The theory of induced representations, of the Eisenstein integral, the c-functions, the Maass–Selberg relations, of wave packets and the Schwartz space proceed very much as in the real case. Already in [1973] he is able to state the Plancherel Formula for the wave packets constructed from a given series of induced representations (Theorem 34). In this Plancherel Formula, the Plancherel measure is determined by a product formula, as it is for real groups. However, it is not explicitly known. But luckily, it is not necessary to know it explicitly in the p-adic case. This is because any p-adic torus A, modulo an open compact subgroup, is discrete. Hence $\hat{A}^0$, the identity component of the Pontrjagin dual of A, is compact. It is a real torus—a product of circles. If M is the Levi component of a parabolic subgroup P, and A the center of M, then a series from which one builds wave packets consists of representations of the form $\pi \otimes \chi$ where π is a square integrable (mod A) unitary representation of M and $\chi \in \hat{A}^0$. In the real case, $\hat{A}^0$ is a vector group, and it is necessary to know that Plancherel measure on $\hat{A}^0$ grows only polynomially at ∞ in order to be able to control the Fourier transform. In fact, for real groups, Harish-Chandra gives an explicit expression for the Plancherel measure, and one can see directly that this expression has moderate growth. But since $\hat{A}^0$ is compact for p-adic groups, the growth of Plancherel measure is not an issue, and no explicit knowledge of it is necessary.

We will close this account by describing what is involved in the passage from the "local" Plancherel Formula of [1973] to the full formula announced in [1977b].

First the notion of constant term described above must be modified so that matrix coefficients of representations which are square integrable but not super-cuspidal have zero constant term. This leads to the notion of the weak constant term.

Let K be a good maximal compact subgroup of G. Let Ξ be, as for real groups, the K-spherical matrix coefficient of the representation unitarily induced from the trivial representation of a minimal parabolic subgroup of G. Let σ be the measure of slow growth on G; roughly σ is the logarithm of Ξ (see [1973]; §14). For each compact open $K_1 \subseteq K$, one defines $\mathcal{C}(G//K_1)$ to be the space of K_1 bi-invariant functions f on G such that

$$|f| \leqslant c(f, k)\Xi(1 + \sigma)^{-k}$$

for an appropriate positive number $c(f, k)$ and for each positive number k. Then $\mathcal{C}(G)$, the *Schwartz space* of G is the union over compact open K_1 of the spaces $\mathcal{C}(G//K_1)$.

Consider $\phi \in \mathcal{C}(G)$. One says ϕ satisfies the *weak inequality* if for some constant c and some integer k one has the estimate

$$|\phi| \leqslant c\Xi(1+\sigma)^k.$$

Such a ϕ is called *tempered*, and the subspace of $\mathcal{C}(G)$ consisting of tempered matrix coefficients is denoted $\mathcal{C}^w(G)$.

Let f be in $\mathcal{C}^w(G)$, and let f_P be its constant term along the parabolic subgroup P. Write

$$f_P = \sum f_{P,\psi}$$

where ψ runs over the exponents of f, and $f_{P,\psi}$ is the component of f in the generalized ψ-eigenspace of the A translates of f, A being the center of a Levi component M of P. The sum of the $f_{P,\psi}$ over those ψ which are unitary is called the *weak constant term* of f along P, and is denoted f_P^w. It is in $\mathcal{C}^w(M)$. A *cusp form* on G is an element of $\mathcal{C}^w(G)$ such that the weak constant terms $(\lambda(g)f)_P^w$ of all left translates of f are zero for all proper parabolic subgroups P of G. The space of cusp forms on G is denoted $^0\mathcal{C}^w(G)$.

There is a notion of constant term for $\mathcal{C}(G)$ that is dual to that for $\mathcal{C}^w(G)$. For $f \in \mathcal{C}(G)$ and P a parabolic subgroup of G, set

$$f^{(P)}(m) = \delta_P(m)^{1/2} \int_N g(mn)\, dn \qquad m \in P/N$$

where N is the unipotent radical of P. Let $^0\mathcal{C}(G)$ denote the subspace of $\mathcal{C}(G)$ consisting of those f such that $(\lambda(g)f)^{(P)} = 0$ for all proper parabolic subgroups.

If G were semisimple rather than reductive, then it is easy to see that $^0\mathcal{C}^w(G) \subseteq {}^0\mathcal{C}(G)$. If G possesses a non-compact center, then one must formulate the relation between $^0\mathcal{C}^w$ and $^0\mathcal{C}$ in a more complicated way, but the situation is in essence the same. So to simplify the discussion we take G semisimple. Then $^0\mathcal{C}^w(G)$ consists of matrix coefficients of the discrete series, while $^0\mathcal{C}(G)$ is the subspace of $\mathcal{C}(G)$ which cannot be obtained by inducing tempered representations from proper parabolic subgroups. Hence to pass from the Plancherel Theorem of [1973], §17, to the Plancherel Formula for G, one needs to show that in fact $^0\mathcal{C}^w(G) = {}^0\mathcal{C}(G)$. This essentially amounts to the statement that (for G semisimple), the representation of G on the span of the left translates of $f \in {}^0\mathcal{C}(G)$ is admissible, i.e., consists of finitely many irreducible summands. This is a consequence of Lemma 4 of [1977b]. A crucial step in the proof of Lemma 4 is Theorem 11, which states that if $\theta \in {}^0\mathcal{C}(G)$, then the map

$$f \to \int_{G/Z} dy^* \int_G f(x)\theta(yxy^{-1})\, dx$$

is well defined and yields a tempered distribution on G. Although the details of this result are not yet published, in flavor it reminds one of the analysis of supercuspidal characters in [1970d].

The form of Harish-Chandra's p-adic Plancherel Theorem is this. The space $\mathcal{C}(G)$ is the orthogonal direct sum of wave packets formed from series of representations induced unitarily from discrete series of (the Levi components of) parabolic subgroups of P. Moreover if two such series of induced representations yield the same subspace of $\mathcal{C}(G)$, then the parabolics from which they are induced are associate, and the representations of the Levi components are conjugate. To complete the analogy with real groups, one needs only to explicitly determine the discrete series; this is the outstanding problem left in p-adic representation theory.

REFERENCES

[Ar] J. Arthur, On some problems suggested by the trace formula, Special Year in Representation Theory, U. of Maryland, Nov. 1982, Springer Lecture Notes, to appear.

[Be] I. N. Bernstein, All reductive p-adic groups are tame, *Fun. Anal. and App.* **8** (1974), 91–93.

[Bo] A. Borel, Automorphic L-functions, in *Automorphic Forms, Representations, and L-functions, Proc. Sym. Pure Math.*, **XXXIII**, Part 2, Amer. Math. Soc. Providence, R.I., 1974, 27–63.

[BW] A. Borel and N. Wallach, Continuous cohomology, discrete subgroups, and representations of reductive groups, *Ann. of Math. Studies* **94**, Princeton University Press, Princeton, 1980.

[Br1] F. Bruhat, Sur les representations induites des groupes de Lie, *Bull. Soc. Math. France*, **84** (1956), 97–205.

[BT] F. Bruhat and J. Tits, Groupes reductifs sur un corps local, I, Donnees radicielles valuees, *Pub. Math. I.H.E.S.* **41** (1972), 5–251.

[CM] W. Casselman and D. Milicic, Asymptotic behavior of matrix coefficients of admissible representations, *Duke Math. J.* **49** (1982), 869–930.

[He] G. Henniart, La conjecture de Langlands locale pour GL(3), *I.H.E.S. Notes* (1982).

[Ho] R. Howe, The Fourier transform and germs of characters, *Math. Ann.* **208** (1974), 305–322.

[H] H. Jacquet, Representations des groupes lineaires p-adiques, Theory of Group Representations and Harmonic Analysis (C.I.M.E., II, Ciclo, Montecatini terme, 1970) Edizione Cremonese, Roma, 1971, 119–220.

[JL] H. Jacquet and R. Langlands, Automorphic forms on GL(2). *Lec. Notes Math.*, **260**, Springer-Verlag, Berlin, New York, 1972.

[Ko] R. E. Kottwitz, Orbital integrals on GL(3), *Am. J. Math*, **102** (1980), 327–384.

[K] P. Kutzko, The Langlands conjecture for GL_2 of a local field, *Ann. Math.* **112** (1980), 381–412.

[KM] P. Kutzko and A. Moy, On the local Langlands Conjecture in prime dimension, preprint.

[L1] R. Langlands, Problems in the theory of automorphic forms, *Lectures in Modern Analysis and Applications, Lecture Notes in Math.*, **170**, Springer-Verlag, New York, 1973, 18–86.

[L2] R. Langlands, On the classification of irreducible representations of real reductive groups, *Notes, I.A.S.*, Princeton, 1973.

[Lu1] G. Lusztig, Irreducible representations of finite classical groups, *Inv. Math.* **43** (1977), 125–175.

[Lu2] G. Lusztig, Some examples of square integrable representations of semisimple p-adic groups, preprint, *I. H. E. S.*, March 1982.

[M] A. Moy, Local constants and the tame Langlands correspondence, Thesis, University of Chicago, 1982.

[MS] A. Moy and P. Sally, Supercuspidal representations of SL_n over a p-adic field: the tame case, preprint.

[Ra] R. Rao, Orbital integrals in reductive groups, *Ann. of Math.* **96** (1972), 505–510.

[Re1] J. Repka, Shalika's germs for p-adic $GL(n)$: the leading term, preprint.

[Re2] J. Repka, Shalika's germs for p-adic $GL(n)$, II: the subregular term, preprint.

[Ro1] J. Rogawski, An application of the building to orbital integrals, *Comp. Math.* **42** (1981), 417–423.

[Ro2] J. Rogawski, Some remarks on Shalika germs, preprint.

[Sh] J. Shalika, A theorem on semisimple p-adic groups, *Ann. Math.* **95** (1972), 226–242.

[Si] A. Silberger, Introduction to Harmonic Analysis on Reductive p-adic Groups, *Math. Notes*, Princeton University Press, Princeton, N.J., 1979.

[T] J. Tunnell, Report on the local Langlands conjecture for GL_2, *Proc. Symp. Pure Math.*, **XXXIII**, Part 2, Amer. Math. Soc., Providence, R.I., 1979, 135–138.

[V] V. Varadarajan, Harmonic Analysis on Real Reductive Groups, *Lecture Notes in Math.*, **576**, Springer-Verlag, New York, 1976.

[Vi] M. F. Vigneras, Caracterisation des integrales orbitales sur un groupe reductif p-adique, *J. Fac. Sci. Tokyo*, Sec. IA, **28**, (1982), 945–961.

[Vo] D. Vogan, Representation of Real Reductive Lie Groups, *Progress in Math.*, **15**, Birkhauser, Boston, Basel, Stuttgart, 1981.

[Wa] N. Wallach, Asymptotic expansions of generalized matrix entries, special year on representation theory, University of Maryland, Vol. I, *Lecture Notes in Mathematics*. Springer-Verlag, Berlin, Heidelberg, New York, Tokyo, 1983.

[W] A. Weil, *Basic Number Theory*, 2nd ed., Grund. Math. Wiss., **144**, Springer-Verlag, Berlin, Heidelberg, New York, 1973.

Permissions

Springer-Verlag would like to thank the original publishers of Harish-Chandra's papers for granting permissions to reprint specific papers in this collection. The following list contains the credit lines for those articles.

[1944a] Reprinted from *Proc. Royal Soc. A*. **183**, ©1944 by The Royal Society of London.
[1944b] Reprinted from *Proc. Royal Soc. A*. **183**, ©1944 by The Royal Society of London.
[1945a] Reprinted from *Proc. Indian Acad. Sci. Sect. A*. **21**, ©1945 by The Indian Academy of Sciences.
[1945b] Reprinted from *Proc. Indian Acad. Sci. Sect. A*. **22**, ©1945 by The Indian Academy of Sciences.
[1946a] Reprinted from *Proc. Royal Soc. A*. **185**, ©1946 by The Royal Society of London.
[1946b] Reprinted from *Proc. Royal Soc. A*. **185**, ©1946 by The Royal Society of London.
[1946c] Reprinted from *Proc. Indian Acad. Sci. Sect. A*. **23**, ©1946 by The Indian Academy of Sciences.
[1946d] Reprinted from *Proc. Royal Soc. A*. **186**, ©1946 by The Royal Society of London.
[1947a] Reprinted from *Proc. Camb. Phil. Soc*. **43**, ©1947 by Cambridge University Press.
[1947b] Reprinted from *Phys. Rev.*, **71**, ©1947 by The American Physical Society.
[1947c] Reprinted from Report of an International Conf. on Fundamental Particles and Low Temperatures held at the Cavendish Laboratory, Cambridge, (1946). Vol I, Fundamental Particles, ©1947 by The Institute of Physics.
[1947d] Reprinted from *Proc. Royal Soc. A*. **189**, ©1947 by The Royal Society of London.
[1948a] Reprinted from *Proc. Royal Soc. A*. **192**, ©1948 by The Royal Society of London.
[1948b] Reprinted from *Phys. Rev.* **74**, ©1948 by The American Physical Society.
[1949a] Reprinted from *Ann. of Math*. **50**, ©1949 by Princeton University Press.
[1949b] Reprinted from *Ann. of Math*. **50**, ©1949 by Princeton University Press.
[1950a] Reprinted from *Proc. Amer. Math. Soc*. **1**, ©1950 by The American Mathematical Society.
[1950b] Reprinted from *Proc. Amer. Math. Soc*. **1**, ©1950 by The American Mathematical Society.
[1950c] Reprinted from *Ann. of Math*. **51**, ©1950 by Princeton University Press.
[1951a] Reprinted from *Trans. Amer. Math. Soc*. **70**, ©1951 by The American Mathematical Society.

Some applications of the Schwartz space of a semisimple Lie group

by Harish-Chandra

Let G be a connected semisimple Lie group with finite center. Then in a paper in Acta [1(a)], I have introduced a space $\mathscr{C}(G)$ of functions on G, which I call the Schwartz space of G. I shall not define it here but recall some of its properties.

1) $C_c^\infty(G) \subset \mathscr{C}(G) \subset C^\infty(G)$. $\mathscr{C}(G)$ is a complete, locally convex Hausdorff space and $C_c^\infty(G)$ is dense in $\mathscr{C}(G)$. Moreover the inclusion mapping of $C_c^\infty(G)$ into $\mathscr{C}(G)$ is continuous.

2) Let $\mathfrak{g}$ be the Lie algebra of G. Then for any $X \in \mathfrak{g}$, the corresponding left- and right-invariant vector-fields L_X and R_X, define continuous endomorphisms of $\mathscr{C}(G)$. $\mathscr{C}(G)$ is stable under left and right translations of G and the corresponding representations of G on $\mathscr{C}(G)$ are continuous.

3) $\mathscr{C}(G) \subset L_2(G)$.

4) $\mathscr{C}(G)$ is closed under convolution.[1]

5) "Wave-packets" lie in $\mathscr{C}(G)$. (This property, which is crucial, is more precisely expressed in Theorem 6.)

Let u be a differential operator on G. Its adjoint u^* is the differential operator given by the relation

$$\int_G uf . g\, dx = \int_G f . u^* g\, dx \qquad\qquad (f, g \in C_c^\infty(G)) \ ,$$

where dx is the Haar measure of G. Let T be a distribution on G. Then the mapping $f \longrightarrow T(u^* f)$ $(f \in C_c^\infty(G))$ is also a distribution which we denote by uT. As usual we identify a locally summable function F on G with the distribution T_F given by

$$T_F(f) = \int F f\, dx \qquad\qquad (f \in C_c^\infty(G))$$

Fix a maximal compact subgroup K of G and let $\mathcal{Z}$ denote the algebra of all differential operators on G which commute with both left and right translations of G. A distribution T on G is said to be $\mathcal{Z}$-finite if the space of all distributions of the form zT $(z \in \mathcal{Z})$ has finite dimension. Similarly it is called K-finite if the left and right translates of T under K, span a vector-space of finite dimension

Theorem 1. <u>Any K-finite and $\mathcal{Z}$-finite function in $L_2(G)$ lies in $\mathcal{C}(G)$.</u>

This shows that $\mathcal{C}(G)$ contains the "wave-packet" corresponding to a point in the discrete spectrum.

By a parabolic subgroup (psgp) P of G, we mean a closed subgroup of G with the following two properties.

1) If $\mathfrak{p}$ is the Lie algebra of P, then $\mathfrak{p}_c$ contains a Borel subalgebra (i.e. a maximal solvable subalgebra) of $\mathfrak{g}_c$.

2) P is the normalizer of $\mathfrak{p}$ in G.

By the radical N of P, we mean the maximal normal subgroup of P such that $\mathrm{Ad}(n)$ is unipotent for every $n \in N$. For any $f \in \mathcal{C}(G)$, put

$$f^P(x) = \int_N f(xn)\,dn \qquad\qquad (x \in G) \ ,$$

where dn is the Haar measure on N. (This integral always exists.) We say that f is a cusp form if $f^P = 0$ for every psgp $P \neq G$. Let $^{\circ}\mathcal{C}(G)$ denote the space of all cusp forms on G. Then $^{\circ}\mathcal{C}(G)$ is a closed subspace of $\mathcal{C}(G)$.

Theorem 2. <u>Every $\mathcal{Z}$-finite function in $\mathcal{C}(G)$ is a cusp form.</u> <u>Conversely $\mathcal{Z}$-finite functions are dense in $^{\circ}\mathcal{C}(G)$.</u>

Put $\mathcal{H} = L_2(G)$ and let $^{\circ}\mathcal{H}$ denote the closure of $^{\circ}\mathcal{C}(G)$ in $\mathcal{H}$. Let λ denote the left-regular representation of G on $\mathcal{H}$.

Theorem 3. <u>$^{\circ}\mathcal{H}$ is an orthogonal sum of invariant and irreducible subspaces.</u> <u>Conversely every irreducible subspace of $\mathcal{H}$ is contained in $^{\circ}\mathcal{H}$. Let $^{\circ}E$ denote the orthogonal projection of $\mathcal{H}$ on $^{\circ}\mathcal{H}$. Then $^{\circ}E\,\mathcal{C}(G) = {}^{\circ}\mathcal{C}(G)$ and</u>

$^{\circ}E$ <u>defines a continuous mapping of</u> $\mathcal{C}(G)$ <u>onto</u> $^{\circ}\mathcal{C}(G)$.

Theorem 4. $^{\circ}\mathcal{C}(G) \neq \{0\}$ <u>if and only if</u> rank G = rank K.

Let τ be a unitary double representation of K on a finite-dimensional Hilbert space V. Let $\mathcal{C}(G, \tau)$ denote the set of all $f \in \mathcal{C}(G) \otimes V$ such that $f(k_1 x k_2) = \tau(k_1) f(x) \tau(k_2)$ $(k_1, k_2 \in K, x \in G)$. We denote by $^{\circ}\mathcal{C}(G, \tau)$ the space of all cusp forms in $\mathcal{C}(G, \tau)$.

Lemma 1. dim $^{\circ}\mathcal{C}(G, \tau) < \infty$.

Let θ denote the Cartan involution of G with respect to K. An abelian subgroup A of G is said to split, if for every $a \in A$, Ad(a) can be diagonalized over $\mathbb{R}$. Let P be a psgp and N its radical. Put $M_1 = P \cap \theta(P)$ and let A be a maximal connected split abelian subgroup lying in the center of M_1. Then A is unique and M_1 is the centralizer of A in G. Let $X(M_1)$ denote the group of all continuous homomorphisms of M_1 into the multiplicative group $\mathbb{R}^{\times}$ of real numbers. Put

$$M = \bigcap_{\chi \in X(M_1)} \ker |\chi|$$

where $|\chi|(m) = |\chi(m)|$ $(\chi \in X(M_1), m \in M_1)$. Then $M_1 = MA$, P = MAN and the corresponding mapping of $M \times A \times N$ into P is a diffeomorphism. We call this the Langlands decomposition of P, and A the split component of P. Let α be the Lie algebra of A. Then the exponential mapping defines a bijection of α on A. M is a reductive group with finitely many connected components and $K_M = K \cap M$ is a maximal compact subgroup of M. Let τ_M denote the restriction of τ on K_M. Then the spaces $\mathcal{C}(M)$, $^{\circ}\mathcal{C}(M)$ and $^{\circ}\mathcal{C}(M, \tau_M)$ can be defined in the same way as above.

Let $\mathfrak{h}$ be a Cartan subalgebra of $\mathfrak{g}$. By $\mathfrak{h}_R$ we mean the set of all points in $\mathfrak{h}$ where every root of $(\mathfrak{g}, \mathfrak{h})$ takes a real value. A psgp P = MAN is called cuspidal if $^{\circ}\mathcal{C}(M) \neq \{0\}$.

Lemma 2. P <u>is cuspidal if and only if there exists a</u> θ-<u>stable Cartan subalgebra</u> $\mathfrak{h}$ <u>of</u> $\mathfrak{g}$ <u>such that</u> $\alpha = \mathfrak{h}_R$.

Let $f \in \mathcal{C}(G)$ and $P = MAN$ a psgp of G. We write $f^P \sim 0$ if[2)]

$$\int_M \text{conj } \phi(m) . f^P(xm) dm = 0$$

for all $x \in G$ and $\phi \in {}^0\mathcal{C}(M)$. Here dm is the Haar measure on M and the integral always exists.

Lemma 3. _Let_ f _be an element in_ $\mathcal{C}(G)$ _such that_ $f^P \sim 0$ _for all cuspidal_ _subgroups_ P _of_ G. _Then_ $f = 0$.

This is the analogue of a theorem of Langlands [1(b), Theorem 4, p. 16].

Two psgps P_1, P_2 are said to be associated, if their split components A_1, A_2 are conjugate under G (or equivalently under K).

Let $\mathfrak{h}_i = \theta(\mathfrak{h}_i)$ $(1 \leq i \leq r)$ be a complete set of Cartan subalgebras, no two of which are conjugate under G. Put $\alpha_i = \mathfrak{h}_{i,R}$ and $A_i = \exp \alpha_i$. Let $\mathcal{C}_i(G)$ denote the set of all $f \in \mathcal{C}(G)$ with the following property. If $P = MAN$ is any cuspidal subgroup of G, then $f^P \sim 0$ unless A is conjugate to A_i under K. Then $\mathcal{C}_i(G)$ is a closed subspace of $\mathcal{C}(G)$.

Theorem 5. $\mathcal{C}(G) = \sum_{1 \leq i \leq r} \mathcal{C}_i(G)$ _where the sum is direct and smooth._

Let E_i denote the projection of $\mathcal{C}(G)$ on $\mathcal{C}_i(G)$ corresponding to the above direct sum. Then smoothness means that E_i are continuous endomorphisms of $\mathcal{C}(G)$. This theorem and its proof are closely related to the Plancherel formula for G.

Let $P = MAN$ be a psgp of G. Then $G = KP$ and therefore any $x \in G$ can be written as $x = kman$ $(k \in K, m \in M, a \in A, n \in N)$. The element a being unique, we define $H_P(x) = \log a \in \alpha$. Also put

$$\rho_P(H) = \frac{1}{2} \text{tr}(\text{ad } H)_{\mathfrak{n}} \qquad (H \in \alpha) \ ,$$

$\mathfrak{n}$ being the Lie algebra of N.

Now suppose P is cuspidal. Then from Lemma 1, $L = {}^0\mathcal{C}(M, \tau_M)$ is a finite-dimensional space. Given $\phi \in L$, we extend it to a function on G as follows:

$$\phi(kman) = \tau(k)\phi(m) \qquad (k \in K,\ m \in M,\ a \in A,\ n \in N)\ .$$

Let α^* denote the space dual to α. For any $\Lambda \in \alpha_c^*$ and $\phi \in L$, define

$$E(P : \phi : \Lambda : x)$$
$$= \int_K \phi(xk)\tau(k^{-1})\exp\{((-1)^{1/2}\Lambda - \rho_P)(H_P(xk))\}dk \qquad (x \in G)\ ,$$

where dk is the normalized Haar measure on K. Then for fixed ϕ and x, this is an entire function of $\Lambda \in \alpha_c^*$. Let $\mathcal{Z}_M$ be the analogue of $\mathcal{Z}$ for M. If ϕ is an eigenfunction of $\mathcal{Z}_M$, then $E(P : \phi : \Lambda)$ is an eigenfunction of $\mathcal{Z}$. We call $E(P : \phi : \Lambda)$ the Eisenstein integral in analogy with the Eisenstein series (see [1(b), p. 29] and [2(a), (b)]).

Let $d\Lambda$ denote the Euclidean measure on α^* and $\mathcal{C}(\alpha^*)$ the usual Schwartz space for α^*.

Theorem 6. <u>There exists a polynomial function</u> $q \neq 0$ <u>on</u> α^*, <u>with the following property. For any</u> $f \in \mathcal{C}(\alpha^*) \otimes L$, <u>the integral</u>

$$E(P : f : x) = \int_{\alpha^*} q(\Lambda)E(P : f(\Lambda) : \Lambda : x)d\Lambda \qquad (x \in G)$$

<u>converges</u>, $E(P : f) \in \mathcal{C}(G,\tau)$ <u>and</u>

$$f \longrightarrow E(P : f)$$

<u>is a continuous mapping of</u> $\mathcal{C}(\alpha^*) \otimes L$ <u>into</u> $\mathcal{C}(G,\tau)$.

Roughly speaking, this theorem says that wave-packets lie in $\mathcal{C}(G)$.

Lemma 4. <u>Let</u> P' <u>be another</u> psgp <u>of</u> G. <u>Then if</u> $f \in \mathcal{C}(\alpha^*) \otimes L$,

$$E(P : f)^{P'} \sim 0$$

<u>unless</u> P <u>and</u> P' <u>are associated.</u>

Fix i $(1 \leq i \leq r)$ such that A is conjugate to A_i under K. (Since P is cuspidal there exists a unique such index i.) Put $\mathcal{C}_i(G,\tau) = \mathcal{C}(G,\tau) \cap (\mathcal{C}_i(G) \otimes V)$. Then Lemma 4 shows that

$E(P : f) \in \mathscr{C}_i(G, \tau)$.

Let $P = MAN$ as above. By a root of (P, A) we mean an element $a \in \mathfrak{a}^*$ such that $[H, X] = a(H)X$ for all $H \in \mathfrak{a}$ and some $X \neq 0$ in $\mathfrak{n}$. Fix a norm on $\mathfrak{a}$ and let a be a variable element in A. We say that $a \xrightarrow[P]{} \infty$ if $|\log a| \longrightarrow \infty$ and there exists a number $\epsilon > 0$ such that $a(\log a) \geq \epsilon |\log a|$ for every root a of (P, A). Put

$$w = w(A) = (\text{Normalizer of } A \text{ in } G)/(\text{Centralizer of } A \text{ in } G)$$

Then w operates on $\mathfrak{a}$ and $\mathfrak{a}^*$ in the usual way.

Let $P = MAN$, $P' = MAN'$ be two psgps with the same split component A. Then if $\Lambda \in \mathfrak{a}^*$ is not too special (i.e. does not lie on a certain finite set of hyperplanes passing through the origin), there exist unique endomorphisms $c_{P'|P}(s : \Lambda)$ $(s \in w)$ of L such that

$$\lim_{a \xrightarrow[P']{} \infty} \left| e^{\rho'(\log a)} E(P : \phi : \Lambda : ma) - \sum_{s \in w} (c_{P'|P}(s : \Lambda)\phi)(m)e^{(-1)^{1/2} s\Lambda(\log a)} \right| = 0$$

for all $m \in M$ and $\phi \in L$. (Here $\rho' = \rho_{P'}$.) Regarded as functions of Λ, $c_{P'|P}(s : \Lambda)$ are meromorphic on $\mathfrak{a}^* + (-1)^{1/2}U$ where U is an open neighborhood of zero in $\mathfrak{a}^*$. But it seems likely that they are actually meromorphic in the whole complex space $\mathfrak{a}_c^*$. Moreover $\det c_{P|P}(1 : \Lambda)$ is not identically zero[3] in Λ. Put

$$E^o(P : \phi : \Lambda) = E(P : c_{P|P}(1 : \Lambda)^{-1}\phi : \Lambda) .$$

Then I believe the following statements are true although the full proofs have not yet been worked out.

1) $E^o(P : \phi : \Lambda)$ is a meromorphic function of $\Lambda \in \mathfrak{a}_c^*$ which is holomorphic on $\mathfrak{a}^*$.

2) It satisfies functional equations very similar to those known for the Eisenstein series (see [1(b), p. 114] and [2(a), (b)]).

Finally, I am inclined to believe that entirely analogous results hold for the corresponding Eisenstein integrals in the $\mathfrak{p}$-adic case as well. But this is, at present, a largely unexplored territory.

The Institute for Advanced Study
    Princeton, N. J.

References

1.   Harish-Chandra, (a) "Discrete series for semisimple Lie groups II,"
     Acta Math., vol. 116 (1966), pp. 1-111.
     (b) "Automorphic forms on semisimple Lie groups," Notes by
     J. G. M. Mars, Lecture Notes in Mathematics, vol. 62 (1968),
     Springer-Verlag.

2.   R. P. Langlands, (a) "On the functional equations satisfied by
     Eisenstein Series," Mimeographed manuscript, 1965
     (unpublished).
     (b) "Eisenstein Series," Algebraic groups and discontinuous
     groups (1966), Amer. Math. Soc., pp. 235-252.

Footnotes

[1] In case of $SL(2, \mathbb{R})$, this fact had been observed by Langlands.

[2] conj c denotes the conjugate of a complex number c.

[3] This was pointed out to me by Langlands.

Reprinted from
Lecture Notes in Mathematics, No. **140**
Springer-Verlag, Berlin-Heidelberg-New York, 1970, pp. 1–7

Eisenstein Series over Finite Fields

HARISH-CHANDRA

1. Introduction

Let me begin by recalling the definition and some properties of the classical Eisenstein Series. Let $G = SL(2, \mathbf{R})$ and $\Gamma = SL(2, \mathbf{Z})$. Then $G = KAU$ where

$$K = SO(2), \quad A = \left\{ a_t = \begin{pmatrix} e^t & 0 \\ 0 & e^{-t} \end{pmatrix}, t \in \mathbf{R} \right\}, \quad U = \left\{ \begin{pmatrix} 1 & u \\ 0 & 1 \end{pmatrix}, u \in \mathbf{R} \right\}.$$

Let P be the normalizer of U in G. Then $P = MAU$ where $M = \{1, -1\}$. Every element x of G can be written uniquely as $x = kau$ ($k \in K$, $a \in A$, $u \in U$). Put $t(x) = t$ where $a = a_t$. For $\lambda \in C$ with* $\Re \lambda < -1$, put

$$E(\lambda : x) = \sum_{\gamma \in \Gamma/\Gamma \cap P} e^{(\lambda - 1)t(x\gamma)} = \tfrac{1}{2} \sum_{\gamma \in \Gamma/\Gamma_\infty} e^{(\lambda - 1)t(x\gamma)}$$

where $\Gamma_\infty = \Gamma \cap U$. If

$$x = \begin{pmatrix} a & b \\ c & d \end{pmatrix}, \quad x_1 = \begin{pmatrix} a \\ c \end{pmatrix} \quad \text{and} \quad |x_1| = (a^2 + c^2)^{\frac{1}{2}},$$

then

$$E(\lambda : x) = \tfrac{1}{2} \sum_{\gamma \in \Gamma/\Gamma_\infty} |(x\gamma)_1|^{\lambda - 1}.$$

For any $\varepsilon > 0$, this series converges absolutely and uniformly if x remains in a compact subset of G and $\Re \lambda \leq -1 - \varepsilon$. Hence the function E is holomorphic in λ for $\Re \lambda < -1$. Moreover $E(\lambda : kx) = E(\lambda : x)$ ($k \in K$) and

$$e^t \int\limits_{U/\Gamma_\infty} E(\lambda : a_t u)\, du = e^{\lambda t} + c(\lambda) e^{-\lambda t} \quad (t \in \mathbf{R}, \Re \lambda < -1).$$

Here $c(\lambda)$ is a holomorphic function of λ for $\Re \lambda < -1$ and the Haar measure du on U is so normalized that the total measure of U/Γ_∞ is 1. The functions c and E have the following properties.

1) c extends to a meromorphic function on C satisfying the functional equation $c(\lambda) c(-\lambda) = 1$. Moreover $|c(\lambda)| = 1$ when $\Re \lambda = 0$.

---

* $\Re \lambda$ denotes the real part of a complex number λ.

2) For any $x \in G$, $E(\lambda:x)$ extends to a meromorphic function of $\lambda \in C$ with the functional equation

$$E(\lambda:x) = c(\lambda) E(-\lambda:x).$$

In fact

$$c(\lambda) = \pi^\lambda \Gamma(-\lambda/2)\zeta(-\lambda) \{\Gamma(\lambda/2)\zeta(\lambda)\}^{-1}$$

where ζ is the Riemann Zeta function.

Similar results hold in other cases. During the past fifteen years, such series have been studied by SELBERG [5]. More recently LANG-LANDS [4(a), (b)] has proved corresponding results for the Eisenstein Series on G/Γ where G is any semisimple algebraic group defined over Q and Γ an arithmetic subgroup of G.

In 1966—1967 I gave some lectures on LANGLANDS' work in Princeton (see [3]) and thereby learnt, what I now call, the philosophy of cusp forms. In the case of G/Γ this philosophy is certainly implicit in SELBERG [5]. Moreover it has been expounded, in a more general context, by GELFAND in his 1962 Stockholm address [2], although I could not understand it then. Actually I realized its full scope only when I tried to relate LANGLANDS' work to my own on harmonic analysis on G. This philosophy works in the following four cases.

1) G/Γ. In fact it originated there.

2) A real semisimple group G.

3) I believe it is also applicable to a reductive $\mathfrak{p}$-adic group G, although this case has not yet been sufficiently investigated.

4) It works for reductive algebraic groups defined over a finite field.

In this lecture we shall be concerned with 4). Our main object is to define the Eisenstein Series and prove their functional equations (see §7, Theorem 3).

2. Bruhat's Lemma and its consequences

We begin by recalling some known facts (see [1]). Let K be a field which will be kept fixed throughout. By a K-group we mean a (linear) algebraic group defined over K. Let G be a connected reductive K-group. By a parabolic subgroup P of G we mean an algebraic subgroup which contains a Borel subgroup of G. We say that P is K-parabolic if it is parabolic and defined over K. Fix a K-parabolic subgroup P and let U denote the unipotent radical of P. Then U is a K-subgroup. By a Levi K-subgroup M of P, we mean a reductive K-subgroup such that the mapping $(m, u) \to mu$ $(m \in M, u \in U)$ defines a K-isomorphism of the algebraic varieties $M \times U$ and P. Such a subgroup M always exists and is connected. Fix M and let A be a maximal K-split torus lying in the center of M. Then A is unique and M is the centralizer of A in G. We call A a split component of P. Let G denote the group of K-rational points of G. For any split component A' of P there exists a unique element $u \in U = U \cap G$ such that* $A' = A^u$. Hence dim A depends only on P. We call it the parabolic rank of P and denote it by prk P.

A cuspidal subgroup P of G is a group of the form $P = G \cap \mathbf{P}$, where $\mathbf{P}$ is a K-parabolic subgroup of G. P determines $\mathbf{P}$ completely. By a split component A of P, we mean a split component of $\mathbf{P}$. We write prk $P = \dim A$ and call (P, A) a cuspidal pair in G. Once A is fixed, we have the corresponding Levi K-decompositions $\mathbf{P} = \mathbf{M}\mathbf{U}$ and $P = MU$ where $M = \mathbf{M} \cap G$. We shall call U the unipotent radical of P.

Let (P_i, A_i) $(i = 1, 2)$ be two cuspidal pairs in G. We write $(P_1, A_1) > (P_2, A_2)$ if $P_1 \supset P_2$ and $A_1 \subset A_2$. A cuspidal pair is called mincuspidal if it is minimal with respect to this partial order. Let $\mathfrak{w}(A_1, A_2)$ denote the set of all bijections $s : A_1 \to A_2$ with the following property. There should exist an element $y \in G$ such that $a^s = a^y$ for all $a \in A_1$. P_1, P_2 are called associated if $\mathfrak{w}(A_1, A_2) \neq \varnothing$ i.e. if A_1 and A_2 are conjugate under G. (The two pairs (P_i, A_i) $i = 1, 2$, are then also called associated.) It is known that $\mathfrak{w}(A_1, A_2)$ is a finite set. Fix $s \in \mathfrak{w}(A_1, A_2)$. We say that $y \in G$ is a representative of s in G if $a^s = a^y$ for all $a \in A_1$. In case $A_1 = A_2 = A$, we write $\mathfrak{w}(A) = \mathfrak{w}(A, A)$. Then $\mathfrak{w}(A)$ is a finite group. The following important result has been proved in [1, p. 100].

Bruhat's Lemma. *Let (P, A) be a mincuspidal pair in G and $P = MU$ the corresponding Levi K-decomposition. For each $s \in \mathfrak{w}(A)$, fix a representative y_s of s in G. Then*

$$G = \bigcup_{s \in \mathfrak{w}(A)} U y_s P$$

where the union is disjoint.

Fix two cuspidal pairs (P_i, A_i) $(i = 1, 2)$ in G and let $P_i = M_i U_i$ be the corresponding decompositions. The following facts are simple consequences of BRUHAT's lemma.

Lemma 1. *$P_2 \backslash G / P_1$ is a finite set. Fix $x \in G$. Then $P_2 \cap U_1^x \subset U_2$ if and only if there exist elements $u_i \in U_i$ $(i = 1, 2)$ such that $A_1^{x u_1} \subset A_2^{u_2}$.*

Corollary. *The following two statements are equivalent.*

1) $P_2 \cap U_1^x \subset U_2$, $U_2^{x^{-1}} \cap P_1 \subset U_1$.

2) *There exist $u_i \in U_i$ $(i = 1, 2)$ such that $A_1^y = A_2$ for $y = u_2 x u_1$.*

---

* We write $x^y = y x y^{-1}$ for $x, y \in G$. If S is any subset of G then $S^y = y S y^{-1}$.

3. The cusp forms

Let S be a finite set. By the standard measure on S, we mean the measure which assigns to each point of S the mass 1. Let $C(S)$ denote the space of all complex-valued functions on S and $[S]$ the number of elements of S.

Now assume that K is a finite field. Then G is a finite group. For any $f \in C(G)$ and any cuspidal subgroup P of G, put

$$f_P(x) = \int_U f(xu)\, du \qquad (x \in G)$$

where du is a Haar measure on the unipotent radical U of P. We say that f is a cusp form if $f_P = 0$ for every cuspidal subgroup $P \neq G$ of G.

Let $^0C(G)$ be the space of all cusp forms. It is easy to see that $^0C(G)$ is stable under both left- and right-translations of G. Hence $^0C(G)$ is a two-sided ideal in the group algebra $C(G)$.

Let λ denote the left-regular representation of G on $C(G)$ and $^0\lambda$ the restriction of λ on $^0C(G)$. Let $\mathscr{E}(G)$ denote the set of all equivalence classes of irreducible (complex) representations of G and $^0\mathscr{E}(G)$ the subset of those classes which occur in the reduction of $^0\lambda$. It is easy to see that an element $\omega \in \mathscr{E}(G)$ lies in $^0\mathscr{E}(G)$ if and only if the character of ω is a cusp form. For any $\omega \in \mathscr{E}(G)$, let $C(G, \omega)$ denote the space of all elements in $C(G)$ which transform under λ according to ω.

Let (P, A) be a cuspidal pair in G and $P = MU$ the corresponding decomposition. Fix $f \in C(G)$. Then we write $f_P \sim 0$ if*

$$\int_M \operatorname{conj} \phi(m) \cdot f_P(xm)\, dm = 0$$

for all $\phi \in {}^0C(M)$ and $x \in G$. (Here dm is the standard measure on M.) It is easy to verify that this definition is independent of the choice of A.

Lemma 2. *Let f be an element in $C(G)$ such that $f_P \sim 0$ for all cuspidal subgroups P of G (including $P = G$). Then $f = 0$.*

This is entirely analogous to a result of LANGLANDS [4(a), p. 3.24].

4. The irreducibility of induced representations

Let (P, A) be a cuspidal pair in G and $P = MU$ the corresponding decomposition. Fix $\omega \in {}^0\mathscr{E}(M)$ and let ω^* denote the class contragredient to ω. Let σ be a right representation of M on a finite-dimensional complex vector space L such that ω^* is the class of σ. (This means that $\operatorname{tr} \sigma(m) = \operatorname{conj} \theta(m)$ $(m \in M)$ where θ is the character of ω.) Let $D(P, \omega)$ denote the space of all functions $f: G \to L$ such that $f(xmu) = f(x)\sigma(m)$

---

* conj c denotes the complex conjugate of a number $c \in C$.

$(x \in G, m \in M, u \in U)$. We get a representation λ of G on $D(P, \omega)$ by setting

$$(\lambda(x)f)(y) = f(x^{-1}y) \qquad (y \in G)$$

for $x \in G$ and $f \in D(P, \omega)$. Let $\Omega(P, \omega)$ denote the class of λ.

Let (P_i, A_i) $(i = 1, 2)$ be two cuspidal pairs and $P_i = M_i U_i$ the corresponding decompositions. Fix $\omega \in {}^0\mathscr{E}(M_1)$, $s \in \mathfrak{w}(A_1, A_2)$ and let y be a representative of s in G. Then $M_2 = M_1^y$. If σ is a representation of M_1 in ω, we define the representation σ^y of M_2 by $\sigma^y(m^y) = \sigma(m)$ $(m \in M_1)$. Let ω^y denote the class of σ^y. Then ω^y depends only on s and we also denote it by ω^s. It is easy to verify that $\omega^s \in {}^0\mathscr{E}(M_2)$.

The following theorem and its corollary play an important role in our discussion.

Theorem 1. *Fix* $\omega_i \in {}^0\mathscr{E}(M_i)$ $(i = 1, 2)$ *and let* $\mathfrak{w}_0$ *be the set of all* $s \in \mathfrak{w}(A_1, A_2)$ *such that* $\omega_1^s = \omega_2$. *Then*

$$I\big(\Omega(P_1, \omega_1), \Omega(P_2, \omega_2)\big) = [\mathfrak{w}_0].$$

Here I *denotes the intertwining number.*

It follows in particular that if P_1, P_2 are not associated, then this intertwining number is zero.

Let (P, A) be a cuspidal pair $(P = M U)$. Fix $\omega \in {}^0\mathscr{E}(M)$ and let $\mathfrak{w}(\omega)$ denote the subgroup of all $s \in \mathfrak{w}(A)$ such that $\omega^s = \omega$. Then $r = [\mathfrak{w}(\omega)]$ is called the ramification index of ω. We say that ω is unramified (in G) if $r = 1$.

Corollary. $\Omega(P_1, \omega_1) = \Omega(P_2, \omega_2)$ *if* $\mathfrak{w}_0 \neq \varnothing$. *Moreover* $\Omega(P_1, \omega_1)$ *is irreducible if* ω_1 *is unramified.*

5. Definition and some properties of Eisenstein Series

Now fix, once for all, a subgroup Γ of G. Let (P, A) be a cuspidal pair $(P = M U)$. We denote by ${}^0\mathscr{D}(P)$ the space of all functions $f \in C(G)$ such that:

1) $f(xy) = f(x)$ for all $x \in G$ and $y \in (\Gamma \cap P) U$.

2) For every $x \in G$, the function $m \rightarrow f(xm)$ $(m \in M)$ lies in ${}^0C(M)$. It is easy to see that the space ${}^0\mathscr{D}(P)$ does not depend on the choice of A. For $f \in {}^0\mathscr{D}(P)$, define

$$E_f(x) = \sum_{\gamma \in \Gamma/\Gamma \cap P} f(x\gamma) \qquad (x \in G).$$

Then $E_f \in C(G/\Gamma)$. For any $g \in C(G/\Gamma)$, put

$$g_P(x) = \int_{U/\Gamma \cap U} g(xu)\, d^0u \qquad (x \in G)$$

where d^0u is the Haar measure on U so normalized that the total measure of $U/\Gamma \cap U$ is 1.

Theorem 2. *Let (P, A) and (P', A') be two cuspidal pairs $(P = MU,$ $P' = M'U')$. Fix $f \in {}^0\mathscr{D}(P)$. Then $(E_f)_{P'} \sim 0$ unless P' is associated to P. Now suppose P and P' are associated. Put $\mathfrak{w} = \mathfrak{w}(A, A')$ and let d^0u' denote the Haar measure on U' so normalized that the total measure of $U'/\Gamma \cap U'$ is 1. Then*

$$\int\limits_{U'/\Gamma \cap U'} E_f(xu')\, d^0u' = \sum_{s \in \mathfrak{w}} \big(c_{P'|P}(s)f\big)(x) \qquad (x \in G).$$

Here $c_{P'|P}(s)$ is a linear transformation of ${}^0\mathscr{D}(P)$ into ${}^0\mathscr{D}(P')$ defined as follows. Put

$$\Gamma(s) = \Gamma \cap (P' y P)$$

where y is a representative of s in G. Then

$$\big(c_{P'|P}(s)f\big)(x) = \int\limits_{U'/U' \cap \Gamma} \sum_{\gamma \in \Gamma(s)/\Gamma \cap P} f(xu'\gamma)\, d^0u'$$

for $f \in {}^0\mathscr{D}(P)$ and $x \in G$.

Fix $\omega \in {}^0\mathscr{E}(M)$ and let $\mathscr{D}(P, \omega)$ be the space of all $f \in {}^0\mathscr{D}(P)$ with the following property. For every $x \in G$, the function $m \to f(xm)$ $(m \in M)$ lies in $C(M, \omega)$. Then $c_{P'|P}(s)$ maps $\mathscr{D}(P, \omega)$ into $\mathscr{D}(P', \omega.)$. In fact $c_{P'|P}(s)$ is the zeta-function of GELFAND [2, §8].

For $f, g \in {}^0\mathscr{D}(P)$, define

$$(f, g)_{G/\Gamma \cap P} = \int\limits_{G/\Gamma \cap P} \operatorname{conj} f(x) \cdot g(x)\, dx$$

where dx is the standard measure on $G/\Gamma \cap P$. In this way ${}^0\mathscr{D}(P)$ becomes a Hilbert space.

Lemma 3. *Let (P_i, A_i) $(i = 1, 2)$ be two associated cuspidal pairs. Then*

$$(f_2, c_{P_2|P_1}(s)f_1)_{G/\Gamma \cap P_2} = (c_{P_1|P_2}(s^{-1})f_2, f_1)_{G/\Gamma \cap P_1}$$

for $f_i \in {}^0\mathscr{D}(P_i)$ $(i = 1, 2)$ and $s \in \mathfrak{w}(A_1, A_2)$.

We regard $C(G/\Gamma)$ as a Hilbert space in the usual way so that

$$(\psi, \phi)_{G/\Gamma} = \int\limits_{G/\Gamma} \operatorname{conj} \psi(x) \cdot \phi(x)\, dx \qquad (\psi, \phi \in C(G/\Gamma))$$

where dx is the standard measure on G/Γ.

Corollary. *Let (P_i, A_i) $(i = 1, 2)$ be two cuspidal pairs in G. Fix $f_i \in {}^0\mathscr{D}(P_i)$ and put*

$$E_{f_i}(x) = \sum_{\gamma \in \Gamma/\Gamma \cap P_i} f_i(x\gamma) \qquad (x \in G).$$

Then

$$(E_{f_2}, E_{f_1})_{G/\Gamma} = \sum_{s \in w(A_1, A_2)} (f_2, c_{P_2 | P_1}(s) f_1)_{G/\Gamma \cap P_2}.$$

In particular,

$$(E_{f_2}, E_{f_1})_{G/\Gamma} = 0$$

if P_1, P_2 are not associated.

We return to the notation of Theorem 2.

Lemma 4. *Let $g \in C(G/\Gamma)$. Then*

$$(E_f, g)_{G/\Gamma} = (f, g_P)_{G/\Gamma \cap P}$$

for all $f \in {}^0\mathscr{D}(P)$.

Combining this with Lemma 2 we get the following corollary.

Corollary. $C(G/\Gamma)$ *is spanned by E_f for all $f \in {}^0\mathscr{D}(P)$ and all cuspidal subgroups P of G.*

6. The series $E(P:\phi)$

Consider a cuspidal pair (P, A) in G with $P = MU$. Fix $\omega \in {}^0\mathscr{E}(M)$ and let $y_i (1 \leq i \leq q)$ be a complete system of representatives for $\Gamma \backslash G / P$ in G. Define

$$(P_i, A_i, M_i, U_i, \omega_i) = (P^{y_i}, A^{y_i}, M^{y_i}, U^{y_i}, \omega^{y_i})$$

and put

$$\mathscr{D}(P, \omega) = \prod_{1 \leq i \leq q} \mathscr{D}(P_i, \omega_i), \quad {}^0\mathscr{D}(P) = \prod_{1 \leq i \leq q} {}^0\mathscr{D}(P_i),$$

where the products are direct. Then ${}^0\mathscr{D}(P)$ is a Hilbert space with the scalar product

$$(\phi, \psi) = \sum_{1 \leq i \leq q} (\phi_i, \psi_i)_{G/\Gamma \cap P_i} \quad (\phi, \psi \in {}^0\mathscr{D}(P)).$$

(Here ϕ_i is the component of ϕ in ${}^0\mathscr{D}(P_i)$, $1 \leq i \leq q$. Similarly for ψ.) Put

$$E(P:\phi:x) = \sum_{1 \leq i \leq q} \sum_{\gamma \in \Gamma/\Gamma \cap P_i} \phi_i(x\gamma) \quad (x \in G)$$

for $\phi \in {}^0\mathscr{D}(P)$.

Lemma 5. *Let $f \in C(G/\Gamma)$ and $\phi \in {}^0\mathscr{D}(P)$. Then*

$$\big(E(P:\phi), f\big)_{G/\Gamma} = \sum_{1 \leq i \leq q} (\phi_i, f_{P_i})_{G/\Gamma \cap P_i}.$$

7. Functional equations of the Eisenstein Series

Let (P_i, A_i) $(1 \leq i \leq r)$ be a set of associated cuspidal pairs in G with $P_i = M_i U_i$. For each i, fix a complete system of representatives $y_{ij} (1 \leq j \leq q_i)$ for $\Gamma \backslash G / P_i$. Then we can define the Hilbert spaces

$^0\mathcal{D}(P_i)$ and $\mathcal{D}(P_i, \omega_i)$ $(\omega_i \in {}^0\mathcal{E}(M_i))$ as in §6. For any $s \in \mathfrak{w}(A_i, A_j)$, define a linear transformation $c_{P_j | P_i}(s)$ of $^0\mathcal{D}(P_i)$ into $^0\mathcal{D}(P_j)$ as follows. If $\phi \in {}^0\mathcal{D}(P_i)$, then

$$\left(c_{P_j | P_i}(s)\phi\right)_l = \sum_{1 \le i \le q_i} c_{P_{jl} | P_{ik}}(y_{jl} \circ s \circ y_{ik}^{-1})\phi_k \quad (1 \le l \le q_j).$$

Here $y_{jl} \circ s \circ y_{ik}^{-1}$ denotes the element $t \in \mathfrak{w}(A_i^{y_{ik}}, A_j^{y_{jl}})$ given by

$$(a^{y_{ik}})^t = (a^s)^{y_{jl}} \quad (a \in A_i).$$

(We note that $(P_{ik}, A_{ik}) = (P_i^{y_{ik}}, A_i^{y_{ik}})$ and similarly for (P_{jl}, A_{jl}).)

Theorem 3. *Fix* i, j, k $(1 \le i, j, k \le r)$ *and* $\omega \in {}^0\mathcal{E}(M_i)$. *Then if* ω *is not ramified in* G,

$$E(P_i : \phi) = E\left(P_j : c_{P_j | P_i}(s)\phi\right)$$

and

$$c_{P_k | P_j}(t)\, c_{P_j | P_i}(s)\phi = c_{P_k | P_i}(ts)\phi$$

for $\phi \in \mathcal{D}(P_i, \omega)$, $s \in \mathfrak{w}(A_i, A_j)$ *and* $t \in \mathfrak{w}(A_j, A_k)$. *Moreover* $c_{P_j | P_i}(s)$ *then defines a unitary transformation of* $\mathcal{D}(P_i, \omega)$ *onto* $\mathcal{D}(P_j, \omega^s)$.

The situation when ω is ramified seems to be complicated. However, not all ramified ω behave badly. For example suppose prk $P_i = 1$ and ω is ramified. Then $\Omega(P_i, \omega) = \Omega_1 + \Omega_2$ where Ω_1, Ω_2 are two distinct elements of $\mathcal{E}(G)$. Let us further assume that the degrees of Ω_1 and Ω_2 are the same. Then one can show that the statements of Theorem 3 continue to hold in this case.

8. Proof of Theorem 3

Let me briefly sketch the proof of Theorem 3.

Lemma 6. *Let* $\phi \in {}^0\mathcal{D}(P_i)$. *Then*

$$\left(E(P_i : \phi)\right)_{P_{jl}} = \sum_{s \in \mathfrak{w}(A_i, A_j)} \left(c_{P_j | P_i}(s)\phi\right)_l$$

for $1 \le l \le q_j$.

This follows from Theorem 2.

We may regard $C(G/\Gamma)$ as a subspace of $C(G)$. For any $\Omega \in \mathcal{E}(G)$, put $C(G/\Gamma, \Omega) = C(G/\Gamma) \cap C(G, \Omega)$. Now let us use the notation of §6 and suppose

$$\Omega(P, \omega) = \sum_{1 \le i \le N} n_i \Omega_i$$

where n_i are positive integers and Ω_i $(1 \le i \le N)$ are distinct elements of $\mathcal{E}(G)$. Put

$$V(P, \omega) = \sum_{1 \le i \le N} C(G/\Gamma, \Omega_i).$$

Lemma 7. $\phi \to E(P:\phi)$ $\big(\phi \in \mathcal{D}(P,\omega)\big)$ *is a surjective mapping of* $\mathcal{D}(P,\omega)$ *onto* $V(P,\omega)$.

Let $D(P,\omega)$ denote the space of all functions $f \in C(G)$ such that:

1) $f(xu) = f(x)$ $(x \in G,\ u \in U)$.

2) For every $x \in G$, the function $m \to f(xm)$ $(m \in M)$ lies in $C(M,\omega)$.

Fix $y \in G$ and for any $f \in D(P,\omega)$, define the function f_y by $f_y(x) = f(xy)$ $(x \in G)$. Then the mapping $f \to f_y$ is a bijection of $D(P,\omega)$ onto $D(P^y,\omega^y)$. Since

$$G = \bigcup_{1 \le i \le q} \Gamma y_i P,$$

it follows easily that the space

$$W = \sum_{1 \le i \le q} \sum_{\gamma \in \Gamma} D(P^{\gamma y_i}, \omega^{\gamma y_i})$$

is stable under both left and right translations of G. Therefore it is clear that

$$W = \sum_{1 \le i \le N} C(G, \Omega_i).$$

For any $f \in W$, define

$$F_f(x) = \sum_{\gamma \in \Gamma} f(x\gamma) \qquad (x \in G).$$

Then $f \to F_f$ is a surjective mapping of W onto

$$\sum_{1 \le i \le N} C(G/\Gamma, \Omega_i) = V(P,\omega).$$

Moreover $F_{f_\gamma} = F_f$ for $\gamma \in \Gamma$. Therefore every element of $V(P,\omega)$ can be written in the form F_f where

$$f \in \sum_{1 \le i \le q} D(P^{y_i}, \omega^{y_i}).$$

The assertion of Lemma 7 is now obvious.

Lemma 8. *Let* ω, ω' *be two distinct elements of* $^0\mathcal{E}(M)$. *Then the spaces* $\mathcal{D}(P,\omega)$ *and* $\mathcal{D}(P,\omega')$ *are mutually orthogonal in* $^0\mathcal{D}(P)$.

We now return to the notation of §7. Fix $\omega \in {}^0\mathcal{E}(M_i)$ and $s \in \mathfrak{w}(A_i, A_j)$. Then $\Omega(P_i,\omega) = \Omega(P_j,\omega^s)$ from the corollary of Lemma 1. Hence

$$V(P_i, \omega) = V(P_j, \omega^s) = V \text{ (say)}.$$

Let $(\phi, \psi)_i$ $(\phi, \psi \in {}^0\mathcal{D}(P_i))$ denote the scalar product in $^0\mathcal{D}(P_i)$.

Lemma 9. *Fix* $s \in \mathfrak{w}(A_i, A_j)$. *Then*

$$(\psi,\ \mathbf{c}_{P_j|P_i}(s)\,\phi)_j = (\mathbf{c}_{P_i|P_j}(s^{-1})\,\psi,\ \phi)_i$$

for $\phi \in {}^0\mathcal{D}(P_i)$ and $\psi \in {}^0\mathcal{D}(P_j)$. Moreover

$$c_{P_j|P_i}(s)\,\phi \in \mathcal{D}(P_j, \omega^s)$$

for $\phi \in \mathcal{D}(P_i, \omega)$. Finally

$$c_{P_i|P_i}(1)\,\phi = \phi \quad (\phi \in {}^0\mathcal{D}(P_i)).$$

Here 1 denotes the unit element of the group $\mathfrak{w}(A_i)$.

This follows from the results of §5.

Now we come to the proof of Theorem 3. Fix $\phi \in \mathcal{D}(P_i, \omega)$ and let

$$f = E(P_i:\phi) - E(P_j:c_{P_j|P_i}(s)\,\phi) \in V.$$

We have to show that $f = 0$. Fix $\psi \in \mathcal{D}(P_j, \omega^s)$. In view of Lemma 7, it is enough to verify that

$$\big(E(P_j:\psi), f\big)_{G/\Gamma} = 0.$$

But

$$
\begin{aligned}
\big(E(P_j:\psi), f\big)_{G/\Gamma} &= \sum_{1 \le l \le q_j} (\psi_l, f_{P_{jl}})_{G/\Gamma \cap P_{jl}} \\
&= \sum_{t' \in \mathfrak{w}(A_i, A_j)} (\psi, c_{P_j|P_i}(t')\,\phi)_j - \sum_{t \in \mathfrak{w}(A_j)} (\psi, c_{P_j|P_j}(t)\,c_{P_j|P_i}(s)\,\phi)_j \\
&= \sum_{t \in \mathfrak{w}(A_j)} (\psi, c_{P_j|P_i}(ts)\,\phi - c_{P_j|P_j}(t)\,c_{P_j|P_i}(s)\,\phi)_j
\end{aligned}
$$

from Lemmas 5 and 6. Now by Lemma 9, $c_{P_j|P_i}(ts)\,\phi$ and $c_{P_j|P_j}(t)\,c_{P_j|P_i}(s)\,\phi$ are both in $\mathcal{D}(P_j, \omega^{ts})$. Since ω is unramified, $\omega^{ts} \ne \omega^s$ unless $t = 1$. Therefore we conclude from Lemmas 8 and 9 that

$$\big(E(P_j:\psi), f\big)_{G/\Gamma} = 0.$$

This proves that $f = 0$.

Now fix s and t as in Theorem 3. Then it follows from the above result that

$$E(P_i:\phi) = E(P_j:c_{P_j|P_i}(s)\,\phi) = E(P_k:c_{P_k|P_j}(t)\,c_{P_j|P_i}(s)\,\phi).$$

Hence

$$f = E(P_i:\phi) - E(P_k:c_{P_k|P_j}(t)\,c_{P_j|P_i}(s)\,\phi) = 0.$$

Now fix $\psi \in \mathcal{D}(P_k:\omega^{ts})$. Then

$$
\begin{aligned}
0 = \big(E(P_k:\psi), f\big)_{G/\Gamma} &= \sum_{1 \le l \le q_k} (\psi_l, f_{P_{kl}})_{G/\Gamma \cap P_{kl}} \\
&= \sum_{u \in \mathfrak{w}(A_i, A_k)} (\psi, c_{P_k|P_i}(u)\,\phi)_k - \sum_{v \in \mathfrak{w}(A_k)} (\psi, c_{P_k|P_k}(v)\,c_{P_k|P_j}(t)\,c_{P_j|P_i}(s)\,\phi)_k \\
&= \sum_{v \in \mathfrak{w}(A_k)} (\psi, c_{P_k|P_i}(vts)\,\phi - c_{P_k|P_k}(v)\,c_{P_k|P_j}(t)\,c_{P_j|P_i}(s)\,\phi)_k
\end{aligned}
$$

from Lemmas 5 and 6. Since ω is unramified, we again conclude from Lemmas 8 and 9 that the right side is equal to

$$(\psi, c_{P_k|P_i}(ts)\,\phi - c_{P_k|P_j}(t)\,c_{P_j|P_i}(s)\,\phi)_k.$$

This implies that

$$c_{P_k|P_i}(ts)\,\phi - c_{P_k|P_j}(t)\,c_{P_j|P_i}(s)\,\phi = 0.$$

On the other hand from Lemma 9,

$$\left(c_{P_j|P_i}(s)\right)^* = c_{P_i|P_j}(s^{-1}) \qquad \left(s\in\mathfrak{w}(A_i,\,A_j)\right),$$

where the star denotes the adjoint. Therefore if ω is unramified it follows from the above proof that

$$\left(c_{P_j|P_i}(s)\right)^* c_{P_j|P_i}(s)\,\phi = c_{P_i|P_j}(s^{-1})\,c_{P_j|P_i}(s)\,\phi = c_{P_i|P_i}(1)\,\phi = \phi$$

for $\phi\in\mathcal{D}(P_i,\omega)$. Since the situation is symmetrical in i and j, it is clear that $c_{P_j|P_i}(s)$ defines a unitary transformation of $\mathcal{D}(P_i,\omega)$ onto $\mathcal{D}(P_j,\omega^s)$. This completes the proof of Theorem 3.

9. A counterexample

Let $G = SL(2)$ and P the group of all upper triangular matrices in G. Then P is a K-parabolic subgroup of G. Let M be the group of all diagonal matrices in G and U the group of all unipotent matrices in P. Then $P = MU$ is a Levi K-decomposition of P and $A = M$ the corresponding split component of P.

Now $G = SL(2, K)$, the pair (P, A) is mincuspidal and prk $P = 1$. Put $\Gamma = U = U\cap G$ and

$$y = \begin{pmatrix} 0 & 1 \\ -1 & 0 \end{pmatrix}.$$

Then $\{1, y\}$ is a complete set of representatives of $\Gamma\backslash G/P$. Put $P_1 = P$, $P_2 = P^y$. Then $P_2 = \bar{P}$ where $\bar{P}$ is the subgroup of all lower triangular matrices in G. The group $\mathfrak{w} = \mathfrak{w}(A)$ consists of just two elements $\{1, s\}$ and y is a representative of s in G.

We observe that

$$^0\mathcal{D}(P) = {}^0\mathcal{D}(P_1) \times {}^0\mathcal{D}(P_2)$$

and $c_{P|P}(1) = 1$. Put $C = c_{P|P}(s)$. Then if $\phi\in {}^0\mathcal{D}(P)$, we have

$$(C\phi)_1 = c_{P|P}(s)\,\phi_1 + c_{P|\bar{P}}(1)\,\phi_2$$

$$(C\phi)_2 = c_{\bar{P}|P}(1)\,\phi_1 + c_{\bar{P}|\bar{P}}(s)\,\phi_2.$$

Moreover $c_{P|P}(s)=0$ since $\Gamma\cap(PyP)=\varnothing$. Hence

$$C=\begin{pmatrix} 0 & , & c_{P|\bar{P}}(1) \\ c_{\bar{P}|P}(1), & c_{\bar{P}|\bar{P}}(s) \end{pmatrix}.$$

On the other hand $\Gamma\cap(P\bar{P})=\Gamma\cap(\bar{P}P)=U=\Gamma$. Hence

$$\bigl(c_{P|\bar{P}}(1)\,\phi_2\bigr)\,(x)=\int_U \phi_2(xu)\,du$$

and

$$\bigl(c_{\bar{P}|P}(1)\,\phi_1\bigr)\,(x)=[\bar{U}]^{-1}\int_{\bar{U}} \phi_2(x\bar{u})\,d\bar{u}\qquad(x\in G),$$

where du and $d\bar{u}$ are the standard measures on U and $\bar{U}$ respectively. ($\bar{U}=U^y$ is the unipotent radical of $\bar{P}$.) Finally

$$\boldsymbol{E}(P:\phi:x)=\phi_1(x)+\int_U \phi_2(xu)\,du$$

and

$$\boldsymbol{E}(P:C\phi:x)=\int_U \phi_2(xu)\,du+[\bar{U}]^{-1}\int_{U\times\bar{U}} \phi_1(xu\bar{u})\,du\,d\bar{u}$$

for $\phi\in{}^0\mathscr{D}(P)$ and $x\in G$. Hence

$$\boldsymbol{E}(P:\phi)=\boldsymbol{E}(P:C\phi)$$

if and only if

$$\phi_1(x)=[\bar{U}]^{-1}\int_{U\times\bar{U}} \phi_1(xu\bar{u})\,du\,d\bar{u}\qquad(x\in G).$$

Similarly the second assertion of Theorem 3 may be written in the form

$$C^2\phi=\phi\qquad\bigl(\phi\in\mathscr{D}(P,\omega)\bigr).$$

But a simple calculation shows that the above condition implies that $c_{\bar{P}|\bar{P}}(s)\,\phi_2=0$ for $\phi_2\in\mathscr{D}(\bar{P},\omega^s)$.

Since P is mincuspidal, $\mathscr{E}(M)={}^0\mathscr{E}(M)$. Moreover $M=A$ where A is the group of all diagonal matrices in G. Hence $\mathscr{E}(M)$ may be identified with the set A^* of all characters of the finite abelian group A. Fix $\chi\in A^*$ and suppose χ is ramified. Then there are two cases.

1) $\chi=1$.

2) $\chi\neq1,\chi^2=1$.

If $\chi=1$, we can take $\phi_2=1$. Then $\phi_2\in\mathscr{D}(\bar{P},\chi^s)$ and

$$\bigl(c_{\bar{P}|\bar{P}}(s)\,\phi_2\bigr)\,(x)=[\bar{U}]^{-1}\int_{\bar{U}}\sum_{\gamma\in U\cap(\bar{P}y\bar{P})}\phi_2(x\bar{u}\gamma)\,d\bar{u}=[U\cap(\bar{P}y\bar{P})]\qquad(x\in G).$$

By Bruhat's lemma G is the disjoint union of $\bar{P}$ and $\bar{P}y\bar{P}$. Since $U\cap\bar{P}=\{1\}$, we conclude that

$$\begin{pmatrix} 1 & 1 \\ 0 & 1 \end{pmatrix}\in U\cap(\bar{P}y\bar{P})$$

and therefore

$$c_{\bar{P}|\bar{P}}(s)\,\phi_2 \neq 0.$$

This means that Theorem 3 does not hold in this case. On the other hand it can be shown that the statements of Theorem 3 remain valid for case 2) above.

10. Concluding remarks

It is clear from the above discussion that the ramified case requires further study. However, in my opinion, a more serious problem is to find an effective method of obtaining the elements of $^0\mathscr{E}(G)$. This is entirely similar to the problem of determining the discrete series in the real or the p-adic case.

Let B be a Cartan subgroup of G defined over K and let $X_K(B)$ denote the group of all K-morphisms of B in $GL(1)$. We say that B is $(K$-$)$anisotropic if $X_K(B) = \{1\}$. Put $B = B \cap G$ and let B^* be the group of all (complex) characters of the finite abelian group B. There seem to be some indications (see TANAKA [6, p. 83]) that in case prk $G = 0$, $^0\mathscr{E}(G)$ is „parameterized" by B^* for the various anisotropic Cartan K-subgroups B of G. This is in fact so for real semisimple Lie groups and the same is believed to be true for the p-adic groups. Thus the construction of the „discrete series" appears to be the central problem in all cases.

References

1. BOREL, A., and J. TITS: Groupes rèductifs. Pub. Math. I.H.E.S., No. 27 (1965).
2. GELFAND, I. M.: Automorphic functions and the theory of representations. Proc. Intern. Congr. Math., 1962, p. 74—85.
3. HARISH-CHANDRA: Automorphic forms on semisimple Lie groups. Notes by J. G. M. MARS, Lecture notes in mathematics, vol. 62. Berlin-Heidelberg-New York: Springer 1968.
4. LANGLANDS, R. P.: (a) On the functional equations satisfied by Eisenstein series. Mimeographed manuscript, 1965 (unpublished). (b) Eisenstein series. Algebraic groups and discontinuous groups (1966). Am. Math. Soc., p. 235—252.
5. SELBERG, A.: Discontinuous groups and harmonic analysis. Proc. Intern. Congr. Math., 1962, p. 177—189.
6. TANAKA, S.: Construction and classification of irreducible representations of special linear group of the second order over a finite field, Osaka J. Math. 4, 65—84 (1967).

HARMONIC ANALYSIS ON SEMISIMPLE LIE GROUPS

BY HARISH-CHANDRA

1. Introduction. Let G be a locally compact group which we assume to be separable and unimodular. Let dx denote the Haar measure on G. If π is a unitary representation of G on a Hilbert space $\mathfrak{H}$ and $f \in L_1(G)$, we write

$$\pi(f) = \int_G f(x)\pi(x)dx.$$

Then $\pi(f)$ is a bounded operator on $\mathfrak{H}$ and

$$\pi(f * g) = \pi(f)\pi(g) \quad (f, g \in L_1(G)),$$

where $f * g$ denotes the convolution of f and g.

Let A be a bounded linear operator on $\mathfrak{H}$. We say that A is of the trace class if the series

$$\sum_i |(\psi_i, A\psi_i)|$$

converges for every orthonormal base $\{\psi_i\}_{i \in J}$ of $\mathfrak{H}$. Moreover if this is so, we define

$$\operatorname{tr} A = \sum_i (\psi_i, A\psi_i).$$

Then $\operatorname{tr} A$ is actually independent of the choice of this base.

Let V_π denote the set of all $f \in L_1(G)$ such that $\pi(f)$ is of the trace class. Then V_π is a linear subspace of $L_1(G)$. Put

$$\Theta_\pi(f) = \operatorname{tr} \pi(f) \qquad (f \in V_\pi).$$

Then Θ_π is a linear function on V_π which we may call the character of π. Of course this concept would be useful only when the space V_π is fairly large.

Let $\mathcal{E}(G)$ denote the set of all equivalence classes of irreducible unitary representations of G. It is easy to see that for any representation π, V_π and Θ_π depend only on the class ω of π. Hence we may de-

A Colloquium Lecture delivered before the 74th Summer meeting of the Society at Eugene, Oregon, on August 26, 1969; received by the editors October 29, 1969.

AMS Subject Classifications. Primary 2265, 2260.

Key Words and Phrases. Characters, invariant eigendistributions, parabolic subgroups, Schwartz space, cusp forms, discrete series, Fourier transform of a Schwartz function, Plancherel formula, p-adic groups.

529

note them by V_ω and Θ_ω respectively. Put

$$V_0 = \bigcap_{\omega \in \mathcal{E}(G)} V_\omega.$$

For any function $f \in L_1(G)$ and $x \in G$, define the function f^x by $f^x(y) = f(x^{-1}yx)$. It is clear that $f^x \in V_0$ whenever $f \in V_0$ and therefore the group G operates on V_0. Now, in order to describe the main problem of harmonic analysis, we have to fix a certain topological vector space V satisfying the following conditions:[1]

(1) $V \subset V_0 \cap C(G)$, where $C(G)$ is the space of all complex-valued continuous functions on G.

(2) V is stable under G and every $x \in G$ defines a continuous endomorphism of V.

(3) For any $\omega \in \mathcal{E}(G)$, the restriction of Θ_ω on V is a continuous linear function on V, which determines ω uniquely.

For any $f \in V$, define $\hat{f}(\omega) = \Theta_\omega(\hat{f})$ $(\omega \in \mathcal{E}(G))$. Then $\hat{f}$ is a complex-valued function on $\mathcal{E}(G)$, which may be regarded as the Fourier transform of f. Let us agree to call a continuous linear function on V, a distribution. For any distribution T and $x \in G$, define the distribution T^x by $T^x(f) = T(f^{x^{-1}})$ $(f \in V)$. We say that T is invariant if $T^x = T$ for all $x \in G$.

The central question in harmonic analysis may now be formulated as follows. *Given an invariant distribution T on G, how to express it as a "linear combination" of the characters Θ_ω $(\omega \in \mathcal{E}(G))$?* In fact, to every such T, we would like to associate a "distribution" $\hat{T}$ on $\mathcal{E}(G)$ in such a way that $T(f) = \hat{T}(\hat{f})$ for all $f \in V$. (Here by a "distribution" on $\mathcal{E}(G)$, we mean some sort of linear functional on a suitable space of functions on $\mathcal{E}(G)$.) If δ is the Dirac measure, so that $\delta(f) = f(1)$ $(f \in V)$, the determination of $\hat{\delta}$ is just the problem of the explicit Plancherel formula for G.

For an arbitrary G, our understanding of $\mathcal{E}(G)$ is still very rudimentary (see [5]) and therefore no serious attempt to attack the above problem can yet be contemplated. However the situation for real semisimple (or reductive) Lie groups seems rather encouraging [4(a), (d)] and there is reason to hope that some substantial results on reductive $\mathfrak{p}$-adic groups may be attainable within a few years. Therefore I shall limit myself to these two cases in these lectures. The other extreme case, when G is a nilpotent or solvable Lie group, has been studied extensively by Dixmier, Kirillov, Moore, and Pukánszky (see [6], [7]).

---

[1] We do not claim here that such a space V necessarily exists. Of course the crucial condition is (3).

Our entire approach to harmonic analysis on reductive groups is based on the philosophy of cusp forms (see [2] and [4(c), (d)]) and the success of this method depends on a good knowledge of the discrete series (see §3) for such groups. In the case of real groups the characters of the discrete series are all known (see Theorem 8) and therefore the theory works reasonably well. But for $\mathfrak{p}$-adic groups our understanding of the supercuspidal representations (§14) is still very poor. Nevertheless the resemblance between the two cases is so striking that one cannot give up the hope that the same philosophy would eventually work in both cases.

2. **Characters in the real case.** Suppose G is a real, connected, semisimple Lie group with finite center. Fix a maximal compact subgroup K of G. If $\omega \in \mathcal{E}(G)$ and $\mathfrak{b} \in \mathcal{E}(K)$, we define $[\omega:\mathfrak{b}]$ as follows. Fix a representation $\pi \in \omega$ and let π_K denote the restriction of π on K. Then $[\omega:\mathfrak{b}]$ is the multiplicity of $\mathfrak{b}$ in π_K. Since K is compact, every irreducible representation of K is finite-dimensional. We denote by $d(\mathfrak{b})$ the degree of a representation in the class $\mathfrak{b}$.

THEOREM 1. *There exists an integer $N \geq 1$ such that*

$$[\omega:\mathfrak{b}] \leq Nd(\mathfrak{b})$$

for all $\omega \in \mathcal{E}(G)$ and $\mathfrak{b} \in \mathcal{E}(K)$.

Let $C_c^\infty(G)$ denote the space of all complex-valued C^∞ functions on G with compact support, taken with its usual topology [9]. The following result is an easy consequence of Theorem 1.

THEOREM 2. *Let π be an irreducible unitary representation of G. Then for any $f \in C_c^\infty(G)$, $\pi(f)$ is of the trace class. Put*

$$\Theta_\pi(f) = \operatorname{tr} \pi(f) \qquad (f \in C_c^\infty(G)).$$

Then Θ_π is a distribution on G in the sense of Schwartz [9].

As we have seen in §1, Θ_π depends only on the class ω of π. Hence we may denote it by Θ_ω.

THEOREM 3. *Let ω_1, ω_2 be two elements in $\mathcal{E}(G)$. Then $\Theta_{\omega_1} = \Theta_{\omega_2}$ if and only if $\omega_1 = \omega_2$.*

This shows that if we take $V = C_c^\infty(G)$, then all the conditions of §1 are fulfilled.

Let u be a differential operator on G. Then its adjoint u^* is the differential operator given by the relation

$$\int_G u^*f \cdot g\,dx = \int_G f \cdot ug\,dx \qquad (f, g \in C_c^\infty(G)).$$

If T is a distribution on G, the mapping

$$f \rightarrow T(u^*f) \qquad (f \in C_c^\infty(G)),$$

is also a distribution which we denote by uT. Let F be a locally summable function on G. Put

$$T_F(f) = \int Ffdx \qquad (f \in C_c^\infty(G)).$$

Then T_F is a distribution on G. We say that $T = F$ if $T = T_F$.

Let $\mathfrak{Z}$ be the algebra of all differential operators on G which commute with both left and right translations of G. Then $\mathfrak{Z}$ is abelian. A distribution T is called $\mathfrak{Z}$-finite if the space of all distributions of the form zT $(z \in \mathfrak{Z})$ has finite dimension. Moreover T is said to be an eigendistribution of $\mathfrak{Z}$, if this space has dimension ≤ 1. Finally we say that T is K-finite, if the left and right translates of T by elements of K, span a finite-dimensional space.

Fix an indeterminate t and let $l = \mathrm{rank}\ G$. For any $x \in G$, let $D(x)$ denote the coefficient of t^l in the polynomial $\det(t+1-\mathrm{Ad}(x))$. Then D is an analytic function on G which is not identically zero. A point $x \in G$ is called regular if $D(x) \neq 0$. Let G' denote the set of all regular points of G. Then G' is an open dense subset of G whose complement is of measure zero.

THEOREM 4. *Let Θ be an invariant and $\mathfrak{Z}$-finite distribution on G. Then there exists an analytic function F on G' such that F is locally summable on G and $\Theta = F$.*

Fix $\omega \in \mathcal{E}(G)$ and let Θ_ω denote, as before, the character of ω. It is easy to see that Θ_ω is an invariant eigendistribution of $\mathfrak{Z}$. Therefore it follows from Theorem 4 that Θ_ω is a function.

3. The discrete series. We return to the assumptions of §1. Let π be an irreducible unitary representation of G on a Hilbert space $\mathfrak{H}$. We say that π is of type L_p $(p \geq 1)$, if there exist nonzero elements $\phi, \psi \in \mathfrak{H}$ such that

$$\int_G |(\phi, \pi(x)\psi)|^p dx < \infty.$$

In case π is of type L_2 (or, as we sometimes say, square-integrable), there exists a positive number $d(\pi)$ such that[2]

---

[2] conj c $(c \in C)$ denotes the complex conjugate of c.

$$\int_G (\phi_1, \pi(x)\psi_1) \cdot \mathrm{conj}(\phi_2, \pi(x)\psi_2) dx = d(\pi)^{-1}(\phi_1, \phi_2) \cdot \mathrm{conj}(\psi_1, \psi_2)$$

for all $\phi_i, \psi_i \in \mathfrak{H}$ ($i = 1, 2$) (see [3] and [5, p. 640]). These are called the Schur orthogonality relations for π and $d(\pi)$ the formal degree of π. Let $\mathcal{E}_p(G)$ denote the set of all classes $\omega \in \mathcal{E}(G)$ which contain representations of type L_p. If $\omega \in \mathcal{E}_2(G)$, we put $d(\omega) = d(\pi)$ for $\pi \in \omega$. The set $\mathcal{E}_2(G)$ is called the discrete series for G.

Now suppose K is an *open and compact* subgroup of G. Normalize dx in such a way that the total measure of K is 1. For any $\omega \in \mathcal{E}(G)$ and $\mathfrak{b} \in \mathcal{E}(K)$, let $[\omega:\mathfrak{b}]$ denote, as before (see §2), the multiplicity of $\mathfrak{b}$ in ω and $d(\mathfrak{b})$ the degree of $\mathfrak{b}$.

THEOREM 5. *Fix* $\mathfrak{b} \in \mathcal{E}(K)$. *Then*

$$d(\mathfrak{b}) \geqq \sum_{\omega \in \mathcal{E}_2(G)} d(\omega)[\omega:\mathfrak{b}].$$

This is a simple consequence of the Schur orthogonality relations.

COROLLARY. *Fix* $\omega \in \mathcal{E}_2(G)$. *Then*

$$[\omega:\mathfrak{b}] \leqq d(\omega)^{-1}d(\mathfrak{b})$$

for all $\mathfrak{b} \in \mathcal{E}(K)$.

4. Characters of the discrete series (real case). Define G as in §2.

THEOREM 6. $\mathcal{E}_2(G) \neq \varnothing$ *if and only if* G *has a compact Cartan subgroup.*

Let us now assume that G has a compact Cartan subgroup B. Then any two such Cartan subgroups are conjugate in G. Let $\mathfrak{g}$ be the Lie algebra of G and $\mathfrak{g}_c$ its complexification. Let G_c be the simply connected complex-analytic group corresponding to $\mathfrak{g}_c$. For simplicity we assume that G is the real analytic subgroup of G_c corresponding to $\mathfrak{g}$.

Let $\mathfrak{b}$ be the Lie algebra of B and $\mathfrak{b}^*$ the dual of $\mathfrak{b}$. Fix some order on the real vector space $(-1)^{1/2}\mathfrak{b}^*$. Let $W = W(\mathfrak{g}/\mathfrak{b})$ denote the Weyl group of[3] $(\mathfrak{g}_c, \mathfrak{b}_c)$ and B^* the character group of B. Then W operates on $\mathfrak{b}$ and B and therefore by duality, it also operates on $\mathfrak{b}^*$ and B^*. Let $\langle b^*, b \rangle$ $(b^* \in B^*, b \in B)$ denote the value of the character b^* at b. Then we have the relations

---

[3] For any finite-dimensional vector space V over R, we denote by V^* its dual and by V_c its complexification. Moreover $V_c^* = (V^*)_c$.

$$(\exp H)^s = \exp(sH), \qquad \langle b^{*s}, b \rangle = \langle b^*, b^{s^{-1}} \rangle$$

for $H \in \mathfrak{b}$ and $s \in W$.

Put

$$\varpi = \prod_{\alpha \in P} H_\alpha$$

where P is the set of all positive roots of $(\mathfrak{g}, \mathfrak{b})$ and H_α the element in $\mathfrak{b}_\theta$, which is dual to α under the Killing form so that

$$\alpha(H) = \mathrm{tr}(\mathrm{ad}\, H \cdot \mathrm{ad}\, H_\alpha) \qquad (H \in \mathfrak{b}).$$

Then ϖ is a polynomial function on $\mathfrak{b}_\theta^*$ and $\varpi^s = \epsilon(s)\varpi$ $(s \in W)$ where $\epsilon(s) = 1$ or -1. Moreover there exists an analytic function Δ on B such that

$$\Delta(\exp H) = \prod_{\alpha \in P} (e^{\alpha(H)/2} - e^{-\alpha(H)/2}) \qquad (H \in \mathfrak{b}).$$

For any $b^* \in B^*$, let $\log b^*$ denote the element $\lambda \in (-1)^{1/2}\mathfrak{b}^*$ such that

$$\langle b^*, \exp H \rangle = e^{\lambda(H)} \qquad (H \in \mathfrak{b}),$$

and put

$$\varpi(b^*) = \varpi(\log b^*) \qquad (b^* \in B^*).$$

We say that b^* is regular if $\varpi(b^*) \neq 0$.

Let $\tilde{B}$ be the normalizer of B in G. Then $W(G/B) = \tilde{B}/B$ may be regarded as a subgroup of W.

THEOREM 7. *Fix $b^* \in B^*$. Then there exists an invariant eigendistribution Θ of $\mathfrak{Z}$ on G such that*[4]
(1) $\sup_{x \in G'} |D(x)|^{1/2} |\Theta(x)| < \infty$,
(2) $\Theta = \Delta^{-1} \sum_{s \in W(G/B)} \epsilon(s) b^{*s}$ *pointwise on $B \cap G'$.*
Moreover Θ is unique if $\varpi(b^*) \neq 0$.

Let $B^{*\prime}$ be the set of all regular elements in B^* and for any $b^* \in B^{*\prime}$, let Θ_{b^*} denote the corresponding distribution of Theorem 7. Put $\epsilon(b^*) = \mathrm{sign}\, \varpi(b^*)$ and $q = \frac{1}{2}\dim G/K$.

THEOREM 8. *Given $b^* \in B^{*\prime}$, there exists a unique class $\omega(b^*) \in \mathcal{E}_2(G)$ such that*

$$(-1)^q \epsilon(b^*) \Theta_{b^*}$$

is the character of $\omega(b^*)$. The mapping $b^* \rightarrow \omega(b^*)$ from $B^{*\prime}$ to $\mathcal{E}_2(G)$ is

---

[4] Here we have to keep in mind Theorem 4.

surjective. Moreover $\omega(b_1^*) = \omega(b_2^*)$ *if and only if* b_1^*, b_2^* *are conjugate under* $W(G/B)$.

Let σ be an automorphism of G such that $\sigma B = B$. Then σ also operates on B^* by the relation

$$\langle \sigma b^*, b \rangle = \langle b^*, \sigma^{-1} b \rangle \qquad (b^* \in B^*, \ b \in B).$$

Define $\epsilon(\sigma) = \pm 1$ by the rule $\varpi^\sigma = \epsilon(\sigma)\varpi$. Also if T is a distribution on G, we define the distribution T^σ by

$$T^\sigma(f) = T(f') \qquad (f \in C_c^\infty(G)),$$

where $f'(x) = f(\sigma x)$ $(x \in G)$.

LEMMA 1. *Let* σ *be an automorphism of* G *such that* $\sigma B = B$. *Then*

$$(\Theta_{b^*})^\sigma = \epsilon(\sigma)\Theta_{\sigma b^*} \qquad (b^* \in B^{*\prime}).$$

Actually it is possible to define Θ_{b^*} for all $b^* \in B^*$ in such a way that both Theorem 7 and Lemma 1 hold for all σ.

5. **Parabolic subgroups.** Let G be as in §2. By a parabolic subgroup (psgp) P of G, we mean a closed subgroup of G with the following two properties.

(1) If $\mathfrak{p}$ is the Lie algebra of P, then $\mathfrak{p}_c$ contains a Borel subalgebra (i.e. a maximal solvable subalgebra) of $\mathfrak{g}_c$.

(2) P is the normalizer of $\mathfrak{p}$ in G.

By the radical N of P, we mean the maximal normal subgroup of P such that $\mathrm{Ad}(n)$ is unipotent for every $n \in N$. An abelian subgroup A of G is said to split if, for every $a \in A$, $\mathrm{Ad}(a)$ can be diagonalized over $\mathbf{R}$. Fix K as in §2 and let θ denote the Cartan involution of G corresponding to K. Put $M_1 = P \cap \theta(P)$ and let A be a maximal connected split abelian subgroup lying in the center of M_1. Then A is unique and M_1 is the centralizer of A in G. Let $X(M_1)$ denote the group of all continuous homomorphisms of M_1 into the multiplicative group $\mathbf{R}^\times$ of real numbers. Put

$$M = \bigcap_{\chi \in X(M_1)} \ker |\chi|,$$

where $|\chi|(m) = |\chi(m)|$ $(m \in M_1)$. Then $M_1 = MA$, $P = MAN$ and the corresponding mapping of $M \times A \times N$ into P is a diffeomorphism. We call this the Langlands decomposition of P and A the split component of P. Let $\mathfrak{a}$ be the Lie algebra of A. Then the exponential mapping defines a bijection of $\mathfrak{a}$ on A and we denote its inverse by log. By the parabolic rank of P we mean the dimension of A and denote it by prk P.

The group M is, in general, neither connected nor semisimple. However it is always reductive. We say that the psgp P is cuspidal if M has a compact Cartan subgroup.

Let $\mathfrak{h}$ be a Cartan subalgebra of $\mathfrak{g}$. We denote by $\mathfrak{h}_R$ the set of all points in $\mathfrak{h}$ where every root of $(\mathfrak{g}, \mathfrak{h})$ takes a real value.

LEMMA 2. *A psgp $P = MAN$ is cuspidal if and only if there exists a θ-stable Cartan subalgebra $\mathfrak{h}$ of $\mathfrak{g}$ such that $\mathfrak{a} = \mathfrak{h}_R$.*

Let $P = MAN$ be a prgp of G. Then $G = KP$. Hence every element $x \in G$ can be written in the form $x = kman$ ($k \in K, m \in M, a \in A, n \in N$). The element a is uniquely determined and we put $H_P(x) = \log a$. Let $\mathfrak{a}^*$ be the dual of $\mathfrak{a}$ and $\mathfrak{n}$ the Lie algebra of N. Then we define $\rho_P \in \mathfrak{a}^*$ by

$$\rho_P(H) = \tfrac{1}{2}\,\mathrm{tr}(\mathrm{ad}\,H)_\mathfrak{n} \qquad (H \in \mathfrak{a}),$$

where $(\mathrm{ad}\,H)_\mathfrak{n}$ denotes the restriction of $\mathrm{ad}\,H$ on $\mathfrak{n}$. By a root of (P, A) (or P) we mean an element $\alpha \in \mathfrak{a}^*$ with the following property. There should exist an element $X \neq 0$ in $\mathfrak{n}$ such that $[H, X] = \alpha(H)X$ for all $H \in \mathfrak{a}$. Let $\mathfrak{a}^+$ be the set of all $H \in \mathfrak{a}$ such that $\alpha(H) \geqq 0$ for every root α of (P, A). We put $A^+ = \exp \mathfrak{a}^+$.

Two psgps P_1, P_2 are said to be associated if their split components A_1, A_2 are conjugate under G (or equivalently under K).

A psgp P is called minimal if it is minimal among all psgps of G. The following three conditions are mutually equivalent.

(1) $P = MAN$ is a minimal psgp.

(2) $M \subset K$.

(3) prk $P = \mathrm{rank}\ G/K$.

Any two minimal psgps are conjugate under K.

6. **The functions Ξ and σ.** Fix a minimal psgp $P = MAN$ of G. Then $G = KAN$ and this is an Iwasawa decomposition of G. Put

$$\Xi(x) = \int_K e^{-\rho(H(xk))}dk \qquad (x \in G)$$

where $H(x) = H_P(x)$, $\rho = \rho_P$ and the Haar measure dk on K is so normalized that the total measure of K is 1.

Define a norm on $\mathfrak{g}$ by setting

$$\|X\|^2 = -\,\mathrm{tr}(\mathrm{ad}\,X\ \mathrm{ad}\,\theta(X)) \qquad (X \in \mathfrak{g}).$$

Then since $G = KAK$, there exists a unique function σ on G such that

(1) $\sigma(k_1 x k_2) = \sigma(x)$ ($k_1, k_2 \in K, x \in G$),

(2) $\sigma(\exp H) = \|H\|$ ($H \in \mathfrak{a}$).

The functions Ξ and σ are both independent of the choice of P and they have the following remarkable properties.

(1) $\int_K \Xi(xky)dk = \Xi(x)\Xi(y)$ $(x,\ y \in G)$.

(2) There exist numbers $c,\ d \geqq 0$ such that

$$1 \leqq \Xi(a)e^{\rho(\log a)} \leqq c(1 + \sigma(a))^d \qquad (a \in A^+).$$

(3) We can choose $r_0 \geqq 0$ such that

$$\int_G \Xi(x)^2(1 + \sigma(x))^{-r_0}dx < \infty.$$

(4) There exists a number $r \geqq 0$ such that[5]

$$\int_G |\ D(x)\ |^{-1/2}\Xi(x)(1 + \sigma(x))^{-r}dx < \infty.$$

Let $\mathfrak{q}$ be any linear subspace of $\mathfrak{g}$. Then we denote by $d\mathfrak{q}$ the Euclidean measure on $\mathfrak{q}$ corresponding to the above Euclidean norm on $\mathfrak{g}$. It is convenient to normalize the Haar measure dx on G as follows. Let da and dn denote the Haar measures on A and N which corespond to $d\mathfrak{a}$ and $d\mathfrak{n}$ respectively under the exponential mapping. Then if $x = kan$ $(k \in K,\ a \in A,\ n \in N)$,

$$dx = e^{2\rho(\log a)}dkdadn$$

where dk is the normalized Haar measure of K. This normalization of dx is independent of the choice of the minimal psgp P.

7. **The Schwartz space and the cusp forms.** Let $\mathfrak{D}$ be the algebra of all differential operators on G. Let $\mathfrak{D}_l$ and $\mathfrak{D}_r$ denote the subalgebras consisting of those $D \in \mathfrak{D}$ which commute with the left and right translations of G respectively. Let $\mathfrak{D}_0$ be the subalgebra of $\mathfrak{D}$ generated by $\mathfrak{D}_l \cup \mathfrak{D}_r$.

For $f \in C^\infty(G)$, $D \in \mathfrak{D}_0$ and $r \geqq 0$, put

$$\nu_{D,r}(f) = \sup_G |Df|\, (1 + \sigma)^r \Xi^{-1}.$$

Let $\mathcal{C}(G)$ denote the space of all $f \in C^\infty(G)$ such that $\nu_{D,r}(f) < \infty$ for all $D \in \mathfrak{D}_0$ and $r \geqq 0$. We topologize $\mathcal{C}(G)$ by means of the set of seminorms $\nu_{D,r}$ $(D \in \mathfrak{D}_0,\ r \geqq 0)$. Then $\mathcal{C}(G)$ is a Hausdorff, locally convex and complete space and we call it the Schwartz space of G. We list here some of the properties of this space.

(1) The inclusion mapping of $C_c^\infty(G)$ into $\mathcal{C}(G)$ is continuous and the image is dense in $\mathcal{C}(G)$.

---

[5] Here D has the same meaning as in §2.

(2) $\mathcal{C}(G) \subset L_2(G)$.

(3) $\mathcal{C}(G)$ is closed under convolution and the mapping $(f,\ g)$ $\rightarrow f * g$ $(f,\ g \in \mathcal{C}(G))$ is continuous.

For $f \in \mathcal{C}(G)$ and $P = MAN$ a psgp of G, define

$$f^P(x) = \int_N f(xn)dn \qquad (x \in G)$$

where dn is the Haar measure on N. (This integral always exists.) We say that f is a cusp form if $f^P = 0$ for every psgp $P \neq G$. Let $^\circ\mathcal{C}(G)$ denote the space of all cusp forms on G. Then $^\circ\mathcal{C}(G)$ is a closed subspace of $\mathcal{C}(G)$.

THEOREM 9. *Every K-finite and $\mathfrak{Z}$-finite function in $L_2(G)$ is a cusp form. Conversely K-finite eigenfunctions of $\mathfrak{Z}$ in $\mathcal{C}(G)$ span a dense subspace of $^\circ\mathcal{C}(G)$.*

Combining this with Theorem 6 we get the following result.

COROLLARY. $^\circ\mathcal{C}(G) \neq \{0\}$ *if and only if* rank $G =$ rank K.

Let τ be a unitary double representation of K on a finite-dimensional Hilbert space V. We denote by $\mathcal{C}(G, \tau)$ the space of all $f \in \mathcal{C}(G) \otimes V$ such that $f(k_1 x k_2) = \tau(k_1)f(x)\tau(k_2)$ $(k_1,\ k_2 \in K,\ x \in G)$. Put

$$^\circ\mathcal{C}(G,\tau) = \mathcal{C}(G,\tau) \cap (^\circ\mathcal{C}(G) \otimes V).$$

THEOREM 10. $\dim {}^\circ\mathcal{C}(G, \tau) < \infty$.

Fix a psgp $P = MAN$ of G. Then M is a reductive group with finitely many connected components. Moreover $K_M = K \cap M$ is a maximal compact subgroup of M. Let τ_M denote the restriction of τ on K_M. Then it is not difficult to define the spaces $\mathcal{C}(M)$, $^\circ\mathcal{C}(M)$ and $^\circ\mathcal{C}(M, \tau_M)$ in the same way as above.

Let $f \in \mathcal{C}(G)$. Then we write $f^P \sim 0$ if

$$\int_M \operatorname{conj} \phi(m) \cdot f^P(xm)dm = 0$$

for all $x \in G$ and $\phi \in {}^\circ\mathcal{C}(M)$. (Here dm is the Haar measure on M and the integral always exists.)

THEOREM 11. *Let f be an element in $\mathcal{C}(G)$ such that $f^P \sim 0$ for all cuspidal subgroups P of G. Then $f = 0$.*

Let $\mathfrak{h}_i$ $(1 \leq i \leq r)$ be a complete set of θ-stable Cartan subalgebras of $\mathfrak{g}$, no two of which are conjugate under G. Put $\mathfrak{a}_i = \mathfrak{h}_{i,R}$ and $A_i = \exp \mathfrak{a}_i$. Let $\mathcal{C}_i(G)$ denote the set of all $f \in \mathcal{C}(G)$ with the following property.

If $P = MAN$ is any cuspidal subgroup of G, then $f^P \sim 0$ unless A is conjugate to A_i under K. Then $\mathcal{C}_i(G)$ is a closed subspace of $\mathcal{C}(G)$.

THEOREM 12. $\mathcal{C}(G) = \sum_{1 \leq i \leq r} \mathcal{C}_i(G)$ *where the sum is direct and smooth.*

Let E_i denote the projection of $\mathcal{C}(G)$ on $\mathcal{C}_i(G)$ corresponding to the above direct sum. Then smoothness means that E_i are continuous endomorphisms of $\mathcal{C}(G)$. Let $\mathfrak{H}_i$ denote the closure of $\mathcal{C}_i(G)$ in $\mathfrak{H} = L_2(G)$. Then $\mathfrak{H}$ is the orthogonal sum of $\mathfrak{H}_i$ $(1 \leq i \leq r)$ and $E_i f$ $(f \in \mathcal{C}(G))$ is, actually, the orthogonal projection of f in $\mathfrak{H}_i$.

A distribution T on G is called tempered if it extends to a continuous linear function on $\mathcal{C}(G)$.

LEMMA 3. *Let Θ be an invariant and $\mathfrak{Z}$-finite distribution on G. Then Θ is tempered if and only if we can choose c, $r \geq 0$ such that*[4,5]

$$| D(x) |^{1/2} | \Theta(x) | \leq c(1 + \sigma(x))^r$$

for all $x \in G'$.

COROLLARY. *Suppose Θ is tempered in the above lemma. Then*

$$\Theta(f) = \int_G f \cdot \Theta \, dx$$

for $f \in \mathcal{C}(G)$.

This follows immediately from (4) of §6. We write

$$(\Theta, f) = \int_G \operatorname{conj} \Theta \cdot f \, dx \qquad (f \in \mathcal{C}(G)).$$

Let f be a continuous function on G. We say that f satisfies the weak inequality if there exist numbers c, $r \geq 0$ such that

$$| f(x) | \leq c \Xi(x)(1 + \sigma(x))^r$$

for all $x \in G$.

LEMMA 4. *Let T be a tempered distribution on G which is both K-finite and $\mathfrak{Z}$-finite. Then T satisfies the weak inequality.*

Here we have to observe that a K-finite and $\mathfrak{Z}$-finite distribution is necessarily an analytic function.

8. **The projection of $\mathcal{C}(G)$ on $°\mathcal{C}(G)$.** We now return to the assumptions and notation of §4. Put $B' = B \cap G'$ and

$$F_f(b) = \Delta(b) \int_G f(xbx^{-1}) \, dx \qquad (b \in B')$$

for $f\in \mathcal{C}(G)$. This integral is always convergent and F_f is a C^∞ function on B'.

Theorem 13. *Let $f\in {}^\circ\mathcal{C}(G)$. Then F_f extends to a C^∞ function on B and*

$$F_f(b) = \sum_{b^*\in B^{*'}} (\Theta_{b^*}, f)\langle b^*, b\rangle \qquad (b\in B).$$

Now

$$\varpi = \prod_{\alpha>0} H_\alpha$$

may be regarded as a differential operator on B. Then, for any $f\in \mathcal{C}(G)$, ϖF_f extends to a continuous function B.

Lemma 5. *There exists a number $c_G>0$ such that*

$$\varpi F_f(1) = (-1)^q\, c_G\, f(1)$$

for all $f\in \mathcal{C}(G)$.

It is possible to determine c_G explicitly as follows. Let $\mathfrak{k}$ be the Lie algebra of K. We assume, as we may, that $\mathfrak{k}\supset\mathfrak{b}$. Let P be the set of all positive roots of $(\mathfrak{g}, \mathfrak{b})$ and P_k the subset of those $\alpha\in P$ which can be regarded as roots of $(\mathfrak{k}, \mathfrak{b})$. Put

$$\varpi_k = \prod_{\alpha\in P_k} H_\alpha, \qquad \rho_k = \frac{1}{2}\sum_{\alpha\in P_k}\alpha$$

and normalize the measure dx as in §6.

Lemma 6. *With the above normalization of the Haar measure of G, we have*[6]

$$c_G = [W(G/B)](2\pi)^q\, 2^{n/2}\, \varpi_k(\rho_k)$$

where

$$n = \dim(G/K) - \mathrm{rank}(G/K).$$

Theorem 14. *For any $f\in \mathcal{C}(G)$ and $b^*\in B^*$, define*

$$f_{b^*}(x) = (-1)^q c_G^{-1}\varpi(b^*)(\Theta_{b^*}, r(x)f) \qquad (x\in G)$$

where r denotes the right-regular representation of G on $\mathcal{C}(G)$. Then

---

[6] $[S]$ denotes the number of elements in a finite set S.

$f_{b*} \in {}^\circ \mathcal{C}(G)$ *and the series*

$$\sum_{b* \in B*} f_{b*}$$

converges absolutely [4(a), p. 8] *in* $\mathcal{C}(G)$. *Let* ${}^\circ f$ *denote its sum. Then* $f \rightarrow {}^\circ f$ *is a continuous projection of* $\mathcal{C}(G)$ *onto* ${}^\circ \mathcal{C}(G)$.

Let ${}^\circ \mathfrak{H}$ denote the closure of ${}^\circ \mathcal{C}(G)$ in $\mathfrak{H} = L_2(G)$ and l the left-regular representation of G on $\mathfrak{H}$.

THEOREM 15. ${}^\circ \mathfrak{H}$ *is the closure of an orthogonal sum of closed subspaces which are invariant and irreducible under* l. *Conversely every irreducible subspace of* $\mathfrak{H}$ *is contained in* ${}^\circ \mathfrak{H}$. *Let* ${}^\circ E$ *denote the orthogonal projection of* $\mathfrak{H}$ *on* ${}^\circ \mathfrak{H}$. *Then* ${}^\circ f = {}^\circ E f$ *for* $f \in \mathcal{C}(G)$.

9. **The space** $\mathcal{Q}(G, \tau)$. Fix G and τ as in §2 and §7. We denote by $\mathcal{Q}(G, \tau)$ the space of all functions $f \in C^\infty(G) \otimes V$ satisfying the following three conditions:

(1) f is τ-spherical i.e. $f(k_1 x k_2) = \tau(k_1) f(x) \tau(k_2)$ $(k_1, k_2 \in K, x \in G)$.

(2) f is $\mathfrak{Z}$-finite.

(3) $|f|$ satisfies the weak inequality.

For a psgp $P = MAN$ of G, put

$$d_P(m) = \big| \det(Ad(m))_\mathfrak{n} \big|^{1/2} \qquad (m \in MA).$$

Let a be a variable element in A. We say that $a \rightarrow_P \infty$ if $\sigma(a) \rightarrow \infty$ and there exists a number $\epsilon > 0$ such that $\alpha(\log a) \geqq \epsilon \sigma(a)$ for every root α of (P, A). Since MA is a reductive group with finitely many connected components and $K_M = K \cap M$ is a maximal compact subgroup of MA, the space $\mathcal{Q}(MA, \tau_M)$ can be defined without difficulty.

THEOREM 16. *For any* $f \in \mathcal{Q}(G, \tau)$, *there exists a unique element* $f_P \in \mathcal{Q}(MA, \tau_M)$ *such that*

$$\lim_{\substack{a \rightarrow \infty \\ P}} \{ d_P(ma) f(ma) - f_P(ma) \} = 0$$

for $m \in MA$ *and* $a \in A$.

Given $f \in \mathcal{Q}(G, \tau)$, we write $f_P \sim 0$ if

$$\int_M (\phi(m), f_P(ma))_V \, dm = 0$$

for all $\phi \in {}^\circ \mathcal{C}(M, \tau_M)$ and $a \in A$. Here the scalar product is in V and the integral is always convergent.

LEMMA 7. *Let* f *be an element in* $\mathcal{Q}(G, \tau)$ *such that* $f_P \sim 0$ *for every* *cuspidal subgroup P of G. Then $f = 0$.*

Here the analogy with Theorem 11 is obvious.

For any integer q, let $\mathcal{C}_q(G, \tau)$ denote the space of all $f \in \mathcal{C}(G, \tau)$ with the following property. If P is a psgp of G, then $f_P \sim 0$ unless prk $P = q$.

THEOREM 17. $\mathcal{C}_0(G, \tau) = {}^\circ\mathcal{C}(G, \tau)$. *Fix $f \neq 0$ in $\mathcal{C}(G, \tau)$ and choose a psgp $P = MAN$ of G with the following two properties:*

(1) $f_P \not\sim 0$.

(2) P is minimal with respect to condition (1).

Then for any fixed $a \in A$, the function $m \to f_P(ma)$ $(m \in M)$ lies in ${}^\circ\mathcal{C}(M, \tau_M)$.

10. **The Eisenstein integral.** Fix a cuspidal subgroup $P = MAN$ of G. Given $\psi \in {}^\circ\mathcal{C}(M, \tau_M)$, we extend it to a function on G as follows:

$$\psi(kman) = \tau(k)\psi(m) \qquad (k \in K, \; m \in M, \; a \in A, \; n \in N).$$

For $\nu \in \mathfrak{a}_c{}^*$ and $\psi \in {}^\circ\mathcal{C}(M, \tau_M)$, define

$$E(P:\psi:\nu:x) = \int_K \psi(xk)\tau(k^{-1}) \exp\left\{((-1)^{1/2}\nu - \rho_P)(H_P(xk))\right\} dk$$

for $x \in G$. We call this the Eisenstein integral. For fixed P and ψ, it is an analytic function of (ν, x) on $\mathfrak{a}_c{}^* \times G$ which is holomorphic in ν.

LEMMA 8. $E(P:\psi:\nu) \in \mathcal{C}(G, \tau)$ *for $\nu \in \mathfrak{a}^*$. If P' is another psgp of G, then*

$$E_{P'}(P:\psi:\nu) \sim 0 \qquad (\nu \in \mathfrak{a}^*)$$

unless P' is associated to P.

Let $N(A)$ and $Z(A)$ respectively denote the normalizer and centralizer of A in G. Then $Z(A) = MA$ and

$$\mathfrak{w} = \mathfrak{w}(A) = N(A)/Z(A)$$

is a finite group which operates on $\mathfrak{a}$ and $\mathfrak{a}^*$ in the usual way. Let $P' = MAN'$ be a psgp with the same split component A. Then if $\nu \in \mathfrak{a}^*$ is not too special (i.e. does not lie on a finite set of hyperplanes passing through the origin), there exist unique endomorphisms $c_{P'|P}(s: \nu)$ $(s \in \mathfrak{w})$ of the finite-dimensional space $L = {}^\circ\mathcal{C}(M, \tau_M)$ (Theorem 10) such that

$$E_{P'}(P:\psi:\nu:ma) = \sum_{s \in \mathfrak{w}} (c_{P'|P}(s:\nu)\psi)(m) \exp((-1)^{1/2}s\nu(\log a))$$

for all $m \in M$, $a \in A$ and $\psi \in L$. Regarded as functions of ν, $c_{P'|P}(s:\nu)$ are meromorphic on $\mathfrak{a}^* + (-1)^{1/2} U$ where U is an open neighborhood of zero in $\mathfrak{a}^*$. However it seems likely that they are actually meromorphic in the whole complex space $\mathfrak{a}_c^*$. Here the analogy with Eisenstein series is quite clear (see [4(b), Chapter V] and [4(d)]).

11. The Fourier transform of a Schwartz function. From now on we shall assume for simplicity that G satisfies the conditions of §4. Let $\mathfrak{a}$ be a θ-stable Cartan subalgebra of $\mathfrak{g}$ and A the corresponding Cartan subgroup of G. Put $A_I = A \cap K$ and $A_R = \exp \mathfrak{a}_R$. Let A^*, A_I^* and A_R^* denote the character groups of A, A_I and A_R respectively. Then we can identify A_R^* with $\mathfrak{a}_R^*$ and therefore A^* with $A_I^* \times \mathfrak{a}_R^*$.

Let $\mathcal{P}(\mathfrak{a}_R)$ denote the set of all psgps P of G with split component A_R. Then $P = M A_R N$ and A_I is a compact Cartan subgroup of M. Let $\mathfrak{Z}_M$ denote the algebra of all differential operators on M which commute with left and right translations of M. Then for every $a^* \in A_I^*$, we have (see §4) a tempered and invariant eigendistribution θ_{a^*} of $\mathfrak{Z}_M$ on M.

Fix $a^* \in A^*$. Then $a^* = (a_0^*, \nu)$ where $a_0^* \in A_I^*$ and $\nu \in \mathfrak{a}_R^*$. We now define a tempered distribution $\Theta_{a^*} = \Theta_{a_0^*, \, \nu}$ on G as follows:

$$\Theta_{a^*}(f) = \theta_{a_0^*}(g_{f, \nu}) \qquad (f \in \mathcal{C}(G)).$$

Here

$$g_{f, \nu}(m) = \int_{A_R \times N} \bar{f}(man) \exp((-1)^{1/2}\nu + \rho)(\log a)\, da\, dn, \qquad (m \in M)$$

$\rho = \rho_P$ and

$$\bar{f}(x) = \int_K \bar{f}(kxk^{-1})\, dk \qquad (x \in G).$$

The Haar measures da and dn are normalized as in §6 by $da = d\mathfrak{a}_R$ and $dn = d\mathfrak{n}$. One can show that $g_{f, \nu} \in \mathcal{C}(M)$ and Θ_{a^*} is a tempered and invariant eigendistribution of $\mathfrak{Z}$ on G. Moreover the definition of Θ_{a^*} is independent of the choice of P in $\mathcal{P}(\mathfrak{a}_R)$.

Let $\tilde{A}$ be the normalizer of A in G. Put $W(G/A) = \tilde{A}/A$. Then $W(G/A)$ is a finite group which operates in the usual way on $\mathfrak{a}$, A and A^*. Let Q be the set of all positive roots of $(\mathfrak{g}, \mathfrak{a})$ (under some order) and Q_I the subset of those $\alpha \in Q$ which vanish identically on $\mathfrak{a}_R$. Put

$$\omega = \prod_{\alpha \in Q} H_\alpha, \qquad \omega_I = \prod_{\alpha \in Q_I} H_\alpha,$$

where H_α denotes, as usual, the element in $\mathfrak{a}_c$ dual to α under the Killing form. Let $\mathfrak{a}_I$ be the Lie algebra of A_I and λ the linear function on $\mathfrak{a}_I$ given by

$$\langle a_0{}^*, \exp H \rangle = e^{\lambda(H)} \qquad (H \in \mathfrak{a}_I).$$

We define $\log a_0{}^* = \lambda$, $\log a^* = \lambda + (-1)^{1/2}\nu$ and $\varpi_I(a^*) = \varpi_I(a_0{}^*) = \varpi_I(\lambda)$, $\varpi(a^*) = \varpi(a_0{}^*{:}\nu) = \varpi(\lambda + (-1)^{1/2}\ \nu)$. Since $\varpi_I{}^2$ is invariant under $W(G/A)$, $\varpi_I{}^s = \epsilon_I(s)\varpi_I$ $(s \in W(G/A))$ where $\epsilon_I(s) = \pm 1$. Let $A_I{}^{*\prime}$ denote the set of all $a_0{}^* \in A_I{}^*$ such that $\varpi_I(a_0{}^*) \neq 0$.

LEMMA 9. $\Theta_{sa*} = \epsilon_I(s)\Theta_{a*}$ *for* $s \in W(G/A)$ *and* $a^* \in A^*$.

This follows from Lemma 1 (§4).

Since $A_I{}^*$ is a discrete group, $A^* = A_I{}^* \times \mathfrak{a}_R{}^*$ is a Lie group. By a polynomial function p on A^*, we mean a function of the form

$$p(a^*) = q(\log a^*) \qquad (a^* \in A^*),$$

where q is a polynomial function on $\mathfrak{a}_c{}^*$. Let $\mathfrak{D}(\mathfrak{a}_R{}^*)$ denote the algebra of all translation-invariant differential operators on $\mathfrak{a}_R{}^*$ and $\mathfrak{D}(A^*)$ the algebra of differential operators on A^* generated by $\mathfrak{D}(\mathfrak{a}_R{}^*)$ and the polynomial functions on A^*. Let $\mathcal{C}(A^*)$ be the space of all functions $f \in C^\infty(A^*)$ such that

$$\mu_D(f) = \sup_{A^*} |Df| < \infty$$

for all $D \in \mathfrak{D}(A^*)$. We topologize $\mathcal{C}(A^*)$ by means of the seminorms μ_D $(D \in \mathfrak{D}(A^*))$. Then $C_c^\infty(A^*)$ is dense in $\mathcal{C}(A^*)$. By a tempered distribution on A^*, we mean a continuous linear function on $\mathcal{C}(A^*)$.

For any $f \in \mathcal{C}(G)$, put (see §7)

$$f_A(a^*) = (\Theta_{a*}, f) \qquad (a^* \in A^*).$$

Then $\mathring{f}_A \in \mathcal{C}(A^*)$ and $f \to \mathring{f}_A$ is a continuous mapping of $\mathcal{C}(G)$ into $\mathcal{C}(A^*)$.

LEMMA 10. *Let* $f \in \mathcal{C}(G)$. *Then*

$$f_A(sa^*) = \epsilon_I(s)f_A(a^*) \qquad (s \in W(G/A),\ a^* \in A^*).$$

This is an immediate consequence of Lemma 9.

Let $A^*(S)$ denote the set of all points $a^* \in A^*$ where $\varpi(a^*) = 0$.

Now fix a complete set $\mathfrak{a}_i$ $(1 \leq i \leq r)$ of θ-stable Cartan subalgebras of $\mathfrak{g}$, no two of which are conjugate under G. Let A_i denote the corresponding Cartan subgroup. Put $\mathring{f}_i = \mathring{f}_{A_i}$ $(f \in \mathcal{C}(G))$ and define $\epsilon_{i,I}(s)$ $(s \in W(G/A_i))$ as above.

THEOREM 18. *Suppose we are given, for each i $(1 \leq i \leq r)$, a tempered distribution T_i on $A_i{}^*$ such that:*[7]

(1) $T_i^s = \epsilon_{i,I}(s) T_i$ $(s \in W(G/A_i))$,

(2) $\sum_{1 \leq i \leq r} T_i(\hat{f}_i) = 0$ *for all* $f \in \mathcal{C}(G)$.

Then[8] $\operatorname{Supp} T_i \subset A_i{}^*(S)$ $(1 \leq i \leq r)$.

We think of $(\hat{f}_i)_{1 \leq i \leq r}$ as the Fourier transform of $f \in \mathcal{C}(G)$.

12. The Plancherel formula for G.

Define $\mathfrak{b}$, B, $W = W(\mathfrak{g}/\mathfrak{b})$ and ϖ as in §4. (We assume that $B \subset K$.) For any $f \in \mathcal{C}(G)$, put

$$*F_f(b) = [W]^{-1} \sum_{s \in W} \varpi F_f(b^s) \qquad (b \in B).$$

Then (see §8) $*F_f$ is a continuous function on B. We now use the notation of Theorem 18. Let $d_i a^*$ denote the Haar measure on $A_i{}^*$.

THEOREM 19. *There exist unique continuous functions C_i on $A_i{}^* \times B$ $(1 \leq i \leq r)$ with the following properties:*

(1) $C_i(a^*: b) = 0$ *if* $\varpi_{i,I}(a^*) = 0$ $(a^* \in A_i{}^*, b \in B)$.

(2) $C_i(sa^*: b) = \epsilon_{i,I}(s) C_i(a^*: b)$ $(s \in W(G/A_i))$.

(3) *We can choose numbers c, $p \geq 0$ such that*[9]

$$\left| C_i(a^*:b) \right| \leq c(1 + \|\log a^*\|)^p$$

for all $a^* \in A_i{}^*$ and $b \in B$.

$$(4) \qquad *F_f(b) = \sum_{1 \leq i \leq r} \int_{A_i^*} C_i(a^*: b) f_i(a^*) d_i a^*$$

for $f \in \mathcal{C}(G)$ and $b \in B$.

The uniqueness is an immediate consequence of Theorem 18. Combining this with Lemma 5, we get the following result.

COROLLARY. $(-1)^q c_G f(1) = \sum_{1 \leq i \leq r} \int_{A_i{}^*} C_i(a^*: 1) \hat{f}_i(a^*) d_i a^*$ *for* $f \in \mathcal{C}(G)$.

This is substantially the Plancherel formula for G.

The proof of Theorem 19 is based on a long induction on the rank of G. It makes essential use of the differential equations connecting the functions F_f and F_{zf} $(z \in \mathfrak{Z}, f \in \mathcal{C}(G))$. The basic idea of utilizing these differential equations to obtain a theorem of the above type,

---

[7] $W(G/A_i)$ operates on the space of distributions on $A_i{}^*$ in the obvious way.

[8] $\operatorname{Supp} T_i$ denotes the support of T_i.

[9] The Euclidean norm on $\mathfrak{g}$ (see §6) defines the structure of a Hilbert space on $\mathfrak{g}_0$ and hence also on $\mathfrak{a}_0$ and $\mathfrak{a}_0{}^*$ for $\mathfrak{a} = \mathfrak{a}_i$.

goes back to R. P. Langlands who had worked out the case of SL(2, R) by this method in 1962.

Fix i ($1 \leq i \leq r$), $a_0^* \in A_{i,I}^*$ and put $\mathfrak{F}_i = \mathfrak{a}_{i,R}^*$. Then the function

$$C_i(a_0^*:v:b) \qquad (v \in \mathfrak{F}_i, \ b \in B)$$

is of class C^∞ on $\mathfrak{F}_i \times B'$ and it is analytic in v. In fact it is a rather simple function which, for a fixed $b \in B'$, is meromorphic on $\mathfrak{F}_{i,c}$.

13. **The relation between the Plancherel measure and the asymptotic behaviour of certain Eisenstein integrals.** Fix i ($1 \leq i \leq r$) and drop it from the notation. Now put

$$q(a_0^*:v) = C_i(a_0^*:v:1) \qquad (a_0 \in A_I^{*\prime}, \ v \in \mathfrak{F}).$$

We wish to obtain a formula for $q(a_0^*: v)$.

For any $v \in \mathfrak{F}_c$, put $v = v_R + (-1)^{1/2} v_I$ ($v_R, v_I \in \mathfrak{F}$). Fix $P = MA_R N$ in $\mathcal{P}(\mathfrak{a}_R)$ and for any $\delta > 0$ let $\mathfrak{F}_c(\delta)$ denote the set of all $v \in \mathfrak{F}_c$ such that $\|v_I\| < \delta$. Then if δ is sufficiently small, we have defined (see §10) the meromorphic function $c_{P|P}(1: v)$ on $\mathfrak{F}_c(\delta)$ with values in the space of endomorphisms of $^\circ \mathcal{C}(M, \tau_M)$. Let $\alpha_1, \cdots, \alpha_l$ be all the distinct simple roots of (P, A_R). Define $H_j \in \mathfrak{a}_R$ by $\alpha_i(H_j) = \delta_{ij}$ ($1 \leq i, \ j \leq l$). Let $\mathfrak{F}_c(P)$ denote the set of all $v \in \mathfrak{F}_c$ such that $v_I(H_j) < 0$ ($1 \leq j \leq l$). Then $c_{P|P}(1: v)$ is actually holomorphic on $\mathfrak{F}_c(P)$. Put $\overline{N} = \theta(N)$. Given $x \in G$, we can write it uniquely in the form $x = kman$ ($k \in K$, $m \in M$, $a \in A$, $n \in N$) where $\theta(m) = m^{-1}$. Define $\kappa(x) = k$ and $\mu(x) = m$.

LEMMA 11. *Put $\rho = \rho_P$ and $H(x) = H_P(x)$ ($x \in G$). Then*

$$(c_{P|P}(1:v)\psi)(m)$$

$$= \int_{\overline{N}} \psi(m\mu(\bar{n})^{-1})\tau(\kappa(\bar{n}))^{-1} \exp[-((-1)^{1/2}v + \rho)(H(\bar{n}))]d_0\bar{n}$$

for $\psi \in {}^\circ \mathcal{C}(M, \tau_M)$, $v \in \mathfrak{F}_c(P)$ and $m \in M$. Here the integral is convergent and the Haar measure $d_0\bar{n}$ on $\overline{N}$ is so normalized that

$$\int_{\overline{N}} e^{-2\rho(H(\bar{n}))}d_0\bar{n} = 1.$$

Let V_0 be the subspace of all $v_0 \in V$ such that $\tau(k)v_0 = v_0\tau(k)$ for all $k \in K$. Fix $a_0^* \in A_I^{*\prime}$, $v_0 \in V_0$ and define $\psi(a_0^*: v_0) = \psi \in {}^\circ \mathcal{C}(M, \tau_M)$ by

$$\psi(m) = \int_{K_M} \theta_{a_0^*}(mk^{-1})\tau(k)v_0 d_0 k \qquad (m \in M),$$

where d_0k is the normalized Haar measure on K_M and $\theta_{a_0}^*$ has the same meaning as in §11. Let $\mathfrak{F}'(a_0^*)$ denote the set of all $\nu \in \mathfrak{F}$ such that $\varpi(a_0^*:\nu) \neq 0$. Then $c_{P|P}(1:\nu)\psi$ is holomorphic in ν for $\nu \in \mathfrak{F}'(a_0^*)$. Put

$$\|\phi\|_M^2 = \int_M |\phi(m)|^2 dm \qquad (\phi \in {}^\circ\mathcal{C}(M, \tau_M)),$$

$q_G = \tfrac{1}{2} \dim G/K$ and $q_M = \tfrac{1}{2} \dim M/K_M$.

THEOREM 20. *There exists a constant $c_0(P) > 0$ (independent of a_0^* and τ) such that*

$$c_0(P)q(a_0^*:\nu)\|c_{P|P}(1:\nu)\psi\|_M^2 = (-1)^{q_G + q_M} \varpi_I(a_0^*)\|\psi\|_M^2$$

for $\nu \in \mathfrak{F}'(a_0^*)$ and $\psi = \psi(a_0^*:v_0)$ $(v_0 \in V_0)$.

Actually it is possible to determine $c_0(P)$ explicitly. Moreover it is not difficult to show that for any given $a_0^* \in A_I^{*'}$, we can choose τ in such a way that $\psi(a_0^*:v_0) \neq 0$ for some $v_0 \in V_0$.

14. The p-adic case. First we recall some standard facts about algebraic groups [1].

Let Ω be a field. By an Ω-group, we mean a (linear) algebraic group defined over Ω. Let $\mathbf{G}$ be a connected and reductive Ω-group. By a parabolic subgroup $\mathbf{P}$ of $\mathbf{G}$, we mean an algebraic subgroup which contains a Borel subgroup of $\mathbf{G}$. We say that $\mathbf{P}$ is Ω-parabolic if it is parabolic and defined over Ω. Let $\mathbf{N}$ denote the (unipotent) radical of $\mathbf{P}$. Then $\mathbf{N}$ is an Ω-subgroup and $\mathbf{P}$ is the normalizer of $\mathbf{N}$ in $\mathbf{G}$.

Now assume that Ω is a p-adic field (i.e. a locally compact field with a discrete valuation). Let G denote the set of all Ω-rational points of $\mathbf{G}$. Then $G \subset GL(n, \Omega)$ for a suitable $n \geq 1$. $GL(n, \Omega)$, being an open subset of a vector space over Ω of dimension n^2, is a locally compact group. Since G is closed in $GL(n, \Omega)$, it is also locally compact. Moreover it is easy to verify that it is separable and unimodular.

By a parabolic subgroup P of G, we mean a subgroup of the form $P = G \cap \mathbf{P}$, where $\mathbf{P}$ is an Ω-parabolic subgroup of $\mathbf{G}$. Then $\mathbf{P}$ is uniquely determined by P. Put $N = G \cap \mathbf{N}$ where $\mathbf{N}$ is the radical of $\mathbf{P}$. Let $C_c(G)$ denote the space of all continuous complex-valued functions on G with compact support. For any $f \in C_c(G)$, define

$$f^P(x) = \int_N f(xn)dn \qquad (x \in G),$$

where dn is the Haar measure of the unimodular group N. f is said to be a cusp form if $f^P = 0$ for all parabolic subgroups $P \neq G$. Let $°C_c(G)$ denote the space of all cusp forms in $C_c(G)$.

Let π be an irreducible unitary representation of G and $\mathfrak{H}_\pi$ the representation space of π. We say that π is supercuspidal if there exist nonzero elements ϕ, ψ in $\mathfrak{H}_\pi$ such that the function $x \to (\phi, \pi(x)\psi)$ $(x \in G)$ lies in $°C_c(G)$. Let $°\mathcal{E}(G)$ denote the set of all classes $\omega \in \mathcal{E}(G)$, which contain supercuspidal representations. It is obvious that $°\mathcal{E}(G) \subset \mathcal{E}_2(G)$.

Let G_0 be an open subset of G and $C_c^\infty(G_0)$ the space of all locally constant complex-valued functions on G_0 with compact support. By a distribution T on G_0, we mean a linear mapping of $C_c^\infty(G_0)$ into $\mathbf{C}$. Let F be a function on G_0 which is locally summable with respect to the Haar measure dx of G. Then we say that $T = F$ if

$$T(f) = \int_G F \cdot f \, dx$$

for all $f \in C_c^\infty(G_0)$.

THEOREM 21. *Let π be a square-integrable representation of G. Then for any $f \in C_c^\infty(G)$, the operator*

$$\pi(f) = \int_G f(x) \pi(x) \, dx$$

is of the trace class.

Since every neighborhood of 1 in G, contains an open compact subgroup, this is an immediate consequence of the corollary of Theorem 5.

Define

$$\Theta_\pi(f) = \operatorname{tr} \pi(f) \qquad (f \in C_c^\infty(G)).$$

Then Θ_π is a distribution on G which depends only on the class ω of π. Hence we may denote it by Θ_ω. Then Θ_ω is invariant and the mapping $\omega \to \Theta_\omega$ $(\omega \in \mathcal{E}_2(G))$ is injective.

THEOREM 22. *Let $\mathfrak{H}_\pi$ denote the representation space of a square-integrable representation π of G. Fix $\phi, \psi \in \mathfrak{H}_\pi$. Then*

$$(\phi, \psi)\Theta_\pi(f) = d(\pi) \int_G (\phi, \pi(x)\pi(f)\pi(x^{-1})\psi) \, dx$$

$$= d(\pi) \int_G dx \int_G f(y)(\phi, \pi(xyx^{-1})\psi) \, dy$$

for $f \in C_c^\infty(G)$.

This is an easy consequence of the Schur orthogonality relations.

The function D and the regular set G' are defined in exactly the same way as in §2. Then D is a regular function on G and G' is an open and dense subset of G whose complement is of measure zero.

THEOREM 23. *Fix* $\omega \in {}^\circ\mathcal{E}(G)$. *Then there exists a locally constant function F on G' such that* $\Theta_\omega = F$ *on* G'.

It seems very likely that F is actually locally summable on G and $\Theta_\omega = F$. In any case, the analogy with the results of §2 is obvious.

For any $\alpha \in \Omega$, we define its absolute value $|\alpha|$ in the usual way so that

$$d(\alpha t) = |\alpha|\, dt$$

for any additive Haar measure dt on Ω.

LEMMA 12. *If Ω has characteristic zero, the function* $|D|^{-1/2}$ *is locally summable on G.*

Put $\Phi = |D|^{1/2} F$ in the notation of Theorem 23. It seems reasonable to expect that Φ remains bounded on G'.

By a Cartan subgroup A of G, we mean a subgroup of the form $A = G \cap \mathbf{A}$ where $\mathbf{A}$ is a maximal Ω-torus in $\mathbf{G}$. Let Φ_A denote the restriction of Φ on $A' = A \cap G'$.

LEMMA 13. *There exists a compact subset C of A such that* $\Phi_A = 0$ *outside C. Moreover if A is compact and Ω has characteristic zero, then* Φ_A *remains bounded on A'.*

When $\mathbf{G} = \mathrm{SL}(2)$ and the characteristic of the residue field of Ω is not 2, the characters of G have been explicitly computed by Sally and Shalika [8] and their results agree with the above statements.

15. **Two problems.** Let me conclude these lectures by discussing two problems (or conjectures) in the $\mathfrak{p}$-adic case.

Fix an open compact subgroup K_0 of G and let ${}^\circ C_c(G//K_0)$ denote the space of all functions in ${}^\circ C_c(G)$ which are constant on double cosets of K_0. Then one would like to show that, in case prk $G = 0$,

$$\dim {}^\circ C_c(G//K_0) < \infty.$$

Here the analogy with Theorem 10 is obvious.

In view of Theorem 5, it would be enough to prove that

$$\inf_{\omega \in {}^\circ\mathcal{E}(G)} d(\omega) > 0.$$

Actually I am inclined to believe that the numbers $d(\omega)$ $(\omega \in \mathcal{E}_2(G))$ are all integers under a suitable normalization of the Haar measure of G.

K being any open compact subgroup of G, an affirmative answer to the above question would imply that

$$\sup_{\omega \in \mathcal{E}(G)} [\omega : \mathfrak{b}] < \infty$$

for $\mathfrak{b} \in \mathcal{E}(K)$, provided Ω has characteristic zero.

In order to state the second problem, we have to recall a recent theorem of Bruhat and Tits. Let (P_0, A_0) be a mincuspidal pair in G and $P_0 = M_0 N_0$ the corresponding Levi decomposition [4(c), §2]. Put $A_0 = G \cap \mathbf{A}_0$ and for any root α of (P_0, A_0), let ξ_α denote the corresponding character of $\mathbf{A}_0$. Let A_0^+ be the set of all points $a \in A_0$ where $|\xi_\alpha(a)| \geqq 1$ for every root α of (P_0, A_0). Define $\mathfrak{w}(A_0)$ as in [4(c), §2].

THEOREM (BRUHAT-TITS). *We can choose an open and compact subgroup K of G with the following properties:*

(1) $G = KP_0$.

(2) $G = KA_0^+ CK$, where C is a finite subset of M_0.

(3) *Every element of* $\mathfrak{w}(A_0)$ *has a representative in K.*

(4) *If $(P, A) \succ (P_0, A_0)$ is a cuspidal pair and $P = MN$ the corresponding Levi decomposition, then $P \cap K = (M \cap K)(N \cap K)$.*

(5) *Put $K_M = K \cap M$ and $*P_0 = M \cap P_0$ in the notation of (4). Then if we replace (G, P_0, A_0, K) by $(M, *P_0, A_0, K_M)$, the above four conditions are again fulfilled.*

·We keep to the notation of the above theorem. Let τ be a unitary representation of K on a finite-dimensional Hilbert space V and let $C_c(G, \tau)$ denote the space of all functions f from G to $\text{End}(V)$ with compact support such that

$$f(k_1 x k_2) = \tau(k_1)f(x)\tau(k_2) \qquad (k_1, k_2 \in K, \ x \in G).$$

Then $C_c(G, \tau)$ is an algebra under convolution. On the other hand, let V_P be the subspace of all $v \in V$ such that $\tau(n)v = v$ for all $n \in N \cap K$. Let E_P denote the orthogonal projection of V on V_P. Put

$$\tau_M(m) = \tau(m)E_P \qquad (m \in K_M).$$

Then τ_M may be regarded as a representation of K_M on V_P and so we can consider the algebra $C_c(M, \tau_M)$.

Let $d_r p$ denote a right-invariant Haar measure on P and define the function δ_P on P by the relation

$$d_r(qp) = \delta_P(q)d_r p \qquad (q \in P).$$

For any $f \in C_c(G, \tau)$, put

$$f^{(P)}(m) = \delta_P(m)^{1/2} \int_N f(mn)E_P dn \qquad (m \in M),$$

where dn is a Haar measure on N. Then $f^{(P)} \in C_c(M, \tau_M)$ and the mapping $\mu_P: f \to f^{(P)}$ is actually a homomorphism of $C_c(G, \tau)$ into $C_c(M, \tau_M)$. We would now like to assert that $C_c(M, \tau_M)$ *is a finite right-module over* $\mu_P(C_c(G, \tau))$. If this is true, much of the theory for real groups [4(a), Part II] can be imitated in the $\mathfrak{p}$-adic case.

<h1 style="text-align:center">REFERENCES</h1>

1. A. Borel and J. Tits, *Groupes réductifs*, Inst. Hautes Études Sci. Publ. Math No. 27 (1965), 55–150. MR **34** #7527.

2. I. M. Gelfand, *Automorphic functions and the theory of representations*, Proc. Internat. Congress Math. (1962) Inst. Mittag-Leffler, Djursholm, 1963, pp. 74–85. MR **28** #1.

3. R. Godement, *Sur les relations d'orthogonalité de V. Bargmann.* I: *Démonstration générale*, C. R. Acad. Sci. Paris **225** (1947), 657–659. MR **9**, 134.

4. Harish-Chandra, (a) *Discrete series for semisimple Lie groups.* II, Acta Math. **116** (1966), 1–111. MR **36** #2745. (b) *Automorphic forms on semisimple Lie groups*, Lecture notes in Math., no. 62, Springer-Verlag, Berlin and New York, 1968. MR **38** #1216. (c) *Eisenstein series over finite fields*, Proc. Stone Jubilee Conf., Chicago, 1968. (d) *Some applications of the Schwartz space of a semisimple Lie group*, Lecture Series, Modern Anal. and Appl., George Washington Univ., 1968.

5. G. W. Mackey, *Infinite-dimensional group representations*, Bull. Amer. Math. Soc. **69** (1963), 628–686. MR **27** #3745.

6. C. C. Moore, *Decomposition of unitary representations defined by discrete subgroups of nilpotent groups*, Ann. of Math. (2) **82** (1965), 146–182. MR **31** #5928.

7. L. Pukánszky, *Leçons sur les représentations des groupes*, Monographies de la Société Mathématique de France, no. 2, Dunod, Paris, 1967. MR **36** #311.

8. P. J. Sally, Jr. and J. A. Shalika, *Characters of the discrete series of representations of* SL(2) *over a local field*, Proc. Nat. Acad. Sci. U.S.A. **61** (1968), 1231–1237.

9. L. Schwartz, *Théorie des distributions.* Tome I, Actualités Sci. Indust., no. 1091, Hermann, Paris, 1950. MR **12**, 31.

INSTITUTE FOR ADVANCED STUDY, PRINCETON, NEW JERSEY 08540

Reprinted from
Bull. Amer. Math. Soc.
76 (1970), 529–551

Harmonic Analysis on Reductive p-adic Groups

HARISH-CHANDRA

Introduction

The object of these lectures is to illustrate, what I like to call the
Lefschetz principle, which, in the context of reductive groups, says that
whatever is true for real groups is also true for $\wp$-adic groups. The theory
of reductive algebraic groups can be divided into four parts arranged according
to their increasing transcendental component.

1) Algebraic theory.

2) Structure theory.

3) Local Fourier Analysis.

4) Global Fourier Analysis on G_A/G_F.

With the work of Chevalley and that of Borel and Tits, the Lefschetz principle
may now be regarded as quite well established for the algebraic theory.
Moreover the recent work of Bruhat and Tits shows that this principle also
holds with regard to the structure theory of real and $\wp$-adic groups. The
results of Mautner, Bruhat, Satake, Gelfand-Graev, Macdonald, and Sally-
Shalika suggest that it is also valid at stage 3). These lectures are meant to
strengthen this conclusion. Finally the work of Jacquet and Langlands on automorphic forms on $GL(2)$, shows that all primes ought to be treated on an equal
footing and this is precisely the content of the Lefschetz principle in the global
case.

These lectures may also be regarded as an attempt to justify a claim
about the philosophy of cusp forms which I made some years ago (see
"Eisenstein series over finite fields," Stone Jubilee volume, Springer). We
shall see how this philosophy can be used to suggest, and occasionally to
prove, results on harmonic analysis on reductive $\wp$-adic groups.

Let Ω be a $\wp$-adic field (i.e. a locally compact field with a discrete
valuation) and let $\underset{\sim}{G}$ be a linear algebraic group defined over Ω. We assume
that $\underset{\sim}{G}$ is connected and reductive. Let G denote the set of all Ω-rational
points in $\underset{\sim}{G}$. Then G is a closed subgroup of $GL(n, \Omega)$ and so it is locally

compact. Also it is separable and unimodular.

Let dx denote the Haar measure on G and $C_c^\infty(G)$ the space of all complex-valued locally constant functions on G with compact support. Let π be an irreducible unitary representation of G on a Hilbert space $\mathcal{H}$. For any $f \in C_c^\infty(G)$, define the operator

$$\pi(f) = \int_G f(x)\pi(x)dx \ .$$

The following two statements are equivalent.

1) Let K be an open compact subgroup of G and $\underline{d}$ an irreducible representation of K. Then the multiplicity of $\underline{d}$ in the restriction of π to K is finite.

2) For any $f \in C_c^\infty(G)$, $\pi(f)$ is of the trace class.

By analogy with the real case, one expects them to be true. Assuming that this is so, put

$$\Theta_\pi(f) = \operatorname{tr} \pi(f)$$

Then Θ_π is a distribution on G which depends only on the class ω of π. Hence we may denote it by Θ_ω. It is easy to show that the mapping $\omega \longmapsto \Theta_\omega$ is injective. If $\ell = \operatorname{rank} \underline{G}$, let $D(x)$ $(x \in G)$ denote the coefficient of t^ℓ in the polynomial $\det(t+1-\operatorname{Ad}(x))$ in the indeterminate t. Let G' be the set of all $x \in G$ where $D(x) \neq 0$. Then G' is an open dense subset of G whose complement is of measure zero. Again by analogy with the real case, one expects that the following statements are true.

3) Θ_ω is a locally summable function on G which is locally constant on G'.

4) The function $\Phi_\omega = |D|^{1/2}\Theta_\omega$ is locally bounded on G.

By a Cartan subgroup A of G, we mean a subgroup of the form $A = \underline{A} \cap G$ where $\underline{A}$ is a maximal Ω-torus in $\underline{G}$. For $f \in C_c^\infty(G)$, define

$$F_f^A(a) = F_f(a) = |D(a)|^{1/2}\int_{G/A} f(xax^{-1})dx^* \qquad (a \in A' = A \cap G') \ ,$$

where dx^* is the invariant measure on G/A. It is easy to see that F_f is locally constant on A'. The Lefschetz principle leads us to expect the following.

5) F_f remains bounded on A. Moreover at any point $a_0 \in A$, F_f has only finitely many distinct limits.
I think it is very important to investigate these limits. (Those familiar with the real case, would see the analogy immediately.) The significant case is when A is compact.

Finally we come to the following question.

6) What is the connection, if any, between the functions F_f^A (for the various Cartan subgroups A of G) and $f(1)$?
Of course the main goal here is the Plancherel formula. However, I hope that a correct understanding of this question would lead us in a natural way to the discrete series for G. (This is exactly what happens in the real case. But the p-adic case seems to be much more difficult here.)

I propose to discuss some of these questions in these lectures and offer some partial answers which seem to be quite encouraging.

These lectures were given at The Institute for Advanced Study during the Fall Term of '69. I am very grateful to Dr. van Dijk for having taken the trouble of preparing these notes. In addition to streamlining the presentation, he has had to work out the details of many proofs and although I have not personally checked them, I trust there are no mistakes.

Harish-Chandra

Reprinted from
Lecture Notes in Mathematics, No. **162**
Springer-Verlag, Berlin-Heidelberg-New York, 1970

<u>ON THE THEORY OF THE EISENSTEIN INTEGRAL</u>

by

<u>Harish-Chandra</u>

The Institute for Advanced Study
Princeton, N. J.

§1. Introduction

Let F be a field and $\underset{\sim}{G}$ a connected, reductive algebraic group defined over F. Let G_F denote the subgroup of all F-rational points of G. If F is a local field (i.e. $F = \underset{\sim}{R}$, $\underset{\sim}{C}$ or a $\mathfrak{p}$-adic field), G_F has a natural locally compact topology. (We write $G_F = G_{\mathfrak{p}}$ when F is a $\mathfrak{p}$-adic field.) On the other hand if F is a finite field, G_F is a finite group. Finally if F is a global field (i.e. a number-field or a function-field in one variable over a finite field), we have the adèle group $G_{\underset{\sim}{A}}$ associated to $\underset{\sim}{G}$. From the point of view of harmonic analysis and the theory of automorphic forms, the study of the following five cases is important.

1) $G_{\underset{\sim}{R}}/\Gamma$ where Γ is an arithmetic subgroup of G,
2) $G_{\underset{\sim}{R}}$,
3) $G_{\underset{\sim}{\mathfrak{p}}}$,
4) G_F where F is a finite field,
5) $G_{\underset{\sim}{A}}/G_F$ where F is a global field.

Actually 5) is the most difficult case and the other four may, in fact, be regarded merely as its several facets. Nevertheless it is useful to pursue their study individually since a knowledge of one case enables us to guess, often quite accurately, analogous results in another case. This similarity from the stand-point of harmonic analysis, is most striking between $G_{\underset{\sim}{R}}$ and $G_{\underset{\sim}{\mathfrak{p}}}$ (see [4(e)]). But in this lecture I propose to bring out the resemblance between $G_{\underset{\sim}{R}}/\Gamma$ and $G_{\underset{\sim}{R}}$.

The theory of Eisenstein Series over $G_{\underset{\sim}{R}}/\Gamma$, which is largely due to Selberg, Gelfand-Piatetsky-Shapiro and Langlands (see [4(a)]), has an exact counterpart in the theory of Eisenstein Integrals over $G_{\underset{\sim}{R}}$. These integrals have functional equations (see §6) which are governed by the coefficients

$c_{P_2|P_1}(s : \nu)$ appearing in their asymptotic expansion. Following Gelfand [2] (see also [4(b), §5]), one is tempted to call these coefficients the local zeta-functions at infinity. Since a similar theory holds for $G_{\mathfrak{p}}$, one gets in this way local zeta-functions at each prime. It seems likely that the global zeta-functions (i.e. the corresponding coefficients appearing in the Eisenstein Series for G_A/G_F) are actually products built out of the local factors. This leads one to expect that the local factors are "elementary" or "Eulerian." Therefore, in particular, $c_{P_2|P_1}(s : \nu)$ should be expressible in terms of gamma factors. Although this conjecture remains unproved, we do obtain a rather simple formula for the "absolute value" of this operator and therefore for the Plancherel measure $\mu_\omega(\nu)d\nu$ (see §13). Moreover there is some evidence (see §14) that the zeros of the function μ_ω are related to the occurrence of the exceptional series (see also [6(a), (b)] and [7]).

In the first few sections of this paper, we restate in a precise form those results of [4(c)] and [4(d)], which are necessary for the understanding of our main theorems.

§2. The assumptions on G

For any Lie group G, we denote by G° the connected component of 1 in G and by $X(G)$ the group of all continuous homomorphisms of G into $R^\times$. Put

$$^\circ G = \bigcap_{\chi \epsilon X(G)} \ker |\chi| \ .$$

Then $G/{}^\circ G$ is an abelian Lie group. By the parabolic rank of G we mean $\dim G/{}^\circ G$ and denote it by prk G. A split component of G is a closed subgroup A of G such that

$$G = {}^\circ G . A , \quad {}^\circ G \cap A = \{1\} \ .$$

Note that A, if it exists, is abelian. In fact $A \simeq G/{}^\circ G$ and therefore dim A = prk G. By a vector subgroup V of G, we mean a closed subgroup which is topologically isomorphic to the additive group of $\underset{\sim}{R}^n$ for some $n \geq 0$.

Let G be a Lie group with Lie algebra $\mathfrak{g}$. Let G_c denote the connected complex adjoint group[1] of $\mathfrak{g}_c$. We make the following assumptions on G.

1) $\mathfrak{g}$ is reductive and $Ad(G) \subset G_c$.

2) Let G_1 denote the analytic subgroup of G corresponding to $\mathfrak{g}_1 = [\mathfrak{g}, \mathfrak{g}]$. Then the center of G_1 is finite.

3) $[G : G^\circ] < \infty$.

Fix a maximal compact subgroup K of G. (Such a subgroup exists and is unique up to conjugacy by G°.) Let C be the center of G°. Fix a maximal vector subgroup C_2 of C. Then $C = C_1 \cdot C_2$ where $C_1 = C \cap K$. Let $\mathfrak{k}$, $\mathcal{L}_1$, $\mathcal{L}_2$ be the Lie algebras of K, C_1, C_2 respectively and θ the Cartan involution of $\mathfrak{g}_1$ with respect to $\mathfrak{k}_1 = \mathfrak{k} \cap \mathfrak{g}_1$. Extend θ to an involution of $\mathfrak{g}$ by setting

$$\theta(X_1 + X_2) = X_1 - X_2 \qquad (X_i \in \mathcal{L}_i, \ i = 1, 2) \ .$$

Let $\mathfrak{p}$ be the set of all points $X \in \mathfrak{g}$ such that $\theta(X) = -X$.

Lemma 1. __The mapping__ $(k, X) \longmapsto k \exp X$ $(k \in K, X \in \mathfrak{p})$ __is an analytic diffeomorphism of__ $K \times \mathfrak{p}$ __onto__ G __and__ θ __extends to an automorphism of__ G __such that__

$$\theta(k \exp X) = k \exp(-X) \qquad (k \in K, X \in \mathfrak{p}) \ .$$

We denote by $\log$ the inverse of the mapping $\exp : \mathfrak{p} \longrightarrow \exp \mathfrak{p}$.

Lemma 2. __There exists a real symmetric bilinear form__ B __on__ $\mathfrak{g}$ __such that:__
1) $B(\theta X, \theta Y) = B(X, Y)$ __and__

$$B([X, Y], Z) + B(Y, [X, Z]) = 0 \qquad (X, Y, Z \in \mathfrak{g}) \ .$$

2) <u>The quadratic form</u>

$$\|X\|^2 = -B(X, \theta X) \qquad (X \in \mathfrak{g})$$

<u>is positive-definite on</u> $\mathfrak{g}$.

We fix K, θ and B as above, once for all and define $\sigma(x) = \|X\|$ for $x = k \exp X$ $(k \in K, X \in \mathfrak{p})$.

§3. Parabolic subgroups[2]

A subalgebra $\mathfrak{q}$ of $\mathfrak{g}$ is called parabolic, if $\mathfrak{q}_c$ contains a Borel subalgebra (i.e. a maximal solvable subalgebra) of $\mathfrak{g}_c$. A subgroup Q of G is called parabolic, if it is the normalizer in G of some parabolic subalgebra $\mathfrak{q}$ of $\mathfrak{g}$. Then Q is closed and its Lie algebra is $\mathfrak{q}$.

Lemma 3. <u>Let</u> Q <u>be a parabolic subgroup</u> (psgp) <u>of</u> G. <u>Then</u> $G = KQ$.

Let $\mathfrak{q}$ be a parabolic subalgebra of $\mathfrak{g}$. By the radical of $\mathfrak{q}$, we mean the maximal ideal $\mathfrak{n}$ of $\mathfrak{q}_1 = \mathfrak{q} \cap [\mathfrak{g}, \mathfrak{g}]$ such that ad X is nilpotent for every $X \in \mathfrak{n}$. If Q is the psgp corresponding to $\mathfrak{q}$ and N the analytic subgroup corresponding to $\mathfrak{n}$, then N is called the radical of Q. Put $M_1 = Q \cap \theta(Q)$ and $M = {}^\circ M_1$ (see §2). Let A be the maximal θ-stable vector subgroup lying in the center of M_1. Then A is a split component both for Q and M_1. We call A <u>the</u> split component of Q. Then $M_1 = MA$ is the centralizer of A in G and

$$Q = MAN ,$$

the mapping $(m, a, n) \longmapsto man$ being an analytic diffeomorphism of $M \times A \times N$ onto Q. We call this the Langlands decomposition of Q.

Let $\mathfrak{m}$, $\mathfrak{a}$ denote the Lie algebra of M, A respectively and put $\mathfrak{m}_1 = \mathfrak{m} + \mathfrak{a}$, $K_M = K \cap M$. Let θ_M and B_M be the restrictions of θ and B respectively on $\mathfrak{m}$. Then if we replace (G, K, θ, B) by

(M, K_M, θ_M, B_M) all the conditions of §2 are fulfilled. The same holds for $(M_1, K_M, \theta_1, B_1)$ where θ_1 and B_1 are the restrictions of θ and B respectively on $\mathfrak{m}_1$.

By a p-pair (Q, A), we mean a psgp Q and its split component A. Then $Q = MAN$. By a root of Q (or (Q, A)), we mean an element a in $\mathfrak{a}^*$ with the following property. Let $\mathfrak{n}_a$ denote the set of all $X \in \mathfrak{n}$ such that $[H, X] = a(H)X$ for all $H \in \mathfrak{a}$. Then $\mathfrak{n}_a \neq \{0\}$. It is clear that prk $Q = \dim A \geq$ prk G.

Let $l =$ prk $Q -$ prk G and let $\Sigma(Q)$ denote the set of all roots of Q. A root $a \in \Sigma(Q)$ is called simple if it cannot be written in the form $a = \beta + \gamma$ with $\beta, \gamma \in \Sigma(Q)$. Let $\Sigma^o(Q)$ be the set of all simple roots. Then[5] $l = [\Sigma^o(Q)]$. Let $a_1, \ldots, a_l$ be all the simple roots. Then they are linearly independent over $\underset{\sim}{R}$ and every $a \in \Sigma(Q)$ can be written in the form

$$a = m_1 a_1 + \ldots + m_l a_l$$

where $m_i \in \underset{\sim}{Z}$ and $m_i \geq 0$ $(1 \leq i \leq l)$.

Fix a subset F of $\Sigma^o(Q)$ and let $\mathfrak{a}_F$ denote the set of all $H \in \mathfrak{a}$ such that $a(H) = 0$ for all $a \in F$. Let Σ_F be the set of all $a \in \Sigma = \Sigma(Q)$ which vanish identically on $\mathfrak{a}_F$. Put

$$\mathfrak{n}_F = \sum_{a \in \Sigma'_F} \mathfrak{n}_a$$

where Σ'_F is the complement of Σ_F in Σ. Then $\mathfrak{n}_F$ is an ideal in $\mathfrak{n}$. Put $N_F = \exp \mathfrak{n}_F$, $A_F = \exp \mathfrak{a}_F$ and let Q_F denote the normalizer of N_F in G. Then (Q_F, A_F) is a p-pair in G and

$$Q_F = M_F A_F N_F$$

where $M_F = {}^o Z(A_F)$ and $Z(A_F) = M_F A_F$ is the centralizer of A_F in G. We write $(Q, A)_F = (Q_F, A_F)$.

Let (Q', A') be any p-pair in G. We write $(Q', A') \succ (Q, A)$ if

$Q' \supset Q$. This implies that $A' \subset A$. Every p-pair $(Q', A') \succ (Q, A)$ is of the form $(Q', A') = (Q, A)_F$ for a unique $F \subset \Sigma^o(Q)$.

Lemma 4. <u>There is a one-one correspondence between psgps</u> P <u>of</u> G <u>which are contained in</u> Q <u>and psgps</u> ${}^*P$ <u>of</u> M. <u>This correspondence is given by the relation</u> ${}^*P = P \cap M$. <u>If</u>

$$P = M'A'N' , \qquad {}^*P = {}^*M{}^*A{}^*N$$

<u>are the corresponding Langlands decompositions, then</u>

$$M' = {}^*M , \quad A' = {}^*A.A , \quad N' = {}^*N.N ,$$

$${}^*A = M \cap A' , \quad {}^*N = M \cap N' .$$

Lemma 5. <u>Any two minimal psgps of</u> G <u>are conjugate under</u> K^o. <u>Let</u> P_1, P_2 <u>and</u> P <u>be three psgps of</u> G. <u>Suppose</u> $P_1 \cap P_2 \supset P$ <u>and</u> P_2 <u>is conjugate to</u> P_1 <u>under</u> G. <u>Then</u> $P_1 = P_2$.

Let (P_i, A_i) $(i = 1, 2)$ be two p-pairs. We denote by $W(A_2 | A_1) = W(\alpha_2 | \alpha_1)$ the set of all linear mappings

$$s : \alpha_1 \longrightarrow \alpha_2$$

satisfying the following condition. There should exist an element $y \in G$ such that $sH = \mathrm{Ad}(y)H$ for all $H \in \alpha_1$. y is then called a representative of s in G. (One can always choose a representative $y \in K$.) We also write $a^s = yay^{-1}$ $(a \in A_1)$.

P_1, P_2 are said to be associated if $W(A_2 | A_1)$ and $W(A_1 | A_2)$ are both nonempty. This is equivalent to saying that A_1 and A_2 are conjugate under G.

Let (P, A) be a p-pair. We write $W(A) = W(A | A)$. Then $W(A)$ is a finite group and in fact

$$W(A) = (\text{Normalizer of } A \text{ in } K)/(\text{Centralizer of } A \text{ in } K) .$$

We call $\mathcal{W}(A)$ the Weyl group of A in G and sometimes denote it by $\mathcal{W}(G/A)$.

Let $P = MAN$ be the Langlands decomposition of P. Define $\rho_P \in \mathcal{a}^*$ by

$$\rho_P(H) = \tfrac{1}{2}\mathrm{tr}(\mathrm{ad}\, H)_{\mathcal{n}} \qquad (H \in \mathcal{a}) ,$$

where $(\mathrm{ad}\, H)_{\mathcal{n}}$ denotes the restriction of $\mathrm{ad}\, H$ on $\mathcal{n}$. Since $G = KP$, every $x \in G$ can be written in the form $x = kman$ ($k \in K$, $m \in M$, $a \in A$, $n \in N$). Here a is uniquely determined and we put $H_P(x) = \log a$. Then $H_P : G \longrightarrow \mathcal{a}$ is an analytic mapping.

Let $d_\ell p$ and $d_r p$ denote the left- and right-invariant Haar measures on P so that $d_r p = d_\ell p^{-1}$. Then

$$d_r p = \delta_P(p) d_\ell p$$

where δ_P is a continuous homomorphism of P into $R_+^\times$. In fact

$$\delta_P(man) = e^{2\rho(\log a)} \qquad (m \in M,\ a \in A,\ n \in N)$$

where $\rho = \rho_P$. Note that $\delta_P = 1$ on $K \cap P = K \cap M$.

Now suppose P is a minimal psgp of G. We extend δ_P to a function on G by setting $\delta_P(kp) = \delta_P(p)$ ($k \in K$, $p \in P$). Now put

$$\Xi_G(x) = \Xi(x) = \int_K \delta_P(xk)^{-1/2} dk$$

where dk is the normalized Haar measure on the compact group K. It follows from Lemma 5 that Ξ is actually independent of the choice of P.

§4. Cusp forms and the space $\mathcal{A}(G, \tau)$

Let $\mathcal{U}$ be the universal enveloping algebra of $\mathcal{g}_c$. We regard elements of $\mathcal{U}$ as left-invariant differential operators on G. There is an obvious anti-isomorphism $g \longmapsto g'$ of $\mathcal{U}$ onto the algebra $\mathcal{U}'$ of right-invariant differential operators on G. If g_1, $g_2 \in \mathcal{U}$, we write

$$f(g_1; x; g_2) = (g_1' g_2 f)(x) \qquad (x \in G)$$

for any $f \in C^\infty(G)$. Put

$$\nu_{g_1, g_2, r}(f) = \sup_{G} |f(g_1 ; x; g_2)| \, \Xi(x)^{-1}(1+\sigma(x))^{r}$$

for $r \geq 0$. Then the Schwartz space $\mathscr{C}(G)$ consists of all functions $f \in C^\infty(G)$ such that

$$\nu_{g_1, g_2, r}(f) < \infty$$

for all $g_1, g_2 \in \mathfrak{G}$ and $r \geq 0$. The set of all seminorms $\nu_{g_1, g_2, r}$ defines the structure of a locally convex Hausdorff space on $\mathscr{C}(G)$ which is complete. Moreover $\mathscr{C}(G)$ is contained in $L_2(G)$.

If $P = MAN$ is a psgp of G and $f \in \mathscr{C}(G)$, we put

$$f^P(x) = \int_N f(xn)dn \qquad\qquad (x \in G) \ ,$$

where dn is the Haar measure on N. (This integral is always convergent.) f is said to be a cusp form if $f^P = 0$ whenever $\mathrm{prk}\, P > 0$. Let $^c\mathscr{C}(G)$ denote the space of all cusp forms. Then $^o\mathscr{C}(G)$ is a closed subspace of $\mathscr{C}(G)$.

Let τ be a unitary double representation of K on a finite-dimensional Hilbert space V. Let $C^\infty(G, \tau)$ denote the subspace of all $f \in C^\infty(G) \otimes V$ such that

$$f(k_1 x k_2) = \tau(k_1)f(x)\tau(k_2) \qquad (k_1, k_2 \in K, x \in G) \ .$$

Put $^o\mathscr{C}(G, \tau) = C^\infty(G, \tau) \cap (^o\mathscr{C}(G) \otimes V)$.

Theorem 1. $\dim {}^o\mathscr{C}(G, \tau) < \infty.$

Let $\mathfrak{Z}$ be the algebra of all differential operators on G which commute with both left and right translations of G. Then $\hat{\mathfrak{Z}}$ is the center of $\mathfrak{G}$ and therefore abelian.

Corollary. <u>Every element in</u> $^{\circ}\mathcal{C}(G,\ \tau)$ <u>is</u> $\mathfrak{Z}$-<u>finite</u>.

A continuous function f on G is said to satisfy the weak inequality, if the're exist numbers $c,\ r \geq 0$ such that

$$|f(x)| \leq c\,\Xi(x)\,(1+\sigma(x))^{r}$$

for all $x \in G$.

Let $\mathcal{A}(G,\ \tau)$ denote the space of all $f \in C^{\infty}(G,\ \tau)$ such that:

1) f is $\mathfrak{Z}$-finite;

2) $|f|$ satisfies the weak inequality.

Let $P = MAN$ be a psgp of G and put

$$\gamma(a) = \inf_{\alpha} \alpha(\log a) \qquad (a \in A)\ ,$$

where α runs over all roots of $(P,\ A)$. a being a variable element of A, we write $a \xrightarrow[P]{} \infty$ if 1) $\sigma(a) \longrightarrow \infty$ and 2) we can choose $\epsilon > 0$ such that $\gamma(a) \geq \epsilon\,\sigma(a)$. Let τ_{M} denote the restriction of τ on $K_{M} = K \cap M$.

Theorem 2. <u>Given</u> $f \in \mathcal{A}(G,\ \tau)$, <u>there exists a unique element</u> $f_{P} \in \mathcal{A}(MA,\ \tau_{M})$ <u>such that</u>

$$\lim_{a \xrightarrow[P]{} \infty} \{\delta_{P}(ma)^{1/2}f(ma) - f_{P}(ma)\} = 0$$

<u>for</u> $m \in MA$ <u>and</u> $a \in A$.

We call f_{P} the constant term of f along P.

Let $^{*}P = {}^{*}M\,{}^{*}A\,{}^{*}N$ be a psgp of M and $P' = M'A'N'$ the psgp of G contained in P, which corresponds to $^{*}P$ according to Lemma 4.

Lemma 6. <u>Fix</u> $f \in \mathcal{A}(G,\ \tau)$, $a \in A$ <u>and put</u>

$$g(m) = f_{P}(ma) \qquad (m \in M)\ .$$

Then $g \in \mathcal{A}(M, \tau_M)$ and

$$g_{*_P}(^*m \cdot {}^*a) = f_{P'}(^*m \cdot {}^*a \cdot a)$$

for $^*m \in {}^*M = M'$ and $^*a \in {}^*A$.

We write $S^x = xSx^{-1}$ $(x \in G)$ for any subset S of G. If $k \in K$, it is clear that $P^k = M^k A^k N^k$ is the Langlands decomposition of the psgp P^k.

Lemma 7. Fix $f \in \mathcal{A}(G, \tau)$ and $k \in K$. Then

$$f_{P^k}(m^k) = \tau(k)f_P(m)\tau(k^{-1})$$

for $m \in MA$.

Let $f \in \mathcal{A}(G, \tau)$. We write $f_P \sim 0$ if

$$\int_M (\phi(m), f_P(ma))_V \, dm = 0$$

for all $\phi \in {}^{\circ}\mathcal{C}(M, \tau_M)$ and $a \in A$. Here the scalar product is in V, dm is the Haar measure on M and the integral is always convergent.

Theorem 3. Suppose $f \in \mathcal{A}(G, \tau)$ and $f_P \sim 0$ for all psgps P of G (including $P = G$). Then $f = 0$.

Theorem 4. Fix $f \neq 0$ in $\mathcal{A}(G, \tau)$ and choose a psgp $P = MAN$ of G such that:

    1) $f_P \not\sim 0$,

    2) P is minimal with respect to condition 1).

Then for any $a \in A$, the function $m \longmapsto f_P(ma)$ lies in $\mathcal{C}(M, \tau_M)$. Let Q be a psgp of G such that $Q \subset P$ and $Q \neq P$. Then $f_Q = 0$.

§5. The Eisenstein Integral and its constant term

Let $P = MAN$ be a psgp of G. Fix $\psi \epsilon {}^{\circ}\mathcal{C}(M, \tau_M)$ and extend it to a function on $G = KP$ as follows:

$$\psi(kman) = \tau(k)\psi(m) \qquad\qquad (k \epsilon K, \ m \epsilon M, \ a \epsilon A, \ n \epsilon N) \ .$$

Put

$$E(P : \psi : \nu : x) = \int_K \psi(xk)\tau(k^{-1})\exp\{((-1)^{1/2}\nu - \rho_P)(H_P(xk))\}dk$$

for $\nu \epsilon \, \mathcal{Q}_c^{\ *}$ and $x \epsilon G$. Then E is an analytic function on $\mathcal{Q}_c^{\ *} \times G$ which, for a fixed ν, is $\mathcal{Z}$-finite and for a fixed x, an entire function of ν.

Lemma 8. <u>Fix</u> $\psi \epsilon {}^{\circ}\mathcal{C}(M, \tau_M)$ <u>and</u> $\nu \epsilon \, \mathcal{Q}^{\ *}$. <u>Then</u> $E(P : \psi : \nu) \epsilon \mathcal{A}(G, \tau)$. <u>Let</u> $P' = M'A'N'$ <u>be another psgp of</u> G. <u>Then</u>

$$E_{P'}(P : \psi : \nu) = 0$$

<u>unless</u> P <u>and</u> P' <u>are associated.</u>

Let $\mathcal{P}(A)$ be the set of all psgps P' of G such that A is the split component of P'. Then $\mathcal{P}(A)$ is a finite set and if N' is the radical of P', it is clear that $P' = MAN'$ is the Langlands decomposition of P'. Put

$$\varpi_P = \prod_{\alpha > 0} H_\alpha$$

where α runs over all roots of (P, A) and H_α is the element of $\mathcal{Q}$ given by

$$B(H, H_\alpha) = \alpha(H) \qquad\qquad (H \epsilon \, \mathcal{Q}) \ .$$

Then ϖ_P may be regarded as a polynomial function on $\mathcal{Q}_c^{\ *}$. Put $\mathcal{W} = \mathcal{W}(A)$ and $L = {}^{\circ}\mathcal{C}(M, \tau_M)$. Then by Theorem 1, $\dim L < \infty$.

Theorem 5. <u>We can choose an open connected neighborhood</u> U <u>of zero in</u> $\mathcal{Q}^{\ *}$ <u>and an integer</u> $r \geq 0$ <u>with the following properties. Fix</u> $P_1, \ P_2 \epsilon \mathcal{P}(A)$

and a point $\nu \in \mathfrak{a}^*$ such that $\varpi_P(\nu) \neq 0$. Then there exist unique elements $c_{P_2|P_1}(s : \nu) \in \text{End } L$ $(s \in W)$ such that

$$E_{P_2}(P_1 : \psi : \nu : ma) = \sum_{s \in W} (c_{P_2|P_1}(s : \nu)\psi)(m)e^{(-1)^{1/2}s\nu(\log a)}$$

for $\psi \in L$, $m \in M$ and $a \in A$. Moreover for any $s \in W$, the function

$$\nu \longmapsto \varpi_P(\nu)^r c_{P_2|P_1}(s : \nu)$$

extends to a holomorphic function of ν on $\mathfrak{a}^* + (-1)^{1/2}U$.

As usual $s\nu(H) = \nu(s^{-1}H)$ for $\nu \in \mathfrak{a}_c^*$ and $H \in \mathfrak{a}_c$.

For $\nu \in \mathfrak{a}_c^*$, define ν_R and ν_I in $\mathfrak{a}^*$ by $\nu = \nu_R + (-1)^{1/2}\nu_I$. Let $\mathcal{F}_c(P)$ denote the set of all $\nu \in \mathfrak{a}_c^*$ such that $\nu_I(H_\alpha) > 0$ for every root α of (P, A). (The definition of $\mathcal{F}_c(P)$ given in [4(d), p. 546] is incorrect.) Put $\overline{P} = \theta(P)$, $\overline{N} = \theta(N)$, $\rho = \rho_P$ and $H(x) = H_P(x)$ $(x \in G)$. Any $x \in G$ can be written uniquely in the form $x = kman$ where $k \in K$, $m \in M \cap \exp \mathfrak{p}$, $a \in A$, $n \in N$ (in the notation of Lemma 1). Put $\kappa(x) = k$, $\mu(x) = m$.

Lemma 9. $c_{\overline{P}|P}(1 : \nu)$ and $c_{P|P}(1 : -\nu)$ extend to holomorphic functions of ν on $\mathcal{F}_c(P)$ and they are given there by the following convergent integrals.

$$(c_{\overline{P}|P}(1 : \nu)\psi)(m) = \int_{\overline{N}} \tau(\kappa(\overline{n}))\psi(\mu(\overline{n})m)e^{((-1)^{1/2}\nu - \rho)(H(\overline{n}))}d\overline{n} \ ,$$

$$(c_{P|P}(1 : -\nu)\psi)(m) = \int_{\overline{N}} \psi(m\mu(\overline{n})^{-1})\tau(\kappa(\overline{n}))^{-1}e^{((-1)^{1/2}\nu - \rho)(H(\overline{n}))}d\overline{n} \ .$$

Here $\psi \in {}^{\circ}\mathcal{C}(M, \tau_M)$, $\nu \in \mathcal{F}_c(P)$, $m \in M$ and the Haar measure $d\overline{n}$ on $\overline{N}$ is so normalized that

$$\int_{\overline{N}} e^{-2\rho(H(\overline{n}))}d\overline{n} = 1 \ .$$

The following consequence of the above integral representation was pointed out to me by R. P. Langlands.

Corollary. $\det c_{P|P}(1:-\nu)$ is not identically zero in ν.

$$\S 6. \quad \text{The functional equations}$$

Put

$$\|\psi\|_M^2 = \int_M |\psi(m)|^2 dm$$

for $\psi \in L = {}^{\circ}\mathcal{C}(M, \tau_M)$. This defines the structure of a Hilbert space on L.

Theorem 6. **Fix** P_1, $P_2 \in \mathcal{P}(A)$. **Then for any** $s \in W$, $c_{P_2|P_1}(s:\nu)$ **extends to a meromorphic function of** ν **on** α_c^*. Put

$$c^{\circ}_{P_2|P_1}(s:\nu) = c_{P_2|P_1}(s:\nu)c_{P_1|P_1}(1:\nu)^{-1}$$

and

$$E^{\circ}(P:\psi:\nu) = E(P: c_{P|P}(1:\nu)^{-1}\psi:\nu) \qquad (\psi \in L, \ P \in \mathcal{P}(A)) \ .$$

Then

    1) $c^{\circ}_{P_2|P_1}(s:\nu)$ **is holomorphic and unitary on** α^*,

    2) $E^{\circ}(P:\psi:\nu)$ **is holomorphic for** $\nu \in \alpha^*$,

    3) $c_{P|P}(1:\nu) = (c_{\bar{P}|\bar{P}}(1:\bar{\nu}))^*$, $\ c_{\bar{P}|P}(1:\nu) = (c_{P|\bar{P}}(1:\bar{\nu}))^*$

where $\bar{\nu} = \nu_R - (-1)^{1/2}\nu_I$ **and the star denotes the adjoint of a linear transformation. Moreover we have the following functional equations.**

    a) $c^{\circ}_{P_3|P_1}(ts:\nu) = c^{\circ}_{P_3|P_2}(t:s\nu)c^{\circ}_{P_2|P_1}(s:\nu)$,

    b) $E^{\circ}(P_1:\psi:\nu) = E^{\circ}(P_2: c^{\circ}_{P_2|P_1}(s:\nu)\psi:s\nu)$

for P_1, P_2, $P_3 \in \mathcal{P}(A)$ **and** $s, t \in W$.

Theorem 7. **Fix** P_1, $P_2 \in \mathcal{P}(A)$ **and suppose** $P' = M'A'N'$ **is a psgp of** G **such that** $P' \supset P_1 \cup P_2$ **and** $\mathrm{prk}\ P' \geq 1$. **Put** ${}^*P_i = M' \cap P_i$ $(i = 1, 2)$, ${}^*A = M' \cap A$ **and let** ${}^*W$ **denote the subgroup of all** $s \in W$ **which leave** A'

pointwise fixed. Then $({}^*P_i, {}^*A)$ $(i = 1, 2)$ _are parabolic pairs in_ M' _and_ ${}^*\mathcal{W}$ _may be identified with the Weyl group of_ ${}^*A$ _in_ M'. _For any_ $\nu \in \mathcal{a}^*$, _let_ ${}^*\nu$ _denote the restriction of_ ν _on_ ${}^*\mathcal{a}$. _Fix_ $\psi \in L$ _and_ $\nu \in \mathcal{a}^*$ _and put_ $f(\nu) = E^o(P_1 : \psi : \nu)$. _Then_

$$f_{P'}(\nu : m'a') = \sum_{s \in {}^*\mathcal{W} \backslash \mathcal{W}} E^o({}^*P_2 : c^o_{P_2|P_1}(s : \nu)\psi : {}^*(s\nu) : m')e^{(-1)^{1/2} s\nu(\log a')}$$

for $m' \in M'$, $a' \in A'$. _Moreover_

$$c^o_{P_2|P_1}(t : \nu) = c^o_{{}^*P_2|{}^*P_1}(t : {}^*\nu)$$

for $t \in {}^*\mathcal{W}$.

Here s runs over a complete system of representatives in $\mathcal{W}$ for ${}^*\mathcal{W}\backslash\mathcal{W}$. Theorems 6 and 7 reduce the determination of $c^o_{P_2|P_1}(s : \nu)$ to that of $c^o_{\bar P|P}(1 : \nu)$ in case $\mathrm{prk}\, P = 1$.

§7. The Maass-Selberg relations and their consequences

Let $\mathcal{A}_P(G, \tau)$ denote the space of all $f \in \mathcal{A}(G, \tau)$ with the following property. If Q is a psgp of G, then $f_Q \sim 0$ unless Q is associated to P. The following theorem plays a decisive role in the proof of Theorems 6 and 7. The case when $\mathrm{prk}\, P = 1$ is especially important.

Theorem 8. _Fix_ $\nu \in \mathcal{a}^*$ _such that_ $\varpi_P(\nu) \neq 0$ _and suppose_ $f \in \mathcal{A}_P(G, \tau)$ _and_ $\phi_{Q,s} \in L$ $(Q \in \mathcal{P}(A),\ s \in \mathcal{W})$ _are given functions satisfying the relation_

$$f_Q(ma) = \sum_{s \in \mathcal{W}} \phi_{Q,s}(m)e^{(-1)^{1/2} s\nu(\log a)} \qquad (m \in M,\ a \in A)$$

for every $Q \in \mathcal{P}(A)$. _Then_

$$\|\phi_{P_1, s_1}\|_M = \|\phi_{P_2, s_2}\|_M$$

<u>for</u> P_1, $P_2 \in \mathcal{P}(A)$ <u>and</u> s_1, $s_2 \in W$.

Corollary. <u>Suppose</u> $\phi_{Q,s} = 0$ <u>for some pair</u> (Q, s). <u>Then</u> $f = 0$.

This theorem is a consequence of, what I call, the Maass-Selberg relations when $\mathrm{prk}\, P = 1$. (These relations are similar to those discussed in [4(a), Chap. IV, §2].) The rest follows by induction on $\mathrm{prk}\, P$. In fact Theorems 6, 7 and 8 are proved together in this induction (cf. [4(a), Chap. V]).

$$\S 8. \quad \underline{\text{The evaluation of } (\phi_a)_\nu^{(P)}}$$

Put $\mathcal{H} = \alpha^*$ and let $\mathcal{H}'$ be the set of all $\nu \in \mathcal{H}$ such that $\varpi_P(\nu) \neq 0$. Let $\mathcal{C}(\mathcal{H})$ denote the Schwartz space on the finite-dimensional vector space $\mathcal{H}$.

Lemma 10. <u>For any</u> $a \in \mathcal{C}(\mathcal{H}) \otimes L$, <u>put</u>

$$\phi_a(x) = \int E(P : a(\nu) : \nu : x)d\nu \qquad (x \in G) \ ,$$

<u>where</u> $d\nu$ <u>denotes the Euclidean measure on</u> $\mathcal{H}$. <u>Suppose</u> a <u>satisfies the following condition.</u> <u>For any</u> $s \in W$ <u>and</u> P_1, $P_2 \in \mathcal{P}(A)$, <u>the function</u>

$$\nu \longmapsto \| c_{P_2|P_1}(s : \nu)a(\nu) \|_M \qquad (\nu \in \mathcal{H}')$$

<u>remains locally bounded on</u> $\mathcal{H}$. <u>Then</u> $\phi_a \in \mathcal{C}(G, \tau)$.

In particular the condition of the lemma is fulfilled if $a \in C_c^\infty(\mathcal{H}') \otimes L$. For any $f \in \mathcal{C}(G, \tau)$, define a function $f^{(P)}$ on MA by

$$f^{(P)}(m) = \delta_P(m)^{1/2} \int_N f(mn)dn \qquad (m \in MA)$$

Then $f^{(P)} \in \mathcal{C}(MA, \tau_M)$ and $f \longmapsto f^{(P)}$ is a continuous mapping of $\mathcal{C}(G, \tau)$ into $\mathcal{C}(MA, \tau_M)$.

Theorem 9. <u>Fix</u> P_1, $P_2 \in \mathcal{P}(A)$ <u>and suppose</u> $a \in \mathcal{C}(\mathcal{H}) \otimes L$ <u>satisfies the</u> <u>condition of Lemma</u> 10. <u>Put</u>

$$\phi_\alpha(x) = \int_{\mathcal{H}} E(P_1 : \alpha(\nu) : \nu : x)d\nu \qquad\qquad (x \in G)$$

and

$$\phi_{P_2, \alpha}(m) = \int_{\mathcal{H}} E_{P_2}(P_1 : \alpha(\nu) : \nu : m)d\nu \qquad (m \in MA) \ .$$

Extend $\phi_{P_2, \alpha}$ to a function on G by the rule

$$\phi_{P_2, \alpha}(kmn) = \tau(k)\phi_{P_2, \alpha}(m) \qquad\qquad (k \in K, \ m \in MA, \ n \in N_2) \ .$$

Then

$$\phi_\alpha^{(\overline{P}_2)}(m) = \int_{\overline{N}_2} e^{-\rho_2(H_2(\overline{n}))} \phi_{P_2, \alpha}(\overline{n}m)d\overline{n} \qquad (m \in MA) \ .$$

Here $P_i = MAN_i$ $(i = 1, 2)$, $\rho_2 = \rho_{P_2}$, $H_2(x) = H_{P_2}(x)$ $(x \in G)$, and all the integrals are convergent.

Put

$$f_\nu^{(P)}(m) = \int_A f^{(P)}(ma)e^{-(-1)^{1/2}\nu(\log a)}da \qquad (m \in M)$$

for $f \in \mathcal{C}(G, \tau)$ and $\nu \in \mathcal{H}$.

Theorem 10. <u>Fix</u> $\alpha \in C_c^\infty(\mathcal{H}')$, $\psi \in L$ <u>and put</u>

$$\phi_\alpha(x) = \int \alpha(\nu)E(P : \psi : \nu : x)d\nu \qquad\qquad (x \in G) \ .$$

<u>Then</u> $\phi_\alpha \in \mathcal{C}(G, \tau)$ <u>and</u>

$$(\phi_\alpha)_\nu^{(P)} = \gamma(P) \sum_{s \in W} \alpha(s^{-1}\nu)(c_{P|\overline{P}}(1 : \nu)c_{\overline{P}|P}(s : s^{-1}\nu)\psi)$$

<u>for</u> $\nu \in \mathcal{H}$. <u>Here</u>

$$\gamma(P) = \int_N e^{-2\rho(H(\theta(n)))}dn$$

<u>and the measures</u> da <u>and</u> $d\nu$ <u>are assumed to be dual to each other.</u>

§9. Normalization of the Haar measures

$\mathcal{q}$ being any linear subspace of $\mathcal{g}$, we denote by $d\mathcal{q}$ the Euclidean measure on $\mathcal{q}$ corresponding to the Euclidean norm on $\mathcal{g}$ defined in Lemma 2. Let $P_o = M_o A_o N_o$ be a minimal psgp of G. Then $G = KA_o N_o$. Put $\rho_o = \rho_{P_o}$ and normalize the Haar measure dx on G in such a way that

$$dx = e^{2\rho_o(\log a_o)} dk\, da_o\, dn_o .$$

Here $x = ka_o n_o$ ($k \in K$, $a_o \in A_o$, $n_o \in N_o$) and da_o and dn_o are the Haar measures on A_o and N_o respectively which correspond to the Euclidean measures on their Lie algebras under the exponential mapping. dk is the normalized Haar measure on K (so that the total measure of K is 1). This normalization of dx is independent of the choice of P_o. We call dx the standard Haar measure on G.

Now let $P = MAN$ as in §8. We can apply the above procedure to M instead of G and thus obtain the standard Haar measure dm on M.

Lemma 11. <u>Let</u> $P = MAN$ <u>be any psgp of</u> G <u>and</u> dx, dm <u>the standard Haar measures on</u> G <u>and</u> M <u>respectively.</u> <u>Let</u> da <u>and</u> dn <u>denote the Haar measures on</u> A <u>and</u> N <u>which correspond to</u> $d\mathcal{a}$ <u>and</u> $d\mathcal{n}$ <u>respectively under the exponential mapping.</u> <u>Then</u>

$$\int_G f(x)dx = \int_{K\times M\times A\times N} f(kman) e^{2\rho(\log a)} dk\, dm\, da\, dn$$

<u>for</u> $f \in C_c(G)$. <u>Here</u> dk <u>is the normalized Haar measure on</u> K <u>and</u> $\rho = \rho_P$.

From now on we shall always assume that the various Haar measures are normalized as in the above lemma.

§10. The space $\mathcal{C}_\omega(G, \tau)$

Let $\mathcal{E}(G)$ be the set of all equivalence classes of irreducible unitary representations of G and $\mathcal{E}_2(G)$ the subset of those classes ω which are square-integrable. For any $\omega \in \mathcal{E}_2(G)$, let $d(\omega)$ denote the formal degree of ω.

Let $G_1 = ZG^o$ where Z is the centralizer of G^o in G. Then G/G_1 is a finite group. Fix $\omega_1 \in \mathcal{E}_2(G_1)$ and let $\mathrm{Ind}\, \omega_1$ denote the class of the representation of G obtained from ω_1 by inducing from G_1 to G. Then $\mathrm{Ind}\, \omega_1$ is irreducible and $\omega_1 \longrightarrow \mathrm{Ind}\, \omega_1$ is a surjective mapping of $\mathcal{E}_2(G_1)$ on $\mathcal{E}_2(G)$. Taking into account the results of [4(f), §41], one can give an explicit formula for $d(\omega)$ $(\omega \in \mathcal{E}_2(G))$.

Now fix $\omega \in \mathcal{E}_2(G)$ and let $\mathcal{H}_\omega$ denote the smallest closed subspace of $L_2(G)$ containing all K-finite matrix coefficients of the class ω. Put $\mathcal{C}_\omega(G) = \mathcal{C}(G) \cap \mathcal{H}_\omega$ and

$$\mathcal{C}_\omega(G,\ \tau) = \mathcal{C}(G,\ \tau) \cap (\mathcal{C}_\omega(G) \otimes V)\ .$$

Then $\mathcal{C}_\omega(G,\ \tau) \subset {}^o\mathcal{C}(G,\ \tau)$.

§11. The characters $\Theta_{\omega,\nu}$

Let $P = MAN$ be a psgp of G. Fix $\omega \in \mathcal{E}_2(M)$, $\nu \in \mathfrak{a}^*$ and define a tempered distribution $\Theta_{\omega,\nu}$ on G as follows:

$$\Theta_{\omega,\nu}(f) = \theta_\omega(g_{f,\nu}) \qquad\qquad (f \in \mathcal{C}(G))\ .$$

Here θ_ω is the character of ω,

$$g_{f,\nu}(m) = \int_{A \times N} \bar{f}(man)\exp\{((-1)^{1/2}\nu + \rho)(\log a)\}da\ dn \qquad (m \in M)$$

and

$$\bar{f}(x) = \int_K f(kxk^{-1})dk\ .$$

Then $\Theta_{\omega,\nu}$ is the character of a unitary representation of G. Let $\Omega(P,\ \nu)$ denote the class of this representation. Then

1) $\Omega(P,\ \nu)$ is irreducible if $\varpi_P(\nu) \neq 0$,

2) $\Omega(P_1, \nu) = \Omega(P_2, \nu)$ for P_1, $P_2 \in \mathcal{P}(A)$.

The group $\mathbf{W} = \mathbf{W}(A)$ operates on $\mathcal{E}_2(M)$ as follows. Fix $s \in \mathbf{W}$ and $\omega \in \mathcal{E}_2(M)$. Choose a representative y in G for s and a representation β of M in the class ω. Put

$$\beta^y(m) = \beta(y^{-1}my) \qquad (m \in M) \ .$$

Since y normalizes M, β^y is a representation of M. We define ω^s to be the class of β^y. It is easy to see that $\omega^s \in \mathcal{E}_2(M)$ and it is independent of the choice of y and β.

Lemma 12. <u>Fix</u> ω_1, $\omega_2 \in \mathcal{E}_2(M)$ <u>and</u> ν_1, $\nu_2 \in \mathfrak{a}^*$. <u>Then</u>

$$\textcircled{H}_{\omega_1, \nu_1} = \textcircled{H}_{\omega_2, \nu_2}$$

<u>if and only if there exists an element</u> $s \in \mathbf{W}$ <u>such that</u>

$$\omega_2 = \omega_1^{\,s}, \ \nu_2 = s\nu_1 \ .$$

The distributions θ_ω and $\textcircled{H}_{\omega, \nu}$ are actually functions. We write[3]

$$(\textcircled{H}_{\omega, \nu}, \ f) = \int_G \mathrm{conj} \, \textcircled{H}_{\omega, \nu} \cdot f dx \qquad (f \in \mathcal{C}(G)) \ ,$$

the integral being convergent. A similar notation is used also for $f \in \mathcal{C}(G, \tau)$.

Lemma 13. <u>Let</u> F <u>denote the projection in</u> V <u>given by</u>

$$Fv = \int_K \tau(k)v\tau(k^{-1})dk \qquad (v \in V) \ .$$

<u>Then</u>

$$(\textcircled{H}_{\omega, \nu}, \ f) = F(\theta_\omega, \ f_\nu^{(P)}) \qquad (f \in \mathcal{C}(G, \tau))$$

<u>for</u> $\omega \in \mathcal{E}_2(M)$ <u>and</u> $\nu \in \mathfrak{a}^*$.

Here

$$(\theta_\omega,\ g) = \int_M \text{conj}\ \theta_\omega \cdot g\, dm$$

for g in $\mathcal{C}(M)$ or $\mathcal{C}(M,\ \tau_M)$.

Fix $\nu_0 \in \mathcal{F}'$. Then $s\nu_0 \neq \nu_0$ for $s \neq 1$ in $\mathcal{W}$. Hence we can choose an open neighborhood U of ν_0 in $\mathcal{F}'$ such that $U \cap sU = \emptyset$ for $s \neq 1$ in $\mathcal{W}$.

Lemma 14. <u>Put</u> $L_\omega = \mathcal{C}_\omega(M,\ \tau_M)$ <u>and suppose</u> $a \in C_c^\infty(U)$ <u>and</u> $\psi \in L_\omega$. <u>Fix</u> $\omega' \in \mathcal{E}_2(M)$ <u>and</u> $\nu \in \mathcal{F}$. <u>Then</u>

$$(\textcircled{H}_{\omega',\nu},\ \phi_a) = 0$$

<u>unless</u> $\nu \in \bigcup_{s \in \mathcal{W}} sU$. <u>Now suppose</u> $\nu \in U$. <u>Then</u>

$$(\textcircled{H}_{\omega',\nu},\ \phi_a) = \begin{cases} \gamma(P)a(\nu)F(\theta_\omega,\ {}^cP|\bar{P}(1:\nu)c_{\bar{P}|P}(1:\nu\psi)) & \underline{\text{if}}\ \omega' = \omega, \\ 0 & \underline{\text{otherwise.}} \end{cases}$$

Here ϕ_a and $\gamma(P)$ have the same meaning as in Theorem 10.

<u>§12. The Plancherel measure μ_ω</u>

Let $\mathcal{C}_A(G)$ denote the set of all $f \in \mathcal{C}(G)$ with the following property. If $P' = M'A'N'$ is any psgp of G, then $f^{P'} \sim 0$ (see [4(d)], p. 538]) unless P' is associated to P. For $f \in \mathcal{C}(G)$, put

$$\hat{f}(\omega : \nu) = (\textcircled{H}_{\omega,\nu},\ f) \qquad (\omega \in \mathcal{E}_2(M),\ \nu \in \mathcal{a}^*)\ .$$

We give $\mathcal{E}_2(M)$ the discrete topology.

Theorem 11.[4)] <u>There exists a unique continuous function</u> μ <u>on</u> $\mathcal{E}_2(M) \times \mathcal{a}^*$ <u>with the following properties:</u>

1) $\mu(\omega^s : s\nu) = \mu(\omega : \nu)$ <u>for</u> $s \in \mathcal{W}$.

2) <u>For any</u> $f \in \mathcal{C}(G)$, <u>the series</u>

$$\sum_{\omega \in \mathcal{E}_2(M)} d(\omega) \int_{\alpha^*} |\hat{f}(\omega : \nu)\mu(\omega : \nu)| \, d\nu$$

is convergent.

3) $f(1) = \displaystyle\sum_{\omega \in \mathcal{E}_2(M)} d(\omega) \int_{\alpha^*} \hat{f}(\omega : \nu)\mu(\omega : \nu) \, d\nu$

for all $f \in \mathcal{C}_A(G)$.

Fix $\omega \in \mathcal{E}_2(M)$ and put $\mu_\omega(\nu) = \mu(\omega : \nu)$.

Lemma 15. μ_ω extends to a meromorphic function on α_c^* which is holomorphic on α^*. Moreover $\mu_\omega(\nu) > 0$ on $\mathcal{H}'$ and[5)]

$$[\mathcal{W}]\mu_\omega(\nu) \cdot \gamma(P)c_{P|\overline{P}}(1 : \nu)c_{\overline{P}|P}(1 : \nu)\psi = \psi$$

for $\nu \in \mathcal{H}'$ and $\psi \in L_\omega$.

§13. Explicit determination of μ_ω

We now make $\mathcal{W}$ operate on L as follows. Fix $s \in \mathcal{W}$ and $\psi \in L$ and let y be a representative of s in K. Then $s\psi$ is the function $m \longmapsto \tau(y)\psi(y^{-1}my)\tau(y^{-1})$ on M. We also put $P^s = P^y$.

Lemma 16. Let P_1, $P_2 \in \mathcal{P}(A)$ and $s, t \in \mathcal{W}$. Then

$$^{s}c_{P_2|P_1}(t : \nu) = c_{P_2^{s}|P_1^{s}}(st : \nu), \quad c_{P_2|P_1}(t : \nu)s^{-1} = c_{P_2|P_1^{s}}(ts^{-1} : s\nu)$$

and

$$^{s}c^{o}_{P_2|P_1}(t : \nu) = c^{o}_{P_2^{s}|P_1^{s}}(st : \nu), \quad c^{o}_{P_2|P_1}(t : \nu)s^{-1} = c^{o}_{P_2|P_1^{s}}(ts^{-1} : s\nu)$$

for $\nu \in \alpha_c^*$.

Fix $P = MAN$ in $\mathcal{P}(A)$ and for any $P_1 \in \mathcal{P}(A)$ ($P_1 = MAN_1$) and $\nu \in \mathcal{H}_c(P)$, define a linear transformation $J_{P_1|P}(\nu)$ on L as follows.

$$(J_{P_1|P}(\nu)\psi)(m) = \int_{\overline{N}\cap N_1} \psi(\overline{n}m)e^{((-1)^{1/2}\nu-\rho)(H(\overline{n}))}d\overline{n} \qquad (\psi \in L, \ m \in M) \ .$$

Here $d\overline{n}$ is the Haar measure on $\overline{N} \cap N_1$ which corresponds, under the exponential mapping, to the Euclidean measure on its Lie algebra. The function ψ is extended on $G = KP$ as before by the rule

$$\psi(kman) = \tau(k)\psi(m) \qquad (k \in K, \ m \in M, \ a \in A, \ n \in N) \ .$$

The above integral is convergent when $\nu \in \mathcal{F}_c(P)$. We note that from Lemma 9,

$$J_{P|P}(\nu) = 1 \ , \quad J_{\overline{P}|P}(\nu) = \gamma(P)c_{\overline{P}|P}(1:\nu) \ ,$$

where $\gamma(P)$ has the same meaning as in Theorem 10.

Let Σ be the set of all roots of (P, A). A root $\alpha \in \Sigma$ is called reduced if $t\alpha \notin \Sigma$ for $0 < t < 1$ $(t \in R)$. Let Φ be the set of all reduced roots. For any $\alpha \in \Phi$, let $\Sigma(\alpha)$ denote the set of all roots in Σ of the form $t\alpha$ $(t \geq 1)$. Put

$$\mathcal{N}_\alpha = \sum_{\beta \in \Sigma(\alpha)} \mathcal{N}(\beta) \ ,$$

where $\mathcal{N}(\beta)$ is the set of all $X \in \mathcal{N}$ such that $[H, X] = \beta(H)X$ for all $H \in \mathcal{A}$. Put $N_\alpha = \exp \mathcal{N}_\alpha$.

Let σ_α denote the hyperplane $\alpha = 0$ on $\mathcal{A}$ and Z_α the centralizer of σ_α in G. Put $M_\alpha = {}^\circ(Z_\alpha)$, $A_\alpha = M_\alpha \cap A$ and $\overline{N}_\alpha = \theta(N_\alpha)$. Then

$$ {}^*P_\alpha = MA_\alpha N_\alpha \ , \quad {}^*\overline{P}_\alpha = MA_\alpha \overline{N}_\alpha$$

are maximal psgps of M_α. Put $\rho_\alpha = \rho_{{}^*P_\alpha}$, $H_\alpha(y) = H_{{}^*P_\alpha}(y)$ $(y \in M_\alpha)$ and define

$$\gamma({}^*P_\alpha) = \int_{\overline{N}_\alpha} e^{-2\rho_\alpha(H_\alpha(\overline{n}))} d_\alpha\overline{n} \ ,$$

where $d_\alpha\overline{n}$ is the Haar measure on $\overline{N}_\alpha$ which corresponds to the Euclidean measure on its Lie algebra.

A point $H \in \mathcal{O}$ is called regular if $\alpha(H) \neq 0$ for every $\alpha \in \Sigma$ and it is called semiregular if there is exactly one root $\alpha \in \Phi$ such that $\alpha(H) = 0$. Let C be the set of all points $H \in \mathcal{O}$ where $\alpha(H) > 0$ for all $\alpha \in \Sigma$. Then we can choose two points H_o, H_1 in $\mathcal{O}$ such that the following conditions hold.

1) $H_o \in C$ and $-H_1 \in C$.

2) Put $H(t) = (1-t)H_o + tH_1$ $(0 \leq t \leq 1)$. Then $H(t)$ is either regular or semiregular.

Let $0 < t_1 < t_2 < \ldots < t_r < 1$ be all the values of t such that $H(t)$ is semiregular. Let α_i be the root in Φ which vanishes at $H(t_i)$ $(1 \leq i \leq r)$. It is clear that $r = [\Phi]$. Put

$$c_i(\nu) = \gamma(^*P_{\alpha_i})c_{*\overline{P}_{\alpha_i} \mid ^*P_{\alpha_i}}(1 : \nu_{\alpha_i}) \qquad (\nu \in \mathcal{H}_c(P))$$

where ν_α is the restriction of ν on $\mathcal{O}_\alpha = R\underset{\sim}{H}_\alpha$ $(\alpha \in \Phi)$.

Lemma 17. $\gamma(P)c_{\overline{P} \mid P}(1 : \nu) = c_r(\nu)c_{r-1}(\nu) \ldots c_1(\nu)$ $(\nu \in \mathcal{H}_c(P))$.

Since both sides are meromorphic on $\mathcal{H}_c$, this relation must hold for all ν. Lemma 17 may be regarded as a generalization of a result of Gindikin and Karpelevič [3].

Let W_α denote the Weyl group of A_α in M_α and $\mu_{\omega, \alpha}$ the function on $\mathcal{O}_\alpha^*$ which corresponds to μ_ω when we replace (G, P) by $(M_\alpha, ^*P_\alpha)$. The following theorem is an immediate consequence of Lemmas 15 and 17.

Theorem 12. $\mu_\omega(\nu) = c \prod_{\alpha \in \Phi} \mu_{\omega, \alpha}(\nu_\alpha)$ $(\omega \in \mathcal{E}_2(M), \nu \in \mathcal{O}^*)$

where

$$c = \gamma(P)[W]^{-1} \prod_{\alpha \in \Phi} [W_\alpha]\gamma(^*P_\alpha)^{-1} .$$

Observe that $\operatorname{prk} M_\alpha = 0$ and $\operatorname{prk} {}^*P_\alpha = 1$. Hence, in order to compute μ_ω, it is enough to consider the case when $\operatorname{prk} G = 0$ and $\operatorname{prk} P = 1$.

Then there are two possibilities.

1) $\mathcal{E}_2(G) = \emptyset$.

2) $\mathcal{E}_2(G) \neq \emptyset$.

The first case is easier and there one can show that μ_ω is a polynomial function on $\mathcal{O}^*$. On the other hand the second case can be dealt with by the method of [4(g), §24]. In this way one obtains an explicit formula for μ_ω.

§14. Relation with the exceptional series

Now we assume that prk $G = 0$ and prk $P = 1$. Fix $\omega \in \mathcal{E}_2(M)$ and put $\mathbf{W} = \mathbf{W}(A)$.

Lemma 18.[6) $\textcircled{H}_{\omega, o}$ is an irreducible character unless the following two con-ditions hold.

1) $\mu_\omega(0) > 0$, $[\mathbf{W}] = 2$ and $\omega^s = \omega$ ($s \in \mathbf{W}$).

2) $\mathcal{E}_2(G) \neq \emptyset$.

Actually 2) is a consequence of 1). Moreover it seems likely that $\textcircled{H}_{\omega, o}$ is the sum of two distinct irreducible characters when these conditions are fulfilled (cf. [4(b), Theorem 1] and [6(b)]).

Lemma 19. Fix $\omega \in \mathcal{E}_2(M)$ such that $\mu_\omega(0) = 0$. Then $[\mathbf{W}] = 2$ and $\omega^s = \omega$ where s is the element of $\mathbf{W}$ other than 1. Put

$$C(\nu)\psi = -c^o_{P|P}(s : (-1)^{1/2}\nu)\psi \qquad (\nu \in \mathcal{O}_c^*, \ \psi \in L_\omega) \ .$$

Then $C(\nu)$ is self-adjoint for $\nu \in \mathcal{O}^*$ and

$$C(\nu)C(-\nu) = 1 \ .$$

Finally $C(0) = 1$.

Let a denote the unique simple root of (P, A). We identify $\mathcal{O}_c^*$ with $\underset{\sim}{C}$ by means of the mapping $\nu \longmapsto \langle \nu, a \rangle$. Let $\delta_o > 0$ be the distance, from the origin, of the nearest pole of C on the real axis. Then $C(\nu)$ is a positive-definite operator for $|\nu| < \delta_0$ ($\nu \in \underset{\sim}{R}$). This enables us to prove the following theorem (cf. [6(a), Theorem 3.3]).

Theorem 13. <u>Fix</u> $\omega \in \mathcal{E}_2(M)$ <u>such that</u> $\mu_\omega(0) = 0$. <u>Then we can choose</u> $\delta > 0$ <u>with the following property.</u> <u>Suppose</u> $\nu \in (-1)^{1/2}\alpha^*$, $|\langle \nu, a \rangle| < \delta$ <u>and</u> $\nu \neq 0$. <u>Then</u> $\textcircled{H}_{\omega,\nu}$ <u>is the character of an irreducible unitary representation of</u> G <u>belonging to the exceptional series</u>.

It would be interesting to extend the above results to the case $\mathrm{prk}\, P > 1$.

References

1. A. Borel and J. Tits, Groupes réductifs, Inst. Hautes Études Sci. Publ. Math. No. 27 (1965), pp. 55-150.

2. I. M. Gelfand, Automorphic functions and the theory of representations, Proc. Internat. Congress Math. (1962), pp. 74-85.

3. S. G. Gindikin and F. I. Karpelevič, Plancherel measure of Riemannian symmetric spaces of nonpositive curvature, Sov. Math. vol. 3 (1962), pp. 962-965.

4. Harish-Chandra, (a) Automorphic forms on semisimple Lie groups, Lecture notes in Math., no. 62, Springer-Verlag (1968).
(b) Eisenstein Series over finite fields, Functional Analysis and Related Fields, pp. 76-88, Springer-Verlag (1970).
(c) Some applications of the Schwartz space of a semisimple Lie group, Lecture notes in Math., no. 140 (1970), pp. 1-7, Springer-Verlag.
(d) Harmonic Analysis on semisimple Lie groups, Bull. Amer. Math. Soc. vol. 78 (1970), pp. 529-551.
(e) Harmonic Analysis on reductive p-adic groups, Lecture notes in Math., no. 162 (1970), Springer-Verlag.
(f) Discrete series for semisimple Lie groups II, Acta Math. vol. 116 (1966), pp. 1-111.
(g) Two theorems on semisimple Lie groups, Ann. of Math. vol. 83 (1966), pp. 74-128.

5. S. Helgason, A duality theory for symmetric spaces with applications to group representations, Advances in Math. vol. 5 (1970), pp. 1-154.

6. A. W. Knapp and E. M. Stein, (a) Existence of complementary series, Problems in Analysis, pp. 249-259, Princeton University Press, 1970.
(b) Singular Integrals and the Principal Series II, Proc. Nat. Acad. Sci. U.S.A. vol. 66 (1970), pp. 13-17.

7. B. Kostant, On the existence and irreducibility of certain series of representations, Bull. Amer. Math. Soc. vol 75 (1969), pp. 627-642.

Footnotes

1) For any finite-dimensional vector space V over $\underset{\sim}{R}$, we denote by V^* its dual and by V_c its complexification. Moreover $V_c^* = (V^*)_c$.

2) See Borel and Tits [1].

3) conjc stands for the complex conjugate of $c \in \underset{\sim}{C}$.

4) Cf. [4(d), §12].

5) [F] denotes the number of elements in a finite set F.

6) The fact that there is a rather direct connection between questions of reducibility and the theory of the Eisenstein Integral was first pointed out to me by J. G. Arthur (cf. his thesis "Harmonic analysis of tempered distributions on semisimple Lie groups of real rank one," Yale, 1970). See also [5].

Reprinted from
Lecture Notes in Mathematics, No. **266**
Springer-Verlag, Berlin-Heidelberg-New York, 1972, pp. 123–149

HARMONIC ANALYSIS ON REDUCTIVE
p-ADIC GROUPS

HARISH-CHANDRA

1. Introduction. The object of this article is to present fresh evidence in support of what I call the Lefschetz principle, which says that whatever is true for real reductive groups is also true for p-adic groups. It is a well-established fact in number theory that global arithmetic problems can often be analyzed into local questions at the individual primes. This is so, for example, in class-field theory, the arithmetic theory of algebras and the theory of quadratic forms. However, the local results at infinity are usually much too simple to provide an adequate insight into the complexity of the problem at finite primes. Therefore, we should regard it as a very fortunate fact that the theory of representations of real groups has a sufficiently rich structure so as to serve as a useful guide in our search for corresponding results for p-adic groups.

The work of Chevalley, Borel, Borel-Tits and Bruhat-Tits has given us a fairly good understanding of the structure of reductive p-adic groups. Harmonic analysis on such groups was started by Mautner and then pursued by Bruhat, Satake, Gelfand-Graev, Sally-Shalika and Macdonald. It was clear from the beginning that the theory of induced representations plays an important role here. In his thesis [2(a)] Bruhat had applied the theory of distributions on real groups to prove the irreducibility of such representations (see also [2(b)]). A few years ago [3(d)] I observed that the same ideas, when combined with the theory of cusp forms, give rather sharp results for finite groups. We shall now adapt this method to the p-adic case.

In their work on $GL(2)$, Jacquet and Langlands [6] introduced the notion of

AMS (MOS) subject classifications (1970). Primary 22E50; Secondary 22E35.

Key words and phrases. Admissible representations, theory of the constant term, supercusp forms, Maass-Selberg relations, functional equations of the Eisenstein integrals, Steinberg character, Plancherel measure.

an admissible representation. They noticed that one could work with such representations in an entirely algebraic fashion. (The analogue in the real case is the representation of the Lie algebra on the space of K-finite vectors.) Jacquet [5] then pushed this theory to $GL(n)$ and obtained the remarkable result that every admissible and irreducible representation of $GL(n)$ is contained in one which is induced from a supercuspidal representation of a parabolic subgroup. Jacquet's ideas enables us to obtain results about the asymptotic behavior of the matrix coefficients of admissible representations and to develop the theory of the constant term.

The whole subject now begins to show a strong resemblance to the theory of automorphic forms [3(c)]. In particular, we get the Maass-Selberg relations (§10) which then imply the functional equations of the Eisenstein integrals (§§11, 12). The c-functions appear in the asymptotic formula for the Eisenstein integral and the Plancherel measure turns out to be directly related to their absolute value (§17). The analogy with the real case (see [3(e)], [3(f)]) is therefore quite exact.

2. Smooth functions. By a totally disconnected (t.d.) space we mean a Hausdorff space X with the following property: Given a point $x \in X$ and a neighborhood U of x in X, we can choose an open and compact subset ω of X such that $x \in \omega \subset U$. Clearly a t.d. space is locally compact.

Let X be a t.d. space and S a set. A mapping $f: X \to S$ is called smooth if it is locally constant. Let V be a complex vector space. By $C^\infty(X:V)$ we mean the space of all smooth functions $f: X \to V$ and by $C_c^\infty(X:V)$ the subspace of those f which have compact support. We omit V in this notation when $V = C$. One can identify $C_c^\infty(X:V)$ with $C_c^\infty(X) \otimes V$ by means of the mapping $i: C_c^\infty(X) \otimes V \to C_c^\infty(X:V)$ defined as follows: If $f \in C_c^\infty(X)$ and $v \in V$, then $i(f \otimes v)$ is the function $x \mapsto f(x) v$ $(x \in X)$ from X to V.

Let μ be a Radon measure on X. Then, for any $f \in C_c^\infty(X:V)$, define

$$\int_X f \, d\mu = \sum_{1 \leqq i \leqq r} \mu(f^{-1}(v_i)) \, v_i$$

where $v_1, \ldots, v_r$ are all the distinct nonzero values of f. (The right side is understood to be zero if $f = 0$.) Moreover if μ has compact support, we define

$$\int_X f \, d\mu = \int_X \varphi f \, d\mu \qquad (f \in C^\infty(X:V))$$

where φ is any function in $C_c^\infty(X)$ such that $\varphi = 1$ on $\operatorname{Supp} \mu$.

Let End V be the algebra of all endomorphisms of V and π a mapping from X to End V such that, for any $v \in V$, the function $x \mapsto \pi(x) v$ from X to V is smooth. If μ is a Radon measure on X with compact support, we define an element $E(\pi, \mu) \in$

End V by

$$E(\pi, \mu)\, v = \int_X \pi(x)\, v\, d\mu \qquad (x \in V).$$

We shall often write $\int_X \pi(x)\, d\mu$ for $E(\pi, \mu)$.

As usual $C(X)$ denotes the space of all (complex-valued) continuous functions f on X and $C_c(X)$ the subspace of those f which have compact support. A distribution on X is a linear mapping of $C_c^\infty(X)$ into $\mathbf{C}$.

3. Admissible representations. Let G be a t.d. group and V a vector space over $\mathbf{C}$. By a representation of G on V, we mean a mapping $\pi: G \to \operatorname{End} V$ such that $\pi(1) = 1$ and $\pi(xy) = \pi(x)\,\pi(y)$ $(x, y \in G)$. A vector $v \in V$ is called π-smooth if the mapping $x \mapsto \pi(x)\, v$ of G into V is smooth. Let V_∞ be the subspace of all smooth vectors. Then V_∞ is π-stable. Let π_∞ denote the restriction of π on V_∞. We say that π is smooth if $V = V_\infty$. Note that π_∞ is always smooth.

Let H be a closed subgroup of G and σ a smooth representation of H on V. Then we define a smooth representation $\pi = \operatorname{Ind}_H^G \sigma$ as follows: Let B denote the space of all smooth functions $\beta: G \to V$ such that

(1) $\beta(hx) = \sigma(h)\,\beta(x)$ $(h \in H,\ x \in G)$,

(2) $\operatorname{Supp} \beta$ is compact mod H.

Then π is the representation of G on B given by $(\pi(y)\,\beta)\,(x) = \beta(xy)$ $(x, y \in G,\ \beta \in B)$.

A representation π of G on V is said to be admissible if

(1) π is smooth,

(2) for any open subgroup H of G, $\dim V_H < \infty$.

Here V_H is the space of all $v \in V$ such that $\pi(h)v = v$ for all $h \in H$.

LEMMA 1. *Let H be a closed subgroup of G and σ an admissible representation of H. Then if G/H is compact, $\operatorname{Ind}_H^G \sigma$ is also admissible.*

Let π be a smooth representation of G on V and V' the (algebraic) dual of V. Then the dual representation π' of G on V' is given by

$$\langle \pi'(x)\, \lambda,\, v \rangle = \langle \lambda,\, \pi(x^{-1})\, v \rangle \qquad (x \in G,\ \lambda \in V',\ v \in V).$$

Put $\tilde{V} = (V')_\infty$ and $\tilde{\pi} = (\pi')_\infty$. Then $\tilde{\pi}$ is a smooth representation which is said to be contragredient to π.

LEMMA 2. *π is admissible if and only if $\tilde{\pi}$ is admissible. Moreover if π is admissible then $(\tilde{\pi})^\sim = \pi$.*

Let π be an admissible representation of G on V. By a matrix coefficient of π, we mean a function on G of the form

$$x \mapsto \langle \tilde{v}, \pi(x)\, v \rangle \qquad (x \in G),$$

where v and $\tilde{v}$ are fixed elements in V and $\tilde{V}$ respectively. Let $\mathscr{A}(\pi)$ denote the space spanned over C by all matrix coefficients of π.

Let $\mathfrak{T} = \mathrm{End}^{\circ} V$ denote the space of all $T \in \mathrm{End}\, V$ such that the mappings $x \mapsto \pi(x)\, T,\ x \mapsto T\pi(x)\ (x \in G)$ are both smooth. Clearly $\mathfrak{T}$ is a subalgebra of $\mathrm{End}\, V$ and $\pi(x)\, T\pi(y) \in \mathfrak{T}$ for $x, y \in G$ and $T \in \mathfrak{T}$. Moreover since π is admissible, $\dim TV < \infty$ and therefore $\mathrm{tr}\, T$ is defined for $T \in \mathfrak{T}$. Put $f_T(x) = \mathrm{tr}\, T\pi(x)\ (x \in G)$.

LEMMA 3. *Let π be an admissible representation of G. Then π is irreducible[1] if and only if the mapping $T \mapsto f_T$ of $\mathfrak{T}$ into $\mathscr{A}(\pi)$ is bijective.*

Let π be a smooth representation of G on V. We say that π is unitary, if there exists a positive-definite hermitian form H on V such that

$$H(\pi(x)\, v_1, \pi(x)\, v_2) = H(v_1, v_2) \qquad (v_1, v_2 \in V;\ x \in G).$$

A double representation π of G on V is a pair (π_1, π_2) where π_1 is a left-representation and π_2 a right-representation of G on V and the action of G on the left commutes with its action on the right. π is called smooth if π_1, π_2 are both smooth. Similarly π is unitary if there exists a pre-hilbertian structure on V with respect to which both π_1 and π_2 are unitary.

Let π_1, π_2 be two representations of G on V_1 and V_2 respectively. We write $\pi_1 \subset \pi_2$ if there exists a linear injection $T : V_1 \to V_2$ such that

$$\pi_2(x)\, T = T\pi_1(x) \qquad (x \in G).$$

Moreover π_1, π_2 are said to be equivalent if T can be taken to be a bijection.

We denote by $\mathscr{E}_c(G)$ the set of all equivalence classes of irreducible admissible representations of G and by $\mathscr{E}(G)$ the subset of those classes which are unitary.

Let $d_l x$ and $d_r x$ respectively denote the left- and right-invariant Haar measures on G. We assume that $d_r x = d_l x^{-1}$. Then

$$d_r x = \delta_G(x)\, d_l x.$$

Here $\delta = \delta_G$ is a homomorphism of G into the multiplicative group of positive real numbers, which is called the module of G. For $f, g \in C_c^{\infty}(G)$ define their convolution $f * g$ by

$$(f * g)(x) = \int f(y)\, g(y^{-1}x)\, d_l y = \int f(xy^{-1})\, g(y)\, d_r y.$$

Under this convolution product, $C_c^{\infty}(G)$ becomes an associative algebra. If π is an admissible representation of G on V, put

---

[1] Here irreducibility is meant in the algebraic sense.

$$\pi(f)=\int f(x)\,\pi(x)\,\delta(x)^{-1/2}\,d_rx=\int f(x)\,\pi(x)\,\delta(x)^{1/2}\,d_lx.$$

Then $\pi(f)\in\mathfrak{X}=\mathrm{End}^\circ V$ and $f\mapsto\pi(f)$ is a homomorphism of $C_c^\infty(G)$ into $\mathfrak{X}$. Put

$$\Theta_\pi(f)=\mathrm{tr}\,\pi(f)\qquad(f\in C_c^\infty(G)).$$

Then Θ_π is a distribution on G which is called the character of π.

Now assume G is unimodular so that $d_lx=d_rx=dx$ (say). Suppose K is an open compact subgroup and P a closed subgroup of G such that $G=KP$. We can normalize the Haar measures dk and d_lp on K and P respectively in such a way that

$$\int \alpha(x)\,dx=\int \alpha(pk)\,d_lp\,dk\qquad(\alpha\in C_c(G))$$

and the total measure of K is 1. Let σ be an admissible representation of P on U. Then $\pi=\mathrm{Ind}_P^G(\delta_P^{1/2}\sigma)$ is an admissible representation of G.

THEOREM 1. *Let θ_σ and Θ_π denote the characters of σ and π respectively. Then*

$$\Theta_\pi(\alpha)=\theta_\sigma(\bar\alpha)\qquad(\alpha\in C_c^\infty(G))$$

where $\bar\alpha(p)=\int_K \alpha(kpk^{-1})\,dk\quad(p\in P).$

Let $\mathfrak{H}$ denote the representation space of π. Then $\mathfrak{H}$ consists of all smooth functions $h:K\to U$ such that $h(pk)=\delta_P(p)^{1/2}\sigma(p)\,h(k)$ for $p\in P\cap K$ and $k\in K$. Now suppose σ is unitary with respect to a suitable pre-hilbertian structure on U. Define

$$\|h\|^2=\int_K |h(k)|^2\,dk\qquad(h\in\mathfrak{H}).$$

Then π is unitary with respect to this norm on $\mathfrak{H}$.

Let Z be a t.d. group which is abelian. By a quasi-character χ of Z we mean a continuous homomorphism of Z into $C^\times$. χ is called a character if $|\chi(z)|=1$ $(z\in Z)$. We denote by $\mathfrak{X}(Z)$ the group of all quasi-characters of Z and by $\hat Z$ the subgroup of all characters.

Let G be a unimodular t.d. group and π an admissible and irreducible representation of G on $\mathfrak{H}$. Let Z_G denote the center of G and Z a closed subgroup of Z_G such that Z_G/Z is compact. Since π is irreducible, there exists an element $\chi_\pi\in\mathfrak{X}(Z)$ such that $\pi(z)=\chi_\pi(z)$ for all $z\in Z$.

THEOREM 2. *Suppose $\chi_\pi\in\hat Z$ and there exists an element $f\neq0$ in $\mathscr{A}(\pi)$ such that*

$$\int_{G/Z} |f(x)|^2 \, dx^* < \infty.$$

Then π is unitary. Fix a pre-hilbertian structure on $\mathfrak{H}$ which is invariant under π. Then there exists a number $d(\pi) > 0$ such that

$$\int_{G/Z} |f_T(x)|^2 \, dx^* = d(\pi)^{-1} \|T\|^2$$

for all $T \in \mathrm{End}^\circ \mathfrak{H}$.

Here $f_T(x) = \mathrm{tr}\, T\pi(x)$ and dx^* is the Haar measure on G/Z. Moreover $\|T\|^2 = \mathrm{tr}(T^*T)$ where T^* is the adjoint of T with respect to the pre-hilbertian structure on $\mathfrak{H}$. (It is easy to see that T^* exists and lies in $\mathrm{End}^\circ \mathfrak{H}$.)

COROLLARY (SCHUR ORTHOGONALITY RELATIONS). *Let S, $T \in \mathrm{End}^\circ \mathfrak{H}$. Then*[2]

$$\int_{G/Z} \mathrm{conj}\, f_S(x) \cdot f_T(x) \, dx^* = d(\pi)^{-1} \mathrm{tr}\, S^*T.$$

Let ω denote the class of π in $\mathscr{E}_c(G)$. We say that π (or ω) is square-integrable, if it satisfies the conditions of Theorem 2. $d(\pi)$ is then called the formal degree of π. Since it depends only on ω, we can also denote it by $d(\omega)$. We shall denote by $\mathscr{E}_2(G)$ the set of all square-integrable elements of $\mathscr{E}(G)$.

Given an open compact subgroup K of G, let $C(G//K)$ denote the space of all functions in $C(G)$ which are constant on double cosets of K. Put $C_c(G//K) = C(G//K) \cap C_c(G)$. Then $H_K = C_c(G//K)$ is an algebra under convolution. An element $f \in C(G)$ is said to be H_K-finite if the space spanned by all functions of the form $h_1 * f * h_2$ $(h_1, h_2 \in H_K)$ has finite dimension. Moreover f is said to be Hecke finite if it is H_K-finite for every open compact subgroup K of G.

Let $\mathscr{A}(G)$ denote the space of all functions $f \in C(G)$ such that
(1) $f \in C(G//K)$ for some open compact subgroup K of G,
(2) f is Hecke finite.

THEOREM 3. *$\mathscr{A}(G) = \bigcup_\pi \mathscr{A}(\pi)$ where π runs over all admissible representations of G.*

4. Intertwining forms. Let G be a t.d. group and π_1, π_2 admissible representations of G on V_1 and V_2 respectively. By an intertwining form (between π_1, π_2) we mean a bilinear form B on $V_1 \times V_2$ such that

---

[2] $\mathrm{conj}\, c$ denotes the conjugate of a complex number c.

$$B(\pi_1(x)\, v_1, \pi_2(x)\, v_2) = B(v_1, v_2) \qquad (x \in G,\ v_i \in V_i,\ i = 1, 2).$$

Let $\mathscr{B}(\pi_1, \pi_2)$ denote the space of all intertwining forms and put

$$I(\pi_1, \pi_2) = \dim \mathscr{B}(\pi_1, \pi_2).$$

Similarly an intertwining operator from π_1 to π_2 is a linear mapping $T : V_1 \to V_2$ such that $T\pi_1(x) = \pi_2(x)\, T$. Let $\mathfrak{T}(\pi_2 \mid \pi_1)$ denote the space of all such operators and put

$$J(\pi_2 \mid \pi_1) = \dim \mathfrak{T}(\pi_2 \mid \pi_1).$$

LEMMA 4.   *Let π_1, π_2 be two admissible representations of G. Then*

$$I(\tilde{\pi}_2, \pi_1) = J(\pi_2 \mid \pi_1).$$

COROLLARY.   *Let π be an admissible unitary representation of G. Then π is irreducible if and only if $I(\tilde{\pi}, \pi) = J(\pi \mid \pi) = 1$.*

Let π be a representation of G on V. A subspace W of V is called π-admissible if W is π-stable and the restriction of π on W is admissible. Moreover π is called quasi-admissible if V is a union of π-admissible subspaces.

5. A theorem of Jacquet.   Let Ω be a p-adic field (i.e., a t.d. field which is not discrete). We shall now use the terminology of [3(g), p. 8]. Let $\mathbf{G}$ be a connected reductive Ω-group and G the subgroup of all Ω-rational points in $\mathbf{G}$. Then G, being a closed subgroup of $GL(n, \Omega)$, is a t.d. group. Moreover since $\mathbf{G}$ is reductive, G is unimodular.

Fix a parabolic pair (p-pair) (P, A) in G and let $P = MN$ be the corresponding Levi decomposition.

LEMMA 5.   *There exists a one-one correspondence between p-pairs $(*P, *A)$ in M and p-pairs (P', A') in G such that $(P, A) \succ (P', A')$. It is given by*

$$(*P, *A) = (M \cap P', A').$$

Moreover if $P' = M'N'$ and $*P = *M*N$ are the corresponding Levi decompositions, then

$$M' = *M, \quad A' = *A, \quad N' = *N \cdot N.$$

π being a representation of G on V, we denote by $V(N)$ the subspace of V spanned by all elements of the form $\pi(n)\, v - v\, (n \in N,\ v \in V)$. Then $V(N)$ is stable under $\pi(P)$. Hence we get a representation of P on $V/V(N)$. Clearly N lies in the kernel of this representation. Since $M \simeq P/N$, we get a representation σ of M on $V/V(N)$. The following theorem of Jacquet plays an important role in harmonic analysis.

THEOREM 4 (JACQUET).   *Suppose π is quasi-admissible. Then the same holds for σ.*

6. The constant term and its applications.   Fix a p-pair (P, A) in G $(P = MN)$. For any $t > 0$, let $A^+(t)$ denote the set of all $a \in A$ such that[3] $|\xi_\alpha(a)|_p \geq t$ for every simple root α of (P, A). (We denote by ξ_α the rational character of A corresponding to α.)

THEOREM 5.[4]   *Fix $f \in \mathscr{A}(G)$. Then there exists exactly one element $f_P \in \mathscr{A}(M)$ with the following property: Given a compact subset ω in M, we can choose $t \geq 1$ such that $\delta_P(ma)^{1/2} f(ma) = f_P(ma)$ for $m \in \omega$ and $a \in A^+(t)$.*

We call f_P the constant term of f along P.

Let $(*P, *A)$ be a p-pair in M and (P', A') the p-pair in G which corresponds to it under Lemma 5.

LEMMA 6.   *Let $f \in \mathscr{A}(G)$. Then $(f_P)_{*P} = f_{P'}$.*

Let $°\mathscr{A}(G)$ denote the space of all $f \in \mathscr{A}(G)$ such that $f_P = 0$ for every p-pair (P, A) in G with $P \neq G$. Let Z be the split component of G. An element $g \in C(G)$ is called a supercusp form if

(1) Supp g is compact mod Z;
(2) let N be the radical of a parabolic subgroup (psgp) $P \neq G$, then

$$\int_N g(xn)\, dn = 0 \qquad (x \in G).$$

(Here dn is the Haar measure of N.)

THEOREM 6.   *Let f be an element in $\mathscr{A}(G)$. Then the following three conditions on f are mutually equivalent:*

(1) $f \in °\mathscr{A}(G)$.
(2) Supp f is compact mod Z.
(3) f is a supercusp form.

Let π be an admissible representation of G. We say that π is supercuspidal if $\mathscr{A}(\pi) \subset °\mathscr{A}(G)$.

LEMMA 7.   *Let π be an admissible representation of G on V. Then the following two conditions are equivalent:*

---

[3] $|\ |_p$ denotes the usual p-adic absolute value.
[4] Cf. [3(f), Theorem 2].

(1) π *is supercuspidal.*
(2) *N being the radical of any psgp $P \neq G$, define $V(N)$ as in §5. Then $V(N) = V$.*

We note that $V(N)$ is also the space of all $v \in V$ with the following property: There exists an open compact subgroup U of N such that

$$\int_U \pi(n)\, v\, dn = 0.$$

Let π be an admissible and irreducible representation of G. By the central exponent of π, we mean the quasi-character $\chi_\pi \in \mathfrak{X}(Z)$ such that $\pi(z) = \chi_\pi(z)$ for $z \in Z$. χ_π depends only on the class ω of π in $\mathscr{E}_c(G)$. Hence we may denote it by χ_ω.

Let $^\circ\mathscr{E}_c(G)$ denote the set of all supercuspidal elements in $\mathscr{E}_c(G)$. Put $^\circ\mathscr{E}(G) = \mathscr{E}(G) \cap {}^\circ\mathscr{E}_c(G)$. It is clear from Theorem 2 that an element $\omega \in {}^\circ\mathscr{E}_c(G)$ lies in $^\circ\mathscr{E}(G)$ if and only if $\chi_\omega \in \hat{Z}$.

Let π be a quasi-admissible representation (§4) of G on V. For any $\chi \in \mathfrak{X}(Z)$ let $V(\chi)$ denote the space of all $v \in V$ with the following property: There exists an integer $n \geq 1$ such that $(\pi(z) - \chi(z))^n v = 0$ for all $z \in Z$. Since π is quasi-admissible, it follows without difficulty that

$$V = \sum_{\chi \in \mathfrak{X}(Z)} V(\chi)$$

where the sum is direct.

Let r denote the right-regular representation of G on $\mathscr{A}(G)$. It follows from the definition of $\mathscr{A}(G)$ that r is quasi-admissible. Hence

$$\mathscr{A}(G) = \sum_{\chi \in \mathfrak{X}(Z)} \mathscr{A}(G, \chi).$$

If $f \in \mathscr{A}(G)$, we denote by f_χ the component of f in $\mathscr{A}(G, \chi)$. Fix a p-pair (P, A) $(P = MN)$ and an element $\eta \in \mathfrak{X}(A)$. Then if $f \in \mathscr{A}(G)$, $f_{P,\eta}$ is the component of f_P in $\mathscr{A}(M, \eta)$.

Let π be an admissible and irreducible representation of G. By an exponent of π (with respect to (P, A)) we mean an element $\chi \in \mathfrak{X}(A)$ such that $f_{P,\chi} \neq 0$ for some $f \in \mathscr{A}(\pi)$. Let $\mathfrak{X}_\pi(P, A)$ denote the set of all such exponents. Then this set depends only on the class ω of π and so we may denote it by $\mathfrak{X}_\omega(P, A)$.

THEOREM 7. *Fix $\omega \in \mathscr{E}_c(G)$ and a p-pair (P, A) in G. Then $\mathfrak{X}_\omega(P, A)$ is a finite set.*

Fix $\chi \in \mathfrak{X}_\omega(P, A)$ and let r denote the right-regular representation of M on $\mathscr{A}(M)$. Then there exists an integer $d \geq 1$ such that $(r(a) - \chi(a))^d f_{P,\chi} = 0$ for all $f \in \mathscr{A}(\pi)$ and $a \in A$. By the multiplicity $d(\chi)$ of χ, we mean the least such integer d.

Moreover we say that χ is simple if $d(\chi) = 1$.

Notice that $\mathfrak{X}_\omega(G, Z) = \{\chi_\omega\}$ where χ_ω is the central exponent of ω. Moreover $\mathfrak{X}_\omega(P, A) = \emptyset$ if $\omega \in {}^\circ\mathscr{E}_c(G)$ and $P \neq G$.

7. The complex structure on ${}^\circ\mathscr{E}_c(M)$.

By a special torus in G, we mean a split component of some psgp of G. Given a special torus A, we denote by $\mathscr{P}(A)$ the set of all psgps P of G such that (P, A) is a p-pair. Let M be the centralizer of A in G. Then if $P \in \mathscr{P}(A)$, $P = MN$ where N is the radical of P. Moreover $\mathscr{P}(A)$ is a finite set.

Let A_1, A_2 be two special tori. By $\mathfrak{w}(A_2 \mid A_1)$ we mean the set of all homomorphisms $s: A_1 \to A_2$ with the following property: There exists an element $y \in G$ such that $a^s = yay^{-1}$ for all $a \in A_1$. y is then called a representative of s in G.

Let A be a special torus. Then $\mathfrak{w}(A \mid A)$ is a group which we denote by $\mathfrak{w}(A)$ or sometimes by $\mathfrak{w}(G/A)$. If M is the centralizer and $\tilde{M}$ the normalizer of A in G, then $\mathfrak{w}(G/A) \simeq \tilde{M}/M$. Hence it is obvious that $\mathfrak{w} = \mathfrak{w}(A)$ operates on $\mathscr{E}_c(M)$. We say that an element $\omega \in \mathscr{E}_c(M)$ is unramified (in G) if $\omega^s \neq \omega$ for $s \neq 1$ in $\mathfrak{w}$.

Let $X(M)$ and $X(A)$ respectively denote the groups of all rational characters of M and A which are defined over Ω. Then the inclusion $i: A \to M$ defines an injective homomorphism $i^*: X(M) \to X(A)$ and $X(A)/i^*(X(M))$ is a finite group (see [1]). Moreover if $l = \dim A$, $X(A)$ is a free abelian group of rank l. We identify $X(M)$ with its image under i^* and put

$$\mathfrak{a}^* = X(A) \otimes R = X(M) \otimes R,$$
$$\mathfrak{a} = \operatorname{Hom}(X(A), R) = \operatorname{Hom}(X(M), R).$$

These are real vector spaces of dimension l which are dual to each other. We denote the corresponding complexifications by $\mathfrak{a}_c^*$ and $\mathfrak{a}_c$.

Define a homomorphism $H_M = H$ of M into $\operatorname{Hom}(X(M), Z)$ as follows. If $\chi \in X(M)$ and $m \in M$, then $\langle \chi, H(m) \rangle \in Z$ is given by[5]

$$|\chi(m)|_\mathfrak{p} = q^{\langle \chi, H(m) \rangle}.$$

Let ${}^\circ M = \ker H$. Then ${}^\circ M$ contains every compact subgroup of M and ${}^\circ MA$ has finite index in M.

If $\omega \in {}^\circ\mathscr{E}_c(M)$ and $v \in \mathfrak{a}_c^*$, we define $\omega_v \in {}^\circ\mathscr{E}_c(M)$ as follows. Fix a representation σ of M in the class ω and put

$$\sigma_v(m) = \sigma(m) \, q^{i\langle v, H(m) \rangle} \qquad (m \in M).$$

Then ω_v is the class of σ_v. The mapping $(v, \omega) \mapsto \omega_v$ defines an action of the additive group $\mathfrak{a}_c^*$ on ${}^\circ\mathscr{E}_c(M)$. Now $L = H(M)$ is a lattice in $\mathfrak{a}$. Let L^* denote the dual lattice consisting of all $\lambda \in \mathfrak{a}^*$ such that

---

[5] q is the number of elements in the residue field of Ω.

$$\langle \lambda, H(m) \rangle \in 2\pi (\log q)^{-1} \mathbf{Z} \qquad (m \in M).$$

It is easy to see that L^* is exactly the stabilizer in $\mathfrak{a}_c^*$ of every $\omega \in {}^\circ \mathscr{E}_c(M)$. Hence the orbit $\mathfrak{o}_c$ of ω under $\mathfrak{a}_c^*$ may be identified with the complex Lie group $\mathfrak{a}_c^*/L^*$. We now introduce a complex structure on ${}^\circ \mathscr{E}_c(M)$ by demanding that each orbit be an open submanifold. It is easy to see that every orbit intersects ${}^\circ \mathscr{E}(M)$.

We shall sometimes call $\mathfrak{a}$ the real Lie algebra of A.

8. Some consequences of Bruhat's theory.   Let (P_i, A_i) $(i = 1, 2)$ be two p-pairs in G and $P_i = M_i N_i$ their Levi decompositions. Let σ_i be an admissible representation of M_i on V_i. Since $M_i \simeq P_i/N_i$, we may regard σ_i as a representation of P_i. Put

$$\pi_i = \mathrm{Ind}_{P_i}^G (\delta_{P_i}^{1/2} \sigma_i).$$

THEOREM 8.   *Assume σ_1, σ_2 are both supercuspidal. Then $I(\pi_1, \pi_2) = 0$ unless A_1 and A_2 are conjugate in G.*

Now suppose $A_1 = A_2$. Then $M_1 = M_2$ and so we can drop the subscripts and write M and A.

THEOREM 9.   *Suppose σ_1, σ_2 are irreducible and supercuspidal representations of M and ω_1, ω_2 the corresponding elements in ${}^\circ \mathscr{E}_c(M)$. Let $\mathfrak{w}_0$ denote the set of all $s \in \mathfrak{w} = \mathfrak{w}(G/A)$ such that $\omega_1^s = \omega_2$. Then[6]*

$$I(\tilde{\pi}_1, \pi_2) \leqq [\mathfrak{w}_0].$$

COROLLARY.   *Fix $\omega \in {}^\circ \mathscr{E}(M)$ and assume that ω is unramified in G. Then if $\sigma \in \omega$ and $P \in \mathscr{P}(A)$, $\pi = \mathrm{Ind}_P^G \delta_P^{1/2} \sigma$ is a unitary, admissible and irreducible representation of G.*

This is proved by combining Lemma 7 with Bruhat's theory (cf. [3(d), Theorem 1]).

9. Induced representations.   Let π be an admissible and irreducible representation of G. A p-pair (P, A) in G $(P = MN)$ is said to be π-minimal if $\mathfrak{X}_\pi(P, A) \neq \emptyset$ but $\mathfrak{X}_\pi(P', A') = \emptyset$ for every p-pair $(P', A') \prec (P, A)$ $(P' \neq P)$. Let $\bar{P}$ denote the unique element in $\mathscr{P}(A)$ such that α is a root of $(\bar{P}, A)$ if and only if $-\alpha$ is a root of (P, A).

THEOREM 10.[7]   *Fix a π-minimal p-pair (P, A) in G and $\chi \in \mathfrak{X}_\pi(P, A)$. Then there exists an irreducible, admissible, supercuspidal representation σ of M such that*

---

[6] $[S]$ stands for the number of elements in a set S.

[7] This is a generalization of a result of Jacquet [5].

(1) χ is the central exponent of σ,
(2) $\pi \subset \mathrm{Ind}_P^G(\delta_P^{1/2}\sigma)$.

COROLLARY. *Suppose $\chi \in \hat{A}$. Then π is unitary.*

We keep to the notation of the above theorem.

THEOREM 11. *If (P', A') is any p-pair in G, then every element of $\mathfrak{X}_\pi(P', A')$ is of the form $\chi \circ s$ where $s \in \mathfrak{w}(A \mid A')$. Moreover if (P', A') is also π-minimal, then A' is conjugate to A in G.*

For any $\chi \in \mathfrak{X}(A)$, put

$$\chi^*(a) = \mathrm{conj}\,\chi(a^{-1}) \qquad (a \in A).$$

Then $\chi \mapsto \chi^*$ is an automorphism of the group $\mathfrak{X}(A)$ and $\hat{A}$ is exactly the set of all fixed points of this automorphism.

LEMMA 8. *Suppose π is unitary and (P, A) is any p-pair in G. Then $\mathfrak{X}_\pi(\bar{P}, A)$ $= \mathfrak{X}_\pi(P, A)^*$.*

Let A be a special torus in G and M its centralizer. Fix $\omega \in {}^\circ\mathscr{E}(M)$ and, for any $P \in \mathscr{P}(A)$, put

$$\pi_P = \mathrm{Ind}_P^G(\delta_P^{1/2}\sigma)$$

where $\sigma \in \omega$. Then π_P is unitary. Let $C(P, \omega)$ denote its class. If ω is unramified in G, $C(P, \omega) \in \mathscr{E}(G)$ (§8).

THEOREM 12. *Suppose $P_1, P_2 \in \mathscr{P}(A)$ and $s \in \mathfrak{w} = \mathfrak{w}(G/A)$. Then*

$$C(P_1, \omega) = C(P_2, \omega^s).$$

Hence we may write $C(\omega)$ (or $C_M^G(\omega)$) instead of $C(P, \omega)$.

THEOREM 13. *Suppose $\omega \in {}^\circ\mathscr{E}(M)$ is unramified in G. Then if $\pi \in C(\omega)$ and $P \in \mathscr{P}(A)$,*

$$\mathfrak{X}_\pi(P, A) = \{\chi \circ s\}_{s \in \mathfrak{w}}$$

where χ is the central exponent of ω. Moreover if (P', A') is any p-pair in G, every exponent $\eta \in \mathfrak{X}_\pi(P', A')$ is simple.

10. Maass-Selberg relations. Fix a maximal split torus A_0 in G and a maximal compact subgroup K of G satisfying the conditions of Bruhat and Tits [3(g), p. 16] with respect to A_0. By a standard torus we mean a special torus contained in A_0. A p-pair (P, A) in G is called semistandard if $A \subset A_0$.

Let τ be a smooth and unitary double representation of K on V. For any $\omega \in \mathscr{E}_c(G)$ put $\mathscr{A}(\omega) = \mathscr{A}(\pi)$ where $\pi \in \omega$. Let $C(G, \tau)$ denote the space of all functions $f: G \to V$ such that $f(k_1 x k_2) = \tau(k_1) f(x) \tau(k_2)$ $(k_1, k_2 \in K, x \in G)$, and $C_c(G, \tau)$ the subspace of all f with compact support. Moreover we denote by $^\circ C_c(G, \tau)$ the space of all $f \in C_c(G, \tau)$ such that

$$\int_N f(xn)\, dn = 0 \qquad (x \in G)$$

where N is the radical of a psgp $P \neq G$. Put

$$\mathscr{A}(G, \tau) = (\mathscr{A}(G) \otimes V) \cap C(G, \tau).$$

If (P, A) is a semistandard p-pair in G $(P = MN)$, let τ_M denote the restriction of τ on $K_M = K \cap M$. Then by Theorem 5 we have the mapping $f \mapsto f_P$ of $\mathscr{A}(G, \tau)$ into $\mathscr{A}(M, \tau_M)$. Fix $f \in \mathscr{A}(G, \tau)$. Then we write $f_P \sim 0$ if

$$\int_M (\varphi(m), f_P(m))\, dm = 0$$

for every $\varphi \in {}^\circ C_c(M, \tau_M)$. (Here dm is the Haar measure of M and the scalar product is in V.)

LEMMA 9. *Let f be an element in $\mathscr{A}(G, \tau)$ such that $f_P \sim 0$ for every semistandard p-pair (P, A) in G (including $(P, A) = (G, Z)$). Then $f = 0$.*

If π is an admissible representation of G and C its equivalence class, we put

$$\mathscr{A}(C, \tau) = \mathscr{A}(\pi, \tau) = C(G, \tau) \cap (\mathscr{A}(\pi) \otimes V).$$

Let M be the centralizer in G of a standard torus A. Fix $\omega \in {}^\circ \mathscr{E}(M)$. Then $\mathscr{A}(\omega, \tau_M)$ has a natural pre-hilbertian structure given by

$$\|g\|^2 = \int_{M/A} |g(m)|^2\, dm^* \qquad (g \in \mathscr{A}(\omega, \tau_M)),$$

where dm^* is the Haar measure on M/A. Put $\mathfrak{w} = \mathfrak{w}(G/A)$.

THEOREM 14. *Let ω be an element in $^\circ \mathscr{E}(M)$ which is unramified in G and put $C(\omega) = C_M^G(\omega)$. Fix $f \in \mathscr{A}(C(\omega), \tau)$. Then the following statements are true:*

(1) If (P', A') is a semistandard p-pair in G, then $f_{P'} \sim 0$ unless A' is conjugate to A in G.

(2) If $P \in \mathscr{P}(A)$, then

$$f_P \in \sum_{s \in \mathfrak{w}} \mathscr{A}(\omega^s, \tau_M)$$

where the sum is direct.

Let $f_{P,s}$ denote the component of f_P in $\mathscr{A}(\omega^s, \tau_M)$.

(3) $\|f_{P_1, s_1}\| = \|f_{P_2, s_2}\|$ for $P_1, P_2 \in \mathscr{P}(A)$ and $s_1, s_2 \in \mathfrak{w}$.

By the Maass-Selberg relations we mean the equalities in (3) (cf. [3(f), §7]).

COROLLARY.   *Fix $P \in \mathscr{P}(A)$ and $s \in \mathfrak{w}$. Then the mapping $f \mapsto f_{P,s}$ of $\mathscr{A}(C(\omega), \tau)$ into $\mathscr{A}(\omega^s, \tau_M)$ is injective.*

11. The functional equations.   We keep to the notation of Theorem 14. If $P_i = MN_i$ $(i = 1, 2)$ are two elements of $\mathscr{P}(A)$, let $V(P_1 \mid P_2)$ denote the subspace of all $v \in V$ such that $\tau(n_1) v \tau(n_2) = v$ $(n_i \in N_i \cap K, i = 1, 2)$. Then $V(P_1 \mid P_2)$ is stable under τ_M. Let $\tau_{P_1 \mid P_2}$ denote the restriction of τ_M on $V(P_1 \mid P_2)$. Put

$$L(\omega, P) = \mathscr{A}(\omega, \tau_{P \mid P}), \qquad \mathscr{L}(\omega, P) = \mathscr{A}(\omega, \tau_{P \mid \bar{P}})$$

for $\omega \in {}^{\circ}\mathscr{E}(M)$ and $P \in \mathscr{P}(A)$. For $\psi \in L(\omega, P)$, define an element $E(P : \psi) \in \mathscr{A}(G, \tau)$ as follows. Extend ψ and δ_P to functions on G by setting

$$\psi(kmn) = \tau(k) \psi(m), \quad \delta_P(kp) = \delta_P(p) \qquad (k \in K, m \in M, n \in N, p \in P)$$

and put

$$E(P : \psi : x) = \int \psi(xk) \tau(k^{-1}) \delta_P(xk)^{-1/2} \, dk \qquad (x \in G)$$

where dk is the Haar measure on K so normalized that the total measure of K is 1.

We shall now assume that ω is unramified in G.

LEMMA 10.   *Fix $P \in \mathscr{P}(A)$. Then $\psi \mapsto E(P : \psi)$ is a bijective mapping of $L(\omega, P)$ onto $\mathscr{A}(C(\omega), \tau)$.*

THEOREM 15.   *Fix $P_1, P_2 \in \mathscr{P}(A)$. Then there exist unique linear mappings $c_{P_2 \mid P_1}(s : \omega)$ $(s \in \mathfrak{w})$ of $L(\omega, P_1)$ into $\mathscr{L}(\omega^s, P_2)$ such that*

$$E_{P_2}(P_1 : \psi) = \sum_{s \in \mathfrak{w}} c_{P_2 \mid P_1}(s : \omega) \psi$$

for all $\psi \in L(\omega, P_1)$. Moreover $c_{P_2 \mid P_1}(s : \omega)$ is bijective and there exists a number $\mu(\omega) > 0$ such that

$$\mu(\omega)\,\|c_{P_2|P_1}(s:\omega)\,\psi\|^2 = \|\psi\|^2$$

for all $P_1, P_2 \in \mathscr{P}(A)$, $s \in \mathfrak{w}$ and $\psi \in L(\omega, P_1)$. Finally $\mu(\omega) = \mu(\omega^s)$ $(s \in \mathfrak{w})$.

Put

$$^{\circ}c_{P_2|P_1}(s:\omega) = c_{P_2|P_2}(1:\omega^s)^{-1}\,c_{P_2|P_1}(s:\omega),$$
$$c^{\circ}_{P_2|P_1}(s:\omega) = c_{P_2|P_1}(s:\omega)\,c_{P_1|P_1}(1:\omega)^{-1}.$$

Then $^{\circ}c_{P_2|P_1}(s:\omega)$ is a unitary bijection of $L(\omega, P_1)$ onto $L(\omega^s, P_2)$. Similarly $c^{\circ}_{P_2|P_1}(s:\omega)$ is a unitary bijection of $\mathscr{L}(\omega, P_1)$ onto $\mathscr{L}(\omega^s, P_2)$.

THEOREM 16.   *We have the functional equations*

$$E(P_2:{^{\circ}c}_{P_2|P_1}(s:\omega)\,\psi) = E(P_1:\psi)\qquad (\psi \in L(\omega, P_1))$$

and

$$^{\circ}c_{P_3|P_1}(st:\omega) = {^{\circ}c}_{P_3|P_2}(s:\omega^t)\,{^{\circ}c}_{P_2|P_1}(t:\omega)$$

for $P_i \in \mathscr{P}(A)$ $(i = 1, 2, 3)$ and $s, t \in \mathfrak{w}$.

Now put

$$E^{\circ}(P:\psi) = E(P:c_{P|P}(1:\omega)^{-1}\,\psi)$$

for $P \in \mathscr{P}(A)$ and $\psi \in \mathscr{L}(\omega, P)$.

THEOREM 17.   *The following functional equations hold:*

$$E^{\circ}(P_2:c^{\circ}_{P_2|P_1}(s:\omega)\,\psi) = E^{\circ}(P_1:\psi)\qquad (\psi \in \mathscr{L}(\omega, P_1))$$

and

$$c^{\circ}_{P_3|P_1}(st:\omega) = c^{\circ}_{P_3|P_2}(s:\omega^t)\,c^{\circ}_{P_2|P_1}(t:\omega)$$

for $P_i \in \mathscr{P}(A)$ $(i = 1, 2, 3)$ and $s, t \in \mathfrak{w}$.

Let P_1, P_2 be two elements in $\mathscr{P}(A)$ and (P', A') a p-pair in G $(P' = M'N')$ such that

$$(P', A') \succ (P_i, A)\qquad (i = 1, 2).$$

Put $*P_i = M' \cap P_i$. Then $(*P_i, A)$ is a p-pair in M' and $\mathscr{L}(\omega, P_1) \subset \mathscr{L}(\omega, *P_1)$. Let $*\mathfrak{w}$ be the subgroup of those elements of $\mathfrak{w}$ which leave A' pointwise fixed. Then $*\mathfrak{w} = \mathfrak{w}(M'/A)$.

THEOREM 18.   *Let $s \in *\mathfrak{w}$. Then $c^{\circ}_{P_2|P_1}(s:\omega) = c^{\circ}_{*P_2|*P_1}(s:\omega)$ on $\mathscr{L}(\omega, P_1)$.*

We now make the group $\mathfrak{w}$ operate on $\mathscr{A}(M, \tau_M)$ as follows. Fix $s \in \mathfrak{w}$ and choose a representative y for s in K. Then if $\varphi \in \mathscr{A}(M, \tau_M)$ we define $s\varphi$ to be the function $m \mapsto \tau(y)\,\varphi(y^{-1}my)\,\tau(y^{-1})\,(m \in M)$. Also if $P \in \mathscr{P}(A)$, we define $P^s = yPy^{-1}$.

LEMMA 11.   *Let* $P_1,\ P_2 \in \mathscr{P}(A)$ *and* $s,\ t \in \mathfrak{w}$. *Then*

$$sc_{P_2 \mid P_1}(t:\omega) = c_{P_2^s \mid P_1}(st:\omega),$$

$$c_{P_2 \mid P_1}(t:\omega)\,s^{-1} = c_{P_2 \mid P_1^s}(ts^{-1}:\omega^s).$$

Similar statements hold for $c^{\circ}_{P_2 \mid P_1}(t:\omega)$ and $^{\circ}c_{P_2 \mid P_1}(t:\omega)$ instead of $c_{P_2 \mid P_1}(t:\omega)$.

12. Extension to the complex domain.   Now assume that $\dim V < \infty$. Then $\dim \mathscr{A}(\omega, \tau_M) < \infty$ for any $\omega \in {}^{\circ}\mathscr{E}(M)$. For $P \in \mathscr{P}(A)$, $\psi \in L(\omega, P)$ and $v \in \mathfrak{a}_c^*$, define

$$E(P:\psi:v:x) = \int_K \psi(xk)\,\tau(k^{-1})\,q^{\langle iv - \varrho_P.\,H_P(xk)\rangle}\,dk \qquad (x \in G).$$

Here ψ is extended to a function on G as in §11, the mapping $H_P : G \to \mathfrak{a}$ is defined by (see §7) $H_P(kmn) = H(m)$ $(k \in K,\ m \in M,\ n \in N)$ and $\varrho_P \in \mathfrak{a}^*$ is given by $q^{\langle \varrho_P,\,H(m)\rangle} = \delta_P(m)^{1/2}$ $(m \in M)$. It is easy to see that $E(P:\psi:v) \in \mathscr{A}(G, \tau)$.

Let $\mathfrak{F}_c'(\omega)$ be the set of all $v \in \mathfrak{a}_c^*$ such that $\chi_{\omega,\,v} \circ s \neq \chi_{\omega,\,v}$ for any $s \neq 1$ in $\mathfrak{w} = \mathfrak{w}(G/A)$. (Here $\chi_{\omega,\,v}$ is the central exponent of ω_v.) Then $\mathfrak{F}_c'(\omega)$ is an open, connected and everywhere dense subset of $\mathfrak{a}_c^*$. We observe that $\mathfrak{w}$ acts on $\mathfrak{a}_c$ and therefore by duality also on $\mathfrak{a}_c^*$.

THEOREM 19.   *Fix* $P_1,\ P_2 \in \mathscr{P}(A)$. *Then, for any* $v \in \mathfrak{F}_c'(\omega)$, *there exist unique linear mappings*

$$c_{P_2 \mid P_1}(s:\omega:v): L(\omega, P_1) \to \mathscr{L}(\omega^s, P_2) \qquad (s \in \mathfrak{w})$$

such that

$$E_{P_2}(P_1:\psi:v:m) = \sum_{s \in \mathfrak{w}} (c_{P_2 \mid P_1}(s:\omega:v)\,\psi)\,(m)\,q^{i\langle sv,\,H(m)\rangle}$$

for $\psi \in L(\omega, P_1)$ *and* $m \in M$. *Moreover the functions* $v \mapsto c_{P_2 \mid P_1}(s:\omega:v)$ *are meromorphic on* $\mathfrak{a}_c^*$.

We have seen in §10 that the space $\mathscr{A}(\omega, \tau_M)$ has a natural hilbertian structure. Let

$$(c_{P_2 \mid P_1}(s:\omega:v))^* : \mathscr{L}(\omega^s, P_2) \to L(\omega, P_1)$$

denote the adjoint of $c_{P_2 \mid P_1}(s:\omega:v)$.

THEOREM 20.   *There exists a complex-valued meromorphic function* $v \to \mu(\omega:v)$ *on* $\mathfrak{a}_c^*$ *such that*

$$\mu(\omega:v)\,(c_{P_2\mid P_1}(s:\omega:\bar v))^*c_{P_2\mid P_1}(s:\omega:v)$$

is the identity on $L(\omega, P_1)$ for all $P_1, P_2\in\mathscr{P}(A)$ and $s\in\mathfrak{w}$. Moreover $\mu(\omega^s:sv)=\mu(\omega:v)$ and $\mu(\omega:v)$ is holomorphic and nonnegative on $\mathfrak{a}^*$.

As usual $\bar v$ stands for the complex conjugate of v with respect to $\mathfrak{a}^*$.

Put $\mathfrak{F}'(\omega)=\mathfrak{a}^*\cap\mathfrak{F}'_c(\omega)$. Then $\mathfrak{F}'(\omega)$ is an open and dense subset of $\mathfrak{a}^*$. Fix $v\in\mathfrak{F}'(\omega)$. Then ω_v is unramified in G, $\mu(\omega:v)=\mu(\omega_v)$ in the notation of §11 and $c_{P_2\mid P_1}(s:\omega:v)$ is bijective. Hence

$$^\circ c_{P_2\mid P_1}(s:\omega:v)=c_{P_2\mid P_2}(1:\omega^s:sv)^{-1}\,c_{P_2\mid P_1}(s:\omega:v),$$

$$c^\circ_{P_2\mid P_1}(s:\omega:v)=c_{P_2\mid P_1}(s:\omega:v)\,c_{P_1\mid P_1}(1:\omega:v)^{-1}$$

are defined as meromorphic functions of v on $\mathfrak{a}^*_c$. They are holomorphic and unitary on $\mathfrak{a}^*$. Put

$$E^\circ(P:\psi:v)=E(P:c_{P\mid P}(1:\omega:v)^{-1}\,\psi:v)\qquad(\psi\in\mathscr{L}(\omega, P)).$$

Then $E^\circ(P:\psi:v)$ is also holomorphic in v on $\mathfrak{a}^*$.

THEOREM 21. *The following functional equations hold:*
(1) $E(P_2:{}^\circ c_{P_2\mid P_1}(s:\omega:v)\,\psi:sv)=E(P_1:\psi:v)\,(\psi\in L(\omega, P_1))$,

(2) $E^\circ(P_2:c^\circ_{P_2\mid P_1}(s:\omega:v)\,\psi:sv)=E^\circ(P_1:\psi:v)\,(\psi\in\mathscr{L}(\omega, P_1))$,

(3) $^\circ c_{P_3\mid P_1}(st:\omega:v)={}^\circ c_{P_3\mid P_2}(s:\omega^t:tv)^\circ c_{P_2\mid P_1}(t:\omega:v)$,

(4) $c^\circ_{P_3\mid P_1}(st:\omega:v)=c^\circ_{P_3\mid P_2}(s:\omega^t:tv)\,c^\circ_{P_2\mid P_1}(t:\omega:v)$,
for $P_i\in\mathscr{P}(A)\,(i=1, 2, 3)$, $s, t\in\mathfrak{w}$ *and* $v\in\mathfrak{a}^*_c$.

Let $\mathfrak{a}_0$ be the real Lie algebra of A_0 (see §10). Fix a positive-definite quadratic form on $\mathfrak{a}_0$ which is invariant under $\mathfrak{w}_0=\mathfrak{w}(G/A_0)$. This defines a scalar product in $\mathfrak{a}_0$ and therefore also in $\mathfrak{a}$. Hence we can speak of the reflexion in $\mathfrak{a}$ corresponding to a given hyperplane. Fix any root α of A, let s_α denote the reflexion corresponding to the hyperplane defined by the equation $\alpha=0$. Fix $P\in\mathscr{P}(A)$. A root α of (P, A) is called reduced if $r\alpha$ is not a root for $0<r<1$ $(r\in\mathbf{R})$. For any $\omega\in{}^\circ\mathscr{E}(M)$, let $\mathfrak{w}(\omega)$ denote the subgroup of all $s\in\mathfrak{w}$ such that $\omega^s=\omega$.

Fix a connected component $\mathfrak{o}_c$ of $^\circ\mathscr{E}_c(M)$ (see §7) and put $\mathfrak{o}=\mathfrak{o}_c\cap{}^\circ\mathscr{E}(M)$. Let $\Sigma_r(\mathfrak{o})$ denote the set of all reduced roots α of (P, A) such that $s_\alpha\in\mathfrak{w}(\omega)$ for some $\omega\in\mathfrak{o}$.

THEOREM 22. *Fix* $P_1, P_2\in\mathscr{P}(A)$, $s\in\mathfrak{w}$, $\omega\in\mathfrak{o}$ *and for every* $\alpha\in\Sigma_r(\mathfrak{o})$ *choose an element* $a_\alpha\in A$. *Then the function*

$$v\mapsto\prod_{\alpha\in\Sigma_r(\mathfrak{o})}(\chi_{\omega, v}(a_\alpha^{s_\alpha})-\chi_{\omega, v}(a_\alpha))\,c_{P_2\mid P_1}(s:\omega:v)$$

is holomorphic on $\mathfrak{a}^*_c$.

We recall that $\chi_{\omega,\,v}$ is the central exponent of ω_v.

13. The product formula for $\mu(\omega:v)$.　Let dh be a Haar measure on a closed and unimodular subgroup H of G. We shall say that dh is normalized if

$$\int_{H\cap K} dh = 1 .$$

Fix $P\in\mathscr{P}(A)$ and choose a complete set of representatives $\{p_i\}_{i\in I}$ for $K\cap P\backslash P$ in P. Since $G=KP$, every element $x\in G$ can be written uniquely in the form $x=kp_i$ where $k\in K$ and $i\in I$. Let $p_i=mn$ $(m\in M,\,n\in N)$. We define $\kappa(x)=k$, $\mu(x)=m$. Then κ and μ are continuous mappings of G into K and M respectively.

Recall that we have a positive-definite scalar product in $\mathfrak{a}$ and hence also in $\mathfrak{a}^*$ (see §12). Let $^{+}\mathfrak{a}^*$ denote the set of all $v\in\mathfrak{a}^*$ such that $\langle v,\alpha\rangle>0$ for every root α of (P,A). If $v\in\mathfrak{a}_c^*$, we write $v=v_R+(-1)^{1/2}v_I$ with v_R and v_I in $\mathfrak{a}^*$. Let $\mathfrak{F}_c(P)$ be the set of all $v\in\mathfrak{a}_c^*$ such that $v_I\in{}^{+}\mathfrak{a}^*$.

THEOREM 23.　*Fix $\omega\in{}^{\circ}\mathscr{E}(M)$, $v\in\mathfrak{F}_c(P)$, $\psi\in\mathscr{A}(\omega,\tau_M)$ and put*

$$J(m)=\int_{\bar N} \tau(\kappa(\bar n))\,\psi(\mu(\bar n)\,m)\,q^{\langle(-1)^{1/2}v-\varrho,\,H(\bar n)\rangle}\,d\bar n \qquad (m\in M),$$

where $\varrho=\varrho_P$, $H=H_P$ and $d\bar n$ is the normalized Haar measure on $\bar N$. Then the above integral is convergent and

$$J=\gamma c_{\bar P\,|\,P}(1:\omega:v)\,\psi \qquad if\ \psi\in L(\omega,P),$$
$$=\gamma(c_{P\,|\,\bar P}(1:\omega:\bar v))^*\psi \qquad if\ \psi\in\mathscr{L}(\omega,P).$$

Here

$$\gamma=\int_{\bar N} q^{-2\langle\varrho,\,H(\bar n)\rangle}\,d\bar n.$$

It turns out that γ is actually independent of $P\in\mathscr{P}(A)$. Hence we may denote it by $\gamma(G/M)$.

Let Φ be the set of all reduced roots of (P,A). For $\alpha\in\Phi$, let A_α be the maximal torus lying in the kernel of ξ_α and M_α the centralizer of A_α in G. Then

$$(^{*}P_\alpha,\,A)=(M_\alpha\cap P,\,A)$$

is a maximal p-pair in M_α and $^{*}P_\alpha=MN_\alpha$ where $N_\alpha=M_\alpha\cap N$. Let $\mu_\alpha(\omega:v)$ have the same meaning for (M_α,A) as $\mu(\omega:v)$ has for (G,A).

THEOREM 24.   *Put $\gamma = \gamma(G/M)$ and $\gamma_\alpha = \gamma(M_\alpha/M)$ $(\alpha \in \Phi)$. Then*

$$\gamma^{-2}\mu(\omega : v) = \prod_{\alpha \in \Phi} \gamma_\alpha^{-2}\mu_\alpha(\omega : v)$$

for all $\omega \in {}^\circ\mathscr{E}(M)$ and $v \in \mathfrak{a}_c^*$.

14. The Schwartz space $\mathscr{C}(G)$.   Fix $P_0 \in \mathscr{P}(A_0)$, extend δ_{P_0} on G as in §11 and put

$$\Xi_G(x) = \Xi(x) = \int_K \delta_{P_0}(xk)^{-1/2}\, dk.$$

Then Ξ is independent of the choice of P_0, it lies in $\mathscr{A}(G)$ and it is constant on double cosets of KZ.

We recall that $G \subset GL(n, \Omega)$. Put $|x| = \max_{i,j}|x_{ij}|_{\mathfrak{p}}$ for any $n \times n$ matrix with coefficients in Ω. Then

$$\|x\| = \max(|x|, |x^{-1}|) \geq 1 \qquad (x \in G).$$

Put $\sigma(x) = \log_q \|x\|$ and $\sigma_*(x) = \inf_{z \in Z} \sigma(xz)$ $(x \in G)$.

THEOREM 25.   *Let A_0^+ be the set of all $a \in A_0$ such that $|\xi_\alpha(a)|_{\mathfrak{p}} \geq 1$ for every root α of (P_0, A_0). Then we can choose an integer $d \geq 0$ and numbers $c_2 \geq c_1 > 0$ such that*

$$c_1 \leq \delta_{P_0}(a)^{1/2}\, \Xi(a) \leq c_2(1 + \sigma_*(a))^d$$

for all $a \in A_0^+$.

COROLLARY 1.   *We can choose $r \geq 0$ such that*

$$\int_G \Xi(x)^2(1 + \sigma(x))^{-r}\, dx < \infty, \qquad \int_{G/Z} \Xi(x)^2(1 + \sigma_*(x))^{-r}\, dx^* < \infty.$$

Let (P, A) be a semistandard p-pair $(P = MN)$. Then A_0 is a maximal split torus contained in M. Hence we obtain the function Ξ_M on M by replacing (G, A_0, K) by (M, A_0, K_M) in the above definition.

COROLLARY 2.   *Fix $r \geq 0$ and $\varepsilon > 0$. Then we can choose $c = c(r, \varepsilon) > 0$ such that*

$$\delta_P(m)^{1/2} \int_{\bar{N}} \Xi(\bar{n}m)\,(1 + \sigma(\bar{n}m))^{-(2d + r + \varepsilon)}\, d\bar{n} \leq c\,\Xi_M(m)\,(1 + \sigma(m))^{-r} \qquad (m \in M).$$

For any open compact subgroup K_0 of G, let $\mathscr{C}_{K_0}(G)$ denote the space of all functions $f \in C(G/\!/K_0)$ such that

$$v_r(f) = \sup_G |f(x)| \, \Xi(x)^{-1}(1 + \sigma(x))^r < \infty$$

for every $r \geqq 0$. We topologize $\mathscr{C}_{K_0}(G)$ by means of the seminorms $\{v_r\}_{r \geqq 0}$. Put

$$\mathscr{C}(G) = \bigcup_{K_0} \mathscr{C}_{K_0}(G)$$

where K_0 runs over all open compact subgroups of G. Let $\mathfrak{S}$ be the collection of all seminorms v on $\mathscr{C}(G)$ with the following property. For any K_0, the restriction of v on $\mathscr{C}_{K_0}(G)$ is continuous. We topologize $\mathscr{C}(G)$ by means of $\mathfrak{S}$. Then $\mathscr{C}(G)$ becomes a complete locally convex Hausdorff space and it is easy to verify that $C_c^\infty(G)$ is a dense subspace of $\mathscr{C}(G)$.

LEMMA 12.   *For $f \in \mathscr{C}(G)$, put*

$$f^{(P)}(m) = \delta_P(m)^{1/2} \int_N f(mn)\, dn \qquad (m \in M).$$

Then this integral converges and $f^{(P)} \in \mathscr{C}(M)$. Moreover $f \mapsto f^{(P)}$ is a continuous mapping of $\mathscr{C}(G)$ into $\mathscr{C}(M)$.

Let $l_0 = \dim Z$ and $\chi_1, \ldots, \chi_{l_0}$ be a set of generators of $X(G)$ (see §7). Define $v(x)$ by

$$q^{v(x)} = \max_{1 \leqq i \leqq l_0} \left(|\chi_i(x)|, |\chi_i(x)|^{-1} \right) \qquad (x \in G).$$

For any $T \geqq 0$, let G_T denote the set of all $x \in G$ such that $v(x) \leqq T$.

THEOREM 26.   *Fix $f \in \mathscr{A}(G)$. Then the following two conditions on f are equivalent.*

(1) For any $T \geqq 0$,

$$\int_{G_T} |f(x)|^2 \, dx < \infty.$$

(2) Given $T \geqq 0$ and $r \geqq 0$, we can choose $c \geqq 0$ such that

$$|f(x)| \leqq c \Xi(x)(1 + \sigma(x))^{-r}$$

for all $x \in G_T$.

A distribution on G is said to be tempered if it extends (uniquely) to a continuous linear function on $\mathscr{C}(G)$.

COROLLARY.   *Let Θ be the character of an element in $\mathscr{E}_2(G)$. Then Θ is tempered.*

Let $\mathscr{E}'(G)$ denote the set of all elements $\omega \in \mathscr{E}(G)$ with the following property. Let (P, A) be any p-pair in G. Then (see §6) $\mathfrak{X}_\omega(P, A) \cap \hat{A} = \emptyset$ unless $P = G$. The following result is a simple consequence of Theorem 26.

THEOREM 27. $\mathscr{E}_2(G) \subset \mathscr{E}'(G)$.

It is now possible to generalize the results of §8 and replace $^\circ\mathscr{E}(M)$ by $\mathscr{E}'(M)$ in Theorem 9. In particular we have the following theorem.

THEOREM 28.   *Fix $\omega \in \mathscr{E}'(M)$ and assume that ω is unramified in G. Then if $\sigma \in \omega$ and $P \in \mathscr{P}(A)$, $\pi = \mathrm{Ind}_P^G \delta_P^{1/2}\sigma$ is a unitary, admissible and irreducible representation of G.*

Fix $C \in \mathscr{E}_2(G)$ and $f \in \mathscr{A}(C)$. Then if (P, A) is a p-pair in G $(P = MN)$, it follows from Theorem 26 that the integral $\int_N f(xn)\,dn$ $(x \in G)$ converges absolutely.

THEOREM 29.   *Suppose $P \neq G$ and $C \in \mathscr{E}_2(G)$. Then*

$$\int_N f(xn)\,dn = 0 \qquad (x \in G)$$

for $f \in \mathscr{A}(C)$.

15. The Steinberg character.   We fix $P_0 \in \mathscr{P}(A_0)$ and call a p-pair (P, A) standard if $(P, A) \succ (P_0, A_0)$. Let $\mathfrak{S}$ be the set of all standard p-pairs. Then $\mathfrak{S}$ is a finite set. For $(P, A) \in \mathfrak{S}$, put

$$\pi_P = \mathrm{Ind}_P^G 1_P$$

where 1_P denotes the trivial representation of P. It follows from Theorem 1 that its character θ_P is a locally summable function on G.

Let l be the (absolute) rank of G and t an indeterminate. We denote by $D_G(x)$ the coefficient of t^l in $\det(t - \mathrm{Ad}(x) + 1)$ $(x \in G)$. As usual let G' be the set of all points $x \in G$ where $D_G(x) \neq 0$. It is easy to see that θ_P is locally constant on G'.

Put

$$\Theta = \sum_{(P, A) \in \mathfrak{S}} (-1)^{\dim A}\theta_P.$$

Then $(-1)^{\dim A_0}\Theta$ is called the Steinberg character of G. Borel and Serre have proved that it is, in fact, the character of an element in $\mathscr{E}_2(G)$.

Let Γ be a Cartan subgroup of G and A_Γ the maximal split torus contained in Γ. Let M_Γ be the centralizer of A_Γ in G and $\mathfrak{g}$ and $\mathfrak{m}_\Gamma$ the Lie algebras (over Ω) of G and M_Γ respectively. Let

$$D_{G/M_\Gamma}(m) = \det(1 - \mathrm{Ad}(m))_{\mathfrak{g}/\mathfrak{m}_\Gamma} \qquad (m \in M_\Gamma).$$

THEOREM 30. *Put $\Gamma' = G' \cap \Gamma$ and $\Phi(\gamma) = |D_{G/M_\Gamma}(\gamma)|^{1/2} \Theta(\gamma)$ $(\gamma \in \Gamma')$. Let $\mathfrak{Q}(\Gamma)$ be the set of all p-pairs (P, A) in G with $A \subset A_\Gamma$. Then $\mathfrak{Q}(\Gamma)$ is a finite set,*

$$\Phi(\gamma) = \sum_{(P, A) \in \mathfrak{Q}(\Gamma)} (-1)^{\dim A} \delta_P(\gamma)^{1/2} |D_{M/M_\Gamma}(\gamma)|_{\mathfrak{p}}^{1/2}$$

and [6]

$$|\Phi(\gamma)| \leq [\mathfrak{Q}(\Gamma)] \qquad (\gamma \in \Gamma').$$

A Cartan subgroup Γ of G is called elliptic if Γ/Z is compact. Moreover a point $x \in G$ is called elliptic if x lies in some elliptic Cartan subgroup. Let G_e denote the set of all elliptic points in G'. Then G_e is open in G. Let φ_e denote the characteristic function of G_e.

It is easy to see that $\Theta = 1$ on G_e and $\int_K \Theta(xk)\, dk = 0$. These two facts, together with Theorem 30, give us the following result (cf. [3(a), Theorem 5]).

THEOREM 31. *There exists a number $c > 0$ such that $\int_K \varphi_e(xk)\, dk \leq c \Xi(x)$ for all $x \in G$.*

COROLLARY 1. *Let $f \in \mathscr{C}(G)$. Then $\int_{G_e} |f(x)|\, dx < \infty$.*

COROLLARY 2. *Let Θ_π be the character of a square-integrable representation π of G. Then Θ_π coincides with a locally summable function on G_e.*

16. A theorem of Howe. Let K_0 be an open compact subgroup of G. For $\mathfrak{d} \in \mathscr{E}(K_0)$, let $\xi_\mathfrak{d}$ denote the character of $\mathfrak{d}$. We extend it to a function on G by defining it to be zero outside K_0.

Let K_1, K_2 be two open compact subgroups of G. Fix $\mathfrak{d}_i \in \mathscr{E}(K_i)$ $(i = 1, 2)$ and a subset ω of G. We say that ω intertwines $\mathfrak{d}_1$ with $\mathfrak{d}_2$ if there exists a function $f \in C(G)$ such that

$$\omega \cap \mathrm{Supp}(\xi_{\mathfrak{d}_1} * f * \xi_{\mathfrak{d}_2}) \neq \emptyset.$$

The following theorem of Howe ([4(a)], [4(b)]) plays an important role in harmonic analysis on G. Its proof has so far been worked out only in the case of characteristic zero. So we assume in this section that $\mathrm{char}\,\Omega = 0$.

THEOREM 32 (HOWE). *Let Γ be a Cartan subgroup of G and ω a compact subset of $\Gamma' = \Gamma \cap G'$. Then there exists a compact open subgroup K_1 of G with the following property. Fix an open compact subgroup K_2 of G and an element $\mathfrak{d}_2 \in \mathscr{E}(K_2)$ and let F denote the set of all $\mathfrak{d}_1 \in \mathscr{E}(K_1)$ such that*

(1) *G intertwines $\mathfrak{d}_1$ with $\mathfrak{d}_2$,*
(2) *ω intertwines $\mathfrak{d}_1$ with itself.*
Then F is a finite set.

COROLLARY. *Let π be an admissible representation of G on V and suppose that V is a finite G-module under π. Then the character of π coincides with a locally constant function on G'.*

Combining this with the results of §15 we get the following lemma.

LEMMA 13. *Let Γ be a Cartan subgroup of G and dx^* the invariant measure on G/Γ. Then for any $f\in\mathscr{C}(G)$ and $\gamma\in\Gamma'=\Gamma\cap G'$, the integral $\int_{G/\Gamma}|f(x\gamma x^{-1})|\,dx^*$ converges. Put*

$$F_f(\gamma)=|D_G(\gamma)|^{1/2}\int_{G/\Gamma} f(x\gamma x^{-1})\,dx^* \qquad (\gamma\in\Gamma')$$

and let Γ'' be the set of all points $\gamma\in\Gamma$ where $D_{M_\Gamma}(\gamma)\neq 0$. Then F_f extends to a locally constant function on Γ''.

One would actually like to prove that

$$\sup_{\gamma\in\Gamma'}|F_f(\gamma)|<\infty.$$

This has been done so far only for $f\in C_c^\infty(G)$ (see [**3**(g), p. 82]). Of course the crucial case is when Γ is elliptic.

17. Wave packets and the Plancherel measure. We return to the notation of §12. Fix $\omega\in{}^\circ\mathscr{E}(M)$, $P\in\mathscr{P}(A)$ and $\psi\in L(\omega, P)$. Then $\mu(\omega:v)$ and $E(P:\psi:v)$ depend only on $v \bmod L^*$ (see §7). We consider the space $C_c^\infty(\mathfrak{a}^*)$ with its usual topology.

LEMMA 14. *Fix $\omega\in{}^\circ\mathscr{E}(M)$, $P\in\mathscr{P}(A)$, $\alpha\in C_c^\infty(\mathfrak{a}^*)\otimes L(\omega, P)$ and put*

$$\varphi_\alpha=\int_{\mathfrak{a}^*} \mu(\omega:v)\,E(P:\alpha(v):v)\,dv$$

where dv is the Euclidean measure on $\mathfrak{a}^*$. Then $\varphi_\alpha\in\mathscr{C}(G,\tau)$ and $\alpha\mapsto\varphi_\alpha$ is a continuous mapping of $C_c^\infty(\mathfrak{a}^*)\otimes L(\omega, P)$ into $\mathscr{C}(G,\tau)$.

Here $\mathscr{C}(G,\tau)=(\mathscr{C}(G)\otimes V)\cap C(G,\tau)$.

LEMMA 15. *Fix $P_1, P_2\in\mathscr{P}(A)$ and put*

$$\varphi_\alpha = \int_{\mathfrak{a}^*} \mu(\omega:v)\, E(P_1:\alpha(v):v)\, dv$$

for $\alpha \in C_c^\infty(\mathfrak{a}^*) \otimes L(\omega, P_1)$. Then

$$\varphi_\alpha^{(P_2)} = \gamma \sum_{s \in \mathfrak{w}} \int_{\mathfrak{a}^*} {}^\circ c_{P_2 \mid P_1}(s:\omega:s^{-1}v)\, \alpha(s^{-1}v)\cdot \chi_v\, dv$$

where $\chi_v(m) = q^{(-1)^{1/2}\langle v, H(m)\rangle}$ $(m \in M)$ and $\gamma = \gamma(G/M)$.

Let $\mathfrak{o}$ be a connected component of ${}^\circ\mathscr{E}(M)$. Fix $P \in \mathscr{P}(A)$ and let $L(\mathfrak{o}, P)$ denote the set of all functions ψ which assign to every $\omega \in \mathfrak{o}$ an element $\psi(\omega) \in L(\omega, P)$ in such a way that

$$\psi(\omega_v) = \psi(\omega)\cdot \chi_v \qquad (v \in \mathfrak{a}^*).$$

Then $L(\mathfrak{o}, P)$ is a complex vector space and in fact the mapping $\psi \mapsto \psi(\omega)$ defines a bijection of $L(\mathfrak{o}, P)$ onto $L(\omega, P)$ for every $\omega \in \mathfrak{o}$.

Put ${}^\circ c_{P_2 \mid P_1}(s:\omega) = {}^\circ c_{P_2 \mid P_1}(s:\omega:0)$ in the notation of §12 (see also §11). Put $\mu(\omega) = \mu(\omega:0)$ and let $d\omega$ denote the unique measure on $\mathfrak{o}$ which is invariant under the action of $\mathfrak{a}^*$ and such that the total measure of $\mathfrak{o}$ is 1. **Then Lemma 15 may be restated in the following form.**

THEOREM 33.   *Fix $P_1, P_2 \in \mathscr{P}(A)$ and put*

$$\varphi_\alpha = \int \mu(\omega)\, E(P_1:\alpha:\omega)\, d\omega$$

for $\alpha \in C^\infty(\mathfrak{o}) \otimes L(\mathfrak{o}, P_1)$. Then $\varphi_\alpha \in \mathscr{C}(G, \tau)$ and

$$\varphi_\alpha^{(P_2)} = \gamma \sum_{s \in \mathfrak{w}} \int_{\mathfrak{o}} {}^\circ c_{P_2 \mid P_1}(s:\omega)\, \alpha(\omega:\omega)\, d\omega.$$

Here $E(P_1:\alpha:\omega) = E(P_1:\alpha(\omega:\omega))$.

For any $\omega \in {}^\circ\mathscr{E}(M)$, let Θ_ω denote the character of the class $C(\omega) = C_M^G(\omega)$ (§9). Then Θ_ω is a tempered distribution.

Let T be a distribution on G. Then we put[2]

$$(T, f) = (T, f)_G = \operatorname{conj} T(g) \qquad (f \in C_c^\infty(G)),$$

where $g = \operatorname{conj} f$. Moreover if U is a finite-dimensional complex vector space, we define $(T, f \otimes u) = (T, f)\, u$ for $f \in C_c^\infty(G)$ and $u \in U$. So if T is tempered, we obtain in this way a continuous mapping $f \mapsto (T, f)$ of $\mathscr{C}(G) \otimes U$ into U.

Let F denote the projection in V given by

$$Fv = \int_K \tau(k)\, v\tau(k^{-1})\, dk \qquad (v \in V).$$

Then if $\omega_0 \in {}^\circ\mathscr{E}(M)$, we have

$$
\begin{aligned}
(\Theta_{\omega_0}, \varphi_\alpha) &= F(\theta_{\omega_0}, \varphi_\alpha^{(P_2)})_M \\
&= \gamma' \sum_{s \in \mathfrak{w}(\mathfrak{o}\,|\,\omega_0)} F(\theta_{\omega_0}, {}^\circ c_{P_2|P_1}(s^{-1}:s\omega_0)\,\alpha(s\omega_0:s\omega_0))_{M/A}
\end{aligned}
$$

in the notation of Theorem 33 where $\gamma' = \gamma[M:{}^\circ MA]^{-1}$ (see §7). Here θ_{ω_0} is the character of ω_0 and $\mathfrak{w}(\mathfrak{o}\,|\,\omega_0)$ is the set of all $s \in \mathfrak{w}$ such that $s\omega_0 = \omega_0^s \in \mathfrak{o}$. Now observe that

$$
\begin{aligned}
(\theta_{\omega_0}, \psi_0)_{M/A} &= d(\omega_0)^{-1}\,\psi_0(1), \\
E(P:\psi_0:1) &= F\psi_0(1) \qquad (P \in \mathscr{P}(A),\ \psi_0 \in L(\omega_0, P))
\end{aligned}
$$

and

$$E(P_2:{}^\circ c_{P_2|P_1}(s^{-1}:s\omega_0)\,\psi) = E(P_1:\psi) \qquad (\psi \in L(s\omega_0, P_1)).$$

From this it follows that

$$d(\omega_0)\,(\Theta_{\omega_0}, \varphi_\alpha) = \gamma' \sum_{s \in \mathfrak{w}(\mathfrak{o}\,|\,\omega_0)} F\alpha(s\omega_0:s\omega_0:1).$$

On the other hand

$$\varphi_\alpha(1) = \int \mu(\omega)\, F\alpha(\omega:\omega:1)\, d\omega.$$

Therefore we obtain the following result (cf. [3(e)], §§12, 13]).

THEOREM 34. — *Put $f = \varphi_\alpha$ and $\hat{f}(\omega) = (\theta_\omega, f)$ $(\omega \in {}^\circ\mathscr{E}(M))$. Then*

$$f(1) = [\mathfrak{w}]^{-1}\,\gamma(G/M)^{-1}\,[M:{}^\circ MA] \int_{{}^\circ\mathscr{E}(M)} d(\omega)\,\mu(\omega)\,\hat{f}(\omega)\, d\omega.$$

Here $d\omega$ is the measure on ${}^\circ\mathscr{E}(M)$ which coincides on each connected component of ${}^\circ\mathscr{E}(M)$ with the measure defined above.

REFERENCES

1. A. Borel and J. Tits, *Groupes réductifs*, Inst. Hautes Etudes Sci. Publ. Math. No. 27 (1965), 55–150. MR **34** #7527.

2. F. Bruhat, (a) *Sur les représentations induites des groupes de Lie,* Bull. Soc. Math. France **84** (1956), 97–205. MR **18**, 907.

(b) *Distributions sur un groupe localement compact et applications à l'étude des représentations des groupes p-adiques,* Bull. Soc. Math. France **89** (1961), 43–75. MR **25** #4354.

3. Harish-Chandra, (a) *Two theorems on semisimple Lie groups,* Ann. of Math. (2) **83** (1966), 74–128. MR **33** #2766.

(b) *Discrete series for semisimple Lie groups.* II. *Explicit determination of the characters,* Acta Math. **116** (1966), 1–111. MR **36**# 2745.

(c) *Automorphic forms on semisimple Lie groups,* Lecture Notes in Math., vol. 62, Springer-Verlag, Berlin and New York, 1968. MR **38**# 1216.

(d) *Eisenstein series over finite fields,* Functional Analysis and Related Fields, Springer-Verlag, Berlin and New York, 1970, pp. 76–88.

(e) *Harmonic analysis on semisimple Lie groups,* Bull. Amer. Math. Soc. **78** (1970), 529–551. MR **41** #1933.

(f) *On the theory of the Eisenstein integral,* Proc. Intern. Conf. on Harmonic Analysis, University of Maryland, College Park, Md., 1971; Lecture Notes in Math., vol. 266, Springer-Verlag, Berlin and New York, 1971.

(g) *Harmonic analysis on reductive p-adic groups,* Lecture Notes in Math., vol. 162, Springer-Verlag, Berlin and New York, 1970.

4. R. E. Howe, (a) *Kirillov theory for compact p-adic groups* (preprint).

(b) *Some qualitative results on the representation theory of GL (n) over a p-adic field,* 1972. (preprint).

5. H. Jacquet, *Représentations des groupes linéaires p-adiques,* Theory of Group Representations and Fourier Analysis (C.I.M.E., II Ciclo, Montecatini Terme, 1970), Edizioni Cremonese, Rome, 1971, pp. 119–220. MR **45**# 453.

6. H. Jacquet and R. P. Langlands, *Automorphic forms on GL* (2), Lecture Notes in Math., vol. 114, Springer-Verlag, Berlin and New York, 1970.

Institute for Advanced Study

Reprinted from
Proceedings of Symposia in Pure Mathematics, Vol. **XXVI**
Amer. Math. Soc., Providence, R.I., 1973, pp. 167–192

Harmonic Analysis on Real Reductive Groups I
The Theory of the Constant Term

Harish-Chandra

Institute for Advanced Study, Princeton, New Jersey 08540

The author is a member of the Editorial Board

Received September 23, 1974

This paper studies the asymptotic behavior of tempered and K-finite eigenfunctions of $\mathfrak{z}$ on a real reductive group.

1. Introduction

The object of this paper is to begin laying the groundwork for the proof of results announced in [17, 18, 19]. In order to allow induction, one has to start with a reductive group G which is not necessarily connected and although the extension of the earlier theory [12, 14, 15] to this case is not difficult, it requires some care. Moreover, I have taken this opportunity to rephrase some of the key results of [15] in terms of parabolic subgroups and the theory of the constant term (e.g., Theor. 21.1 and Lemma 25.1). This makes the analogy with the theory of automorphic forms [16] much more transparent (e.g., Lemma 21.1 and Theor. 28.1) and, in particular, the Eisenstein integral (Sect. 19) appears as a natural counterpart of the Eisenstein series. This analogy will be further developed in later papers when we come to the study of the constant term of the Eisenstein integral and the c-function (see [19]).

In order to obtain explicit formulas for the Plancherel measure and the c-functions, it is important to fix the normalization of the various invariant measures appearing in the discussion. This is done in a coherent way in Sections 7, 8 and the formulation of Theorem 9.1, which will play an important role later, depends on this fact. Another outcome is the evaluation of the constant c (Theor. 37.1) which appears in the Plancherel formula (see Cor. 1 of Lemma 27.5).

Almost all the work presented here was done several years ago and in fact I gave lectures on it in Princeton during the fall of 1968.

104

However various unforeseen circumstances have delayed its publication for so long.

Our method for cross reference is as follows. Lemma 23.3 means Lemma 3 of Section 23. But we omit the number of the section when referring to lemmas or theorems of the current section.

2. Definition of a Split Component

Let $\mathbf{R}^\times$ denote, as usual, the multiplicative group of $\mathbf{R}$. For any Lie group G, we denote by $X(G)$, the group of all (continuous) homomorphisms of G into $\mathbf{R}^\times$ and by G^0 the connected component of 1 in G. By a vector subgroup of G, we mean a closed subgroup A such that A is topologically isomorphic to the additive group of $\mathbf{R}^n$ for some $n \geqslant 0$. For $\chi \in X(G)$, let $|\chi|$ denote the homomorphism $x \to |\chi(x)|$ $(x \in G)$. Put

$$^0G = \bigcap_{\chi \in X(G)} \ker |\chi|.$$

Then ^0G is a closed normal subgroup of G and $G/^0G$ is an abelian Lie group. We call $\dim G/^0G$ the parabolic rank of G and denote it by $\mathrm{prk}\, G$. Notice that ^0G is unimodular. Moreover if G is abelian and $[G : G^0] < \infty$, then ^0G is the maximal compact subgroup of G. Hence if A is a maximal vector subgroup of G, then $G/^0G \simeq A$ and therefore $\mathrm{prk}\, G = \dim A$ in this case.

By a split component of G, we mean a vector subgroup A of G such that $G = {}^0GA$ and $^0G \cap A = \{1\}$. Note that if such a subgroup A exists, then $A \simeq G/^0G$ and therefore $\dim A = \mathrm{prk}\, G$.

Notation. If V is a (real or complex) vector space of finite dimension, we denote its dual by V^*. Moreover if V is real, we denote its complexification by V_c. The Lie algebra of a Lie group will be denoted by the corresponding l.c. german letter unless explicitly stated otherwise.

3. The Assumptions on G

Let $\mathfrak{g}$ be the Lie algebra of G and G_c the connected complex adjoint group of $\mathfrak{g}_c$. We now make the following assumptions on G.

(1) $\mathfrak{g}$ is reductive and $\mathrm{Ad}(G) \subset G_c$.

(2)   Let G_1 be the analytic subgroup of G corresponding to $\mathfrak{g}_1 = [\mathfrak{g}, \mathfrak{g}]$. Then the center of G_1 is finite.

(3) $[G : G^0] < \infty$.

As was pointed out to me by A. Borel, condition (3) is equivalent to the following (see [22, Sect. 3]).

(4)   There exists a compact subgroup K of G such that $K \cap G^0$ is a maximal compact subgroup of G^0 and K meets every connected component of G.

Fix K as in (4) and let $\mathfrak{k}$ denote the Lie algebra of K. If C is the center of G^0, then it follows from (2) that $C = C_1 C_2$ where $C_1 = C \cap K$ and C_2 is a maximal vector subgroup of C. Let θ denote the Cartan involution of $\mathfrak{g}_1 = [\mathfrak{g}, \mathfrak{g}]$ with respect to $\mathfrak{k}_1 = \mathfrak{k} \cap \mathfrak{g}_1$. Now $\mathfrak{g} = \mathfrak{g}_1 + \mathfrak{c}_1 + \mathfrak{c}_2$ where $\mathfrak{c}_i$ is the Lie algebra of C_i ($i = 1, 2$). Extend θ to an involution of $\mathfrak{g}$ by setting

$$\theta(X_1 + X_2) = X_1 - X_2 \qquad (X_i \in \mathfrak{c}_i , i = 1, 2).$$

Let $\mathfrak{p}$ be the subspace of all $X \in \mathfrak{g}$ such that $\theta(X) = -X$.

LEMMA 1.   *The mapping $(k, X) \to k \exp X$ ($k \in K, X \in \mathfrak{p}$) defines an analytic diffeomorphism of $K \times \mathfrak{p}$ onto G. If K_1 is any compact subgroup of G such that $K_1 \supset K \cap G^0$, then $K_1 \subset K$. Finally θ extends to an automorphism of G such that*

$$\theta(k \exp X) = k \exp(-X) \qquad (k \in K, X \in \mathfrak{p}).$$

This is well known (see [22, Sect. 3]).

LEMMA 2.   *There exists a real symmetric bilinear form B on $\mathfrak{g}$ such that*:

(1) $B([X, Y], Z) + B(Y, [X, Z]) = 0 \ (X, Y, Z \in \mathfrak{g})$,

(2)   *The quadratic form*

$$\| X \|^2 = -B(X, \theta(X)) \qquad (X \in \mathfrak{g})$$

is positive-definite on $\mathfrak{g}$.

(3)   *B is invariant under θ i.e.,*

$$B(\theta X, \theta Y) = B(X, Y) \qquad (X, Y \in \mathfrak{g}).$$

We have seen that $\mathfrak{g} = \mathfrak{g}_1 + \mathfrak{c}_1 + \mathfrak{c}_2$ where the sum is direct. Let Q_i be a positive-definite quadratic form on $\mathfrak{c}_i$ $(i = 1, 2)$. Put

$$Q(Y + X_1 + X_2) = \operatorname{tr}(\operatorname{ad} Y)^2 - Q_1(X_1) + Q_2(X_2),$$

for $Y \in \mathfrak{g}_1$, $X_i \in \mathfrak{c}_i$ $(i = 1, 2)$. Let B denote the bilinear form on $\mathfrak{g}$ corresponding to Q. Then B satisfies all the conditions of the lemma. Extend B to a bilinear form on $\mathfrak{g}_c$. Then it is clear that B is invariant under G_c. Therefore since $\operatorname{Ad}(G) \subset G_c$, we conclude that B is also invariant under G.

We fix K, θ, B as above, once for all.

Let Z_G denote the center of G and $\mathfrak{z}$ its Lie algebra. Put $\mathfrak{c} = \mathfrak{z} \cap \mathfrak{p}$ and $C = \exp \mathfrak{c}$.

LEMMA 3. $^0G = KG_1$ and C is a split component of G. Moreover $C = Z_G \cap \exp \mathfrak{p}$.

It is clear that $^0G \supset KG_1$. Also G_1 being a closed normal subgroup of G, the same holds for $G_2 = KG_1$. It follows from Lemma 1 that C is a vector subgroup of G, $G_2 \cap C = \{1\}$ and $G = G_2 C$. Hence $G/G_2 \simeq C$. Since C is a vector group, it is clear that $^0C = \{1\}$. This shows that $^0G = G_2$. Fix $X \in \mathfrak{p}$ such that $\exp X \in Z_G$. Then it follows from [12, Cor. 2, p. 480] that $X \in \mathfrak{c}$. This proves that $C = Z_G \cap \exp \mathfrak{p}$.

Remark. We note that C is the unique maximal vector subgroup of Z_G which is θ-stable. We shall call C *the* split component of G.

It is obvious that ^0G is stable under θ. Let $^0\mathfrak{g}$ denote its Lie algebra and let ^0B and $^0\theta$ denote the restrictions of B and θ, respectively, on $^0\mathfrak{g}$.

COROLLARY. *^0G satisfies the same conditions as those imposed on G. Moreover if we replace (G, K, θ, B) by $(^0G, K, {}^0\theta, {}^0B)$, all the above conditions are again fulfilled.*

This is obvious since $^0\mathfrak{g}$ is the orthogonal complement of $\mathfrak{c}$ in $\mathfrak{g}$ with respect to the euclidean norm defined above.

By a Cartan subgroup of G, we mean the centralizer in G of a Cartan subalgebra of $\mathfrak{g}$.

LEMMA 4. *Suppose* $\operatorname{rank} \mathfrak{g} = \operatorname{rank} \mathfrak{k}$ *and let* $\mathfrak{b}$ *be a Cartan subalgebra of* $\mathfrak{k}$ *and* B *the corresponding Cartan subgroup of* G. *Let* Z *denote the centralizer of* $\mathfrak{g}$ *in* G. *Then* $B = ZB^0 \subset K$.

We may obviously assume for the proof that $\mathfrak{g}$ is semisimple. Let ρ denote the adjoint representation of G. Then $\rho(G) \subset G_c$. Let B_c be the centralizer of $\mathfrak{b}$ in G_c. Then $B_c = \exp_c \mathfrak{b}_c$ where $\exp_c$ denotes the exponential mapping of $\mathfrak{g}_c$ into G_c. It is clear that $\exp_c \mathfrak{b} = \rho(B^0)$ is the maximal compact subgroup of B_c. This means that $\rho(B) = \rho(B^0)$ and therefore $B = ZB^0$. Moreover $B \subset K$ from [12, Cor. 2, p. 480].

4. Decomposition of Parabolic Subgroups

A subalgebra $\mathfrak{q}$ of $\mathfrak{g}$ is called parabolic if $\mathfrak{q}_c$ contains a Borel subalgebra (i.e., a maximal solvable subalgebra) of $\mathfrak{g}_c$. Suppose $\mathfrak{q}$ is parabolic. Then the normalizer of $\mathfrak{q}$ in $\mathfrak{g}$ coincides with $\mathfrak{q}$. A subgroup Q of G is called parabolic if Q is the normalizer (in G) of some parabolic subalgebra $\mathfrak{q}$ of $\mathfrak{g}$. If this is so, it is clear that Q is closed and $\mathfrak{q}$ is its Lie algebra.

LEMMA 1. *Let Q be a parabolic subgroup (psgp) of G. Then $G = KQ$.*

This is well known when G is connected and follows in the general case from the fact that $G = KG^0$.

An element $H \in \mathfrak{g}$ is said to split (in $\mathfrak{g}$) if $\operatorname{ad} H$ is diagonalizable over $\mathbf{R}$. An abelian subalgebra $\mathfrak{a}$ of $\mathfrak{g}$ is said to split (in $\mathfrak{g}$) if every element of $\mathfrak{a}$ splits.

LEMMA 2. *Let $\mathfrak{a}$ be a split abelian subalgebra containing the center of $\mathfrak{g}$. Then we can choose $x \in G^0$ such that[1] $\mathfrak{a}^x$ is θ-stable.*

This is well known.

Let $\mathfrak{a}$ be a split abelian subalgebra of $\mathfrak{g}$. For any $\lambda \in \mathfrak{a}^*$, let $\mathfrak{g}_\lambda$ denote the set of all $X \in \mathfrak{g}$ such that $[H, X] = \lambda(H)X$ for all $H \in \mathfrak{a}$. By a root of $(\mathfrak{g}, \mathfrak{a})$, we mean an element $\alpha \neq 0$ in $\mathfrak{a}^*$ such that $\mathfrak{g}_\alpha \neq \{0\}$. It is known that whenever α is a root, the same holds for $-\alpha$.

Let $\mathfrak{q}$ be a parabolic subalgebra of $\mathfrak{g}$. By the radical of $\mathfrak{q}$, we mean the maximal ideal $\mathfrak{n}$ of $\mathfrak{q} \cap \mathfrak{g}_1$ such that $\operatorname{ad} X$ is nilpotent for every $X \in \mathfrak{n}$. If Q is the psgp of G corresponding to $\mathfrak{q}$ and N the analytic subgroup of Q corresponding to $\mathfrak{n}$, then N is called the radical of Q.

The following two lemmas are well known.

---

[1] We write $y^x = xyx^{-1}$ and $X^x = x \cdot X = \operatorname{Ad}(x)X$ for $x, y \in G$ and $X \in \mathfrak{g}_c$.

LEMMA 3. *Let $\mathfrak{n}$ be the radical of a parabolic subalgebra $\mathfrak{q}$. Then there exists a subalgebra $\mathfrak{z}$ of $\mathfrak{q}$ such that $\mathfrak{q} = \mathfrak{z} + \mathfrak{n}$, the sum being direct.*

LEMMA 4. *$\mathfrak{z}$ is reductive in $\mathfrak{g}$ and* rank $\mathfrak{z} =$ rank $\mathfrak{g}$. *An element H of $\mathfrak{z}$ splits in $\mathfrak{g}$ iff it splits in $\mathfrak{z}$. Let $\mathfrak{a}$ be a maximal split abelian algebra lying in the center of $\mathfrak{z}$. Then $\mathfrak{a}$ is unique and $\mathfrak{z}$ is the centralizer of $\mathfrak{a}$ in $\mathfrak{g}$.*

Choose $\mathfrak{z}$ and $\mathfrak{a}$ as above. By Lemma 2, we can choose $x \in G$ such that $\mathfrak{a}^x$ is θ-stable. Since $G = KQ$, we can write $x = kq$ ($k \in K$, $q \in Q$). Then $\mathfrak{a}^q$ is θ-stable and the same holds for $\mathfrak{z}^q$. Replacing $(\mathfrak{z}, \mathfrak{a})$ by $(\mathfrak{z}^q, \mathfrak{a}^q)$, we may assume that $\mathfrak{z}$ and $\mathfrak{a}$ are both θ-stable. By considering the roots of $(\mathfrak{g}, \mathfrak{a})$, it is clear that

$$\mathfrak{g} = \theta(\mathfrak{n}) + \mathfrak{z} + \mathfrak{n},$$

where the sum is direct. Hence $\mathfrak{z} = \mathfrak{q} \cap \theta(\mathfrak{q})$ and we get the following result.

LEMMA 5. *Put $\mathfrak{m}_1 = \mathfrak{q} \cap \theta(\mathfrak{q})$. Then $\mathfrak{m}_1$ is reductive in $\mathfrak{g}$ and $\mathfrak{q} = \mathfrak{m}_1 + \mathfrak{n}$ where the sum is direct.*

Let $\mathfrak{a}_1$ be the maximal split abelian subalgebra lying in the center of $\mathfrak{m}_1$. Let M_1 denote the centralizer of $\mathfrak{a}_1$ in G. Then clearly M_1 and $\mathfrak{a}_1$ are θ-stable.

LEMMA 6. *$M_1 \cap N = \{1\}$, $Q = M_1 N$, $\theta(N) \cap Q = \{1\}$ and $M_1 = Q \cap \theta(Q)$.*

Let X be an element of $\mathfrak{n}$ such that $\exp X \in M_1$. Then X centralizes $\mathfrak{a}_1$ and therefore $X \in \mathfrak{m}_1 \cap \mathfrak{n} = \{0\}$. This proves that $M_1 \cap N = \{1\}$. Similarly if $\exp \theta(X) \in Q$ ($X \in \mathfrak{n}$), it follows that $\exp \operatorname{ad} \theta(X)$ maps $\mathfrak{m}_1$ into $\mathfrak{q} = \mathfrak{m}_1 + \mathfrak{n}$. But then it is obvious that $X = 0$.

Since $\operatorname{Ad}(G) \subset G_c$, we may, in order to verify that $Q = M_1 N$, assume that $G \subset G_c$. Let Q_c denote the normalizer of $\mathfrak{q}_c$ in G_c. Then it is known that Q_c is connected. Since $\mathfrak{q} = \mathfrak{m}_1 + \mathfrak{n}$ and $Q = G \cap Q_c$, it is obvious that $Q = M_1 N$.

Finally since $\theta(N) \cap Q = \{1\}$, it is clear that $Q \cap \theta(Q) = M_1$.

LEMMA 7. *M_1 satisfies the conditions of Section 3.*

Let M_{1c} denote the centralizer of $\mathfrak{a}_{1c}$ in G_c. Then M_{1c} is connected and since $\operatorname{Ad}(G) \subset G_c$, it is obvious that $\operatorname{Ad}(M_1) \subset M_{1c}$. Put $\mathfrak{m}_2 = [\mathfrak{m}_1, \mathfrak{m}_1]$ and let M_2 and G_1 denote the analytic subgroups of G corresponding to $\mathfrak{m}_2$ and $\mathfrak{g}_1 = [\mathfrak{g}, \mathfrak{g}]$, respectively. Also let ρ denote the adjoint representation of G_1. Then $\rho(M_2)$ is a connected semi-simple linear Lie group and so its center is finite. Therefore since $\ker \rho$ is finite, we conclude that M_2 has finite center. Finally we conclude from [6, Lemma 15, p. 679] that

$$[M_1 : M_1^0] = [M_1 : M_1 \cap G^0][M_1 \cap G^0 : M_1^0]$$

$$\leqslant [G : G^0][M_1 \cap G^0 : M_1^0] < \infty.$$

Put $K_1 = K \cap M_1$ and let B_1 and θ_1 denote the restrictions of B and θ, respectively, on $\mathfrak{m}_1$.

LEMMA 8. *If we replace (G, K, θ, B) by $(M_1, K_1, \theta_1, B_1)$ then all the conditions of Section 3 are fulfilled.*

Since M_1 is θ-stable, $\mathfrak{m}_1 = \mathfrak{m}_1 \cap \mathfrak{k} + \mathfrak{m}_1 \cap \mathfrak{p}$. We have only to verify that $M_1 = K_1 \exp(\mathfrak{m}_1 \cap \mathfrak{p})$. Fix $m \in M_1$. Then $m = k \exp X$ where $k \in K$, $X \in \mathfrak{p}$. Since $\mathfrak{a}_1$ is θ-stable, we conclude from [12, Cor. 2, p. 480] that $k \in K_1$ and $X \in \mathfrak{m}_1 \cap \mathfrak{p}$.

Let A denote the split component of M_1 (see the remark after Lemma 3.3). Then A is the maximal θ-stable vector group lying in the center of M_1 and $M_1 = MA$ where $M = {}^0M_1$ and $M \cap A = \{1\}$. Clearly M and A are closed θ-stable subgroups of G. Let $\mathfrak{m}$ and $\mathfrak{a}$ denote the corresponding Lie algebras. Then $\mathfrak{a} = \mathfrak{a}_1 \cap \mathfrak{p}$. Put $K_M = K \cap M$ and let B_M and θ_M denote the restrictions of B and θ, respectively, on $\mathfrak{m}$. It is clear that $K_M = K_1$.

LEMMA 9. *If we replace (G, K, θ, B) by (M, K_M, θ_M, B_M), then all the conditions of Section 3 are fulfilled.*

This follows immediately from Lemma 8 and the corollary of Lemma 3.3.

LEMMA 10. *The mapping $(m, a, n) \to man$ defines an analytic diffeomorphism of $M \times A \times N$ onto Q. Moreover ${}^0Q = MN$, A is a split component of Q and MA is the centralizer of A in G.*

Since $Q = M_1 N = MAN$, the first statement is an easy consequence of Lemma 6. Moreover it is clear that $\mathfrak{a}_1 \cap \mathfrak{k}$, since it splits, must lie in the center of $\mathfrak{g}$. Therefore $\mathfrak{a}$ and $\mathfrak{a}_1$ have the same centralizer in G. Also $[\mathfrak{a}, \mathfrak{n}] = [\mathfrak{a}_1, \mathfrak{n}] = \mathfrak{n}$ and therefore $N \subset {}^0Q$. On the other hand $Q/N \simeq M_1$ and so it is obvious that ${}^0Q = {}^0M_1 N = MN$ and A is a split component of Q.

We shall call A *the* split component of Q and refer to the decompositions

$$Q = MAN, \mathfrak{q} = \mathfrak{m} + \mathfrak{a} + \mathfrak{n}$$

as the Langlands decompositions of Q and $\mathfrak{q}$, respectively.

LEMMA 11. *M meets every connected component of G.*

Extend $\mathfrak{a}$ to a maximal abelian subspace $\mathfrak{a}_0$ of $\mathfrak{p}$. It is known that every maximal abelian subspace of $\mathfrak{p}$ is conjugate to $\mathfrak{a}_0$ under K^0. Fix $k \in K$. Then we can choose $k_1 \in K^0$ such that $\mathfrak{a}_0{}^k = \mathfrak{a}_0^{k_1}$. Then $k_1^{-1}k$ normalizes $\mathfrak{a}_0$. Hence (see Sect. 5) we can choose $k_2 \in K^0$ such that

$$m = k_2^{-1}k_1^{-1}k$$

centralizes $\mathfrak{a}_0$. Put $k_0 = k_1 k_2$. Then $k_0 \in K^0$, $m \in MA$ and $m = k_0^{-1}k$. This shows that MA meets every connected component of K. Since $G = KG^0$ and A is connected, the required result is now obvious.

5. WEYL GROUPS

In this section we recall some known facts about Weyl groups (see [1]).

Let $\mathfrak{a}_0$ be a maximal abelian subspace of $\mathfrak{p}$. Consider the set $\mathfrak{a}_0'$ of all $H \in \mathfrak{a}_0$ such that $\alpha(H) \neq 0$ for every root α of $(\mathfrak{g}, \mathfrak{a}_0)$. Fix a connected component $\mathfrak{a}_0^+$ of $\mathfrak{a}_0'$.

Put

$$\mathfrak{w} = N(\mathfrak{a}_0)/Z(\mathfrak{a}_0),$$

where $N(\mathfrak{a}_0)$ is the normalizer and $Z(\mathfrak{a}_0)$ the centralizer of $\mathfrak{a}_0$ in G. Then $\mathfrak{w}$ is a finite group which operates on $\mathfrak{a}_0$ in the usual way. Extend $\mathfrak{a}_0$ to a Cartan subalgebra $\mathfrak{h}$ of $\mathfrak{g}$ and let W denote the Weyl group of $(\mathfrak{g}_c, \mathfrak{h}_c)$.

LEMMA 1. *Two elements of $\mathfrak{a}_0$ are conjugate under G_c iff they are conjugate under any one of the three groups*

$$W, \mathfrak{w}, N(\mathfrak{a}_0) \cap K^0.$$

Moreover given $H \in \mathfrak{a}_0$, there exists a unique element $H_0 \in \mathrm{Cl}\, \mathfrak{a}_0^+$ which is conjugate to H under $\mathfrak{w}$.

COROLLARY 1. *If two elements in $\mathfrak{p}$ are conjugate under G_c then they are also conjugate under K^0.*

Since $\mathfrak{p} = \bigcup_{k \in K^0} \mathfrak{a}_0{}^k$, this is obvious.

COROLLARY 2. *Let $\mathfrak{a}$ be a linear subspace of $\mathfrak{a}_0$ and x an element in G_c such that $\mathfrak{a}^x \subset \mathfrak{a}_0$. Then we can choose $s \in \mathfrak{w}$ such that $\mathrm{Ad}(x)H = sH$ for all $H \in \mathfrak{a}$.*

We can select an element $H_0 \in \mathfrak{a}$ with the following property. If α is a root of $(\mathfrak{g}, \mathfrak{a}_0)$ and $\alpha(H_0) = 0$, then α vanishes identically on $\mathfrak{a}$. Let M_c be the centralizer of H_0 in G_c. Then M_c is connected and it is also the centralizer of $\mathfrak{a}$ in G_c. By Lemma 1 we can choose $s \in \mathfrak{w}$ such that $\mathrm{Ad}(x)\, H_0 = sH_0$. Then if y is a representative of s in $N(\mathfrak{a}_0)$, it follows that $y^{-1}x \in M_c$ and therefore

$$\mathrm{Ad}(x)H = \mathrm{Ad}(y)H = sH$$

for all $H \in \mathfrak{a}$.

Let $\mathfrak{a}_1$, $\mathfrak{a}_2$ be two abelian subspaces of $\mathfrak{p}$. We denote by $\mathfrak{w}(\mathfrak{a}_2 \mid \mathfrak{a}_1)$ the set of all linear mappings s of $\mathfrak{a}_1$ into $\mathfrak{a}_2$ with the following property. There exists an element $x \in G$ such that $\mathrm{Ad}(x) = s$ on $\mathfrak{a}_1$.

COROLLARY 3. $\mathfrak{w}(\mathfrak{a}_2 \mid \mathfrak{a}_1)$ *is a finite set.*

We can choose $k_i \in K^0$ ($i = 1, 2$) such that $\mathfrak{a}_i^{k_i} \subset \mathfrak{a}_0$. Hence without loss of generality, we may assume that $\mathfrak{a}_i \subset \mathfrak{a}_0$. Then our assertion follows from Corollary 2 above.

Let $\mathfrak{a}$ be a Cartan subalgebra of $\mathfrak{g}$ and A the corresponding Cartan subgroup of G. We denote by $W(\mathfrak{g}/\mathfrak{a})$ the Weyl group of $(\mathfrak{g}_c, \mathfrak{a}_c)$. Let $\tilde{A}$ be the normalizer of $\mathfrak{a}$ in G. We define $W(G/A) = \tilde{A}/A$. Then $W(G/A)$ may be regarded as a subgroup of $W(\mathfrak{g}/\mathfrak{a})$.

6. PARABOLIC PAIRS[2]

By a parabolic pair (or p-pair) in G, we mean a pair (Q, A) where Q is a psgp of G and A the split component of Q. Let $Q = MAN$ and $\mathfrak{q} = \mathfrak{m} + \mathfrak{a} + \mathfrak{n}$ be the Langlands decompositions of Q and its Lie algebra $\mathfrak{q}$. By a root of Q or (Q, A) we mean a root α of $(\mathfrak{g}, \mathfrak{a})$ such that $\mathfrak{g}_\alpha \subset \mathfrak{n}$ (see Sect. 4 for the definition of $\mathfrak{g}_\alpha$). Let $\Sigma(Q)$ denote the set of all roots of Q. We write $\mathfrak{n}_\alpha = \mathfrak{g}_\alpha$ for $\alpha \in \Sigma(Q)$. A root α is called simple if it cannot be written in the form $\alpha = \beta + \gamma$ with $\beta, \gamma \in \Sigma(Q)$.

---

[2] We recapitulate some well-known facts in this section (see [1]).

If $\ell = \operatorname{prk} Q - \operatorname{prk} G$, there exist exactly ℓ distinct simple root $\alpha_1, \ldots, \alpha_\ell$. They are linearly independent over $\mathbf{R}$ and every $\alpha \in \Sigma(Q)$ can be written (uniquely) in the form $\alpha = m_1\alpha_1 + \cdots + m_\ell\alpha_\ell$ where m_i are nonnegative integers. We denote by $\Sigma^0(Q)$ the set of simple roots of Q.

Fix a subset F of $\Sigma^0(Q)$ and let $\mathfrak{a}_F$ denote the subspace of all $H \in \mathfrak{a}$ such that $\alpha(H) = 0$ for all $\alpha \in F$. Let Σ_F be the set of those $\alpha \in \Sigma = \Sigma(Q)$ which vanish identically on $\mathfrak{a}_F$. We denote by Σ_F' the complement of Σ_F in Σ and put

$$\mathfrak{n}_F = \sum_{\alpha \in \Sigma_F'} \mathfrak{n}_\alpha, \quad \mathfrak{q}_F = \mathfrak{z}_F + \mathfrak{n}_F,$$

where $\mathfrak{z}_F$ is the centralizer of $\mathfrak{a}_F$ in $\mathfrak{g}$. Then $\mathfrak{q}_F$ is a parabolic subalgebra of $\mathfrak{g}$ and the corresponding psgp Q_F has the Langlands decomposition

$$Q_F = M_F A_F N_F,$$

where $A_F = \exp \mathfrak{a}_F$, $N_F = \exp \mathfrak{n}_F$. Moreover

$$Q_F \supset Q, \; M_F \supset M, \; A_F \subset A, \; N_F \subset N.$$

Given any psgp Q' of G containing Q, there exists a unique subset F of $\Sigma^0(Q)$ such that $Q' = Q_F$. We write $(Q, A)_F = (Q_F, A_F)$.

LEMMA 1. *There is a one-one correspondence between* psgps P *of* G *which are contained in* Q *and* psgps $*P$ *of* M. *This correspondence is given by the relation* $*P = P \cap M$. *If*

$$P = M'A'N', \qquad P* = *M *A *N$$

are the corresponding Langlands decompositions, then

$$M' = *M, \qquad A' = *A \cdot A, \qquad N' = *N \cdot N,$$

$$*A = M \cap A', \qquad *N = M \cap N'.$$

If (P_i, A_i) $(i = 1, 2)$ are two p-pairs in G, we write $(P_1, A_1) \succ (P_2, A_2)$ if $P_1 \supset P_2$. This implies that $A_1 \subset A_2$.

A psgp P is called minimal if it is minimal in the set of all psgps of G. Similarly it is called maximal if $P \neq G$ and P is maximal in the set of all psgps $Q \neq G$.

LEMMA 2. *Any two minimal* psgps *of* G *are conjugate under* K^0. *Let* P_1, P_2 *and* Q *be three* psgps *of* G. *Suppose* $P_1 \cap P_2 \supset Q$ *and* P_1 *is conjugate to* P_2 *under* G. *Then* $P_1 = P_2$.

LEMMA 3. *Let $P = MAN$ be a psgp of G. Then the following three conditions are equivalent.*

(1)   *P is a minimal psgp.*

(2)   *$M \subset K$.*

(3)   *prk $P = $ rank G/K.*

A vector subgroup A of G is called special if $A \subset \exp \mathfrak{p}$ and there exists a psgp P of G with A as a split component. If A is a special vector subgroup, we denote by $\mathscr{P}(A)$ the set of all psgps P of G such that (P, A) is a p-pair. Then $P = MAN$ where M is independent of the choice of P in $\mathscr{P}(A)$. Moreover $\mathscr{P}(A)$ is a finite set.

Let $P = MAN$ be a psgp of G. Since $G = KP$, every element $x \in G$ can be written uniquely in the form

$$x = kman \quad (k \in K, m \in M_{\mathfrak{p}}, a \in A, n \in N),$$

where $M_{\mathfrak{p}} = M \cap \exp \mathfrak{p}$. Put $H_P(x) = \log a \in \mathfrak{a}$. Then H_P is an analytic mapping of G into $\mathfrak{a}$. Define $\rho_P \in \mathfrak{a}^*$ by

$$\rho_P(H) = (1/2)\, \mathrm{tr}(\mathrm{ad}\, H)_{\mathfrak{n}} \quad (H \in \mathfrak{a}).$$

Let $\mathfrak{a}^+$ denote the set of all $H \in \mathfrak{a}$ such that $\alpha(H) > 0$ for every root α of (P, A). Put $A^+ = \exp \mathfrak{a}^+$. We note that $\mathfrak{a}^+ = \mathfrak{a}$ and $A^+ = A$ in case $P = G$.

Two psgps are said to be associated if their split components are conjugate in G or equivalently, under K. By the class of a psgp P we mean the set of all psgps associated to P.

7. NORMALIZATION OF MEASURES

Let E denote any affine subspace of $\mathfrak{g}$. Then the Euclidean norm on $\mathfrak{g}$ defines a Euclidean measure dE on E (see [10]). Now suppose L is a Lie subgroup of G and $\mathfrak{l}$ its Lie algebra. Let dL be a left-invariant Haar measure on L. We can choose an open neighborhood $\mathfrak{l}_0$ of zero in $\mathfrak{l}$ such that the exponential mapping defines a diffeomorphism of $\mathfrak{l}_0$ onto an open subset $L_0 = \exp \mathfrak{l}_0$ of L. We normalize dL in such a way that

$$\int_{L_0} f\, dL = \int_{\mathfrak{l}_0} F\, d\mathfrak{l} \quad (f \in C_c(L_0)),$$

where

$$F(X) = f(\exp X)\,|\det((1 - e^{-\mathrm{ad}X})/\mathrm{ad}\,X)_\mathfrak{l}|\qquad (X \in \mathfrak{l}_0).$$

This fixes dL uniquely and its definition is independent of the choice of $\mathfrak{l}_0$. If L is compact, we put

$$v(L) = \int_L dL,$$

and call $v(L)^{-1}\,dL$ the normalized Haar measure of L.

Let $P = MAN$ be a minimal psgp of G. Then $M \subset K$ and therefore

$$G = KP = KAN.$$

Put $da = dA$, $dn = dN$ and let dk denote the normalized Haar measure of K. We define the Haar measure dx on G by

$$dx = e^{2\rho(\log a)}dk\,da\,dn\qquad (x = kan),$$

where $\rho = \rho_P$. Since any two minimal psgps are conjugate under K, this measure is independent of the choice of P. We call it the standard Haar measure on G.

Let $P = MAN$ be any psgp of G. Replacing (G, K, θ, B) by (M, K_M, θ_M, B_M) (see Lemma 4.9), we get the standard measure dm on M.

LEMMA 1. *Let $P = MAN$ be a psgp of G. Put $da = dA$, $dn = dN$ and let dm denote the standard measure on M and dk the normalized Haar measure on K. Then*

$$\int_G f(x)\,dx = \int_{K \times M \times A \times N} f(kman)e^{2\rho(\log a)}dk\,dm\,da\,dn$$

for $f \in C_c(G)$. Here $\rho = \rho_P$ and dx is the standard Haar measure on G.

Let $*P = *M*A*N$ be a minimal psgp of M and $P' = M'A'N'$ the corresponding psgp of G contained in P (Lemma 6.1). Then P' is a minimal psgp of G and it follows from the definition dx that

$$\int_G f(x)\,dx = \int_{K \times A' \times N'} f(ka'n')e^{2\rho(\log a')}da'\,dn',$$

where $\rho' = \rho_{P'}$, $da' = dA'$ and $dn' = dN'$. Our assertion is now an immediate consequence of Lemma 6.1.

8. Definition of F_f

Let $\mathfrak{a}$ be a Cartan subalgebra of $\mathfrak{g}$ and A the corresponding Cartan subgroup of G. For any root α of $(\mathfrak{g}, \mathfrak{a})$ we denote by ξ_α the corresponding quasicharacter of A. Now assume that $\mathfrak{a}$ is θ-stable. Put $\mathfrak{a}_I = \mathfrak{a} \cap \mathfrak{k}$ and $\mathfrak{a}_R = \mathfrak{a} \cap \mathfrak{p}$. Then $A = A_I A_R$ where $A_I = A \cap K$ and $A_R = \exp \mathfrak{a}_R$. Moreover A_R is a special vector subgroup of G. Let $\mathfrak{z}(\mathfrak{a}_R) = \mathfrak{z}$ denote the centralizer of $\mathfrak{a}_R$ in $\mathfrak{g}$ and P_I the set of all positive roots of $(\mathfrak{z}, \mathfrak{a})$ (under some order). Put

$$\Delta_+(a) = |\det(1 - \mathrm{Ad}(a^{-1}))_{\mathfrak{g}/\mathfrak{z}}|^{1/2} \qquad (a \in A),$$

$$\Delta_I(H) = \prod_{\beta \in P_1} (e^{\beta(H)/2} - e^{-\beta(H)/2}) \qquad (H \in \mathfrak{a}).$$

Let $Z(G : \mathfrak{a}) = Z(\mathfrak{a})$ denote the centralizer of $\mathfrak{z}$ in K. Let $P \in \mathscr{P}(A_R)$. Then $P = M A_R N$ where M is independent of the choice of P.

LEMMA 1. $A_I = Z(\mathfrak{a})\, A_I^0$.

Let $\mathfrak{m}$ be the Lie algebra of M. Then A_I is the Cartan subgroup of M corresponding to $\mathfrak{a}_I$ and $Z(\mathfrak{a})$ is the centralizer of $\mathfrak{m}$ in K_M. Hence our assertion follows from Lemma 3.4.

Let dx denote the standard measure on G and dx^* the invariant measure on G/A_R normalized so that $dx = dx^*\, dA_R$. For any $f \in C_c^\infty(G)$ and $\zeta \in Z(\mathfrak{a})$, put

$$F_f^{\mathfrak{a}}(\zeta : H) = \Delta_I(H)\, \Delta_+(\zeta \exp H) \int_{G/A_R} f((\zeta \exp H)^{x^*})\, dx^*$$

for $H \in \mathfrak{a}'(\zeta)$. Here $\mathfrak{a}'(\zeta)$ is the set of all $H \in \mathfrak{a}$ such that

$$\xi_\alpha(\zeta \exp H) \neq 1,$$

for every root α of $(\mathfrak{g}, \mathfrak{a})$. Let $\mathfrak{a}'$ be the set of all $H \in \mathfrak{a}$ such that $\alpha(H) \notin 2\pi(-1)^{1/2}\, \mathbf{Z}$ for every singular imaginary root α of $(\mathfrak{g}, \mathfrak{a})$. Then $\mathfrak{a}' = \mathfrak{a}_I' + \mathfrak{a}_R$ where $\mathfrak{a}_I' = \mathfrak{a}' \cap \mathfrak{a}_I$.

LEMMA 2.   *Fix $f \in C_c^\infty(G)$ and $\zeta \in Z(\mathfrak{a})$. Then $F_f(\zeta : H)$ extends to a C^∞ function of $H \in \mathfrak{a}'$. If $P = M A_R N$ is a psgp of G, then*

$$F_f^{\mathfrak{a}}(\zeta : H) = \Delta_I(H) e^{\rho(H_2)} \int_{M \times N} f((\zeta \exp H_1)^m \cdot \exp H_2 \cdot n)\, dm\, dn$$

for $H \in \mathfrak{a}'$. Here $H = H_1 + H_2$ ($H_1 \in \mathfrak{a}_I$, $H_2 \in \mathfrak{a}_R$), $\rho = \rho_P$ and

$$\tilde{f}(x) = \int_K f(kxk^{-1})\, dk \qquad (x \in G).$$

The measures dk, dm, da, dn have the same meaning as in Lemma 7.1.

Since $G = KP = KNMA_R$, it is clear that $G/A_R = (KNM)^*$. Moreover it follows without difficulty from Lemma 7.1 that

$$\int_{G/A_R} F(x^*)\, dx^* = \int_{K \times M \times N} F((knm)^*)\, dk\, dm\, dn \qquad (F \in C_c(G^*)).$$

Our statements are now immediate consequences of [13, Cor. p. 93].

Sometimes we shall write $F_f^{G/\mathfrak{a}}$ or $F_f^{G/A}$ instead of $F_f^{\mathfrak{a}}$.

Let $\mathfrak{a}$ and $\mathfrak{b}$ be two θ-stable Cartan subalgebras of $\mathfrak{g}$ such that $\mathfrak{a}_R \supset \mathfrak{b}_R$. Then it is clear from the definition that $Z(\mathfrak{a}) \supset Z(\mathfrak{b})$. Fix a p-pair (P, B_R) ($P = MB_R N$) in G. Then $*A = A \cap M$ is a Cartan subgroup of M and $*A_R = A_R \cap M$. Choose a psgp $*P$ of M with the split component $*A_R$ so that $*P = *M \, *A_R \, *N$. Put $N' = *N \cdot N$. Then (Lemma 6.1)

$$P' = *MA_R N'$$

is a psgp of G. It is clear that $\mathfrak{a}_R \subset \mathfrak{z}(\mathfrak{a}_R) \subset \mathfrak{z}(\mathfrak{b}_R) = \mathfrak{b}_R + \mathfrak{m}$. Hence $\mathfrak{a}_R = \mathfrak{b}_R + *\mathfrak{a}_R$. Let $\mathfrak{z}(*\mathfrak{a}_R)$ denote the centralizer of $*\mathfrak{a}_R$ in $\mathfrak{m}$. Then $\mathfrak{z}(\mathfrak{a}_R) = \mathfrak{z}(*\mathfrak{a}_R) + \mathfrak{b}_R$ and therefore

$$Z(G : \mathfrak{a}) = Z(M : *\mathfrak{a}).$$

Put $\rho = \rho_P$.

LEMMA 3. *Fix $f \in C_c^\infty(G)$ and put*

$$g(H_2 : m) = g_{H_2}(m) = e^{\rho(H_2)} \int_N \tilde{f}(m \exp H_2 \cdot n)\, dn$$

for $H_2 \in \mathfrak{b}_R$ and $m \in M$. Then $g \in C_c^\infty(\mathfrak{b}_R \times M)$ and

$$F_f^{\mathfrak{a}}(\zeta : H_1 + H_2) = F_{g_{H_2}}^{M/*\mathfrak{a}}(\zeta : H_1)$$

for $\zeta \in Z(\mathfrak{a})$, $H_1 \in *\mathfrak{a}'$ and $H_2 \in \mathfrak{b}_R$.

This follows by applying Lemma 2 to (G, A) and $(M, *A)$.

If λ is a linear function on $\mathfrak{a}_c$, we define $H_\lambda \in \mathfrak{a}_c$ by the condition

$$B(H, H_\lambda) = \lambda(H) \qquad (H \in \mathfrak{a}_c).$$

Given $\lambda, \mu \in \mathfrak{a}_c{}^*$, we put

$$\langle \lambda, \mu \rangle = B(H_\lambda, H_\mu).$$

In particular if λ takes only real values on $\mathfrak{a}_R + (-1)^{1/2}\, \mathfrak{a}_I$, then $\langle \lambda, \lambda \rangle \geqslant 0$. Put $|\lambda| = \langle \lambda, \lambda \rangle^{1/2} \geqslant 0$ in that case.

9. The Relation between $F_f^{\mathfrak{a}}$ and $F_f^{\mathfrak{b}}$

Let $\mathfrak{b}$ be a θ-stable Cartan subalgebra of $\mathfrak{g}$ and H_0 a point in $\mathfrak{b}$ such that:

(1)   There exists exactly one positive imaginary root β of $(\mathfrak{g}, \mathfrak{b})$ such that $\beta(H_0) \in 2\pi(-1)^{1/2}\, \mathbf{Z}$.

(2) β is singular.

Put $B_R = \exp \mathfrak{b}_R$, $b_0 = \exp H_0$ and let $\mathfrak{z}$ denote the centralizer of $b_0 B_R$ in $\mathfrak{g}$. Then $\mathfrak{l} = [\mathfrak{z}, \mathfrak{z}]$ is a simple Lie algebra of dimension 3 of noncompact type. We can choose a base (H', X', Y') for $\mathfrak{l}$ over $\mathbf{R}$ such that

$$[H', X'] = 2X', [H', Y'] = -2Y', [X', Y'] = H',$$

θ maps (H', X', Y') into $(-H', -Y', -X')$, $\mathfrak{b} = \sigma_\beta + \mathbf{R}(X' - Y')$ and $\beta(X' - Y') = -2(-1)^{1/2}$. Here σ_β is the set of all $H \in \mathfrak{b}$ with $\beta(H) = 0$. Put

$$\mathfrak{a} = \sigma_\beta + \mathbf{R}H'.$$

Then $\mathfrak{a}$ is also a θ-stable Cartan subalgebra of $\mathfrak{g}$. Put

$$\nu = \exp\{(-1)^{1/2}\, (\pi/4)\, \mathrm{ad}(X' + Y')\}.$$

Then ν is an automorphism of $\mathfrak{z}_c$ and $\mathfrak{b}_c{}^\nu = \mathfrak{a}_c$ (see [10, Sect. 7]).

Let $P_I(\mathfrak{a})$ and $P_I(\mathfrak{b})$ denote the sets of positive imaginary roots of $(\mathfrak{g}, \mathfrak{a})$ and $(\mathfrak{g}, \mathfrak{b})$, respectively (under some given orders). We assume that $\beta \in P_I(\mathfrak{b})$ and put

$$\Delta_{\mathfrak{b}, I, \beta} = \prod_{\gamma \neq \beta} (e^{\gamma(H)/2} - e^{-\gamma(H)/2}, \qquad (H \in \mathfrak{b}),$$

where γ runs over all roots in $P_I(\mathfrak{b})$ different from β. Similarly put

$$\Delta_{\mathfrak{a}, I}(H) = \prod_{\alpha \in P_I(\mathfrak{a})} (e^{\alpha(H)/2} - e^{-\alpha(H)/2}) \qquad (H \in \mathfrak{a}).$$

Since $\mathfrak{b}_R = \sigma_\beta \cap \mathfrak{p} \subset \mathfrak{a}_R$, it is clear that $Z(\mathfrak{b}) \subset Z(\mathfrak{a})$.

THEOREM 1. *Fix $\zeta \in Z(\mathfrak{b})$, $u \in S(\mathfrak{b}_c)$ and $f \in C_c^\infty(G)$. Then*[3]

$$F_f^{\mathfrak{b}}(\zeta : H_0 ; \partial(u))^+ - F_f^{\mathfrak{b}}(\zeta : H_0 ; \partial(u))^-$$

$$= -2\pi(-1)^{1/2} \mid \beta \mid^{-1} \epsilon(H_0) F_f^{\mathfrak{a}}(\zeta_0 : H_0'; \partial(u^v)).$$

Here $H_0' = H_0 - \beta(H_0) H_\beta \mid \beta \mid^{-2}$, $\zeta_0 = \zeta \exp(\beta(H_0) \mid \beta \mid^{-2} H_\beta) \in Z(\mathfrak{a})$,

$$F_f^{\mathfrak{b}}(\zeta : H_0 ; \partial(u))^{\pm} = \lim_{\phi \to \pm 0} F_f(\zeta : H_0 + \phi(X' - Y'); \partial(u))$$

and

$$\epsilon(H_0) = \operatorname{sign}(e^{\beta(H_0)/2} \Delta_{\mathfrak{b},I,\beta}(H_0) \cdot \Delta_{\mathfrak{a},I}(H_0')^{-1}).$$

We observe that $H_0' \in \sigma_\beta$, $\beta(H_0) H_\beta \in I \cap \mathfrak{k}$ and

$$\mid \beta \mid^{-2} H_\beta = (1/2)(-1)^{1/2} (X' - Y').$$

Therefore $b_1 = \exp(\beta(H_0) \mid \beta \mid^{-2} H_\beta)$ centralizes $\mathfrak{z}$. We claim that $a = \exp \pi(X' - Y') \in Z(\mathfrak{a})$. Let L_c be the simply connected complex analytic group corresponding to I_c. Then $L_c \simeq SL(2, \mathbf{C})$ and therefore it is easy to verify that

$$\exp \pi(X' - Y') = \exp\{\pi(-1)^{1/2} H'\}$$

in L_c. Therefore

$$\operatorname{Ad}(a) = \exp\{\pi(-1)^{1/2} \operatorname{ad} H'\}$$

in G_c. As before let $\mathfrak{z}(\mathfrak{a}_R)$ denote the centralizer of $\mathfrak{a}_R$ in $\mathfrak{z}$. Then since $H' \in \mathfrak{a}_R$ and $X' - Y' \in \mathfrak{k}$, it is obvious that $a \in Z(\mathfrak{a})$. Since $\beta(H_0) \in 2\pi(-1)^{1/2} \mathbf{Z}$, b_1 is a power of a and therefore it lies in $Z(\mathfrak{a})$. Hence $\zeta_0 = \zeta b_1 \in Z(\mathfrak{a})$ and

$$\zeta \exp H_0 = \zeta_0 \exp H_0'.$$

Now we shall prove the theorem by induction on $\dim G$. Without loss of generality we may assume that $\operatorname{prk} G = 0$. First suppose $\mathfrak{b}_R \neq \{0\}$. Then, using the notation of Lemma 8.3, we have

$$F_f^{\mathfrak{b}}(\zeta : H_1 + H_2) = F_{{}^\sigma H_2}^{M/\mathfrak{b}_I}(\zeta : H_1) \qquad (H_1 \in \mathfrak{b}_I', H_2 \in \mathfrak{b}_R),$$

and

$$F_f^{\mathfrak{a}}(\zeta_0 : H_1 + H_2) = F_{{}^\sigma H_2}^{M/{}^*\mathfrak{a}}(\zeta_0 : H_1) \ (H_1 \in {}^*\mathfrak{a}', H_2 \in \mathfrak{b}_R).$$

---

[3] We use here the standard notation (see [10, p. 561]).

Without loss of generality, we may assume that $u = u_1 u_2$ where $u_1 \in S(\mathfrak{b}_{I,c})$, and $u_2 \in S(\mathfrak{b}_{R,c})$. Put $H(\phi) = H_0 + \phi(X' - Y')$ where $\phi \neq 0$ and $|\phi|$ is small ($\phi \in \mathbf{R}$). Then

$$F_f^{\mathfrak{b}}(\zeta : H(\phi); \partial(u)) = F_{g_0}^{M/\mathfrak{b}_I}(\zeta : H_1(\phi); \partial(u_1)),$$

where

$$g_0(m) = g(H_{02} ; \partial(u_2) : m) \qquad (m \in M),$$

and $H_1(\phi) = H_{01} + \phi(X' - Y')$. Here H_{01}, H_{02} are the projections of H_0 in $\mathfrak{b}_I$ and $\mathfrak{b}_R$, respectively. On the other hand v leaves σ_β fixed and $\mathfrak{a}_I + \mathfrak{b}_R \subset \sigma_\beta$. Hence u_2 and H_0' are both left fixed by v. Moreover if γ is an imaginary root of $(\mathfrak{g}, \mathfrak{a})$, then $H_\gamma = v^{-1} H_\gamma \in \sigma_{\beta,c}$. Let $\gamma_0 = v^{-1}\gamma$ be the corresponding root of $(\mathfrak{g}, \mathfrak{b})$. Then $\langle \gamma_0 , \beta \rangle = 0$ and therefore

$$\gamma(H_0') = \gamma_0(H_0') = \gamma_0(H_0).$$

Moreover $\gamma_0 \neq \pm\beta$ and hence $\gamma(H_0') \notin 2\pi(-1)^{1/2}\, \mathbf{Z}$. Therefore

$$F_f^{\mathfrak{a}}(\zeta_0 : H_0'; \partial(u^v)) = F_{g_0}^{M/*\mathfrak{a}}(\zeta_0 : H_{01}' ; \partial(u_1^v)),$$

where H_{01}' is the projection of H_0' in $*\mathfrak{a}$ so that $H_0' = H_{01}' + H_{02}$. Since $\dim M < \dim G$, Theorem 9.1 is true for $(M, *\mathfrak{a}, \mathfrak{b}_I , g_0)$ in place of $(G, \mathfrak{a}, \mathfrak{b}, f)$ by induction hypothesis. The required statement for (G, f) is then an immediate consequence of the above relations.

So we may now assume that $\mathfrak{b}_R = \{0\}$. Then $Z(\mathfrak{b})$ is the centralizer of $\mathfrak{g}$ in G. Without loss of generality we may assume that v takes positive roots of $(\mathfrak{g}, \mathfrak{b})$ into positive roots of $(\mathfrak{g}, \mathfrak{a})$ and a root γ of $(\mathfrak{g}, \mathfrak{b})$ is positive if $\langle \gamma, \beta \rangle > 0$.

Put $b = \zeta \exp H_0$ and let G_b denote the centralizer of b in G. Then $\mathfrak{z} = \sigma_\beta + I$ is the centralizer of b in $\mathfrak{g}$ as before. Put $\Xi = G_b^0$ and

$$\Delta(H) = \prod_{\gamma \in P} (e^{\gamma(H)/2} - e^{-\gamma(H)/2}) \qquad (H \in \mathfrak{b}),$$

where P is the set of all positive roots of $(\mathfrak{g}, \mathfrak{b})$. Then there exists a unique holomorphic function D on $\mathfrak{z}_c$ such that

(1) D is invariant under Ξ,

(2) $\Delta(H_0 + H) = \beta(H) D(H)$ for $H \in \mathfrak{b}$.

(This follows by applying Lemma 39 of [12] to $(\mathfrak{z}, \mathfrak{b})$.) Fix an open neighborhood $\mathfrak{z}_0$ of zero in $\mathfrak{z}$ such that:

(1) $\mathfrak{z}_0$ is completely invariant under Ξ.

(2)   The exponential mapping (from $\mathfrak{z}$ to $\mathcal{Z}$) is regular and univalent on $\mathfrak{z}_0$.

This is possible (see [12, p. 472]). Select a C^∞ function v on $\mathfrak{z}$ such that:

(1) v is invariant under $\mathcal{Z}$,
(2)   Supp $v \subset \mathfrak{z}_0$,
(3) $v = 1$ around zero.

That this is possible can be seen either directly or by [11, Lemma 45, p. 46].

Fix an open and relatively compact subset Ω of G and suppose $f \in C_c^\infty(\Omega)$. Then if $H \in \mathfrak{b}$ and $\Delta(H_0 + H) \neq 0$,

$$F_f^{\mathfrak{b}}(\zeta : H_0 + H) = \Delta(H_0 + H) \int_G f(xb \exp H \cdot x^{-1})\, dx$$

$$= \Delta(H_0 + H) \int_{G/\mathcal{Z}} d\bar{x} \int_{\mathcal{Z}} f(xb \exp(H^y) \cdot x^{-1})\, dy,$$

where the invariant measures $d\bar{x}$ and dy on $\overline{G} = G/\mathcal{Z}$ and $\mathcal{Z}$ are so normalized that $dx = d\bar{x}\, dy$.

Let $\mathfrak{a}_0$ and $\mathfrak{b}_0$ be open and relatively compact neighborhoods of zero in $\mathfrak{a}$ and $\mathfrak{b}$ such that:

(1) $v = 1$ on $\mathfrak{a}_0 \cup \mathfrak{b}_0$,
(2) $\Delta_\beta(H_0 + H) \neq 0$ for $H \in \mathfrak{b}_0$,
(3) $\Delta_{\mathfrak{a},l}(H_0' + H) \neq 0$ for $H \in \mathfrak{a}_0$.

(Here $\Delta_\beta = \Delta_{\mathfrak{b},l,\beta}$.) Assuming that $\mathfrak{a}_0$, $\mathfrak{b}_0$ are sufficiently small, we can choose a compact set ω in $\overline{G}$ with the following property (see [7, Theor. 1, p. 736]). If x is an element of G such that

$$x(b \exp H)x^{-1} \in \Omega,$$

for some $H \in \mathfrak{a}_0 \cup \mathfrak{b}_0$, then $\bar{x} \in \omega$. Choose a function $\Gamma \in C_c^\infty(G)$ such that

$$\int_{\mathcal{Z}} \Gamma(xy)\, dy = 1$$

whenever $\bar{x} \in \omega$ $(x \in G)$. Put

$$g_f(Z) = v(Z)\, D(Z) \int_G \Gamma(x)\, f(x(b \exp Z)x^{-1})\, dx \qquad (Z \in \mathfrak{z}).$$

For any $g \in C_c^{\infty}(\mathfrak{z})$, put

$$\psi_g{}^{\mathfrak{b}}(H) = \beta(H) \int_{\Xi} g(H^y) \, dy \qquad (H \in \mathfrak{b}_0 \,, \beta(H) \neq 0),$$

$$\psi_g{}^{\mathfrak{a}}(H) = |\,\alpha(H)| \int_{\Xi/A_R} g(H^y) \, dy^* \qquad (H \in \mathfrak{a}_0 \,, \alpha(H) \neq 0).$$

Here $\alpha = \beta^y$ and the invariant measure dy^* on Ξ/A_R is so normalized that $dy = dy^* \, dA_R$.

LEMMA 1. *For any $u \in S(\mathfrak{b}_c)$ and $g \in C_c^{\infty}(\mathfrak{z})$,*

$$\psi_g{}^{\mathfrak{b}}(0; \partial(u))^+ - \psi_g{}^{\mathfrak{b}}(0; \partial(u))^- = -2\pi(-1)^{1/2} \,|\,\beta\,|^{-1} \psi_g(0; \partial(u^y)).$$

Here (see [10, Theor. 2])

$$\psi_g{}^{\mathfrak{b}}(0; \partial(u))^{\pm} = \lim_{\phi \to \pm 0} \psi_g{}^{\mathfrak{b}}(\phi(X' - Y'); \partial(u)).$$

Assuming this lemma, let us finish the proof of Theorem 1. Put $g = g_f$. Then

$$F_f{}^{\mathfrak{b}}(\zeta : H_0 + H) = \psi_g{}^{\mathfrak{b}}(H) \qquad (H \in \mathfrak{b}_0 \,, \beta(H) \neq 0).$$

Hence

$$F_f{}^{\mathfrak{b}}(\zeta : H_0 \,; \partial(u))^+ - F_f{}^{\mathfrak{b}}(\zeta : H_0 \,; \partial(u))^- = -2\pi(-1)^{1/2} \,|\,\beta\,|^{-1} \psi_g{}^{\mathfrak{a}}(0; \partial(u^y)).$$

We have seen above that $\zeta_0 \in Z(\mathfrak{a})$ and

$$b = \zeta \exp H_0 = \zeta_0 \exp H_0'.$$

Now consider

$$F_f{}^{\mathfrak{a}}(\zeta_0 : H_0' + H) = \varDelta_{\mathfrak{a},I}(H_0' + H) \,|\det(1 - \mathrm{Ad}(b \exp H)^{-1})_{\mathfrak{g}/\mathfrak{z}(\mathfrak{a}_R)}\,|^{1/2}$$

$$\times \int_{G/A_R} f((b \exp H)^x) \, dx^* \qquad (H \in \mathfrak{a}_0).$$

But

$$\int_{G/A_R} f((b \exp H)^x) \, dx^* = \int_{G/\Xi} d\bar{x} \int_{\Xi/A_R} f(x(b \exp H^y)x^{-1}) \, dy^*$$

$$= \int_{G} \Gamma(x) \, dx \int_{\Xi/A_R} f(x(b \exp H^y)x^{-1}) \, dy^* \quad (H \in \mathfrak{a}_0).$$

Lemma 2.

$$| \alpha(H)|\, D(H) = \epsilon_0 \Delta_{\mathfrak{a},I}(H_0' + H)\, |\det(1 - \mathrm{Ad}(b\exp H)^{-1})_{\mathfrak{g}/\mathfrak{z}(\mathfrak{a}_R)}|^{1/2},$$

for $H \in \mathfrak{a}_0$. *Here*

$$\epsilon_0 = \mathrm{sign}\{e^{\beta(H_0)/2}\Delta_{\mathfrak{b},\beta}(H_0) \cdot \Delta_{\mathfrak{a},I}(H_0')^{-1}\}.$$

Assuming this lemma, we conclude that

$$F_f{}^\alpha(\zeta_0 : H_0' + H) = \epsilon_0 \psi_g{}^\alpha(H) \qquad (H \in \mathfrak{a}_0).$$

The statement of Theorem 1 is now an immediate consequence of
Lemma 1.

We shall now prove Lemma 2. Since $\alpha = \beta^\nu$,

$$\alpha(H)\, D(H) = \beta(\nu^{-1}H)\, D(\nu^{-1}H) = \Delta(H_0 + \nu^{-1}H) = \Delta_{\mathfrak{a}}(\nu H_0 + H).$$

We claim that

$$\mathrm{Ad}(b) = \exp \mathrm{ad}\, H_0 = \exp \mathrm{ad}(\nu H_0).$$

Since $b = \zeta \exp H_0$ and ζ centralizes $\mathfrak{g}$, the first assertion is obvious.
Moreover since b centralizes $\mathfrak{z}$,

$$\mathrm{Ad}(b) = \nu\, \mathrm{Ad}(b)\nu^{-1} = \exp \mathrm{ad}(\nu H_0).$$

On the other hand

$$H_0 = H_0' + \beta(H_0)\, |\beta|^{-2}H_\beta$$

and $H_0' \in \sigma_\beta$. Therefore

$$\nu H_0 = H_0' + \beta(H_0)\, |\beta|^{-2}H_\alpha.$$

Hence

$$\alpha(H)\, D(H) = \Delta_{\mathfrak{a}}(\nu H_0 + H)$$
$$= \Delta_{\mathfrak{a},I}(H_0' + H) \prod_{\gamma \in P_+} (e^{\gamma(\nu H_0 + H)/2} - e^{-\gamma(\nu H_0 + H)/2}),$$

where P_+ is the set of all positive roots of $(\mathfrak{g}, \mathfrak{a})$ which do not vanish
identically on $\mathfrak{a}_R$. Put

$$\rho_+ = \frac{1}{2} \sum_{\gamma \in P_+} \gamma.$$

We recall that a root δ of $(\mathfrak{g}, \mathfrak{b})$ is positive if $\langle \delta, \beta \rangle > 0$. This implies that a root γ of $(\mathfrak{g}, \mathfrak{a})$ is positive if $\gamma(H_\alpha) > 0$. Hence if $\gamma \in P_+$ then $-\theta\gamma$ is also in P_+ and $\gamma \neq -\theta\gamma$ unless $\gamma = \alpha$. (Here we use the fact that $\mathfrak{a}_R = \mathbf{R}H_\alpha$.) Since $H_0' \in \sigma_\beta$, we conclude that $\rho_+(H_0') = 0$. Therefore

$$\prod_{\gamma \in P_+} \left(e^{\gamma(\nu H_0 + H)/2} - e^{-\gamma(\nu H_0 + H)/2} \right)$$

$$= e^{\beta(H_0)|\alpha|^{-2}\langle \rho_+, \alpha \rangle} \prod_{\gamma \in P_+} \left(e^{\gamma(H)/2} - \xi_\gamma(b^{-1}) e^{-\gamma(H)/2} \right),$$

since $\mathrm{Ad}(b) = \exp \mathrm{ad}\, \nu H_0$. Note that $\xi_\alpha(b^{-1}) = 1$ since b centralizes $\mathfrak{z}$. Moreover

$$\xi_{-\theta\gamma}(a) = \xi_\gamma(\theta a)^{-1} = \mathrm{conj}\, \xi_\gamma(a)$$

for $a \in A$ and any root γ of $(\mathfrak{g}, \mathfrak{a})$. This holds in particular for $b = \zeta_0 \exp H_0' \in A$. Therefore since

$$e^{\alpha(H)/2} - e^{-\alpha(H)/2} = \mathrm{sign}\, \alpha(H) \cdot \mid e^{\alpha(H)/2} - e^{-\alpha(H)/2} \mid \qquad (H \in \mathfrak{a}),$$

we find that

$$\mid \det(1 - \mathrm{Ad}(b \exp H)^{-1})_{\mathfrak{g}/\mathfrak{z}(\mathfrak{a}_R)} \mid^{1/2}$$

$$= \prod_{\gamma \in P_+} \mid e^{\gamma(H)/2} - \xi_\gamma(b^{-1}) e^{-\gamma(H)/2} \mid$$

$$= \mathrm{sign}\, \alpha(H) \cdot \prod_{\gamma \in P_+} \left(e^{\gamma(H)/2} - \xi_\gamma(b^{-1}) e^{-\gamma(H)/2} \right) \qquad (H \in \mathfrak{a}).$$

This shows that

$$\Delta_\mathfrak{a}(\nu H_0 + H) = \Delta_{\mathfrak{a}, I}(H_0' + H) e^{\beta(H_0)|\alpha|^{-2}\langle \rho_+, \alpha \rangle} \cdot \mathrm{sign}\, \alpha(H)$$

$$\cdot \mid \det(1 - \mathrm{Ad}(b \exp H)^{-1})_{\mathfrak{g}/\mathfrak{z}(\mathfrak{a}_R)} \mid^{1/2}$$

for $H \in \mathfrak{a}_0$.

Now put

$$\rho = \frac{1}{2} \sum_{\gamma > 0} \gamma,$$

where γ runs over all positive roots of $(\mathfrak{g}, \mathfrak{a})$. Then it is clear that $\langle \rho, \alpha \rangle = \langle \rho_+, \alpha \rangle$. Moreover

$$2\langle \rho, \alpha \rangle \mid \alpha \mid^{-2} \in \mathbf{Z}.$$

Since $\beta(H_0) \in 2\pi(-1)^{1/2}\,\mathbf{Z}$, it follows that

$$e^{\beta(H_0)\langle\rho_+,\alpha\rangle|\alpha|^{-2}} = \epsilon_1,$$

where $\epsilon_1 = \pm 1$. Therefore

$$\alpha(H)D(H) = \Delta_\alpha(\nu H_0 + H),$$
$$= \epsilon_1\,\mathrm{sign}\,\alpha(H)\cdot\Delta_{\alpha,I}(H_0' + H)\,|\det(1 - \mathrm{Ad}(b\exp H)^{-1})_{\mathfrak{g}/\mathfrak{z}(\mathfrak{a}_R)}|^{1/2}$$

for $H \in \mathfrak{a}_0$. But

$$\Delta_\alpha(\nu H_0 + H) = \Delta(H_0 + \nu^{-1}H) = \Delta_{\mathfrak{b},\beta}(H_0 + \nu^{-1}H)e^{\beta(H_0)/2}\,(e^{\alpha(H)/2} - e^{-\alpha(H)/2})$$

since $e^{\beta(H_0)/2} = \pm 1$ and $\alpha = \beta^\nu$. Therefore

$$\Delta_{\mathfrak{b},\beta}(H_0 + \nu^{-1}H)e^{\beta(H_0)/2} = \epsilon_1\Delta_{\alpha,I}(H_0' + H)\prod_{\substack{\gamma\in P_+ \\ \gamma\neq\alpha}}|\,e^{\gamma(H)/2} - \xi_\gamma(b^{-1})e^{-\gamma(H)/2}\,|.$$

Letting $H \to 0$ $(H \in \mathfrak{a}_0,\ \alpha(H) \neq 0)$, we get

$$\Delta_{\mathfrak{b},\beta}(H_0)e^{\beta(H_0)/2} = \epsilon_1\Delta_{\alpha,I}(H_0')\prod_{\substack{\alpha\in P_+ \\ \gamma\neq\alpha\bullet}}|\,1 - \xi_\gamma(b^{-1})|.$$

Hence

$$\epsilon_1 = \mathrm{sign}\{e^{\beta(H_0)/2}\Delta_{\mathfrak{b},\beta}(H_0)\,\Delta_{\alpha,I}(H_0')^{-1}\}.$$

So now it remains to prove Lemma 1. It follows from [10, Theor. 2, p. 561] that it is enough to consider the case when $u = 1$. Let L be the analytic subgroup of Ξ corresponding to I. Since $\sigma_\beta = \mathfrak{a}_I$, it is clear that $\Xi = LA_I^0$. Moreover Ξ, being the connected component of 1 in the centralizer of b in G, is closed and θ-stable. Let Z_Ξ denote the center of Ξ and A the Cartan subgroup of G corresponding to $\mathfrak{a}$. Since $\mathfrak{a} \subset \mathfrak{z}$, it is obvious that $Z_\Xi \subset \Xi \cap A = (\Xi \cap A_I)\cdot A_R$. On the other hand Z_Ξ is also θ-stable and $\mathfrak{a}_I = \sigma_\beta$ is the center of $\mathfrak{z}$. Hence we conclude that $Z_\Xi \subset \Xi \cap A_I$ and therefore Z_Ξ is compact. On the other hand $I = [\mathfrak{z}, \mathfrak{z}]$. Hence it follows easily (see [13, p. 126]) that L is closed in Ξ and the center of L is finite. Moreover $\mathfrak{a}_R = \mathbf{R}H'$ and so we can identify $\mathbf{R}$ with A_R under the mapping $t \to \exp tH'$ $(t \in \mathbf{R})$. Since $H' = 2\,|\,\alpha\,|^{-2}\,H_\alpha$, it follows that

$$\|H'\| = 2\,|\,\alpha\,|^{-1} = 2\,|\,\beta\,|^{-1},$$

and therefore $dA_R = 2\,|\,\beta\,|^{-1}\,dt$. So the required result follows immediately from Lemma 9A.2.

We now return to the notation of Theorem 1. Fix $n \in \mathbf{Z}$ and let H_0 vary in $\mathfrak{b}$ in such a way that $\beta(H_0) = 2\pi(-1)^{1/2} n$ and $\gamma(H_0) \notin 2\pi(-1)^{1/2} \mathbf{Z}$ for any root $\gamma \neq \beta$ in $P_I(\mathfrak{b})$.

LEMMA 3. *Under the above conditions $\epsilon(H_0)$ remains constant.*

The above statement is clearly independent of the choice of positive roots. Hence we may choose systems $P(\mathfrak{a})$ and $P(\mathfrak{b})$ of positive roots for $(\mathfrak{g}, \mathfrak{a})$ and $(\mathfrak{g}, \mathfrak{b})$, respectively, in such a way that the following conditions hold.

(1) A root γ of $(\mathfrak{g}, \mathfrak{b})$ is positive if $\langle \gamma, \beta \rangle > 0$,

(2) $P(\mathfrak{a}) = P(\mathfrak{b})^\nu$.

LEMMA 4. *Under the above conditions*

$$\epsilon(H_0) = e^{\beta(H_0)\langle \rho_I, \beta \rangle |\beta|^{-2}},$$

where

$$\rho_I = \frac{1}{2} \sum_{\gamma \in P_I(\mathfrak{b})} \gamma.$$

Let P_0 be the set of all roots $\gamma \in P_I(\mathfrak{b})$ such that $\langle \gamma, \beta \rangle = 0$. Since $\mathfrak{a}_R = \mathfrak{b}_R + \mathbf{R}H'$, it is clear that $P_0{}^\nu = P_I(\mathfrak{a})$. Now let $\gamma \in P_0$. Then

$$\gamma(H_0) = \gamma(H_0') = \gamma^\nu(\nu H_0') = \gamma^\nu(H_0'),$$

since $\langle \gamma, \beta \rangle = 0$. Hence if P' is the complement of $P_0 \cup \{\beta\}$ in $P_I(\mathfrak{b})$, it follows from Theorem 9.1 that

$$\epsilon(H_0) = (-1)^n \operatorname{sign} \prod_{\gamma \in P'} (e^{\gamma(H_0)/2} - e^{-\gamma(H_0)/2}).$$

As usual let s_β denote the Weyl reflexion corresponding to β. Then if $\gamma \in P'$, it is obvious that $\gamma' = -s_\beta \gamma$ is also in P' and $\gamma' \neq \gamma$. Moreover

$$\gamma'(H_0) = -\gamma(H_0) + 2\pi(-1)^{1/2} n n_\gamma,$$

where

$$n_\gamma = 2\langle \gamma, \beta \rangle |\beta|^{-2} \in \mathbf{Z}.$$

Hence

$$e^{\gamma'(H_0)/2} - e^{-\gamma'(H_0)/2} = (-1)^{n n_\gamma} \operatorname{conj}(e^{\gamma(H_0)/2} - e^{-\gamma(H_0)/2}).$$

Since $n_\gamma = n_{\gamma'}$, we conclude that

$$\epsilon(H_0) = (-1)^{n(1+m)},$$

where

$$m = \frac{1}{2} \sum_{\gamma \in P'} n_\gamma .$$

Since $P_I(\mathfrak{b}) = P_0 \cup \{\beta\} \cup P'$, it is clear that

$$2\langle \rho_I, \beta \rangle \, |\beta|^{-2} = 1 + m$$

and so our assertion follows.

9A. *Some computations on* $SL(2, \mathbf{R})$

Let $G = SL(2, \mathbf{R})$ and $\mathfrak{g} = sl(2, \mathbf{R})$. As usual put

$$H = \begin{pmatrix} 1 & 0 \\ 0 & -1 \end{pmatrix}, \qquad X = \begin{pmatrix} 0 & 1 \\ 0 & 0 \end{pmatrix}, \qquad Y = \begin{pmatrix} 0 & 0 \\ 1 & 0 \end{pmatrix},$$

and let K, A, N be the one-parameter subgroups of G corresponding to $(X - Y)$, H and X, respectively. Then $G = KAN$ is an Iwasawa decomposition of G. Put

$$a_t = \exp tH = \begin{pmatrix} e^t & 0 \\ 0 & e^t \end{pmatrix}, \qquad k_\theta = \exp \theta(X - Y) = \begin{pmatrix} \cos \theta & \sin \theta \\ -\sin \theta & \cos \theta \end{pmatrix}$$

$$n_s = \exp sX = \begin{pmatrix} 1 & s \\ 0 & 1 \end{pmatrix} \qquad (t, \theta, s \in \mathbf{R}).$$

Identify A and N with $\mathbf{R}$ under the mappings $t \to a_t$ and $s \to n_s$. Then $da = dt$, $dn = ds$ are the Haar measures on A and N, respectively. Similarly K may be identified with $\mathbf{R}/2\pi\mathbf{Z}$ under the mapping $\theta \to k_\theta$. Let $dk = d\theta/2\pi$ be the normalized Haar measure on K. Then

$$dx = e^{2t}(d\theta/2\pi) \, dt \, ds \qquad (x = k_\theta a_t n_s)$$

is a Haar measure on G.

LEMMA 1.

$$\int_G f(x) \, dx = \pi \int_0^\infty (e^{2T} - e^{-2T}) \, dT \int_{K \times K} f(k a_T k') \, dk \, dk'$$

for $f \in C_c(G)$.

We know from general theory [3, Lemma 22] that the two sides are equal up to a positive factor c which is independent of f. Now put $f(x) = F(\| x \|^2)$ $(x \in G)$ where $F \in C_c(R)$ and

$$\| x \|^2 = a^2 + b^2 + c^2 + d^2$$

for $x = \begin{pmatrix} a & b \\ c & d \end{pmatrix}$ in G. Then

$$\| k_\theta a_t n_s \|^2 = (e^t - e^{-t})^2 + s^2 e^{2t} + 2.$$

Put $g(t) = F(t^2 + 2)$. Then

$$\int_G f(x)\, dx = \frac{1}{2\pi} \int_0^{2\pi} d\theta \int_{-\infty}^{\infty} e^{2t}\, dt \int_{-\infty}^{\infty} f(k_\theta a_t n_s)\, ds$$

$$= \int_{-\infty}^{\infty} e^{2t}\, dt \int_{-\infty}^{\infty} g((e^t - \bar{e}^t)^2 + s^2 e^{2t})\, ds.$$

Let $y = se^t$ and $u = (e^t - \bar{e}^t)$. Then

$$\int_G f(x)\, dx = \int_{-\infty}^{\infty} e^t\, dt \int_{-\infty}^{\infty} g((e^t - \bar{e}^t)^2 + y^2)\, dy$$

$$= \int_0^{\infty} (e^t + \bar{e}^t)\, dt \int_{-\infty}^{\infty} g((e^t - \bar{e}^t)^2 + y^2)\, dy,$$

$$= \frac{1}{2} \int_{-\infty}^{\infty} du \int_{-\infty}^{\infty} g(u^2 + y^2)\, dy = \pi \int_0^{\infty} g(r^2)\, r dr,$$

where $r = (u^2 + y^2)^{1/2}$.

On the other hand

$$\| k a_T k' \|^2 = (e^T - \bar{e}^T)^2 + 2.$$

Hence if we put $r = (e^T - \bar{e}^T)$, we have $(e^{2T} - \bar{e}^{2T})\, dT = r\, dr$ and the right side of our lemma becomes

$$\pi \int_0^{\infty} g(r^2)\, r dr.$$

So if we assume that $g \geqslant 0$ and $g(1) = 1$, we conclude that $c = 1$.

Now $\mathfrak{a} = \mathbf{R}H$ and $\mathfrak{b} = \mathbf{R}(X - Y)$ are two Cartan subalgebras of $\mathfrak{g}$. Normalize the invariant measure dx^* on G/A in such a way that $dx = dx^*\, da$ and put

$$\psi_f{}^{\mathfrak{a}}(t) = 2\, | t | \int_{G/A} f(t H^x)\, dx^* \qquad (t \in \mathbf{R},\ t \neq 0)$$

for $f \in C_c^\infty(\mathfrak{g})$. Then it is easy to verify that

$$\psi_f{}^a(t) = \int_{-\infty}^{\infty} \tilde{f}(tH + sX)\, ds,$$

where

$$\tilde{f}(Z) = \int_K f(Z^k)\, dk \qquad (Z \in \mathfrak{g}).$$

Similarly put

$$\psi_f{}^b(\theta) = -2(-1)^{1/2}\theta \int_G f(\theta(X - Y)^x)\, dx \qquad (\theta \neq 0, \theta \in \mathbf{R}).$$

Then we conclude from Lemma 1 that

$$\psi_f{}^b(\theta) = -2\pi\theta(-1)^{1/2} \int_0^{\infty} \tilde{f}(\theta(e^{2t}X - \bar{e}^{2t}Y))\,(e^{2t} - \bar{e}^{2t})\, dt.$$

Hence we get the following result from [10, Lemma 2].

LEMMA 2. $\psi_f(+0) - \psi_f(-0) = -(-1)^{1/2}\,\pi\psi_f(0)$ for $f \in C_c^\infty(\mathfrak{g})$.

10. PROPERTIES OF THE FUNCTION Ξ

Since $G = K \exp \mathfrak{p}$, we can define functions σ and Ξ on G as in [15, Sect. 7] by

$$\sigma(k \exp X) = \| X \|, \ \Xi(k \exp X) = \Xi(\exp X) \qquad (k \in K, X \in \mathfrak{p}).$$

Let $P_0 = M_0 A_0 N_0$ be any minimal psgp of G. Then

$$\Xi(x) = \int_K e^{-\rho_0(H_0(xk))}\, dk \qquad (x \in G),$$

where $\rho_0 = \rho_{P_0}$, $H_0(x) = H_{P_0}(x)$ and dk is the normalized Haar measure on K. Let Z be the split component of G and $A_0{}^+$ the set of all $a \in A$ such that $\alpha(\log a) \geqslant 0$ for every root α of (P_0, A_0).

LEMMA 1.   *The function $\Xi = \Xi_G$ has the following properties.*

(1) $\Xi(x) = \Xi(x^{-1}) = \Xi(xz) > 0$ *for $x \in G$ and $z \in KZ$.*

(2)   *There exist numbers $c > 0$ and $d \geqslant 0$ such that*

$$1 \leqslant e^{\rho_0(\log a)}\Xi(a) \leqslant c(1 + \sigma(a))^d$$

for all $a \in A_0^{+}$.

(3)   *Given a compact set ω in G, we can choose a number $c = c(\omega) > 0$ such that*

$$\Xi(y_1 x y_2) \leqslant c\Xi(x)$$

for all $x \in G$ and $y_1 , y_2 \in \omega$.

(4) $\int_K \Xi(xky)\, dk = \Xi(x)\, \Xi(y)\ (x, y) \in G).$

(5)   *There exists a number $r \geqslant 0$ such that*

$$\int_G \Xi(x)^2 (1 + \sigma(x))^{-r}\, dx < \infty.$$

(6) $\sigma(xy) \leqslant \sigma(x) + \sigma(y)$ *for $x, y \in G$ and*

$$\sigma(ma) \geqslant \max(\sigma(m), \sigma(a)) \qquad (m \in M,\, a \in A),$$

where $P = MAN$ is a psgp of G.

These facts are well known (see [8], [15, Lemma 11], and [20, p. 399]). The second assertion of (6) follows from the fact that $\mathfrak{m} \cap \mathfrak{p}$ and $\mathfrak{a}$ are orthogonal. We shall call (4) the doubling principle.

Let $P = MAN$ be a psgp of G. Let Ξ_M denote the function on M corresponding to Ξ when the pair (G, K) is replaced by (M, K_M). Put $\rho = \rho_P$.

LEMMA 2.   *Fix $r > r' \geqslant 0$ and define d as in (2) of Lemma 1. Then we can choose a number $c = c(r, r') > 0$ such that*

$$e^{\rho(\log a)} \int_N (1 + \sigma(man))^{-(r+2d)}\Xi(man)\, dn \leqslant c(1 + \sigma(ma))^{-r'}\Xi_M(m)$$

for all $m \in M$ and $a \in A$. Moreover there exists a number $c_0 \geqslant 1$ such that

$$1 + \sigma(ma) \leqslant c_0(1 + \sigma(man)) \qquad (m \in M,\, a \in A,\, n \in N).$$

This is a slight generalization of Lemma 21 of [15]. We shall give a proof of this lemma in Section 31.

COROLLARY 1. *We can choose $c' > 0$ such that*

$$e^{\rho(\log a)} \int_{\bar{N}} (1 + \sigma(\bar{n}am))^{-(r+2d)} \Xi(\bar{n}am) \, d\bar{n} \leqslant c'(1 + \sigma(am))^{-r'} \Xi_M(m)$$

$$(m \in M, a \in A),$$

where $\bar{N} = \theta(N)$.

This follows by applying the above lemma to $\theta(P) = \bar{N}AM$ (see [15, p. 28]).

COROLLARY 2. *Let ω be a compact subset of G. Then if $r > 2d$, the integral*

$$\int_N \Xi(xny)(1 + \sigma(xny))^{-r} \, dn$$

converges uniformly for $x, y \in \omega$.

Define $c = c(\omega)$ as in (3) of Lemma 1. Then

$$\Xi(xny)(1 + \sigma(xny))^{-r} \leqslant c(1 + \sigma(x))^r \, (1 + \sigma(y))^r \, \Xi(n)(1 + \sigma(n))^{-r}$$

for $x, y \in \omega$ and $n \in N$. Hence our assertion follows immediately from Lemma 2.

11. CENTRAL $\mathfrak{Z}$-FINITE DISTRIBUTIONS

Put $\ell = \operatorname{rank} G = \operatorname{rank} \mathfrak{g}$ and let $D(x)$ be the coefficient of t^ℓ in $\det(t + 1 - \operatorname{Ad}(x))$ $(x \in G)$ where t is an indeterminate. As usual, G' denotes the set of all regular points in G, i.e., those x in G where $D(x) \neq 0$.

An element $x \in G$ is called semisimple, if $\operatorname{Ad}(x)$ is semisimple. Put $\mathcal{N}_G = \exp \mathcal{N} \subset G$, where $\mathcal{N}$ is the set of all nilpotent elements in $\mathfrak{g}$ (see [11, Sect. 3]). Since $\operatorname{Ad}(G) \subset G_c$, it follows without difficulty that the statements of [12, Sect. 3] remain true (even though G is not assumed to be connected).

Let $\mathfrak{G}$ be the universal enveloping algebra of $\mathfrak{g}_c$. We regard elements of $\mathfrak{G}$ as left-invariant differential operators on G. Let $\mathfrak{Z}$ be the algebra of all differential operators on G which commute with both left and right translations of G. Since $\operatorname{Ad}(G) \subset G_c$, $\mathfrak{Z}$ is just the center of $\mathfrak{G}$.

Notation. Let λ denote the canonical bijection of $S(\mathfrak{g}_c)$ onto $\mathfrak{G}$ [2, p. 192]. For any linear subspace $\mathfrak{l}$ of $\mathfrak{g}_c$, define $\mathfrak{G}(\mathfrak{l}) = \lambda(S(\mathfrak{l}))$. If $\mathfrak{m}$ is a subalgebra of $\mathfrak{g}$, we usually denote $\mathfrak{G}(\mathfrak{m}_c)$ by the corresponding capital german letter $\mathfrak{M}$. Moreover suppose $\mathfrak{m}$ is reductive and rank $\mathfrak{g} = $ rank $\mathfrak{m}$. Then $\gamma_{\mathfrak{g}/\mathfrak{m}}$ denotes the usual isomorphism of $\mathfrak{Z}$ into the center of $\mathfrak{M}$ [12, Sect. 12].

Given a distribution T on G and $x \in G$, define the distribution T^x by

$$T^x(f) = T(f') \qquad (f \in C_c^\infty(G)),$$

where

$$f'(y) = f(y^x) \qquad (y \in G).$$

We say that T is central (or invariant) if $T = T^x$ for all $x \in G$. Similarly T is said to be $\mathfrak{Z}$-finite, if the space of all distributions of the form zT $(z \in \mathfrak{Z})$ has finite dimension. The proof of [12, Theor. 2, Sect. 15] now goes through in the present case without any substantial change. In particular we have the following result.

THEOREM 1.   *Let Θ be a central and $\mathfrak{Z}$-finite distribution on G. Then Θ is a locally summable function which is analytic on G'.*

Let A be a Cartan subgroup of G and $\mathfrak{a}$ its Lie algebra. Put

$$'\Delta(a) = \prod_{\alpha \in P} (1 - \xi_\alpha(a^{-1})) \qquad (a \in A),$$

where P is the set of all positive roots of $(\mathfrak{g}, \mathfrak{a})$ (under some given order). Let P_R be the set of all real roots in P and $A'(R)$ the set of all points $a \in A$ such that $\xi_\alpha(a) \neq 1$ for every $\alpha \in P_R$. Put $A' = A \cap G'$.

LEMMA 1.   *Θ being as above, put*

$$\Phi(a) = '\Delta(a)\,\Theta(a) \qquad (a \in A').$$

Then Φ extends to an analytic function on $A'(R)$.

This is proved in the same way as [12, Lemma 31].

12. AN INEQUALITY FOR Θ

Since $[G : G^0] < \infty$, we can define the Schwartz space $\mathscr{C}(G)$ as in [15, Sect. 9]. The following theorem is a slight extension of [15, Sect. 19, Theor. 7].

THEOREM 1.   *Let Θ be a central and $\mathfrak{Z}$-finite distribution on G.*

Then Θ is tempered if and only if there exist numbers $c, r \geq 0$ such that

$$| D(x)|^{1/2} | \Theta(x)| \leq c(1 + \sigma(x))^r$$

for all $x \in G'$.

We first verify the necessity of our condition.

LEMMA 1. *Let A be a Cartan subgroup of G. Then we can choose $c > 0$ such that*

$$1 + \sigma(a) \leq c(1 + \sigma(a^x))$$

for all $a \in A$ and $x \in G$.

Without loss of generality we may assume that A is θ-stable. Fix $P \in \mathscr{P}(A_R)$ so that $P = MA_R N$. By Lemma 10.2 we can choose $c_0 \geq 1$ such that

$$1 + \sigma(ma_2) \leq c_0(1 + \sigma(ma_2 n)),$$

for $m \in M$, $a_2 \in A_R$ and $n \in N$. Now let $a = a_1 a_2$ $(a_1 \in A_I, a_2 \in A_R)$. Since $G = KP$ and $A_I \subset M$, we can, for a given $x \in G$, choose $m \in M$, $n \in N$ such that $\sigma(a^x) = \sigma(ma_2 n)$. Hence

$$1 + \sigma(ma_2) \leq c_0(1 + \sigma(a^x)).$$

Let $*P = *M *A *N$ be a minimal psgp of M and

$$P_0 = *MA_0 N_0,$$

the corresponding psgp of G contained in P (Lemma 6.1). Then $A_0 = *A \cdot A_R$. We can choose $k_1, k_2 \in K \cap *M$ and $h \in *A$ such that $m = k_1 h k_2$. Then $\sigma(ma_2) = \sigma(ha_2)$. On the other hand $*\mathfrak{a}$ and $\mathfrak{a}_R$ are orthogonal. Therefore it is clear that

$$\sigma(h_1 h_2) \geq \max(\sigma(h_1), \sigma(h_2)),$$

for $h_1 \in *A$, $h_2 \in A_R$. Therefore since $a_1 \in A_I \subset K$, it follows that

$$\sigma(ma_2) = \sigma(ha_2) \geq \sigma(a_2) = \sigma(a).$$

Therefore

$$1 + \sigma(a) \leq c_0(1 + \sigma(a^x)).$$

We can choose a finite number of θ-stable Cartan subgroups A_i $(1 \leqslant i \leqslant p)$ of G such that G' is the disjoint union of

$$G_i = \bigcup_{x \in G} x A_i' x^{-1} \qquad (1 \leqslant i \leqslant p),$$

where $A_i' = A_i \cap G'$. Therefore, in view of Lemma 1, it would be enough to prove the inequality of Theorem 1 on $A' = A \cap G'$ for a fixed θ-stable Cartan subgroup A.

Let $\mathfrak{a}$ denote the Lie algebra of A. Fix a point $a_0 \in A_I$ and choose an open, relatively compact, convex neighborhood $\mathfrak{a}_I{}^0$ of zero in $\mathfrak{a}_I$ such that:

(1)　$|(1 - \xi_\alpha(a_0 \exp H)| \geqslant (1/2)|(1 - \xi_\alpha(a_0))|$ and,

$$|(1 - e^{\alpha(H)})| \geqslant (1/2)| \alpha(H)|$$

for every root α of $(\mathfrak{g}, \mathfrak{a})$ and $H \in \mathfrak{a}_I{}^0$.

(2)　The mapping $H \to a_0 \exp H$ defines a diffeomorphism of $\mathfrak{a}_I{}^0$ on an open neighborhood U of a_0 in A_I.

Since A_I is compact, it would be sufficient to verify the inequality on $A' \cap (U A_R)$.

Let $\mathfrak{a}_R'$ be the set of all $H \in \mathfrak{a}_R$ such that $\alpha(H) \neq 0$ for every root α of $(\mathfrak{g}, \mathfrak{a})$ which does not vanish identically on $\mathfrak{a}_R$. Also let $\mathfrak{a}_I^{0'}$ denote the set of all points $H \in \mathfrak{a}_I{}^0$ where $\alpha(H) \neq 0$ for every imaginary root α of $(\mathfrak{g}, \mathfrak{a})$ such that $\xi_\alpha(a_0) = 1$. Then both $\mathfrak{a}_I^{0'}$ and $\mathfrak{a}_R'$ have only finite number of connected components. Put

$$U' = a_0 \exp \mathfrak{a}_I^{0'}, \; A_R' = \exp \mathfrak{a}_R'.$$

Then $U' A_R'$ is contained in A' and it is dense in $U A_R$. Fix connected components $\mathfrak{a}_I{}^+$ and $\mathfrak{a}_R{}^+$ of $\mathfrak{a}_R^{0'}$ and $\mathfrak{a}_R'$, respectively, and put $A^+ = U^+ \cdot A_R{}^+$ where

$$U^+ = a_0 \exp \mathfrak{a}_I{}^+, \; A_R{}^+ = \exp \mathfrak{a}_R{}^+.$$

Then it would be enough to verify the inequality on A^+.

Fix a point $H^0 \in \mathfrak{a}_R{}^+$ and introduce an order so that a root α of $(\mathfrak{g}, \mathfrak{a})$ is positive if $\alpha(H^0) > 0$. Let P be the set of all positive roots under this order. Since $\mathrm{Ad}(G) \subset G_c$, we can choose $H_0 \in \mathfrak{a}_c$ such that

$$\mathrm{Ad}(a_0) = \exp \mathrm{ad} H_0.$$

Then if $a = a_0 \exp H$ $(H \in \mathfrak{a})$, we have

$$D(a) = \prod_{\alpha \in P} (1 - \xi_\alpha(a)^{-1})(1 - \xi_\alpha(a)),$$

$$= (-1)^p\, e^{2\rho(H_0 + H)} \prod_{\alpha \in P} (1 - \xi_\alpha(a^{-1}))^2,$$

where p is the number of roots in P and

$$\rho = \frac{1}{2} \sum_{\alpha \in P} \alpha.$$

Put

$$\xi_\rho(a_0 \exp H) = e^{\rho(H_0 + H)}$$

for $H \in \mathfrak{a}_I{}^0 + \mathfrak{a}_R$. Since $D(a) \in \mathbf{R}$, it is obvious that

$$|\, D(a)|^{1/2} = \epsilon(a)^{-1}\, \xi_\rho(a) \prod_{\alpha \in P} (1 - \xi_\alpha(a^{-1}))$$

for $a \in A'' = A' \cap (UA_R)$. Here ϵ is a locally constant function on A'' such that $\epsilon^4 = 1$.

For any root α of $(\mathfrak{g}, \mathfrak{a})$, let η_α denote the function $(1 - \xi_\alpha^{-1})^{-1}$ on A' and let $\mathscr{R}$ be the ring of analytic functions on A' generated over $\mathbf{C}$ by 1 and η_α for all α.

LEMMA 2. *Put* $\eta = \prod_{\alpha \in P} \eta_\alpha$. *Then*

$$|\, D\,|^{-1/2} = \epsilon \xi_\rho^{-1} \eta$$

on A''.

This is obvious.

Let A_0 be the center of A and $\tilde{A}$ the normalizer of A in G. Put $W_A = \tilde{A}/A_0$ and let $x \to x^*$ denote the natural projection of G on $G^* = G/A_0$. Then W_A is a finite group and a^s, sH and x^*s $(s \in W_A,\ a \in A,\ H \in \mathfrak{a},\ x^* \in G^*)$ are defined as usual (see [15, p. 40]).

Put $G_A = (A')^G$ and normalize the invariant measure dx^* on G^* in such a way that

$$\int_G f(x)\, dx = \int_{G^*} dx^* \int_{A_0} f(xa)\, da \qquad (f \in C_c(G)),$$

where da is the Haar measure on A. Fix a function $\alpha^* \in C_c^\infty(G^*)$ such that $\alpha^*(x^*) = \alpha^*(x^*s)$ $(x^* \in G^*, s \in W_A)$ and

$$\int_{G^*} \alpha^*(x^*)\, dx^* = 1.$$

Then for any $\beta \in C_c^\infty(A')$, we define $f_\beta \in C_c^\infty(G_A)$ as follows.

$$f_\beta(a^x) = \alpha^*(x^*)\, |D(a)|^{-1/2} \sum_{s \in W_A} \beta(a^s) \qquad (a \in A',\ x \in G).$$

Put

$$\Phi(a) = |D(a)|^{1/2}\, \Theta(a) \qquad (a \in A').$$

Then it follows from [15, Lemma 91] that

$$\Theta(f_\beta) = \int_A \Phi\beta\, da \qquad (\beta \in C_c^\infty(A')).$$

Fix a compact set $C = C^{-1}$ in G such that $\mathrm{Supp}\ \alpha^* \subset C^*$. For any $a \in A$, we write $a = a_1 a_2$ $(a_1 \in A_I,\ a_2 \in A_R)$.

LEMMA 3. *Given* $g \in \mathfrak{G}$, *we can choose a finite set of elements* $u_i \in \mathfrak{S}(\mathfrak{a}_c)$ *and* $\eta_i \in \mathscr{R}$ $(1 \leqslant i \leqslant q)$ *such that*

$$|f_\beta(a^x; g)| \leqslant e^{-\rho(\log a_2)} \sum_{1 \leqslant i \leqslant q} \sum_{s \in W_A} |\eta_i(a)\, \beta^s(a; u_i)|$$

for $\beta \in C_c^\infty(A')$, $a \in A''$ *and* $x \in C$.

This is proved in the same way as [15, Lemma 31] except that now we use Lemma 2.

COROLLARY. *We can choose* $m \geqslant 0$, $\eta_i \in \mathscr{R}$ *and* $u_i \in \mathfrak{S}(\mathfrak{a}_c)$ $(1 \leqslant i \leqslant p)$ *such that*

$$\left| \int_A \beta\Phi\, da \right| \leqslant \sum_{1 \leqslant i \leqslant p} \sup_{x \in A^+} (1 + \sigma(a))^m\, |\eta_i(a)\, \beta(a; u_i)|$$

for all $\beta \in C_c^\infty(A^+)$.

This is proved in the same way as [15, Lemma 32].
It follows from Lemmas 2 and 11.1 that

$$\Phi(a) = \epsilon(a)^{-1}\, \xi_\rho(a)\, \Phi_0(a) \qquad (a \in A^+),$$

where Φ_0 is an analytic function on $a_0 A_I{}^0 A_R{}^+$. Since Θ is $\mathfrak{z}$-finite, it follows from [4, Theor. 2] that Φ_0 is $\mathfrak{S}(\mathfrak{a}_c)$-finite. Therefore we conclude from the above corollary (see [15, p. 43]) that we can choose c, $r \geqslant 0$ such that

$$| \Phi(a)| \leqslant c(1 + \sigma(a))^r \qquad (a \in A^+).$$

This proves the necessity of the inequality in Theorem 1. The sufficiency is an immediate consequence of Corollary 1 of Lemma 13.1 below.

13. Convergence of an Integral

LEMMA 1. *There exists a number $r \geqslant 0$ such that*

$$\int_G | D(x)|^{-1/2}\, \Xi(x)(1 + \sigma(x))^{-r}\, dx < \infty.$$

We proceed by induction on dim G. First assume that prk $G \geqslant 1$. Let $G = MA$ be the Langlands decomposition of G. Then

$$\sigma(ma) \geqslant \max(\sigma(m), \sigma(a)) \qquad (m \in M, a \in A).$$

Therefore

$$\int_G | D(x)|^{-1/2}\, \Xi(x)(1 + \sigma(x))^{-r}\, dx$$

$$\leqslant \int_M | D(m)|^{-1/2}\, \Xi(m)(1 + \sigma(m))^{-r/2}\, dm \int_A (1 + \sigma(a))^{-r/2}\, da < \infty.$$

by induction hypothesis, if r is large enough.

So now assume that prk $G = 0$. Fix a θ-stable Cartan subgroup A of G and put

$$G_A = \bigcup_{x \in G} xA'x^{-1},$$

where $A' = A \cap G'$. It would be enough (see [12, p. 505]) to show that

$$\int_{G_A} | D(x)|^{-1/2}\, \Xi(x)(1 + \sigma(x))^{-r}\, dx < \infty$$

for some $r \geqslant 0$.

Let dx^* denote the invariant measure on $G^* = G/A_R$. Then

$$\int_{G_A} |D(x)|^{-1/2}\, \Xi(x)(1 + \sigma(x))^{-r}\, dx$$

$$= \int_A |D(a)|^{1/2}\, da \int_{G^*} \Xi(a^x)(1 + \sigma(a^x))^{-r}\, dx^* \qquad (a \in A').$$

Fix a psgp $P \in \mathscr{P}(A_R)$. Then $P = MA_R N$. Define D_M on M corresponding to the function D on G. Then if $a = a_1 a_2\ (a_1 \in A_I,\ a_2 \in A_R)$, we conclude from [13, Cor. p. 93] that

$$|D(a)|^{1/2} \int_{G^*} \Xi(a^x)(1 + \sigma(a^x))^{-r}\, dx^*$$

$$= |D_M(a_1)|^{1/2}\, e^{\rho(\log a_2)} \int_{M \times N} \Xi(a^m \cdot n)(1 + \sigma(a^m \cdot n))^{-r}\, dm\, dn \qquad (a \in A').$$

Therefore

$$\int_A |D(a)|^{1/2}\, da \int_{G^*} \Xi(a^x)(1 + \sigma(a^x))^{-r}\, dx^*$$

$$\leqslant \int_{A_R} e^{\rho(\log a_2)}\, da_2 \int_M |D_M(m)|^{-1/2}\, dm \int_N \Xi(ma_2 n)(1 + \sigma(ma_2 n))^{-r}\, dn,$$

where $\rho = \rho_P$ and da_2 is the Haar measure on A_R.

Now first assume that $A_R \neq \{1\}$. Then since $\operatorname{prk} G = 0$, $\dim(MA_R) < \dim G$. Hence by induction hypothesis, we can choose $r' \geqslant 0$ such that

$$\int_{M \times A_R} |D_M(m)|^{-1/2}\, \Xi_M(m)(1 + \sigma(ma_2))^{-r'}\, dm\, da_2 < \infty.$$

Then if $r > r' + 2d$, it follows from Lemma 10.2 that

$$\int_A |D(a)|^{1/2}\, da \int_{G^*} \Xi(a^x)(1 + \sigma(a^x))^{-r}\, dx^* < \infty.$$

So now we may suppose that $A_R = \{1\}$ and therefore A is compact. Then the required result follows from [15, Theor. 5, p. 32]. This completes the proof of Lemma 1.

COROLLARY 1. *Let T be a locally summable function on G. Suppose that there exist numbers $c,\ m \geqslant 0$ such that*

$$|D|^{1/2}\,|T| \leqslant c(1 + \sigma)^m$$

almost everywhere on G. Then T is tempered as a distribution and

$$T(f) = \int_G T(x) f(x) \, dx$$

for all $f \in \mathscr{C}(G)$.

This is obvious from Lemma 1.

COROLLARY 2. *Let Θ be a central and $\mathfrak{z}$-finite distribution on G which is tempered. Then*

$$\Theta(f) = \int_G \Theta(x) f(x) \, dx \qquad (f \in \mathscr{C}(G)).$$

This follows from Theorem 12.1 and Corollary 1 above.

14. SOME ESTIMATES AND THEIR APPLICATIONS

LEMMA 1. *Let Θ be a locally summable function on G such that $\Theta(y^k) = \Theta(y)$ $(k \in K, y \in G)$ and*

$$c = \int_G |\Theta(y)| \, \Xi(y)(1 + \sigma(y))^{-r} \, dy < \infty$$

for some $r \geqslant 0$. Then

$$\int_G |\Theta(y)| \, \Xi(yx)(1 + \sigma(yx))^{-r} \, dy \leqslant c\Xi(x)(1 + \sigma(x))^{r}$$

for all $x \in G$.

Since

$$1 + \sigma(yx) \geqslant (1 + \sigma(y)) (1 + \sigma(x))^{-1},$$

it is clear that

$$\int |\Theta(y)| \, \Xi(yx)(1 + \sigma(yx))^{-r} \, dy \leqslant (1 + \sigma(x))^{r} \int |\Theta(y)| \, \Xi(yx)(1 + \sigma(y))^{-r} \, dy.$$

But since $\Theta(y) = \Theta(y^k)$ $(k \in K)$, we conclude that

$$\int |\Theta(y)| \, \Xi(yx)(1 + \sigma(y))^{-r} \, dy$$

$$= \int |\Theta(y)| \, (1 + \sigma(y))^{-r} \, dy \int_K \Xi(ykx) \, dk,$$

$$= c\Xi(x),$$

from the doubling principle. Our assertion is now obvious.

Put

$$\varXi_r = \varXi(1 + \sigma)^r$$

for $r \in \mathbf{R}$. Fix $r_0 \leqslant 0$ such that

$$c_0 = \int_G \varXi_{r_0}(x)^2 \, dx < \infty$$

(see Lemma 10.1).

LEMMA 2. *Let r_1, r_2 be two real numbers such that $|r_1| + r_2 \leqslant 2r_0$.
Then*

$$\int_G \varXi_{r_1}(xy^{-1}) \, \varXi_{r_2}(y) \, dy \leqslant c_0 \varXi_{r_1}(x) \qquad (x \in G).$$

It is clear that

$$(1 + \sigma(xy^{-1}))^{r_1} \leqslant (1 + \sigma(x))^{r_1} (1 + \sigma(y))^{|r_1|}.$$

Therefore

$$\varXi_{r_1}(xy^{-1}) \, \varXi_{r_2}(y) \leqslant (1 + \sigma(x))^{r_1} \varXi(xy^{-1}) \, \varXi_{r_3}(y),$$

where $r_3 = |r_1| + r_2$. But

$$\int \varXi(xy^{-1}) \, \varXi_{r_3}(y) \, dy = \int \varXi_{r_3}(y) \, dy \int_K \varXi(xky^{-1}) \, dk$$

$$= \varXi(x) \int \varXi_{r_3}(y) \, \varXi(y) \, dy \leqslant c_0 \varXi(x).$$

From this our assertion follows immediately.

COROLLARY. *Let $f, g \in \mathscr{C}(G)$. Then the convolution product $f * g$
lies in $\mathscr{C}(G)$. Moreover $(f, g) \to f * g$ is a continuous mapping of
$\mathscr{C}(G) \times \mathscr{C}(G)$ into $\mathscr{C}(G)$.*

This is an easy consequence of the above lemma.

A distribution T on G is said to be K-finite if both left and right
translates of T by elements of K span a finite-dimensional space.
We recall that a K-finite and $\mathfrak{z}$-finite distribution is an analytic
function [13, Lemma 33].

A continuous function f on G is said to satisfy the weak inequality if there exist numbers $c, r \geqslant 0$ such that

$$|f| \leqslant c\Xi(1 + \sigma)^r$$

THEOREM 1. *Let T be a tempered distribution on G which is K-finite and $\mathfrak{Z}$-finite. Then T is an analytic function and it satisfies the weak inequality.*

We note that both Theorem 9 and Lemma 65 of [15] are special cases of the above theorem.

By [15, Theor. 1] we can choose $\alpha_1, \alpha_2 \in C_c^\infty(G)$ such that $T = \alpha_1 * T = T * \alpha_2$. Put $\beta_i(x) = \alpha_i(x^{-1})$ $(x \in G, \ i = 1, 2)$. Then $T(f) = T(f')$ where $f' = \beta_1 * f * \beta_2$ $(f \in C_c^\infty(G))$. It is clear that given $g_1, g_2 \in \mathfrak{G}$ and $r \geqslant 0$, we can choose $c \geqslant 0$ such that

$$\sup_{x \in G} |f'(g_1 : x; g_2)| \, (1 + \sigma(x))^r \, \Xi(x)^{-1} \leqslant c \sup_{x \in G} |f(x)| \, (1 + \sigma(x))^r \, \Xi(x)^{-1}$$

for all $f \in C_c^\infty(G)$. Since T is tempered and $T(f) = T(f')$, we conclude that there exist numbers $c, r \geqslant 0$ such that

$$| T(f)| \leqslant c \sup |f| \, \Xi^{-1}(1 + \sigma)^r \qquad (f \in C_c^\infty(G)).$$

Clearly this implies that

$$\int_G | T(x)| \, \Xi(x)(1 + \sigma(x))^{-r} \, dx \leqslant c.$$

On the other hand $T * \alpha_2 = T$. Hence

$$T(y) = \int T(yx) \beta_2(x) \, dx.$$

Therefore

$$| T(y)| \leqslant c_1 \int | T(yx)| \, \Xi(x)(1 + \sigma(x))^{-r} \, dx$$

where

$$c_1 = \sup | \beta_2 | \, \Xi^{-1}(1 + \sigma)^r.$$

This shows that

$$| T(y)| \leqslant c_1 \int | T(x)| \, \Xi(y^{-1}x)(1 + \sigma(y^{-1}x))^{-r} \, dx,$$

$$\leqslant c_1(1 + \sigma(y))^r \int | T(x)| \, \Xi(y^{-1}x)(1 + \sigma(x))^{-r} \, dx.$$

But since T is K-finite, we can choose a finite subset F of $\mathscr{E}(K)$ such that[4]

$$T(y) = \int_K \alpha_F(k)\, T(k^{-1}y)\, dk.$$

Hence

$$|T(y)| \leqslant c_2(1 + \sigma(y))^r \int_{G \times K} |T(x)|\, \Xi(y^{-1}kx)(1 + \sigma(x))^{-r}\, dx\, dk$$

$$= c_3(1 + \sigma(y))^r\, \Xi(y) \qquad (y \in G),$$

from the doubling principle. Here $c_2 = c_1 \sup |\alpha_F|$ and $c_3 = c_2 c$.

15. The Space $\mathscr{C}(G, V)$

Let V be a complex locally convex space and $\mathscr{S}(V)$ the set of all continuous seminorms on V. Often it is convenient to write $|v|_{\mathbf{s}} = \mathbf{s}(v)$ for $\mathbf{s} \in \mathscr{S}(V)$ and $v \in V$.

Now suppose V is Hausdorff and complete and π is a representation of G on V [15, Sect. 2]. We say that π is differentiable if (1) every element of V is differentiable under π and (2) for any $g \in \mathfrak{G}$, $\pi(g)$ is a continuous endomorphism of V.

Let G_0 be an open set in G and $C^\infty(G_0, V)$ the space of all C^∞ functions from G_0 to V. Put $\mathfrak{G}^{(2)} = \mathfrak{G} \otimes \mathfrak{G}$ and regard it as an associative algebra under the multiplication given by

$$(g_1 \otimes g_2) \cdot (g_3 \otimes g_4) = g_3 g_1 \otimes g_2 g_4 \qquad (g_i \in \mathfrak{G}, 1 \leqslant i \leqslant 4).$$

Then $C^\infty(G_0, V)$ becomes a left $\mathfrak{G}^{(2)}$-module as follows. Let $f \in C^\infty(G_0, V)$ and $g_i \in \mathfrak{G}$ $(i = 1, 2)$. Then $f' = (g_1 \otimes g_2)f$ is given by (see [12, p. 459])

$$f'(x) = f(g_1 : x; g_2) \qquad (x \in G_0).$$

For $\mathbf{s} \in \mathscr{S}(V)$, $D \in \mathfrak{G}^{(2)}$ and $r \in \mathbf{R}$, put

$$\mathbf{s}_{D,r}(f) = \sup_G |Df|_{\mathbf{s}}\, \Xi^{-1}(1 + \sigma)^{-r} \qquad (f \in C^\infty(G_0, V)).$$

Let $\mathscr{C}(G_0, V)$ denote the space of all $f \in C^\infty(G_0, V)$ such that $\mathbf{s}_{D,r}(f) < \infty$ for all $\mathbf{s}$, D and r. We introduce the structure of a locally convex space in $\mathscr{C}(G_0, V)$ by means of the seminorms $\mathbf{s}_{D,r}$ ($\mathbf{s} \in \mathscr{S}(V)$, $D \in \mathfrak{G}^{(2)}$, $r \in \mathbf{R}$).

---

[4] $\alpha_F = \Sigma_{\mathfrak{d} \in F}\, \alpha_{\mathfrak{d}}$ in the notation of [15, Sect. 5].

Let M be a differentiable manifold (not necessarily connected) and $C_c^\infty(M, V)$ the space of all C^∞-functions from M to V with compact support. We topologize it in the usual way. If $\Phi = \{D_i\}_{i\in I}$ is a scattered family of differential operators on M and $\mathbf{s} \in \mathscr{S}(V)$, then we put

$$\mathbf{s}_\Phi(f) = \sum_{i\in I} \sup |D_i f|_\mathbf{s} \qquad (f \in C_c^\infty(M, V)).$$

The seminorms $\mathbf{s}_\Phi$ define the structure of a locally convex space in $C_c^\infty(M, V)$. One proves as usual that the spaces $C_c^\infty(M, V)$ and $\mathscr{C}(G_0, V)$ are Hausdorff and complete. Moreover the inclusion mapping $C_c^\infty(G_0, V)$ into $\mathscr{C}(G_0, V)$ is continuous.

LEMMA 1. $C_c^\infty(G, V)$ *is dense in* $\mathscr{C}(G, V)$.

This can be proved as in [15, Sect. 13] but the following proof is better. For any $t > 0$, let G_t denote the set of all $x \in G$ with $\sigma(x) < t$ and let ξ_t be the characteristic function of G_t. Fix $\alpha \in C_c^\infty(G)$ such that

$$\int_G \alpha(x)\, dx = 1$$

and put

$$u_t =: \alpha * (1 - \xi_t) * \alpha = 1 - \alpha * \xi_t * \alpha.$$

Choose $a > 0$ such that $\operatorname{Supp} \alpha \subset G_a$.

LEMMA 2.

$$u_t(x) = \begin{cases} 0 & \textit{if} \;\; \sigma(x) \leqslant t - 2a, \\ 1 & \textit{if} \;\; \sigma(x) \geqslant t + 2a. \end{cases}$$

Moreover

$$|u_t(g_1 : x; g_2)| \leqslant \int_{G\times G} |\alpha(g_1 : y_1)\, \alpha(y_2 ; g_2)|\, dy_1\, dy_2$$

for $x \in G$ *and* $g_1, g_2 \in \mathfrak{G}$.

This follows from the definition of u_t.
Now fix $f \in \mathscr{C}(G, V)$ and put

$$f_t = (1 - u_t)f = (\alpha * \xi_t * \alpha)f.$$

Then $f_t \in C_c^\infty(G, V)$ and one proves easily from the above lemma (see [15, p. 27]) that $f_t \to f$ in $\mathscr{C}(G, V)$ as $t \to \infty$.

LEMMA 3. *K-finite elements are dense in both* $C_c^\infty(G, V)$ *and* $\mathscr{C}(G, V)$.

This is proved in the same way as [15, Lemmas 9 and 16].
For $f \in \mathscr{C}(G)$ and $v \in V$, let $f \otimes v$ denote the function

$$x \to f(x)v$$

from G to V. Then $f \otimes v \in \mathscr{C}(G, V)$. In this way we get an injection
of $\mathscr{C}(G) \otimes V$ into $\mathscr{C}(G, V)$ and we may thus regard $\mathscr{C}(G) \otimes V$ as
a subspace of $\mathscr{C}(G, V)$. Similarly $C_c^\infty(G) \otimes V \subset C_c^\infty(G, V)$.

LEMMA 4. $C_c^\infty(G) \otimes V$ *is dense in* $\mathscr{C}(G, V)$.

First suppose G is compact. Then $G = K$ and $\mathscr{C}(G, V) = C^\infty(K, V)$.
It is obvious that every K-finite element of $C^\infty(K, V)$ is in $C^\infty(K) \otimes V$.
Hence our assertion follows in this case from Lemma 3.
On the other hand we have the following general result.

LEMMA 5. *If M is a differentiable manifold, then* $C_c^\infty(M) \otimes V$ *is
dense in* $C_c^\infty(M, V)$.

Assuming this it follows from Lemma 1 that $C_c^\infty(G) \otimes V$ is dense
in $\mathscr{C}(G, V)$.
Let J denote the compact Lie group $\mathbf{R}/\mathbf{Z}$ and I the open subset of J
obtained by removing the zero element of J. Then I may be identified
in the natural way with the open interval $0 < t < 1$ $(t \in \mathbf{R})$. Let J^n
and I^n denote the corresponding n-fold Cartesian products. Then by
the result proved above, $C^\infty(J^n) \otimes V$ is dense in $C^\infty(J^n, V)$. From
this we conclude immediately that $C_c^\infty(I^n) \otimes V$ is dense in $C_c^\infty(I^n, V)$.
Using a partition of unity, it is now clear that $C_c^\infty(M) \otimes V$ is dense
in $C_c^\infty(M, V)$. (This proof was pointed out to me by L. Hörmander.)

16. The Mapping $f \to f^{(P)}$

Let $P = MAN$ be a psgp of G. If $\mathbf{s} \in \mathscr{S}(V)$ and $f \in \mathscr{C}(G, V)$, it
follows from Lemma 10.2 that

$$\int_N |f(xn)|_{\mathbf{s}}\, dn < \infty \qquad (x \in G).$$

Therefore since V is complete, we can define

$$f^P(x) = \int_N f(xn)\, dn \qquad (x \in G).$$

Using Corollary 2 of Lemma 10.2, it is easy to verify that f^P lies in $C^\infty(G, V)$ and one can differentiate inside the integral sign.

Put $\mathfrak{z} = \mathfrak{m} + \mathfrak{a}$ and let $\mu_P = \gamma_{\mathfrak{g}/\mathfrak{z}}$ denote the usual isomorphism of $\mathfrak{z}$ into $\mathfrak{z}_M\mathfrak{A}$ [12, p. 474]. ($\mathfrak{z}_M$ is the center of $\mathfrak{M}$.)

LEMMA 1. *For $f \in \mathscr{C}(G, V)$, put*

$$f^{(P)}(ma) = e^{\rho(\log a)}f^P(ma) \qquad (m \in M, \ a \in A),$$

where $\rho = \rho_P$. Then $f^{(P)} \in \mathscr{C}(MA, V)$ and $f \to f^{(P)}$ is a continuous mapping of $\mathscr{C}(G, V)$ into $\mathscr{C}(MA, V)$. Moreover

$$(\mathfrak{z}f)^{(P)} = \mu_P(\mathfrak{z})f^{(P)}$$

for $\mathfrak{z} \in \mathfrak{z}$.

This is a simple consequence of Lemma 10.2 and the definition of μ_P (see [15, Sect. 15]).

Let $*P = *M \, *A \, *N$ be a psgp of M and $P' = M'A'N'$ the corresponding psgp of G so that $P' \subset P$ (see Lemma 6.1). Fix $a \in A$ and put $g(m) = f^{(P)}(ma)$ $(m \in M)$. It is clear that $g \in \mathscr{C}(M, V)$.

LEMMA 2. $g^{(*P)}(*m \, *a) = f^{(P')}(*m \, *a \cdot a)$ *for* $*m \in *M, \ *a \in *A$.

Since $M' = *M$, $A' = *A \cdot A$ and $N' = *N \cdot N$, this is an easy consequence of the definitions.

17. THE FUNCTION $'F_f$

Let A be a θ-stable Cartan subgroup of G. Let P_I denote the set of all positive imaginary roots of $(\mathfrak{g}, \mathfrak{a})$ (under some order). Put

$$'\Delta_I(a) = \prod_{\alpha \in P_1} (1 - \xi_\alpha(a^{-1})) \qquad (a \in A),$$

and

$$'F_f{}^A(a) = 'F_f(a) = '\Delta_I(a)\,\Delta_+(a) \int_{G/A_R} f(a^x)\, dx^* \qquad (a \in A \cap G'),$$

for $f \in \mathscr{C}(G, V)$, where Δ_+ is defined as in Section 8. Since $G = KG^0$, it follows from [15, Sect. 17, Theor. 5] that this integral is well defined.

THEOREM 1. *Let $A'(I)$ be the set of all $a \in A$ such that $\xi_\alpha(a) \neq 1$ for every singular imaginary root α of $(\mathfrak{g}, \mathfrak{a})$. Then $f \to 'F_f$ defines a continuous mapping of $\mathscr{C}(G, V)$ into $\mathscr{C}(A'(I), V)$.*

This means, in particular, that for any $f \in \mathscr{C}(G, V)$, the function $'F_f$ has a (unique) continuous extension on $A'(I)$ which lies in $\mathscr{C}(A'(I), V)$. (Here $A'(I)$ is to be regarded an open subset of A and the space $\mathscr{C}(A'(I), V)$ is then defined as in Sect. 15.)

We proceed by induction on $\dim G$. First assume $\mathfrak{a}_R \neq \{0\}$. Fix $P \in \mathscr{P}(A_R)$. Then $P = MA_RN$. Put

$$\tilde{f}(x) = \int_K f(kxk^{-1}) \, dk \qquad (x \in G),$$

and $g_f = \tilde{f}^{(P)}$. Then we conclude easily (see Lemma 8.2 and [15, Theor. 5]) that

$$'F_f(a) = '\Delta_I(a) \int_M g_f(a^m) \, dm.$$

Now if $\mathfrak{a}_R$ does not lie in the center of $\mathfrak{g}$, $\dim(MA_R) < \dim G$. Since $f \to \tilde{f}$ is obviously a continuous endomorphism of $\mathscr{C}(G, V)$, our assertion follows by applying the induction hypothesis to MA_R.

The case when $\mathfrak{a}_R \neq \{0\}$ and $\mathfrak{a}_R$ lies in the center of $\mathfrak{g}$, presents no trouble (see [15, p. 36]). So we may now assume that $\mathfrak{a}_R = \{0\}$. We write B instead of A, in this case.

Let $\mathscr{C}_0 = \mathscr{C}_0(G, V)$ denote the space $C_c^\infty(G, V)$ with the topology induced on it from $\mathscr{C}(G, V)$. For any $\mathbf{s} \in \mathscr{S}(V)$ and $f \in \mathscr{C}(G, V)$, put

$$\nu_\mathbf{s}(f) = \int_B | \, '\Delta(b) \, 'F_f(b)|_\mathbf{s} \, db,$$

where $'\Delta = '\Delta_I$. It follows from [15, Theor. 5], that $\nu_\mathbf{s}$ is a continuous seminorm on $\mathscr{C}(G, V)$. On the other hand, it follows from elementary arguments that for $f \in \mathscr{C}_0$, $'F_f$ is a C^∞-function on $B' = B \cap G'$.

LEMMA 2.　*Given $u \in \mathfrak{S}(\mathfrak{b}_c)$ and $\mathbf{s} \in \mathscr{S}(V)$, we can choose a finite number of elements $z_1, ..., z_N$ in $\mathfrak{Z}$ such that*

$$\sup_{b \in B'} | \, 'F_f(b; u)|_\mathbf{s} \leqslant \sum_{1 \leqslant i \leqslant N} \nu_s(z_i f)$$

for all $f \in \mathscr{C}_0$.

This is proved in the same way as [12, Lemma 43].

Applying Hahn–Banach theorem, we conclude from [12, Lemma 40] that $'F_f$ extends to a C^∞-function on $B'(I)$ ($f \in \mathscr{C}_0$). Since $\mathscr{C}_0$ is dense in $\mathscr{C}(G, V)$ and $\mathscr{C}(B'(I), V)$ is complete, the assertion of Theorem 1 is now obvious from [15, Theor. 5], if we recall that $B \subset K$.

It is now clear that we can define $F_f{}^\mathfrak{a}$ as in Section 8 for $f \in \mathscr{C}(G, V)$.

LEMMA 3. *The statement of Theorem 9.1 remains true for* $f \in \mathscr{C}(G, V)$.

This is obvious from Theorem 1.
Put

$$\varpi = \varpi_{\mathfrak{g}/\mathfrak{a}} = \prod_{\alpha > 0} H_\alpha ,$$

where α runs over all positive roots of $(\mathfrak{g}, \mathfrak{a})$. Define

$$\rho_I = \frac{1}{2} \sum_{\alpha \in P_I} \alpha$$

and let $u \to u'$ denote the automorphism of $\mathfrak{A}$ such that $H' = H + \rho_I(H)$ $(H \in \mathfrak{a})$. Finally put $\gamma = \gamma_{\mathfrak{g}/\mathfrak{a}}$ (see Sect. 11).

LEMMA 4. *For any* $f \in \mathscr{C}(G, V)$, $\varpi'\, 'F_f{}^A$ *extends to a continuous function on* A. *Moreover* $'F_{zf}^A = \gamma(z)'\, 'F_f{}^A$ $(z \in \mathfrak{Z})$.
This is obvious from Theorem 1 and [12, Sect. 22].

LEMMA 5. *Suppose* $\mathfrak{a}$ *is fundamental in* $\mathfrak{g}$. *Then there exists a constant* $c \neq 0$ *such that*

$$'F_f{}^A(1; \varpi') = cf(1)$$

for all $f \in \mathscr{C}(G, V)$.

This is clear (see [15, p. 47]).

18. CUSP FORMS

Let $\mathscr{P}$ be the set of all psgps P of G with $\operatorname{prk} P \geqslant 1$. We denote by $^0\mathscr{C}(G, V)$ the space of all $f \in \mathscr{C}(G, V)$ such that $f^P = 0$ for all $P \in \mathscr{P}$. Then $^0\mathscr{C}(G, V)$ is a closed subspace of $\mathscr{C}(G, V)$ which is stable under left and right translations of G. An element $f \in \mathscr{C}(G, V)$ is called a cusp form if $f \in {}^0\mathscr{C}(G, V)$.

THEOREM 1. *Every* $\mathfrak{Z}$-*finite function in* $\mathscr{C}(G, V)$ *is a cusp form. Moreover* $^0\mathscr{C}(G, V) = \{0\}$ *unless* $\operatorname{rank} G = \operatorname{rank} K$.

Let f be a $\mathfrak{Z}$-finite function in $\mathscr{C}(G, V)$. Fix $P = MAN$ in $\mathscr{P}$. Since $\mathfrak{Z}_M\mathfrak{A}$ is a finite module over $\mu_P(\mathfrak{Z})$, we conclude from

Lemma 16.1 that $f^{(P)}$ is $\mathfrak{Z}_M\mathfrak{A}$-finite. Fix $m \in M$. Then the function

$$g : a \to f^{(P)}(ma) \qquad (a \in A)$$

is $\mathfrak{A}$-finite and it lies in $\mathscr{C}(A, V)$ (Lemma 16.1). But $\dim A =$ prk $P \geqslant 1$ and so it follows by standard arguments [15, p. 76] that $g = 0$. This shows that $f^{(P)} = 0$ and therefore $f^{(P)}(1) = 0$. Replacing f by a left translate of f, we conclude that $f^P = 0$. This proves that f is a cusp form.

Fix a θ-stable fundamental Cartan subgroup A of G. If rank $G >$ rank K, then $\dim A_R \geqslant 1$. Fix a psgp $P = MA_RN$ in $\mathscr{P}(A_R)$. Then if $f \in {}^0\mathscr{C}(G, V)$, $f^P = 0$. Hence we conclude (see Lemma 8.3 and Sect. 17) that $'F_f{}^A = 0$. But then it follows from Lemma 17.5 that $f(1) = 0$. Replacing f by a left translate of f, we conclude that $f = 0$. This completes the proof of Theorem 1.

Now assume that rank $G =$ rank K. Fix a compact Cartan subgroup B of G $(B \subset K)$. Let B^* denote the set of all irreducible characters of B and $\langle b^*, b \rangle$ the value of b^* at b $(b^* \in B^*, b \in B)$. Let $\log b^*$ denote the linear function μ on $\mathfrak{b}$ given by

$$\langle b^*, \exp H \rangle = d(b^*)e^{\mu(H)} \qquad (H \in \mathfrak{b}),$$

where $d(b^*) = \langle b^*, 1 \rangle$ is the degree of b^*. We put $\lambda(b^*) = \mu + \rho$ and

$$\varpi(b^*) = \varpi(\lambda(b^*)).$$

Here $\rho = \rho_I$ (see Sect. 17).

LEMMA 2.　*Let $f \in {}^0\mathscr{C}(G, V)$. Then $'F_f = {}'F_f{}^B$ lies in $\mathscr{C}(B, V)$ and*

$$'F_f(b) = \sum_{b^* \in B^*} (b^*, {}'F_f)\langle b^*, b \rangle \qquad (b \in B).$$

Here

$$(b^*, {}'F_f) = \int_B \mathrm{conj}\langle b^*, b \rangle \cdot {}'F_f(b) \, db,$$

and db is the normalized Haar measure on the compact group B.

To prove the first statement we apply Lemma 17.3 and observe that $F_f{}^\mathfrak{a} = 0$ since f is a cusp form. Since $'F_f$ is a class-function on the compact group B, the rest is obvious.

Corollary.

$$'F_f(b; \varpi') = \sum_{b^* \in B^*} \varpi(b^*)(b^*, \, 'F_f)\langle b^*, b \rangle$$

for $f \in {}^0\mathscr{C}(G, V)$ and $b \in B$.

This follows immediately by differentiation.

Let $B^{*\prime}$ be the set of all $b^* \in B^*$ such that $\varpi(b^*) \neq 0$. Also let L' be the set of all $\lambda(b^*)$ ($b^* \in B^{*\prime}$). For any $\lambda \in L'$, let $\mathscr{C}_\lambda(G, V)$ denote the set of all $f \in \mathscr{C}(G, V)$ such that

$$zf = \gamma(z : \lambda)f \qquad (z \in \mathfrak{Z}),$$

where $\gamma = \gamma_{\mathfrak{g}/\mathfrak{b}}$ and $\gamma(z : \lambda)$ is the value of $\gamma(z)$ at λ, regarded as a polynomial function on the dual of $\mathfrak{b}_c$.

THEOREM 3. *Every $\mathfrak{Z}$-finite function in $\mathscr{C}(G, V)$ can be written as a finite sum of functions in $\mathscr{C}_\lambda(G, V)$ ($\lambda \in L'$).*

It will be enough to prove the following lemma.

LEMMA 4. *Let χ be a homomorphism of $\mathfrak{Z}$ into $\mathbf{C}$, N an integer $\geqslant 1$ and f an element in $\mathscr{C}(G, V)$ such that*

$$(z - \chi(z))^N f = 0 \qquad (z \in \mathfrak{Z}).$$

Then we can choose $\lambda \in L'$ such that $f \in \mathscr{C}_\lambda(G, V)$.

We know that

$$'F_{zf} = \gamma(z)' \cdot {}'F_f \qquad (z \in \mathfrak{Z}).$$

Hence we conclude from the corollary of Lemma 2 that

$$0 = \sum_{b^* \in B^*} \varpi(b^*)(b^*, \, 'F_f)\langle b^*, b \rangle \, (\gamma(z : b^*) - \chi(z))^N,$$

for $b \in B$. Here $\gamma(z : b^*) = \gamma(z : \lambda(b^*))$. Put $\chi_{b^*}(z) = \gamma(z : b^*)$. Then it is clear from the above relation that

$$(b^*, \, 'F_f) = 0 \qquad (b^* \in B^{*\prime}),$$

unless $\chi_{b^*} = \chi$. Therefore

$$0 = \sum_{b^* \in B^*} \varpi(b^*)(b^*, \, 'F_f)\langle b^*, b \rangle \, (\gamma(z : b^*) - \chi(z))$$

for $z \in \mathfrak{Z}$ and $b \in B$. Now fix $z \in \mathfrak{Z}$ and put $f_0 = (z - \chi(z))f$. Then it is clear that

$$\varpi' \cdot {}'F_{f_0} = 0,$$

and therefore $f_0(1) = 0$ (Lemma 17.5). Replacing f by a left translate of f, we conclude that $f_0 = 0$. This proves that

$$zf = \chi(z)f \qquad (z \in \mathfrak{Z}).$$

Now suppose $\chi \neq \chi_{b*}$ for every $b* \in B*'$. Then we conclude as above that

$$\varpi' \cdot {}'F_f = 0,$$

and therefore $f(1) = 0$. Again replacing f by a left translate of f, we conclude that $f = 0$. This proves the lemma.

19. Definition of the Eisenstein Integral

By a double representation τ of K on V, we mean a pair (τ_1, τ_2) where τ_1 is a left representation and τ_2 a right representation of K on V (see [15, Sect. 2]) and $\tau_1(k_1)$ commutes with $\tau_2(k_2)$ $(k_1, k_2 \in K)$ as linear transformations on V. We write $\dim \tau = \dim V$ and say that τ is differentiable if both τ_1, τ_2 are differentiable (see Sect. 15). Often it is convenient to drop subscripts and write $\tau(k_1) \, v\tau(k_2)$ instead of $\tau_1(k_1) \, v\tau_2(k_2)$ $(k_1, k_2 \in K, v \in V)$. A continuous function $f \colon G \to V$ is said to be τ-spherical (or of type τ) if

$$f(k_1 x k_2) = \tau(k_1) f(x) \tau(k_2) \qquad (k_1, k_2 \in K, x \in G).$$

We shall assume from now on that τ is differentiable. Let $C^\infty(G, \tau)$ and $\mathscr{C}(G, \tau)$ be the spaces of all τ-spherical functions in $C^\infty(G, V)$ and $\mathscr{C}(G, V)$ respectively. Put ${}^0\mathscr{C}(G, \tau) = \mathscr{C}(G, \tau) \cap {}^0\mathscr{C}(G, V)$.

Let $P = MAN$ be a psgp of G. We denote by τ_M the restriction of τ on $K_M = K \cap M$. Given $\phi \in C^\infty(M, \tau_M)$, we extend it to a function on G by setting

$$\phi(kman) = \tau(k) \phi(m) \qquad (k \in K, m \in M, a \in A, n \in N).$$

Since $K \cap P = K_M$, this extension is well defined and is a C^∞-function. For any $\nu \in \mathfrak{a}_c{}^*$, put

$$E(P : \phi : \nu : x) = E(\phi : \nu : x)$$

$$= \int_K \phi(xk) \, \tau(k^{-1}) \exp\{((-1)^{1/2}\nu - \rho_P) (H_P(xk))\} \, dk$$

for $x \in G$ and $\phi \in C^\infty(M, \tau_M)$. For a fixed ϕ, it is a C^∞-function from $\mathfrak{a}_c^* \times G$ to V, which is holomorphic in ν. We call it the Eisenstein integral. (The important case is when $\phi \in {}^0\mathscr{C}(M, \tau_M)$.)

Put $\mu = \mu_P$ in the notation of Lemma 16.1. Then for any $z \in \mathfrak{Z}$, $\mu(z) \in \mathfrak{Z}_M \mathfrak{A} \simeq \mathfrak{Z}_M \otimes \mathfrak{A}$. Hence we may regard $\mu(z)$ as a polynomial function on $\mathfrak{a}_c^*$ with values in $\mathfrak{Z}_M$. Let $\mu(z : \nu)$ denote the value of this function at $\nu \in \mathfrak{a}_c^*$.

LEMMA 1. $E(\phi : \nu : x; z) = E(\mu(z : (-1)^{1/2} \nu) \phi : \nu : x)$ *for* $\phi \in$ $C^\infty(M, \tau_M)$, $\nu \in \mathfrak{a}_c^*$, $x \in G$ *and* $z \in \mathfrak{Z}$.

This is a simple consequence of the definition of μ and $E(\phi : \nu)$.

It follows from this lemma that if ϕ is an eigenfunction of $\mathfrak{Z}_M$, then $E(\phi : \nu)$ is an eigenfunction of $\mathfrak{Z}$.

20. A CHARACTERISTIC PROPERTY OF E

We say that τ is unitary if there exists a positive-definite, continuous hermitian form H on V which is invariant under τ. Put $(u, v) = H(u, v)$ and $|u| = (u, u)^{1/2}$ $(u, v \in V)$. This defines a prehilbertian structure on V.

Assume τ is unitary, $P = MAN$ is a psgp of G and $f \in \mathscr{C}(G, \tau)$. Then we write $f^P \sim 0$ if

$$\int_M (\phi(m), f^P(ma)) \, dm = 0$$

for all $\phi \in {}^0\mathscr{C}(M, \tau_M)$ and $a \in A$. Here the scalar product is in V and it follows from Lemma 16.1 that the integral is well defined.

LEMMA 1.   *Let* f *be an element in* $\mathscr{C}(G, \tau)$ *such that* $f^P \sim 0$ *for every* psgp P *of* G. *Then* $f = 0$.

This follows by an easy induction on $\dim G$ if we make use of Lemma 16.2.

LEMMA 2.   *Fix a* psgp $P = MAN$ *of* G *and* $f \in \mathscr{C}(G, \tau)$. *Then* $f^P \sim 0$ *if and only if*

$$\int_G (E(P : \phi : \nu : x), f(x)) \, dx = 0$$

for all $\phi \in {}^0\mathscr{C}(M, \tau_M)$ *and* $\nu \in \mathfrak{a}^*$.

Put $\rho = \rho_P$. Then it is clear that

$$\int_G |E(P:\phi:\nu:x)|\,|f(x)|\,dx$$

$$= \int_{M\times A\times N} |\phi(m)|\,|f(man)|\,e^{\rho(\log a)}\,dm\,da\,dn < \infty$$

from Lemma 10.2. Therefore we conclude that

$$\int_G (E(P:\phi:\nu:x), f(x))\,dx$$

$$= \int_{M\times A} (\phi(m), f^{(P)}(ma))e^{-(-1)^{1/2}\nu(\log a)}\,dm\,da.$$

Since $f^{(P)} \in \mathscr{C}(MA, \tau_M)$ (Lemma 16.1) our assertion is obvious.

21. The Constant Term

A function $f: G \to V$ is said to satisfy the weak inequality if for every $\mathbf{s} \in \mathscr{S}(V)$, we can choose $r \geqslant 0$ such that

$$\sup_{x \in G} |f(x)|_{\mathbf{s}}\, \Xi(x)^{-1}(1 + \sigma(x))^{-r} < \infty.$$

Given V and τ as in Section 19, we denote by $\mathscr{A}(G, \tau)$ the space of all $f \in C^\infty(G, \tau)$ satisfying the following two conditions.

(1) f is $\mathfrak{z}$-finite.
(2) For every $D \in \mathfrak{G}^{(2)}$, Df satisfies the weak inequality.

Remark. If $\dim \tau < \infty$, it follows from [15, Theor. 1] that it is enough to assume the weak inequality for f itself (see [15, Lemma 48]).
Let $P = MAN$ be a psgp. Put

$$d_P(x) = e^{\rho(H(x))} \qquad (x \in G),$$

where $\rho = \rho_P$ and $H(x) = H_P(x)$. Let a be a variable element in A. We say that $a \to_P \infty$ if there exists a number $\epsilon > 0$ such that (1) $\alpha(\log a) \geqslant \epsilon\sigma(a)$ and (2) $\alpha(\log a) \to +\infty$ for every root α of (P, A). (If $P = G$, the statement $a \to_P \infty$ means that a varies freely in A.)

THEOREM 1. *Given $f \in \mathscr{A}(G, \tau)$, there exists a unique element $f_P \in \mathscr{A}(MA, \tau_M)$ such that*

$$\lim_{\substack{a \to \infty \\ P}} \{d_P(ma)\, f(ma) - f_P(ma)\} = 0$$

for $m \in MA$.

In particular $f_G = f$. We call f_P the constant term of f along P.

Before giving a proof, we state some important properties of the mapping $f \mapsto f_P$.

Let $*P = *M\, *A\, *N$ be a psgp of M and $P' = M'A'N'$ the corresponding psgp of G so that (Lemma 6.1)

$$M' = *M, \ A' = *A \cdot A, \ N' = *N \cdot N.$$

LEMMA 1. *Fix $f \in \mathscr{A}(G, \tau)$, $a \in A$ and put*

$$g(m) = f_P(ma) \qquad (m \in M).$$

Then $g \in \mathscr{A}(M, \tau_M)$ and

$$g_{*P}(*m) = f_{P'}(*ma) \qquad (*m \in *M*A).$$

Moreover

$$(zf)_P = \mu_P(z) f_P \qquad (z \in \mathfrak{Z}),$$

and

$$f_{P^k}(m^k) = \tau(k)\, f_P(m)\, \tau(k^{-1}) \qquad (m \in MA, \ k \in K).$$

Here the analogy with the results of Section 16 is obvious.

Let U, V, W be three locally convex Hausdorff spaces. By a pairing of U, V into W we mean a continuous bilinear mapping $(u, v) \mapsto u \cdot v$ of $U \times V$ into W.

LEMMA 2. *A bilinear mapping of $U \times V$ into W is continuous if and only if the following condition holds. Given $\mathbf{s} \in \mathscr{S}(W)$, we can choose $\mathbf{s_1} \in \mathscr{S}(U)$, $\mathbf{s_2} \in \mathscr{S}(V)$ such that*

$$\mathbf{s}(u \cdot v) \leqslant \mathbf{s_1}(u)\, \mathbf{s_2}(v)$$

for all $u \in U$, $v \in V$.

This is immediate.

Now suppose U, V, W are complete and we are given a pairing of U, V into W. Moreover suppose we are also given differentiable double representations τ_U, τ_V, τ_W of K on U, V, W, respectively.

We assume that these representations are compatible, which means that

$$\tau_W(k)(u \cdot v) = (\tau_U(k)u) \cdot v, \quad u\tau_U(k) \cdot v = u \cdot \tau_V(k)v,$$

$$(u \cdot v)\,\tau_W(k) = u \cdot v\tau_V(k)$$

for $k \in K$, $u \in U$, $v \in V$. Fix $\alpha \in \mathscr{C}(G, \tau_U)$, $f \in \mathscr{A}(G, \tau_V)$ and define $\phi = \alpha * f$ by the integral

$$\phi(x) = \int_G \alpha(xy^{-1}) \cdot f(y)\,dy \qquad (x \in G).$$

It follows immediately from Lemma 14.2 that $\phi \in \mathscr{A}(G, \tau_W)$.

THEOREM 2. *Let* $P = MAN$ *be a* psgp *of* G *and suppose* $\alpha \in \mathscr{C}(G, \tau_U)$ *and* $f \in \mathscr{A}(G, \tau_V)$. *Then* $\alpha * f \in \mathscr{A}(G, \tau_W)$ *and*

$$(\alpha \underset{G}{*} f)_P = \alpha^{(P)} \underset{MA}{*} f_P \,.$$

Here the convolution on the left is over G and on the right over MA. Since $\alpha^{(P)} \in \mathscr{C}(MA, (\tau_U)_M)$ (Lemma 16.1) and $f_P \in \mathscr{A}(MA, (\tau_V)_M)$, the convolution $\alpha^{(P)} * f_P$ is well defined.

The uniqueness in Theorem 1 is an immediate consequence of the following simple result.

LEMMA 3. *Fix* $g \in \mathscr{A}(MA, \tau_M)$, $m \in MA$, $H \in \mathfrak{a}$ *and suppose that* $g(m \exp tH) \to 0$ *weakly in* V *as* $t \to +\infty$. *Then* $g(m) = 0$.

Fix a continuous linear function λ on V. It would be enough to show that $\lambda(g(m)) = 0$. Put

$$F(t) = \lambda(g(m \exp tH)) \qquad (t \in \mathbf{R}).$$

Since g is $\mathfrak{A}$-finite, F satisfies a differential equation of the form

$$\sum_{0 \leqslant j \leqslant p} c_j d^j F/dt^j = 0,$$

where $c_j \in \mathbf{C}$ and $c_p = 1$. Moreover we can choose $c, q \geqslant 0$ such that

$$|F(t)| \leqslant c(1 + |t|)^q \qquad (t \in \mathbf{R}).$$

This follows from the weak inequality. Therefore we conclude that

$$F(t) = \sum_{1 \leqslant i \leqslant r} p_i(t) e^{(-1)^{1/2} \nu_i t},$$

where p_i are polynomials and $v_i \in \mathbf{R}$. We may assume that $v_1, \ldots, v_r$ are all distinct. Since $F(t) \to 0$ as $t \to +\infty$, we conclude (see [8, Sect. 15]) that $p_1 = p_2 = \cdots = p_r = 0$. This proves that $\lambda(g(m)) = F(0) = 0$.

22. The Differential Equations for $\Phi(f)$

We shall now begin the proof of Theorem 21.1.

For any $\mathbf{s} \in \mathscr{S}(V)$, $r \in \mathbf{R}$ and $D \in \mathfrak{G}^{(2)}$, define (as in Sect. 15)

$$\mathbf{s}_{D,r}(f) = \sup_G |Df| \cdot \varXi^{-1}(1 + \sigma)^{-r} \qquad (f \in C^\infty(G, V)).$$

If F is a finite subset of $\mathfrak{G}^{(2)}$, put

$$\mathbf{s}_{F,r}(f) = \sum_{D \in F} \mathbf{s}_{D,r}(f).$$

Fix an ideal $\mathfrak{U}$ in $\mathfrak{Z}$ such that $\dim \mathfrak{Z}/\mathfrak{U} < \infty$ and let $\mathscr{A}(G, \tau, \mathfrak{U})$ denote the set of all $f \in \mathscr{A}(G, \tau)$ such that $uf = 0$ for all $u \in \mathfrak{U}$.

Let $\mathscr{S}^0(V)$ denote the set of all $\mathbf{s} \in \mathscr{S}(V)$ such that

$$|\tau(k_1)\, v\tau(k_2)|_\mathbf{s} = |v|_\mathbf{s}$$

for all $k_1, k_2 \in K$ and $v \in V$. For any $s \in \mathscr{S}(V)$, define

$$\mathbf{s}^0(v) = \sup_{k_1, k_2 \in K} |\tau(k_1)\, v\tau(k_2)|_\mathbf{s} \qquad (v \in V).$$

Since K is compact, it is clear (see [15, p. 5]) that $\mathbf{s}^0 \in \mathscr{S}^0(V)$. Therefore the set $\mathscr{S}^0(V)$ of seminorms is equivalent to $\mathscr{S}(V)$.

Fix a psgp $P = MAN$ of G and a minimal psgp $P_0 = M_0 A_0 N_0$ contained in P. Then $A \subset A_0 \subset M_1 = MA$. Put

$$M_1^+ = K_1 Cl(A_0^+)\, K_1 \subset M_1,$$

where $K_1 = K_M = K \cap M$ and A_0^+ is defined as in Section 6. Since any two such subgroups P_0 are conjugate under K_1 (Lemmas 6.1 and 6.2), the definition of M_1^+ is actually independent of the choice of P_0. Choose $c_0, r_0 \geqslant 0$ such that

$$d_P(m)\, \varXi(m) \leqslant c_0 \varXi_M(m)(1 + \sigma(m))^{r_0}$$

for $m \in M_1^+$ (see [15, p. 63]) and put

$$\beta_P(H) = \inf_\alpha \alpha(H) \qquad (H \in \mathfrak{a}),$$

where α runs over all roots of (P, A). (If $P = G$, we define $\beta_G = 0$.) Moreover let

$$|(x, y)| = (1 + \sigma(x))(1 + \sigma(y)) \qquad (x, y \in G).$$

LEMMA 1. *Fix* $g_1, g_2 \in \mathfrak{G}$ *and* $X \in \mathfrak{n}$. *Then we can choose a finite subset F of* $\mathfrak{G}^{(2)}$ *with the following property:*

$$d_P(ma)\, \{|\, f(g_1 X \,;\, ma; g_2)|_{\mathbf{s}} + |\, f(g_1 \,;\, ma; \theta(X)g_2)|_{\mathbf{s}}\}$$
$$\leqslant \mathbf{s}_{F,r}(f)\, \Xi_M(m)\, |(m, a)|^{r_0 + r} e^{-\beta_P(\log a)}$$

and

$$d_P(m)\, |\, f(g_1 \,;\, m; g_2)|_s \leqslant \mathbf{s}_{F,r}(f)\, \Xi_M(m)(1 + \sigma(m))^{r_0 + r}$$

for all $\mathbf{s} \in \mathscr{S}^0(V)$, $f \in \mathscr{A}(G, \tau)$, $m \in M_1^+$, $a \in Cl\, A^+$ *and* $r \geqslant 0$.

We can write $m = k_1^{-1} a_0 k_2$ $(k_1, k_2 \in K_1,\ a_0 \in Cl\, A_0^+)$. Then $ma = k_1^{-1} h k_2$ where $h = a_0 a \in Cl\, A_0^+$. Therefore

$$f(g_1 X \,;\, ma; g_2) = \tau(k_1^{-1})\, f(g_1^{k_1} \,;\, h; X^{h^{-1} k_1} \cdot g^{k_2})\, \tau(k_2),$$

and

$$f(g_1 \,;\, ma; Y g_2) = \tau(k_1^{-1})\, f(g_1^{k_1} \cdot Y^{h k_2} \,;\, h; g_2^{k_2})\, \tau(k_2),$$

where $Y = \theta(X)$. The proof now goes through in the same way as in [15, Sect. 28].

Put $\mathfrak{Z}_1 = \mathfrak{Z}_M \mathfrak{A}$ and $\mathfrak{U}_1 = \mathfrak{Z}_1 \mu(\mathfrak{U})$ where $\mu = \mu_P$ (see Sect. 16). Since $\mathfrak{Z}_1$ is a finite module over $\mu(\mathfrak{Z})$ [12, Lemma 21], dim $\mathfrak{Z}_1/\mathfrak{U}_1 < \infty$. We regard $\mathfrak{Z}_1^* = \mathfrak{Z}_1/\mathfrak{U}_1$ as a left $\mathfrak{Z}_1$-module. Let $^*\mathfrak{Z}_1$ be the space dual to $\mathfrak{Z}_1^*$. Then $^*\mathfrak{Z}_1$ becomes a right $\mathfrak{Z}_1$-module in the usual way. Put $\mathbf{V} = V \otimes {}^*\mathfrak{Z}_1$. Since dim $^*\mathfrak{Z}_1 < \infty$, V is a complete locally convex space. Fix elements $\eta_1 = 1, \eta_2, \ldots, \eta_p$ in $\mathfrak{Z}_1$ such that their images $(\eta_1^*, \ldots, \eta_p^*)$ form a base for $\mathfrak{Z}_1^*$. Let $(^*\eta_1, \ldots, {}^*\eta_p)$ denote the dual base for $^*\mathfrak{Z}_1$. We regard $^*\mathfrak{Z}_1$ as a Hilbert space with this as an orthonormal base. If $\mathbf{s} \in \mathscr{S}(V)$ and

$$\mathbf{v} = \sum_{1 \leqslant i \leqslant p} v_i \otimes {}^*\eta_i \qquad (v_i \in V),$$

we put

$$\mathbf{s}(\mathbf{v}) = |\,\mathbf{v}\,|_{\mathbf{s}} = \left(\sum_i |\, v_i\,|_{\mathbf{s}}^2\right)^{1/2}.$$

By making K act trivially on $*\mathfrak{Z}_1$, we get a double representation τ of K on V. We define a right representation Γ of $\mathfrak{Z}_1$ on $\mathbf{V}$ by

$$(v \otimes {}^*\eta)\, \Gamma(\zeta) = v \otimes {}^*\eta \cdot \zeta \qquad (v \in V,\ {}^*\eta \in {}^*\mathfrak{Z}_1,\ \zeta \in \mathfrak{Z}_1).$$

For any $f \in \mathscr{A}(G, \tau, \mathfrak{U})$, define

$$\Phi(f : m) = \sum_i \phi_f(m; \eta_i) \otimes {}^*\eta_i \qquad (m \in M_1),$$

where

$$\phi_f(m) = d_P(m)\, f(m).$$

Then $\Phi(f)$ lies in $C^\infty(M_1, \tau_M)$.

Fix $\zeta \in \mathfrak{Z}_1$ and consider $\Phi(f : m; \zeta)$. There exist unique complex numbers c_{ij} such that

$$u_i(\zeta) = \zeta \eta_i - \sum_{1 \leqslant j \leqslant p} c_{ij} \eta_j \in \mathfrak{U}_1 \qquad (1 \leqslant i \leqslant p).$$

Put

$$\Psi_\zeta(f : m) = \sum_{1 \leqslant i \leqslant p} \phi_f(m; u_i(\zeta)) \otimes {}^*\eta_i \qquad (m \in M_1).$$

(We observe that $\mathfrak{U}_1 = \mathfrak{U}$ and $\Psi_\zeta = 0$ in case $P = G$.)

LEMMA 2. $\Phi(f : m; \zeta) = \Phi(f : m)\, \Gamma(\zeta) + \Psi_\zeta(f : m)$ for all $f \in \mathscr{A}(G, \tau, \mathfrak{U})$, $m \in M_1$ and $\zeta \in \mathfrak{Z}_1$.

COROLLARY.

$$\Phi(f : m \exp TH) = \Phi(f : m)e^{T\Gamma(H)} + \int_0^T \Psi_H(f : m \exp tH)e^{(T-t)\Gamma(H)}\, dt$$

for $f \in \mathscr{A}(G, \tau, \mathfrak{U})$, $m \in M_1$, $H \in \mathfrak{a}$ and $T \in \mathbf{R}$.

The proof here is the same as in [15, Sect. 27].
Put $|(x, X)| = (1 + \sigma(x))(1 + \| X \|)$ for $x \in G$ and $X \in \mathfrak{g}$.

LEMMA 3. Fix $\zeta \in \mathfrak{Z}_1$ and $v_1, v_2 \in \mathfrak{M}_1$. Then we can choose a finite subset F of $\mathfrak{G}^{(2)}$ such that

$$| \Psi_\zeta(f : v_1 ; m \exp H; v_2)|_\mathfrak{s} \leqslant \mathbf{s}_{F,\tau}(f)\, \Xi_M(m)\, |(m, H)|^{r+r_0} e^{-\beta_P(H)}$$

and

$$| \Phi(f : v_1 ; m \exp H; v_2)|_{\mathbf{s}} \leqslant \mathbf{s}_{F,r}(f)\, \Xi_M(m)\, |\,(m, H)|^{r+r_0}$$

for all $f \in \mathscr{A}(G, \tau, \mathfrak{U})$, $\mathbf{s} \in \mathscr{S}^0(V)$, $m \in M_1{}^+$, $H \in Cl\, \mathfrak{a}^+$ and $r \geqslant 0$.

This follows from Lemma 1 and the definition of μ (see [15, p. 64]).

We recall that $*\mathfrak{Z}_1$ is a right $\mathfrak{Z}_1$-module. For any linear function λ on $\mathfrak{a}_c$, let $*\mathfrak{Z}_1(\lambda)$ denote the space of all $*\eta \in *\mathfrak{Z}_1$ such that

$$*\eta \cdot (H - \lambda(H))^p = 0 \qquad (H \in \mathfrak{a}),$$

where $p = \dim *\mathfrak{Z}_1$. Let Q be the set of all λ such that $*\mathfrak{Z}_1(\lambda) \neq 0$. For any $\lambda \in Q$, let E_λ denote the projection of $*\mathfrak{Z}_1$ on $*\mathfrak{Z}_1(\lambda)$ corresponding to the direct sum

$$*\mathfrak{Z}_1 = \sum_{\lambda \in Q} *\mathfrak{Z}_1(\lambda).$$

If T is a linear transformation in $*\mathfrak{Z}_1$, we denote by $\| T \|$ its Hilbert–Schmidt norm. Also we write T on the right and denote the linear transformation $1 \otimes T$ in $\mathbf{V}$ again by T. Then it is easy to verify that

$$\mathbf{s}(\mathbf{v} \cdot T) \leqslant \mathbf{s}(\mathbf{v}) \| T \|$$

for $\mathbf{s} \in \mathscr{S}(V)$ and $\mathbf{v} \in \mathbf{V}$.

Since $\mathfrak{Z}_1$ is abelian, it is clear that E_λ commutes with $\Gamma(\zeta)$ ($\lambda \in Q$, $\zeta \in \mathfrak{Z}_1$). Put

$$\Phi_\lambda(f : m) = \Phi(f : m)E_\lambda .$$

LEMMA 4.   *Fix $\lambda \in Q$, $H \in \mathfrak{a}$ and $f \in \mathscr{A}(G, \tau, \mathfrak{U})$. Then*

$$\Phi_\lambda(f : m \exp TH) = \Phi_\lambda(f : m)e^{T\Gamma(H)} + \int_0^T \Psi_H(f : m \exp tH)E_\lambda e^{(T-t)\Gamma(H)}\, dt$$

for $m \in M_1$ and $T \in \mathbf{R}$.

This is obvious from the corollary of Lemma 2.
Put

$$E_\lambda(H) = E_\lambda \exp\{\Gamma(H) - \lambda(H)\} \qquad (H \in \mathfrak{a}).$$

Since $E_\lambda(\Gamma(H) - \lambda(H))$ is nilpotent, the following result is obvious.

LEMMA 5. *We can choose $c \geqslant 0$ such that*

$$\| E_\lambda(H) \| \leqslant c(1 + \| H \|)^p$$

for $\lambda \in Q$ and $H \in \mathfrak{a}$.

We say that a variable element x in G remains bounded if it stays within a compact set.

LEMMA 6. *Fix v_1, $v_2 \in \mathfrak{M}_1$, $\lambda \in Q$, $H \in \mathfrak{a}^+$ and suppose[5]*

$$\Re\lambda(H) + \beta_P(H) > 0.$$

Then for any $f \in \mathscr{A}(G, \tau, \mathfrak{U})$ and $\mathbf{s} \in \mathscr{S}^0(V)$, the integral

$$\int_0^\infty |\, \Psi_H(f : v_1 \,;\, m \exp tH; v_2)|_s \, \| E_\lambda e^{-t\Gamma(H)} \| \, dt \qquad (m \in M_1)$$

converges uniformly provided m remains bounded.

This follows immediately from Lemmas 3 and 5 and [15, Lemma 54].

Fix f, λ and H as above and put

$$\Phi_{\lambda,\infty}(f : m : H) = \lim_{T \to +\infty} \Phi_\lambda(f : m \exp TH)e^{-T\Gamma(H)} \qquad (m \in M_1).$$

It follows from Lemmas 4 and 6 that this limit exists and is a C^∞-function from M_1 to V. In fact

$$\begin{aligned}
\Phi_{\lambda,\infty}(f : v_1 \,;\, m; v_2 : H) \\
= \Phi_\lambda(f : v_1 \,;\, m; v_2) + \int_0^\infty \Psi_H(f : v_1 \,;\, m \exp tH; v_2)E_\lambda e^{-t\Gamma(H)} \, dt
\end{aligned}$$

for v_1, $v_2 \in \mathfrak{M}_1$.

LEMMA 7. *Fix $\lambda \in Q$ and $H \in \mathfrak{a}^+$ and suppose $\Re\lambda(H) > 0$. Then*

$$\Phi_{\lambda,\infty}(f : m : H) = 0 \qquad (m \in M_1).$$

Fix $\mathbf{s} \in \mathscr{S}^0(V)$. Then

$$| \Phi_\lambda(f : m \exp TH)e^{-T\Gamma(H)} |_\mathbf{s}$$

$$\leqslant | \Phi(f : m \exp TH)|_\mathbf{s} \, \| E_\lambda(-TH) \| \, e^{-T\Re\lambda(H)} \to 0$$

as $T \to +\infty$ from Lemmas 3 and 5. This proves our assertion.

---

[5] $\Re c$ denotes the real part of a complex number c.

LEMMA 8.　*Let H_1, $H_2 \in \mathfrak{a}^+$, $\lambda \in \underset{\sim}{Q}$ and suppose*

$$\lambda(H_i) + \beta_P(H_i) > 0 \qquad (i = 1, 2).$$

Then

$$\Phi_{\lambda,\infty}(f : m : H_1) = \Phi_{\lambda,\infty}(f : m : H_2) \qquad (m \in M_1).$$

Put $m_1 = m \exp T_1 H$ $(T_1 \geqslant 0)$ for a fixed $m \in M_1$. Moreover f being kept fixed, we suppress it from the notation whenever convenient. Then

$$\Phi_\lambda(m_1 \exp T_2 H_2)e^{-T_2 \Gamma(H_2)} = \Phi_\lambda(m_1) + \int_0^{T_2} \Psi_{H_2}(m_1 \exp t_2 H_2)E_\lambda e^{-t_2 \Gamma(H_2)}\, dt_2 .$$

Hence

$$\Phi_\lambda(m \exp(T_1 H_1 + T_2 H_2))e^{-\Gamma(T_1 H_1 + T_2 H_2)}$$

$$= \Phi_\lambda(m \exp T_1 H_1)e^{-T_1 \Gamma(H_1)}$$

$$+ \int_0^{T_2} \Psi_{H_2}(m \exp(T_1 H_1 + t_2 H_2))E_\lambda e^{-\Gamma(T_1 H_1 + t_2 H_2)}\, dt_2$$

for $T_2 \geqslant 0$. But if $\mathbf{s} \in \mathscr{S}^0(V)$, it follows from Lemmas 3 and 5 that

$$\int_0^\infty |\Psi_{H_2}(m \exp(T_1 H_1 + t_2 H_2))|_\mathbf{s} \, \| E_\lambda(-(T_1 H_1 + t_2 H_2)) \| \, e^{-\Re\lambda(T_1 H_1 + t_2 H_2)}\, dt_2$$

tends to zero as $T_1 \to +\infty$. Therefore

$$\lim_{T_1, T_2 \to +\infty} \Phi_\lambda(m \exp(T_1 H_1 + T_2 H_2))e^{-\Gamma(T_1 H_1 + T_2 H_2)} = \Phi_{\lambda,\infty}(m : H_1).$$

But since the left side is symmetrical in H_1, H_2, our assertion follows.

We decompose Q into three disjoint sets Q^+, Q^0 and Q^- as follows. Let $\lambda \in Q$. Then

  (1) $\lambda \in \underset{\sim}{Q}^+$ if $\Re\lambda(H) > 0$ for some $H \in \mathfrak{a}^+$.

  (2) $\lambda \in \underset{\sim}{Q}^0$ if $\Re\lambda(H) = 0$ for all $H \in \mathfrak{a}$.

  (3) $\lambda \in \underset{\sim}{Q}^-$ if $\lambda \notin Q^+ \cup Q^0$.

We observe that if $P = G$, then $\mathfrak{a}^+ = \mathfrak{a}$, $Q^- = \varnothing$ and

$$\Phi(m \exp H) = \Phi(m)e^{\Gamma(H)} \qquad (m \in M_1, H \in \mathfrak{a}).$$

Hence we conclude from Lemma 7 that $\Phi_\lambda = 0$ for $\lambda \in Q^+$ in this case.

Let $(\alpha_1 ,..., \alpha_l)$ be the set of simple roots of (P, A). Then if $\lambda \in Q^-$, it is clear that we can choose $c_i \geqslant 0$ $(1 \leqslant i \leqslant l)$ such that

$$\lambda(H) = - \sum_i c_i \alpha_i(H) \qquad (H \in \mathfrak{a}).$$

Moreover since $\lambda \notin Q^0$, it is obvious that

$$c = \sum_i c_i > 0.$$

Therefore we can choose δ $(0 < \delta \leqslant 1/2)$ such that

$$\Re\lambda(H) \leqslant -\delta\beta_P(H),$$

for all $\lambda \in Q^-$ and $H \in Cl\,\mathfrak{a}^+$.

LEMMA 9. *Fix $\lambda \in Q^+$ and $H \in \mathfrak{a}^+$ such that $\Re\lambda(H) + \beta_P(H) > 0$. Then $\Phi_{\lambda,\infty}(f : m : H) = 0$ and*

$$\Phi_\lambda(f : m \exp TH) = - \int_T^\infty \Psi_H(f : m \exp tH)E_\lambda e^{-(t-T)\Gamma(H)}\, dt$$

for $m \in M_1$ and $T \in \mathbf{R}$.

Since $\lambda \in Q^+$, we can choose $H_0 \in \mathfrak{a}^+$ such that $\Re\lambda(H_0) > 0$. Then

$$\Phi_{\lambda,\infty}(m : H) = \Phi_{\lambda,\infty}(m : H_0) = 0$$

from Lemmas 8 and 7. Therefore

$$0 = \Phi_{\lambda,\infty}(m : H) = \Phi_\lambda(m) + \int_0^\infty \Psi_H(m \exp tH)E_\lambda e^{-t\Gamma(H)}\, dt$$

for $m \in M_1$. The second assertion follows by replacing m by $m \exp TH$ in this relation.

COROLLARY. *Suppose $\Re\lambda(H) + (1/2)\beta_P(H) \geqslant 0$ in the above lemma. Then for any $\mathbf{s} \in \mathscr{S}(V)$ and $v_1 , v_2 \in \mathfrak{M}_1$,*

$$| \Phi_\lambda(f : v_1 : m \exp TH; v_2)|_\mathbf{s}$$

$$\leqslant e^{-T\beta_P(H)/2} \int_T^\infty | \Psi_H(f : v_1 : m \exp tH; v_2)|_\mathbf{s}\, \| E_\lambda((T - t)H)\|\, e^{t\beta_P(H)/2}\, dt$$

for $m \in M_1$ and $T \geqslant 0$.

It is clear that

$$| \Phi_\lambda(v_1 \,;\, m \exp TH; v_2)|_\mathbf{s}$$
$$\leqslant \int_T^\infty | \Psi_H(v_1 \,;\, m \exp tH; v_2)|_\mathbf{s} \, \| E_\lambda((T-t)H)\| \, e^{-(t-T)\Re\lambda(H)} \, dt.$$

Therefore since $t \geqslant T$ and $-\Re\lambda(H) \leqslant (1/2)\,\beta_P(H)$, our assertion follows.

LEMMA 10. *Suppose $\lambda \in Q$, $H \in \mathfrak{a}^+$ and $\Re\lambda(H) + (1/2)\,\beta_P(H) \leqslant 0$. Then*

$$| \Phi_\lambda(f : v_1 \,;\, m \exp TH; v_2)|_\mathbf{s}$$
$$\leqslant e^{-T\beta_P(H)/2}\{| \Phi(f : v_1 \,;\, m; v_2)|_\mathbf{s} \, \| E_\lambda(TH)\|$$
$$+ \int_0^\infty | \Psi_H(f : v_1 \,;\, m \exp tH; v_2)|_\mathbf{s} \, \| E_\lambda((T-t)H)\| \, e^{t\beta_P(H)/2} \, dt\}$$

for $v_1 , v_2 \in \mathfrak{M}_1$, $m \in M_1$, $\mathbf{s} \in \mathscr{S}(V)$ and $T \geqslant 0$.

This follows from Lemma 4.

LEMMA 11. *Let $\lambda \in Q^-$ and $H \in \mathfrak{a}^+$. Then*

$$| \Phi_\lambda(f : v_1 \,;\, m \exp TH; v_2)|_\mathbf{s}$$
$$\leqslant e^{-T\delta\beta_P(H)} \{| \Phi(f : v_1 \,;\, m; v_2)|_\mathbf{s} \, \| E_\lambda(TH)\|$$
$$+ \int_0^\infty | \Psi_H(f : v_1 \,;\, m \exp tH; v_2)|_\mathbf{s} \, \| E_\lambda((T-t)H)\| \, e^{t\beta_P(H)/2} \, dt\}$$

for $v_1 , v_2 \in \mathfrak{M}_1$, $m \in M_1$, $\mathbf{s} \in \mathscr{S}(V)$ and $T \geqslant 0$.

This is proved in the same way as Lemma 10. We have only to observe that $\Re\lambda(H) \leqslant -\delta\beta_P(H)$.

It follows from Lemma 8 that if $\lambda \in Q^0$ and $H \in \mathfrak{a}^+$, then $\Phi_{\lambda, \infty}(f : m : H)$ $(m \in M_1)$ is actually independent of H. (One should observe that this remains true when $P = G$). Hence we may denote it by $\Phi_{\lambda, \infty}(f : m)$.

LEMMA 12. *Let $\lambda \in Q^0$ and $H \in \mathfrak{a}^+$. Then*

$$\Phi_\lambda(f : v_1 \,;\, m \exp TH; v_2) - \Phi_{\lambda\infty}(f : v_1 \,;\, m \exp TH; v_2)|_\mathbf{s}$$
$$\leqslant e^{-T\beta_P(H)/2} \int_0^\infty | \Psi_H(f : v_1 \,;\, m \exp tH; v_2)|_\mathbf{s} \, e^{t\beta_P(H)/2} \| E_\lambda((T-t)H)\| \, dt$$

for $v_1 , v_2 \in \mathfrak{M}_1$, $m \in M_1$, $T \geqslant 0$ and $\mathbf{s} \in \mathscr{S}(V)$.

This follows from the fact that

$$\Phi_{\lambda\infty}(v_1 : m \exp TH; v_2) = \Phi_\lambda(v_1 : m \exp TH; v_2)$$

$$+ \int_T^\infty \Psi_H(v_1 : m \exp tH; v_2)E_\lambda e^{-(t-T)\Gamma(H)}\, dt.$$

Define $\Phi_{\lambda\infty}(f : m) = 0\ (m \in M_1)$ if $\lambda \in Q^+ \cup Q^-$.

THEOREM 1. *Let $\lambda \in Q$ and $H \in \mathfrak{a}^+$. Then*

$$|\Phi_\lambda(f : v_1 : m \exp TH; v_2) - \Phi_{\lambda\infty}(f : v_1 : m \exp TH; v_2)|_\mathbf{s}$$

$$\leqslant e^{-T\delta\beta_P(H)} \Big\{ |\Phi(f : v_1 : m; v_2)|_\mathbf{s}\, \|E_\lambda(TH)\|$$

$$+ \int_0^\infty |\Psi_H(f : v_1 : m \exp tH; v_2)|_\mathbf{s}\, \|E_\lambda((T - t)H\|\, e^{t\beta_P(H)/2}\, dt \Big\}$$

for $v_1, v_2 \in \mathfrak{M}_1$, $m \in M_1$, $T \geqslant 0$ and $\mathbf{s} \in \mathscr{S}(V)$. (In case $P = G$, the right side should be replaced by zero.)

First suppose $P = G$. Then $Q^- = \varnothing$ and $\Phi_\lambda = 0$ for $\lambda \in Q^+$. Hence our result follows from Lemma 12 in this case.

So now assume that $P \neq G$ and recall that $\delta \leqslant 1/2$. Therefore if $\lambda \in Q^0 \cup Q^-$, our assertion follows from Lemmas 11 and 12. On the other hand if $\lambda \in Q^+$, we can apply Lemma 10 and the Corollary to Lemma 9.

23. The Function $\Theta(f)$

Now we assume that $P \neq G$. Fix a compact set Ω in $\mathfrak{a}^+$ and choose $\epsilon_0 > 0$ such that $\beta_P(H) \geqslant 2\epsilon_0$ for all $H \in \Omega$. Put $\epsilon = \delta\epsilon_0$.

LEMMA 1. *Given $v_1, v_2 \in \mathfrak{M}_1$, we can choose a finite subset F of $\mathfrak{G}^{(2)}$ such that*

$$|\Phi_\lambda(f : v_1 : m \exp TH; v_2) - \Phi_{\lambda\infty}(f : v_1 : m \exp TH; v_2)|_\mathbf{s}$$

$$\leqslant \mathbf{s}_{F,r}(f)e^{-\epsilon T}\, \Xi_M(m)(1 + \sigma(m))^{r_0+r}$$

for all $f \in \mathscr{A}(G, \tau, \mathfrak{U})$, $\mathbf{s} \in \mathscr{S}^0(V)$, $\lambda \in Q$, $m \in M_1^+$, $H \in \Omega$, $T \geqslant 0$ and $r \geqslant 0$.

This follows immediately from Theorem 22.1 and Lemmas 22.3 and 22.5.

On the other hand we have the following result.

LEMMA 2. *For any* $\lambda \in Q$ *and* $f \in \mathscr{A}(G, \tau, \mathfrak{U})$, $\Phi_{\lambda,\infty}(f)$ *lies in* $\mathscr{A}(M_1, \tau_M)$ *and*

$$\Phi_{\lambda\infty}(f : m; \zeta) = \Phi_{\lambda\infty}(f : m)\,\Gamma(\zeta) \qquad (m \in M_1, \zeta \in \mathfrak{Z}_1).$$

Fix $f \in \mathscr{A}(G, \tau, \mathfrak{U})$. We may assume that $\lambda \in Q^0$ for otherwise $\Phi_{\lambda\infty} = 0$. It is obvious that E_λ and $\Gamma(\zeta)$ commute with the action of K on $\mathbf{V}$. Hence Φ_λ and therefore also $\Phi_{\lambda\infty}$ is of type τ_M.

Fix $\zeta \in \mathfrak{Z}_1$ and $H \in \mathfrak{a}^+$. Then

$$\Phi_{\lambda\infty}(m; \zeta) = \lim_{T \to +\infty} \Phi_\lambda(m \exp TH; \zeta)e^{-T\Gamma(H)} \qquad (m \in M_1).$$

But

$$\Phi_\lambda(m \exp TH; \zeta) = \Phi_\lambda(m \exp TH)\,\Gamma(\zeta) + \Psi_\zeta(m \exp TH)E_\lambda$$

from Lemma 22.2 and

$$\lim_{T \to +\infty} \Psi_\zeta(m \exp TH)E_\lambda e^{-T\Gamma(H)} = 0$$

from Lemmas 22.3, 22.5 and [15, Lemma 54]. Therefore

$$\Phi_{\lambda\infty}(m; \zeta) = \lim_{T \to +\infty} \Phi_\lambda(m \exp TH)e^{-T\Gamma(H)}\Gamma(\zeta),$$
$$= \Phi_{\lambda\infty}(m)\,\Gamma(\zeta).$$

Now put $T = 0$ in Lemma 1 and apply Lemma 22.3. Then it follows without difficulty (see [15, p. 69]) that $\Phi_{\lambda\infty} \in \mathscr{A}(M_1, \tau_M)$.

COROLLARY. $\Phi_{\lambda\infty}(m \exp H) = \Phi_{\lambda\infty}(m)\,E_\lambda(H)\,e^{\lambda(H)}$ *for* $m \in M_1$, $H \in \mathfrak{a}$.

Since $\mathfrak{a} \subset \mathfrak{Z}_1$, it follows from the above lemma that

$$\Phi_{\lambda\infty}(m; H) = \Phi_{\lambda\infty}(m)\,\Gamma(H) \qquad (m \in M_1, H \in \mathfrak{a}).$$

Now fix m and H. Then

$$d\{\Phi_{\lambda\infty}(m \exp tH)e^{-t\Gamma(H)}\}/dt = 0 \qquad (t \in \mathbf{R}),$$

and therefore

$$\Phi_{\lambda\infty}(m \exp tH) = \Phi_{\lambda\infty}(m)e^{t\Gamma(H)}.$$

So if we put $t = 1$ and observe that $\Phi_{\lambda\infty}(m) = \Phi_{\lambda\infty}(m) E_\lambda$, our assertion follows.

Put

$$\Theta(f) = \sum_{\lambda \in Q} \Phi_{\lambda\infty}(f) \qquad (f \in \mathscr{A}(G, \tau, \mathfrak{U})).$$

Then

$$\Theta(f) = \sum_{1 \leqslant i \leqslant p} \theta_i(f) \otimes {}^*\eta_i \,,$$

where $\theta_i(f) \in \mathscr{A}(M_1, \tau_M)$. Since $\eta_1 = 1$, it is clear from Lemma 1 that $f_P = \theta_1(f)$ satisfies the conditions of Theorem 21.1. (If $P = G$, we take $f_G = f$.) This proves the existence of f_P and thus completes the proof of Theorem 21.1.

We can now summarize our results as follows.

LEMMA 3. *Put*

$$\Theta(f) = \sum_{\lambda \in Q} \Phi_{\lambda\infty}(f)$$

for $f \in \mathscr{A}(G, \tau, \mathfrak{U})$. Then $\Theta(f)$ has the following properties:

(1) $\Theta(f) \in \mathscr{A}(M_1, \tau_M)$.

(2) $\Theta(f : m; \zeta) = \Theta(f : m) \, \Gamma(\zeta) \, (m \in M_1, \zeta \in \mathfrak{Z}_1)$.

(3) $\Theta(f : m \exp H) = \Theta(m) \, e^{\Gamma(H)} \, (m \in M_1, H \in \mathfrak{a})$.

(4) *Given $v_1, v_2 \in \mathfrak{M}_1$, we can choose a finite subset F of $\mathfrak{G}^{(2)}$* *such that*

$$| \Phi(f : v_1 ; m \exp TH; v_2) - \Theta(f : v_1 ; m \exp TH; v_2)|_s$$

$$\leqslant \mathbf{s}_{F,r}(f) e^{-\epsilon T} \, \Xi_M(m)(1 + \sigma(m))^{r_0 + r}$$

for all $f \in \mathscr{A}(G, \tau, \mathfrak{U})$, $\mathbf{s} \in \mathscr{S}^0(V)$, $m \in M_1^+$, $H \in \Omega$ and $T, r \geqslant 0$.

(5) $\Theta(f) = \sum_{1 \leqslant i \leqslant p} \eta_i f_P \otimes {}^*\eta_i$.

(6) $\Theta(zf) = \mu(z) \, \Theta(f) = \Theta(f) \, \Gamma(\mu(z)) \, (z \in \mathfrak{Z})$.

The first four statements have been proved above. Put $d(m) = d_P(m)$ $(m \in M_1)$ and define the automorphism $\eta \to \eta'$ of $\mathfrak{M}_1$ by

$$\eta' = d^{-1} \eta \circ d \qquad (\eta \in \mathfrak{M}_1).$$

Since $f_P = \theta_1(f)$, the following result is obvious from (4).

LEMMA 4. *Given $v_1, v_2 \in \mathfrak{M}_1$, we can choose a finite subset F of $\mathfrak{G}^{(2)}$ such that*

$$| d_P(m \exp TH) f(v_1' ; m \exp TH; v_2') - f_P(v_1 ; m \exp TH; v_2)|_\mathfrak{s}$$
$$\leqslant \mathfrak{s}_{F,r}(f) e^{-\epsilon T} \, \Xi_M(m)(1 + \sigma(m))^{r_0+r}$$

for all $f \in \mathscr{A}(G, \tau, \mathfrak{U})$, $\mathbf{s} \in \mathscr{S}^0(V)$, $m \in M_1{}^+$, $H \in \Omega$ and $T, r \geqslant 0$.

Now

$$\Phi(f : m) = d_P(m) \sum_i f(m; \eta_i') \otimes {}^*\eta_i \qquad (m \in M_1).$$

Therefore part (5) of Lemma 3 follows from (4) and Lemma 4. To prove (6), fix $z \in \mathfrak{Z}$. Then (see [15, p. 110])

$$g = z - \mu(z)' \in \bar{\mathfrak{n}}_\alpha \mathfrak{G},$$

where $\bar{\mathfrak{n}} = \theta(\mathfrak{n})$. Hence

$$f(m; z) = f(m; \mu(z)') + f(m; g) \qquad (m \in M_1),$$

for $f \in \mathscr{A}(G, \tau, \mathfrak{U})$. Now fix $m \in M_1$, $H \in \Omega$ and put $m_T = m \exp TH$. Then

$$d_P(m_T) f(m_T ; \mu(z)') - f_P(m_T ; \mu(z)) \to 0$$

from Lemma 4 and

$$d_P(m_T) f(m_T ; g) \to 0$$

from Lemma 22.1 as $T \to +\infty$. This proves that $(zf)_P = \mu(z) f_P$ and therefore part (6) of Lemma 3 follows from (5) and (2). This completes the proof of Lemma 3.

It follows from the Corollary of Lemma 2 that

$$\Theta(m \exp H) = \sum_{\lambda \in Q^0} \Phi_{\lambda\infty}(m) E_\lambda(H) e^{\lambda(H)} \qquad (m \in M_1, H \in \mathfrak{a}).$$

Since $f_P = \theta_1(f)$, the following lemma is obvious.

LEMMA 5. *Fix $f \in \mathscr{A}(G, \tau)$. Then we can choose distinct elements $\lambda_1, ..., \lambda_q \in \mathfrak{a}^*$, linearly independent polynomial functions $p_1, ..., p_s$ on $\mathfrak{a}_c$ and $\phi_{ij} \in \mathscr{A}(M, \tau_M)$ $(1 \leqslant i \leqslant q, 1 \leqslant j \leqslant s)$ such that*

$$f_P(ma) = \sum_{1 \leqslant i \leqslant q} \sum_{1 \leqslant j \leqslant s} \phi_{ij}(m) \, p_j(\log a) e^{(-1)^{1/2}\lambda_i(\log a)}$$

for $m \in M$ and $a \in A$.

We now come to the proof of Lemma 21.1. We may obviously assume that $P \neq G$, $*P \neq M$ and $P' \supset P_0$. Since $\mathfrak{a}' = *\mathfrak{a} + \mathfrak{a}$, we can select $*H_0 \in *\mathfrak{a}^+$, $H_0 \in \mathfrak{a}^+$ such that $*H_0 + (1/2) H_0 \in \mathfrak{a}'^+$. Choose compact neighborhoods $*\Omega$ and Ω of $*H_0$ and H_0 in $*\mathfrak{a}^+$ and $\mathfrak{a}^+$, respectively, such that

$$*\Omega + \tfrac{1}{2}\Omega \subset \mathfrak{a}'^+.$$

Fix $m' \in M'A'$, $*H \in *\Omega$, $H \in \Omega$ and put $H' = *H + H$, $m_t' = m' \exp tH'$. Then $H' \in \mathfrak{a}'^+$ and

$$d_P(m_t') f(m_t') - f_{P'}(m_t') \to 0$$

as $t \to +\infty$, by the definition of $f_{P'}$. On the other hand $M'A' \subset MA$ and we can choose $t_0 \geqslant 0$ such that

$$m_t = m' \exp t(*H + (1/2)H) \in (M'A')^+ \subset M_1^+$$

for $t \geqslant t_0$. Moreover $m_t' = m_t \exp tH/2$. Fix $\mathbf{s} \in \mathscr{S}^0(V)$. Then by Lemma 4, we can choose $c, r \geqslant 0$ and $\epsilon > 0$ such that

$$|\, d_P(m_t') f(m_t') - f_P(m_t')|_\mathbf{s} \leqslant c\Xi_M(m' \exp t\, *H)(1 + \sigma(m_t))^r e^{-\epsilon t}$$

for $t \geqslant t_0$. Now fix $a \in A$ as in Lemma 21.1 and let $m' = *ma$ where $*m \in *M \cdot *A$. Then

$$d_{P'}(m_t') = d_P(m_t') d_{*P}(*m_t),$$

where $*m_t = *m \exp t\, *H$. Therefore

$$|\, d_{P'}(m_t') f(m_t') - d_{*P}(*m_t) f_P(m_t')|_\mathbf{s} \leqslant c d_{*P}(*m_t)\, \Xi_M(*m_t)(1 + \sigma(m_t))^r e^{-\epsilon t} \to 0$$

as $t \to +\infty$. (This is an easy consequence of (2) of Lemma 10.1.) This shows that

$$f_{P'}(m_t') - d_{*P}(*m_t) f_P(m_t') \to 0.$$

On the other hand

$$f_P(m_t') = \sum_{i,j} \phi_{ij}(*m_t)\, p_j(\log a + tH) e^{(-1)^{1/2}\lambda_i(\log a + tH)}$$

from Lemma 5. Moreover, again by Lemma 4, we can choose c_1, $t_1 \geqslant 0$ and $\epsilon_1 > 0$ such that

$$|\, d_{*P}(*m_t) \phi_{ij}(*m_t) - \phi_{ij, *P}(*m_t)|_\mathbf{s} \leqslant c_1 e^{-\epsilon_1 t}$$

for $t \geqslant t_1$ and all i, j. Hence we conclude from Lemma 21.3 that

$$f_{P'}(m') = \sum_{i,j} \phi_{ij,*P}(*m)\, p_{ij}(H(m'))e^{(-1)^{1/2}\lambda_i(H(m'))},$$

$$= g_{*P}(*m),$$

since $H(m') = H_P(m') = \log a$. This proves the first statement of Lemma 21.1. The second has already been proved above and the third is obvious from the uniqueness in Theorem 21.1.

24. Proof of Theorem 21.2

We shall now begin the proof of Theorem 21.2. We may assume that $P \neq G$. Put $\bar{\mathfrak{n}} = \theta(\mathfrak{n})$ and $\mathfrak{P} = \mathfrak{S}(\bar{\mathfrak{p}}_c)$ where $\bar{\mathfrak{p}} = \mathfrak{m} + \mathfrak{a} + \bar{\mathfrak{n}}$. Then $\bar{\mathfrak{n}}_c\mathfrak{P}$ is a two-sided ideal in $\mathfrak{P}$ and

$$\mathfrak{P} = \mathfrak{M}_1 + \bar{\mathfrak{n}}_c\mathfrak{P},$$

where the sum is direct. Define the automorphism $v \to v'$ of $\mathfrak{M}_1$ as in Section 23 and let $v \to {}'v$ denote its inverse. We can extend these uniquely to homomorphisms $q \to q'$ and $q \to {}'q$ of $\mathfrak{P}$ into $\mathfrak{M}_1$ with kernel $\bar{\mathfrak{n}}_c\mathfrak{P}$.

For any $f \in \mathscr{A}(G, \tau)$, we extend f_P to a function on G as follows:

$$f_P(kmn) = \tau(k)\, f_P(m) \qquad (k \in K,\ m \in M_1,\ n \in N).$$

LEMMA 1. *Let Ω be a compact set in G and fix $f \in \mathscr{A}(G, \tau)$, $q \in \mathfrak{P}$, $g \in \bar{\mathfrak{n}}_c\mathfrak{G}$ and $\mathbf{s} \in \mathscr{S}^0(V)$. Then*

$$\lim_{\substack{a \to \infty \\ P}} \sup_{x \in \Omega} |\, d_P(xa)\, f(xa;\, q) - f_P(xa;\, {}'q)|_{\mathbf{s}} = 0$$

and

$$\lim_{\substack{a \to \infty \\ P}} \sup_{x \in \Omega} |\, d_P(xa)\, f(xa;\, g)|_{\mathbf{s}} = 0.$$

Let $x = kmn$ ($k \in K$, $m \in M_1$, $n \in N$). Since Ω is compact, there exist compact sets Ω_1, Ω_N in M_1, N, respectively, such that $m \in \Omega_1$ and $n \in \Omega_N$ for $x \in \Omega$. Let $v \in \mathfrak{M}_1$. Then

$| d_P(xa)\, f(xa;\, v) - f_P(xa;\, {}'v)|_{\mathbf{s}}$

$\qquad \leqslant d_P(ma)\, |f(mna;\, v) - f(ma;\, v)|_{\mathbf{s}} + |\, d_P(ma)\, f(ma;\, v) - f_P(ma;\, {}'v)|_{\mathbf{s}}\,.$

Put $n_t = \exp tX$ $(t \in \mathbf{R})$ where $X = \log n \in \mathfrak{n}$. Then

$$f(mna; v) - f(ma; v) = \int_0^1 f(mn_t a; X^{a^{-1}} \cdot v) \, dt.$$

Since m, n_t $(0 \leqslant t \leqslant 1)$ and X remain bounded (as x varies in Ω), it follows from standard arguments (see [15, Lemma 49]) that we can choose c, $r \geqslant 0$ such that

$$| f(mna; v) - f(ma; v)|_{\mathbf{s}} \leqslant ce^{-\beta_P(\log a)} \, \Xi(a)(1 + \sigma(a))^r$$

for $x \in \Omega$ and $a \in A^+$. Since

$$e^{-\beta_P(\log a)} d_P(a) \, \Xi(a)(1 + \sigma(a))^r \to 0$$

as $a \to_P \infty$, we conclude that

$$| f(mna; v) - f(ma; v)|_{\mathbf{s}} \to 0$$

as $a \to_P \infty$ uniformly for $m \in \Omega_1$, $n \in \Omega_N$. On the other hand

$$\lim_{\substack{a \to \infty \\ P}} \sup_{m \in \Omega_1} | \, d_P(ma) f(ma; v) - f_P(ma; {}'v)|_{\mathbf{s}} = 0$$

from Lemma 23.4. This shows that the first assertion of our lemma is true when $q \in \mathfrak{M}_1$.

On the other hand if $X \in \bar{\mathfrak{n}}$ and $g_0 \in \mathfrak{G}$, then

$$f(xa; Xg_0) = f(X^{xa} ; xa; g_0)$$

and $X^a \to 0$ as $a \to_P \infty$. By the same argument as used above, this implies that

$$\lim_{\substack{a \to \infty \\ P}} \sup_{x \in \Omega} | \, d_P(xa) f(xa; Xg_0)|_{\mathbf{s}} = 0.$$

Since $\mathfrak{P} = \mathfrak{M}_1 + \bar{\mathfrak{n}}_c \mathfrak{P}$ the rest is obvious.

We now use the notation of Theorem 21.2.

COROLLARY.   *Suppose* $\alpha \in C_c^\infty(G, \tau_U)$ *and* $f \in \mathscr{A}(G, \tau_V)$. *Then*

$$(\alpha * f)_P = \alpha^{(P)} * f_P .$$

Let $\phi = \alpha * f$. Then

$$\phi(x) = \int_G \alpha(y^{-1}) \cdot f(yx)\, dy = \int_{N \times M_1} \alpha((nm)^{-1}) \cdot f(nmx)\, dn\, dm,$$

$$= \int_{M_1 \times N} \alpha(mn) \cdot f(n^{-1}m^{-1}x)\, dm\, dn.$$

Therefore if $m_0 \in M_1$ and $a \in A$,

$$d_P(m_0 a)\, \phi(m_0 a) = \int_{M_1 \times N} d_P(m)\, \alpha(mn) \cdot d_P(m^{-1}m_0 a)\, f(n^{-1}m^{-1}m_0 a)\, dm\, dn.$$

Now put $\psi = \alpha^{(P)} * f_P$. Then

$$\psi(m_0 a) = \int_{M_1} \alpha^{(P)}(m) \cdot f_P(m^{-1}m_0 a)\, dm,$$

$$= \int_{M_1 \times N} d_P(m)\, \alpha(mn) \cdot f_P(m^{-1}m_0 a)\, dm\, dn.$$

Therefore

$$d_P(m_0 a)\, \phi(m_0 a) - \psi(m_0 a)$$

$$= \int_{M_1 \times N} d_P(m)\, \alpha(mn) \cdot \{d_P(m^{-1}m_0 a)\, f(n^{-1}m^{-1}m_0 a) - f_P(n^{-1}m^{-1}m_0 a)\}\, dm\, dn$$

$$\to 0$$

as $a \to_P \infty$ from Lemma 1. This proves that $\psi = \phi_P$.

Fix $f \in \mathscr{A}(G, \tau_V)$ and for any $\alpha \in \mathscr{C}(G, \tau_U)$, put $f_\alpha = \alpha * f$.

LEMMA 2. *Fix* $\mathbf{s} \in \mathscr{S}(W)$ *and a finite subset* F *of* $\mathfrak{G}^{(2)}$. *Then we can choose a number* $r \geqslant 0$ *and a continuous seminorm* ν *on* $\mathscr{C}(G, U)$ *such that*

$$\mathbf{s}_{F,r}(f_\alpha) \leqslant \nu(\alpha)$$

for all $\alpha \in \mathscr{C}(G, \tau_U)$.

It is enough to prove this when F consists of a single element $g_1 \otimes g_2$ ($g_1, g_2 \in \mathfrak{G}$). It follows from Lemma 14.2 that

$$f_\alpha(g_1 ; x ; g_2) = \int \alpha(g_1 ; y) \cdot f(y^{-1}x ; g_2)\, dy.$$

By Lemma 21.2 we can choose $\mathbf{s}_1 \in \mathscr{S}(U)$, $\mathbf{s}_2 \in \mathscr{S}(V)$ such that

$$\mathbf{s}(u \cdot v) \leqslant \mathbf{s}_1(u)\,\mathbf{s}_2(v) \qquad (u \in U, v \in V).$$

Also fix $r \geqslant 0$ such that

$$\sup_{x \in G} |f(x; g_2)|_{\mathbf{s}}\, \Xi(x)^{-1}(1 + \sigma(x))^{-r} < \infty.$$

Then by Lemma 14.2 we can choose $c, r_1 \geqslant 0$ such that

$$\int_G |\alpha(g_1 : y)|_{\mathbf{s}_1}\, |f(y^{-1}x; g_2)|_{\mathbf{s}_2}\, dy \leqslant \nu(\alpha)\, \Xi(x)(1 + \sigma(x))^r$$

for all $\alpha \in \mathscr{C}(G, \tau_U)$ and $x \in G$. Here

$$\nu(\alpha) = c \sup_{y \in G} |\alpha(g_1 : y)|_{\mathbf{s}_1}\, \Xi(y)^{-1}(1 + \sigma(y))^{r_1}.$$

This implies that

$$|f_\alpha(g_1 : x; g_2)|_{\mathbf{s}} \leqslant \nu(\alpha)\, \Xi(x)(1 + \sigma(x))^r.$$

Since ν is a continuous seminorm on $\mathscr{C}(G, U)$, the lemma is proved. We observe that $\mathscr{C}(G, \tau_U)$ is a closed subspace of $\mathscr{C}(G, U)$.

COROLLARY 1. *Suppose* $\alpha \to 0$ *in* $\mathscr{C}(G, \tau_U)$. *Then*

$$\mathbf{s}_{F,r}(f_\alpha) \to 0$$

for all r sufficiently large and positive.

This is obvious.

COROLLARY 2. *Suppose* $\alpha \to 0$ *in* $\mathscr{C}(G, \tau_U)$. *Then* $(f_\alpha)_P \to 0$ *pointwise on MA.*

Let $\mathfrak{U}$ be the set of all $z \in \mathfrak{Z}$ such that $zf = 0$. Then $\mathfrak{U}$ is an ideal in $\mathfrak{Z}$ of finite codimension and $f_\alpha \in \mathscr{A}(G, \tau_W, \mathfrak{U})$ for all $\alpha \in \mathscr{C}(G, \tau_U)$. Now fix $m \in M_1^+$ and let $\alpha \to 0$. Then it follows from the above Corollary and (4) of Lemma 23.3 that

$$|\Phi(f_\alpha : m) - \Theta(f_\alpha : m)|_{\mathbf{s}} \to 0.$$

On the other hand it is obvious from Lemma 22.1 and Corollary 1 above that

$$|\Phi(f_\alpha : m)|_{\mathbf{s}} \to 0.$$

Hence we conclude that $\Theta(f_\alpha : m) \to 0$.

Now fix $m_0 \in M_1$. Then we can choose $H \in \mathfrak{a}^+$ such that $m = m_0 \exp H \in M_1^+$. But by Lemma 23.3,

$$\Theta(f_\alpha : m_0) = \Theta(f_\alpha : m)e^{-\Gamma(H)},$$

and so we conclude from the above result that

$$\Theta(f_\alpha : m_0) \to 0.$$

In view of (5) of Lemma 23.3, this implies our assertion.

Now we come to the proof of Theorem 21.2. It follows easily from Lemma 15.1 that $C_c^\infty(G, \tau_U)$ is dense in $\mathscr{C}(G, \tau_U)$. Fix $\alpha \in \mathscr{C}(G, \tau_U)$ and let β be a variable element in $C_c^\infty(G, \tau_U)$ such that $\beta \to \alpha$. Then by Corollary 2 of Lemma 2,

$$(f_\beta)_P \to (f_\alpha)_P$$

pointwise on MA. Put $\phi_\alpha = (f_\alpha)_P$. Then we conclude from the Corollary of Lemma 1 that

$$\phi_\alpha(m_0) = \lim_{\beta \to \alpha} \int_{M_1} \beta^{(P)}(m) \cdot f_P(m^{-1}m_0) \, dm \qquad (m_0 \in M_1).$$

Now $f_P \in \mathscr{A}(M_1, (\tau_V)_M)$ and by Lemma 16.1,

$$\beta^{(P)} \to \alpha^{(P)}$$

in $\mathscr{C}(M_1, (\tau_U)_M)$. Therefore

$$\phi_\alpha(m_0) = \int_{M_1} \alpha^{(P)}(m) \cdot f_P(m^{-1}m_0) \, dm$$

from Lemma 14.2. This shows that $(f_\alpha)_P = \alpha^{(P)} * f_P$ and so Theorem 21.2 is proved.

Remark. If $\alpha \in \mathscr{C}(G, \tau_V)$ and $f \in \mathscr{A}(G, \tau_U)$, one can show that

$$(f * \alpha)_P = f_P * \alpha^{(\bar{P})},$$

where $\bar{P} = \theta(P)$. Put $\check{f}(x) = f(x^{-1})$. Then it is easy to see that

$$(\check{f})_P(m) = f_P(m^{-1}) \qquad (m \in MA),$$

and our result follows from Theorem 21.2 by a formal argument.

25. More About Cusp Forms

Lemma 1. *Suppose f is an element in $\mathscr{A}(G, \tau)$ such that $f_P = 0$ for every psgp P of G with $\operatorname{prk} P \geqslant 1$. Then $f \in {}^0\!\mathscr{C}(G, \tau)$.*

If $\operatorname{prk} G > 0$, then $f = f_G = 0$. Hence we may assume that $\operatorname{prk} G = 0$. In view of Theorem 18.1 it would be enough to show that $f \in \mathscr{C}(G, \tau)$. But this is an easy consequence of Lemma 23.4.

Now assume τ is unitary (see Sect. 20). Then if $f \in \mathscr{A}(G, \tau)$, we write $f_P \sim 0$ if

$$\int_M (\phi(m), f_P(ma))\, dm = 0$$

for all $\phi \in {}^0\!\mathscr{C}(M, \tau_M)$ and $a \in A$. (Since $f_P \in \mathscr{A}(MA, \tau_M)$, it follows from Lemma 14.2 that this integral is well defined.)

Lemma 2. *Let f be an element in $\mathscr{A}(G, \tau)$ such that $f_P \sim 0$ for every psgp P of G. Then $f = 0$.*

This follows by an easy induction on $\dim G$ if we make use of Lemmas 1 and 21.1.

Corollary. *Let $\mathscr{A}_0(G, \tau)$ be the space of all $f \in \mathscr{A}(G, \tau)$ such that $f_P \sim 0$ for all psgps P of G with $\operatorname{prk} P \geqslant 1$. Then $\mathscr{A}_0(G, \tau)$ is the space of all $\mathfrak{Z}$-finite functions in ${}^0\!\mathscr{C}(G, \tau)$.*

We may again assume that $\operatorname{prk} G = 0$. Fix a psgp $P \neq G$. Then we conclude from Lemmas 2 and 21.1 that $f_P = 0$. Therefore $f \in {}^0\!\mathscr{C}(G, \tau)$ from Lemma 1. The rest is obvious.

Theorem 1. *Fix $f \neq 0$ in $\mathscr{A}(G, \tau)$ and choose a psgp $P = MAN$ of G with the following two properties:*

(1) $f_P \nsim 0$.

(2) *P is minimal with respect to condition (1).*

Then for any $a \in A$, the function $m \mapsto f_P(ma)$ $(m \in M)$ lies in ${}^0\!\mathscr{C}(M, \tau_M)$.

Fix a and put $g(m) = f_P(ma)$ $(m \in M)$. Then if follows from Lemma 21.1 that $g \in \mathscr{A}_0(M, \tau_M)$. But then $g \in {}^0\!\mathscr{C}(M, \tau_M)$ from the Corollary of Lemma 2.

26. An Example

Let U be a complete, locally convex, Hausdorff space. Put $V = C^\infty(K \times K, U)$. Then V, with its usual topology, is also complete. Define a double representation τ of K on V as follows. If $v \in V$ and $k \in K$, then $v_1 = \tau(k)v$ and $v_2 = v\tau(k)$ are given by

$$v_1(k_1 : k_2) = v(k_1 k : k_2), \quad v_2(k_1 : k_2) = v(k_1 : kk_2) \qquad (k_1, k_2 \in K).$$

It is easy to verify that τ is differentiable. We can therefore consider the spaces $\mathscr{C}(G, \tau)$ and $\mathscr{A}(G, \tau)$.

Put $\mathbf{G} = K \times G \times K$ and $\mathbf{K} = K \times K \times K$. Then $\mathbf{K}$ is a maximal compact subgroup of $\mathbf{G}$ and the mapping

$$\boldsymbol{\theta} : (k_1, x, k_2) \mapsto (k_1, \theta(x), k_2) \qquad (k_1, k_2 \in K, x \in G)$$

is an involution of $\mathbf{G}$. Define a bilinear form $\mathbf{B}$ on $\mathfrak{k} \oplus \mathfrak{g} \oplus \mathfrak{k}$ by

$$\mathbf{B}(\mathbf{X}, \mathbf{Y}) = B(Z_1, Z_3) + B(X, Y) + B(Z_2, Z_4),$$

where $\mathbf{X} = (Z_1, X, Z_2), \mathbf{Y} = (Z_3, Y, Z_4)$ $(X, Y \in \mathfrak{g}, Z_i \in \mathfrak{k}, 1 \leqslant i \leqslant 4)$. Then if we replace (G, K, θ, B) by $(\mathbf{G}, \mathbf{K}, \boldsymbol{\theta}, \mathbf{B})$ all the conditions of Section 3 are fulfilled. Therefore we can define the space $\mathscr{C}(\mathbf{G}, U)$.

For $f \in \mathscr{C}(\mathbf{G}, U)$, let $\mathbf{f}$ denote the function from G to V defined as follows. Fix $x \in G$. Then $\mathbf{f}(x) = v$ where

$$v(k_1 : k_2) = f(k_1 : x : k_2) \qquad (k_1, k_2 \in K).$$

It is easy to verify that $f \to \mathbf{f}$ is a topological linear isomorphism of $\mathscr{C}(\mathbf{G}, U)$ onto $\mathscr{C}(G, V)$.

Let j denote the mapping of $\mathbf{G}$ onto G given by

$$j(k_1, x, k_2) = k_1 x k_2 \qquad (k_1, k_2 \in K, x \in G).$$

If $f \in \mathscr{C}(G, U)$, then $f \circ j \in \mathscr{C}(\mathbf{G}, U)$ and $f \mapsto f \circ j$ is a topological mapping of $\mathscr{C}(G, U)$ onto a closed subspace of $\mathscr{C}(\mathbf{G}, U)$. We may therefore identify $\mathscr{C}(G, U)$ with its image under this mapping.

LEMMA 1.   *For any $f \in \mathscr{C}(G, U)$, define $\mathbf{f} \in \mathscr{C}(G, V)$ by*

$$\mathbf{f}(x) : (k_1, k_2) \mapsto f(k_1 x k_2) \qquad (k_1, k_2 \in K)$$

for $x \in G$. Then $f \mapsto \mathbf{f}$ is a topological mapping of $\mathscr{C}(G, U)$ onto $\mathscr{C}(G, \tau)$.

Now take $U = \mathbf{C}$. Then $V = C^\infty(K \times K)$. For v_1, $v_2 \in V$, define $v = v_1 \cdot v_2 \in V$ by

$$v(k_1 : k_2) = \int_K v_1(k_1 : k) \, v_2(k^{-1} : k_2) \, dk.$$

Under this multiplication, V becomes an associative algebra. We can now take $U = W = V$ and $\tau_U = \tau_W = \tau$ in the set up of Section 21. Then $\mathscr{C}(G, \tau)$ becomes an algebra under convolution.

For $f \in C^\infty(G)$ define $\mathbf{f} \in C^\infty(G, V)$ by

$$\mathbf{f}(x) : (k_1, k_2) \mapsto f(k_1 x k_2) \qquad (k_1, k_2 \in K, x \in G).$$

Then $f \mapsto \mathbf{f}$ is a bijection of $C^\infty(G)$ on $C^\infty(G, \tau)$.

Let $\mathscr{A}(G)$ denote the space of all functions $f \in C^\infty(G)$ satisfying the following two conditions:

(1) f is $\mathfrak{Z}$-finite

(2) Df satisfies the weak inequality for every $D \in \mathfrak{G}^{(2)}$.

It is easy to verify that the mapping $f \mapsto \mathbf{f}$ defines a bijection of $\mathscr{A}(G)$ on $\mathscr{A}(G, \tau)$.

LEMMA 2.   *Suppose* $\alpha, \beta \in \mathscr{C}(G)$ *and* $\gamma = \alpha * \beta$. *Then* $\gamma = \alpha * \beta$ *in* $\mathscr{C}(G, \tau)$. *Similarly if* $f \in \mathscr{A}(G)$ *and* $g = \alpha * f$, $h = f * \beta$, *then* $\mathbf{g} = \alpha * \mathbf{f}, \mathbf{h} = \mathbf{f} * \beta$ *in* $\mathscr{A}(G, \tau)$.

This follows immediately from the definitions.

Let $P = MAN$ be a psgp of G. Then $\mathbf{P} = K \times P \times K$ is a psgp of $\mathbf{G}$ and $\mathbf{P} = \mathbf{M} \cdot A \cdot N$ is its Langlands decomposition. Here $\mathbf{M} = K \times M \times K$ and we have identified G with a subgroup of $\mathbf{G}$ by means of the mapping $x \to (1, x, 1)$ $(x \in G)$. It is easy to see that $\mathscr{C}(M, \tau_M)$ may be identified with the space of all $\phi \in \mathscr{C}(\mathbf{M}) = \mathscr{C}(K \times M \times K)$ such that

$$\phi(k : k_1 m k_2 : k') = \phi(k k_1 : m : k_2 k'),$$

for $k, k' \in K$, $m \in M$ and $k_1, k_2 \in K_M$.

Fix $\phi \in \mathscr{C}(M, \tau_M)$, $v \in \mathfrak{a}_c{}^*$ and consider the Eisenstein integral $E(P : \phi : v) = E(\phi : v)$ (Sect. 19). Extend ϕ to a function on $K \times G \times K$ by the rule

$$\phi(k_1 : kman : k_2) = \phi(k_1 k : m : k_2),$$

for $k_1, k_2, k \in K$, $m \in M$, $a \in A$, $n \in N$. Also we may regard $E(\phi : v)$

as a function on $K \times G \times K$. Then we get the following formula from the definition of $E(\phi : \nu)$.

LEMMA 3.

$$E(P : \phi : \nu : k_1 : x : k_2)$$

$$= \int_K \phi(k_1 : xk : k^{-1}k_2) \exp\{((-1)^{1/2} \nu - \rho_P)(H_P(xk))\} \, dk$$

for $x \in G$, k_1, $k_2 \in K$, $\phi \in \mathscr{C}(M, \tau_M)$ and $\nu \in \mathfrak{a}_c{}^*$.

27. HARMONIC ANALYSIS ON THE SPACE OF CUSP FORMS

We assume in this section that rank $G = $ rank K. Fix a Cartan subalgebra $\mathfrak{b}$ of $\mathfrak{k}$ and let B be the centralizer of $\mathfrak{b}$ in G. Then B is a Cartan subgroup of G and it is compact. Let Z be the centralizer of G^0 in G. Then $B = ZB^0$ (Lemma 3.4).

Let $\tilde{B}$ denote the normalizer of $\mathfrak{b}$ in G. We put $W(G/B) = \tilde{B}/B$. Then $W(G/B)$ may be regarded as a subgroup of $W(\mathfrak{g}/\mathfrak{b})$ (see Sect. 5).

Let P be the set of all positive roots of $(\mathfrak{g}, \mathfrak{b})$ (under some fixed order). Put $\varpi = \varpi_{\mathfrak{g}/\mathfrak{b}}$ (see Sect. 17) and

$$\rho = \frac{1}{2} \sum_{\alpha \in P} \alpha, \quad \Delta(H) = \prod_{\alpha \in P} (e^{\alpha(H)/2} - e^{-\alpha(H)/2}) \qquad (H \in \mathfrak{b}).$$

Define $\lambda(b^*) = \log b^* + \rho$ and $\varpi(b^*) = \varpi(\lambda(b^*))$ $(b^* \in B^*)$ as in Section 18. We say that b^* is singular or regular according as $\varpi(b^*) = 0$ or not. As in Section 18, let $B^{*\prime}$ denote the set of all regular elements of B^*.

Let $\mathscr{A}$ be the group of all automorphisms σ of G such that $\sigma B = B$. Fix $\sigma \in \mathscr{A}$. Then σ defines an automorphism of $\mathfrak{b}$ and hence also of $S(\mathfrak{b}_c)$. Define $\epsilon(\sigma) = \pm 1$ by $\varpi^\sigma = \epsilon(\sigma)\varpi$.

Now $\mathscr{A}$ operates on $\mathfrak{b}$ and hence by duality also on $\mathfrak{b}_c{}^*$. For any root α of $(\mathfrak{g}, \mathfrak{b})$, define $X_\alpha \neq 0$ in $\mathfrak{g}_c$ as usual [10, Sect. 4] and let ξ_α denote the abelian (i.e., one-dimensional) character of B given by

$$\mathrm{Ad}(b)X_\alpha = \xi_\alpha(b)X_\alpha \qquad (b \in B).$$

Let $\mathscr{L}$ be the additive subgroup of $\mathfrak{b}_c{}^*$ generated by all the roots of $(\mathfrak{g}, \mathfrak{b})$. Then if $B_0{}^*$ is the group of all abelian characters of B, we can extend the mapping $\alpha \mapsto \xi_\alpha$ uniquely to a homomorphism $\gamma \to \xi_\gamma$ $(\gamma \in \mathscr{L})$ of $\mathscr{L}$ into $B_0{}^*$.

Now we make $\mathscr{A}$ operate on B^* as follows. Fix $\sigma \in \mathscr{A}$. It is well known that $\sigma\rho - \rho \in \mathscr{L}$. We define σb^* ($b^* \in B^*$) by

$$\langle \sigma b^*, b \rangle = \langle b^*, \sigma^{-1}b \rangle \xi_{\sigma\rho - \rho}(b) \qquad (b \in B).$$

Then $\lambda(\sigma b^*) = \sigma\lambda(b^*)$. Let $\mathscr{A}_0$ be the subgroup of those $\sigma \in \mathscr{A}$, for which there exists an element $b \in B$ such that $\sigma x = bxb^{-1}$ for all $x \in G$. If $\sigma \in \mathscr{A}_0$, it is clear that σ leaves B^* pointwise fixed. Hence $\mathscr{A}/\mathscr{A}_0$ operates on B^*. (Note that $\mathscr{A}_0$ is normal in $\mathscr{A}$.) Since $W(G/B) = \tilde{B}/B$ may be regarded as a subgroup of $\mathscr{A}/\mathscr{A}_0$, $W(G/B)$ also operates on B^*.

Fix $b^* \in B^*$ and put $\lambda = \lambda(b^*)$. Then in [13, Sect. 8] we have defined an invariant tempered eigendistribution Θ_λ of $\mathfrak{z}$ on G^0 such that

$$\Delta(H)\,\Theta_\lambda(\exp H) = \sum_{s \in W(G^0/B^0)} \epsilon(s)e^{s\lambda(H)} \qquad (H \in \mathfrak{b}').$$

Here $\mathfrak{b}'$ is the set of all points $H \in \mathfrak{b}$ such that $\Delta(H) \neq 0$.

Put $G_1 = ZG^0$ and let y_i ($1 \leqslant i \leqslant r$) be a complete set of representatives for G/G_1 in G. Define a locally summable function Θ_{b^*} on G as follows. Supp $\Theta_{b^*} \subset G_1$ and

$$\Theta_{b^*}(zx) = \sum_{1 \leqslant i \leqslant r} \langle b^*, z^{y_i} \rangle \, \Theta_\lambda(x^{y_i}) \qquad (z \in Z, x \in G^0).$$

It is easy to check that Θ_{b^*} is a well-defined invariant function which is a tempered eigendistribution of $\mathfrak{z}$ on G. Moreover its definition is independent of the choice of y_i ($1 \leqslant i \leqslant r$).

Put

$$'\Delta(b) = \prod_{\alpha \in P} (1 - \xi_\alpha(b^{-1})) \qquad (b \in B).$$

LEMMA 1. *Let $\sigma \in \mathscr{A}$. Then*

$$'\Delta(\sigma b) = \epsilon(\sigma) \, '\Delta(b)\xi_{\rho - \sigma^{-1}\rho}(b) \qquad (b \in B).$$

This is obvious since

$$\Delta(\sigma H) = \epsilon(\sigma)\,\Delta(H) \qquad (H \in \mathfrak{b}).$$

LEMMA 2. *Fix $b^* \in B^*$. Then*

$$'\Delta(b)\,\Theta_{b^*}(b) = \sum_{s \in W(G/B)} \epsilon(s)\langle sb^*, b \rangle \qquad (b \in B \cap G').$$

We know that $G = KG^0$. Fix $k \in K$. Then $\mathfrak{b}^k$ is a Cartan subalgebra of $\mathfrak{k}$. Hence we can choose $k_0 \in K^0$ such that $\mathfrak{b}^k = \mathfrak{b}^{k_0}$ and therefore $k_0^{-1}k \in \tilde{B}$. This shows that $G = \tilde{B}G^0$ and therefore $G/G_1 \simeq \tilde{B}/\tilde{B} \cap G_1$. Moreover $W(G/B) = \tilde{B}/B$ and $\tilde{B} \cap G_1/B \simeq W(G^0/B^0)$. Our assertion is an immediate consequence of this fact and the definition of Θ_{b^*}.

Put $q = (1/2) \dim G/K$ and $\epsilon(b^*) = \text{sign } \varpi(b^*)$ $(b^* \in B^{*\prime})$.

THEOREM 1. *Fix $b^* \in B^{*\prime}$. Then $(-1)^q \epsilon(b^*) \Theta_{b^*}$ is the character of an irreducible, unitary and square-integrable representation of G.*

Put $\lambda = \lambda(b^*)$ and $\epsilon(\lambda) = \text{sign } \varpi(\lambda) = \epsilon(b^*)$. Fix an irreducible, unitary and square-integrable representation π^0 of G^0 on a Hilbert space $\mathfrak{H}^0$ such that $(-1)^q \epsilon(\lambda) \Theta_\lambda$ is the character of π^0 [15, Theor. 16]. Also let μ_0 denote an irreducible unitary representation of B on a finite-dimensional Hilbert space V with the character b^*. Now $B = ZB^0$ and B^0 lies in the center of B. Hence if μ is the restriction of μ_0 on Z, it follows by Schur's lemma, that μ is also irreducible. We can define an irreducible representation π_1 of $G_1 = ZG^0$ on $\mathfrak{H}_1 = V \otimes \mathfrak{H}^0$ by

$$\pi_1(zx) = \mu(z) \otimes \pi^0(x) \qquad (z \in Z, \, x \in G^0).$$

This is possible since there exists a character χ of the abelian group $Z \cap G^0$ such that

$$\mu(z) = \chi(z)\,\mu(1), \quad \pi^0(z) = \chi(z)\,\pi^0(1) \qquad (z \in Z \cap G^0).$$

Let π be the representation of G obtained from π_1 by inducing from G_1 to G. It is obvious that $(-1)^q \epsilon(b^*) \Theta_{b^*}$ is the character of π. We shall now verify that π is irreducible. We have seen above that we can choose $y_i \in \tilde{B}$ $(1 \leqslant i \leqslant r)$. Then if $\mathfrak{H}$ is the representation space of π, we may identify $\mathfrak{H}_1$ with a closed subspace of $\mathfrak{H}$ and write

$$\mathfrak{H} = \sum_{1 \leqslant i \leqslant r} \mathfrak{H}_i \,,$$

where the sum is orthogonal and $\mathfrak{H}_i = \pi(y_i^{-1})\,\mathfrak{H}_1$. (We assume that $y_1 = 1$.) Let π_i denote the representation of G_1 on $\mathfrak{H}_i$ obtained from π by restriction. If Θ_i is the character of π_i, it is clear that

$$\Theta_i(x) = \Theta(x^{y_i}) \qquad (x \in G_1).$$

(Here $\Theta = \Theta_1$.) Let s_i be the element of $W(G/B) \simeq \tilde{B}/B$ defined by y_i. Then, as we have seen above, s_i $(1 \leqslant i \leqslant r)$ are a complete set of representatives for $W(G/B)/W(G^0/B^0)$. Therefore it follows from [15, Theor. 16] that $\Theta_i \neq \Theta_j$ for $i \neq j$. This shows that the representations π_i $(1 \leqslant i \leqslant r)$ are all irreducible and inequivalent and so it is clear that π is irreducible.

Since $[G : G^0] < \infty$, it is obvious that π is square-integrable.

LEMMA 3. *Let $\sigma \in \mathscr{A}$. Then*

$$(\Theta_{b*})^\sigma = \epsilon(\sigma)\Theta_{\sigma b*} \qquad (b* \in B*).$$

Fix $b* \in B*$ and put $\lambda = \lambda(b*)$. Let σ_0 denote the restriction of σ on G^0. Then one sees easily that it would be enough to verify that

$$\Theta_\lambda^{\sigma_0} = \epsilon(\sigma)\Theta_{\sigma\lambda} \, .$$

Hence we may assume that G is connected. Let $\tilde{G}$ denote the simply connected covering group of G and ϕ the natural projection of $\tilde{G}$ on G. Let $Z = \ker \phi$. Then Z is a finitely generated discrete abelian subgroup of $\tilde{G}$. For any integer $N \geqslant 1$, let Z^N denote the subgroup of Z consisting of all elements of the form z^N $(z \in Z)$. Then Z/Z^N is a finite group and $G_1 - \tilde{G}/Z^2$ is acceptable [12, Sect. 18]. Let $\tilde\sigma$ denote the unique automorphism of $\tilde{G}$ such that $\phi \circ \tilde\sigma = \sigma \circ \phi$. Then Z is stable under $\tilde\sigma$ and therefore it defines an automorphism σ_1 of G_1. It would clearly be enough to verify our lemma for (G_1 , σ_1) instead of (G, σ). Hence we may assume that G is connected and acceptable.

Put $L = \log B*$ (see Sect. 18). Then L is a lattice in $(-1)^{1/2}\,\mathfrak{b}*$ and $\rho \in L$. Let L' be the set of all $\lambda \in L$ such that $\varpi(\lambda) \neq 0$. Fix $\lambda \in L'$. It follows from [14, Theor. 3] that

$$\Theta_\lambda{}^\sigma = \epsilon(\sigma)\Theta_{\sigma\lambda} \, .$$

Now let us use the notation of [14, Lemma 59]. Then

$$\Delta_A(h_1 h_2)\, \Theta_\lambda(h_1 h_2)$$
$$= \sum_{t \in \mathfrak{w} \backslash W_G} \epsilon(t) \sum_{s \in W(A^+)} \epsilon(s)\, c(s : t\mathfrak{F}^+ : A^+) \cdot \xi_{t\lambda}(h_1) \exp(s(t\lambda)^y (\log h_2))$$

for $h_1 \in A_I{}^+$, $h_2 \in \in A_R{}^+$ and $\lambda \in L \cap \mathfrak{F}^+$. Put $\tau = \sigma^{-1}$. Then τA is another Cartan subgroup of G. Note that τ defines an automorphism of $\mathfrak{g}_c$ and therefore also of G_c. Hence τy is defined and

$$\mathrm{Ad}(\tau y)\mathfrak{b}_\alpha = \tau\mathfrak{a}_\alpha \, .$$

Without loss of generality, we may assume that

$$\Delta_{\tau A}(\tau a) = \Delta_A(a) \qquad (a \in A).$$

Then applying [14, Lemma 59] to τA instead of A, we get

$$\Delta_A(h_1 h_2)\, \Theta_\lambda(\tau(h_1 h_2))$$
$$= \sum_{t \in \mathfrak{w} \backslash W_G} \epsilon(t) \sum_{s \in W(A^+)} \epsilon(s)\, c(s^\tau : t^\tau \mathfrak{F}^+ : \tau A^+) \cdot \xi_{t\sigma\lambda}(h_1)\, \exp(s(t\sigma\lambda)^y\,(\log h_2))$$

for $h_1 \in A_I{}^+$ and $h_2 \in A_R{}^+$. Here s^τ $(s \in W(A^+))$ is the element of $W(\tau A^+)$ defined by

$$s^\tau(\tau H) = \tau(sH) \qquad (H \in \mathfrak{a}),$$

and t^τ $(t \in W_G)$ is defined similarly. But since $\Theta_\lambda{}^\sigma = \epsilon(\sigma)\, \Theta_{\sigma\lambda}$ for every $\lambda \in L'$, we conclude that

$$c(s^\tau : t^\tau \mathfrak{F}^+ : \tau A^+) = \epsilon(\sigma)\, c(s : t\sigma\mathfrak{F}^+ : A^+)$$

for $s \in W(A^+)$ and $t \in W_G$.

Now fix a connected component $\mathfrak{F}^+$ of $\mathfrak{F}'$ and for any $\lambda \in L \cap Cl\, \mathfrak{F}^+$, define $\Theta_{\lambda,\mathfrak{F}^+} = \Theta_\lambda{}^+$ as in [14, Sect. 22]. Then it follows from the above result and [14, Lemma 59] that

$$(\Theta_{\lambda,\mathfrak{F}^+})^\sigma = \epsilon(\sigma)\Theta_{\sigma\lambda,\sigma\mathfrak{F}^+}.$$

Hence we conclude from the definition of Θ_λ [13, Sect. 8] that

$$\Theta_\lambda{}^\sigma = \epsilon(\sigma)\Theta_{\sigma\lambda} \qquad (\lambda \in L).$$

This proves Lemma 3.

Consider the space $\mathscr{C}(G, V)$ of Section 15 and for any $f \in \mathscr{C}(G, V)$, put

$$'F_f(b) = '\Delta(b) \int_G f(b^x)\, dx \qquad (b \in B \cap G').$$

Let B' be the set of $b \in B$ such that $\xi_\alpha(b) \neq 1$ for every singular root α of $(\mathfrak{g}, \mathfrak{b})$. Then by Theorem 17.1, $'F_f \in \mathscr{C}(B', V)$.

LEMMA 4. *For any $f \in {}^0\mathscr{C}(G, V)$ and $b^* \in B^*$,*

$$(\Theta_{b^*}, f) = (b^*, 'F_f)$$

in the notation of Lemma 18.2.

Let A be a θ-stable Cartan subgroup of G which is not compact.

Since f is a cusp form, it follows from Lemma 8.3 that $'F_f{}^A = 0$. Hence

$$(\Theta_{b^*}, f) = [W(G/B)]^{-1} \int_B \text{conj}('\varDelta(b)\, \Theta_{b^*}(b)) \cdot {}'F_f(b)\, db = (b^*, {}'F_f)$$

from Lemma 2.

COROLLARY. $'F_f(b) = \sum_{b^* \in B^*} (\Theta_{b^*}, f)\langle b^*, b \rangle$ $(b \in B)$ for $f \in {}^0\mathscr{C}(G, V)$.

This follows from Lemma 18.2.

Let P_k denote the set of all compact positive roots of $(\mathfrak{g}, \mathfrak{b})$. Put

$$\rho_k = \frac{1}{2} \sum_{\alpha \in P_k} \alpha,\ \varpi_k = \prod_{\alpha \in P_k} H_\alpha,$$

and $\varpi' = e^{-\rho}\varpi \circ e^{\rho} = \prod_{\alpha \in P}(H_\alpha + \rho(H_\alpha))$ as in Section 17.

LEMMA 5. $'F_f(1; \varpi) = (-1)^q\, c_G f(1)$ $(f \in \mathscr{C}(G, V))$, where

$$c_G = [W(G^0/B^0)]\, (2\pi)^q 2^{\nu/2}\, \varpi_k(\rho_k)$$

and

$$\nu = \dim G/K - \text{rank } G/K.$$

In view of Lemma 17.5 we have only to determine the precise value of c_G. This is done in Theorem 37.1.

COROLLARY 1.[6] $f(x) = c_G^{-1} \sum_{b^* \in B^*} (-1)^q\, d(b^*)\, \varpi(b^*)(\Theta_{b^*}, r(x)f)$ $(x \in G)$ for $f \in {}^0\mathscr{C}(G, V)$.

This follows from the Corollary of Lemma 4.

COROLLARY 2. Let f be a function in ${}^0\mathscr{C}(G, V)$ which is left K-finite. Then f is $\mathfrak{z}$-finite.

Fix a finite subset $F \subset \mathscr{E}(K)$ and put[4]

$$\Theta_{b^*, F} = \alpha_F * \Theta_{b^*} = \Theta_{b^*} * \alpha_F.$$

Let $B^{*\prime}(F)$ denote the set of all $b^* \in B^{*\prime}$ such that $\Theta_{b^*, F} \neq 0$. Then $B^{*\prime}(F)$ is a finite set [15, Lemma 70]. Now as f is left K-finite, we can choose F so that $\alpha_F * f = f$. Then

$$(\Theta_{b^*}, r(x)f) = (\Theta_{b^* F}, r(x)f),$$

[6] As usual $r(x)f$ is the function $y \to f(yx)$ $(y \in G)$.

and therefore

$$f(x) = (-1)^q c_G^{-1} \sum_{b^* \in B^{*\prime}(F)} \varpi(b^*)\, d(b^*)(\Theta_{b^*},\, r(x)f).$$

Since $B^{*\prime}(F)$ is a finite set and Θ_{b^*} is an eigendistribution of $\mathfrak{Z}$, it is now clear that f is $\mathfrak{Z}$-finite.

COROLLARY 3. *Let f be a function in $^0\mathscr{C}(G, V)$ with compact support. Then if G is not compact, $f = 0$.*

For otherwise suppose $f \neq 0$. We can choose a finite subset F of $\mathscr{E}(K)$ such that $g = \alpha_F * f \neq 0$. Then g is a left K-finite function in $^0\mathscr{C}(G, V)$ and therefore by Corollary 2, it is $\mathfrak{Z}$-finite. This implies [13, Lemma 33] that g is analytic. But g has compact support and G^0 is not compact since $[G : G^0] < \infty$. Therefore $g = 0$ and this contradiction proves the corollary.

Since K-finite functions are dense in $^0\mathscr{C}(G, V)$ [15, Lemma 4], we get the following result from Theorem 18.3.

THEOREM 2. *The space*

$$\sum_{\lambda \in L'} \mathscr{C}_\lambda(G, V)$$

is dense in $^0\mathscr{C}(G, V)$.

Put $\mathscr{C}_\lambda(G, \tau) = \mathscr{C}_\lambda(G, V) \cap \mathscr{C}(G, \tau)$ in the notation of Section 19.

COROLLARY. *$\sum_{\lambda \in L'} \mathscr{C}_\lambda(G, \tau)$ is dense in $^0\mathscr{C}(G, \tau)$.*

This is obvious from the above theorem.

Let G_0 be an open subgroup of G. Put $B_0 = B \cap G_0$ and let B_0^* denote, as usual, the set of all irreducible characters of B_0. Let σ and σ_0 be irreducible representations of B and B_0, respectively, and b^* and b_0^* the corresponding characters. Let $[b^* : b_0^*]$ denote the multiplicity of σ_0 in the restriction of σ on B_0. Also let $\Theta_{b_0^*}$ denote the central distribution on G_0 defined above.

LEMMA 6. *Let dx and d_0x denote the standard Haar measures on G and G_0, respectively. Then*

$$\int_{G_0} \Theta_{b_0^*}(x)\, f(x)\, d_0x = \sum_{b^* \in B^*} [b^* : b_0^*] \int_G \Theta_{b^*}(x)\, f(x)\, dx$$

for $b_0{}^* \in B_0{}^*$ and $f \in \mathscr{C}(G)$. Here

$$\bar{f}(x) = \int_K f(kxk^{-1})\, dk,$$

dk being the normalized Haar measure on K.

Let $\lambda = \lambda(b_0{}^*)$. It is clear that $[b^* : b_0{}^*] = 0$ unless $\lambda(b^*) = \lambda$. Now fix b^* such that $\lambda(b^*) = \lambda$. Also fix $f \in \mathscr{C}(G)$ and put

$$g(z) = \int_{G^0} \bar{f}(zx)\, \Theta_\lambda(x)\, d^0x \qquad (z \in Z),$$

where d^0x is the standard Haar measure on G^0. Then g is a C^∞ central function on Z. Moreover it is easy to verify from the definitions of $\Theta_{b_0{}^*}$ and Θ_{b^*} that

$$\int_{G_0} \Theta_{b_0{}^*}(x)\, \bar{f}(x)\, d_0x = \int_{Z_0} \langle b_0{}^*, z \rangle\, g(z)\, d_0z,$$

$$\int_{G} \Theta_{b^*}(x)\, f(x)\, dx = \int_{Z} \langle b^*, z \rangle\, g(z)\, dz,$$

where d^0z and dz are the normalized Haar measures on $Z_0 = B_0 \cap Z$ and Z, respectively.

Let σ_0 be an irreducible representation of B_0 with the character $b_0{}^*$. Then

$$\sum_{b^* \in B^*} [b^* : b_0{}^*]b^*$$

is the character of $\sigma = \mathrm{Ind}_{B_0}^{B}\, \sigma_0$. Let τ, τ_0 be the restrictions of σ, σ_0 on Z and Z_0, respectively. Since $B = ZB^0$, $B_0 = Z_0B^0$ and B^0 lies in the center of B, it is obvious that

$$\tau = \mathrm{Ind}_{Z_0}^{Z}\, \tau_0\,.$$

Let χ denote the character of τ. Then

$$\chi(z) = \sum_{b^* \in B^*} [b^* : b_0{}^*]\langle b^*, z \rangle \qquad (z \in Z).$$

Since g is a class function on Z, it follows from the definition of the induced representation that

$$\int_{Z} g(z)\, \chi(z)\, dz = \mathrm{tr} \int_{Z} g(z)\, \tau(z)\, dz = \mathrm{tr} \int_{Z_0} g(z)\, \tau_0(z)\, d_0z$$

$$= \int_{Z_0} g(z)\langle b_0{}^*, z \rangle\, d_0z,$$

and this is equivalent to the assertion of the lemma.

THEOREM 3. *Suppose* $\dim \tau < \infty$. *Then* ${}^{0}\mathscr{C}(G, \tau)$ *also has finite dimension.*

Since $\dim \tau < \infty$, we can choose a finite subset F of $\mathscr{E}(K)$ such that $f = \alpha_F * f = f * \alpha_F$ for all $f \in \mathscr{C}(G, \tau)$. Hence, as we have seen during the proof of Corollary 2 of Lemma 5,

$$f(x) = (-1)^q c_G^{-1} \sum_{b^* \in B^{*\prime}(F)} \varpi(b^*)\, d(b^*)(\Theta_{b^*, F}\,,\, r(x)f)$$

for all $f \in {}^{0}\mathscr{C}(G, \tau)$. Now $B^{*\prime}(F)$ is a finite set and for a fixed $b^* \in B^{*\prime}(F)$ the functions of the form

$$(r(x^{-1})\Theta_{b^*, F}) * \alpha_F \qquad (x \in G),$$

span a finite-dimensional subspace of $L_2(G)$ [15, p. 89]. Since $\dim V = \dim \tau < \infty$, it follows from the above formula that

$$\dim {}^{0}\mathscr{C}(G, \tau) < \infty.$$

COROLLARY. $\mathscr{A}_0(G, \tau) = {}^{0}\mathscr{C}(G, \tau)$ *if* τ *is unitary and* $\dim \tau < \infty$.

It is obvious from the above theorem that every element in ${}^{0}\mathscr{C}(G, \tau)$ is $\mathfrak{Z}$-finite. Hence our assertion follows from the corollary of Lemma 25.2.

28. FINITENESS OF $\dim \mathscr{A}(G, \tau, \mathfrak{U})$

We now drop the condition that $\operatorname{rank} G = \operatorname{rank} K$. Fix an ideal $\mathfrak{U}$ in $\mathfrak{Z}$ of finite codimension and define $\mathscr{A}(G, \tau, \mathfrak{U})$ as in Section 22.

THEOREM 1. *Suppose* $\dim \tau < \infty$. *Then* $\dim \mathscr{A}(G, \tau, \mathfrak{U})$ *is also finite.*

We proceed by induction on $\dim G$. Let $\mathscr{P}$ denote the set of all psgps P of G with $\operatorname{prk} P \geqslant 1$. Fix $P = MAN$ in $\mathscr{P}$ and for any $f \in \mathscr{A}(G, \tau, \mathfrak{U})$, define $\Theta(f) \in \mathscr{A}(MA, \tau_M, \mathfrak{U}_1)$ as in Lemma 23.3. Then

$$\Theta(f : ma) = \Theta(f : m)e^{\Gamma(\log a)} \qquad (m \in M,\, a \in A).$$

Put $\mathfrak{U}_M = \mathfrak{U}_1 \cap \mathfrak{Z}_M$. Then $\mathfrak{U}_M$ is an ideal in $\mathfrak{Z}_M$ of finite codimension. Let $\Theta_0(f)$ denote the restriction of $\Theta(f)$ on M. Then $\Theta_0(f) \in \mathscr{A}(M, \tau_M, \mathfrak{U}_M)$ and $\Theta(f) = 0$ if and only if $\Theta_0(f) = 0$.

Since $\dim M < \dim G$ and $\dim \tau_M < \infty$, it follows by induction hypothesis that

$$\dim \mathscr{A}(M, \tau_M, \mathfrak{U}_M) < \infty.$$

Let $\mathscr{B}(P)$ denote the set of all $f \in \mathscr{A} = \mathscr{A}(G, \tau, \mathfrak{U})$ such that $\Theta(f) = 0$. Then it is clear that

$$\dim \mathscr{A}/\mathscr{B}(P) \leqslant \dim \mathscr{A}(M, \tau_M, \mathfrak{U}_M) < \infty.$$

Moreover it follows from the fifth assertion of Lemma 23.3 that $f_P = 0$ for $f \in \mathscr{B}(P)$.

Let $P_1, \ldots, P_r$ be a complete set of elements in $\mathscr{P}$ no two of which are conjugate under K. Put

$$\mathscr{B} = \bigcap_{1 \leqslant i \leqslant r} \mathscr{B}(P_i).$$

Then

$$\dim \mathscr{A}/\mathscr{B} \leqslant \sum_{1 \leqslant i \leqslant r} \dim \mathscr{A}/\mathscr{B}(P_i) < \infty.$$

On the other hand $f \in \mathscr{B}$ implies that $f_P = 0$ for all $P \in \mathscr{P}$ (Lemma 21.1). Without loss of generality we may assume that τ is unitary. Then we conclude from the corollary of Lemma 25.2 that

$$\mathscr{B} \subset \mathscr{A}_0(G, \tau) \subset \mathscr{C}(G, \tau),$$

and therefore from Theorems 18.1 and 27.3 that $\dim \mathscr{B} < \infty$. This proves that $\dim \mathscr{A} < \infty$.

29. The Class of Parabolic Subgroups Attached to a Regular Character of $\mathfrak{Z}$

By a character of $\mathfrak{Z}$ we mean a homomorphism χ of $\mathfrak{Z}$ into $\mathbf{C}$ such that $\chi(1) = 1$. Let $\mathfrak{X}$ be the set of all characters of $\mathfrak{Z}$. For any $\chi \in \mathfrak{X}$, let $\mathscr{A}(G, \tau, \chi)$ denote the space of all $f \in \mathscr{A}(G, \tau)$ such that $zf = \chi(z)f$ for all $z \in \mathfrak{Z}$.

Let $\mathfrak{h}$ be a Cartan subalgebra of $\mathfrak{g}$. Put $\gamma = \gamma_{\mathfrak{g}/\mathfrak{h}}$ and $\varpi = \varpi_{\mathfrak{g}/\mathfrak{h}}$. For any $\Lambda \in \mathfrak{h}_c^*$, set

$$\chi_\Lambda^{\mathfrak{h}}(z) = \gamma(z : \Lambda) \qquad (z \in \mathfrak{Z}).$$

Then $\chi_\Lambda^{\mathfrak{h}} \in \mathfrak{X}$. Conversely for any $\chi \in \mathfrak{X}$, we can choose $\Lambda \in \mathfrak{h}_c^*$ such that $\chi = \chi_\Lambda{}^{\mathfrak{h}}$. Moreover Λ is unique up to an operation of $W(\mathfrak{g}/\mathfrak{h})$. We say that χ is regular if $\varpi(\Lambda) \neq 0$. This definition is independent of the choice of $\mathfrak{h}$ and Λ. Let $\mathfrak{X}'$ denote the set of all regular elements in $\mathfrak{X}$.

Now suppose $\mathfrak{h}$ is θ-stable. For any $\Lambda \in \mathfrak{h}_c{}^*$, we define two elements $\bar{\Lambda}$ and $\theta\Lambda$ in $\mathfrak{h}_c{}^*$ as follows:

$$\bar{\Lambda}(H) = \operatorname{conj} \Lambda(H) \qquad (H \in \mathfrak{h} \cap \mathfrak{p} + (-1)^{1/2}\,\mathfrak{h} \cap \mathfrak{k}),$$

$$\theta\Lambda(H) = \Lambda(\theta H) \qquad (H \in \mathfrak{h}).$$

Then the condition $\theta\Lambda = \bar{\Lambda}$ is equivalent to saying that Λ takes only pure imaginary values on $\mathfrak{h}$.

We assume that τ is unitary (see Sect. 20).

THEOREM 1. *Fix $\chi \in \mathfrak{X}'$ and let $P = MAN$ be a psgp of G such that $f_P \not\sim 0$ for some $f \in \mathscr{A}(G, \tau, \chi)$. Then the class of P is uniquely determined by χ.*

Since $f_P \not\sim 0$, we can choose $\phi \in {}^{0}\mathscr{C}(M, \tau_M)$ such that the function

$$J(a) = \int_M (\phi(m), f_P(ma))\, dm \qquad (a \in A),$$

is not identically zero. It follows from Theorem 18.1 that rank $M =$ rank K_M and therefore we can extend $\mathfrak{a}$ to a θ-stable Cartan subalgebra $\mathfrak{h}$ of $\mathfrak{g}$ such that $\mathfrak{a} = \mathfrak{h} \cap \mathfrak{p}$ and $\mathfrak{b} = \mathfrak{h} \cap \mathfrak{k}$ is a Cartan subalgebra of $\mathfrak{m}$. Moreover, in view of Theorem 27.2, we may assume that ϕ is an eigenfunction of $\mathfrak{Z}_M$. Put $\gamma_1 = \gamma_{\mathfrak{m}/\mathfrak{b}}$ and let λ be a linear function on $\mathfrak{b}_c$ such that

$$\zeta\phi = \gamma_1(\zeta : \lambda)\phi \qquad (\zeta \in \mathfrak{Z}_M).$$

Since $\phi \neq 0$, we conclude from Theorem 18.3 that λ takes only pure imaginary values on $\mathfrak{b}$ and $\varpi_{\mathfrak{m}/\mathfrak{b}}(\lambda) \neq 0$. Define an anti-automorphism $g \mapsto g^\dagger$ of $\mathfrak{G}$ by $g^\dagger = \eta(g^*)$ $(g \in \mathfrak{G})$ where g^* is the adjoint of g and η the conjugation of $\mathfrak{g}_c$ with respect to $\mathfrak{g}$. Then $\mathfrak{Z}_{M^\dagger} = \mathfrak{Z}_M$ and [15, p. 81]

$$\gamma_1(\zeta^\dagger : \lambda) = \operatorname{conj} \gamma_1(\zeta : \lambda) \qquad (\zeta \in \mathfrak{Z}_M).$$

Hence it follows that

$$\int_M (\phi(m), f_P(ma; \zeta))\, dm = \int_M (\phi(m; \zeta^\dagger), f_P(ma))\, dm = \gamma_1(\zeta : \lambda)\, J(a)$$

$$(\zeta \in \mathfrak{Z}_M,\, a \in A).$$

On the other hand

$$\chi(z)f_P = (zf)_P = \mu_P(z)f_P \qquad (z \in \mathfrak{Z})$$

from Lemma 21.1. Let α_λ denote the homomorphism of $\mathfrak{Z}_M\mathfrak{A} \simeq \mathfrak{Z}_M \otimes \mathfrak{A}$ into $\mathfrak{A}$, given by $\alpha_\lambda(\zeta u) = \gamma_1(\zeta : \lambda)u$ ($\zeta \in \mathfrak{Z}_M$, $u \in \mathfrak{A}$). Put

$$q_\lambda(z) = \alpha_\lambda(\mu_P(z)) \in \mathfrak{A} \qquad (z \in \mathfrak{Z}).$$

Then it is clear that

$$\chi(z)\,J(a) = \int_M (\phi(m),\, f_P(ma;\, \mu_P(z)))\, dm$$

$$= \int_M (\phi(m),\, f_P(ma;\, q_\lambda(z)))\, dm = J(a;\, q_\lambda(z)) \qquad (a \in A,\, z \in \mathfrak{Z}).$$

On the other hand it follows from Lemma 23.5 that

$$J = \sum_{1 \leqslant i \leqslant r} J_i\,,$$

with

$$J_i(a) = p_i(\log a)e^{(-1)^{1/2}\nu_i(\log a)} \qquad (a \in A).$$

Here $\nu_1, \ldots, \nu_r$ are distinct real-valued linear functions and p_i complex-valued polynomial functions on $\mathfrak{a}$. Since $J \neq 0$, we may assume that $J_1 \neq 0$. Moreover since $\nu_1, \nu_2, \ldots, \nu_r$ are distinct, we can conclude that

$$q_\lambda(z)J_1 = \chi(z)J_1 \qquad (z \in \mathfrak{Z}).$$

Put $p = p_1$ and $\nu = \nu_1$. Then $p \neq 0$. Let $d = d^0 p$. Then it is obvious that

$$J_1(a;\, q_\lambda(z)) = e^{(-1)^{1/2}\nu(\log a)}\, \{q_\lambda(z : (-1)^{1/2}\nu)\, p(\log a) + p_0(\log a)\},$$

where p_0 is a polynomial function on $\mathfrak{a}$ of degree less than d. (As usual, $q_\lambda(z : (-1)^{1/2}\nu)$ denotes the value of $q_\lambda(z)$, regarded as a polynomial on $\mathfrak{a}_c{}^*$, at $(-1)^{1/2}\nu$.) Therefore we conclude immediately that

$$\chi(z) = q_\lambda(z : (-1)^{1/2}\nu) = \chi_\Lambda{}^\flat(z) \qquad (z \in \mathfrak{Z}),$$

where Λ is the linear function on $\mathfrak{h}_c$ which coincides with $(-1)^{1/2}\nu$ on $\mathfrak{a}$ and λ on $\mathfrak{b}$. Note that $\bar{\Lambda} = \theta\Lambda$.

So far we have not used the fact that χ is regular. Hence we can state our result as follows.

LEMMA 1. *Fix $\chi \in \mathfrak{X}$ and let $f \neq 0$ be an element in $\mathscr{A}(G, \tau, \chi)$. Choose a psgp $P = MAN$ such that $f_P \nsim 0$. Then we can extend $\mathfrak{a}$ to a θ-stable Cartan subalgebra $\mathfrak{h}$ of $\mathfrak{g}$ and choose a linear function Λ on $\mathfrak{h}_c$ such that:*

(1) $\mathfrak{a} = \mathfrak{h}_R$, $\mathfrak{h}_I \subset \mathfrak{m}$,

(2) $\chi = \chi_\Lambda{}^\mathfrak{h}$,

(3) $\bar{\Lambda} = \theta\Lambda$ *and* $\Lambda(H_\alpha) \neq 0$ *for every root α of $(\mathfrak{g}, \mathfrak{h})$ which vanishes identically on $\mathfrak{a}$.*

In order to complete the proof of Theorem 1, it is sufficient to verify the following result.

LEMMA 2. *Let $\mathfrak{h}_i$ $(i = 1, 2)$ be two θ-stable Cartan subalgebras of $\mathfrak{g}$ and Λ_i a linear function on $\mathfrak{h}_i$ such that $\bar{\Lambda}_i = \theta\Lambda_i$. Suppose*

(1) Λ_1 *is nonsingular, i.e.,* $\varpi_{\mathfrak{g}/\mathfrak{h}_1}(\Lambda_1) \neq 0$,

(2) $\gamma_{\mathfrak{g}/\mathfrak{h}_1}(z : \Lambda_1) = \gamma_{\mathfrak{g}/\mathfrak{h}_2}(z : \Lambda_2)$ *for all* $z \in \mathfrak{Z}$.
Then $\mathfrak{h}_1$, $\mathfrak{h}_2$ are conjugate under K.

We shall give a proof in Section 33.

For the following lemma we drop the condition that τ be unitary.

LEMMA 3. *Let $\mathfrak{b}$ be a θ-stable Cartan subalgebra of $\mathfrak{g}$ and μ a linear function on $\mathfrak{b}$. Put $\chi = \chi_\mu{}^\mathfrak{b}$. Then if $\mathscr{A}(G, \tau, \chi) \neq 0$, we can choose $s \in W(\mathfrak{g}/\mathfrak{b})$ such that $\bar{\mu} = s(\theta\mu)$.*

Fix $f \neq 0$ in $\mathscr{A}(G, \tau, \chi)$ and choose a psgp $P = MAN$ such that:

(1) $f_P \neq 0$,

(2) P is minimal among all psgps which satisfy condition (1). It follows from Lemmas 21.1 and 25.1 that for any $a \in A$, the function $m \mapsto f_P(ma)$ $(m \in M)$ lies in $^0\mathscr{C}(M, \tau_M)$. Hence we conclude from Theorems 18.1 and 18.3 that we can extend $\mathfrak{a}$ to a θ-stable Cartan subalgebra $\mathfrak{h}$ of $\mathfrak{g}$ and choose a linear function Λ on $\mathfrak{h}$ such that the three conditions of Lemma 1 are fulfilled. Since

$$\chi_\mu{}^\mathfrak{b} = \chi = \chi_\Lambda{}^\mathfrak{h},$$

we can choose $x \in G_c$ such that $\mathfrak{b}_c = \mathfrak{h}_c{}^x$ and $\mu = \Lambda^x$. Then $\bar{\mu} = (\bar{\Lambda})^x$. Moreover since $\bar{\Lambda} = \theta\Lambda$, it follows that $\theta\mu = (\bar{\Lambda})^y$ where $y = \theta x \theta^{-1} \in G_c$. Hence

$$\chi_{\bar{\mu}}{}^\mathfrak{b} = \chi_{\bar{\Lambda}}{}^\mathfrak{h} = \chi_{\theta\mu}^\mathfrak{b}.$$

This implies that $\bar{\mu} = s(\theta\mu)$ for some $s \in W(\mathfrak{g}, \mathfrak{b})$.

30. Some Inequalities

Fix a minimal psgp $P_0 = M_0 A_0 N_0$ of G and let $P = MAN$ be any psgp such that $P \supset P_0$. Put $\rho_0 = \rho_{P_0}$, $\rho = \rho_P$ and

$$H_0(x) = H_{P_0}(x), \ H(x) = H_P(x) \qquad (x \in G).$$

Any $x \in G$ can be expressed uniquely in the form $x = kman$ where $k \in K$, $m \in \exp(\mathfrak{m} \cap \mathfrak{p})$, $a \in A$, $n \in N$. We write $\mu(x) = m$.

LEMMA 1. *Let $d_1 k$ denote the normalized Haar measure on $K_1 = K \cap M$. Then*

$$\int_{K_1} e^{-\rho_0(H_0(xk))} d_1 k = e^{-\rho(H(x))} \Xi_M(\mu(x)) \qquad (x \in G).$$

Let $*P = *M\,*A\,*N$ be the minimal psgp of M corresponding to P_0 (Lemma 6.1). Let us write $*H$ and $*\rho$ instead of H_{*P} and ρ_{*P}, respectively. Then it is easy to see that

$$H_0(xk) = *H(\mu(x)k) + H(x)$$

for $x \in G$ and $k \in K_1$. Hence

$$\rho_0(H_0(xk)) = *\rho(*H(\mu(x)k)) + \rho(H(x)) \qquad (k \in K_1),$$

and from this our assertion follows immediately.

COROLLARY. *Let dk denote the normalized Haar measure on K. Then*

$$\int_K e^{-\rho(H(xk))} \Xi_M(\mu(xk))\, dk = \Xi(x) \qquad (x \in G).$$

This is obvious from the above lemma and the definition of Ξ (Sect. 10).

Put $\overline{N}_0 = \theta(N_0)$, $\overline{N} = \theta(N)$ and

$$\beta(H) = \inf_\alpha \alpha(H) \qquad (H \in \mathfrak{a}),$$

where α runs over all roots of (P, A).

LEMMA 2. $\rho(H(\bar{n})) \geqslant 0$ *and* $\rho_0(H_0(\bar{n})) \geqslant 0$ *for* $\bar{n} \in \overline{N}_0$. *Moreover if* $a \in A$ *and* $\beta(\log a) \geqslant 0$, *then*

$$e^{\rho(H(\bar{n}^a))} \leqslant 1 + \exp\{-\tfrac{1}{2}\beta(\log a) + \rho(H(\bar{n}))\} \qquad (\bar{n} \in \overline{N}_0),$$

and

$$e^{\rho_0(H_0(\bar{n}^a))} \leqslant 1 + \exp\{-\beta(\log a) + \rho_0(H_0(\bar{n}))\} \qquad (\bar{n} \in \bar{N}).$$

Except for some minor changes the proof of this lemma is the same as that of [15, Lemma 85]. Hence we sketch it only briefly. We may assume that G is semisimple and agree to standard conventions [8, p. 244].

Extend $\mathfrak{a}_0$ to a Cartan subalgebra $\mathfrak{h}$ of $\mathfrak{g}$. Let $\mathfrak{F}$ be the set of all real-valued linear functions on $\mathfrak{h} \cap \mathfrak{p} + (-1)^{1/2} \mathfrak{h} \cap \mathfrak{k}$. Introduce compatible orders in $\mathfrak{F}$ and $\mathfrak{a}^*$ such that $\alpha > 0$ for every root α of (P_0, A_0). Let Q be the set of all positive roots of $(\mathfrak{g}, \mathfrak{h})$ under this order and J the set of all $\lambda \in \mathfrak{F}$ such that $2\langle \lambda, \alpha \rangle / \langle \alpha, \alpha \rangle$ is an integer $\geqslant 0$ for all $\alpha \in Q$. Then for each $\lambda \in J$, we have an irreducible finite-dimensional representation π_λ of G on V_λ with the highest weight λ. Let v_λ denote a unit vector in V_λ belonging to the highest weight λ.

LEMMA 3. *Fix $\lambda \in J$ and let U be the subspace consisting of all $v \in V_\lambda$ such that $\pi_\lambda(H)v = \lambda(H)v$ for all $H \in \mathfrak{a}$. Then U is invariant and irreducible under $\pi_\lambda(\mathfrak{m})$. Moreover $\dim U = 1$ if and only if $\langle \lambda, \alpha \rangle = 0$ for every root α of $(\mathfrak{m}_1, \mathfrak{h})$.*

The proof is the same as that of [15, Lemma 86].

Let Q_1 denote the set of all positive roots of $(\mathfrak{m}_1, \mathfrak{h})$ and J_0 the set of all $\lambda \in J$ such that $\langle \lambda, \alpha \rangle = 0$ for all $\alpha \in Q_1$.

LEMMA 4. *Fix $\lambda \in J_0$. Then $\lambda(H(\bar{n})) \geqslant 0$ and*

$$\exp \lambda(H(\bar{n}^a)) \leqslant 1 + \exp\{-\beta(\log a) + \lambda(H(\bar{n}))\}$$

for $\bar{n} \in \bar{N}_0$ and $a \in A$ provided $\beta(\log a) \geqslant 0$.

This is proved in the same way as [15, Lemma 87].

Fix $\Lambda \in J$, put $(\pi, V) = (\pi_\Lambda, V_\Lambda)$ and define U as in Lemma 3. Let E be the orthogonal projection of V on U and $\| T \|$ the Hilbert–Schmidt norm of an endomorphism T of V.

LEMMA 5. *Fix $a \in A$ such that $\beta(\log a) \geqslant 0$. Then*

$$\exp \Lambda(H_0(\bar{n}^a)) \leqslant 1 + \exp\{-\beta(\log a) + \Lambda(H_0(\bar{n}))\},$$

$$\| \pi(\bar{n}^a)E \| \leqslant \| E \| + e^{-\beta(\log a)} \| \pi(\bar{n})E \|$$

for $\bar{n} \in \bar{N}$.

The proof is the same as for [15, Lemma 88].

Now put $\varLambda = 1/2 \sum_{\alpha \in Q} \alpha$, $W = W(\mathfrak{g}/\mathfrak{h})$ and $W_1 = W(\mathfrak{m}_1/\mathfrak{h})$. Then $\varLambda \in J$ and $W_1 \subset W$. Let s be the unique element of W_1 which maps every root in Q_1 into a negative root. Put $\lambda = \varLambda + s\varLambda$. Then $\lambda \in J_0$, $\varLambda = \rho_0$ on $\mathfrak{a}_0$ and $\lambda = 2\rho$ on $\mathfrak{a}$. The statements of Lemma 2 now follow from Lemmas 4 and 5.

31. Convergence of an Integral

Fix $d \geqslant 0$ such that (Lemma 10.1)

$$\sup_{a \in A_0^+} \varXi(a) e^{\rho_0(\log a)}(1 + \sigma(a))^{-d} < \infty.$$

LEMMA 1. *For any* $\epsilon > 0$,

$$\int_{\bar{N}} e^{-\rho_0(H_0(\bar{n}))} \{1 + \rho_0(H_0(\bar{n}))\}^{-(d+\epsilon)} d\bar{n} < \infty.$$

The proof is the same as before [15, Lemma 89].

From here on the proof of Lemma 10.2 goes through in the same way as that of [15, Lemma 21] (see [15, pp. 106–109]).

32. Some Further Results on Convergence

We shall now derive a consequence of Lemma 31.1 which will be needed in a subsequent paper. We keep to the notation of Section 30 and define d as in Section 31. Recall that $\rho(H(\bar{n})) \geqslant 0$ for $\bar{n} \in \bar{N}$ (Lemma 30.2).

LEMMA 1. *Fix* $r > d$. *Then*

$$\int_{\bar{N}} e^{-\rho(H(\bar{n}))} \varXi_M(\mu(\bar{n})) \{(1 + \rho(H(\bar{n}))) (1 + \sigma(\mu(\bar{n})))\}^{-r} d\bar{n} < \infty.$$

Let $\bar{n} = k_0 man$ ($k_0 \in K$, $m = \mu(\bar{n})$, $a \in A$, $n \in N$). Then

$$H_0(\bar{n}) = H_0(m) + H(\bar{n}),$$

and therefore

$$\rho_0(H_0(\bar{n})) = \rho(H(\bar{n})) + \rho_0(H_0(m))$$
$$\leqslant \rho(H(\bar{n})) + c\sigma(m),$$

where c is a number $\geqslant 1$ independent of $\bar{n}$. Now replace $\bar{n}$ by $\bar{n}^k$ ($k \in K_1 = K \cap M$). Then

$$\bar{n}^k = k_0{}^k m^k a n^k.$$

Hence

$$\rho_0(H_0(\bar{n}^k)) = \rho(H(\bar{n})) + \rho_0(H_0(m^k)).$$

Therefore

$$\int_{K_1} e^{-\rho_0(H_0(\bar{n}^k))}\{1 + \rho_0(H_0(\bar{n}^k))\}^{-r}\, d_1 k$$

$$\geqslant e^{-\rho(H(\bar{n}))}\{1 + \rho(H(\bar{n}))\}^{-r} c^{-r}(1 + \sigma(m))^{-r} \int_{K_1} e^{-\rho_0(H_0(m^k))}\, d_1 k.$$

Taking Lemma 30.1 into account, we conclude that

$$\int_{\bar{N}} e^{-\rho(H(\bar{n}))} \Xi_M(\mu(\bar{n})) \{(1 + \rho(H(\bar{n})))\,(1 + \sigma(\mu(\bar{n})))\}^{-r}\, d\bar{n}$$

$$\leqslant c^r \int_{\bar{N}} d\bar{n} \int_{K_1} e^{-\rho_0(H_0(\bar{n}^k))}\{1 + \rho_0(H_0(\bar{n}^k))\}^{-r}\, d_1 k$$

$$= c^r \int_{\bar{N}} e^{-\rho_0(H_0(\bar{n}))}\{1 + \rho_0(H_0(\bar{n}))\}^{-r}\, d\bar{n} < \infty$$

from Lemma 31.1.

LEMMA 2.   *We can choose $c \geqslant 1$ such that*

$$1 + \sigma(\mu(\bar{n})) \leqslant c(1 + \rho(H(\bar{n}))) \qquad (\bar{n} \in \bar{N}).$$

Let G_1 be the analytic subgroup of G corresponding to $\mathfrak{g}_1 = [\mathfrak{g}, \mathfrak{g}]$. For any linear transformation T of $\mathfrak{g}$, let $\| T \|$ denote the Hilbert–Schmidt norm of T and put

$$\sigma_1(x) = \log \| Ad(x)\| \qquad (x \in G_1).$$

Then $\sigma_1(x) \geqslant 0$ and it is easy to verify that we can choose $c_1 > 0$ such that

$$1 + \sigma(x) \leqslant c_1(1 + \sigma_1(x))$$

for all $x \in G_1$.

Since $2\rho \in J_0$, we can take $(\pi, v) = (\pi_{2\rho}, v_{2\rho})$ in the notation of Section 30. Then

$$e^{2\rho(H(\bar{n}))} = | \pi(\bar{n})v | \qquad (\bar{n} \in \bar{N}).$$

Put

$$q(X) = \mid \pi(\exp X)v \mid^2 \qquad (X \in \bar{\mathfrak{n}}).$$

Then q is a polynomial function on $\bar{\mathfrak{n}}$, $q(X) \geqslant 1$ and $q(X) \to +\infty$ as $\|X\| \to \infty$. Hence we conclude from the Seidenberg–Tarski theorem [21, p. 276] that there exist numbers c_2, $\delta > 0$ such that

$$q(X) \geqslant c_2 \|X\|^\delta,$$

for all $X \in \bar{\mathfrak{n}}$. Now $\mathrm{Ad}(\exp X) = e^{\mathrm{ad}X}$ is a polynomial in $\mathrm{ad}\, X$. Hence it is clear that we can choose c_3, $c_4 > 0$ such that

$$\| e^{\mathrm{ad}X} \| \leqslant c_3(q(X))^{c_4} \qquad (X \in \bar{\mathfrak{n}}).$$

Since $\bar{\mathfrak{n}} \subset \mathfrak{g}_1$ and $1 + \sigma(\mu(\bar{n})) \leqslant c_0(1 + \sigma(\bar{n}))$ $(\bar{n} \in \bar{N})$ from Lemma 10.2, the statement of the lemma is now obvious.

Corollary. *Fix* $\epsilon > 0$. *Then*

$$\int_{\bar{N}} e^{-(1+\epsilon)\rho(H(\bar{n}))} \Xi_M(\mu(\bar{n}))\, d\bar{n} < \infty.$$

Since $\rho(H(\bar{n})) \geqslant 0$, this is obvious from Lemmas 1 and 2.

33. Proof of Lemma 29.2

Let $\mathfrak{h}_i$ $(i = 1, 2)$ be two θ-stable Cartan subalgebras of $\mathfrak{g}$. Put $W_i = W(\mathfrak{g}/\mathfrak{h}_i)$ and

$$\pi_i = \prod_{\alpha \in P_i} \alpha \qquad (i = 1, 2),$$

where P_i is the set of all positive roots of $(\mathfrak{g}, \mathfrak{h}_i)$ under some order.

Lemma 1. *Fix* $x \in G_c$ *such that* $\mathfrak{h}_{2c} = x\mathfrak{h}_{1c}$. *Let* H_0 *be an element in* $\mathfrak{h}_2 \cap \mathfrak{p}$ *and* $\mathfrak{z}$ *the centralizer of* H_0 *in* $\mathfrak{g}$. *We denote by* $\mathfrak{a}$ *the intersection of* $\mathfrak{p}$ *with the center of* $\mathfrak{z}$. *Suppose* $\lambda \in (\mathfrak{h}_1 \cap \mathfrak{k})_c$ *and* $v \in (\mathfrak{h}_1 \cap \mathfrak{p})_c$ *are two elements such that* $\pi_1(\lambda + v) \neq 0$ *and*

$$x(\lambda - v) = s \cdot \theta x(\lambda + v)$$

for some $s \in W_2$ *which leaves* H_0 *fixed. Then* $\mathfrak{a} \subset x(\mathfrak{h}_1 \cap \mathfrak{p})$.

Choose $y \in G_c$ such that $y = s$ on $\mathfrak{h}_{2c}$ and put $\theta^x = x\theta x^{-1}$. Then $\theta^x \mathfrak{h}_{2c} = \mathfrak{h}_{2c}$ and

$$\theta^x x(\lambda + \nu) = x\theta(\lambda + \nu) = x(\lambda - \nu),$$
$$= y\theta x(\lambda + \nu).$$

Therefore the linear transformation

$$z = \theta^{-1} y^{-1} \theta^x$$

leaves $x(\lambda + \nu)$ fixed. But $\theta^{-1} y^{-1} x\theta \in G_c$ and therefore $z \in G_c$. Since $\pi_2(x(\lambda + \nu)) \neq 0$, we conclude that z leaves $\mathfrak{h}_{2c}$ pointwise fixed.

On the other hand the centralizer of H_0 in G_c is connected and $\mathfrak{a} \subset \mathfrak{h}_2$. Since y leaves H_0 fixed and $\mathfrak{a}$ lies in the center of $\mathfrak{z}$, it follows that both y and z leave $\mathfrak{a}$ pointwise fixed. But $y = \theta^x z^{-1} \theta$. Hence if $H \in \mathfrak{a}$,

$$H = yH = -\theta^x H = -x\theta(x^{-1}H).$$

This shows that $\theta(x^{-1}H) = -x^{-1}H$ and therefore $x^{-1}H \in (\mathfrak{h}_1 \cap \mathfrak{p})_c$. Since roots of $(\mathfrak{g}, \mathfrak{h}_i)$ take real values on $\mathfrak{h}_i \cap \mathfrak{p}$, we now conclude easily that $\mathfrak{a} \subset x(\mathfrak{h}_1 \cap \mathfrak{p})$.

For the proof of Lemma 29.2, we may assume that $\mathfrak{g}$ is semisimple and identify $\mathfrak{h}_{ic}$ with its dual by means of the bilinear form B (Sect. 3). Then $\Lambda_1 = \lambda + (-1)^{1/2} \nu$ where $\lambda \in (-1)^{1/2} \mathfrak{h}_1 \cap \mathfrak{k}$ and $\nu \in \mathfrak{h}_2 \cap \mathfrak{p}$. Moreover $\pi_1(\lambda + (-1)^{1/2} \nu) \neq 0$. Let U be the real analytic subgroup of G_c corresponding to $\mathfrak{u} = \mathfrak{k} + (-1)^{1/2} \mathfrak{p}$ which is a compact form of $\mathfrak{g}_c$. Put

$$\mathfrak{b}_i = \mathfrak{u} \cap \mathfrak{h}_{ic} = \mathfrak{h}_i \cap \mathfrak{k} + (-1)^{1/2} \mathfrak{h}_i \cap \mathfrak{p}.$$

Then $\mathfrak{b}_1, \mathfrak{b}_2$ are Cartan subalgebras of U and $\Lambda_i \in (-1)^{1/2} \mathfrak{b}_i$. Moreover it follows from the second condition of Lemma 29.2 that we can choose $x \in U$ such that $\Lambda_2 = x\Lambda_1$. Since Λ_1 is nonsingular this implies that $\mathfrak{b}_2 = x\mathfrak{b}_1$. Moreover λ and ν lie in $(-1)^{1/2} \mathfrak{b}_1$ and therefore $x\lambda$ and $x\nu$ are in $(-1)^{1/2} \mathfrak{b}_2$. This implies that

$$\bar{\Lambda}_2 = x\lambda - (-1)^{1/2} x\nu.$$

But $\bar{\Lambda}_2 = \theta\Lambda_2$. Therefore

$$x(\lambda - (-1)^{1/\rho}\nu) = \theta x(\lambda + (-1)^{1/2}\nu).$$

Now fix $H_0 \in \mathfrak{h}_2 \cap \mathfrak{p}$ such that $\alpha(H) \neq 0$ for every root α of $(\mathfrak{g}, \mathfrak{h}_2)$ which does not vanish identically on $\mathfrak{h}_2 \cap \mathfrak{p}$. Then Lemma 1 is applicable with $\mathfrak{a} = \mathfrak{h}_2 \cap \mathfrak{p}$ and $s = 1$. Hence we conclude that $\mathfrak{h}_2 \cap \mathfrak{p} \subset x(\mathfrak{h}_1 \cap \mathfrak{p})$. Interchanging $\mathfrak{h}_1$, $\mathfrak{h}_2$ and replacing x by x^{-1}, we deduce that

$$\mathfrak{h}_2 \cap \mathfrak{p} = x(\mathfrak{h}_1 \cap \mathfrak{p}),$$

and therefore by Corollary 1 of Lemma 5.1 we can choose $k \in K^0$ such that $x^{-1}H_0 = \mathrm{Ad}(k^{-1})\,H_0$. Then $y = \mathrm{Ad}(k)\,x^{-1}$ leaves H_0 fixed and therefore as we have seen during the proof of Lemma 1, it centralizes $\mathfrak{a} = \mathfrak{h}_2 \cap \mathfrak{p}$. Hence

$$(\mathfrak{h}_1 \cap \mathfrak{p})^k = yx(\mathfrak{h}_1 \cap \mathfrak{p}) = \mathfrak{h}_2 \cap \mathfrak{p}$$

and therefore in order to prove Lemma 29.2 it is enough to verify the following result.

LEMMA 2. *Let $\mathfrak{h}_1$, $\mathfrak{h}_2$ be two θ-stable Cartan subalgebras of $\mathfrak{g}$ such that $\mathfrak{h}_1 \cap \mathfrak{p} \subset \mathfrak{h}_2 \cap \mathfrak{p}$. Let L denote the centralizer of $\mathfrak{h}_1 \cap \mathfrak{p}$ in K. Then we can choose $k \in L^0$ such that*

$$(\mathfrak{h}_1 \cap \mathfrak{k})^k \supset \mathfrak{h}_2 \cap \mathfrak{k}.$$

Let $\mathfrak{z}$ denote the centralizer of $\mathfrak{h}_1 \cap \mathfrak{p}$ in $\mathfrak{g}$. Since $\mathfrak{h}_1$ is a Cartan subalgebra of $\mathfrak{z}$, it is clear that $\mathfrak{h}_1 \cap \mathfrak{k}$ is a Cartan subalgebra of $\mathfrak{l} = \mathfrak{z} \cap \mathfrak{k}$. Now $\mathfrak{l}$ is the Lie algebra of L and $\mathfrak{h}_2 \cap \mathfrak{k}$ is an abelian subalgebra of $\mathfrak{l}$. Therefore our assertion follows from the fact that L is compact.

34. A Limit Formula

For any linear subspace E of $\mathfrak{g}$, put

$$d_0E = (2\pi)^{-(\dim E)/2}\,dE$$

in the notation of Section 7. (Here $\pi = 3.14\ldots$.) Let B_E denote the restriction of the bilinear form B (see Sect. 3) on E and suppose B_E is nondegenerate. Let $S(E_c)$ be the symmetric algebra over E_c and $P(E_c)$ the algebra of all polynomial functions on E_c. We identify E with its dual by means of B_E. Then $S(E_c) = P(E_c)$ and therefore for any $p \in P(E_c)$, we obtain a differential operator $\partial(p)$ on E [9, Sect. 3].

Let $\mathfrak{a}$ be a θ-stable Cartan subalgebra of $\mathfrak{g}$ and Q the set of all positive roots of $(\mathfrak{g}, \mathfrak{a})$ (under some order). Put

$$\pi = \prod_{\alpha \in Q} \alpha, \qquad \pi_R = \prod_{\alpha \in Q_R} \alpha,$$

where Q_R is the set of all real roots in Q [10, Sect. 4]. Let $\mathfrak{a}'$ be the set of all $H \in \mathfrak{a}$ where $\pi(H) \neq 0$. Put

$$\mathfrak{g}_\mathfrak{a} = (\mathfrak{a}')^G.$$

Then $\mathfrak{g}_\mathfrak{a}$ is an open subset of $\mathfrak{g}$.

We now use the notation of Section 8. Normalize the invariant measure $d_0 x^*$ on G/A_R in such a way that

$$\int_{\mathfrak{g}_\mathfrak{a}} f(X)\, d_0 X = \int_\mathfrak{a} |\pi(H)|^2\, d_0 H \int_{G/A_R} f(x \cdot H)\, d_0 x^*$$

for $f \in C_c(\mathfrak{g})$. Here $d_0 X = d_0 \mathfrak{g}$ and $d_0 H = d_0 \mathfrak{a}$.

Let $\mathscr{C}(\mathfrak{g})$ be the Schwartz space of $\mathfrak{g}$. For $f \in \mathscr{C}(\mathfrak{g})$, define a function $\phi_f^{G/\mathfrak{a}} = \phi_f^\mathfrak{a} = \phi_f$ on $\mathfrak{a}'$ by

$$\phi_f(H) = \epsilon_R(H)\, \pi(H) \int_{G/A_R} f(x \cdot H)\, d_0 x^* \qquad (H \in \mathfrak{a}'),$$

where $\epsilon_R(H) = \operatorname{sign} \pi_R(H)$.

LEMMA 1. *Suppose $\mathfrak{a}$ is fundamental in $\mathfrak{g}$. Then*

$$\phi_f(0; \partial(\pi)) = (-1)^q f(0)$$

for all $f \in \mathscr{C}(\mathfrak{g})$. Here

$$q = (1/2)\,\{\dim G/K - \operatorname{rank} G + \operatorname{rank} K\}.$$

Although this lemma is known [23, p. 211], we shall give a simplified proof.

35. Relation between Fourier Transforms

Let $\mathscr{C}(E)$ denote the Schwartz space of a linear subspace E of $\mathfrak{g}$. Put $\bar{E} = \theta(E)$. Then the bilinear form B is nondegenerate on $\bar{E} \times E$. For $f \in \mathscr{C}(E)$ define $\hat{f} \in \mathscr{C}(\bar{E})$ by

$$\hat{f}(\bar{X}) = \int_E e^{(-1)^{1/2} B(\bar{X}, X)} f(X)\, d_0 E \qquad (\bar{X} \in \bar{E}).$$

Then

$$f(0) = \int_E \hat{f} d_0 \bar{E},$$

Note that if E is θ-stable then B_E is nondegenerate.

Let $P = MAN$ be a psgp of G. Put $\mathfrak{m}_1 = \mathfrak{m} + \mathfrak{a}$. Then $\mathfrak{m}_1 = \theta(\mathfrak{m}_1)$. For $f \in \mathscr{C}(\mathfrak{g})$, define a function $f_P \in \mathscr{C}(\mathfrak{m}_1)$ as follows:

$$f_P(Y) = \int \bar{f}(Y + Z)\, d_0 Z \qquad (Y \in \mathfrak{m}_1),$$

where $d_0 Z = d_0 \mathfrak{n}$ and

$$\bar{f}(X) = \int_K f(k \cdot X)\, dk \qquad (X \in \mathfrak{g}),$$

dk being the normalized Haar measure on K.

LEMMA 1. *Let P be a* psgp *of G. Then*

$$(\bar{f})_P = (f_P)^{\wedge}$$

for $f \in \mathscr{C}(\mathfrak{g})$.

Put $\bar{\mathfrak{n}} = \theta(\mathfrak{n})$. Then $\mathfrak{g} = \bar{\mathfrak{n}} \mid \mathfrak{m}_1 \mid \mathfrak{n}$ where the sum is orthogonal. Let $X_i = \bar{Z}_i + Y_i + Z_i$ where $\bar{Z}_i \in \bar{\mathfrak{n}}$, $Y_i \in \mathfrak{m}_1$, $Z \in \mathfrak{n}$ $(i = 1, 2)$. Then

$$B(X_1, X_2) = B(\bar{Z}_1, Z_2) + B(Y_1, Y_2) + B(\bar{Z}_2, Z_1).$$

Moreover if $X = \bar{Z} + Y + Z$ and $d_0 X = d_0 \mathfrak{g}$, $d_0 \bar{Z} = d_0 \bar{\mathfrak{n}}$, $d_0 Y = d_0 \mathfrak{m}_1$, $d_0 Z = d_0 \mathfrak{n}$, then it is obvious that

$$d_0 X = d_0 \bar{Z} d_0\, Y d_0 Z.$$

Since B is G-invariant, it is enough to consider the case when $f = \bar{f}$. Then if $g = \overset{\circ}{f}$,

$$g_P(Y_1) = \int g(Y_1 + Z_1)\, d_0 Z_1 \qquad (Y_1 \in \mathfrak{m}_1),$$

and

$$g(Y_1 + Z_1) = \int e^{(-1)^{1/2} B(Y_1 + Z_1, X_2)} f(X_2)\, d_0 X_2\,.$$

Fix Y_1 and put

$$h(\bar{Z}_2) = \int e^{(-1)^{1/2} B(Y_1, Y_2)} f(\bar{Z}_2 + Y_2 + Z_2)\, d_0 Y_2\, d_0 Z_2$$

for $\bar{Z}_2 \in \bar{\mathfrak{n}}$. Then $h \in \mathscr{C}(\bar{\mathfrak{n}})$ and

$$g_P(Y_1) = \int_{\mathfrak{n}} h(Z_1)\, d_0 Z_1 = h(0) = (f_P)^{\wedge}(Y_1).$$

This proves the lemma.

36. Proof of Lemma 34.1

We return to the notation of Section 34. Fix $P \in \mathscr{P}(A_R)$ $(P = MA_R N)$ and put $M_1 = MA_R$.

Lemma 1. *There exists a constant $c \neq 0$ such that*

$$\phi_f^{G/\mathfrak{a}} = c\phi_{f_P}^{M_1/\mathfrak{a}}$$

for all $f \in \mathscr{C}(\mathfrak{g})$.

This is proved in the same way as Lemma 8.2.

Lemma 2. *Suppose $\mathfrak{a}$ is fundamental and $f \in C_c^{\infty}(\mathfrak{g}_{\mathfrak{a}})$. Then $\phi_f^{\mathfrak{b}} = 0$ for any θ-stable Cartan subalgebra $\mathfrak{b}$ of $\mathfrak{g}$ which is not fundamental. Moreover $\phi_f^{\mathfrak{a}} \in \mathscr{C}(\mathfrak{a})$.*

Fix $P \in \mathscr{P}(B_R)$ $(P = MB_R N)$ and put $\mathfrak{m}_1 = \mathfrak{m} + \mathfrak{b}_R$. Then $\dim \mathfrak{b}_R > \dim \mathfrak{a}_R$ since $\mathfrak{b}$ is not fundamental. Let $\mathfrak{h}$ be any Cartan subalgebra of $\mathfrak{m}_1$. Then $\mathfrak{h} \supset \mathfrak{b}_R$ and therefore $\mathfrak{h}$ cannot be conjugate to $\mathfrak{a}$ under G. This shows that

$$\mathfrak{g}_{\mathfrak{a}} \cap (\mathfrak{m}_1)^G = \varnothing.$$

Let $\mathfrak{g}'$ be the set of all regular elements of $\mathfrak{g}$. Then a simple argument [5, Lemma 16] shows that

$$\mathfrak{g}' \cap (\mathfrak{m}_1 + \mathfrak{n}) \subset \mathfrak{m}_1^N.$$

Hence

$$\mathfrak{g}_{\mathfrak{a}} \cap (\mathfrak{m}_1 + \mathfrak{n})^K \subset \mathfrak{g}_{\mathfrak{a}} \cap \mathfrak{m}_1^G = \varnothing$$

and therefore $f_P = 0$. But then we conclude from Lemmas 35.1 and 1 that

$$\phi_f^{\mathfrak{b}} = c\phi_{(f_P)_{\wedge}}^{M_1/\mathfrak{b}} = 0.$$

The rest follows from [5, Theor. 3] and [10, Theor. 2].

Fix $g \in C_c^\infty(\mathfrak{g}_\mathfrak{a})$ such that

$$\int_\mathfrak{g} g \, d_0 X = 1.$$

Put $f = \hat{g}$. Then $f(0) = 1$. Since $\mathfrak{a}$ is fundamental, $\mathfrak{a}_I$ is a Cartan subalgebra of $\mathfrak{k}$. Hence $\dim \mathfrak{k} \equiv \dim \mathfrak{a}_I \bmod 2$. Moreover $\dim \mathfrak{g} \equiv \dim \mathfrak{a} \bmod 2$. Hence

$$\dim \mathfrak{k} \equiv \dim \mathfrak{a}_I \bmod 2,$$

$$\dim \mathfrak{p} \equiv \dim \mathfrak{a}_R \bmod 2.$$

Let ω denote the polynomial function on $\mathfrak{g}$ given by $\omega(X) = B(X, X)$ $(X \in \mathfrak{g})$. Put[7] $n = \dim \mathfrak{g}$, $\ell = \dim \mathfrak{a}$, $m = [n/2]$, $k = [\ell/2]$. Assume $\ell \geqslant 2$ and define

$$\xi_\mathfrak{g} = \xi_{p,q} \qquad (p = \dim \mathfrak{p}, \ q = \dim \mathfrak{k}),$$

$$\xi_\mathfrak{a} = \xi_{p,q} \qquad (p = \dim \mathfrak{a}_R, \ q = \dim \mathfrak{a}_I),$$

where $\xi_{p,q} = \xi$ has the same meaning as in [10, Sect. 12]. Put

$$\Xi_\mathfrak{g}(X) = \xi_\mathfrak{g}(\omega(X)) \qquad (X \in \mathfrak{g}),$$
$$\Xi_\mathfrak{a}(H) = \xi_\mathfrak{a}(\omega(H)) \qquad (H \in \mathfrak{a}).$$

Then

$$1 = f(0) = \int_\mathfrak{g} \Xi_\mathfrak{g} \partial(\omega^m) f \, dX = (2\pi)^{n/2} \int_\mathfrak{g} \Xi_\mathfrak{g} \partial(\omega^m) f \, d_0 X,$$

where $dX = d\mathfrak{g}$ and $d_0 X = d_0 \mathfrak{g}$.

Let Q_I and Q_c, respectively, be the sets of all imaginary and complex roots in Q. We may assume that if $\alpha \in Q_c$ then $-\theta\alpha$ is also in Q_c. Since $\mathfrak{a}$ is fundamental, Q_R is empty. Let r, r_I and r_c, respectively, be the number of roots in Q, Q_I and Q_c. Then

$$\operatorname{conj} \pi(H) = (-1)^{r_I} \pi(H) = (-1)^r \pi(H) \qquad (H \in \mathfrak{a}),$$

since $r - r_I = r_c \equiv 0 \bmod 2$. Hence

$$1 = (2\pi)^{n/2}(-1)^r \int_\mathfrak{a} \pi(H) \, \Xi_\mathfrak{g}(H) \, \phi_{f_m}(H) \, d_0 H,$$

where $f_m = \partial(\omega^m) f$. Let ω_1 be the restriction of ω on $\mathfrak{a}$. Then

$$\phi_{f_m} = \partial(\omega_1^m) \phi_f$$

[7] As usual $[r]$ denotes the greatest integer not exceeding r.

from [5, Theor. 1]. Moreover since $n = \ell + 2r$, it is obvious that $m = k + r$. Therefore a simple calculation [10, p. 572] shows that

$$\pi \circ \partial(\omega_1{}^m) = \partial(\omega_1{}^k) \circ \eta,$$

where η is a polynomial differential operator on $\mathfrak{a}$ whose local expression at zero is given by

$$\eta_0 = (-1)^r 2^r \left(\Gamma(m)/\Gamma(k)\right) \partial(\pi).$$

On the other hand

$$(1/2)(\dim \mathfrak{p} - \dim \mathfrak{a}_R) = (1/2)(\dim G/K - \operatorname{rank} G + \operatorname{rank} K) = q,$$

and one verifies directly that

$$\xi_\mathfrak{g} = \left((-1)^q/2^{2r}\pi^r\right) \left(\Gamma(k)/\Gamma(m)\right)\xi_\mathfrak{a} .$$

Hence

$$1 = (2\pi)^r(-1)^r \int \pi(H)\, \Xi_\mathfrak{g}(H)\, \phi_f(H;\, \partial(\omega_1{}^m))\, dH$$

$$= (-1)^{r+q}\, 2^{-r}\, \frac{\Gamma(k)}{\Gamma(m)} \int \Xi_\mathfrak{a}\partial(\omega_1{}^k)(\eta\phi_f)\, dH,$$

where $dH = d\mathfrak{a}$. But

$$\partial(\omega_1{}^k)\Xi_\mathfrak{a} = \delta_\mathfrak{a} ,$$

where $\delta_\mathfrak{a}$ is the Dirac measure on $\mathfrak{a}$ at the origin. Hence

$$1 = (-1)^{r+q}2^{-r}\left(\Gamma(k)/\Gamma(m)\right)\phi_f(0;\, \eta_0),$$
$$= (-1)^q\, \phi_f(0;\, \partial(\pi)).$$

In view of [10, Lemma 17], this proves Lemma 34.1 when $\ell \geqslant 2$. The case $\ell = 1$ can be handled by considering the direct sum of $\mathfrak{g}$ with a one-dimensional Lie algebra and thus reducing it to the case $\ell = 2$.

37. Evaluation of the Constant c

Let us use the notation of Section 7 and define a positive constant v_G by

$$dG = v_G\, dx,$$

where dx is the standard Haar measure on G. Let $\tilde{A}$ be the normalizer of $\mathfrak{a}$ in G. Put $W(G/A) = \tilde{A}/A$ and define dx^* as in Section 8.

LEMMA 1. $dx^* = (2\pi)^r\, v_G^{-1} v(A_I)[W(G/A)]\, d_0 x^*$, **where** $r = (1/2)\dim \mathfrak{g}/\mathfrak{a}$.

Put $A' = A \cap G'$ and $G_A = (A')^G$. Since $A^0 = A_I{}^0 A_R$ lies in the center of A, we conclude from [15, Lemma 91] that

$$\int_{G_A} f(x)\, dx = [\tilde{A} : A^0]^{-1} \int_A v(a)\, d_0 a \int_{G/A_R} f(a^x)\, dx^*$$

for $f \in C_c(G)$. Here

$$v(a) = |\det(\mathrm{Ad}(a^{-1}) - 1)_{\mathfrak{g}/\mathfrak{a}}| \qquad (a \in A),$$

and

$$d_0 a = v(A_I{}^0)^{-1}\, dA.$$

Put $da = v(A_I)^{-1}\, dA$. Then

$$\int_{G_A} f(x)\, dx = [W(G/A)]^{-1} \int_A v(a)\, da \int_{G/A_R} f(a^x)\, dx^*,$$

and this is equivalent to

$$\int_{G_A} f d_0 G = (2\pi)^{-r}\, v_G [W(G/A)]^{-1}\, v(A_I)^{-1} \int_A v(a)\, d_0 A \int_{G/A_R} f(a^x)\, dx^*,$$

where $d_0 G = (2\pi)^{-n/2}\, dG$ and $d_0 A = (2\pi)^{-\ell/2}\, dA$ $(n = \dim G, \ell = \dim A)$. Clearly this relation implies the assertion of the lemma.

LEMMA 2. $v_G = 2^{-\nu/2} v(K)$ *where* $\nu = \dim G/K - \mathrm{rank}\, G/K$.

Let $P_0 = M_0 A_0 N_0$ be a minimal psgp of G. Then

$$G = K A_0 N_0,$$

and this is an Iwasawa decomposition of G. Hence

$$\mathfrak{g} = \mathfrak{k} + \mathfrak{a}_0 + \mathfrak{n}_0,$$

where the sum is direct. Moreover

$$\nu = \dim \mathfrak{g}/\mathfrak{k} - \dim \mathfrak{a}_0 = \dim \mathfrak{n}_0.$$

Let $E_t = (1 + \theta)/2$ and $E_\mathfrak{p} = (1 - \theta)/2$ denote the orthogonal projections of $\mathfrak{g}$ on $\mathfrak{k}$ and $\mathfrak{p}$, respectively. Put

$$(X, Y) = -B(\theta(X), Y) \qquad (X, Y \in \mathfrak{g}),$$

and choose an orthonormal base Z_i $(1 \leqslant i \leqslant \nu)$ for $\mathfrak{n}_0$. Then a simple calculation shows that

$$(E_t Z_i, E_t Z_j) = (E_\mathfrak{p} Z_i, E_\mathfrak{p} Z_j) = \delta_{ij}/2 \qquad (1 \leqslant i, j \leqslant \nu).$$

Hence $\sqrt{2}\, E_\mathfrak{p} Z_i$ $(1 \leqslant i \leqslant \nu)$ is an orthonormal base for $E_\mathfrak{p} \mathfrak{n}_0$ and $\sqrt{2}\, E_t Z_i$ $(1 \leqslant i \leqslant \nu)$ an orthonormal base for $E_t \mathfrak{n}_0$. Now

$$\mathfrak{p} = \mathfrak{a}_0 + E_\mathfrak{p} \mathfrak{n}_0 , \quad \mathfrak{k} = \mathfrak{m}_0 + E_t \mathfrak{n}_0 ,$$

where both sums are orthogonal. Moreover

$$\sum_{1 \leqslant i \leqslant \nu} x_i Z_i + \sum_{1 \leqslant i \leqslant \nu} y_i \sqrt{2}\, E_t Z_i = \sum_i x_i{}' \sqrt{2}\, E_\mathfrak{p} Z_i + \sum_i y_i{}' \sqrt{2}\, E_t Z_i ,$$

where

$$x_i{}' = 2^{-1/2} x_i , \; y_i{}' = 2^{-1/2} x_i + y_i \qquad (1 \leqslant i \leqslant \nu),$$

and $x_i , y_i \in \mathbf{R}$. This shows that

$$d\mathfrak{g} = 2^{-\nu/2}\, d\mathfrak{k}\, d\mathfrak{a}_0\, d\mathfrak{n}_0$$

corresponding to the direct sum $\mathfrak{g} = \mathfrak{k} + \mathfrak{a}_0 + \mathfrak{n}_0$. Since this implies that

$$dG = 2^{-\nu/2}\, v(K)\, dx,$$

our assertion follows.

COROLLARY. $dx^* = (2\pi)^r\, 2^{\nu/2}(v(A_l)/v(K))[W(G/A)]\, d_0 x^*.$

Now suppose $\mathfrak{a}$ is fundamental and let $(-1)^q\, c_G$ denote the constant c of Lemma 17.5. (Here q has the same meaning as in Lemma 34.1.)

LEMMA 3. $c_G = (2\pi)^r\, 2^{\nu/2}(v(A_l)/v(K))[W(G/A)].$

This follows easily from the above corollary and Lemma 34.1 [7, p. 759].

By choosing $f \in C_c^\infty(G^0)$ with $f(1) \neq 0$, it becomes clear that $c_G = c_{G^0}$.

Let Q_k denote the set of all compact roots [10, p. 555] in Q. Since $\mathfrak{a}$

is fundamental, $\mathfrak{a}_I$ is a Cartan subalgebra of $\mathfrak{k}$ and Q_k may be regarded as the set of all positive roots of $(\mathfrak{k}, \mathfrak{a}_I)$. Put

$$\varpi_k = \prod_{\alpha \in Q_k} \alpha, \quad \rho_k = \frac{1}{2} \sum_{\alpha \in Q_k} \alpha,$$

and let r_k denote the number of roots in Q_k.

LEMMA 4. *Suppose K is connected. Then*

$$(v(A_I)/v(K)) = (2\pi)^{-r_k} \varpi_k(\rho_k).$$

We apply Lemma 3 to (K, A_I) in place of (G, A). Then

$$c_K = (2\pi)^{r_k} (v(A_I)/v(K)) [W(K/A_I)].$$

Let 1_K be the function on K which is identically equal to 1. Then substituting $f = 1_K$ in Lemma 17.5, we find that

$$[W(K/A_I)] \varpi_k(\rho_k) = c_K.$$

Comparing the two expressions for c_K, we get the required result. Now

$$r - r_k = (1/2)\{\dim \mathfrak{g}/\mathfrak{a} - \dim \mathfrak{k}/\mathfrak{a}_I\},$$

$$= (1/2)\{\dim \mathfrak{g}/\mathfrak{k} - \dim \mathfrak{a} + \dim \mathfrak{a}_I\} = q.$$

Also $G^0 \cap K = K^0$, $K^0 \cap A_I = A_I^0$ and $G^0 \cap A = A^0$. Therefore

$$c_G = c_{G^0} = (2\pi)^r 2^{\nu/2} (v(A_I^0)/v(K^0)) [W(G^0/A^0)],$$

$$= (2\pi)^q 2^{\nu/2} \varpi_k(\rho_k) [W(G^0/A^0)].$$

Hence we have obtained the following theorem.

THEOREM 1. *The constant c of Lemma 17.5 is given by the following formula*

$$c = (-1)^q (2\pi)^q 2^{\nu/2} \varpi_k(\rho_k) [W(G^0/A^0)],$$

where

$$q = (1/2)\{\dim G/K - \operatorname{rank} G + \operatorname{rank} K\},$$

and

$$\nu = \dim G/K - \operatorname{rank} G/K.$$

References

1. A. Borel and J. Tits, Groupes réductifs, *Inst. Hautes Études Sci. Publ. Math.* **27** (1965), 55–150.
2. Harish-Chandra, Representations of a semisimple Lie group on a Banach space I, *Trans. Amer. Math. Soc.* **75** (1953), 185–243.
3. Harish-Chandra, Representations of semisimple Lie groups VI, *Amer. J. Math.* **78** (1956), 564–628.
4. Harish-Chandra, The characters of semisimple Lie groups, *Trans. Amer. Math. Soc.* **83** (1956), 98–163.
5. Harish-Chandra, Fourier transforms on a semisimple Lie algebra I, *Amer. J. Math.* **79** (1957), 193–257.
6. Harish-Chandra, Fourier transforms on a semisimple Lie algebra II, *Amer. J. Math.* **79** (1957), 653–686.
7. Harish-Chandra, A formula for semisimple Lie groups, *Amer. J. Math.* **79** (1957), 733–760.
8. Harish-Chandra, Spherical functions on a semisimple Lie group I, *Amer. J. Math.* **80** (1958), 241–310.
9. Harish-Chandra, Invariant differential operators and distributions on a semisimple Lie algebra, *Amer. J. Math.* **86** (1964), 534–564.
10. Harish-Chandra, Some results on an invariant integral on a semisimple Lie algebra, *Ann. of Math.* **80** (1964), 551–593.
11. Harish-Chandra, Invariant eigendistributions on a semisimple Lie algebra, *Inst. Hautes Études Sci. Publ. Math.* **27** (1965), 5–54.
12. Harish-Chandra, Invariant eigendistributions on a semisimple Lie group, *Trans. Amer. Math. Soc.* **119** (1965), 457–508.
13. Harish-Chandra, Two theorems on semisimple Lie groups, *Ann. of Math.* **83** (1966), 74–128.
14. Harish-Chandra, Discrete series for semisimple Lie groups I, *Acta Math.* **113** (1965), 241–318.
15. Harish-Chandra, Discrete series for semisimple Lie groups II, *Acta Math.* **116** (1966), 1–111.
16. Harish-Chandra, Automorphic forms on semisimple Lie groups, *in* "Lecture Notes in Math.," Vol. 62, Springer-Verlag, 1968.
17. Harish-Chandra, Some applications of the Schwartz space of a semisimple Lie group, *in* "Lecture Notes in Math.," Vol. 140, Springer-Verlag, 1970.
18. Harish-Chandra, Harmonic analysis on semisimple Lie groups, *Bull. Amer. Math. Soc.* **78** (1970), 529–551.
19. Harish-Chandra, On the theory of the Eisenstein integral, *in* "Lecture Notes in Math.," Vol. 266, pp. 123–149, Springer-Verlag, 1971.
20. S. Helgason, "Differential Geometry and Symmetric Spaces," Academic Press, New York, 1962.
21. L. Hörmander, "Linear Partial Differential Operators," Springer-Verlag, New York, 1969.
22. G. D. Mostow, Self-adjoint groups, *Ann. of Math.* **62** (1955), 44–55.
23. G. Warner, "Harmonic Analysis on Semi-Simple Lie Groups II," Springer-Verlag, New York, 1972.

Printed by the St Catherine Press Ltd., Tempelhof 37, Bruges, Belgium

Inventiones math. 36, 1 – 55 (1976)

Inventiones
mathematicae
© by Springer-Verlag 1976

Harmonic Analysis on Real Reductive Groups. II

Wave-Packets in the Schwartz Space

Harish-Chandra (Princeton)

To Jean-Pierre Serre

§ 1. Introduction

The theory of the constant term, which has been developed in [1(e)] will now be applied to construct wave-packets in the Schwartz space of a reductive group G. Keeping to the notation of [1(e)], let A be the split component of a θ-stable Cartan subgroup of G. Fix a psgp $P_1 = MAN_1$ with the split component A and let τ be a unitary double representation of K on a finite-dimensional Hilbert space V. Then $L = {}^\circ\mathscr{C}(M, \tau_M)$ also has finite dimension [1(e), Theorem 27.3]. Put $\mathfrak{F} = \mathfrak{a}^*$ and consider the Eisenstein integral

$$\phi_v = E(P_1 : \psi : v) \quad (v \in \mathfrak{F})$$

for a given $\psi \in L$. We compute the constant term ϕ_{v, P_2} of ϕ_v along a psgp $P_2 \in \mathscr{P}(A)$ (Theorem 18.1). The expression for ϕ_{v, P_2} involves certain endomorphisms $c_{P_2 | P_1}(s : v)$ $(s \in \mathfrak{w}(\mathfrak{a}))$ of L. We shall see later that these c-functions can be extended to meromorphic functions of v on the whole complex space $\mathfrak{F}_c$.

Let $\mathfrak{F}'$ be the set of all regular elements in $\mathfrak{F}$. Fix $\alpha \in C_c^\infty(\mathfrak{F}')$ and put

$$\phi_\alpha = \int_{\mathfrak{F}} \alpha(v) \phi_v \, dv$$

where dv is the Euclidean measure on $\mathfrak{F}$. Then $\phi_\alpha \in \mathscr{C}(G, \tau)$ (Theorem 13.1). Now fix $P_2 \in \mathscr{P}(A)$ and $m \in MA$ and consider the distribution

$$\alpha \rightarrow \phi_\alpha^{(P_2)}(m)$$

on $\mathfrak{F}'$. It turns out that this distribution is actually a function which can be written quite simply in terms of the c-functions (Theorem 19.2).

Theorems 13.1, 13.2 and 18.1 contain the main results of this paper. They may be regarded as generalizations of the corresponding results on spherical functions obtained in [1(a, b)]. In fact here we have combined the methods of [1(a, b)] with those of [1(d)] and our success depends in an essential way on the systematic use of the weak inequality.

As far as possible, we shall keep to the notation of [1(e)] and therefore any undefined symbols should be given the same meaning as in [1(e)].

Most of the work presented here was done some years ago and I have given lectures on it on various occasions.

§ 2. Recapitulation of Some Algebraic Results

Let $(P, A) \succ (P_0, A_0)$ be two p-pairs in G such that (P_0, A_0) is minimal. Then $P = MAN$, $P_0 = M_0 A_0 N_0$. Extend $\mathfrak{a}_0$ to a Cartan subalgebra $\mathfrak{h}_0$ of $\mathfrak{g}$. Then $\mathfrak{h}_0$ is θ-stable and $\mathfrak{a}_0 = \mathfrak{h}_0 \cap \mathfrak{p}$. Put $W_0 = W(\mathfrak{g}/\mathfrak{h}_0)$ and let W_1 be the subgroup of those elements of W_0 which leave $\mathfrak{a}$ pointwise fixed. Put $S = \mathfrak{S}(\mathfrak{h}_{0c}) = S(\mathfrak{h}_{0c})$ and let J and J_1 denote the algebras of invariants of W_0 and W_1 respectively in S. Let $s_1, s_2, \ldots, s_q$ $(q = [W_0 : W_1])$ be a complete system of representatives for $W_1 \backslash W_0$ so that

$$W_0 = \bigcup_{1 \leq i \leq q} W_1 s_i.$$

Select homogeneous elements $u_1 = 1, u_2, \ldots, u_q$ in J_1 such that [1(a), Lemma 8]

$$J_1 = \sum_{1 \leq i \leq q} J u_i.$$

Fix a system of positive roots for $(\mathfrak{g}, \mathfrak{h}_0)$ and put

$$\varpi_0 = \varpi_{\mathfrak{g}/\mathfrak{h}_0}, \qquad \varpi_1 = \varpi_{\mathfrak{m}_1/\mathfrak{h}_0}, \qquad \varpi_{01} = \varpi_{\mathfrak{g}/\mathfrak{m}_1},$$

where $\mathfrak{m}_1 = \mathfrak{m} + \mathfrak{a}$. Then $\varpi_0 = \varpi_{01} \varpi_1$. Define $u^j \in C(J_1)$ by

$$\mathrm{tr}_{J_1/J}(u_i u^j) = \delta_i^j \qquad (1 \leq i, \ j \leq q)$$

and put $\tau^j = \varpi_{01} u^j$. Then [1(a), Lemma 12] $\tau^j \in J_1$.

Every element of S may be regarded as a polynomial function on $\mathfrak{h}_{0c}^*$. For $p \in J_1$ and $\Lambda \in \mathfrak{h}_{0c}^*$, define

$$f_\Lambda = \sum_{1 \leq i \leq q} \tau^i(\Lambda) u_i,$$
$$v^j(p : \Lambda) = \mathrm{tr}_{J_1/J}\{(p - p(\Lambda)) f_\Lambda u^j\} \qquad (1 \leq j \leq q).$$

Then it is clear that $v^j(p : \Lambda) \in J$ and, for p fixed, $\Lambda \mapsto v^j(p : \Lambda)$ is a polynomial mapping of $\mathfrak{h}_{0c}^*$ into J. Let S_Λ denote the set of all $p \in S$ such that $p(\Lambda) = 0$. Put $J_\Lambda = J \cap S_\Lambda$. Then it is obvious that $J_{s\Lambda} = J_\Lambda$ $(s \in W_0)$.

Identify $\mathfrak{h}_0$ with its dual by means of the bilinear form B. We call an element $u \in J_1$ harmonic if $\partial(p) u = 0$ for all $p \in J \cap S_0$ in the notation of [1(c), § 3]. Then it is easy to conclude from [1(c), Lemma 4] that $u_1, \ldots, u_q$ may be so chosen as to span the space U of all harmonic elements in J_1. Moreover $J_1 = U + J_1 J_\Lambda$ where the sum is direct [1(a), p. 256]. The following lemma enables us to diagonalize the action of J_1 on $J_1/J_1 J_\Lambda \simeq U$.

Lemma 1. *Fix* $p \in J_1$, $\Lambda \in \mathfrak{h}_{0c}^*$ *and put* $\Lambda_i = s_i \Lambda$ $(1 \leq i \leq q)$. *Then*

1) $v^j(p : \Lambda_i) \in J_\Lambda$,

2) $(p - p(\Lambda_i)) f_{\Lambda_i} = \sum_{1 \leq j \leq q} v^j(p : \Lambda_i) u_j$,

3) $\sum_{1 \leq k \leq q} \varepsilon(s_k) \varpi_1(\Lambda_k) f_{\Lambda_k} = \varpi_0(\Lambda)$

for $1 \leq i \leq q$. *Here* $\varepsilon(s) = \pm 1$ *is defined as usual by* $\varpi_0^s = \varepsilon(s) \varpi_0$ $(s \in W_0)$.

We know from [1(a), Lemma 15] that

$$(p - p(\Lambda)) f_\Lambda \in S J_\Lambda \cap J_1 = J_\Lambda J_1 .$$

Hence the first two statements are obvious. Both sides of 3) being polynomial in Λ, it is sufficient to consider the case when $\varpi_0(\Lambda) \neq 0$. Then the rational function u^j is defined at Λ_k and

$$\sum_k \varepsilon(s_k) \varpi_1(\Lambda_k) f_{\Lambda_k} = \varpi_0(\Lambda) \sum_{i, k} u^i(\Lambda_k) u_i .$$

But since

$$\sum_k (u^i)^{s_k^{-1}} = \operatorname{tr}_{J_1/J} u^i = \delta_1^i ,$$

we conclude that

$$\sum_{i, k} u^i(\Lambda_k) u_i = 1$$

and this proves 3).

§ 3. Further Algebraic Results

Let $\mathfrak{h}$ be a θ-stable Cartan subalgebra of $\mathfrak{g}$. Then $\mathfrak{h} = \mathfrak{h}_I + \mathfrak{h}_R$ as usual [1(e), § 8]. If $\lambda \in (\mathfrak{h}_I)_c^*$, $\nu \in (\mathfrak{h}_R)_c^*$, we extend them to linear functions on $\mathfrak{h}_c$ by defining $\lambda = 0$ on $\mathfrak{h}_R$ and $\nu = 0$ on $\mathfrak{h}_I$. In this way $\mathfrak{h}_c^*$ becomes the direct sum of $(\mathfrak{h}_I)_c^*$ and $(\mathfrak{h}_R)_c^*$.

An element $\lambda \in (\mathfrak{h}_I)_c^*$ is called singular if $\lambda(H_\alpha) = 0$ for some imaginary root α of $(\mathfrak{g}, \mathfrak{h})$. Otherwise we call it regular. Put $\mathfrak{F} = \mathfrak{h}_R^*$ and

$$\varpi = \varpi_{\mathfrak{g}/\mathfrak{h}} = \prod_{\alpha > 0} H_\alpha$$

where α runs over all positive roots of $(\mathfrak{g}, \mathfrak{h})$ (under some fixed order). Fix a regular element $\lambda \in (-1)^{1/2} \mathfrak{h}_I^*$ and let $\mathfrak{F}_c'(\lambda)$ denote the set of all $\nu \in \mathfrak{F}_c$ such that

$$\varpi(\lambda + (-1)^{1/2} \nu) \neq 0 .$$

Put $\mathfrak{F}'(\lambda) = \mathfrak{F} \cap \mathfrak{F}_c'(\lambda)$. Then $\mathfrak{F}'(\lambda)$ is an open and dense subset of $\mathfrak{F}$.

Now we use the notation of § 2. Fix $k_0 \in K$ such that $\mathfrak{h}_R^{k_0} \subset \mathfrak{a}_0$. Let $\mathfrak{z}$ denote the centralizer of $\mathfrak{h}_R^{k_0}$ in $\mathfrak{g}$. Then $\mathfrak{h}^{k_0}$ and $\mathfrak{h}_0$ are two Cartan subalgebras of $\mathfrak{z}$. Hence we can choose $y_0 \in G_c$ such that y_0 centralizes $\mathfrak{h}_R^{k_0}$ and $\mathfrak{h}_c^y = \mathfrak{h}_{0c}$ where $y = y_0 \operatorname{Ad}(k_0)$. Put $\Lambda_\nu = (\lambda + (-1)^{1/2} \nu)^y$ for $\nu \in \mathfrak{F}_c$. (Here we have identified $\mathfrak{h}_c$ with its dual by means of the restriction of the bilinear form B on $\mathfrak{h}_c$.) Then if $v\in\mathfrak{F}'_c(\lambda)$, it is clear that $\varpi_0(\Lambda_v)\neq 0$ and therefore the rational functions u^i are defined at Λ_v.

Fix an element $\Lambda\in\mathfrak{h}^*_{0c}$ and let $W_0(\Lambda)$ be the subgroup of all $s\in W_0$ which leave Λ fixed. Let p_0 be the set of all positive roots of $(\mathfrak{g},\mathfrak{h}_0)$ and $p_0(\Lambda)$ the set of those $\alpha\in p_0$ for which $\Lambda(H_\alpha)\neq 0$. Put

$$\varpi_{0,\Lambda}=\prod_{\alpha\in p_0(\Lambda)}H_\alpha.$$

Let $J(\Lambda)$ be the algebra of all invariants of $W_0(\Lambda)$ in S.

Lemma 1. *Let v be an element in $\mathbf{C}(S)$ such that* $\operatorname{tr}_{S/J}(uv)\in J$ *for all* $u\in S$. *Then* $\varpi_{0,\Lambda}\operatorname{tr}_{S/J(\Lambda)}(v)\in S$.

Put $v'=\operatorname{tr}_{S/J(\Lambda)}v$. Then if $u\in J(\Lambda)$, it is clear that

$$\operatorname{tr}_{J(\Lambda)/J}(v'u)=\operatorname{tr}_{S/J}(vu)\in J.$$

Hence we conclude from [1(a), Lemma 12] that $\varpi_{0,\Lambda}v'\in S$.

Now put

$$U=\sum_{1\leq i\leq q}\mathbf{C}u_i$$

and $\varpi_{s,\lambda}=\varpi_{0,\Lambda}$ where $\Lambda=s\lambda^y$ $(s\in W_0)$. Define a rational mapping e_s $(s\in W_0)$ of $\mathfrak{F}_c$ into U by

$$e_s(v)=\sum_{1\leq j\leq q}u^j(s\Lambda_v)u_j \qquad (v\in\mathfrak{F}'_c(\lambda)).$$

Since $u^j\in\mathbf{C}(J_1)$, it is clear that $e_{ts}=e_s$ $(t\in W_1)$.

Put $W_0(s,\lambda)=W_0(s\lambda^y)$.

Lemma 2. *Fix* $s\in W_0$. *Then the mapping*

$$v\mapsto\varpi_{s,\lambda}(s\Lambda_v)\sum_{t\in W_0(s,\lambda)}e_{ts}(v)$$

is a polynomial mapping of $\mathfrak{F}_c$ into U.

Let $u\in S$ and put $u'=\operatorname{tr}_{S/J_1}u$. Then $u'\in J_1$ and it is obvious that

$$\operatorname{tr}_{S/J}(u^j u)=\operatorname{tr}_{J_1/J}(u^j u')\in J \qquad (1\leq j\leq q).$$

Hence we conclude from Lemma 1 that

$$\varpi_{0,\Lambda}\operatorname{tr}_{S/J(\Lambda)}u^j\in S$$

where $\Lambda=s\lambda^y$. Since $\mathbf{C}(J(\Lambda))$ is the fixed field of $W_0(\Lambda)=W_0(s,\lambda)$ in $\mathbf{C}(S)$, it follows that

$$\operatorname{tr}_{S/J(\Lambda)}u^j=\sum_{t\in W_0(s,\lambda)}(u^j)^t.$$

Hence the mapping

$$v \mapsto \varpi_{s,\lambda}(s\Lambda_v) \sum_{t \in W_0(s,\lambda)} \sum_j u^j (ts\Lambda_v) u_j$$
$$= \varpi_{s,\lambda}(s\Lambda_v) \sum_{t \in W_0(s,\lambda)} e_{ts}(v) \qquad (v \in \mathfrak{F}'_c(\lambda))$$

extends to a polynomial mapping of $\mathfrak{F}_c$ into U.

Let $p(\lambda)$ be the set of all positive roots α of $(\mathfrak{g}, \mathfrak{h})$ such that $\lambda(H_\alpha) \neq 0$. Put

$$\varpi_\lambda = \prod_{\alpha \in p(\lambda)} H_\alpha.$$

Lemma 3. *Fix* $s \in W_0$ *and* $v \in \mathfrak{F}$. *Then*

$$|\varpi_{s,\lambda}(s\Lambda_v)| \geq |\varpi_{s,\lambda}(s\lambda^v)| = |\varpi_\lambda(\lambda)| > 0.$$

This is obvious from the definitions.
Now put $e_i = e_{s_i}$ and

$$_i e = [W_1 \cap W_0(s_i,\lambda)]^{-1} \sum_{t \in W_0(s_i,\lambda)} e_{ts_i} \qquad (1 \leq i \leq q).$$

Let Q denote the set $\{1, 2, \ldots, q\}$. It is clear that $_i e = {}_j e$ if $s_i \lambda^v = s_j \lambda^v$ $(i, j \in Q)$. Choose a maximal subset $^\circ Q$ of Q such that $s_i \lambda^v \neq s_j \lambda^v$ for $i \neq j$ in $^\circ Q$.

Lemma 4. *Fix* $i \in Q$. *Then* $_i e$ *is a rational mapping of* $\mathfrak{F}_c$ *into* U *which is everywhere defined on* $\mathfrak{F}$. *Moreover the mapping*

$$v \mapsto \varpi_{s_i,\lambda}(s_i\Lambda_v) {}_i e(v) \qquad (v \in \mathfrak{F}'_c(\lambda))$$

extends to a polynomial mapping from $\mathfrak{F}_c$ *into* U. *Finally*

$$\sum_{i \in {}^\circ Q} {}_i e = 1.$$

The first two statements follow from Lemmas 2 and 3. Moreover since $\mathrm{tr}_{J_1/J} u^j = \delta_1^j$, it is clear that

$$\sum_{1 \leq i \leq q} e_i = 1.$$

The third statement is an immediate consequence of this fact.
Put

$$v_{ij}(p:v) = \mathrm{tr}_{J_1/J} \{p - p(s_i\Lambda_v) u^j e_i(v)\} \qquad (v \in \mathfrak{F}'_c(\lambda),\ 1 \leq i,\ j \leq q)$$

for $p \in J_1$. Then $v_{ij}(p:v) \in J$.

Lemma 5. *Fix* $v \in \mathfrak{F}'_c(\lambda)$ *and* $p \in J_1$. *Then*
1) $v_{ij}(p:v) \in J_{\Lambda_v}$,
2) $(p - p(s_i\Lambda_v)) e_i(v) = \sum_{1 \leq k \leq q} v_{ik}(p:v) u_k$,
3) $\sum_{1 \leq k \leq q} e_k(v) = 1$,

for $1 \leq i,\ j \leq q$.

This follows immediately from Lemma 2.1.

We know from [1 (a), p. 256] that $J_1 = U + J_1 J_\mu$ for $\mu \in \mathfrak{h}_{0c}^*$, the sum being direct. Hence for any $v \in \mathfrak{F}_c$, we can define a representation Γ_v of J_1 on U as follows. For $p \in J_1$, $\Gamma_v(p)$ is the linear transformation on U given by

$$\Gamma_v(p) u \equiv p u \bmod J_1 J_{\Lambda_v} \qquad (u \in U).$$

Corollary 1. *Fix* $v \in \mathfrak{F}_c'(\lambda)$. *Then*

$$\Gamma_v(p) e_i(v) = p(s_i \Lambda_v) e_i(v), \qquad \Gamma_v(e_i(v)) e_j(v) = \delta_{ij} e_j(v)$$

for $p \in J_1$ *and* $1 \leq i, j \leq q$. *Moreover*

$$U = \sum_{1 \leq i \leq q} C e_i(v).$$

This follows from Lemma 5 if we note that [1 (a), p. 259]

$$e_i(v : s_j \Lambda_v) = \sum_k u^k(s_i \Lambda_v) u_k(s_j \Lambda_v) = \delta_{ij}.$$

Corollary 2. $\Gamma_v(p\, e_i(v)) = p(s_i \Lambda_v) \Gamma_v(e_i(v))$ *and*

$$\Gamma_v(e_i(v) e_j(v)) = \delta_{ij} \Gamma(e_j(v))$$

for $1 \leq i, j \leq q$ *and* $v \in \mathfrak{F}_c'(\lambda)$.

This is obvious from Corollary 1 above.

Corollary 3. *For any* $p \in J_1$, $v \mapsto \Gamma_v(p)$ *is a polynomial mapping of* $\mathfrak{F}_c$ *into* End U.

Put $p_i^j = \mathrm{tr}_{J_{1/J}}(p u_i u^j) \in J$. It would be enough to verify that

$$\Gamma_v(p) u_i = \sum_j p_i^j(\Lambda_v) u_j \qquad (v \in \mathfrak{F}_c).$$

By Corollary 1 above, the left side is a rational function of v. Hence it would be sufficient to prove this for $v \in \mathfrak{F}_c'(\lambda)$. Fix $v \in \mathfrak{F}_c'(\lambda)$. Then

$$\Gamma_v(p) u_i = \Gamma_v(p u_i) 1 = \sum_k \Gamma_v(p u_i) e_k(v)$$

$$= \sum_k p(s_k \Lambda_v) u_i(s_k \Lambda_v) e_k(v) \qquad \text{from Corollary 1,}$$

$$= \sum_{k, j} p(s_k \Lambda_v) u_i(s_k \Lambda_v) u^j(s_k \Lambda_v) u_j.$$

But [1 (a), p. 258]

$$\sum_k (p u_i u^j)^{s_k^{-1}} = \mathrm{tr}_{J_{1/J}}(p u_i u^j) = p_i^j$$

and therefore the required statement is obvious.

Corollary 4. *Let* $p \in J_1$. *Then*

$$\prod_{1 \leq i \leq q} \{\Gamma_v(p) - p(s_i \Lambda_v)\} = 0 \qquad (v \in \mathfrak{F}_c).$$

If $v\in\mathfrak{F}'_c(\lambda)$, then $e_i(v)$ $(1\leq i\leq v)$ is a base for U and so our statement is obvious from Corollary 1. The rest follows from Corollary 3.

Corollary 5. *Fix* $i\in Q$. *Then*

$$\prod_{t\in W_0(s_i,\lambda)} (\Gamma_v(p) - p(t\,s_i\,\Lambda_v))\,\Gamma_v({}_ie(v)) = 0$$

for $v\in\mathfrak{F}$.

This is proved in the same way by taking Lemma 4 into account.

§ 4. Application to Differential Operators

We keep to the notation of §§ 2, 3. Put $\gamma_0=\gamma_{\mathfrak{g}/\mathfrak{h}_0}$ and $\gamma_1=\gamma_{\mathfrak{m}_1/\mathfrak{h}_0}$ (see [1 (e), § 11]) where $\mathfrak{m}_1=\mathfrak{m}+\mathfrak{a}$ as in § 2. Also define $\mathfrak{M}_1=\mathfrak{M}\mathfrak{A}$ and $\mathfrak{Z}_1=\mathfrak{Z}_M\mathfrak{A}$. (As usual $\mathfrak{Z}_M$ is the center of $\mathfrak{M}$.) Finally put

$$\eta_i(v)=\gamma_1^{-1}(f_{s_i\Lambda_v})\in\mathfrak{Z}_1$$

$$z_{ij}(\zeta:v)=\gamma_0^{-1}(v^j(\gamma_1(\zeta):s_i\Lambda_v))\in\mathfrak{Z} \quad (1\leq i,\,j\leq q)$$

for $\zeta\in\mathfrak{Z}_1$ and $v\in\mathfrak{F}_c$ in the notation of Lemma 2.1. (Here $\Lambda_v=(\lambda+(-1)^{1/2}\,v)^y$ as in § 3.) Then for fixed i,j and ζ, $v\mapsto\eta_i(v)$ and $v\mapsto z_{ij}(\zeta:v)$ are polynomial mappings of $\mathfrak{F}_c$ into $\mathfrak{Z}_1$ and $\mathfrak{Z}$ respectively.

Put $\gamma=\gamma_{\mathfrak{g}/\mathfrak{h}}$ and $\mu=\gamma_{\mathfrak{g}/\mathfrak{m}_1}$ so that $\gamma_0=\gamma_1\circ\mu$ [1 (e), § 11].

Lemma 1. *Define* $w_j=\gamma_1^{-1}(u_j)\in\mathfrak{Z}_1$. *Then*

1) $\gamma(z_{ij}(\zeta:v):\lambda+(-1)^{1/2}\,v)=0$,

2) $\zeta\eta_i(v)-\gamma_i(\zeta:s_i\Lambda_v)\eta_i(v)=\displaystyle\sum_{1\leq j\leq q}\mu(z_{ij}(\zeta:v))w_j$

for $\zeta\in\mathfrak{Z}_1$, $v\in\mathfrak{F}_c$ *and* $1\leq i,\,j\leq q$.

This follows from 1) and 2) of Lemma 2.1.

Put $d(m)=d_P(m)$ [1 (e), § 21] for $m\in M_1=MA$ and define $v'=d^{-1}\,v\circ d$ $(v\in\mathfrak{M}_1)$ as usual [1 (d), § 45]. Let

$$g_i(\zeta:v)= -\sum_{1\leq j\leq q}\{z_{ij}(\zeta:v)-\mu(z_{ij}(\zeta:v))'\}\,w_j' \quad (1\leq i\leq q)$$

for $\zeta\in\mathfrak{Z}_1$ and $v\in\mathfrak{F}_c$.

Corollary. $g_i(\zeta:v)\in\theta(\mathfrak{n})\mathfrak{G}\,\mathfrak{n}$ *and*

$$\zeta'\eta_i(v)'-\gamma_1(\zeta:s_i\Lambda_v)\eta_i(v)'=\sum_j z_{ij}(\zeta:v)w_j'+g_i(\zeta:v) \quad (1\leq i\leq q)$$

for $\zeta\in\mathfrak{Z}_1$ *and* $v\in\mathfrak{F}_c$. *Moreover for* i *and* ζ *fixed,* $v\mapsto g_i(\zeta:v)$ *is a polynomial mapping of* $\mathfrak{F}_c$ *into* $\theta(\mathfrak{n})\mathfrak{G}\,\mathfrak{n}$.

This is obvious from the above lemma if we recall [1 (d), p. 110] that

$$z-\mu(z)'\in\theta(\mathfrak{n})\mathfrak{G}\,\mathfrak{n} \quad (z\in\mathfrak{Z}).$$

§ 5. The Basic Differential Equations

Let V be a complete, locally convex, Hausdorff space and τ a differentiable double representation of K on V [1 (e), § 19]. Fix $v \in \mathfrak{F}_c$ and let ϕ be an element in $C^\infty(G, \tau)$ [1 (e), § 19] such that

$$z\phi = \gamma(z : \lambda + (-1)^{1/2} v)\phi \qquad (z \in \mathfrak{Z}).$$

Put

$$\phi_i(m) = d_P(m)\phi(m; \eta_i(v)') \qquad (m \in M_1).$$

Lemma 1. *Let $m \in M_1$. Then*

$$\varpi_0(\Lambda_v) d_P(m)\phi(m) = \sum_{1 \leq i \leq q} \varepsilon(s_i)\varpi_1(s_i\Lambda_v)\phi_i(m)$$

and

$$\phi_i(m; \zeta) = \gamma_1(\zeta : s_i\Lambda_v)\phi_i(m) + d_P(m)\phi(m; g_i(\zeta : v)) \qquad (1 \leq i \leq q)$$

for $\zeta \in \mathfrak{Z}_1$.

This follows from the corollary of Lemma 4.1.

Let α be a root of (P_0, A_0). Fix $X \in \mathfrak{n}_0$ such that $[H, X] = \alpha(H)X$ for all $H \in \mathfrak{a}_0$.

Lemma 2. *Let $g_1, g_2 \in \mathfrak{G}$ and $h \in A_0$. Then*

$$\phi(g_1 ; h; \theta(X)g_2) = e^{-\alpha(\log h)}\phi(g_1\theta(X); h; g_2)$$

and

$$\phi(g_1 X; h; g_2) = e^{-\alpha(\log h)}\phi(g_1; h; X g_2).$$

This is obvious.

Define

$$\psi_{i,\zeta}(m) = d_P(m)\phi(m; g_i(\zeta : v)) \qquad (1 \leq i \leq q, \; m \in M_1)$$

for $\zeta \in \mathfrak{Z}_1$. It is clear that $\psi_{i,\zeta}$ depends linearly on ζ. Since $\mathfrak{a} \subset \mathfrak{Z}_1$, the following result is an immediate consequence of Lemma 1.

Lemma 3.

$$\phi_i(m \exp TH)e^{-Ts_i\Lambda_v(H)} = \phi_i(m) + \int_0^T \psi_{i,H}(m \exp tH)e^{-ts_i\Lambda_v(H)}dt \qquad (1 \leq i \leq q)$$

for $m \in M_1$, $H \in \mathfrak{a}$ and $T \in \mathbf{R}$.

§ 6. Asymptotic Behavior of Eigenfunctions

For $v \in \mathfrak{F}$, let $\mathscr{A}(G, \tau, \lambda, v) = \mathscr{A}(\lambda, v) = \mathscr{A}(v)$ denote the space of all $\phi \in \mathscr{A}(G, \tau)$ [1 (e), § 21] such that

$$z\phi = \gamma(z : \lambda + (-1)^{1/2} v)\phi \qquad (z \in \mathfrak{Z}).$$

Fix $v \in \mathfrak{F}$, $\phi \in \mathscr{A}(v)$ and let us use the notation of § 5. Our object is to study the asymptotic behavior of ϕ_i. Put $M_1^+ = K_1 \cdot Cl(A_0^+) \cdot K_1$ as in [1 (e), § 22] where $K_1 =$

$K_M = K \cap M$. The following lemma is proved in the same way as [1 (e), Lemma 22.1].

Lemma 1. *Fix* $\zeta \in \mathfrak{Z}_1$, v_1, $v_2 \in \mathfrak{M}_1$ *and* $s \in \mathscr{S}(V)$. *Then we can choose numbers* $c, r \geq 0$ *such that*

$$|\psi_{i,\zeta}(v_1; m \exp H; v_2)|_s \leq c \Xi_M(m)\,|(m, H)|^r\, e^{-\beta_P(H)}$$

for $m \in M_1^+$ *and* $H \in Cl\,\mathfrak{a}^+$.

Here the notation is the same as in [1 (e), Lemma 22.3].

Let λ_i $(i \in Q)$ denote the restriction of $s_i \lambda^v$ on $\mathfrak{a}$. We decompose Q into three disjoint sets Q^+, Q^o and Q^- as follows. An element $i \in Q$ lies in Q^+ if $\lambda_i(H) > 0$ for some $H \in \mathfrak{a}^+$, $i \in Q^o$ if $\lambda_i = 0$ and $i \in Q^-$ if $\lambda_i(H) < 0$ for all $H \in \mathfrak{a}^+$. Define

$$\phi_{i\,\infty}(m) = \lim_{T \to +\infty} \phi_i(m \exp TH)\, e^{-Ts_i \Lambda_v(H)} \qquad (m \in M_1)$$

for $i \in Q^o$ and $H \in \mathfrak{a}^+$. One proves as in [1 (e), § 22] that this limit exists and is independent of the choice of H. Moreover $\phi_{i\infty} \in \mathscr{A}(M_1, \tau_M)$. Define $\phi_{i\infty} = 0$ for $i \in Q^+ \cup Q^-$.

Choose a number δ $(0 < \delta \leq \tfrac{1}{2})$ such that

$$\lambda_i(H) \leq -\delta\,\beta_P(H)$$

for all $i \in Q^-$ and $H \in \mathfrak{a}^+$. We have seen in [1 (e), § 22] that this is possible.

Lemma 2. *Let* $i \in Q$. *Then* $\phi_{i\infty} = 0$ *unless* $i \in Q^o$. *Moreover* $\phi_{i\infty} \in \mathscr{A}(M_1, \tau_M)$ *and*

$$\zeta\,\phi_{i\,\infty} = \gamma_1(\zeta : s_i \Lambda_v)\,\phi_{i\,\infty} \qquad (\zeta \in \mathfrak{Z}_1).$$

Finally

$$|\phi_i(v_1; m \exp TH; v_2) - \phi_{i\,\infty}(v_1; m \exp TH; v_2)|_s$$

$$\leq e^{-T\delta\beta_P(H)}\left\{|\phi_i(v_1; m; v_2)|_s + \int_0^\infty |\psi_{i,H}(v_1; m \exp tH; v_2)|_s\, e^{t\beta_P(H)/2}\right\}$$

for $v_1, v_2 \in \mathfrak{M}_1$, $m \in M_1$, $H \in \mathfrak{a}^+$, $T \geq 0$ *and* $s \in \mathscr{S}(V)$. *(In case* $P = G$, *the right side should be replaced by zero.)*

This is proved in the same way as [1 (e), Theorem 22.1].

Lemma 3. *Fix* i $(1 \leq i \leq q)$ *and suppose* $v \in \mathfrak{F}'(\lambda)$. *Then* $\phi_{i\,\infty} = 0$ *unless* $s_i^{-1} \mathfrak{a} \subset \mathfrak{h}_R^{k_0}$.

Suppose $\phi_{i\,\infty} \neq 0$. Clearly $\mathfrak{h}_0$ is a θ-stable Cartan subalgebra of $\mathfrak{M}_1$. Hence by [1 (e), Lemma 29.3] we can choose $s \in W(\mathfrak{m}_1/\mathfrak{h}_0)$ such that

$$s_i(\lambda - (-1)^{1/2} v)^y = s\,\theta\, s_i(\lambda + (-1)^{1/2} v)^y.$$

Choose $x \in G_c$ such that $x\,y^{-1} = s_i$ on $\mathfrak{h}_0$. Then

$$(\lambda - (-1)^{1/2} v)^x = s\,\theta(\lambda + (-1)^{1/2} v)^x$$

and $x \cdot \mathfrak{h}_c = \mathfrak{h}_{0c}$ since $y \cdot \mathfrak{h}_c = \mathfrak{h}_{0c}$ (see § 3). Fix $H_0 \in \mathfrak{a}^+$. Then we conclude from [1(e), Lemma 33.1] that

$$\mathfrak{a} \subset x \cdot \mathfrak{h}_R = s_i \mathfrak{h}_R^y = s_i \mathfrak{h}_R^{k_0}.$$

This proves the lemma.

§ 7. The Functions $\phi_{P,s}$

Let $P = MAN$ be a psgp of G. Given $k \in K$, let s denote the restriction of $\mathrm{Ad}(k)$ on $\mathfrak{a}$. Then s determines the coset $k K_M$ completely. Hence if H is any subgroup of G which is normalized by K_M, we can define $H^s = H^k = kHk^{-1}$. In particular $P^s = M^s A^s N^s$. For any $\phi \in \mathscr{A}(MA, \tau_M)$, we define $\phi^k = \phi^s \in \mathscr{A}((MA)^s, \tau_{M^s})$ by

$$\phi^s(m^k) = \tau(k)\phi(m)\tau(k^{-1}) \qquad (m \in MA).$$

It is easy to see that ϕ^s depends only on s. Similarly we define

$$\zeta^s = \zeta^k = \mathrm{Ad}(k)\zeta \quad (\zeta \in \mathfrak{Z}_M \mathfrak{U}), \qquad a^s = a^k \ (a \in A).$$

If $\mathfrak{h}$ is a Cartan subalgebra on $\mathfrak{g}$, sometimes it will be convenient to write $\gamma_{G/\mathfrak{h}}$ instead of $\gamma_{\mathfrak{g}/\mathfrak{h}}$.

Let $P' = M'A'N'$ be another psgp of G. Then we have [1(e), § 5] the finite set $\mathfrak{w}(\mathfrak{a}'|\mathfrak{a})$ of linear injections of $\mathfrak{a}$ into $\mathfrak{a}'$. For every $s \in \mathfrak{w}(\mathfrak{a}'|\mathfrak{a})$ we can choose $k \in K$ such that $\mathrm{Ad}(k) = s$ on $\mathfrak{a}$ [1(e), § 5]. Put $\mathfrak{w}(\mathfrak{a}) = \mathfrak{w}(\mathfrak{a}|\mathfrak{a})$. Then $\mathfrak{w}(\mathfrak{a})$ is a group of linear transformations in $\mathfrak{a}$.

Fix λ as in § 6.

Theorem 1. *Suppose* $v \in \mathfrak{F}'(\lambda)$ *and* $\phi \in \mathscr{A}(G, \tau, \lambda, v)$ *in the notation of* § 6. *Put* $\mathfrak{w} = \mathfrak{w}(\mathfrak{h}_R|\mathfrak{a})$. *Then there exist unique functions* $\phi_{P,s} \in \mathscr{A}(M_1, \tau_M)$ $(s \in \mathfrak{w})$ *with the following two properties.*

1) $\phi_P(m) = \sum_{s \in \mathfrak{w}} \phi_{P,s}(m) \qquad (m \in M_1),$

2) $\zeta \phi_{P,s} = \gamma_{M_1^s/\mathfrak{h}}(\zeta^s : \lambda + (-1)^{1/2} v)\phi_{P,s} \qquad (\zeta \in \mathfrak{Z}_1, \ s \in \mathfrak{w}).$

Here ϕ_P is the constant term of ϕ along P [1(e), § 21].

Corollary. $\phi_{P,s}(ma) = \phi_{P,s}(m)e^{(-1)^{1/2} v (\log a^s)}(m \in M_1, a \in A, s \in \mathfrak{w}).$

Since $\mathfrak{a} \subset \mathfrak{Z}_1$, the corollary is obvious from the second statement of the theorem. First we prove the following lemma.

Lemma 1. *Given* $s \in \mathfrak{w}(\mathfrak{h}_R|\mathfrak{a})$, *there exists a unique index* i $(1 \leq i \leq q)$ *such that* $sH = \mathrm{Ad}(k_0^{-1}) s_i^{-1} H$ *for all* $H \in \mathfrak{a}$.

Choose a representative $k \in K$ for s. (This means that $\mathrm{Ad}(k) = s$ on $\mathfrak{a}$.) Then

$$(\mathfrak{a})^{k_0 k} \subset \mathfrak{h}_R^{k_0} \subset \mathfrak{a}_0.$$

Hence we can choose $t \in W_0 = W(\mathfrak{g}/\mathfrak{h}_0)$ such that $\mathrm{Ad}(k_0 k) = t^{-1}$ on $\mathfrak{a}$. Clearly the coset $W_1 t$ is uniquely determined by this condition. Hence there exists a unique i such that $W_1 t = W_1 s_i$. This s_i satisfies our condition.

Lemma 2. *Let s and i be related as in Lemma* 1. *Then*

$$\gamma_1(\zeta: s_i \Lambda_v) = \gamma_{M^s_1/\mathfrak{h}}(\zeta^s: \lambda + (-1)^{1/2} v) \quad (\zeta \in \mathfrak{Z}_1).$$

Choose $y_i \in G_c$ such that $y_i = s_i$ on $\mathfrak{h}_0$ and define k as in the proof of Lemma 1. Then it is clear that

$$m_1 = y_i \, \mathrm{Ad}(k_0 k) \in M_{1c}$$

where M_{1c} is the centralizer of $\mathfrak{a}$ in G_c. Now $s_i \Lambda_v = (\lambda + (-1)^{1/2} v)^{y_i y}$ and

$$y_i y = m_1 \, \mathrm{Ad}(k_0 k)^{-1} y.$$

Moreover $\mathrm{Ad}(k_0^{-1}) y$ centralizes $\mathfrak{h}_R$ (see § 3) and $\mathfrak{h}_R \supset s\mathfrak{a} = \mathfrak{a}^k$. Hence

$$m_2 = \mathrm{Ad}(k_0 k)^{-1} y \, \mathrm{Ad}(k) \in M_{1c}.$$

Put

$$m = m_1 \, m_2 = y_i y \, \mathrm{Ad}(k) \in M_{1c}$$

so that $y_i y = m \, \mathrm{Ad}(k^{-1})$. Since (§ 3)

$$(y_i y)^{-1} \mathfrak{h}_{0c} = y^{-1} \mathfrak{h}_{0c} = \mathfrak{h}_c$$

it follows that $m^{-1} \mathfrak{h}_{0c} = \mathfrak{h}_c^{k^{-1}}$. Therefore

$$\gamma_1(\zeta: s_i \Lambda_v) = \gamma_{M_1/\mathfrak{h}_0}(\zeta: (\lambda + (-1)^{1/2} v)^{y_i y})$$

$$= \gamma_{M_1/\mathfrak{h}^{k^{-1}}}(\zeta: (\lambda + (-1)^{1/2} v)^{k^{-1}}) = \gamma_{M^s_1/\mathfrak{h}}(\zeta^s: \lambda + (-1)^{1/2} v).$$

We now come to the proof of Theorem 1. Since $\varpi_0(\Lambda_v) \neq 0$, it is clear that $s\Lambda_v \neq t\Lambda_v$ for $s \neq t$ in W_0. Hence $s_i \Lambda_v$ and $s_j \Lambda_v$ cannot be conjugate under W_1 unless $i = j$. Put

$$\chi_s(\zeta) = \gamma_{M^s_1/\mathfrak{h}}(\zeta^s: \lambda + (-1)^{1/2} v) \quad (s \in \mathfrak{w}, \zeta \in \mathfrak{Z}_1).$$

Then it follows from Lemma 2 that $\chi_s \neq \chi_t$ if $s \neq t$ in $\mathfrak{w}$. The uniqueness of $\phi_{P,s}$ is now obvious. On the other hand if s and i are related by Lemma 1 and we set

$$\phi_{P,s} = \varpi_{01}(s_i \Lambda_v)^{-1} \phi_{i\infty},$$

it follows from Lemmas 5.1 and 6.2 that all the conditions of Theorem 1 are fulfilled and this completes the proof.

We state the above result as a lemma for later reference.

Lemma 3. *Suppose s and i are related as in Lemma* 1. *Then*

$$\phi_{P,s} = \varpi_{01}(s_i \Lambda_v)^{-1} \phi_{i\infty}.$$

Let $(P', A') \prec (P, A)$ be another p-pair in G and put $^*P = P' \cap (MA)$. Then $(^*P, A')$ is a p-pair in M_1. For any $s \in \mathfrak{w}(\mathfrak{h}_R | \mathfrak{a})$, let $\mathfrak{w}_s(\mathfrak{h}_R | \mathfrak{a}')$ be the set of all $t \in \mathfrak{w}(\mathfrak{h}_R | \mathfrak{a}')$ such that $t = s$ on $\mathfrak{a}$. (We note that $\mathfrak{a} \subset \mathfrak{a}'$.)

Fix $s \in \mathfrak{w}(\mathfrak{h}_R | \mathfrak{a})$ and choose a representative $k \in K$ for s. Put

$$\psi = (\phi_{P,s})^s \in \mathscr{A}((MA)^s, \tau_{M^s}).$$

Then

$$\zeta\,\psi = \gamma_{M_1^s/\mathfrak{h}}(\zeta : \lambda + (-1)^{1/2}\,v)\,\psi \qquad (\zeta \in \mathfrak{Z}_1^s).$$

and $*P^k = (*P)^k$ is a psgp of M_1^s with split component $(A')^k$.

Lemma 4. *For any* $t \in \mathfrak{w}_s(\mathfrak{h}_R|\mathfrak{a}')$,

$$(\phi_{P',t})^t = (\psi_{*P^k, t\circ k^{-1}})^{t\circ k^{-1}}.$$

Here $t\circ k^{-1}$ denotes the mapping $H \mapsto t(\mathrm{Ad}(k^{-1})H)$ $(H \in (\mathfrak{a}')^k)$ of $\mathfrak{a}'^k$ into $\mathfrak{h}_R$. We know [1(e), Lemma 21.1] that

$$(\phi_{P,s})_{*P} = (\psi_{*P^k})^{k^{-1}}.$$

Let $\mathfrak{w}_0(\mathfrak{h}_R|\mathfrak{a}'^k)$ denote the set of all $t' \in \mathfrak{w}(\mathfrak{h}_R|\mathfrak{a}'^k)$ such that $t' = \mathrm{Ad}(m^k)$ on $\mathfrak{a}'^k$ for some $m \in M_1$. Then it is easy to verify that $t \mapsto t\circ k^{-1}$ is a bijection of $\mathfrak{w}_s(\mathfrak{h}_R|\mathfrak{a}')$ on $\mathfrak{w}_0(\mathfrak{h}_R|\mathfrak{a}'^k)$. Therefore by applying Theorem 1 to (M_1^s, ψ) in place of (G, ϕ), we conclude that

$$\psi_{*P^k} = \sum_{t \in \mathfrak{w}_s(\mathfrak{h}_R|\mathfrak{a}')} \psi_{*P^k, t\circ k^{-1}}.$$

Now put $M_1' = M'A'$, $\mathfrak{Z}_1' = \mathfrak{Z}_{M'}\,\mathfrak{A}'$,

$$\chi_t(\eta) = \gamma_{(M_1')^t/\mathfrak{h}}(\eta^t : \lambda + (-1)^{1/2}\,v) \qquad (\eta \in \mathfrak{Z}_1')$$

and

$$\Psi(t) = (\psi_{*P^k, t\circ k^{-1}})^{k^{-1}}$$

for $t \in \mathfrak{w}_s(\mathfrak{h}_R|\mathfrak{a}')$. Then

$$\eta\,\Psi(t) = \chi_t(\eta)\,\Psi(t) \qquad (\eta \in \mathfrak{Z}_1').$$

On the other hand $\phi_{P'} = (\phi_P)_{*P}$ [1(e), Lemma 21.1]. Hence

$$\phi_{P'} = \sum_{s \in \mathfrak{w}(\mathfrak{h}_R|\mathfrak{a})} (\phi_{P,s})_{*P}.$$

For every $s \in \mathfrak{w}(\mathfrak{h}_R|\mathfrak{a})$, choose a representative k_s in K and define

$$\Psi(s, t) = (\psi_{*P^k, t\circ k^{-1}})^{k^{-1}} \qquad (t \in \mathfrak{w}_s(\mathfrak{h}_R|\mathfrak{a}')),$$

with $\psi = (\phi_{P,s})^s$ and $k = k_s$. Then, by the above result,

$$(\phi_{P,s})_{*P} = \sum_{t \in \mathfrak{w}_s(\mathfrak{h}_R|\mathfrak{a}')} \Psi(s, t)$$

and

$$\eta\,\Psi(s, t) = \chi_t(\eta)\,\Psi(s, t) \qquad (\eta \in \mathfrak{Z}_1')$$

for $s \in \mathfrak{w}(\mathfrak{h}_R|\mathfrak{a})$ and $t \in \mathfrak{w}_s(\mathfrak{h}_R|\mathfrak{a}')$. Hence

$$\phi_{P'} = \sum_{s \in \mathfrak{w}(\mathfrak{h}_R|\mathfrak{a})} (\phi_{P,s})_{*P}$$

$$= \sum_{s \in \mathfrak{w}(\mathfrak{h}_R|\mathfrak{a})} \sum_{t \in \mathfrak{w}_s(\mathfrak{h}_R|\mathfrak{a}')} \Psi(s, t).$$

It is now obvious from Theorem 1 that

$$\phi_{P',t} = \Psi(s, t)$$

for $t \in \mathfrak{w}_s(\mathfrak{h}_R|\mathfrak{a}')$ and the statement of the lemma follows immediately.

We define the space $^{\circ}\mathscr{C}(M, \tau_M)$ as in [1(e), §19].

Lemma 5. *Fix* $s \in \mathfrak{w}(\mathfrak{h}_R|\mathfrak{a})$ *and let* f *denote the restriction of* $\phi_{P,s}$ *on* M. *Then if* prk $P = \dim \mathfrak{h}_R$, $f \in {^{\circ}\mathscr{C}}(M, \tau_M)$.

Let $*P = *M *A *N$ be a psgp of M with prk $*P \geq 1$. Then by [1(e), Lemma 25.1], it is enough to verify that $f_{*P} = 0$. Let $P' = M' A' N'$ be the psgp of G corresponding to $*P$ [1(e), Lemma 6.1] so that $(P', A') \prec (P, A)$. Then

$$\text{prk } P' = \text{prk } *P + \text{prk } P > \dim \mathfrak{h}_R$$

and therefore $\mathfrak{w}(\mathfrak{h}_R|\mathfrak{a}')$ is empty. Fix a representative $k \in K$ for s and put $\psi = (\phi_{P,s})^s$, $Q = (P' \cap M_1)^k$. Then Q is a psgp of $M_1 = MA$ and it follows from the proof of Lemma 4 that $\psi_Q = 0$. Since [1(e), Lemma 21.1]

$$f_{*P} = (\psi_Q)^{k^{-1}}$$

on $*M *A$, we conclude that $f_{*P} = 0$.

§ 8. Functions of Type $II(\lambda)$

Now, instead of keeping v fixed, we shall allow it to vary in $\mathfrak{F}$. Note that $\mathfrak{h}_R$, being a subspace of $\mathfrak{g}$, has a Euclidean norm. Hence, by duality, the same holds for $\mathfrak{F}$. Put

$$|(v, x)| = (1 + |v|)(1 + \sigma(x))$$

for $(v, x) \in \mathfrak{F} \times G$. Let $\mathfrak{D} = \mathfrak{D}(\mathfrak{F}_c)$ denote the algebra of polynomial differential operators on $\mathfrak{F}$ (or $\mathfrak{F}_c$) [1(c), §3]. Put $\tilde{\mathfrak{G}} = \mathfrak{D} \otimes \mathfrak{G}^{(2)}$ [1(e), §15]. Let ϕ be a C^{∞} function from $\mathfrak{F} \times G$ to V. For $D \in \tilde{\mathfrak{G}}$, $s \in \mathscr{S}(V)$ and $r \geq 0$, put

$$s_{D,r}(\phi) = \sup_{\mathfrak{F} \times G} |D\phi|_s \, \Xi^{-1} \, |(v, x)|^{-r}$$

in the notation of [1(e), §15]. If F is a finite subset of $\tilde{\mathfrak{G}}$, we set

$$s_{F,r}(\phi) = \sum_{D \in F} s_{D,r}(\phi).$$

A function $\phi: \mathfrak{F} \times G \to V$ will be said to be of type $II(\lambda)$ if the following conditions hold.

1) ϕ is of class C^{∞}.

2) For any $v \in \mathfrak{F}$, the function $\phi_v = \phi(v)$ is a τ-spherical function on G and

$$z \phi_v = \gamma_{\mathfrak{g}/\mathfrak{h}}(z : \lambda + (-1)^{1/2} v) \phi_v \qquad (z \in \mathfrak{Z}).$$

3) For any $D \in \tilde{\mathfrak{G}}$ and $s \in \mathscr{S}(V)$, we can choose a number $r \geq 0$ such that $s_{D,r}(\phi) < \infty$.

Fix a function ϕ of type $II(\lambda)$ and let us use the notation of §5. Then ϕ_i and $\psi_{i,\zeta}$ ($\zeta \in \mathfrak{Z}_1$) are now functions on $\mathfrak{F} \times M_1$. Put

$$|(v, x, X)| = (1 + |v|)(1 + \sigma(x))(1 + \|X\|)$$

for $(v, x, X) \in \mathfrak{F} \times G \times \mathfrak{g}$.

Lemma 1. *Fix* $\zeta \in \mathfrak{Z}_1$, $v_1, v_2 \in \mathfrak{M}_1$, $p \in S(\mathfrak{F}_c)$ *and* $s \in \mathscr{S}(V)$. *Then we can choose* c, $r \geq 0$ *such that*

$$|\psi_{i,\zeta}(v; \partial(p): v_1 \,; m \exp H; v_2)|_s \leq c \, \Xi_M(m) \, |(v, m, H)|^r \, e^{-\beta_P(H)}$$

for $m \in M_1^+$, $H \in Cl\,\mathfrak{a}^+$, $v \in \mathfrak{F}$ *and* $1 \leq i \leq q$.

The proof is the same as for Lemma 6.1.

It follows without difficulty from the above estimates that $\phi_{i\infty}$, regarded as functions on $\mathfrak{F} \times M_1$, are of class C^∞. In fact we have the following analogue of Lemma 6.2.

Lemma 2. 1) $\phi_{i\infty}(v: m; \zeta) = \gamma_1(\zeta: s_i \Lambda_v) \phi_{i\infty}(v: m) \, (\zeta \in \mathfrak{Z}_1)$.

Given $v_1, v_2 \in \mathfrak{M}_1$, $p \in S(\mathfrak{F}_c)$ *and* $s \in \mathscr{S}(V)$, *we can choose* $c, r \geq 0$ *such that*

2) $|\phi_{i\infty}(v; \partial(p): v_1 \,; m; v_2)|_s \leq c \, \Xi_M(m) \, |(v, m)|^r$.

Finally

3) $|\phi_i(v: v_1 \,; m \exp TH; v_2) - \phi_{i\infty}(v: v_1 \,; m \exp TH; v_2)|_s$

$$\leq e^{-T\delta\beta_P(H)} \left\{ |\phi_i(v: v_1 \,; m; v_2|_s + \int_0^\infty |\psi_{i,H}(v: v_1 \,; m \exp tH; v_2)|_s \, e^{t\beta_P(H)/2} \, dt \right\}$$

for $H \in \mathfrak{a}^+$, $T \geq 0$.

Here $i \in Q$, $v \in \mathfrak{F}$, $m \in M_1$ *and the right side in* 3) *is to be replaced by zero in case* $P = G$.

We have only to comment on the proof of 2). Put

$$\phi_i^o(v: m: H) = \phi_i(v: m \exp H) \, e^{-s_i \Lambda_v(H)},$$

$$\psi_{i,\zeta}^o(v: m: H) = \psi_{i,\zeta}(v: m \exp H) \, e^{-s_i \Lambda_v(H)} \qquad (\zeta \in \mathfrak{Z}_1),$$

for $v \in \mathfrak{F}$, $m \in M_1$ and $H \in \mathfrak{a}$. Then

$$\phi_i^o(v: m: TH) = \phi_i(v: m) + \int_0^T \psi_{i,H}^o(v: m: tH) \, dt \qquad (T \in \mathbf{R})$$

from Lemma 5.3. Now if $i \in Q^o$, it follows from Lemma 1 that

$$\phi_{i\infty}(v; \partial(p): v_1 \,; m; v_2) = \phi_i(v; \partial(p): v_1 \,; m; v_2)$$

$$+ \int_0^\infty \psi_{i,H}^o(v; \partial(p): v_1 \,; m; v_2: tH) \, dt$$

for $v_1, v_2 \in \mathfrak{M}_1$ and $p \in S(\mathfrak{F}_c)$. Now fix p. Then it is obvious that

$$\partial(p) \circ e^{-s_i \Lambda_v(H)} = e^{-s_i \Lambda_v(H)} \partial(p_H) \qquad (H \in \mathfrak{a})$$

where $H \mapsto p_H$ is a polynomial mapping of $\mathfrak{a}$ in $S(\mathfrak{F}_c)$. Hence 2) is an easy consequence of Lemma 1 and standard arguments [1(d), p. 69]. (We recall that by Lemma 6.2 $\phi_{i\infty} = 0$ unless $i \in Q^o$.)

Put $\phi_P(v) = (\phi_v)_P$ $(v \in \mathfrak{F})$ and $\phi_{P,s}(v) = (\phi_v)_{P,s}$ for $v \in \mathfrak{F}'(\lambda)$ and $s \in \mathfrak{w}(\mathfrak{h}_R | \mathfrak{a})$.

Lemma 3. *Suppose* $v \in \mathfrak{F}'(\lambda)$. *Then*

$$\phi_P(v) = \sum_{i \in Q^o} \varpi_{01}(s_i \Lambda_v)^{-1} \phi_{i\infty}(v) = \sum_{s \in \mathfrak{w}(\mathfrak{h}_R | \mathfrak{a})} \phi_{P,s}(v).$$

This is obvious from the results of §7.

§9. Functions of Type $II'(\lambda)$

Let $\mathscr{P}$ be the set of all psgps of G. We keep to the notation of §8.

Let ϕ be a function from $\mathfrak{F} \times G$ to V. We say that ϕ is of type $II'(\lambda)$, if it is of type $II(\lambda)$ and the following additional condition holds. Given $P = MAN$ in $\mathscr{P}$ and $s \in \mathfrak{w}(\mathfrak{h}_R | \mathfrak{a})$, the function $(\phi_{P,s})^s$ on $\mathfrak{F}'(\lambda) \times (MA)^s$ extends (uniquely) to a function of type $II(\lambda)$ on $\mathfrak{F} \times (MA)^s$.

Lemma 1. *Suppose* ϕ *is of type* $II'(\lambda)$ *on* $\mathfrak{F} \times G$. *Then for any* $P = MAN$ *in* $\mathscr{P}$ *and* $s \in \mathfrak{w}(\mathfrak{h}_R | \mathfrak{a})$, $(\phi_{P,s})^s$ *is of type* $II'(\lambda)$ *on* $\mathfrak{F} \times (MA)^s$.

This is an immediate consequence of Lemma 7.4.

Theorem 1. *Suppose* ϕ *is a function of type* $II(\lambda)$ *on* $\mathfrak{F} \times G$. *Define*

$$\psi(v:x) = \varpi(\lambda + (-1)^{1/2} v) \phi(v:x) \quad (v \in \mathfrak{F}, x \in G).$$

Then ψ *is of type* $II'(\lambda)$.

This is an immediate consequence of Lemmas 7.3 and 8.2.

§10. Continuity of ϕ_P

Fix a function ϕ of type $II(\lambda)$ on $\mathfrak{F} \times G$ and a psgp $P = MAN$ of G. We intend to show that ϕ_P is a continuous function on $\mathfrak{F} \times MA$. So we may assume that $P \neq G$.

We use the notation of §3. Let U^* be the space dual to U and $(u_1^*, \ldots, u_q^*)$ the base for U^* dual to $(u_1, \ldots, u_q)$. For any $v \in \mathfrak{F}$, we have defined in §3 a representation Γ_v of J_1 on U. The corresponding (right-)representation Γ_v^* on U^* is given by

$$\langle u^* \Gamma_v^*(p), u \rangle = \langle u^*, \Gamma_v(p) u \rangle \quad (p \in J_1, u \in U, u^* \in U^*).$$

Define γ_1 as in §4 and put $\eta_i = \gamma_1^{-1}(u_i) \in \mathfrak{Z}_1$ $(1 \leq i \leq q)$.

We regard U^* as a Hilbert space with $(u_1^*, \ldots, u_q^*)$ as an orthonormal base. Put $\mathbf{V} = V \otimes U^*$. Then by letting K act trivially on U^*, we get a double representation τ of K on $\mathbf{V}$. Put

$$\Gamma_v(\zeta) = 1 \otimes \Gamma_v^*(\gamma_1(\zeta)) \quad (\zeta \in \mathfrak{Z}_1).$$

Then Γ_v is a right-representation of $\mathfrak{Z}_1$ on $\mathbf{V}$ which commutes with τ.

We now proceed in the same way as in [1(e), §22]. If $s \in \mathscr{S}(V)$ and

$$\mathbf{v} = \sum_{1 \leq i \leq q} v_i \otimes u_i^* \qquad (v_i \in V),$$

we put

$$\mathbf{s}(\mathbf{v}) = |\mathbf{v}|_\mathbf{s} = (\sum_i |v_i|_\mathbf{s}^2)^{1/2}.$$

Let $\|T\|$ denote the Hilbert-Schmidt norm of a linear transformation T on U^*. (We write T on the right.) Then it is easy to verify that

$$\mathbf{s}(\mathbf{v} \cdot (1 \otimes T)) \leq \mathbf{s}(\mathbf{v}) \|T\| \qquad (s \in \mathscr{S}(V), \mathbf{v} \in \mathbf{V}).$$

Now define a C^∞ function Φ from $\mathfrak{F} \times M_1$ to $\mathbf{V}$ by

$$\Phi(v:m) = d(m) \sum_{1 \leq i \leq q} \phi(v:m; \eta_i') \otimes u_i^* \qquad (v \in \mathfrak{F}, m \in M_1).$$

Here $M_1 = MA$, $d(m) = d_P(m)\,(m \in M_1)$ and $v' = d^{-1} v \circ d$ for $v \in \mathfrak{M}_1$ as in §4. Fix $\zeta \in \mathfrak{Z}_1$ and consider $\Phi(v:m; \zeta)$. Put $p = \gamma_1(\zeta) \in J_1$. Then

$$p u_i = \Gamma_v(p) u_i + \sum_{1 \leq j \leq q} v_{ij}(p:v) u_j \qquad (1 \leq i \leq q)$$

where

$$v_{ij}(p:v) = \mathrm{tr}_{J_1/J} \{u^j(p u_i - \Gamma_v(p) u_i)\} \in J_{A_v}$$

from the definition of $\Gamma_v(p)$. Define γ_0 and μ as in §4 and put

$$z_{ij}(\zeta:v) = \gamma_0^{-1}(v_{ij}(p:v)) \in \mathfrak{Z}.$$

Then it is clear that

$$\gamma(z_{ij}(\zeta:v): \lambda + (-1)^{1/2} v) = 0$$

and [1(d), p. 110]

$$g_{ij}(\zeta:v) = z_{ij}(\zeta:v) - \mu(z_{ij}(\zeta:v))' \in \theta(\mathfrak{n}) \, \mathfrak{G} \, \mathfrak{n}.$$

Put

$$g_i(\zeta:v) = - \sum_{1 \leq j \leq q} g_{ij}(\zeta:v) \eta_j'.$$

Then $g_i(\zeta:v)$ is linear in ζ and for fixed i and ζ, $v \mapsto g_i(\zeta:v)$ is a polynomial mapping of $\mathfrak{F}$ into $\theta(\mathfrak{n}) \, \mathfrak{G} \, \mathfrak{n}$ by Corollary 3 of Lemma 3.5.

Lemma 1. *Fix* $\zeta \in \mathfrak{Z}_1$ *and put*

$$\Psi_\zeta(v:m) = d(m) \sum_{1 \leq i \leq q} \phi(v:m; g_i(\zeta:v)) \otimes u_i^* \qquad (v \in \mathfrak{F}, m \in M_1).$$

Then

$$\Phi(v:m; \zeta) = \Phi(v:m) \Gamma_v(\zeta) + \Psi_\zeta(v:m)$$

for $v \in \mathfrak{F}$ *and* $m \in M_1$.

Let $p = \gamma_1(\zeta)$. Then

$$\sum_i u_i \otimes u_i^* \, \Gamma_v^*(p) = \sum_i \Gamma_v(p) \, u_i \otimes u_i^* = \sum_i p \, u_i \otimes u_i^* - \sum_{i,j} v_{ij}(p:v) \, u_j \otimes u_i^*$$

in $J_1 \otimes U^*$. Therefore since $\gamma_1(\mu(z_{ij}(\zeta:v))) = v_{ij}(p:v)$ and $z_{ij}(\zeta:v)\,\phi(v) = 0$, we conclude that

$$\Phi(v:m)\,\Gamma_v(\zeta) = d(m) \sum_i \phi(v:m; \zeta' \eta_i') \otimes u_i^* - d(m) \sum_i \phi(v:m; g_i(\zeta:v)) \otimes u_i^*$$

and this implies our assertion.

Lemma 2. *Let $H \in \mathfrak{a}$. Then*

$$\Phi(v:m \exp TH)\, e^{-T\Gamma_v(H)} = \Phi(v:m) + \int_0^T \Psi_H(v:m \exp tH)\, e^{-t\Gamma_v(H)}\, dt$$

for $v \in \mathfrak{F}$, $m \in M_1$ and $T \in \mathbf{R}$.

Since $\mathfrak{a} \subset \mathfrak{Z}_1$, this is an immediate consequence of Lemma 1.

Put

$$E_i(v) = \Gamma_v^*({}_i e(v)) \qquad (v \in \mathfrak{F}, i \in Q)$$

in the notation of Lemma 3.4. Then it is clear that E_i is a C^∞ function from $\mathfrak{F}$ to End U^* and

$$\sum_{i \in {}^\circ Q} E_i(v) = 1.$$

Moreover it is easy to verify from Corollary 2 of Lemma 3.5 that

$$E_i(v)\, E_j(v) = \delta_{ij}\, E_j(v) \qquad (i, j \in {}^\circ Q).$$

Put $\mathbf{E}_i(v) = 1 \otimes E_i(v)$. Since J_1 is an abelian algebra, it is obvious that $\mathbf{E}_i(v)$ commutes with $\Gamma_v(\zeta)\,(\zeta \in \mathfrak{Z}_1)$ and the operations of K on $\mathbf{V}$. Put

$$\Phi_i(v) = \Phi(v)\,\mathbf{E}_i(v) \qquad (v \in \mathfrak{F}).$$

Then the following result is immediate.

Lemma 3. *Let $H \in \mathfrak{a}$. Then*

$$\Phi_i(v:m \exp TH)\, e^{-T\Gamma_v(H)} = \Phi_i(v:m) + \int_0^T \Psi_H(v:m \exp tH)\, \mathbf{E}_i(v)\, e^{-t\Gamma_v(H)}\, dt$$

for $v \in \mathfrak{F}$, $m \in M_1$, $T \in \mathbf{R}$ and $i \in Q$.

Let $\lambda_i\,(i \in Q)$ denote the restriction of $s_i\,\lambda^v$ on $\mathfrak{a}$ as in §6.

Lemma 4. *Put*

$$\Gamma_i(v:H) = E_i(v)\, e^{\Gamma_v(H) - \lambda_i(H)}$$

for $i \in Q$, $v \in \mathfrak{F}$, $H \in \mathfrak{a}$. Then we can choose c_0, $r_0 \geq 0$ such that

$$\|\Gamma_i(v:H)\| \leq c_0 (1 + \|H\|)^{r_0} (1 + |v|)^{r_0} \qquad (i \in Q, H \in \mathfrak{a}, v \in \mathfrak{F}).$$

Put

$$F_i(v:H) = E_i(v)(\Gamma_v^*(H) - \lambda_i(H))$$

and fix $H \in \mathfrak{a}$. We claim that all eigenvalues of $F_i(v:H)$ are purely imaginary. Since $F_i(v:H)$ is a continuous function of $v \in \mathfrak{F}$, it would be enough to verify this for $v \in \mathfrak{F}'(\lambda)$. But this follows from Corollary 1 of Lemma 3.5 since

$$\Gamma_v(H)\, e_j(v) = s_j \Lambda_v(H)\, e_j(v) \qquad (1 \leq j \leq q).$$

Now

$$\Gamma_i(v:H) = E_i(v)\, e^{F_i(v:H)}.$$

Since

$$v \longmapsto \varpi_{s_i,\lambda}(s_i \Lambda_v)\, E_i(v)$$

is a polynomial mapping (Lemma 3.4) and

$$|\varpi_{s_i,\lambda}(s_i \Lambda_v)| \geq |\varpi_\lambda(\lambda)| > 0$$

(Lemma 3.3), the required result follows from [1(a), Lemma 60].

Lemma 5. *Fix* $\zeta \in \mathfrak{Z}_1$, v_1, $v_2 \in \mathfrak{M}_1$ *and* $s \in \mathscr{S}(V)$. *Then we can choose* c, $r \geq 0$ *such that*

$$|\Psi_\zeta(v:v_1 \,;m\exp H\,;v_2)|_s \leq c\, \Xi_M(m)\, |(v,m,H)|^r\, e^{-\beta_P(H)}$$

for $m \in M_1^+$, $H \in Cl\,\mathfrak{a}^+$ *and* $v \in \mathfrak{F}$.

We recall that $v \longmapsto g_j(\zeta:v)$ $(1 \leq j \leq q)$ are polynomial mappings of $\mathfrak{F}$ into $\theta(\mathfrak{n}) \mathfrak{G}\, \mathfrak{n}$. Therefore our assertion follows without difficulty from Lemma 5.2.

Define Q^+, Q^o and Q^- as in §6. Then (see §6) we can choose δ $(0 < \delta \leq \frac{1}{2})$ such that

$$\lambda_i(H) \leq -\delta \beta_P(H)$$

for all $i \in Q^-$ and $H \in \mathfrak{a}^+$.

Fix $i \in Q^o$, v_1, $v_2 \in \mathfrak{M}_1$, $s \in \mathscr{S}(V)$ and $H \in \mathfrak{a}^+$. Then it follows from Lemmas 4 and 5 that the integral

$$\int_0^\infty |\Psi_H(v:v_1\,;m\exp tH\,;v_2)|_s\, \|\Gamma_i(v:-tH)\|\, dt$$

converges uniformly as v and m vary within compact subsets of $\mathfrak{F}$ and M_1 respectively. Put

$$\Phi_{i\infty}(v:m) = \lim_{t \to +\infty} \Phi_i(v:m\exp tH)\, e^{-t\Gamma_v(H)} \qquad (v \in \mathfrak{F}, m \in M_1).$$

Then, from Lemma 3, this limit exists and we prove as in [1(e), §22] that it is independent of $H \in \mathfrak{a}^+$. Moreover $\Phi_{i\infty}$ is a continuous function from $\mathfrak{F} \times M_1$ to $\mathbf{V}$ which is differentiable in $m \in M_1$. In fact

$$\Phi_{i\infty}(v:v_1\,;m\,;v_2) = \lim_{t \to +\infty} \Phi_i(v:v_1\,;m\exp tH\,;v_2)\, e^{-t\Gamma_v(H)}$$

for v_1, $v_2 \in \mathfrak{M}_1$ and $H \in \mathfrak{a}^+$.

Define $\Phi_{i\infty} = 0$ for $i \in Q^+ \cup Q^-$.

Lemma 6. *Fix $i \in Q$. Then*

$$\Phi_{i\infty}(v:m;\zeta) = \Phi_{i\infty}(v:m)\,\Gamma_v(\zeta) \qquad (\zeta \in \mathfrak{Z}_1)$$

and

$$|\Phi_i(v:v_1;m\exp TH;v_2) - \Phi_{i\infty}(v:v_1;m\exp TH;v_2)|_s$$

$$\leq e^{-\delta T \beta_P(H)} \left\{ |\Phi(v:v_1;m;v_2)|_s \,\|\Gamma_i(v:TH)\| \right.$$

$$\left. + \int_0^\infty |\Psi_H(v:v_1;m\exp tH;v_2)|_s \,\|\Gamma_i(v:(T-t)H)\|\, e^{t\beta_P(H)/2}\, dt \right\}$$

for $v_1, v_2 \in \mathfrak{M}_1$, $m \in M_1$, $H \in \mathfrak{a}^+$, $v \in \mathfrak{F}$, $s \in \mathscr{S}(V)$ and $T \geq 0$.

This is proved in the same way as [1(e), Theorem 22.1].
Put $^oQ^o = {}^oQ \cap Q^o$. Since

$$\sum_{i \in {}^oQ} E_i(v) = 1,$$

we get the following corollary.

Corollary.

$$\left|\Phi(v:v_1;m\exp TH;v_2) - \sum_{i \in {}^oQ^o} \Phi_{i\infty}(v:v_1;m\exp TH;v_2)\right|_s$$

$$\leq e^{-\delta T \beta_P(H)} \sum_{i \in {}^oQ} \left\{ |\Phi(v:v_1;m;v_2)|_s \,\|\Gamma_i(v:TH)\| \right.$$

$$\left. + \int_0^\infty |\Psi_H(v:v_1;m\exp tH;v_2)|_s \,\|\Gamma_i(v:(T-t)H)\|\, e^{t\beta_P(H)/2}\, dt \right\}.$$

Define functions ψ_i $(i \in Q)$ from $\mathfrak{F} \times M_1$ to V by the formula

$$\sum_{i \in {}^oQ^o} \Phi_{i\infty} = \sum_{i \in Q} \psi_i \otimes u_i^*.$$

Since $u_1 = 1$, it is clear from the above results and the definition of ϕ_P [1(e), Theorem 21.1] that $\psi_1 = \phi_P$. The following result is now obvious from Lemma 6.

Lemma 7. *Fix $v_1, v_2 \in \mathfrak{M}_1$. Then the function $(v, m) \mapsto \phi_P(v:v_1;m;v_2)$ is continuous on $\mathfrak{F} \times M_1$. Moreover for each $s \in \mathscr{S}(V)$, we can choose $c, r \geq 0$ such that*

$$|\phi_P(v:v_1;m;v_2)|_s \leq c\, \Xi_M(m)\, |(v, m)|^r$$

for $v \in \mathfrak{F}$ and $m \in M_1$.

Corollary. *Suppose ϕ is of type $II'(\lambda)$. Then*

$$\phi_P = \sum_{s \in \mathfrak{w}(\mathfrak{h}_R \,|\, \mathfrak{a})} \phi_{P,s}$$

on $\mathfrak{F} \times M_1$.

By Lemma 8.3 the equality holds on $\mathfrak{F}'(\lambda) \times M_1$. But since both sides are continuous, it must hold on $\mathfrak{F} \times M_1$.

We recall that $P \neq G$. Fix a compact subset Ω of $\mathfrak{a}^+$ and choose $\varepsilon_0 > 0$ such that $\beta_P(H) \geq 2\varepsilon_0$ for all $H \in \Omega$. Put $\varepsilon = \delta\varepsilon_0$. Then the following result is an easy consequence of the corollary of Lemma 6.

Lemma 8. *Given* $v_1, v_2 \in \mathfrak{M}_1$ *and* $s \in \mathcal{S}(V)$, *we can choose* $c, r \geq 0$ *such that*

$$|d_P(m \exp TH)\, \phi(v : v_1' ; m \exp TH ; v_2') - \phi_P(v : v_1 ; m \exp TH ; v_2)|_{\mathbf{s}}$$

$$\leq c\, e^{-\varepsilon T}\, \Xi_M(m)\, |(v, m)|^r$$

for $v \in \mathfrak{F}$, $m \in M_1^+$, $H \in \Omega$ *and* $T \geq 0$.

§ 11. A Criterion for a Function to be of Type *II′*(λ)

We assume in this section that τ is a unitary [1(e), § 20]. Let $\mathscr{P}(\mathfrak{h}_R)$ denote the set of all psgps $P = MAN$ of G such that $\mathfrak{a} = \mathfrak{h}_R$. Clearly M is independent of $P \in \mathscr{P}(\mathfrak{h}_R)$.

Let ϕ be a function on $\mathfrak{F} \times G$ of type $II(\lambda)$. Put $\mathfrak{a} = \mathfrak{h}_R$ and fix $P \in \mathscr{P}(\mathfrak{a})$, $s \in \mathfrak{w}(\mathfrak{a})$ and $v \in \mathfrak{F}'(\lambda)$ $(P = MAN)$. Then the function $m \mapsto \phi_{P,s}(v : m)$ $(m \in M)$ lies in $^o\mathscr{C}(M, \tau_M)$ (Lemma 7.5). We observe that $^o\mathscr{C}(M, \tau_M)$, being a closed subspace of $\mathscr{C}(M, \tau_M)$ [1(e), § 18], is a locally convex space.

Lemma 1. *Let* ϕ *be a function on* $\mathfrak{F} \times G$ *of type* $II(\lambda)$ *and* $P' = M' A' N'$ *a psgp of* G. *Then* $\phi_{P'}(v) \sim 0$ $(v \in \mathfrak{F})$ *unless* $\mathfrak{a}'$ *is a conjugate to* $\mathfrak{a}$ *under* K.

Fix $a' \in A'$, $f \in {}^o\mathscr{C}(M', \tau_{M'})$ and assume that $\mathfrak{a}'$ is not conjugate to under K. Then it follows from [1(e), Theorem 29.1] that

$$\int_{M'} (f(m'),\ \phi_{P'}(v : m' a'))\, dm' = 0$$

for $v \in \mathfrak{F}'(\lambda)$. On the other hand, it is obvious from Lemma 10.7 that the left side is a continuous function of $v \in \mathfrak{F}$. Hence $\phi_{P'}(v) \sim 0$ for all $v \in \mathfrak{F}$.

Corollary. *Fix* $v \in \mathfrak{F}$ *and suppose* $\phi_P(v) = 0$ *for all* $P \in \mathscr{P}(\mathfrak{a})$. *Then* $\phi(v) = 0$.

This is an immediate consequence of [1(e), Lemma 25.2] and the above result.

Theorem 1. *Let* ϕ *be as above and* S *a collection of continuous seminorms on* $^o\mathscr{C}(M, \tau_M)$. *We assume that* $f \in {}^o\mathscr{C}(M, \tau_M)$ *and* $\mathrm{s}(f) = 0$ *for all* $s \in S$, *implies that* $f = 0$. *Then, in order that* ϕ *be of type* $II'(\lambda)$, *it is necessary and sufficient that the following condition holds. For* $P \in \mathscr{P}(\mathfrak{a})$ *and* $s \in \mathfrak{w}(\mathfrak{a})$, *let* $f_{P,s}(v)$ *denote the restriction of* $\phi_{P,s}(v)$ *on* M $(v \in \mathfrak{F}'(\lambda))$. *Then* $\mathrm{s}(f_{P,s}(v))$ *should remain locally bounded on* $\mathfrak{F}$ *for every* $P \in \mathscr{P}(\mathfrak{a})$, $s \in \mathfrak{w}(\mathfrak{a})$ *and* $s \in S$.

For example we can take S to consist of the single element s given by

$$\mathrm{s}(f) = \|f\|_M \qquad (f \in {}^o\mathscr{C}(M, \tau_M)),$$

where

$$\|f\|_M^2 = \int_M |f(m)|^2\, dm.$$

We first need a simple result.

Lemma 2. *Let $H_0 \neq 0$ be a point in $\mathfrak{a}$ and ϕ a function of type $II(\lambda)$ on $\mathfrak{F} \times G$ such that $\phi(v) = 0$ whenever $v(H_0) = 0$ $(v \in \mathfrak{F})$. Then the function*

$$\psi(v : x) = v(H_0)^{-1} \phi(v : x) \qquad (v \in \mathfrak{F}, x \in G)$$

is also of type $II(\lambda)$.

This follows from Lemma 22.1.

Now we come to the proof of Theorem 1. If ϕ is of type $II'(\lambda)$, then for fixed $P \in \mathscr{P}(\mathfrak{a})$ and $s \in \mathfrak{w}(\mathfrak{a})$, $f_{P,s}$ defines a C^∞ mapping from $\mathfrak{F}$ to $^\circ\mathscr{C}(M, \tau_M)$ (see Lemma 12.1 below). Hence our condition is certainly necessary. So it remains to verify that it is sufficient.

Put

$$\psi(v : x) = \varpi(\lambda + (-1)^{1/2} v) \phi(v : x) \qquad (v \in \mathfrak{F}, x \in G).$$

Then by Theorem 9.1, ψ is of type $II'(\lambda)$. Let p be the set of all positive roots of $(\mathfrak{g}, \mathfrak{h})$, $p(\lambda)$ the subset of those $\alpha \in p$ for which $\lambda(H_\alpha) \neq 0$ and $p'(\lambda)$ the complement of $p(\lambda)$ in p. Put

$$\varpi_\lambda = \prod_{\alpha \in p(\lambda)} H_\alpha, \qquad \varpi'_\lambda = \prod_{\alpha \in p'(\lambda)} H_\alpha.$$

Then $\varpi = \varpi_\lambda \cdot \varpi'_\lambda$ and

$$|\varpi_\lambda(\lambda + (-1)^{1/2} v)| \geq |\varpi_\lambda(\lambda)| > 0 \qquad (v \in \mathfrak{F}).$$

Hence it follows without difficulty that

$$\psi'(v : x) = \varpi_\lambda(\lambda + (-1)^{1/2} v)^{-1} \psi(v : x)$$
$$= \varpi'_\lambda(\lambda + (-1)^{1/2} v) \phi(v : x)$$

is a function of type $II'(\lambda)$. Since λ is a regular element in $(-1)^{1/2} \mathfrak{h}_I^*$ (see § 3), it is clear that we can choose elements $H_i \neq 0$ $(1 \leq i \leq r)$ in $\mathfrak{a}$ and a complex number $c \neq 0$ such that

$$\varpi'_\lambda(\lambda + (-1)^{1/2} v) = c \prod_{1 \leq i \leq r} v(H_i) \qquad (v \in \mathfrak{F}).$$

Hence it is enough to prove the following result.

Lemma 3. *Put*

$$Q(v) = \prod_{1 \leq i \leq r} v(H_i) \qquad (v \in \mathfrak{F})$$

where $H_i \neq 0$ are elements in $\mathfrak{a}$. Suppose ϕ satisfies the condition of Theorem 1 and

$$\psi(v : x) = Q(v) \phi(v : x) \qquad (v \in \mathfrak{F}, x \in G)$$

is a function of type $II'(\lambda)$. Then ϕ is also of type $II'(\lambda)$.

By induction we are reduced to the case $r = 1$. Fix a $\cdot$psgp $P' = M' A' N'$ and $t \in \mathfrak{w}(\mathfrak{a}|\mathfrak{a}')$. Then

$$\psi_{P',t}(v) = v(H_1) \phi_{P',t}(v) \qquad (v \in \mathfrak{F}'(\lambda)).$$

We have to verify that $(\phi_{P',t})^t$ is of type $II(\lambda)$. Since ψ is of type $II'(\lambda)$, we know from Lemma 9.1 that $(\psi_{P',t})^t$ is also of type $II'(\lambda)$. Hence in view of Lemma 2, it would be enough to verify that $\psi_{P',t}(v)=0$ whenever $v(H_1)=0$.

Now fix $P\in\mathscr{P}(\mathfrak{a})$ and $s\in\mathfrak{w}(\mathfrak{a})$. Then

$$\psi_{P,s}(v)=v(H_1)\phi_{P,s}(v)\qquad(v\in\mathfrak{F}'(\lambda)).$$

Let $g(v)$ denote the restriction of $\psi_{P,s}(v)$ on M. Then we conclude from Lemma 12.1 below that $v\mapsto g(v)$ is a continuous mapping from $\mathfrak{F}$ into $^o\mathscr{C}(M,\tau_M)$.

Fix a point $v_0\in\mathfrak{F}$ such that $v_0(H_1)=0$. Let v be a variable point in $\mathfrak{F}'(\lambda)$ which tends to v_0. Then if $s\in S$,

$$s(g(v_0))=\lim_v s(g(v))=\lim_v |v(H_1)|\, s(f_{P,s}(v))=0$$

by our assumption on ϕ. Hence $g(v_0)=0$ and this implies (Corollary of Theorem 7.1) that $\psi_{P,s}(v_0)=0$. But then we conclude from Lemma 7.4 and the corollary of Lemma 1 that $\psi_{P',t}(v_0)=0$.

This proves Lemma 3 and therefore also Theorem 1.

§ 12. An Auxiliary Result

Let $G=MA$ be the Langlands decomposition of G and assume $\mathfrak{a}=\mathfrak{h}_R$. Let ϕ be a function of type $II(\lambda)$ on $\mathfrak{F}\times G$ and ψ its restriction on $\mathfrak{F}\times M$. Then we know from Lemma 7.5 that $\psi(v)\in{}^o\mathscr{C}(M,\tau_M)$ for $v\in\mathfrak{F}$. (We note that $\mathfrak{F}'(\lambda)=\mathfrak{F}$ and $\mathfrak{w}(\mathfrak{a})=\{1\}$ in this case.)

Lemma 1. $v\mapsto\psi(v)$ *is a* C^∞ *mapping of* $\mathfrak{F}$ *into* $^o\mathscr{C}(M,\tau_M)$.

This is an immediate consequence of the following lemma.

Lemma 2. *Suppose* $\mathfrak{h}_R=\{0\}$. *Fix* $g_1,g_2\in\mathfrak{G}$ *and* $r_0\geqq0$. *Then we can choose a finite subset* F *of* $\mathfrak{G}^{(2)}$ *with the following property. Given* $r\geqq0$ *and* $s\in\mathscr{S}(V)$, *we can choose a number* $c>0$ *such that*

$$|\phi(g_1;x;g_2)|_s\,\Xi(x)^{-1}(1+\sigma(x))^{r_0}\leqq c\,\mathbf{s}_{F,r}(\phi)\qquad(x\in G)$$

for all functions ϕ *on* G *of type* $II(\lambda)$.

Let $P=MAN$ be a psgp of G $(P\neq G)$. Since $\mathfrak{h}_R=\{0\}$, $\mathfrak{w}(\mathfrak{h}_R|\mathfrak{a})=\emptyset$ and $\mathfrak{F}'(\lambda)=\mathfrak{F}=\{0\}$. Therefore $\phi_P=0$ by Lemma 8.3. Moreover

$$z\phi=\gamma(z:\lambda)\phi\qquad(z\in\mathfrak{Z}).$$

Therefore Lemma 23.4 of [1 (e)] is applicable. Fix a minimal p-pair (P_0,A_0) in G. Then $G=K\cdot Cl\,A_0^+\cdot K$ and there are only a finite number of p-pairs $(P,A)\succ(P_0,A_0)$. Our assertion is an easy consequence of these facts.

§ 13. Statement of the Two Main Theorems

We keep to the notation of § 8. For $D \in \tilde{\mathfrak{G}}$, $s \in \mathscr{S}(V)$ and $r \geq 0$, define

$$^o\mathbf{s}_{D,r}(f) = \sup_{\mathfrak{F} \times G} |Df|_\mathbf{s} \, \Xi^{-1}(1+\sigma)^{-r} \quad (f \in C^\infty(\mathfrak{F} \times G, V)).$$

Similarly if F is any finite subset of $\tilde{\mathfrak{G}}$, we write

$$^o\mathbf{s}_{F,r}(f) = \sum_{D \in F} {}^o\mathbf{s}_{D,r}(f).$$

A function $\phi \colon \mathfrak{F} \times G \to V$ will be said to be of type $I(\lambda)$ if:

1) ϕ is of type $II(\lambda)$.

2) For any $D \in \tilde{\mathfrak{G}}$ and $\mathbf{s} \in \mathscr{S}(V)$, we can choose $r \geq 0$ such that $^o\mathbf{s}_{D,r}(\phi) < \infty$.

Moreover we say that ϕ is of type $I'(\lambda)$ if it is both of type $I(\lambda)$ and type $II'(\lambda)$.

Let $\mathscr{E}(I'(\lambda))$ denote the space of all functions of type $I'(\lambda)$ and dv the Euclidean measure on $\mathfrak{F}$.

Theorem 1. *For* $\phi \in \mathscr{E}(I'(\lambda))$, *define*

$$j_\phi(x) = \int_{\mathfrak{F}} \phi(v : x) \, dv \quad (x \in G).$$

Then $j_\phi \in \mathscr{C}(G, \tau)$. *Fix* $g_1, g_2 \in \mathfrak{G}$ *and* $r_0 \geq 0$. *Then we can choose a finite subset F of $\mathfrak{G}$ with the following property. Given $r \geq 0$ and $\mathbf{s} \in \mathscr{S}(V)$, there exists a number $c > 0$ such that*

$$|j_\phi(g_1 ; x ; g_2)|_s \leq c \, {}^o\mathbf{s}_{F,r}(\phi) \, \Xi(x)(1 + \sigma(x))^{-r_0} \quad (x \in G)$$

for all $\phi \in \mathscr{E}(I'(\lambda))$.

Define the Schwartz space $\mathscr{C}(\mathfrak{F})$ as usual.

Corollary. *Fix a function ϕ on $\mathfrak{F} \times G$ of type $II'(\lambda)$ and define*

$$\phi_\alpha(x) = \int_{\mathfrak{F}} \alpha(v) \phi(v : x) \, dv \quad (x \in G)$$

for $\alpha \in \mathscr{C}(\mathfrak{F})$. Then $\alpha \mapsto \phi_\alpha$ is a continuous mapping of $\mathscr{C}(\mathfrak{F})$ into $\mathscr{C}(G, \tau)$ and

$$\phi_\alpha(g_1 ; x ; g_2) = \int_{\mathfrak{F}} \alpha(v) \phi(v : g_1 ; x ; g_2) \, dv \quad (x \in G)$$

for $g_1, g_2 \in \mathfrak{G}$ and $\alpha \in \mathscr{C}(\mathfrak{F})$.

This is an immediate consequence of Theorem 1.

Fix ϕ as in the above corollary. Then if $P = MAN$ is a psgp of G and $\alpha \in \mathscr{C}(\mathfrak{F})$, it follows from Lemma 9.1 and the corollary of Lemma 10.7 that the function

$$\phi_{P,\alpha}(m) = \int_{\mathfrak{F}} \alpha(v) \phi_P(v : m) \, dv \quad (m \in MA)$$

lies in $\mathscr{C}(MA, \tau_M)$. Extend it to a function on G by setting

$$\phi_{P,\alpha}(kmn) = \tau(k) \phi_{P,\alpha}(m) \quad (k \in K, \ m \in MA, \ n \in N).$$

Put $\bar{P}=\theta(P)$, $\bar{N}=\theta(N)$, $\rho=\rho_P$, $H(x)=H_P(x)$ $(x\in G)$ and define $\phi_\alpha^{(P)}$ as in [1(e), Lemma 16.1].

Theorem 2. *Let $d\bar{n}$ denote the Haar measure on $\bar{N}$. Then*

$$\phi_\alpha^{(P)}(m)=d_P(m)\int_{\bar{N}}\phi_\alpha(\bar{n}m)d\bar{n}=\int_{\bar{N}}e^{-\rho(H(\bar{n}))}\phi_{P,\alpha}(\bar{n}m)d\bar{n}$$

for $m\in MA$ and $\alpha\in\mathscr{C}(\mathfrak{F})$.

This is a generalization of [1(b), Theorem 4, p. 610]. (It is part of the assertion of the theorem that the above integrals are well defined.)

In view of the corollary of Lemma 10.7, the following result is obvious.

Corollary. $\phi_\alpha^{(P)}=0$ *unless* $\mathfrak{a}^k\subset\mathfrak{h}_R$ *for some* $k\in K$.

The above two theorems contain the main results of this paper. The significance of Theorem 2 may be explained as follows. Extend d_P and $\phi_P(v)$ to functions on G as in [1(e), §24]. Then Theorem 2 asserts that

$$\int_{\bar{N}}d_P(\bar{n})^{-1}d\bar{n}\int_{\mathfrak{F}}\alpha(v)d_P(\bar{n}m)\phi(v:\bar{n}m)dv$$
$$=\int_{\bar{N}}d_P(\bar{n})^{-1}d\bar{n}\int_{\mathfrak{F}}\alpha(v)\phi_P(v:\bar{n}m)dv \qquad (m\in MA)$$

for $\alpha\in\mathscr{C}(\mathfrak{F})$. This shows that the integral on the left remains unchanged when we replace $d_P\phi(v)$ by its asymptotic value $\phi_P(v)$ [1(e), Lemma 24.1].

§ 14. Some Preparation

Put $\mathfrak{D}=\mathfrak{D}(\mathfrak{F}_c)$, $\mathscr{E}'=\mathscr{E}(I'(\lambda))$ and let $\mathscr{E}=\mathscr{E}(I(\lambda))$ denote the space of all functions on $\mathfrak{F}\times G$ of type $I(\lambda)$. It is obviously enough to prove the statement of Theorem 13.1 for $s\in\mathscr{S}^o(V)$ [1(e), §22].

Let $\mathbf{R}_+$ denote the set of all real numbers $r\geq 0$. In order to avoid tedious repetitions, we agree to the following conventions. The variables $r,\mathbf{s}$ and v shall range freely over $\mathbf{R}_+$, $\mathscr{S}^o(V)$ and $\mathfrak{F}$ respectively unless explicitly mentioned otherwise. Let Y be any set and f,g two functions from $\mathbf{R}_+\times\mathscr{S}^o(V)\times\mathfrak{F}\times Y$ to $\mathbf{R}_+\cup\{\infty\}$. Then we write

$$f(r,\mathbf{s},v,y)\prec g(r,\mathbf{s},v,y) \qquad (y\in Y),$$

if for any given r and $\mathbf{s}$ we can choose a real number $c(r,\mathbf{s})>0$ such that

$$f(r,\mathbf{s},v,y)\leq c(r,\mathbf{s})g(r,\mathbf{s},v,y)$$

for all $v\in\mathfrak{F}$ and $y\in Y$. Finally the letter F will always stand for a finite set. Thus $F\subset Y$ means that F is a finite subset of Y.

We now use the notation of § 5 and fix numbers $c_0, d_0\geq 0$ such that

$$d_P(m)\,\Xi(m)\leq c_0\,\Xi_M(m)(1+\sigma(m))^{d_0} \qquad (m\in M_1^+).$$

Lemma 1. *Fix* $\zeta \in \mathfrak{Z}_1$, $v_1, v_2 \in \mathfrak{M}_1$ *and* $D \in \mathfrak{D}$. *Then we can choose* $F \subset \tilde{\mathfrak{G}}$ *such that*

$$|\psi_{i,\zeta}(v; D:v_1;m; v_2)|_s \leq {}^{o}\mathbf{s}_{F,r}(\phi)\, \Xi_M(m)\,|(m, H)|^{d_0+r} e^{-\beta P(H)}$$

for $\phi \in \mathscr{E}$, $m \in M_1^+$, $H \in Cl\,\mathfrak{a}^+$ *and* $1 \leq i \leq q$.

Here $\psi_{i,\zeta}$ is the function defined in §5 corresponding to ϕ. This lemma is proved in the same way as [1(e), Lemma 22.3].

Lemma 2. *Given* $D \in \mathfrak{D}$ *and* $v_1, v_2 \in \mathfrak{M}_1$, *we can choose* $F \subset \tilde{\mathfrak{G}}$ *such that*

$$|\phi_{i\infty}(v; D:v_1;m; v_2)|_s \prec {}^{o}\mathbf{s}_{F,r}(\phi)\, \Xi_M(m)(1+\sigma(m))^{d_0+r} \qquad (i \in Q^o)$$

for $m \in M_1$ *and* $\phi \in \mathscr{E}$.

We use the notation of the proof of Lemma 8.2. Fix $H \in \mathfrak{a}^+$. Then

$$\phi_{i\infty}(v; D:v_1;m; v_2)$$
$$= \phi_i(v; D:v_1;m; v_2) + \int_0^\infty \psi_{i,H}{}^o(v; D:v_1;m; v_2:tH)\,dt$$

and our assertion follows without difficulty.

Now suppose $\phi \in \mathscr{E}'$. Then for any $\mathfrak{s} \in \mathfrak{w}(\mathfrak{h}_R|\mathfrak{a})$, $\phi_{P,\mathfrak{s}}$ extends to a C^∞ function on $\mathfrak{F} \times M_1$.

Lemma 3. *Given* $D \in \mathfrak{D}$ *and* $v_1, v_2 \in \mathfrak{M}_1$, *we can choose* $F \subset \tilde{\mathfrak{G}}$ *such that*

$$|\phi_{P,\mathfrak{s}}(v; D:v_1;m; v_2)|_{\mathbf{s}} \prec {}^{o}\mathbf{s}_{F,r}(\phi)\, \Xi_M(m)(1+\sigma(m))^{d_0+r}$$

for $m \in M_1$, $\phi \in \mathscr{E}'$ *and* $\mathfrak{s} \in \mathfrak{w}(\mathfrak{h}_R|\mathfrak{a})$.

In view of Lemmas 3.3 and 7.3, this is an immediate consequence of Lemmas 2 and 22.2.

Corollary. *If* $\phi \in \mathscr{E}'$, *then for any* $\mathfrak{s} \in \mathfrak{w}(\mathfrak{h}_R|\mathfrak{a})$, $(\phi_{P,\mathfrak{s}})^{\mathfrak{s}}$ *is a function of type* $I'(\lambda)$ *on* $\mathfrak{F} \times M_1^{\mathfrak{s}}$.

This follows from Lemmas 3 and 9.1.

Now let us use the notation of Lemma 10.5.

Lemma 4. *Fix* $\zeta \in \mathfrak{Z}_1$, $v_1, v_2 \in \mathfrak{M}_1$ *and* $r_1 \geq 0$. *Then we can choose* $F \subset \tilde{\mathfrak{G}}$ *such that*

$$|\Psi_\zeta(v:v_1;m\exp H; v_2)|_{\mathbf{s}}(1+|v|)^{r_1} \leq {}^{o}\mathbf{s}_{F,r}(\phi)\, \Xi_M(m)\,|(m, H)|^{d_0+r} e^{-\beta P(H)}$$

for $m \in M_1^+$, $H \in Cl\,\mathfrak{a}^+$, $\phi \in \mathscr{E}$.

As before this follows from Lemma 5.2.

Now assume that $P \neq G$. Fix a compact set Ω in $\mathfrak{a}^+$ and choose $\varepsilon_0 > 0$ such that $\beta_P(H) \geq 2\varepsilon_0$ for all $H \in \Omega$. Select δ $(0 < \delta \leq \frac{1}{2})$ as in §10 and put $\varepsilon = \delta\varepsilon_0$.

Lemma 5. *Given* $v_1, v_2 \in \mathfrak{M}_1$ *and* $r_1 \geq 0$, *we can choose* $F \subset \tilde{\mathfrak{G}}$ *such that*

$$|d_P(m\exp TH)\phi(v:v_1';m\exp TH; v_2') - \phi_P(v:v_1;m\exp TH; v_2)|_{\mathbf{s}}(1+|v|)^{r_1}$$
$$\prec {}^{o}\mathbf{s}_{F,r}(\phi)\, \Xi_M(m)(1+\sigma(m))^{d_0+r} e^{-\varepsilon T}$$

for $m \in M_1^+$, $H \in \Omega$, $T \geq 0$ *and* $\phi \in \mathscr{E}$.

This is proved in the same way as Lemma 10.8.

Now fix $r_1 \geq 0$ such that

$$\int_{\mathfrak{F}} (1+|v|)^{-r_1} \, dv < \infty.$$

If $\phi \in \mathscr{E}'$, we know (Corollary of Lemma 10.7) that

$$\phi_P = \sum_{s \in \mathfrak{w}(\mathfrak{h}_R|\mathfrak{a})} \phi_{P,s}.$$

Put $j(\phi:x) = j_\phi(x) \ (x \in G)$ and

$$j(\phi_{P,s}:m) = \int_{\mathfrak{F}} \phi_{P,s}(v:m) \, dv \qquad (m \in M_1)$$

for $s \in \mathfrak{w}(\mathfrak{h}_R|\mathfrak{a})$ and $\phi \in \mathscr{E}'$.

Corollary.

$$\left| d_P(m \exp TH) j(\phi:v_1'; m \exp TH; v_2') - \sum_{s \in \mathfrak{w}(\mathfrak{h}_R|\mathfrak{a})} j(\phi_{P,s}:v_1; m \exp TH; v_2) \right|_{\mathbf{s}}$$
$$\prec {}^0 \mathbf{s}_{F,r}(\phi) \, \Xi_M(m)(1+\sigma(m))^{d_0+r} e^{-\varepsilon T}$$

for $m \in M_1^+$, $H \in \Omega$, $T \geq 0$ and $\phi \in \mathscr{E}'$.

This follows immediately from Lemma 5.

§ 15. Proof of Theorem 13.1

We now come to the proof of Theorem 13.1. It is clearly enough to prove the second part of the theorem.

We proceed by induction on dim G. First assume that prk $G > 0$ and let $G = MA$ be the Langlands decomposition of G. Then $\mathfrak{h}_R = \mathfrak{m} \cap \mathfrak{h}_R + \mathfrak{a}$ where the sum is direct. Let $\mathfrak{F}_1$ and $\mathfrak{F}_2$ be the subspace consisting of all $v \in \mathfrak{F}$ which vanish identically on $\mathfrak{m} \cap \mathfrak{h}_R$ and $\mathfrak{a}$ respectively. Then $\mathfrak{F} = \mathfrak{F}_1 + \mathfrak{F}_2$ where the sum is direct. We note that $\mathfrak{D}_i = \mathfrak{D}(\mathfrak{F}_{ic}) \subset \mathfrak{D} = \mathfrak{D}(\mathfrak{F}_c)$ [1 (c), p. 540]. Let dv_i denote the Euclidean measure on $\mathfrak{F}_i$ so normalized that $dv = dv_1 \, dv_2$ $(v = v_1 + v_2, \ v_i \in \mathfrak{F}_i, \ i = 1, 2)$. Since $\mathfrak{a} \subset \mathfrak{Z}$, it follows from our assumptions that

$$\phi(v_1 + v_2 : ma) = \phi(v_1 + v_2 : m) e^{(-1)^{1/2} v_1 (\log a)} \qquad (m \in M, \ a \in A)$$

for $\phi \in \mathscr{E}'$ and $v_i \in \mathfrak{F}_i$. Fix $v_1, v_2 \in \mathfrak{M}$ and $u \in \mathfrak{A}$. Then

$$j_\phi(v_1 : ma; v_2 u) = \int \phi(v_1 + v_2 : v_1 : m; v_2) u((-1)^{1/2} v_1) e^{(-1)^{1/2} v_1 (\log a)} \, dv_1 \, dv_2.$$

(We regard u as a polynomial function on $\mathfrak{F}_{1c}$ in the right side.) Now fix $r_0 \geq 0$. Then we can choose $p \in S(\mathfrak{F}_{1c})$ such that

$$p((-1)^{1/2} H) \geq (1 + \|H\|)^{r_0} \qquad (H \in \mathfrak{a}).$$

Also we can select a polynomial function p_1 on $\mathfrak{F}_1$ such that $p_1 \geq 1$ on $\mathfrak{F}_1$ and

$$\int_{\mathfrak{F}_1} p_1^{-1} \, dv_1 < \infty.$$

Hence it is obvious that there exists an element $D_1 \in \mathfrak{D}_1$ such that

$$|j_\phi(v_1;ma;v_2u)|_\mathbf{s}(1+\sigma(a))^{r_0} \leq \sup_{v_1 \in \mathfrak{F}_1} \Big| \int_{\mathfrak{F}_2} \phi(v_1+v_2;D_1:v_1;m;v_2)\,dv_2 \Big|_\mathbf{s}$$

for $m \in M$, $a \in A$ and $\phi \in \mathscr{E}'$.

On the other hand $\dim M < \dim G$ and so the induction hypothesis is applicable to M. Let $\mathscr{E}'_M$ be the space of all functions ψ on $\mathfrak{F}_2 \times M$ of type $I'(\lambda)$. Then we can choose a finite subset F_2 of $\mathfrak{M} = \mathfrak{D}_2 \otimes \mathfrak{M}^{(2)}$ such that

$$\Big| \int_{\mathfrak{F}_2} \psi(v_2:v_1;m;v_2)\,dv_2 \Big|_\mathbf{s} \prec^\circ \mathbf{s}_{F_2,r}(\psi)\, \Xi(m)(1+\sigma(m))^{-r_0}$$

for $m \in M$ and $\psi \in \mathscr{E}'_M$.

We regard $\mathfrak{M}$ as a subalgebra of $\mathfrak{G} = \mathfrak{D} \otimes \mathfrak{G}^{(2)}$. Let F denote the subset of $\mathfrak{G}$ consisting of all elements of the form $D_2 D_1$ $(D_2 \in F_2)$. Fix $\phi \in \mathscr{E}'$, $v_1 \in \mathfrak{F}_1$ and put

$$\psi(v_2:m) = \phi(v_1+v_2;D_1:m) \qquad (v_2 \in \mathfrak{F}_2,\ m \in M).$$

Then $\psi \in \mathscr{E}'_M$ and so we conclude from the above result that

$$|j_\phi(v_1;ma;v_2u)|_\mathbf{s} \prec^\circ \mathbf{s}_{F,r}(\phi)\, \Xi(m)(1+\sigma(m))^{-r_0}(1+\sigma(a))^{-r_0}$$

for $m \in M$, $a \in A$ and $\phi \in \mathscr{E}'$. This obviously implies Theorem 13.1 in this case.

So now suppose $\operatorname{prk} G = 0$. The case $G = K$ being trivial, we may assume that G is not compact. Fix a minimal p-pair (P_0, A_0) in G and let S^+ be the set of all $H \in Cl\,\mathfrak{a}_0^+$ with $\|H\| = 1$. Fix $H_0 \in S^+$ and let F_0 be the set of all simple roots of (P_0, A_0) which vanish at H_0. Put $(P,A) = (P_0, A_0)_{F_0}$. Then $H_0 \in \mathfrak{a}^+$. Fix a compact neighborhood Ω_0 of H_0 in S^+ such that

$$\alpha(H) \geq \alpha(H_0)/2 \qquad (H \in \Omega_0)$$

for every root α of (P_0, A_0). Put $\varepsilon_0 = \beta_P(H_0)/4$ and $\varepsilon = \delta \varepsilon_0$ where δ is defined as in § 10. Since

$$\exp tH = m_t \exp(tH_0/2) \qquad (H \in \Omega_0,\ t \geq 0),$$

where $m_t = \exp t(H - \frac{1}{2}H_0) \in Cl\,A_0^+ \subset M_1^+$, we get the following result from the corollary of Lemma 14.5.

Lemma 1. *Given* $v_1, v_2 \in \mathfrak{M}_1$, *we can choose* $F \subset \mathfrak{G}$ *such that*

$$|d_P(\exp tH)j(\phi:v_1';\exp tH;v_2') - \sum_{s \in \mathfrak{w}(\mathfrak{h}_R|\mathfrak{a})} j(\phi_{P,s}:v_1;\exp tH;v_2)|_\mathbf{s}$$

$$\prec^\circ \mathbf{s}_{F,r}(\phi)\,\Xi_M(\exp tH)(1+t)^{d_0+r}e^{-\varepsilon t}$$

for $H \in \Omega_0$, $t \geq 0$ *and* $\phi \in \mathscr{E}'$.

On the other hand since $\operatorname{prk} G = 0$ and $H_0 \neq 0$, it is clear that $\dim M_1 < \dim G$. Moreover for $s \in \mathfrak{w}(\mathfrak{h}_R|\mathfrak{a})$ and $\phi \in \mathscr{E}'$, $(\phi_{P,s})^s$ is a function of type $I'(\lambda)$ on $\mathfrak{F} \times M_1^s$ (Corollary of Lemma 14.3). Hence if we take into account Lemma 14.3 and apply the induction hypothesis to M_1^s, we get the following result immediately.

Lemma 2. *Fix $v_1, v_2 \in \mathfrak{M}_1$ and $r_0 \geq 0$. Then we can choose $F \subset \tilde{\mathfrak{G}}$ such that*

$$|j(\phi_{P,s} : v_1 ; m; v_2)|_s \prec {}^0\mathbf{s}_{F,r}(\phi) \, \Xi_M(m)(1 + \sigma(m))^{-r_0}$$

for $s \in \mathfrak{w}(\mathfrak{h}_R | \mathfrak{a})$, $m \in M_1$ and $\phi \in \mathscr{E}'$.

Combining this with Lemma 1 and standard inequalities relating Ξ and Ξ_M, we get the following result.

Lemma 3. *Given $v_1, v_2 \in \mathfrak{M}_1$ and $r_0 \geq 0$, we can choose $F \subset \tilde{\mathfrak{G}}$ such that*

$$|j(\phi : v_1 ; \exp tH; v_2)|_s \prec {}^0\mathbf{s}_{F,r}(\phi) \, \Xi(\exp tH)(1 + t)^{-r_0}$$

for $H \in \Omega_0$, $t \geq 0$ and $\phi \in \mathscr{E}'$.

On the other hand the following result is an immediate consequence of Lemma 5.2.

Lemma 4. *Fix $D \in \mathfrak{D}$, $g_i \in \mathfrak{G}$ $(1 \leq i \leq 4)$ such that $g_1 \in \mathfrak{G}\mathfrak{n}$ and $g_4 \in \theta(\mathfrak{n})\,\mathfrak{G}$. Then we can choose $F \subset \tilde{\mathfrak{G}}$ such that*

$$\sum_{i=1,3} |\phi(v; D : g_i ; \exp tH; g_{i+1})|_s \leq {}^0\mathbf{s}_{F,r}(\phi) \, \Xi(\exp tH)(1 + t)^r e^{-2\varepsilon_0 t}$$

for $\phi \in \mathscr{E}$, $H \in \Omega_0$ and $t \geq 0$.

Now fix a polynomial function p on $\mathfrak{F}$ such that $p \geq 1$ on $\mathfrak{F}$ and

$$\int_{\mathfrak{F}} p^{-1} \, dv < \infty.$$

Then taking $D = p$ in the above lemma, we get the following corollary.

Corollary. *Let g_i $(1 \leq i \leq 4)$ be as above. Then we can choose $F \subset \tilde{\mathfrak{G}}$ such that*

$$\sum_{i=1,3} |j(\phi : g_i ; \exp tH; g_{i+1})|_s \leq {}^0\mathbf{s}_{F,r}(\phi) \, \Xi(\exp tH)(1 + t)^r e^{-2\varepsilon_0 t}$$

for $\phi \in \mathscr{E}$, $H \in \Omega_0$ and $t \geq 0$.

Now fix $g_1, g_2 \in \mathfrak{G}$. Since $G = K \cdot ClA_0^+ \cdot K$, S^+ is compact and

$$j_\phi(g_1 : k_1^{-1} a k_2 ; g_2) = \tau(k_1^{-1}) j_\phi(g_1^{k_1} ; a; g_2^{k_2}) \tau(k_2)$$

$(k_1, k_2 \in K, a \in A_0, \phi \in \mathscr{E}')$, in order to prove Theorem 13.1, it would be enough to verify the following lemma.

Lemma 5. *Fix $g_1, g_2 \in \mathfrak{G}$ and $H_0 \in S^+$. Then we can choose a neighborhood Ω_0 of H_0 in S^+ satisfying the following condition. Given $r_0 \geq 0$, there exists $F \subset \tilde{\mathfrak{G}}$ such that*

$$|j(\phi : h_1 ; \exp tH; g_2)|_s \prec {}^0\mathbf{s}_{F,r}(\phi) \, \Xi(\exp tH)(1 + t)^{-r_0}$$

for $\phi \in \mathscr{E}'$, $H \in \Omega_0$ and $t \geq 0$.

Since

$$\mathfrak{G} = \mathfrak{K}\mathfrak{M}_1 \mathfrak{N} = \theta(\mathfrak{N}) \mathfrak{M}_1 \mathfrak{K}$$

and τ is differentiable, we may without loss of generality assume that $g_1 \in \mathfrak{M}_1 \mathfrak{N}$ and $g_2 \in \theta(\mathfrak{N}) \mathfrak{M}_1$. Then we can choose $v_i \in \mathfrak{M}_1 (i = 1, 2)$ such that

$$g_1 - v_1 \in \mathfrak{G} \mathfrak{n}, \qquad g_2 - v_2 \in \theta(\mathfrak{n}) \mathfrak{G}.$$

Our assertion now follows immediately from Lemma 3 and the corollary of Lemma 4.

This completes the proof of Theorem 13.1.

§ 16. Proof of Theorem 13.2

We shall now begin preparation for the proof of Theorem 13.2. Fix a function ϕ on $\mathfrak{F} \times G$ of type $II'(\lambda)$. We use the notation of § 10 and assume, as we may, that $P \neq G$. We also agree to the convention that the variables $v, \bar{n}$ and m shall range freely over $\mathfrak{F}, \bar{N}$ and M_1 respectively unless explicitly stated otherwise. Put

$$\Phi(v: \bar{n}: m) = d_P(m) \sum_{1 \leq i \leq q} \phi(v: \bar{n}m; \eta_i') \otimes u_i^*$$

and consider the obvious pairing [1(e), § 21] of $V \otimes U^*$, U into V given by

$$\langle v \otimes u^*, u \rangle = \langle u^*, u \rangle v \qquad (v \in V, u^* \in U^*, u \in U).$$

For any $\mathbf{b} \in \mathscr{C}(\mathfrak{F}, U) = \mathscr{C}(\mathfrak{F}) \otimes U$, define

$$\Phi(\mathbf{b}: \bar{n}: m) = \int_{\mathfrak{F}} \langle \Phi(v: \bar{n}: m), \mathbf{b}(v) \rangle \, dv$$

and put

$$F_{\mathbf{b}}(m) = F(\mathbf{b}: m) = \int_{\bar{N}} \Phi(\mathbf{b}: \bar{n}: m) \, d\bar{n}.$$

It follows from the corollary of Theorem 13.1 and [1(e), § 16] that this integral is well defined and in fact we have the following result.

Lemma 1. $\mathbf{b} \to F_{\mathbf{b}}$ *is a continuous mapping of* $\mathscr{C}(\mathfrak{F}, U)$ *into* $\mathscr{C}(M_1, \tau_M)$.

For $\zeta \in \mathfrak{Z}_1$, define $g_i(\zeta: v)$ $(1 \leq i \leq q)$ as in § 10 and put

$$\Psi_\zeta(v: \bar{n}: m) = d_P(m) \sum_{1 \leq i \leq q} \phi(v: \bar{n}m; g_i(\zeta: v)) \otimes u_i^*.$$

Lemma 2. *Let* $\zeta \in \mathfrak{Z}_1$. *Then*

$$\Phi(v: \bar{n}: m; \zeta) = \Phi(v: \bar{n}: m) \, \Gamma_v(\zeta) + \Psi_\zeta(v: \bar{n}: m).$$

This is proved in the same way as Lemma 10.1.
Now put

$$\Psi_\zeta(\mathbf{b}: \bar{n}: m) = \int \langle \Psi_\zeta(v: \bar{n}: m), \mathbf{b}(v) \rangle \, dv$$

for $\zeta \in \mathfrak{Z}_1$ and $\mathbf{b} \in \mathscr{C}(\mathfrak{F}, U)$.

Lemma 3. *Let* $\zeta \in \mathfrak{Z}_1$ *and* $\mathbf{b} \in \mathscr{C}(\mathfrak{F}, U)$. *Then*

$$\int_{\bar{N}} \Psi_\zeta(\mathbf{b}: \bar{n}: m) \, d\bar{n} = 0.$$

We know (see §10) that $v \mapsto g_i(\zeta : v)$ is a polynomial mapping of $\mathfrak{F}$ into $\bar{n}\,\mathfrak{G}$. Therefore (Corollary of Theorem 13.1) the above integral is defined and it would be enough to verify the following result.

Lemma 4. *Fix* $X \in \bar{n}$, $g \in \mathfrak{G}$ *and* $b \in \mathscr{C}(\mathfrak{F})$. *Then*

$$\int_{\bar{N}} d\bar{n} \int_{\mathfrak{F}} b(v)\, \phi(v : \bar{n}m; X g)\, dv = 0.$$

Put

$$\psi(x) = \int_{\mathfrak{F}} b(v)\, \phi(v : x; g)\, dv \qquad (x \in G).$$

Then $\psi \in \mathscr{C}(G, V)$ (Corollary of Theorem 13.1). Let

$$f(x) = \int_{\bar{N}} \psi(\bar{n}x)\, d\bar{n} \qquad (x \in G).$$

Then $[1(e), §16]\, f \in C^\infty(G, V)$ and

$$f(x; X) = \int_{\bar{N}} \psi(\bar{n}x; X)\, d\bar{n}.$$

Therefore since $f(\bar{n}x) = f(x)$ and $X^m \in \bar{n}$, we conclude that

$$f(m; X) = f(X^m ; m) = 0.$$

This proves the lemma.

For any $b \in \mathscr{C}(\mathfrak{F}, U)$ and $\zeta \in \mathfrak{Z}_1$, let $\Gamma(\zeta)\mathbf{b}$ denote the function $v \mapsto \Gamma_v(\gamma_1(\zeta))\, \mathbf{b}(v)$ from $\mathfrak{F}$ to U in the notation of §10. It is clear from Corollary 3 of Lemma 3.5 that for a fixed ζ, $\mathbf{b} \mapsto \Gamma(\zeta)b$ is a continuous endomorphism of $\mathscr{C}(\mathfrak{F}, U)$.

Lemma 5. *Let* $\mathbf{b} \in \mathscr{C}(\mathfrak{F}, U)$ *and* $\zeta \in \mathfrak{Z}_1$. *Then*

$$F(\mathbf{b} : m; \zeta) = F(\Gamma(\zeta)\, \mathbf{b} : m).$$

This is an immediate consequence of Lemmas 2 and 3.
Define $^\circ Q$ and $_i e\, (i \in {}^\circ Q)$ as in Lemma 3.4 and put

$$_i\mathbf{b}(v) = \Gamma_v(_i e(v))\, \mathbf{b}(v)$$

for $\mathbf{b} \in \mathscr{C}(\mathfrak{F}, U)$. Then it is clear from Lemmas 3.3, 3.4 and Corollary 3 of Lemma 3.5 that $\mathbf{b} \mapsto {}_i\mathbf{b}$ is a continuous endomorphism of $\mathscr{C}(\mathfrak{F}, U)$ and

$$\mathbf{b} = \sum_{i \in {}^\circ Q} {}_i\mathbf{b}.$$

Put $^\circ Q^\circ = {}^\circ Q \cap Q^\circ$ as in §10 and define

$$e^\circ(v) = \sum_{i \in {}^\circ Q^\circ} {}_i e(v),$$

$$\mathbf{b}^\circ = \sum_{i \in {}^\circ Q^\circ} {}_i\mathbf{b} \qquad (\mathbf{b} \in \mathscr{C}(\mathfrak{F}, U)).$$

Lemma 6. *Let* $\mathbf{b} \in \mathscr{C}(\mathfrak{F}, U)$ *and* $i \in {}^\circ Q$. *Then* $F(_i\mathbf{b} : m) = 0$ *unless* $i \in {}^\circ Q^\circ$. *Hence* $F(\mathbf{b} : m) = F(\mathbf{b}^\circ : m)$.

Put

$$F(\mathbf{b} : m : \mu) = \int_{\mathfrak{a}} F(\mathbf{b} : m \exp H)\, e^{-(-1)^{1/2} \mu(H)}\, dH$$

for $\mathbf{b} \in \mathscr{C}(\mathfrak{F}, U)$ and $\mu \in \mathfrak{a}^*$. (Here dH denotes the Euclidean measure on $\mathfrak{a}$ and $\mathfrak{a}^*$ the dual of $\mathfrak{a}$.) It follows from Lemma 1 that for (m, μ) fixed,

$$\mathbf{b} \to F(\mathbf{b}: m: \mu)$$

is a continuous mapping of $\mathscr{C}(\mathfrak{F}, U)$ into V. Moreover we conclude from Lemma 5 that

$$F(\Gamma(H)\mathbf{b}: m: \mu) = (-1)^{1/2} \mu(H) F(\mathbf{b}: m: \mu)$$

for $H \in \mathfrak{a}$.

Now fix $m_0 \in M_1$, $\mu \in \mathfrak{a}^*$, $i \in {}^{\circ}Q$ and put

$$T(\mathbf{b}) = F({}_i\mathbf{b}: m_0: \mu) \qquad (\mathbf{b} \in \mathscr{C}(\mathfrak{F}, U)).$$

Since $\dim U < \infty$, T may be regarded as a tempered distribution on $\mathfrak{F}$ with values in $V \otimes U^*$ (i.e. a continuous linear mapping of $\mathscr{C}(\mathfrak{F})$ into $V \otimes U^*$).

Lemma 7. *Fix* $H \in \mathfrak{a}$, $\mathbf{b} \in \mathscr{C}(\mathfrak{F}, U)$ *and put*

$$\mathbf{b}'(v) = \prod_{t \in W_0(s_i, \lambda)} \{(-1)^{1/2} \mu(H) - t\, s_i\, \Lambda_v(H)\} \cdot {}_i\mathbf{b}(v)$$

in the notation of §3. *Then* $T(\mathbf{b}') = 0$.

It follows from what we have seen above that

$$T(\Gamma(H)\mathbf{b}) = (-1)^{1/2} \mu(H) T(\mathbf{b}) \qquad (H \in \mathfrak{a}, \mathbf{b} \in \mathscr{C}(\mathfrak{F}, U)).$$

Hence our assertion is an immediate consequence of Corollary 5 of Lemma 3.5.

Now suppose $T \neq 0$. Then if $v_0 \in \mathrm{Supp}\, T$ $(v_0 \in \mathfrak{F})$, it follows from Lemma 7 that

$$\prod_{t \in W_0(s_i, \lambda)} \{(-1)^{1/2} \mu(H) - t\, s_i\, \Lambda_{v_0}(H)\} = 0$$

for all $H \in \mathfrak{a}$. Since $\Re\, t\, s_i\, \Lambda_{v_0}(H) = s_i\, \lambda^y(H)$, this implies that $i \in {}^{\circ}Q^{\circ}$. Therefore if $i \notin {}^{\circ}Q^{\circ}$, we conclude that $F({}_i\mathbf{b}: m: \mu) = 0$ for all $m \in M_1$ and $\mu \in \mathfrak{a}^*$. The statement of Lemma 6 now follows immediately by Fourier transform.

Now introduce the structure of a Hilbert space on U so that $(u_1, \ldots, u_q)$ becomes an orthonormal base. Moreover for any $E \in \mathrm{End}\, U$, let $\|E\|$ denote the Hilbert-Schmidt norm of E.

Lemma 8. *Put*

$$E(H: v) = e^{\Gamma_v(H)} \Gamma_v(e^{\circ}(v)) \qquad (H \in \mathfrak{a}).$$

Then for a given $D \in \mathfrak{D}(\mathfrak{F}_c)$, *we can choose* $c, r \geq 0$ *such that*

$$\|E(H: v; D)\| \leq c(1 + |v|)^r (1 + \|H\|)^r$$

for all $H \in \mathfrak{a}$.

Set

$$p(v) = \prod_{i \in {}^{\circ}Q^{\circ}} \varpi_{s_i, \lambda}(s_i\, \Lambda_v)$$

in the notation of §3. Then p is a polynomial function on $\mathfrak{F}$ and by Lemma 3.3,

$$|p(v)| \geq |\varpi_\lambda(\lambda)|^{q_0} > 0$$

where q_0 is the number of elements in $^\circ Q^\circ$. Put $E^\circ(v) = \Gamma_v(e^\circ(v))$. Then for a fixed $H \in \mathfrak{a}$, $v \mapsto \Gamma_v(H)$ and $v \mapsto p(v) E^\circ(v)$ are polynomial mappings of $\mathfrak{F}$ into End U (see §3). Moreover

$$E(H : v) = e^{\Gamma_v(H) E^\circ(v)} \cdot E^\circ(v)$$

and all eigenvalues of $\Gamma_v(H) E^\circ(v)$ are pure imaginary (Corollaries 2 and 5 of Lemma 3.5). Hence our assertion follows without difficulty from [1(a), Lemma 60].

Now fix $H_0 \in \mathfrak{a}$, $\mathbf{a} \in \mathscr{C}(\mathfrak{F}, U)$ and for any $t \in \mathbf{R}$, put

$$\mathbf{a}_t(v) = E(-tH_0 : v)\,\mathbf{a}(v).$$

Then it follows from Lemma 8 that $t \mapsto \mathbf{a}_t$ is a C^∞ function from $\mathbf{R}$ to $\mathscr{C}(\mathfrak{F}, U)$.

Lemma 9. *Fix* $m \in M_1$ *and* $\mu \in \mathfrak{a}^*$. *Then*

$$F(\mathbf{a} : m : \mu) = F(\mathbf{a}^\circ : m : \mu) = e^{(-1)^{1/2} t\mu(H_0)}\, F(\mathbf{a}_t : m : \mu)$$

for $t \in \mathbf{R}$.

Put

$$T(\mathbf{b}) = F(\mathbf{b} : m : \mu) \qquad (\mathbf{b} \in \mathscr{C}(\mathfrak{F}, U)).$$

Then, as we have seen above, T is a continuous linear mapping of $\mathscr{C}(\mathfrak{F}, U)$ into V and

$$T(\Gamma(H)\,\mathbf{b}) = (-1)^{1/2}\, \mu(H)\, T(\mathbf{b}) \qquad (H \in \mathfrak{a}).$$

Now let

$$f(t) = T(\mathbf{a}_t) \qquad (t \in \mathbf{R}).$$

It follows from the definition of $\mathbf{a}_t$ that

$$d\mathbf{a}_t/dt = -\Gamma(H_0)\,\mathbf{a}_t$$

and therefore

$$df/dt = -(-1)^{1/2}\,\mu(H_0)f.$$

This implies that

$$f(t) = e^{-(-1)^{1/2} t\mu(H_0)} f(0),$$

which is equivalent to the required result, if we take Lemma 6 into account.

Now assume that $H_0 \in \mathfrak{a}^+$. Then it is clear from Lemma 2 that

$$d\Phi(\mathbf{a}_t : \bar{n} : m \exp tH_0)/dt$$

$$= -\Phi(\Gamma(H_0)\,\mathbf{a}_t : \bar{n} : m \exp tH_0) + \Phi(\mathbf{a}_t : \bar{n} : m \exp tH_0 ; H_0)$$

$$= \Psi_{H_0}(\mathbf{a}_t : \bar{n} : m \exp tH_0) \qquad (t \in \mathbf{R}).$$

Put $\Psi = \Psi_{H_0}$ and for any $\mathbf{b} \in \mathscr{C}(\mathfrak{F}, U)$ and $\alpha \in C_c^\infty(\bar{N})$, define

$$\Phi(\mathbf{b} : \alpha : \bar{n}_0 : m) = \int_{\bar{N}} \alpha(\bar{n})\, \Phi(\mathbf{b} : \bar{n}_0 \bar{n} : m)\, d\bar{n},$$

$$\Psi(\mathbf{b} : \alpha : \bar{n}_0 : m) = \int_{\bar{N}} \alpha(\bar{n})\, \Psi(\mathbf{b} : \bar{n}_0 n : m)\, d\bar{n}$$

for $\bar{n}_0 \in \bar{N}$. Then the following result is obvious.

Lemma 10. $d\Phi(\mathbf{a}_t: \alpha: \bar{n}: m \ \exp tH_0)/dt = \Psi(\mathbf{a}_t: \alpha: \bar{n}: m \exp tH_0)$ *for* $\alpha \in C_c^\infty(\bar{N})$ *and* $t \in \mathbf{R}$.

Let us now put

$$\phi_b(\bar{n}: x) = \phi_b(\bar{n}x) \qquad (x \in G)$$

for $b \in \mathscr{C}(\mathfrak{F})$, and define

$$\phi_b(\alpha: \bar{n}_0: x) = \int_{\bar{N}} \alpha(\bar{n}) \, \phi_b(\bar{n}_0 \, \bar{n}: x) \, d\bar{n} \qquad (\bar{n}_0 \in \bar{N}, x \in G)$$

for $\alpha \in C_c^\infty(\bar{N})$. Then if $X \in \bar{\mathfrak{n}}$ and $g \in \mathfrak{G}$, it is clear that

$$\phi_b(\bar{n}m; X g) = \phi_b(\bar{n}; X^m: m; g).$$

Since $X^m \in \bar{\mathfrak{n}}$, it follows that

$$\phi_b(\alpha: \bar{n}: m; X g) = -\phi_b(X^m \alpha: \bar{n}: m; g).$$

Put $\beta_P(H_0) = 2\varepsilon$ so that $\varepsilon > 0$.

Lemma 11. *Fix* $m_0 \in M_1$, $\alpha \in C_c^\infty(\bar{N})$, $X \in \bar{\mathfrak{n}}$, $g \in \mathfrak{G}$, $\mathbf{s} \in \mathscr{S}(V)$ *and* $r_0 \geqq 0$. *Then we can choose a continuous seminorm* $\mathbf{t}$ *on* $\mathscr{C}(\mathfrak{F})$ *such that*

$$|\phi_b(\alpha: \bar{n}: m_t: X g)|_\mathbf{s} \leqq \mathbf{t}(b) \, e^{-2\varepsilon t} \int_\omega \Xi_{r_0}(\bar{n}\bar{n}_0 m_t) d\bar{n}_0$$

for $t \geqq 0$ *and* $b \in \mathscr{C}(\mathfrak{F})$. *Here* $m_t = m_0 \exp tH_0$,

$$\Xi_{r_0}(x) = \Xi(x)(1 + \sigma(x))^{-r_0} \qquad (x \in G)$$

and $\omega = \mathrm{Supp}\,\alpha$.

This follows from the corollary of Theorem 13.1 and the above remarks.

Corollary. *We can choose* $c \geqq 0$ *such that*

$$|\Psi(\mathbf{a}_t: \alpha: \bar{n}: m_t)|_\mathbf{s} \leqq c \, e^{-\varepsilon t} \, d_P(m_t) \int_\omega \Xi_{r_0}(\bar{n}\bar{n}_0 m_t) \, d\bar{n}_0$$

for $t \geqq 0$.

This follows from the corollary of Theorem 13.1 and the above remarks.

Now fix $\alpha \in C_c^\infty(\bar{N})$. Then it follows from the above corollary and [1(e), §10] that

$$\int_0^\infty |\Psi(\mathbf{a}_t: \alpha: \bar{n}: m \exp tH_0)|_\mathbf{s} \, dt < \infty$$

for $\mathbf{s} \in \mathscr{S}(V)$. Put

$$\Phi_\infty(\mathbf{a}: \alpha: \bar{n}: m) = \Phi(\mathbf{a}^o: \alpha: \bar{n}: m) + \int_0^\infty \Psi(\mathbf{a}_t: \alpha: \bar{n}: m \exp tH_0) \, dt.$$

Then it follows from Lemma 10 that

$$\Phi_\infty(\mathbf{a}: \alpha: \bar{n}: m) = \lim_{t \to +\infty} \Phi(\mathbf{a}_t: \alpha: \bar{n}: m \exp tH_0).$$

Lemma 12. *Fix* $\alpha \in C_c^\infty(\bar{N})$ *such that* $\int_{\bar{N}} \alpha(\bar{n}) \, d\bar{n} = 1$. *Then*

$$F(\mathbf{a}: m) = \int_{\bar{N}} \Phi_\infty(\mathbf{a}: \alpha: \bar{n}: m) \, d\bar{n}.$$

It follows from [1 (e), §10] and the corollary of Lemma 11 that

$$\int\limits_{N} d\bar{n} \int\limits_{0}^{\infty} |\Psi(\mathbf{a}_t: \alpha: \bar{n}: m \exp tH_0)|_{\mathbf{s}}\, dt < \infty$$

for $s \in \mathscr{S}(V)$. Therefore we conclude from the corollary of Theorem 13.1 that

$$\int\limits_{N} |\Phi_{\infty}(\mathbf{a}: \alpha: \bar{n}: m)|_{\mathbf{s}}\, d\bar{n} < \infty.$$

On the other hand it is clear from Lemma 3 that

$$\int\limits_{N} \Psi(\mathbf{a}_t: \alpha: \bar{n}: m \exp tH_0)\, d\bar{n} = 0.$$

Therefore by Fubini's theorem we obtain

$$\int\limits_{N} \Phi_{\infty}(\mathbf{a}: \alpha: \bar{n}: m)\, d\bar{n} = \int\limits_{N} \Phi(\mathbf{a}^o: \alpha: \bar{n}: m)\, d\bar{n}$$

$$= F(\mathbf{a}^o: m) = F(\mathbf{a}: m)$$

from Lemma 6.

Now put

$$\Phi^o(v: \bar{n}: m) = \Phi(v: \bar{n}: m)\, E^{o*}(v),$$

$$\Psi^o(v: \bar{n}: m) = \Psi(v: \bar{n}: m)\, E^{o*}(v)$$

where $E^{o*}(v) = 1 \otimes \Gamma_v^*(e^o(v))$. Then

$$\Phi(\mathbf{a}_t: \bar{n}: m \exp tH_0) = \int\limits_{\mathfrak{F}} \langle \Phi(v: \bar{n}: m \exp tH_0), \mathbf{a}_t(v) \rangle\, dv$$

$$= \int\limits_{\mathfrak{F}} \langle \Phi^o(v: \bar{n}: m \exp tH_0)\, e^{-t\Gamma_v(H_0)}, \mathbf{a}(v) \rangle\, dv.$$

Lemma 13. *Fix* $x \in G$, $X \in \bar{\mathfrak{n}}$, $g \in \mathfrak{G}$ *and* $s \in \mathscr{S}(V)$. *Then we can choose* $c, r \geq 0$ *such that*

$$|\phi(v: x \exp tH_0; X g)|_{\mathbf{s}} \leq c(1 + |v|)^r\, e^{-2\varepsilon t}\, \Xi(x \exp tH_0)(1 + t)^r$$

for $t \geq 0$.

This follows immediately from the fact that

$$\phi(v: x_t; X g) = \phi(v: \mathrm{Ad}(x_t) X : x_t; g)$$

where $x_t = x \exp tH_0$.

Corollary. *Fix* $\bar{n} \in \bar{N}$, $m \in M_1$ *and* $s \in \mathscr{S}(V)$. *Then we can choose* $c, r \geq 0$ *such that*

$$|\Psi^o(v: \bar{n}: m \exp tH_0)\, e^{-t\Gamma_v(H_0)}|_{\mathbf{s}} \leq c\, e^{-\varepsilon t}(1 + |v|)^r$$

for $t \geq 0$.

This is an immediate consequence of Lemmas 13 and 8.

On the other hand it follows from Lemma 2 that

$$\Phi^o(v: \bar{n}: m \exp TH_0)\, e^{-T\Gamma_v(H_0)}$$

$$= \Phi^o(v: \bar{n}: m) + \int\limits_{0}^{T} \Psi^o(v: \bar{n}: m \exp tH_0)\, e^{-t\Gamma_v(H_0)}\, dt.$$

Moreover we conclude from the above corollary that

$$\int_0^\infty \left| \Psi^o(v:\bar n: m \exp tH_0)\, e^{-t\Gamma_v(H_0)} \right|_{\mathbf s} dt < \infty$$

for $\mathbf s \in \mathscr{S}(V)$. Therefore if we put

$$\Phi_\infty(v:\bar n: m) = \Phi^o(v:\bar n: m) + \int_0^\infty \Psi^o(v:\bar n: m \exp tH_0)\, e^{-t\Gamma_v(H_0)}\, dt,$$

it follows that

$$\left| \Phi_\infty(v:\bar n: m) - \Phi^o(v:\bar n: m \exp TH_0)\, e^{-T\Gamma_v(H_0)} \right|_{\mathbf s}$$

$$\leq \int_T^\infty \left| \Psi^o(v:\bar n: m \exp tH_0)\, e^{-t\Gamma_v(H_0)} \right|_{\mathbf s} dt$$

for $T \geq 0$ and $\mathbf s \in \mathscr{S}(V)$. Hence we get the following result from the corollary of Lemma 13.

Lemma 14. *Fix $\bar n \in \bar N$, $m \in M$ and put*

$$\Phi_\infty(v:\bar n: m) = \lim_{t \to +\infty} \Phi^o(v:\bar n: m \exp tH_0)\, e^{-t\Gamma_v(H_0)}.$$

Then for any $\mathbf s \in \mathscr{S}(V)$, we can choose $c, r \geq 0$ such that

$$\left| \Phi_\infty(v:\bar n: m) - \Phi^o(v:\bar n: m \exp TH_0)\, e^{-T\Gamma_v(H_0)} \right|_{\mathbf s} \leq c\, e^{-\varepsilon T}(1+|v|)^r$$

for $T \geq 0$.

Now define

$$\Phi_\infty(\mathbf a:\bar n: m) = \lim_{t \to +\infty} \Phi(\mathbf a_t:\bar n: m \exp tH_0).$$

It is clear that this limit exists and in fact

$$\Phi_\infty(\mathbf a:\bar n: m) = \lim_{t \to +\infty} \int_{\mathfrak F} \langle \Phi^o(v:\bar n: m \exp tH_0)\, e^{-t\Gamma_v(H_0)}, \mathbf a(v) \rangle\, dv$$

$$= \int_{\mathfrak F} \langle \Phi_\infty(v:\bar n: m), \mathbf a(v) \rangle\, dv.$$

On the other hand, let us put

$$\Phi_\infty(v: m) = \Phi_\infty(v: 1: m)$$

$$= \lim_{t \to +\infty} \Phi(v: m \exp tH_0)\, E^{o*}(v)\, e^{-t\Gamma_v(H_0)}.$$

Extend this to a function on $\mathfrak F \times G$ by setting

$$\Phi_\infty(v: kmn) = \tau(k)\, \Phi_\infty(v: m) \qquad (k \in K,\, m \in M_1,\, n \in N).$$

Lemma 15. $\Phi_\infty(v:\bar n: m) = e^{-\rho(H(\bar n))}\, \Phi_\infty(v: \bar n m).$

It is obvious from Lemma 14 that for fixed $\bar n$ and m, $\Phi(v:\bar n: m)$ is a continuous function of v. Therefore, in view of its definition, the same holds for $\Phi_\infty(v: \bar n m)$. Hence it would be enough to verify the above relation for $v \in \mathfrak F'(\lambda)$.

Fix $v \in \mathfrak F'(\lambda)$ and let $e_i^*(v)$ $(1 \leq i \leq q)$ be the base of U^* dual to $e_i(v)$ $(1 \leq i \leq q)$. Then

$$\sum_i u_i \otimes u_i^* = \sum_i e_i(v) \otimes e_i^*(v).$$

Hence

$$\sum_i u_i \otimes u_i^* \, \Gamma_v^*(e^o(v)) \, e^{-t\Gamma_v^*(H_0)} = \sum_i e^{-t\Gamma_v(H_0)} \, \Gamma_v(e^o(v)) \, e_i(v) \otimes e_i^*(v)$$

$$= \sum_{i \in Q^o} e^{-t s_i \Lambda_v(H_0)} \, e_i(v) \otimes e_i^*(v)$$

from Corollary 1 of Lemma 3.5. Hence

$$\Phi^o(v: \bar{n}: m_t) \, e^{-t\Gamma_v(H_0)}$$

$$= d_P(m_t) \sum_{i \in Q^o} \varpi_{01}(s_i \Lambda_v)^{-1} \, e^{-t s_i \Lambda_v(H_0)} \, \phi(v: \bar{n}m_t; \eta_i(v)') \otimes e_i^*(v),$$

where $m_t = m \exp t H_0$ and ϖ_{01}, $\eta_i(v)$ have the same meaning as in §2 and §4 respectively.

Now fix $i \in Q^o$. Then (see §6)

$$\phi_{i\infty}(v: m) = \lim_{t \to +\infty} d_P(m_t) \, \phi(v: m_t; \eta_i(v)') \, e^{-t s_i \Lambda_v(H_0)}$$

and

$$\phi_{i\infty}(v: m \exp H) = \phi_{i\infty}(v: m) \, e^{s_i \Lambda(H)} \qquad (H \in \mathfrak{a})$$

from Lemma 6.2. Therefore

$$\lim_{t \to +\infty} \{ d_P(m_t) \, \phi(v: m_t; \eta_i(v)') - \phi_{i\infty}(v: m_t) \} = 0$$

and we conclude from [1 (e), Lemmas 21.3 and 24.1] that

$$\phi_P(v: m; \eta_i(v)) = \phi_{i\infty}(v: m).$$

Extend $\phi_{i\infty}(v)$ to a function on G by setting

$$\phi_{i\infty}(v: kmn) = \tau(k) \, \phi_{i\infty}(v: m) \qquad (k \in K, \, m \in M_1, \, n \in N).$$

Then we conclude from [1 (e), Lemma 24.1] that

$$\lim_{t \to +\infty} \{ d_P(x_t) \, \phi(v: x_t; \eta_i(v)') - \phi_{i\infty}(v: x_t) \} = 0.$$

Here x is a fixed element in G and $x_t = x \exp t H_0$. But this implies that

$$\lim_{t \to +\infty} d_P(x_t) \, \phi(v: x_t; \eta_i(v)') \, e^{-t s_i \Lambda_v(H_0)} = \phi_{i\infty}(v: x)$$

and therefore

$$\Phi_\infty(v: \bar{n}: m) = \lim_{t \to +\infty} \Phi^o(v: \bar{n}: m_t) \, e^{-t\Gamma_v(H_0)}$$

$$= e^{-\rho(H(\bar{n}))} \sum_{i \in Q^o} \varpi_{01}(s_i \Lambda_v)^{-1} \, \phi_{i\infty}(v: \bar{n}m) \otimes e_i^*(v).$$

The assertion of the lemma is now obvious from the definition of $\Phi_\infty(v: m)$.

Lemma 16. $e_j^*(v) = \sum_{1 \leq i \leq q} u_i(s_j \Lambda_v) \, u_i^* \quad (1 \leq j \leq q)$ for $v \in \mathfrak{F}_c'(\lambda)$.

By Corollary 1 of Lemma 3.5

$$u_i = \Gamma_v(u_i) 1 = \sum_{1 \leq j \leq q} \Gamma_v(u_i) \, e_j(v)$$

$$= \sum_j u_i(s_i \Lambda_v) \, e_j(v).$$

Therefore

$$\sum_i u_i \otimes u_i^* = \sum_{i,j} u_i(s_j \Lambda_v)\, e_j(v) \otimes u_i^*.$$

But since

$$\sum_i u_i \otimes u_i^* = \sum_j e_j(v) \otimes e_j^*(v),$$

our assertion is now obvious.

Corollary. *Let* $v \in \mathfrak{F}'(\lambda)$. *Then*

$$\Phi_\infty(v:m) = \sum_{i \in Q^o} \sum_{1 \le j \le q} \varpi_{01}(s_i \Lambda_v)^{-1}\, \phi_{i\infty}(v:m) \otimes u_j(s_i \Lambda_v)\, u_j^*.$$

This follows immediately from Lemma 16 and what we have seen above.
Now put

$$\Phi_\infty(\mathbf{a}:m) = \int_{\mathfrak{F}} \langle \Phi_\infty(v:m), \mathbf{a}(v) \rangle \, dv.$$

Then it follows from Lemmas 7.3, 8.2 and the corollary of Theorem 13.1 that
$\Phi_\infty(\mathbf{a}) \in \mathscr{C}(M_1, \tau_M)$. Hence we conclude from [1(c), Lemma 32.1] that

$$\int_N e^{-\rho(H(\bar n))} |\Phi_\infty(\mathbf{a}:\bar n m)|_s \, d\bar n < \infty \qquad (s \in \mathscr{S}(V)),$$

provided $\Phi_\infty(\mathbf{a})$ is extended to a function on G in the usual way so that

$$\Phi_\infty(\mathbf{a}:kmn) = \tau(k)\, \Phi_\infty(\mathbf{a}:m) \qquad (k \in K, m \in M_1, n \in N).$$

Now put, as before,

$$\Phi_\infty(\mathbf{a}:\bar n:m) = \int_{\mathfrak{F}} \langle \Phi_\infty(v:\bar n:m), \mathbf{a}(v) \rangle \, dv.$$

Then it follows from Lemma 15 that

$$\Phi_\infty(\mathbf{a}:\bar n:m) = e^{-\rho(H(\bar n))}\, \Phi_\infty(\mathbf{a}:\bar n m)$$

and therefore from Lemma 12 that

$$F(\mathbf{a}:m) = \int_N e^{-\rho(H(\bar n))}\, \Phi_\infty(\mathbf{a}:\bar n m)\, d\bar n.$$

Substituting the definition of $F(\mathbf{a})$ we obtain the following result.

Lemma 17. *Let* $\mathbf{a} \in \mathscr{C}(\mathfrak{F}, U)$. *Then*

$$\int_N \Phi(\mathbf{a}:\bar n:m)\, d\bar n = \int_N e^{-\rho(H(\bar n))}\, \Phi_\infty(\mathbf{a}:\bar n m)\, d\bar n.$$

In order to prove Theorem 13.2 we take $\mathbf{a}(v) = \alpha(v)\, u_1$. Then we claim that

$$\langle \Phi_\infty(v:m), \mathbf{a}(v) \rangle = \alpha(v)\, \phi_P(v:m).$$

Since both sides are continuous in v, it is sufficient to verify this for $v \in \mathfrak{F}'(\lambda)$. But
$u_1 = 1$ and so this is an immediate consequence of Lemma 7.3 and the corollary
of Lemma 16. The statement of Theorem 13.2 is now obvious from Lemma 17.

§ 17. Application to Eisenstein Integrals

Let U be an open subset of $\mathfrak{F}_c$. A function $f: U \times G \to V$ will be said to be of type $H \times C^\infty$ if 1) it is of class C^∞ on $U \times G$ and 2) for all $x \in G$ the function $v \to f(v: x)$ from U to V is holomorphic.

Fix a psgp $P_1 = MAN_1$ in $\mathscr{P}(\mathfrak{h}_R)$. Then for any $\psi \in C^\infty(M, \tau_M)$, we consider the Eisenstein integral $E(P_1: \psi)$ [1(e), §9]. Clearly it is a function of type $H \times C^\infty$ on $\mathfrak{F}_c \times G$.

Put

$$\mathbf{s}_\delta(\psi) = \sup_M |\delta \psi|_\mathbf{s} \, \Xi_M^{-1}$$

for $\mathbf{s} \in \mathscr{S}(V)$, $\delta \in \mathfrak{M}^{(2)} = \mathfrak{M} \otimes \mathfrak{M}$ and $\psi \in C^\infty(M, V)$. Moreover let

$$\mathbf{s}_F(\psi) = \sum_{\delta \in F} \mathbf{s}_\delta(\psi)$$

for any finite subset F of $\mathfrak{M}^{(2)}$. If $v \in \mathfrak{F}_c$, define v_R and v_I in $\mathfrak{F}$ by $v = v_R + (-1)^{1/2} v_I$. Then it is easy to see that we can choose $c_0 \geq 0$ such that

$$|\mathfrak{R}(-1)^{1/2} v(H_{P_1}(x))| \leq c_0 |v_I| \, \sigma(x) \qquad (v \in \mathfrak{F}_c, \ x \in G).$$

Extend the norm on $\mathfrak{F}_c$ by setting

$$|v|^2 = |v_R|^2 + |v_I|^2$$

and put

$$|(v, x)| = (1 + |v|)(1 + \sigma(x)) \qquad (v \in \mathfrak{F}_c, \ x \in G).$$

Lemma 1. *Fix* $g_1, g_2 \in \mathfrak{G}$ *and* $D \in \mathfrak{D}(\mathfrak{F}_c)$. *Then we can choose* $r \geq 0$ *and a finite subset* F *of* $\mathfrak{M}^{(2)}$ *with the following property. For any* $\mathbf{s} \in \mathscr{S}(V)$, *there exists a number* $c > 0$ *such that*

$$|E(P_1: \psi: v; D: g_1 : x; g_2)|_\mathbf{s} \leq c \, \mathbf{s}_F(\psi) \, \Xi(x) \, |(v, x)|^r \exp \{c_0 |v_I| \, \sigma(x)\}$$

for all $\psi \in C^\infty(M, \tau_M)$, $v \in \mathfrak{F}_c$ *and* $x \in G$.

It is enough to consider the case $D = 1$. The general result would follow from this if we fix x, consider the complex polycylinder with center v and radius $(1 + \sigma(x))^{-1}$ and apply the Cauchy integral formula.

We drop the subscript and write $P = P_1$, $N = N_1$. Put

$$\psi_v(x) = \psi(x) \exp \{((-1)^{1/2} v - \rho)(H(x))\} \qquad (x \in G)$$

in the usual notation [1(e), §19] where $\rho = \rho_P$ and $H(x) = H_P(x)$. Then it is obvious that

$$|E(P: \psi: v: g_1 : x; g_2)|_\mathbf{s} \leq \int_K |\psi_v(g_1 : xk; g_2^k)|_\mathbf{s} \, dk$$

for $\mathbf{s} \in \mathscr{S}^0(V)$ [1(e), §22]. But if $x = kman$ ($k \in K$, $m \in M$, $a \in A$, $n \in N$), it is clear that

$$\psi_v(g_1 : kman; g_2) = \tau(k) \psi_v(g_1^{k^{-1}} ; man; g_2).$$

Moreover

$$\mathfrak{G} = \mathfrak{G} \, \mathfrak{n} + \mathfrak{K} \mathfrak{M}_1 = \mathfrak{G} \, \mathfrak{n} + \mathfrak{M}_1 \, \mathfrak{K}.$$

Therefore for given $g_1, g_2 \in \mathfrak{G}$, we can choose $r \geq 0$ and $u_i, v_i \in \mathfrak{M}$ $(1 \leq i \leq p)$ such that

$$|\psi_v(g_1 : kman; g_2)|_\mathbf{s} \leq \sum_{1 \leq i \leq p} |\psi(u_i \; m; v_i)|_\mathbf{s} (1 + |v|)^r \, e^{-(v_I + \rho)(\log a)}$$

for all $v \in \mathfrak{F}_c$, $\mathbf{s} \in \mathscr{S}^\circ(V)$ and $(k, m, a, n) \in K \times M \times A \times N$. The required result now follows immediately from [1(e), Corollary of Lemma 30.1].

Let $\psi \neq 0$ be an eigenfunction of $\mathfrak{Z}_M$ in $\mathscr{C}(M, \tau_M)$. Then [1(e), Theorem 18.3] there exists a regular element $\lambda \in (-1)^{1/2} \, \mathfrak{h}_I^*$ such that

$$\zeta \psi = \gamma_{\mathfrak{m}/\mathfrak{h}_I}(\zeta : \lambda) \psi \qquad (\zeta \in \mathfrak{Z}_M).$$

Put $\phi = E(P_1 : \psi)$. Then it is obvious from Lemma 1 and [1(e), Lemma 19.1] that ϕ defines a function of type $II(\lambda)$ (see §8) on $\mathfrak{F} \times G$.

Let $P = MAN$ be another psgp in $\mathscr{P}(\mathfrak{h}_R)$. We shall now investigate the behavior of

$$d_P(a) \, \phi(v : ma) \qquad (m \in M, a \in A)$$

as $a \xrightarrow{P} \infty$. The case $P = G$ being trivial, we assume that $\mathfrak{a} = \mathfrak{h}_R$ does not lie in the center of $\mathfrak{g}$.

We now use the notation of §5 and put

$$|(v, x, H)| = |(v, x)| (1 + \|H\|)$$

for $v \in \mathfrak{F}_c$, $x \in G$ and $H \in \mathfrak{a}$. Note that $\mathfrak{a} = \mathfrak{h}_R \subset \mathfrak{a}_0$ and therefore we may assume that $k_0 = 1$ (see §3). Then y centralizes $\mathfrak{a}$ and therefore $\Lambda_v = \lambda^y + (-1)^{1/2} v$.

Lemma 2. *Fix* $\zeta \in \mathfrak{Z}_1$ *and* $v_1, v_2 \in \mathfrak{M}$. *Then we can choose* $r \geq 0$ *and for each* $\mathbf{s} \in \mathscr{S}(V)$ *a number* $c(\mathbf{s}) \geq 0$ *such that*

$$|\psi_{i, \zeta}(v : v_1 \; m \exp H; v_2)|_\mathbf{s}$$
$$\leq c(\mathbf{s}) \, \Xi_M(m) \, |(v, m, H)|^r \, e^{-\beta_P(H)} \exp \{c_0 |v_I| (\sigma(m) + \|H\|)\}$$

for $v \in \mathfrak{F}_c$, $m \in M_1^+$, $H \in Cl \, \mathfrak{a}^+$ *and* $1 \leq i \leq q$.

This is proved in the same way as Lemma 6.1.

Define $\lambda_i (i \in Q)$ and Q° as in §6. Fix two positive numbers ε, δ and an element $H_0 \in \mathfrak{a}^+$ with $\|H_0\| = 1$. Let $\mathfrak{F}_c(\delta)$ denote the set of all $v \in \mathfrak{F}_c$ with $|v_I| < \delta$. By choosing ε, δ sufficiently small, we can assume that:

1) $\beta_P(H_0) \geq 4\varepsilon$,

2) $|\lambda_i(H_0)| \geq 3\varepsilon$ if $\lambda_i(H_0) \neq 0$,

3) $|s_i \, v_I(H_0)| + c_0 |v_I| \leq \varepsilon$

for $i \in Q$ and $v \in \mathfrak{F}_c(\delta)$. Put

$$\psi_i^\circ(v : m : t) = \psi_{i, H_0}(v : m \exp tH_0) \, e^{-ts_i \Lambda_v(H_0)}.$$

Fix $\mathbf{s} \in \mathscr{S}(V)$ and $v \in \mathfrak{M}_1$. Then it follows from Lemma 2 that if $\lambda_i(H_0) \geq 0$, the integral

$$\int_0^\infty |\psi_i^\circ(v : m; v : t)|_\mathbf{s} \, dt$$

converges uniformly as (v, m) varies within a compact subset of $\mathfrak{F}_c(\delta) \times M_1$. Hence by Lemma 5.3, we can define

$$\phi_{i\infty}(v: m) = \lim_{t \to +\infty} \phi_i(v: m \exp tH_0)\, e^{-ts_i \Lambda_v(H_0)}$$

for $(v, m) \in \mathfrak{F}_c(\delta) \times M_1$. Then $\phi_{i\infty}$ is a function of type $H \times C^\infty$.

Lemma 3. *Fix i such that $\lambda_i(H_0) \geq 0$. Then*

$$\phi_{i\infty}(v: m; \zeta) = \gamma_1(\zeta: s_i \Lambda_v)\, \phi_{i\infty}(v: m) \qquad (\zeta \in \mathfrak{Z}_1)$$

for $v \in \mathfrak{F}_c(\delta)$ and $m \in M_1$. Moreover $\phi_{i\infty} = 0$ unless $i \in Q^o$ and $s_i^{-1} \mathfrak{a} = \mathfrak{a}$.

If $\lambda_i(H_0) > 0$, it is clear that

$$\Re s_i \Lambda_v(H_0) - c_0 |v_I| > 0$$

for $v \in \mathfrak{F}_c(\delta)$ and therefore $\phi_{i\infty} = 0$ from Lemma 1. So now assume that $\lambda_i(H_0) = 0$. Then it follows easily from Lemmas 6.2 and 6.3 that our statement is true if $v \in \mathfrak{F}$. The rest is obvious by holomorphy.

Corollary. $\phi_{i\infty}(v: m \exp H) = \phi_{i\infty}(v: m)\, e^{s_i \Lambda_v(H)}$ *for $m \in M_1$, $H \in \mathfrak{a}$ and $v \in \mathfrak{F}_c(\delta)$.*

This is obvious from Lemma 3.
Define $\phi_{i\infty} = 0$ if $\lambda_i(H_0) < 0$.

Lemma 4. *Fix $v \in \mathfrak{F}_c(\delta)$, $m \in M_1$. Then*

$$|\phi_i(v: m \exp TH_0) - \phi_{i\infty}(v: m \exp TH_0)|_{\mathbf{s}}$$

$$\leqq e^{-2\varepsilon T} \left\{ |\phi_i(v: m)|_{\mathbf{s}} + \int_0^\infty |\psi_{i, H_0}(v: m \exp tH_0)|_{\mathbf{s}}\, e^{2\varepsilon t}\, dt \right\}$$

for $\mathbf{s} \in \mathscr{S}(V)$, $T \geqq 0$ and $i \in Q$.

Put $m_t = m \exp tH_0$ $(t \in \mathbf{R})$ and first suppose $\lambda_i(H_0) \geqq 0$. Then

$$\phi_{i\infty}(v: m_T) = \phi_i(v: m_T) + \int_T^\infty \psi_{i, H_0}(v: m_t)\, e^{-(t-T)s_i \Lambda_v(H_0)}\, dt$$

from Lemma 5.3. Moreover

$$\Re s_i \Lambda_v(H_0) = \lambda_i(H_0) - s_i v_I(H_0) \geqq -\varepsilon.$$

Hence

$$|\phi_{i\infty}(v: m_T) - \phi_i(v: m_T)|_{\mathbf{s}} \leqq \int_T^\infty |\psi_{i, H_0}(v: m_t)|\, e^{\varepsilon(t-T)}\, dt$$

and this implies the required inequality.
Now suppose $\lambda_i(H_0) < 0$. Then $\phi_{i\infty} = 0$ and

$$\phi_i(v: m_T) = \phi_i(v: m)\, e^{T s_i \Lambda_v(H_0)} + \int_0^T \psi_{i, H_0}(v: m_t)\, e^{(T-t)s_i \Lambda_v(H_0)}\, dt$$

from Lemma 5.3. But

$$\Re s_i \Lambda_v(H_0) = \lambda_i(H_0) - s_i v_I(H_0) \leqq -2\varepsilon$$

and therefore

$$|\phi_i(v: m \exp T H_0)|_{\mathbf{s}} \leq e^{-2\varepsilon T}\left\{|\phi_i(v: m)|_{\mathbf{s}} + \int_0^\infty |\psi_{i, H_0}(v: m \exp t H_0)|_{\mathbf{s}}\, e^{2\varepsilon t}\, dt\right\}.$$

This proves the lemma.

Let $\mathfrak{F}'_c(\delta, \lambda)$ denote the set of all $v \in \mathfrak{F}_c(\delta)$ where $\varpi(\lambda + (-1)^{1/2} v) \neq 0$, so that (see §3)

$$\mathfrak{F}'_c(\delta, \lambda) = \mathfrak{F}_c(\delta) \cap \mathfrak{F}'_c(\lambda).$$

For any $s \in \mathfrak{w} = \mathfrak{w}(\mathfrak{a})$, there exists a unique index $i \in Q$ such that $s = s_i^{-1}$ on $\mathfrak{a}$ (Lemma 7.1). Define

$$\phi_{P, s}(v: m) = \varpi_{01}(s_i \Lambda_v)^{-1}\, \phi_{i\infty}(v: m)$$

for $v \in \mathfrak{F}'_c(\delta, \lambda)$, $m \in M_1$. Note that

$$s_i \Lambda_v(H) = \Lambda_v(sH) = \lambda^y(sH) + (-1)^{1/2} v(sH)$$
$$= (-1)^{1/2} v(sH) \quad (H \in \mathfrak{a})$$

since y centralizes $\mathfrak{a}$ and $\lambda = 0$ on $\mathfrak{a} = \mathfrak{h}_R$. This shows that $i \in Q^o$. Therefore the following result is obvious from Lemmas 2 and 4, Corollary of Lemma 3 and Lemma 5.1.

Lemma 5. *Let $v \in \mathfrak{F}'_c(\delta, \lambda)$, $m \in M_1$ and $s \in \mathscr{S}(V)$. Then*

$$\lim_{t \to +\infty} e^{\varepsilon t}\left|d_P(m_t)\, \phi(v: m_t) - \sum_{s \in \mathfrak{w}} \phi_{P, s}(v: m)\, e^{(-1)^{1/2} t v(s H_0)}\right|_{\mathbf{s}} = 0$$

where $m_t = m \exp t H_0$.

Corollary. *Fix $v \in \mathfrak{F}'_c(\delta, \lambda)$ and $s_0 \in \mathfrak{w}$ and suppose $v_I(s_0 H_0) < v_I(s H_0)$ for every $s \neq s_0$ in $\mathfrak{w}$. Then*

$$\lim_{t \to +\infty} d_P(m_t)\, \phi(v: m_t)\, e^{-(-1)^{1/2} t v(s_0 H_0)} = \phi_{P, s_0}(v: m)$$

for $m \in M_1$.

Since $|v_I(s_0 H_0)| \leq \varepsilon$, this follows from Lemma 5 if we observe that

$$\Re(-1)^{1/2}\{v(s H_0) - v(s_0 H_0)\} = v_I(s_0 H_0) - v_I(s H_0) < 0$$

for $s \neq s_0$.

§ 18. The *c*-Functions

Now assume that $\dim \tau < \infty$ and τ is unitary. Put

$$L = {}^o\mathscr{C}(M, \tau_M).$$

Then by [1 (e), Theorem 27.9], $\dim L < \infty$. Let $|\cdot|$ denote the norm in the finite-dimensional Hilbert space V. Put

$$\|\psi\|^2 = \int_M |\psi(m)|^2\, dm$$

for $\psi \in L$. This defines the structure of a Hilbert space on L.

Let $\mathscr{E}_2(M)$ be the discrete series of M (i.e. the set of all equivalence classes of irreducible, square-integrable representations of M). For $\omega \in \mathscr{E}_2(M)$, put

$$L(\omega) = L \cap (\mathfrak{H}_\omega \otimes V)$$

where $\mathfrak{H}_\omega$ is the smallest closed subspace of $L_2(M)$ containing all the matrix coefficients of ω. Then

$$L = \sum_\omega L(\omega)$$

where the sum is orthogonal.

We keep to the notation of §17 and put $\mathfrak{a} = \mathfrak{h}_R$ and $\mathfrak{w} = \mathfrak{w}(\mathfrak{a})$. Fix $P \in \mathscr{P}(\mathfrak{a})$ and define

$$\pi(v) = \prod_{1 \leq i \leq r} \langle \alpha_i, v \rangle^{m_i} \qquad (v \in \mathfrak{F}_c)$$

where $\alpha_1, \ldots, \alpha_r$ are all the distinct roots of (P, A) and m_i the multiplicity of α_i. As usual $\langle \alpha_1, v \rangle = \alpha_i(H_v)$. Let $\mathfrak{F}'_c$ be the set of all $v \in \mathfrak{F}_c$ where $\pi(v) \neq 0$. Clearly $\mathfrak{F}'_c$ is independent of the choice of P in $\mathscr{P}(\mathfrak{a})$. Put $\mathfrak{F}' = \mathfrak{F} \cap \mathfrak{F}'_c$ and $\mathfrak{F}'_c(\delta) = \mathfrak{F}_c(\delta) \cap \mathfrak{F}'_c$ for $\delta > 0$.

Theorem 1. *Fix* $v \in \mathfrak{F}'$ *and* $P_1, P_2 \in \mathscr{P}(\mathfrak{a})$. *Then there exist unique elements* $c_{P_2 | P_1}(s: v) \in \mathrm{End}\, L$ $(s \in \mathfrak{w})$ *such that*

$$E_{P_2}(P_1: \psi: v: ma) = \sum_{s \in \mathfrak{w}} (c_{P_2 | P_1}(s: v) \psi)(m)\, e^{(-1)^{1/2} sv(\log a)}$$

for $\psi \in L$, $m \in M$ *and* $a \in A$. *Moreover we can choose* $\delta > 0$ *such that for every* $s \in \mathfrak{w}$, $\pi(v)\, c_{P_2 | P_1}(s: v)$ *extends to a holomorphic function of* v *on* $\mathfrak{F}_c(\delta)$.

Fix $v \in \mathfrak{F}'$. Then $sv \neq v$ for $s \neq 1$ in $\mathfrak{w}$ (Lemma 22.3). Hence the uniqueness is obvious. So now we have to prove existence. Fix $\omega \in \mathscr{E}_2(M)$ such that $L(\omega) \neq \{0\}$. It is enough to define $c_{P_2 | P_1}(s: v)$ on $L(\omega)$. By [1(e), Theorem 18.3] there exists a regular element $\lambda \in (-1)^{1/2} \mathfrak{h}_I^*$ such that

$$\zeta \psi = \gamma_{m/\mathfrak{h}_I}(\zeta: \lambda) \psi \qquad (\zeta \in \mathfrak{Z}_M)$$

for all $\psi \in L(\omega)$. Now fix $\psi \in L(\omega)$ and put $\phi = E(P_1: \psi: v)$. It is easy to verify that $\mathfrak{F}'(\lambda) \supset \mathfrak{F}'$ and therefore by Theorem 7.1

$$\phi_{P_2} = \sum_{s \in \mathfrak{w}} \phi_{P_2, s}.$$

Moreover by Lemma 7.5 the functions

$$m \longmapsto \phi_{P_2, s}(m) \qquad (m \in M)$$

are in L. Now define

$$c_{P_2 | P_1}(s^{-1}: v) \psi = \phi_{P_2, s} \qquad (s \in \mathfrak{w}).$$

Then the first statement of the theorem follows from Theorem 7.1 and its corollary.

For any linear function μ on V and $m \in M$, put

$$\mu_m(\psi) = \mu(\psi(m)) \qquad (\psi \in L).$$

Then μ_m is a linear function on L. For a given $\psi \in L$, the condition $\mu_m(\psi)=0$ for all μ and m, implies that $\psi=0$. Hence we can choose a base $(\Lambda_1, \ldots, \Lambda_n)$ for the space dual to L, consisting of linear functions of the form μ_m. Let $(\psi_1, \ldots, \psi_n)$ be the dual base for L. For each i, choose $m_i \in M$ and a linear function μ_i on V such that $\Lambda_i(\psi)=\mu_i(\psi(m_i))$ for $\psi \in L$. Then

$$\psi = \sum_i \Lambda_i(\psi)\,\psi_i = \sum_i \mu_i(\psi(m_i))\,\psi_i \qquad (\psi \in L).$$

Now fix ω and $\psi \in L(\omega)$ as above and put

$$\psi_s(v) = \varpi(\lambda + (-1)^{1/2}\,v)\,\phi_{P_2,s}(v)$$

for $v \in \mathfrak{F}_c(\delta)$ in the notation of §17 where

$$\phi(v) = E(P_1 : \psi : v).$$

Then for a fixed $s \in \mathfrak{w}$, the function

$$(v, m) \longmapsto \psi_s(v : m)$$

on $\mathfrak{F}_c(\delta) \times M$ is of class $H \times C^\infty$. Moreover $\psi_s(v) \in L$ for $v \in \mathfrak{F}'$. Hence

$$\psi_s(v : m) = \sum_i \mu_i(\psi_s(v : m_i))\,\psi_i(m) \qquad (m \in M)$$

for $v \in \mathfrak{F}'$. Therefore by holomorphy this relation holds for all $v \in \mathfrak{F}_c(\delta)$. This shows that $\psi_s(v) \in L$ and $v \to \psi_s(v)$ is a holomorphic mapping from $\mathfrak{F}_c(\delta)$ to L. The second statement of Theorem 1 is now obvious.

We observe that $\mathfrak{w}$ operates on L. For if $s \in \mathfrak{w}$ and $\psi \in L$, then $s\psi = \psi^s$ (see §7) is also in L. Clearly the sets $\mathfrak{F}_c(\delta)$ and $\mathfrak{F}'$ are also stable under $\mathfrak{w}$.

Lemma 1. *Let* $P_1, P_2 \in \mathscr{P}(\mathfrak{a})$ *and* $s, t \in \mathfrak{w}$. *Then*

$$s\,c_{P_2|P_1}(t : v) = c_{P_2^s|P_1}(st : v)$$

$$c_{P_2|P_1}(t : v)\,s^{-1} = c_{P_2|P_1^s}(ts^{-1} : sv)$$

for $v \in \mathfrak{F}_c(\delta)$.

It is enough to prove this for v in $\mathfrak{F}'$. Fix $\psi \in L$, $v \in \mathfrak{F}'$ and put $\phi = E(P_1 : \psi : v)$. Then it follows from [1(e), Lemma 21.1] that

$$(\phi_{P_2})^s = \phi_{P_2^s}$$

and the first assertion is an immediate consequence of this fact.

Similarly the second statement is an easy consequence of the following lemma.

Lemma 2. *Fix* $P \in \mathscr{P}(\mathfrak{a})$ *and* $s \in \mathfrak{w}$. *Then*

$$E(P : \psi : v) = E(P^s : s\psi : sv)$$

for $\psi \in L$ *and* $v \in \mathfrak{F}_c$.

For $f, g \in C^\infty(G, \tau)$ and $\alpha, \beta \in C^\infty(M, \tau_M)$, put

$$(f, g)_G = \int_G (f(x), g(x))\,dx,$$

$$(\alpha, \beta)_M = \int_M (\alpha(m), \beta(m))\,dm,$$

provided the integrals are absolutely convergent. Moreover for $f \in C_c^\infty(G, \tau)$, define $f_v^{(P)} \in C_c^\infty(M, \tau_M)$ $(v \in \mathfrak{F}_c)$ by

$$f_v^{(P)}(m) = \int_A f^{(P)}(ma)\, e^{-(-1)^{1/2} v(\log a)}\, da \qquad (m \in M)$$

in the notation of [1(e), §16]. Then it is clear that

$$(E(P: \psi: v), f)_G = (\psi, f_v^{(P)})_M$$

for $\psi \in L$ and $v \in \mathfrak{F}$. Similarly

$$(E(P^s: s\psi: sv), f)_G = (s\psi, f_{sv}^{(P^s)})_M.$$

However it is easy to verify that

$$f_{sv}^{(P^s)} = s(f_v^{(P)})$$

and therefore

$$(E(P: \psi: v), f)_G = (E(P^s: s\psi: sv), f)_G$$

for all $f \in C_c^\infty(G, \tau)$. The statement of Lemma 2 is now obvious.

Lemma 1 shows that it is sufficient to investigate the functions $c_{P_2 | P_1}(1: v)$ for $P_1, P_2 \in \mathscr{P}(\mathfrak{a})$.

Lemma 3. *Fix* $P \in \mathscr{P}(\mathfrak{a})$, $\psi \in L$, $v \in \mathfrak{F}$ *and let* $P' = M' A' N'$ *be a psgp of* G. *Then*

$$E_{P'}(P: \psi: v) \sim 0$$

unless A' *is conjugate to* A *under* K.

We may assume, without loss of generality, that $\psi \in L(\omega)$ for some $\omega \in \mathscr{E}_2(M)$. Then our assertion follows from Lemmas 11.1. and 17.1.

§ 19. Some Integral Formulas

Fix $P \in \mathscr{P}(\mathfrak{a})$ and let $\mathfrak{F}_c(P)$ denote the set of all $v \in \mathfrak{F}_c$ such that $\langle \alpha, v_I \rangle > 0$ for every root α of (P, A). Put $\rho = \rho_P$ and $H(x) = H_P(x)$ $(x \in G)$. Every $x \in G$ can be written uniquely in the form $x = kman$ where $k \in K$, $m \in M \cap \exp \mathfrak{p}$, $a \in A$, $n \in N$. Put $k = \kappa(x)$ and $m = \mu(x)$. As usual let $\bar{P} = \theta(P)$ and $\bar{N} = \theta(N)$.

Theorem 1. $c_{\bar{P}|P}(1: v)$ *and* $c_{P|P}(1: -v)$ *extend to holomorphic functions of* v *on* $\mathfrak{F}_c(P)$ *and they are given by the following integrals.*

$$(c_{\bar{P}|P}(1: v)\psi)(m) = \int_{\bar{N}} \tau(\kappa(\bar{n}))\, \psi(\mu(\bar{n})\, m)\, e^{((-1)^{1/2} v - \rho)(H(\bar{n}))}\, d\bar{n},$$

$$(c_{P|P}(1: -v)\psi)(m) = \int_{\bar{N}} \psi(m\mu(\bar{n})^{-1})\, \tau(\kappa(\bar{n}))^{-1}\, e^{((-1)^{1/2} v - \rho)(H(\bar{n}))}\, d\bar{n}.$$

Here $\psi \in L$, $v \in \mathfrak{F}_c(P)$, $m \in M$ *and the Haar measure* $d\bar{n}$ *on* $\bar{N}$ *is so normalized that*

$$\int_{\bar{N}} e^{-2\rho(H(\bar{n}))}\, d\bar{n} = 1.$$

We need some preparation. Observe that $G = KP$ and $\bar{N}P$ is an open dense subset of G whose complement is of Haar measure zero. Let $d_l p$ and $d_r p$ denote the left- and right-invariant Haar measures respectively on P so that $d_r p = d_l p^{-1}$. Then $d_r p = \delta(p) d_l p$ where δ is a homomorphism of P into $\mathbf{R}_+^\times$. We can normalize the Haar measures dx and $d\bar{n}$ on G and $\bar{N}$ respectively in such a way that

$$\int_G f(x)\, dx = \int_{\bar{N} \times P} f(\bar{n}\, p)\, d\bar{n}\, d_r p = \int_{K \times P} f(k\, p)\, dk\, d_r p$$

for $f \in C_c(G)$. Put

$$\bar{f}(\bar{x}) = \int_P f(x\, p)\, d_l p \qquad (x \in G)$$

where $x \mapsto \bar{x}$ is the natural projection of G on $\bar{G} = G/P$. Note that

$$\bar{G} = \bar{K} = K/K \cap P = K/K_M$$

and put $\bar{f}(k) = \bar{f}(\bar{k})$ $(k \in K)$. Then

$$\int_K \bar{f}(k)\, dk = \int f(k\, p)\, dk\, d_l p = \int f(k\, p)\, \delta(p)^{-1}\, dk\, d_r p.$$

Since $K \cap P$ lies in the kernel of δ, we can extend δ on G by defining $\delta(k\, p) = \delta(p)$ $(k \in K, p \in P)$. Then $\delta(y\, p) = \delta(y)\, \delta(p)$ for $y \in G$, $p \in P$ and therefore

$$\int_K \bar{f}(k)\, dk = \int f(x)\, \delta(x)^{-1}\, dx - \int f(\bar{n}\, p)\, \delta(\bar{n}\, p)^{-1}\, d\bar{n}\, d_r p$$

$$= \int f(\bar{n}\, p)\, \delta(\bar{n})^{-1}\, d\bar{n}\, d_l p.$$

On the other hand $\bar{N} \cap P = \{1\}$ and so we may identify $\bar{N}$ with its image under the projection of G on $\bar{G}$. Then the above relation becomes

$$\int_K \bar{f}(k)\, dk = \int_{\bar{N}} \bar{f}(\bar{n})\, \delta(\bar{n})^{-1}\, d\bar{n}.$$

But since $f \mapsto \bar{f}$ is a surjective mapping of $C_c(G)$ on $C(G/P)$, we have obtained the following result.

Lemma 1. *We can normalize the Haar measure $d\bar{n}$ in such a way that*

$$\int_K \phi(k)\, dk = \int_{\bar{N}} \phi(\bar{n})\, \delta(\bar{n})^{-1}\, d\bar{n}$$

for all $\phi \in C(G/P) = C(K/K_M)$.

It is easy to verify that

$$\delta(x) = e^{2\rho(H(x))} \qquad (x \in G).$$

Hence taking $\phi = 1$ in the above lemma we get the following result.

Corollary. *Under the above normalization of $d\bar{n}$ we have*

$$\int_{\bar{N}} e^{-2\rho(H(\bar{n}))}\, d\bar{n} = 1.$$

Now we come to the proof of Theorem 1. It follows from [1 (e), Corollary of Lemma 32.2] that the two integrals converge uniformly when ν varies in a compact subset of $\mathfrak{F}_c(P)$. Therefore (see the proof of Theorem 18.1), it would be enough to verify the two equations for $\nu \in \mathfrak{F}_c(P) \cap \mathfrak{F}_c'(\delta)$. We prove only the first since the proof of the second is quite similar.

Fix $\psi \in L$, $v \in \mathfrak{F}_c(P) \cap \mathfrak{F}'_c(\delta)$ and put

$$\phi = E(P: \psi: v).$$

Then

$$\phi(x) = \int \psi(x \kappa(\bar{n})) \, \tau(\kappa(\bar{n}))^{-1} \exp \{((-1)^{1/2} v - \rho)(H(x \kappa(\bar{n}))) - 2\rho(H(\bar{n}))\} \, d\bar{n}$$

for $x \in G$, from Lemma 1. Now

$$\bar{n} = \kappa(\bar{n}) \, \mu(\bar{n}) \exp H(\bar{n}) \cdot n$$

where $n \in N$. Hence if $m \in M_1 = MA$,

$$\psi(m^{-1} \kappa(\bar{n})) = \psi(m^{-1} \bar{n} \mu(\bar{n})^{-1}),$$
$$H(m^{-1} \kappa(\bar{n})) = H(m^{-1} \bar{n}) - H(\bar{n}).$$

Take $x = m^{-1}$, replace $\bar{n}$ by $\bar{n}^m$ inside the integral and observe that

$$d\bar{n}^m = e^{-2\rho(H(m))} \, d\bar{n}.$$

Then we obtain

$$e^{v_+(H(m))} \phi(m^{-1}) = \int_{\bar{N}} \psi(\bar{n} m^{-1} \mu(\bar{n}^m)^{-1}) \, \tau(\kappa(\bar{n}^m))^{-1} \, e^{v_-(H(\bar{n})) - v_+(H(\bar{n}^m))} \, d\bar{n}$$

where $v_- = (-1)^{1/2} v - \rho$ and $v_+ = (-1)^{1/2} v + \rho$. On the other hand, we can choose $c > 0$ such that

$$|\psi(m)| \leq c \, \Xi_M(m)$$

for all $m \in M_1$. Now let $m = m_0^{-1} a$ where $m_0 \in M$ and $a \in A$. Keep m_0 fixed and let $a \underset{P}{\to} \infty$. Then

$$H(\bar{n}^m) = H(m_0^{-1} \bar{n}^a) = H(m_0^{-1} \kappa(\bar{n}^a)) + H(\bar{n}^a).$$

Hence $H(\bar{n}^m) - H(\bar{n}^a)$ remains bounded. Moreover

$$|\psi(\bar{n} m^{-1} \mu(\bar{n}^m)^{-1})| = |\psi(\mu(\bar{n}) \, m_0 \, \mu(m_0^{-1} \bar{n}^a m_0)^{-1})|.$$

Now

$$\bar{n}^a \in \kappa(\bar{n}^a) \, \mu(\bar{n}^a) \, AN.$$

Hence

$$m_0^{-1} \bar{n}^a m_0 \in m_0^{-1} \kappa(\bar{n}^a) m_0 \cdot \mu(\bar{n}^a)^{m_0^{-1}} \cdot AN$$

and therefore

$$\mu(m_0^{-1} \bar{n}^a m_0) \in K_M \cdot \mu(m_0^{-1} \kappa(\bar{n}^a) m_0) \, \mu(\bar{n}^a)^{m_0^{-1}}.$$

This shows that

$$m_0 \, \mu(m_0^{-1} \bar{n}^a m_0) \, m_0^{-1} \in C \mu(\bar{n}^a)$$

where C is a compact subset of M. Hence

$$\mu(\bar{n}) \, m_0 \, \mu(m_0^{-1} \bar{n}^a m_0)^{-1} \in \mu(\bar{n}) \, \mu(\bar{n}^a)^{-1} \, C^{-1} \, m_0.$$

Therefore we can choose $c_1 > 0$ such that

$$|\psi(\bar{n}m^{-1}\mu(\bar{n}^m)^{-1})| \leqq c_1\, \Xi_M(\mu(\bar{n})\,\mu(\bar{n}^a)^{-1})$$

for all $\bar{n} \in \bar{N}$ and $a \in A$. By Lemma 20.1, we can take the limit inside the integral and conclude that

$$\lim_{a \overrightarrow{P} \infty} e^{v+(\log a)}\, \phi(m_0\, a^{-1}) = \int_{\bar{N}} \psi(\bar{n}m_0)\, e^{v-(H(\bar{n}))}\, d\bar{n}$$
$$= \int_{\bar{N}} \tau(\kappa(\bar{n}))\, \psi(\mu(\bar{n})\, m_0)\, e^{v-(H(\bar{n}))}\, d\bar{n}.$$

The required result now follows from the corollary of Lemma 17.5.

We shall now derive some consequences of Theorem 1.

Lemma 2. *Fix* $\omega \in \mathscr{E}_2(M)$. *Then* $L(\omega)$ *is stable under* $c_{\bar{P}|P}(1:v)$ *and* $c_{P|P}(1:v)$.

Since $\mathscr{C}_\omega(M)$ is stable under both left and right translations of M, this is obvious from Theorem 1.

The following result was pointed out to me by Langlands.

Lemma 3. $\det c_{P|P}(1:v)$ *is not identically zero.*

Put

$$c(t) = \int_{\bar{N}} e^{-t\rho(H(\bar{n}))}\, d\bar{n} \qquad (t \geqq 2)$$

and

$$\alpha_t(\bar{n}) = c(t)^{-1}\, e^{-t\rho(H(\bar{n}))} \qquad (\bar{n} \in \bar{N}).$$

The proof is based on the following simple fact.

Lemma 4. *Let* f *be a continuous function on* $\bar{N}$ *which is integrable with respect to* $d\bar{n}$. *Then*

$$\lim_{t \to +\infty} \int_{\bar{N}} \alpha_t\, f\, d\bar{n} = f(1).$$

We shall prove this in §21.

Now fix $v \in \mathfrak{F}_c(P)$ and put

$$v_t = v + (-1)^{1/2}\, t\rho, \qquad C(t) = c(t)^{-1}\, c_{P|P}(1:-v_t) \qquad (t \geqq 2).$$

Then $v_t \in \mathfrak{F}_c(P)$ and $C(t) \in \operatorname{End} L$. Fix $\psi \in L$. Then it follows from Theorem 1 that

$$(C(t)\,\psi)(m) = \int \psi(m\,\mu(\bar{n})^{-1})\, \tau(\kappa(\bar{n}))^{-1}\, e^{((-1)^{1/2}\,v-\rho)(H(\bar{n}))}\, \alpha_t(\bar{n})\, d\bar{n}.$$

Hence

$$\lim_{t \to +\infty} C(t)\,\psi = \psi$$

from Lemma 3. This proves that $C(t) \to 1$ and therefore $\det C(t) \to 1$. Hence $\det C(t) \neq 0$ for t sufficiently large.

Combining Theorem 1 with Theorem 13.2, we can now obtain the following result.

Theorem 2. *Fix* $\psi \in L$, $\alpha \in C_c^\infty(\mathfrak{F}')$, $P_1, P_2 \in \mathscr{P}(\mathfrak{a})$ *and put*

$$\phi_\alpha(x) = \int_{\mathfrak{F}} \alpha(v)\, E(P_1 : \psi : v : x)\, dv \qquad (x \in G).$$

Then $\phi_\alpha \in \mathscr{C}(G, \tau)$ *and*

$$\phi_\alpha^{(P_2)}(ma) = \gamma(P_2) \int_{\mathfrak{F}} e^{(-1)^{1/2} v(\log a)} \sum_{s \in \mathfrak{w}} \alpha(s^{-1} v)(c_{\bar{P}_2|P_2}(1 : v)\, c_{P_2|P_1}(s : s^{-1} v)\, \psi)(m)\, dv$$

for $m \in M$, $a \in A$. *Here*

$$\gamma(P_2) = \int_{N_2} e^{-2\rho(H(\bar{n}))}\, d\bar{n},$$

the integrand having the same meaning as in Theorem 1 for $P = P_2$.

There is no loss of generality in assuming that $\psi \in L(\omega)$ for some $\omega \in \mathscr{E}_2(M)$. Put

$$\phi(v : x) = E(P_1 : \psi : v : x) \qquad (v \in \mathfrak{F}, \, x \in G).$$

Then it follows from Lemma 17.1 that ϕ is a function on $\mathfrak{F} \times G$ of type $II(\lambda)$ for a suitable $\lambda \in (-1)^{1/2}\, \mathfrak{h}^*$ (see the proof of Theorem 18.1). Therefore since Supp $\alpha \subset \mathfrak{F}'$, it follows from Theorem 18.1 that the function

$$(v, x) \mapsto \alpha(v)\, \phi(v : x)$$

is of type $I'(\lambda)$ (§ 13). Hence we conclude from Theorem 13.1 that $\phi_\alpha \in \mathscr{C}(G, \tau)$.

Now put $P = P_2$ and let us use the notation of Theorem 13.2. Since $\rho(H(\bar{n})) \geq 0$, it is clear from this that

$$\phi_\alpha^{(P)}(m) = \lim_{\varepsilon \to 0} \int_{\bar{N}} e^{-(1+\varepsilon)\rho(H(\bar{n}))}\, \phi_{P,\alpha}(\bar{n}m)\, d\bar{n} \qquad (m \in M_1).$$

(Here $\varepsilon > 0$.) But

$$\phi_{P,\alpha}(\bar{n}m) = \int_{\mathfrak{F}} \alpha(v)\, \tau(\kappa(\bar{n}))\, E_P(P_1 : \psi : v : \mu(\bar{n})\, m \exp H(\bar{n}))\, dv.$$

Fix $\varepsilon > 0$ and put $v_\varepsilon = v + (-1)^{1/2} \varepsilon \rho$ for $v \in \mathfrak{F}$. Then $v_\varepsilon \in \mathfrak{F}_c(P)$ and we conclude from [1(e), Corollary of Lemma 32.2] and Theorems 1 and 18.1 that

$$\int_{\bar{N}} e^{-(1+\varepsilon)\rho(H(\bar{n}))}\, \phi_{P,\alpha}(\bar{n}ma)\, d\bar{n}$$

$$= \gamma(P) \int_{\mathfrak{F}} \alpha(v) \sum_{s \in \mathfrak{w}} (c_{\bar{P}|P}(1 : (sv)_\varepsilon)\, c_{P|P_1}(s : v)\, \psi)(m)\, e^{(-1)^{1/2} sv(H(a))}\, dv$$

for $m \in M$ and $a \in A$. But $c_{\bar{P}|P}(1 : v)$ is holomorphic on $\mathfrak{F}_c'(\delta)$. Therefore since Supp $\alpha \subset \mathfrak{F}'$, we obtain by making $\varepsilon \to 0$ that

$$\phi_\alpha^{(P)}(ma) = \sum_{s \in \mathfrak{w}} \gamma(P) \int_{\mathfrak{F}} \alpha(v)(c_{\bar{P}|P}(1 : sv)\, c_{P|P_1}(s : v)\, \psi)(m)\, e^{(-1)^{1/2} sv(H(a))}\, dv$$

and this is equivalent to the required result.

§ 20. A Result on Uniform Convergence

Let $P = MAN$ be a psgp of G. Define ρ, $H(x)$, $\mu(x)$ $(x \in G)$, A^+ and $\bar{N}$ as usual.

Lemma 1. *Fix* $v \in \mathfrak{a}^*$ *such that* $\langle v, \alpha \rangle > 0$ *for every root* α *of* (P, A) *and put* $v_+ = v + \rho$, $v_- = v - \rho$. *Then the integral*

$$\int_{\bar{N}} e^{-v_+(H(\bar{n})) + v_-(H(\bar{n}^a))} \, \Xi_M(\mu(\bar{n}) \, \mu(\bar{n}^a)^{-1}) \, d\bar{n}$$

converges uniformly for $a \in A^+$.

The present form of this lemma is due to Langlands [2, Lemma 3.12]. My original formulation was more complicated.

We first need an auxiliary result. Let $P_0 = M_0 A_0 N_0$ be a minimal psgp of G contained in P and let us use the notation of [1(e), § 30].

Lemma 2. *Let* x, $y \in G$. *Then*

$$\Xi(xy^{-1}) = \int_{\bar{N}_0} e^{-\rho_0(H_0(x\bar{n}_0) + H_0(y\bar{n}_0))} \, d\bar{n}_0,$$

where the Haar measure $d\bar{n}_0$ *on* $\bar{N}_0$ *is so normalized that*

$$\int_{\bar{N}_0} e^{-2\rho_0(H_0(\bar{n}_0))} \, d\bar{n}_0 = 1.$$

Let $\kappa_0(x)$ $(x \in G)$ denote the component of x in K corresponding to the Iwasawa decomposition $G = KA_0 N_0$. Put $k_y = \kappa_0(yk)$ $(k \in K)$. Then $k \mapsto k_y$ is a diffeomorphism of K and [1(a), p. 281]

$$e^{2\rho_0(H_0(yk))} \, dk_y = dk.$$

Now

$$\Xi(xy^{-1}) = \int_K e^{-\rho_0(H_0(xy^{-1}k))} \, dk.$$

Replacing k by k_y and observing that

$$H_0(xy^{-1}k_y) = H_0(xk) - H_0(yk),$$

we get

$$\Xi(xy^{-1}) = \int_K e^{-\rho_0(H_0(xk) + H_0(yk))} \, dk$$

and the required result now follows from Lemma 19.1.

Let $^*P = \,^*M\,^*A\,^*N$ be the minimal psgp of M corresponding to P_0 [1(e), Lemma 6.1] so that $^*P = M \cap P_0$. Put $^*\bar{N} = \theta(^*N)$. Then $\bar{N}$ is a normal subgroup of $\bar{N}^0$ and the mapping

$$(\bar{n}, \,^*\bar{n}) \mapsto \bar{n}_0 = \bar{n} \cdot \,^*\bar{n}$$

defines a diffeomorphism of $\bar{N} \times \,^*\bar{N}$ onto $\bar{N}^0$. Let $d\bar{n}$ and $d^*\bar{n}$ denote the corresponding Haar measures. Then $d\bar{n} \cdot d^*\bar{n} = c \, d\bar{n}_0$ where c is a positive constant.

Let us now use the notations of [1(e), § 30].

Lemma 3. *We can normalize $d\bar{n}$ and $d*\bar{n}$ in such a way that*

$$\int_N e^{-2\rho(H(\bar{n}))}\, d\bar{n} = \int_{*N} e^{-2*\rho(*H(*\bar{n}))}\, d*\bar{n} = 1.$$

Then $d\bar{n}_0 = d\bar{n}\, d*\bar{n}$ where $d\bar{n}_0$ is normalized as in Lemma 2.

The proof of the first part is the same as that of the corollary of Lemma 19.1. Since $d\bar{n}\, d*\bar{n} = c\, d\bar{n}_0$, we have

$$c = \int_{*N} d*\bar{n} \int_N e^{-2\rho_0(H_0(\bar{n}*\bar{n}))}\, d\bar{n}.$$

Fix $*\bar{n} \in *N$. Then $*\bar{n} = kan$ ($k \in K_M$, $a \in *A$, $n \in *N$) and

$$H_0(\bar{n}*\bar{n}) = H_0(\bar{n}k) + \log a = H_0(k^{-1}\bar{n}k) + *H(*\bar{n}).$$

But since K_M normalizes $\bar{N}$, we conclude that

$$\int_N e^{-2\rho_0(H_0(\bar{n}*\bar{n}))}\, d\bar{n} = e^{-2*\rho(*H(*\bar{n}))} \int_N e^{-2\rho_0(H_0(\bar{n}))}\, d\bar{n}.$$

On the other hand

$$H_0(\bar{n}*k) = H(\bar{n}) + *H(\mu(\bar{n})*k)$$

for $*k \in K_M$. Hence if $d*k$ is the normalized Haar measure on K_M, we conclude from [1(a), Corollary p. 261] that

$$\int_{K_M} e^{-2\rho_0(H_0(\bar{n}*k))}\, d*k = e^{-2\rho(H(\bar{n}))} \int_{K_M} e^{-2*\rho(*H(\mu(\bar{n})*k))}\, d*k$$
$$= e^{-2\rho(H(\bar{n}))}.$$

Therefore

$$\int_N e^{-2\rho_0(H_0(\bar{n}))}\, d\bar{n} = \int_N d\bar{n} \int_{K_M} e^{-2\rho_0(H_0(*k^{-1}\bar{n}*k))}\, d*k$$
$$= \int_N e^{-2\rho(H(\bar{n}))}\, d\bar{n} = 1$$

and this proves that

$$c = \int_{*N} e^{-2*\rho(*H(*\bar{n}))}\, d*\bar{n} = 1.$$

Corollary. $\Xi_M(m_1 m_2^{-1}) = \int_{*N} e^{-*\rho(*H(m_1*\bar{n}) + *H(m_2*\bar{n}))}\, d*\bar{n}$ *for $m_1, m_2 \in M$.*

This follows by applying Lemma 2 to $(M, *P)$ in place of (G, P_0). Now we come to the proof of Lemma 1. Fix $0 < \varepsilon \leq 1$ such that

$$\langle \rho_0 - \varepsilon v, \alpha_0 \rangle \geq 0$$

for every root α_0 of (P_0, A_0). Note that

$$-v_+(H(\bar{n})) + v_-(H(\bar{n}^a)) = v(H(\bar{n}^a) - H(\bar{n})) - \rho(H(\bar{n}^a) + H(\bar{n}))$$

and it follows from [1(e), Lemma 30.4] that we can choose $c \geq 0$ such that

$$v(H(\bar{n}^a) - H(\bar{n})) \leq c$$

for all $\bar{n} \in \bar{N}$ and $a \in A^+$. Put $v' = \varepsilon v$. Then

$$-v_+(H(\bar{n})) + v_-(H(\bar{n}^a)) \leqq (1-\varepsilon) c - v'_+(H(\bar{n})) + v'_-(H(\bar{n}^a)).$$

Hence it would be enough to prove Lemma 1 for εv instead of v.

So we may now assume that

$$\langle \rho_0 - v, \alpha_0 \rangle \geqq 0$$

for every root α_0 of (P_0, A_0). Let $\bar{n}_0 = \bar{n} \cdot {}^*\bar{n}$ where $\bar{n} \in \bar{N}$ and ${}^*\bar{n} \in {}^*\bar{N}$. Then

$$H_0(\bar{n}_0) = H(\bar{n}) + {}^*H(\mu(\bar{n}) \cdot {}^*\bar{n}),$$
$$H_0(\bar{n}_0^a) = H(\bar{n}^a) + {}^*H(\mu(\bar{n}^a) \cdot {}^*\bar{n}) \qquad (a \in A).$$

Therefore

$$\begin{aligned}
(v - \rho_0)(H_0(\bar{n}_0^a)) &- (v + \rho_0)(H_0(\bar{n}_0)) \\
&= v_-(H(\bar{n}^a)) - v_+(H(\bar{n})) \\
&\quad - {}^*\rho({}^*H(\mu(\bar{n}) {}^*\bar{n})) - {}^*\rho({}^*H(\mu(\bar{n}^a) {}^*\bar{n})).
\end{aligned}$$

ω being a measurable subset of $\bar{N}$, put $\omega_0 = \omega \cdot {}^*\bar{N}$. Then integrating both sides, we get

$$\begin{aligned}
\int_{\omega_0} e^{(v - \rho_0)(H_0(\bar{n}_0^a)) - (v + \rho_0)(H_0(\bar{n}_0))} \, d\bar{n}_0 \\
= \int_\omega e^{v_-(H(\bar{n}^a)) - v_+(H(\bar{n}))} \, \Xi_M(\mu(\bar{n}) \, \mu(\bar{n}^a)^{-1}) \, d\bar{n} = I_\omega(a) \qquad \text{(say)}
\end{aligned}$$

from the corollary of Lemma 3. On the other hand $M = K_M \cdot {}^*A \cdot {}^*N$ is an Iwasawa decomposition of M. Hence

$$\bar{n}_0^a = \bar{n}^a \cdot {}^*\bar{n} = \bar{n}^a \cdot {}^*k \cdot {}^*a \cdot {}^*n$$

where ${}^*k \in K_M$, ${}^*a \in {}^*A$, ${}^*n \in {}^*N$. Since M normalizes $\bar{N}$, it is clear that

$$H_0(\bar{n}_0^a) = H_0(\bar{n}') + H_0({}^*a)$$

where $\bar{n}' = {}^*k^{-1} \cdot \bar{n}^a \cdot {}^*k \in \bar{N}$. Hence we conclude from [1(a), Lemma 43] that

$$(\rho_0 - v)(H_0(\bar{n}^a)) \geqq (\rho_0 - v)(H_0({}^*a)) = {}^*\rho({}^*H({}^*\bar{n})).$$

Therefore

$$\begin{aligned}
(v - \rho_0)(H_0(\bar{n}_0^a)) &- (v + \rho_0)(H_0(\bar{n}_0)) \\
&\leqq -{}^*\rho({}^*H({}^*\bar{n})) - v_+(H(\bar{n})) - {}^*\rho({}^*H(\mu(\bar{n}) {}^*\bar{n})).
\end{aligned}$$

Integrating both sides on ω_0 and applying Lemma 3 and its corollary, we find that

$$I_\omega(a) \leqq \int_\omega e^{-v_+(H(\bar{n}))} \, \Xi_M(\mu(\bar{n})) \, d\bar{n} \qquad (a \in A).$$

Now choose $\varepsilon > 0$ so small that $\langle v, \alpha \rangle \geqq \varepsilon \langle \rho, \alpha \rangle$ for every root α of (P, A). Then

$$v^+(H(\bar{n})) \geqq (1 + \varepsilon) \rho(H(\bar{n})) \qquad (\bar{n} \in \bar{N})$$

from [1(e), Lemma 30.4]. On the other hand

$$\int_{\bar{N}} e^{-(1+\varepsilon)\rho(H(\bar{n}))} \, \Xi_M(\mu(\bar{n})) \, d\bar{n} < \infty$$

from [1(e), Corollary of Lemma 32.2]. Therefore the assertion of Lemma 1 is now obvious.

§ 21. Proof of Lemma 19.4

For $T \geq 0$, let $\bar{N}(T)$ denote the set of all points $\bar{n} \in \bar{N}$ such that $\rho(H(\bar{n})) \leq T$. Then $\bar{N}(T)$ is a compact set and $\bar{N}(0) = \{1\}$. Let $(\alpha_1, \ldots, \alpha_l)$ be the system of simple roots of (P, A). Then

$$2\rho = m_1 \alpha_1 + \cdots + m_l \alpha_l$$

where m_i are positive integers. Put $m = m_1 + \cdots + m_l$.

Lemma 1. *There exists a number $c > 0$ such that*

$$\int_{\bar{N}(\varepsilon)} d\bar{n} \geq c \varepsilon^{2m}$$

for $0 < \varepsilon \leq 1$.

Put

$$\beta(a) = \inf_{1 \leq i \leq l} \alpha_i(\log a)/2 \quad (a \in A^+).$$

Then

$$\rho(H(\bar{n}^a)) \leq \log(1 + e^{1 - \beta(a)})$$

for $\bar{n} \in \bar{N}(1)$ and $a \in A^+$ from [1(e), Lemma 30.2]. Fix $\varepsilon\,(0 < \varepsilon \leq 1)$ and choose $a \in A$ such that

$$\alpha_i(\log a) = 2(1 - \log \varepsilon) \quad (1 \leq i \leq l).$$

Then $a \in A^+$ and

$$1 - \beta(a) = \log \varepsilon.$$

Hence

$$\rho(H(\bar{n}^a)) \leq \log(1 + \varepsilon) \leq \varepsilon$$

for $\bar{n} \in \bar{N}(1)$. Therefore

$$\int_{\bar{N}(\varepsilon)} d\bar{n} \geq \int_{(\bar{N}(1))^a} d\bar{n} = e^{-2\rho(\log a)} c_0$$

where

$$c_0 = \int_{\bar{N}(1)} d\bar{n} > 0.$$

But

$$2\rho(\log a) = m\alpha_i(\log a) = 2m(1 - \log \varepsilon).$$

Hence

$$\int_{\bar{N}(\varepsilon)} d\bar{n} \geq c \varepsilon^{2m}$$

where $c = c_0 e^{-2m} > 0$.

Now we come to the proof of Lemma 19.4. Fix ε $(0<\varepsilon\leq1)$ and let $\bar{N}_r(\varepsilon)$ denote the complement of $\bar{N}((r-1)\varepsilon)$ in $\bar{N}(r\varepsilon)$ $(r\geq1)$. Then if $t\geq2$,

$$\int_{\bar{N}_r(\varepsilon)} e^{-t\rho(H(\bar{n}))}\,d\bar{n}\geq e^{-r\varepsilon t}\int_{\bar{N}_r(\varepsilon)} d\bar{n}=e^{-r\varepsilon t}(\mu(r\varepsilon)-\mu((r-1)\varepsilon))$$

where

$$\mu(T)=\int_{\bar{N}(T)} d\bar{n}\quad(T\geq0).$$

Therefore

$$c(t)=\int_{\bar{N}} e^{-t\rho(H(\bar{n}))}\,d\bar{n}\geq\sum_{r\geq1} e^{-r\varepsilon t}(\mu(r\varepsilon)-\mu((r-1)\varepsilon)).$$

On the other hand

$$\mu(T)=\int_{\bar{N}(T)} d\bar{n}\leq e^{2T}\int_{\bar{N}} e^{-2\rho(H(\bar{n}))}\,d\bar{n}=c(2)\,e^{2T}.$$

Hence if $t>2$,

$$e^{-r\varepsilon t}\mu(r\varepsilon)\to0$$

as $r\to+\infty$. Therefore

$$c(t)\geq\sum_{r\geq1}\mu(r\varepsilon)\,e^{-r\varepsilon t}(1-e^{-\varepsilon t})$$
$$\geq\mu(\varepsilon)\,e^{-\varepsilon t}(1-e^{-\varepsilon t}).$$

Now take $\varepsilon=t^{-1}$. Then it follows from Lemma 1 that

$$c(t)\geq\mu(t^{-1})\,e^{-1}(1-e^{-1})\geq c_0\,t^{-2m}\quad(t>2),$$

where c_0 is a positive constant independent of t.

Now let U be any open neighborhood of 1 in $\bar{N}$. We have to show that

$$\int_{^cU}\alpha_t(\bar{n})\,d\bar{n}\to0$$

as $t\to+\infty$. (As usual ^cU denotes the complement of U.) Fix ε $(0<\varepsilon\leq1)$ such that $\bar{N}(\varepsilon)\subset U$. Then if $t>2$,

$$\int_{^cU}\alpha_t(\bar{n})\,d\bar{n}\leq\int_{^c\bar{N}(\varepsilon)}\alpha_t(\bar{n})\,d\bar{n}=c(t)^{-1}\int_{^c\bar{N}(\varepsilon)} e^{-t\rho(H(\bar{n}))}\,d\bar{n}.$$

But $c(t)^{-1}\leq c_0^{-1}\,t^{2m}$ and

$$\int_{^c\bar{N}(\varepsilon)} e^{-t\rho(H(\bar{n}))}\,dt\leq e^{-(t-2)\varepsilon}\int_{^c\bar{N}(\varepsilon)} e^{-2\rho(H(\bar{n}))}\,d\bar{n}$$
$$\leq c(2)\,e^{-(t-2)\varepsilon}.$$

Therefore

$$\int_{^cU}\alpha_t(\bar{n})\,d\bar{n}\leq c_1\,t^{2m}\,e^{-t\varepsilon}\to0$$

as $t\to+\infty$, where $c_1=c_0^{-1}\,c(2)\,e^{2\varepsilon}$. This proves Lemma 19.4.

§ 22. Appendix

Let $x=(x_1, \ldots, x_n)$ denote a variable point in $E=\mathbf{R}^n$. Put $D_i=\partial/\partial x_i$ and $D^\alpha = D_1^{\alpha_1} \ldots D_n^{\alpha_n}$ for a multi-index $\alpha=(\alpha_1, \ldots, \alpha_n)$. We write $|\alpha|=\alpha_1+\alpha_2+\cdots+\alpha_n$, $|x|=\max_i |x_i|$ and denote by M the set of all multi-indices.

Let V and $\mathscr{S}(V)$ be as before (§ 6).

Lemma 1. *Let f be an element in $C^\infty(E, V)$ such that $f=0$ on the hyperplane $x_1=0$. Then $f=x_1 g$ where*

$$g(x)=\int_0^1 f_1(x_1 t, x_2, \ldots, x_n)\, dt$$

and $f_1=D_1 f$. Hence $g\in C^\infty(E, V)$ and

$$|D^\alpha g(x)|_\mathbf{s} \leq \sup_{|y|\leq|x|} |D^\alpha f_1(y)|_\mathbf{s}$$

for all $x\in E$, $\alpha\in M$ and $\mathbf{s}\in\mathscr{S}(V)$.

This is obvious.

Let $p\neq 0$ be the product of N real linear forms on E and E' the set of all points $x\in E$ where $p(x)\neq 0$. A function f from E' to V is said to be locally bounded (on E), if for every compact set ω in E and $\mathbf{s}\in\mathscr{S}(V)$, $|f(x)|_\mathbf{s}$ remains bounded for $x\in\omega\cap E'$.

For $\alpha\in M$, $r\geq 0$ and $\mathbf{s}\in\mathscr{S}(V)$, put

$$\mathbf{s}_{\alpha, r}(f)=\sup_E (1+|x|)^r\, |D^\alpha f|_\mathbf{s} \qquad (f\in C^\infty(E, V)).$$

If F is a finite subset of M, put

$$\mathbf{s}_{F, r}(f)=\sum_{\alpha\in F} \mathbf{s}_{\alpha, r}(f).$$

Let $\mathscr{C}(E, V)$ denote the set of all functions $f\in C^\infty(E, V)$ such that $\mathbf{s}_{\alpha, r}(f)<\infty$ for all $\alpha\in M$ and $r\geq 0$.

Lemma 2. *Fix $\alpha\in M$ and let F denote the set of all $\beta\in M$ such that $|\beta|\leq|\alpha|+N$. Then for every $r\geq 0$, we can choose a number $c_r\geq 1$ with the following property. Suppose $f\in\mathscr{C}(E, V)$ and $p^{-1}f$ is locally bounded. Then $f=pg$ where $g\in\mathscr{C}(E, V)$ and*

$$\mathbf{s}_{\alpha, r}(g)\leq c_r\, \mathbf{s}_{F, r}(f)$$

for all $\mathbf{s}\in\mathscr{S}(V)$.

By an easy induction we are reduced to the case $N=1$. Hence we may assume that $p=x_1$. Then $f=x_1 g$ in the notation of Lemma 1. Let E_1 and E_2 be the sets of points $x\in E$ where $|x_1|\leq 1$ and $|x_1|\geq 1$ respectively. Then if $x\in E_1$, we have

$$1+|x|\leq 2(1+\max_{i\geq 2} |x_i|)$$

and therefore

$$(1+|x|)^r\, |D^\alpha g(x)|_\mathbf{s}\leq 2^r \sup_{|y|\leq|x|} |D^\alpha f_1(y)|(1+|y|)^r.$$

This means that

$$\sup_{E_1} (1+|x|)^r \, |D^\alpha g(x)|_{\mathbf{s}} \leq 2^r \, \mathbf{s}_{\beta,r}(f)$$

where $\beta = (\alpha_1 + 1, \alpha_2, \ldots, \alpha_n)$.

On the other hand $g = x_1^{-1} f$ on E_2. Since $|x_1| \geq 1$, it follows directly by differentiation that

$$|D^\alpha g(x)|_{\mathbf{s}} \leq \alpha_1! \sum_{0 \leq m \leq \alpha_1} |D_1^m D^\beta f(x)|_{\mathbf{s}}$$

on E_2 where $\beta = (0, \alpha_2, \ldots, \alpha_n)$. Therefore since $E = E_1 \cup E_2$ the required result is obvious.

Let us now use the notation of Theorem 18.1.

Lemma 3. *Let H be an element in $\mathfrak{a}$ such that $\alpha(H) \neq 0$ for every root α of $(\mathfrak{g}, \mathfrak{a})$. Then $sH \neq H$ for every $s \neq 1$ in $\mathfrak{w}$.*

Extend $\mathfrak{a}$ to a maximal abelian subspace $\mathfrak{a}_0$ of $\mathfrak{p}$ and put $\mathfrak{w}_0 = \mathfrak{w}(\mathfrak{a}_0)$. Let Q be the set of all roots of $(\mathfrak{g}, \mathfrak{a}_0)$ which vanish at H. Then if $\beta \in Q$, it is clear that $\beta = 0$ on $\mathfrak{a}$.

Let $\mathfrak{w}_1$ be the stabilizer of H in $\mathfrak{w}_0$. Then $\mathfrak{w}_1$ is the subgroup of $\mathfrak{w}_0$ generated by the Weyl reflexions s_β for $\beta \in Q$. Hence every element of $\mathfrak{w}_1$ leaves $\mathfrak{a}$ fixed pointwise.

Now suppose $sH = H$ for some $s \in \mathfrak{w}$. We can choose $s_0 \in \mathfrak{w}_0$ such that $s_0 = s$ on $\mathfrak{a}$. But then $s_0 \in \mathfrak{w}_1$ and hence $s_0 = 1$ on $\mathfrak{a}$. This proves that $s = 1$.

References

1. Harish-Chandra: (a) Spherical functions on a semisimple Lie group I. Amer. J. Math. **80**, 241–310 (1958)

   (b) Spherical functions on a semisimple Lie group II. Amer. J. Math. **80**, 553–613 (1958)

   (c) Invariant differential operators and distributions on a semisimple Lie algebra. Amer. J. Math. **86**, 534–564 (1964)

   (d) Discrete series for semisimple Lie groups II. Acta Math. **116**, 1–111 (1966)

   (e) Harmonic analysis on real reductive groups I. J. Functional Analysis **19**, 104–204 (1975)

2. Langlands, R. P.: On the classification of irreducible representations of real algebraic groups, 1973, preprint

Received July 28, 1975

Harish-Chandra
The Institute for Advanced Study
School of Mathematics
Princeton, N. J. 08540
USA

Annals of Mathematics **104** (1976), 117–201

Harmonic analysis on real reductive groups III. The Maass-Selberg relations and the Plancherel formula

By HARISH-CHANDRA

TABLE OF CONTENTS

Part IV. Commuting algebras of induced representations

1. Introduction

This paper is devoted to a deeper study of the Eisenstein integrals, their functional equations and the relationship of the Plancherel measure to their asymptotic behavior. We have seen in [2 (j), Thm. 18.1] that the constant term of an Eisenstein integral is a finite sum of plane waves whose amplitudes are given by the corresponding c-functions. It is an important principle that the absolute values of these several c-functions are always equal to each other and the Plancherel measure μ can be written down very simply in terms of their common value (Lemma 13.4). The situation is therefore exactly similar to what was found some years ago in the case of the symmetric space G/K [2 (b), p. 611].

The c-functions are intimately related to the intertwining operators for induced representations which I denote by j. The j's have a natural product formula, which can be transferred to the c's and finally to μ. This reduces the computation of μ to the special case of maximal parabolic subgroups and thus enables us to obtain an explicit formula for μ (§ 36). It turns out that μ, c and j can all be extended to meromorphic functions on the complex space $\mathfrak{F}_c$ which parameterizes the Eisenstein integrals. Moreover μ is holomorphic and nonnegative on the real subspace $\mathfrak{F}$. Let $\mathfrak{F}'$ denote the set of all regular points of $\mathfrak{F}$. Then for $\nu \in \mathfrak{F}'$, $\mu(\nu) > 0$ and the corresponding induced representation π_ν of G is irreducible. The decomposition properties of π_ν, when ν is singular, are intimately related to the behavior of μ at ν. For example if $\dim \mathfrak{F} = 1$ and $\mu(0) = 0$, then π_0 is irreducible.

This paper is divided into four parts. After introducing the intertwining operators, we show that the Eisenstein integral $E(P:\psi:\nu)$ is essentially a matrix coefficient of the corresponding induced representation $\pi_{P,\nu}$ (Theorem 7.1). This establishes a close relationship between the c's and the j's and enables us to continue the j's analytically to the real subspace $\mathfrak{F}$ (§ 11). We introduce the function μ_ω on $\mathfrak{F}$ by means of the formula

$$\mu_\omega(\nu)j_{P|\bar{P}}(\nu)j_{\bar{P}|P}(\nu) = 1 \qquad\qquad (\nu \in \mathfrak{F}')$$

and show that μ_ω is holomorphic on $\mathfrak{F}$ and $\mu_\omega(\nu) > 0$ for $\nu \in \mathfrak{F}'$. This is first done under the assumption prk $P = 1$ (§ 12). The general case then follows from the product formula for $j_{\bar{P}|P}$ (§ 5). In particular we obtain the relation (§ 13)

$$\mu_\omega = \prod_{\alpha \in \Sigma(P)} \mu_{\omega,\alpha} \,.$$

The irreducibility of $\pi_{P,\nu}$ ($\nu \in \mathfrak{F}'$) is a consequence of the fact that $\mu_\omega(\nu) > 0$ (§ 13).

Part II is devoted to a proof of the Maass-Selberg relations (§ 14) which assert, roughly speaking, that all the plane waves, appearing in the constant term of a suitable eigenfunction, have the same intensity. So if one of them is missing, we can actually conclude that the eigenfunction itself is zero. The functional equations of the Eisenstein integrals (§ 17) are an immediate consequence of this fact. These equations involve certain linear transformations ${}^0c_{Q|P}(s\colon \nu)$ which are rational functions of $\nu \in \mathfrak{F}_c$ (§ 19).

In Part III we come to the Plancherel formula and the determination of the Plancherel measure. Our first task is to compute $(\Theta_{\omega,\nu}, \phi_\alpha)$ (see Theorem 22.1) where $\Theta_{\omega,\nu}$ is the character of the induced representation $\pi_{P,\nu}(\nu \in \mathfrak{F})$ and ϕ_α the wave-packet given by

$$\phi_\alpha = \int_{\mathfrak{F}} \mu_\omega(\nu)\alpha(\nu)E(P\colon \psi\colon \nu)d\nu \qquad\qquad \left(\alpha \in C_c^\infty(\mathfrak{F})\right).$$

Let Γ be the Cartan subgroup of G associated to P. Then this calculation leads to a simple formula for $'F_{\phi_\alpha}^\Gamma$ (Lemma 24.1). In the special case when Γ is fundamental in G, one immediately obtains a formula for μ_ω by applying [2 (i), Lemma 17.5] to ϕ_α. This shows, in particular, that $\mu_\omega(\nu)$ is a polynomial function of ν in this case (Theorem 24.1).

Theorem 25.1 asserts that μ_ω extends to a meromorphic function on $\mathfrak{F}_c$ and this function is holomorphic on a tube domain around $\mathfrak{F}$ where it has at most polynomial growth. This theorem has the following two important consequences. Firstly it enables us to prove that the c- and j-functions also extend to meromorphic functions on $\mathfrak{F}_c$. Secondly it implies that the wave-packets ϕ_α corresponding to $\alpha \in \mathcal{C}(\mathfrak{F})$ lie in the Schwartz space of G. The second result is crucial for the proof of the Plancherel formula (§ 27).

On account of the product formula for μ_ω, it is enough, for the proof of Theorem 25.1, to consider the case when prk $P = 1$. By the earlier result μ_ω is a polynomial if rank $G >$ rank K. Hence the important case is when rank $G =$ rank K. Theorem 25.1 is proved by an explicit determination of μ_ω in this case (§ 36). The method used here is basically the same as that of [2 (d), § 24].

Theorem 38.1 contains the main result of Part IV. We use it to obtain an upper bound for the self intertwining number of an induced representation (corollary of Lemma 40.2) and, as an application, we show that every representation in the fundamental series is irreducible (Theorem 41.1).

The work presented in this paper has been carried out over a period of several years and a brief account of some parts of it has appeared in two short notes [2 (g), (h)]. Moreover I have given lectures on it in Princeton during the spring of 1971 and again during the academic year 1974–1975. For recent work by other authors on similar questions see [1], [4 (b)], [9].

As far as possible we shall keep to the notation of [2 (i), (j)] and therefore any undefined symbols should be given the same meaning as there.

Historical note. Intertwining operators have been studied intensively for twenty years by Bruhat [6], Kunze and Stein [5], Shiffman [8] and Knapp and Stein [4]. The product formula for μ was first obtained by Gindikin and Karpelevič [7] in the case of G/K. The corresponding results of this paper may therefore be regarded as natural generalizations of this earlier work.

Part I. The c-, j- and μ-functions

2. Some elementary results on integrals

Fix a special vector subgroup A of G [2 (i), p. 114]. A root of $(\mathfrak{g}, \mathfrak{a})$ is called reduced if $r\alpha$ is not a root for $0 < r < 1$ ($r \in \mathbf{R}$). If $P \in \mathscr{P}(A)$, we denote by $\Sigma(P)$ the set of all reduced roots of (P, A). If $P_1, P_2 \in \mathscr{P}(A)$, put $\Sigma(P_2 \mid P_1) = \Sigma(\bar{P}_2) \cap \Sigma(P_1)$ and[1] $d(P_1, P_2) = [\Sigma(P_2 \mid P_1)]$ where $\bar{P}_2 = \theta(P_2)$ as usual. Then d is a metric on the finite set $\mathscr{P}(A)$ and the following lemma is an easy consequence of the definition.

LEMMA 1. *Fix P_1, P_2 and P in $\mathscr{P}(A)$. Then the following conditions are equivalent.*

(1) $d(P_1, P_2) = d(P_1, P) + d(P, P_2)$,
(2) $\Sigma(P_2 \mid P_1) \supset \Sigma(P \mid P_1)$,
(3) $\Sigma(P_2 \mid P_1)$ *is the disjoint union of* $\Sigma(P_2 \mid P)$ *and* $\Sigma(P \mid P_1)$.

We say that P lies between P_1 and P_2 if (1) holds. Moreover P_1, P_2 are said to be adjacent if $d(P_1, P_2) = 1$.

Define

$$I_{P_2 \mid P_1}(\nu) = \int_{N_2 \cap \bar{N}_1} \exp\{-(\nu + \rho_1)(H_1(n))\}dn \qquad (\nu \in \mathfrak{a}^*) .$$

Here $P_i \in \mathscr{P}(A)$ ($P_i = MAN_i$, $i = 1, 2$), $\bar{N}_1 = \theta(N_1)$, and $\rho_1 = \rho_{P_1}$, $H_1(x) = H_{P_1}(x)$ ($x \in G$). Moreover $dn = d(N_2 \cap \bar{N}_1)$ [2 (i), § 7]. It is obvious that

---

[1] $[S]$ denotes the number of elements in a set S.

$$0 < I_{P_2|P_1}(\nu) \leqq \infty$$

and $I_{P_1|P_1}(\nu) = 1$.

LEMMA 2. *Let P_1, P_2 and P be three elements in $\mathscr{P}(A)$ such that P lies between P_1 and P_2. Then*

$$I_{P_2|P_1}(\nu) = I_{P_2|P}(\nu)I_{P|P_1}(\nu)$$

for $\nu \in \mathfrak{a}^*$.

Here the method of proof is the same as in [7].

First we need some preparation. Given $H_0 \in \mathfrak{g}$ such that $\theta(H_0) = -H_0$, we associate to it a psgp Q of G as follows. Let $\mathfrak{q}$ be the subspace spanned by all $X \in \mathfrak{g}$ such that $[H_0, X] = \lambda X$ for some $\lambda \geqq 0$ ($\lambda \in \mathbf{R}$). Then $\mathfrak{q}$ is a parabolic subalgebra of $\mathfrak{g}$ and Q is the normalizer of $\mathfrak{q}$ in G. We say that Q is the psgp of G determined by H_0.

Fix $P \in \mathscr{P}(A)$ and $\alpha \in \Sigma(P)$. Let σ_α denote the hyperplane $\alpha = 0$ in $\mathfrak{a}$ and let Z_α be the centralizer of σ_α in G. Put $M_\alpha = {}^0Z_\alpha$ and $*P_\alpha = M_\alpha \cap P$. For $\lambda \in \mathfrak{a}^*$, let $\mathfrak{g}_\lambda$ denote, as usual, the space of all $X \subset \mathfrak{g}$ such that $[H, X] = \lambda(H)X$ for all $H \in \mathfrak{a}$.

LEMMA 3. *M_α is a reductive group satisfying the conditions of [2(i), §3] and $*P_\alpha$ is a psgp of M_α. Moreover the Langlands decomposition of $*P_\alpha$ is given by*

$$*P_\alpha = MA_\alpha N_\alpha$$

where $\mathfrak{a}_\alpha = \mathbf{R}H_\alpha$ and

$$\mathfrak{n}_\alpha = \sum_{k \geqq 1} \mathfrak{g}_{k\alpha} \, .$$

Fix an element $H_0 \in \sigma_\alpha$ such that every root of $(\mathfrak{g}, \mathfrak{a})$ which vanishes at H_0 is a multiple of α. Let $Q = M_Q A_Q N_Q$ be the psgp of G determined by H_0. Then it is clear that $Z_\alpha = M_Q A_Q$ and $M_\alpha = M_Q$. Therefore our first assertion follows from [2(i), Lem. 4.9].

Now $\mathfrak{z}_\alpha$ is the centralizer of H_0 in $\mathfrak{g}$. Hence $\mathfrak{z}_\alpha \supset \mathfrak{m} + \mathfrak{a}$ and $\mathfrak{z}_\alpha \cap \mathfrak{n} = \mathfrak{n}_\alpha$. This implies that

$$\mathfrak{z}_\alpha = \theta(\mathfrak{n}_\alpha) + \mathfrak{m} + \mathfrak{a} + \mathfrak{n}_\alpha \, .$$

Since $\mathfrak{a} = \sigma_\alpha + \mathbf{R}H_\alpha$ and σ_α lies in the center of $\mathfrak{z}_\alpha$, we conclude that

$$\mathfrak{m}_\alpha = \theta(\mathfrak{n}_\alpha) + \mathfrak{m} + \mathfrak{a}_\alpha + \mathfrak{n}_\alpha \, .$$

This shows that $\mathfrak{m} + \mathfrak{a}_\alpha + \mathfrak{n}_\alpha$ is a parabolic subalgebra of $\mathfrak{m}_\alpha$. Since it is obviously the Lie algebra of $M_\alpha \cap P$, it follows that $*P_\alpha$ is a psgp of M_α and its Langlands decomposition is $MA_\alpha N_\alpha$.

Let P_1, P_2 be two distinct elements in $\mathscr{P}(A)$ and α an element in $\Sigma(P_2 \mid P_1)$.

Then $\alpha \in \Sigma(P_1)$ and $-\alpha \in \Sigma(P_2)$. Hence if $*P = M_\alpha \cap P_1$, it follows from Lemma 3 that

$$M_\alpha \cap P_2 = {}^*\bar{P} .$$

Now suppose P_1, P_2 are adjacent. Then $\Sigma(P_2 \mid P_1) = \{\alpha\}$ where α is a simple root of P_1. Put

$$(P', A') = (P_1, A)_{\{\alpha\}}$$

in the notation of [2(i), § 6]. Then $P' = M'A'N'$ and it is obvious that $M' = M_\alpha$. Since $\dim A = \dim A' + 1$, it follows that $*P$ is a maximal psgp of M'.

LEMMA 4. *Let P_1, P_2 be two adjacent elements in $\mathscr{P}(A)$ and α the unique root in $\Sigma(P_2 \mid P_1)$. Put $(P', A') = (P_1, A)_{\{\alpha\}}$ $(P' = M'A'N')$ and $*P = M' \cap P_1$ $(*P = {}^*M \cdot {}^*A \cdot {}^*N)$. Then*

$$I_{P_2 \mid P_1}(\nu) = I_{*\bar{P} \mid *P}({}^*\nu) \qquad (\nu \in \mathfrak{a}^*)$$

where $*\nu$ is the restriction of ν on $*\mathfrak{a} = \mathbf{R}H_\alpha$.

We observe that $*M = M$, $*N = N_\alpha$ and

$$\mathfrak{n}_2 \cap \theta(\mathfrak{n}_1) = \theta(\mathfrak{n}_\alpha) .$$

Hence $N_2 \cap \bar{N}_1 = {}^*\bar{N}$. Put $*K = K \cap M'$. Then $M' = {}^*K \cdot {}^*P$. Put $*H(m') = H_{*P}(m')$ for $m' \in M'$. Then it is obvious that

$$H_1(m') = {}^*H(m') \in {}^*\mathfrak{a} = \mathbf{R}H_\alpha$$

for $m' \in M'$. Put $*\rho = \rho_{*P}$. Since $\mathfrak{n}_1 = \mathfrak{n}' + \mathfrak{n}_\alpha$, it follows that $\rho_1 = {}^*\rho$ on $*\mathfrak{a}$. Therefore since $*\bar{N} \subset M'$, we conclude that

$$I_{P_2 \mid P_1}(\nu) = \int_{N_2 \cap \bar{N}_1} \exp\{-(\nu + \rho_1)(H_1(n))\}dn$$

$$= \int_{*\bar{N}} \exp\{-(\nu + \rho_1)({}^*H(n))\}dn = I_{*\bar{P} \mid *P}({}^*\nu) .$$

Now we come to the proof of Lemma 2. If $d(P_2, P_1) = 0$, our statement is true since both sides are 1. So we assume that $d(P_2, P_1) \geq 1$. First assume that $d(P_2, P) = 1$. Put $(P', A') = (P, A)_{\{\alpha\}}$ $(P' = M'A'N')$ where α is the unique root in $\Sigma(P_2 \mid P)$. Put $*P = M' \cap P$ $(*P = {}^*M \cdot {}^*A \cdot {}^*N)$. Then

$$M' \cap P_2 = {}^*\bar{P} , \quad *M = M , \quad *N = N_\alpha \quad \text{and} \quad *\mathfrak{a} = \mathbf{R}H_\alpha .$$

Moreover

$$I_{P_2 \mid P}(\nu) = I_{*\bar{P} \mid *P}({}^*\nu)$$

from Lemma 4, where $*\nu$ is the restriction of ν on $*\mathfrak{a}$.

Now $P = MAN$, $N = N' \cdot N_\alpha$, $N_2 = \bar{N}_\alpha N'$ and $N_\alpha \cup \bar{N}_\alpha \subset M'$. Since

$\alpha \in \Sigma(P_2 \mid P) \subset \Sigma(P_2 \mid P_1)$, it follows that

$$N \cap \bar{N}_1 = N' \cap \bar{N}_1, \; N_2 \cap \bar{N}_1 = \bar{N}_\alpha(N' \cap \bar{N}_1) = {}^*\bar{N}(N' \cap \bar{N}_1) \,.$$

Let $d\mu$ denote the invariant measure on $N'/N' \cap N_1$ such that

$$d\mu \cdot d(N' \cap N_1) = dN' \,.$$

We observe that $N' = (N' \cap \bar{N}_1)(N' \cap N_1)$ and the projection of N' on $N'/N' \cap N_1$ defines a homeomorphism of $N' \cap \bar{N}_1$ onto $N'/N' \cap N_1$. Moreover $d(N' \cap \bar{N}_1)$ corresponds to $d\mu$ under this homeomorphism. Hence

$$I_{P_2 \mid P_1}(\nu) = \int_{N_2 \cap \bar{N}_1} \exp\{-(\nu + \rho_1)(H_1(n))\} dn$$

$$= \int_{{}^*\bar{N} \times N'/N' \cap N_1} \exp\{-(\nu + \rho_1)(H_1({}^*\bar{n} \cdot n'))\} d^*\bar{n}\, d\mu \,.$$

Here $d^*\bar{n} = d^*\bar{N}$. Fix ${}^*\bar{n} \in {}^*\bar{N}$ and put ${}^*K = K \cap M'$. Since $M' = {}^*K \cdot {}^*P$, we have ${}^*\bar{n} = {}^*k\,{}^*p$ where ${}^*k \in {}^*K$, ${}^*p \in {}^*P$. Then

$$H_1({}^*\bar{n}n') = H_1({}^*pn') = H_1({}^*pn' \cdot {}^*p^{-1}) + H_1({}^*p)$$
$$= H_1({}^*pn' \cdot {}^*p^{-1}) + {}^*H({}^*p)$$

for $n' \in N'$. Here ${}^*H(m') = H_{*P}(m')$ $(m' \in M')$ as before and we have to use the fact that ${}^*N = N_\alpha \subset N_1$. Moreover ${}^*P = M \cdot {}^*AN_\alpha \subset M' \cap P_1$. Therefore ${}^*p$ normalizes both N' and N_1. Hence

$$\int_{N'/N' \cap N_1} \exp\{-(\nu + \rho_1)(H_1({}^*\bar{n}n'))\} d\mu$$

$$= \exp\{-(\nu + \rho_1)({}^*H({}^*p))\} \delta_{N'/N' \cap N_1}({}^*p^{-1}) \int_{N'/N' \cap N_1} \exp\{-(\nu + \rho_1)(H_1(n'))\} d\mu$$

where

$$\delta_{N'/N' \cap N_1}({}^*p^{-1}) = \left| \det\left(\mathrm{Ad}({}^*p^{-1})\right)_{\mathfrak{n}'/\mathfrak{n}' \cap \mathfrak{n}_1} \right| \,.$$

Let ${}^*p = {}^*m \cdot {}^*a \cdot {}^*n$ $({}^*m \in {}^*M, \; {}^*a \in {}^*A, \; {}^*n \in {}^*N)$. Then ${}^*H({}^*p) = \log {}^*a = {}^*H({}^*\bar{n})$ and

$$\delta_{N'/N' \cap N_1}({}^*p^{-1}) = \det\left(\mathrm{Ad}({}^*a^{-1})\right)_{\mathfrak{n}' \cap \bar{\mathfrak{n}}_1} \,.$$

Define ρ^+ and ρ^- in $\mathfrak{a}^*$ as follows.

$$\rho^+(H) = \tfrac{1}{2}\mathrm{tr}(\mathrm{ad}\,H)_{\mathfrak{n} \cap \mathfrak{n}_1} \,,$$
$$\rho^-(H) = \tfrac{1}{2}\mathrm{tr}(\mathrm{ad}\,H)_{\bar{\mathfrak{n}} \cap \mathfrak{n}_1} \qquad (H \in \mathfrak{a}) \,.$$

Since $\mathfrak{n}_1 = \mathfrak{n} \cap \mathfrak{n}_1 + \bar{\mathfrak{n}} \cap \mathfrak{n}_1$ and $\mathfrak{n} = \mathfrak{n} \cap \mathfrak{n}_1 + \mathfrak{n} \cap \bar{\mathfrak{n}}_1$, it follows that

$$\rho_1 = \rho^+ + \rho^- \,, \qquad \rho = \rho^+ - \rho^-$$

where $\rho = \rho_P$. Therefore since $\mathfrak{n}' \cap \bar{\mathfrak{n}}_1 = \mathfrak{n} \cap \bar{\mathfrak{n}}_1$, we get

$$\delta_{N'/N' \cap N_1}({}^*p^{-1}) = \det\left(\mathrm{Ad}({}^*a^{-1})\right)_{\mathfrak{n} \cap \bar{\mathfrak{n}}_1}$$
$$= \det\left(\mathrm{Ad}({}^*a)\right)_{\bar{\mathfrak{n}} \cap \mathfrak{n}_1} = \exp\{2\rho^-({}^*H({}^*p))\} \,.$$

Hence

$$\exp\{-(\nu + \rho_1)(^*H(^*p))\}\delta_{N'/N'\cap N_1}(^*p^{-1}) = \exp\{-(\nu + \rho)(^*H(^*\bar{n}))\}$$

and this shows that

$$I_{P_2|P_1}(\nu) = \int_{^*\bar{N}} \exp\{-(\nu + \rho)(^*H(^*\bar{n}))\}d^*\bar{n}\cdot\int_{N'/N'\cap N_1} \exp\{-(\nu + \rho_1)(H_1(n'))\}d\mu\ .$$

But since $d\mu$ corresponds to $d(N' \cap \bar{N}_1)$ and $N' \cap \bar{N}_1 = N \cap \bar{N}_1$, we have

$$\int_{N'/N'\cap N_1} \exp\{-(\nu + \rho_1)(H_1(n'))\}d\mu = I_{P|P_1}(\nu)\ .$$

This proves that

$$\begin{aligned} I_{P_2|P_1}(\nu) &= I_{^*\bar{P}|^*P}(^*\nu)I_{P|P_1}(\nu) \\ &= I_{P_2|P}(\nu)I_{P|P_1}(\nu)\ . \end{aligned}$$

The above result enables us to prove the following lemma by induction on $d(P_2, P_1)$.

LEMMA 5. *For* $\alpha \in \Sigma(P_2|P_1)$, *define* $^*P_\alpha = MA_\alpha N_\alpha$ *as in Lemma* 3. *Then*

$$I_{P_2|P_1}(\nu) = \prod_{\alpha \in \Sigma(P_2|P_1)} I_{^*\bar{P}_\alpha|^*P_\alpha}(\nu_\alpha) \qquad\qquad (\nu \in \mathfrak{a}^*)$$

where ν_α *is the restriction of* ν *on* $\mathbf{R}H_\alpha$.

It is clear that Lemma 2 is an immediate consequence of Lemma 5.

For $P \in \mathscr{P}(A)$, put

$$\gamma(P) = I_{\bar{P}|P}(\rho_P) = \int_{\bar{N}} \exp\{-2\rho_P(H_P(\bar{n}))\}d\bar{n}\ .$$

We know from [2(j), Lem. 20.3] that $\gamma(P) < \infty$. If α is a reduced root of $(\mathfrak{g}, \mathfrak{a})$ define

$$c_\alpha = \inf_P \langle \rho_P, \alpha \rangle$$

where P runs over all elements in $\mathscr{P}(A)$ such that $\alpha \in \Sigma(P)$.

COROLLARY 1. *Let* $P_1, P_2 \in \mathscr{P}(A)$ *and* $\nu \in \mathfrak{a}^*$. *Then* $I_{P_2|P_1}(\nu)$ *depends only on the values of* $\langle \nu, \alpha \rangle$ *for* $\alpha \in \Sigma(P_2|P_1)$. *Moreover* $I_{P_2|P_1}(\nu) < \infty$ *if*

$$\langle \nu, \alpha \rangle \geqq c_\alpha$$

for all $\alpha \in \Sigma(P_2|P_1)$.

The first part is obvious from Lemma 5. Now fix $\alpha \in \Sigma(P_2|P_1)$ and suppose there exists an element $P \in \mathscr{P}(A)$ such that $\alpha \in \Sigma(P)$ and

$$\langle \nu, \alpha \rangle \geqq \langle \rho_P, \alpha \rangle\ .$$

Put $\mu = \rho_P$. Then since

$$I_{\bar{P}|P}(\mu) = \gamma(P) < \infty$$

and $\alpha \in \Sigma(P) = \Sigma(\bar{P}|P)$, we conclude from Lemma 5 that

$$I_{*\bar{P}_\alpha|*P_\alpha}(\mu_\alpha) < \infty .$$

Hence

$$I_{*\bar{P}_\alpha|*P_\alpha}(\nu_\alpha) \leqq I_{*\bar{P}_\alpha|*P_\alpha}(\mu_\alpha) < \infty .$$

Our assertion now follows from Lemma 5.

For any reduced root α of $(\mathfrak{g}, \mathfrak{a})$, define

$$\rho_\alpha = \tfrac{1}{2} \sum_{k \geq 1} m_k \cdot k\alpha$$

where $m_k = \dim \mathfrak{g}_{k\alpha}$ in the notation of Lemma 3. Then $\rho_\alpha = \rho_{*P_\alpha}$. Put

$$\gamma_\alpha = \gamma(*P_\alpha) = \int_{\bar{N}_\alpha} \exp\{-2\rho_\alpha(H_\alpha(\bar{n}))\}d\bar{n}$$

where $H_\alpha(y) = H_{*P_\alpha}(y)$ $(y \in M_\alpha)$.

COROLLARY 2. *In order that* $I_{P_2|P_1}(\nu) < \infty$, *it is sufficient that*

$$\langle \nu, \alpha \rangle \geqq \langle \rho_\alpha, \alpha \rangle$$

for all $\alpha \in \Sigma(P_2 \mid P_1)$.

This is obvious from Lemma 5 since $\gamma_\alpha < \infty$.

Put

$$\gamma_{P_2|P_1} = \prod_{\alpha \in \Sigma(P_2|P_1)} \gamma_\alpha$$

for $P_1, P_2 \in \mathscr{P}(A)$. We note that if $P \in \mathscr{P}(A)$ $(P = MAN)$, then

$$H_{\bar{P}}(\theta(\bar{n})) = -H_P(\bar{n}) \qquad\qquad (\bar{n} \in \bar{N})$$

and therefore $\gamma(P) = \gamma(\bar{P})$. In particular $\gamma_{-\alpha} = \gamma_\alpha$ and this shows that $\gamma_{\bar{P}|P}$ is independent of $P \in \mathscr{P}(A)$. Moreover we have the following result.

LEMMA 6. $\gamma(P)$ *is independent of* $P \in \mathscr{P}(A)$.

Fix $P \in \mathscr{P}(A)$ and let (P_0, A_0) be a minimal p-pair in G such that $A_0) \prec (P, A)$. Put $*P = M \cap P_0$. Then $*P$ is a minimal psgp of M and by [2(j), Lem. 20.3]

$$\gamma(P_0) = \gamma(*P)\gamma(P) .$$

Since any two minimal psgps of G are conjugate under K, it is clear that $\gamma(P_0)$ is independent of $P_0 \in \mathscr{P}(A_0)$. Similarly for $\gamma(*P)$. The statement of the lemma is now obvious.

3. A lemma of Arthur

Let π be a representation of G on a complex, locally convex and complete Hausdorff space V. As usual let $\mathscr{E}(K)$ be the set of all equivalence classes of irreducible (finite-dimensional) representations of K. For any $\mathfrak{b} \in \mathscr{E}(K)$, let $\xi_\mathfrak{b}$ denote the character of $\mathfrak{b}$ and put $\alpha_\mathfrak{b} = d(\mathfrak{b}) \operatorname{conj} \xi_\mathfrak{b}$ where

$d(\mathfrak{b}) = \xi_\mathfrak{b}(1)$ is the degree of $\mathfrak{b}$. For a finite subset F of $\mathfrak{E}(K)$, put

$$\alpha_F = \sum_{\mathfrak{b} \in F} \alpha_\mathfrak{b}$$

and consider the operator

$$E_F = \int_K \alpha_F(k)\pi(k)dk$$

on V. (Here dk is the normalized Haar measure on K.) Put $V_F = E_F V$. We say that π is admissible if $\dim V_F < \infty$ for every F.

Suppose π is admissible. Put $\mathfrak{X}_F = \mathrm{End}\, V_F$. We extend $T \in \mathfrak{X}_F$ to a continuous linear transformation on V by setting

$$Tv = TE_F v \qquad\qquad (v \in V).$$

Then $\mathfrak{X}_{F'} \subset \mathfrak{X}_F$ if $F' \subset F$. Put

$$\mathrm{End}^0\, V = \bigcup_F \mathfrak{X}_F$$

where F runs over all finite subsets of $\mathfrak{E}(K)$. Then $\mathrm{End}^0\, V$ is a subalgebra of the algebra of all continuous endomorphisms of V.

Put $\mathfrak{L} = C_c^\infty(G)$ and

$$f_F = \alpha_F *_K f *_K \alpha_F \qquad\qquad (f \in \mathfrak{L})$$

where the convolution is over K. Let $\mathfrak{L}_F$ be the image of $\mathfrak{L}$ under the mapping $f \mapsto f_F$. $\mathfrak{L}$ is an algebra under convolution (over G) and $\mathfrak{L}_F$ is a subalgebra of $\mathfrak{L}$. Define

$$\pi(f) = \int_G f(x)\pi(x)dx \qquad\qquad (f \in \mathfrak{L})$$

where dx is the standard Haar measure on G [2(i), §7]. Then it is obvious that

$$E_F\pi(f)E_F = \pi(f_F).$$

Therefore the mapping $f \mapsto \pi(f)$ $(f \in \mathfrak{L}_F)$ may be regarded as a representation of $\mathfrak{L}_F$ on V_F. We denote this representation by π_F.

For $T \in \mathrm{End}^0\, V$, the operator $T\pi(x)$ $(x \in G)$ is of finite rank. Put

$$f_T(x) = \mathrm{tr}\, T\pi(x) \qquad\qquad (x \in G).$$

The following lemma, which I owe to J. Arthur, is very simple but it will be crucial for us.

LEMMA 1. *Let π be an admissible representation of G on V. Fix a finite subset F of $\mathfrak{E}(K)$ and for $T \in \mathrm{End}\, V_F$ put $f_T(x) = \mathrm{tr}\, T\pi(x)\,(x \in G)$. Then the following three statements are equivalent.*

(1) *The mapping $T \mapsto f_T$ is injective on $\mathrm{End}\, V_F$.*

(2) $\pi(\mathfrak{L}_F) = \mathrm{End}\, V_F.$

(3) π_F *is irreducible.*

Put $\mathfrak{X} = \operatorname{End} V_F$. Then $\mathfrak{X}$ is a simple algebra and obviously (2) and (3) are equivalent by Burnside's theorem. Now the bilinear form

$$(S, T) \longmapsto \operatorname{tr} ST \qquad\qquad (S, T \in \mathfrak{X})$$

is nondegenerate on $\mathfrak{X}$. Let $\mathfrak{X}_0$ be the space of all $T \in \mathfrak{X}$ such that

$$\operatorname{tr} T\pi(\beta) = 0$$

for all $\beta \in \mathfrak{L}_F$. Then $\mathfrak{X}_0 = \{0\}$ if and only if $\pi(\mathfrak{L}_F) = \mathfrak{X}$. Hence in order to prove that (1) is equivalent to (2), it is sufficient to verify that $\mathfrak{X}_0$ is exactly the set of all $T \in \mathfrak{X}$ such that $f_T = 0$.

Fix $T \in \mathfrak{X}$ and $\beta \in \mathfrak{L}$. Then it is obvious that

$$\operatorname{tr} T\pi(\beta_F) = \operatorname{tr} T\pi(\beta) = \int \beta(x) f_T(x) dx \ .$$

Since f_T is a continuous function on G, it is clear that $f_T = 0$ if and only if $\operatorname{tr} T\pi(\beta) = 0$ for all $\beta \in \mathfrak{L}$. This shows that $f_T = 0$ if and only if $T \in \mathfrak{X}_0$.

4. Induced representations

Let A be a special vector subgroup of G. Fix $P \in \mathscr{P}(A)\,(P = MAN)$ and let σ be a unitary and admissible representation of M on a Hilbert space U. Put $K_M = K \cap M$ and let $\mathfrak{H}$ denote the space of the representation

$$\pi_K = \operatorname{Ind}_{K_M}^K (\sigma \mid K_M) \ .$$

Then $\mathfrak{H}$ is the Hilbert space consisting of all measurable functions $f: K \to U$ such that:[2]

(1) $f(mk) = \sigma(m) f(k)$ for all $m \in K_M$ and $k \in K$.

(2) $\|f\|^2 = \displaystyle\int_K |f(k)|^2\, dk < \infty$.

Then $\pi_K(k_0) f \ (k_0 \in K)$ is defined to be the function

$$k \longmapsto f(kk_0) \qquad\qquad (k \in K) \ .$$

Clearly π_K is a unitary representation of K on $\mathfrak{H}$. Fix $\mathfrak{b} \in \mathscr{E}(K)$ and for $\delta \in \mathscr{E}(K_M)$, let $[\mathfrak{b}\!:\!\delta]$ and $[\sigma\!:\!\delta]$ denote the multiplicities of δ in $\mathfrak{b}$ and σ respectively. Since σ is admissible, $[\sigma\!:\!\delta] < \infty$ and we conclude from the Frobenius reciprocity theorem that

$$\dim \mathfrak{H}_\mathfrak{b} = d(\mathfrak{b}) \sum_{\delta \in \mathscr{E}(K_M)} [\mathfrak{b}\!:\!\delta][\sigma\!:\!\delta] < \infty \ .$$

This shows that π_K is admissible.

Let $T \in \operatorname{End} \mathfrak{H}_F$ where F is a finite subset of $\mathscr{E}(K)$. Then since π_K is unitary, its adjoint T^* is also in $\operatorname{End} \mathfrak{H}_F$. Therefore $\operatorname{End}^0 \mathfrak{H}$ is stable under

---

[2] We do not distinguish between two such functions which coincide almost everywhere.

the mapping $T \mapsto T^*$. The same holds for $\mathrm{End}^0 U$, since σ is unitary.

Put $\mathfrak{F} = \mathfrak{a}^*$. Then for $\nu \in \mathfrak{F}_c$, we define a representation $\pi = \pi_{P,\nu}$ of G on $\mathfrak{H}$ as follows. Put

$$\sigma_\nu(man) = \sigma(m) \exp\{(-1)^{1/2}\nu(\log a)\} \qquad (m \in M,\, a \in A,\, n \in N)\,.$$

Then σ_ν is a representation of P on U. Let $\mathfrak{H}_{P,\nu}$ be the Hilbert space of all measurable functions $f\colon G \to U$ such that:[2]

(1) $f(px) = \delta_P(p)^{1/2}\sigma_\nu(p)f(x)$ for all $p \in P$ and $x \in G$.

(2) $\|f\|^2 = \displaystyle\int_K |f(k)|^2 dk < \infty$.

Here δ_P is the module of P[2(j), § 19]. Since $G = PK$, it is clear that $\|f\| = 0$ implies $f = 0$. For $f \in \mathfrak{H}_{P,\nu}$, let h_f denote the restriction of f on K. Then $f \mapsto h_f$ is a unitary isomorphism of $\mathfrak{H}_{P,\nu}$ onto $\mathfrak{H}$ and we shall identify these two spaces under this mapping whenever it is convenient to do so. If $f \in \mathfrak{H}_{P,\nu} = \mathfrak{H}$ and $y \in G$, then $\pi(y)f$ is the function $x \mapsto f(xy)$ on G. It is obvious that $\pi|K = \pi_K$ and therefore π is admissible.

LEMMA 1. *Let* $\alpha \in C_c(G)$ *and* $h \in \mathfrak{H}$. *Then* $h' = \pi_{P,\nu}(\alpha)h$ *is given by*

$$h'(k_2) = \int_K \kappa_\alpha(k_2\colon k_1)h(k_1^{-1})dk_1 \qquad (k_2 \in K)\,,$$

where

$$\kappa_\alpha(k_2\colon k_1) = \int_{M \times A \times N} \alpha(k_2^{-1}man\,k_1^{-1})\sigma(m)\exp\{((-1)^{1/2}\nu + \rho)(\log a)\}dmdadn$$

for $k_1, k_2 \in K$. (*Here* $\rho = \rho_P$.)

We observe that $\kappa_\alpha(k_2\colon k_1)$ is a bounded operator on U and, for a fixed $u \in U$, the mapping

$$(k_2,\, k_1) \longmapsto \kappa_\alpha(k_2\colon k_1)u$$

of $K \times K$ into U is continuous.

It follows from the definition of π that

$$h'(k_2) = \int \alpha(x)h(k_2x)dx = \int \alpha(k_2^{-1}x)h(x)dx$$

$$= \int \alpha(k_2^{-1}man\,k_1)\delta_P(a)^{1/2}\sigma_\nu(ma)h(k_1)dk_1dmdadn$$

$$= \int_K \kappa_\alpha(k_2\colon k_1)h(k_1^{-1})dk_1\,.$$

COROLLARY 1. *Fix* $h_1, h_2 \in \mathfrak{H}$ *and* $\alpha \in C_c(G)$. *Then the function*

$$\nu \longmapsto \big(h_2,\, \pi_{P,\nu}(\alpha)h_1\big)$$

---

[2] We do not distinguish between two such functions which coincide almost everywhere.

is holomorphic on $\mathfrak{F}_c$.

Since α has compact support, this is obvious from the above lemma.

COROLLARY 2. *Fix $h \in \mathfrak{H}$, $x \in G$ and define*

$$h'(k) = \exp\{-((-1)^{1/2}\nu + \rho)(H(y))\}\sigma(\mu(y))^{-1}h(\kappa(y)^{-1}k) \qquad (k \in K)$$

where $y = kx^{-1}k^{-1}$ and the notation is as in [2(j), Thm. 19.1]. Then $h' = \pi_{P,\nu}(x)h$.

Since $kx = y^{-1}k$, this follows from the definition of π.

For $\alpha \in C_c(G)$, put $\tilde{\alpha}(x) = \mathrm{conj}\,\alpha(x^{-1})$ $(x \in G)$. Then it is easy to verify that

$$\kappa_\alpha(k_2\colon k_1)^* = \int_{M \times A \times N} \tilde{\alpha}(k_1 man\, k_2)\sigma(m)\exp\{((-1)^{1/2}\bar{\nu} + \rho)(\log a)\}dm\,da\,dn\,,$$

where the bar denotes conjugation in $\mathfrak{F}_c$ with respect to $\mathfrak{F}$. Therefore

$$\big(h_2,\, \pi_{P,\nu}(\alpha)h_1\big) = \big(\pi_{P,\bar{\nu}}(\tilde{\alpha})h_2,\, h_1\big)$$

for $h_1,\, h_2 \in \mathfrak{H}$. In particular this shows that $\pi_{P,\nu}$ is unitary if $\nu \in \mathfrak{F}$. Moreover we have the following result.

COROLLARY 3. *Suppose $\nu \in \mathfrak{F}$. Then*

$$\kappa_\alpha(k_2\colon k_1)^* = \kappa_{\tilde{\alpha}}(k_1^{-1}\colon k_2^{-1})$$

for $\alpha \in C_c(G)$ and $k_1,\, k_2 \in K$.

We write

$$\pi_{P,\nu} = \mathrm{Ind}_P^G \delta_P^{1/2}\sigma_\nu$$

since $\pi_{P,\nu}$ is obtained by inducing to G from the representation $\delta_P^{1/2}\sigma_\nu$ of P on U.

LEMMA 2. *Let $\alpha,\, \beta \in C_c(G)$. Then*

$$\kappa_\alpha(m_2 k_2\colon k_1 m_1) = \sigma(m_2)\kappa_\alpha(k_2\colon k_1)\sigma(m_1) \qquad (m_1,\, m_2 \in K_M)$$

and

$$\kappa_{\alpha*\beta}(k_2\colon k_1) = \int_K \kappa_\alpha(k_2\colon k)\kappa_\beta(k^{-1}\colon k_2)dk$$

for $k_1,\, k_2 \in K$.

The first statement is obvious and the second follows from the fact that $\pi(\alpha*\beta) = \pi(\alpha)\pi(\beta)$.

5. Intertwining operators

Let $P_i \in \mathscr{P}(A)$ $(P_i = MAN_i,\ i = 1,\, 2)$ and $\nu \in \mathfrak{F}_c$. Then we define, at first formally, a linear mapping

$$J_{P_2|P_1}(\nu): \mathfrak{H}_{P_1,\nu} \longrightarrow \mathfrak{H}_{P_2,\nu}$$

as follows. If $h \in \mathfrak{H}_{P_1,\nu}$, then $h' = J_{P_2|P_1}(\nu)h$ is given by

$$h'(x) = \gamma_{P_2|P_1}^{-1} \int_{N_2 \cap \bar{N}_1} h(\bar{n}x)d\bar{n}$$

$$= \gamma_{P_2|P_1}^{-1} \int_{N_1 \cap N_2 \backslash N_2} h(nx)d\mu \qquad\qquad (x \in G) .$$

Here $d\bar{n} = d(N_2 \cap \bar{N}_1)$ and the invariant measure $d\mu$ on $N_1 \cap N_2\backslash N_2$ is so normalized that

$$dN_2 = d(N_1 \cap N_2) \cdot d\mu .$$

The constant $\gamma_{P_2|P_1}$ has been defined in Section 2.

Fix $\bar{n} \in N_2 \cap \bar{N}_1$. Then we can write

$$\bar{n}^{-1} = k_0 m_0 a_0 n_0 \qquad\qquad (k_0 \in K, \ m_0 \in M, \ a_0 \in A, \ n_0 \in N_1)$$

and since σ is unitary, we have[3]

$$\left\{ \int_K | h(\bar{n}k) |^2 dk \right\}^{1/2} \leqq \| h \| \exp \{(\nu_I - \rho_I)(\log a_0)\} .$$

Therefore the following lemma is obvious.

LEMMA 1. *Suppose ν is an element in $\mathfrak{F}_c$ such that*

$$I_{P_2|P_1}(-\nu_I) < \infty .$$

Then the above integral defines a bounded operator $J_{P_2|P_1}(\nu)$. In fact

$$\| J_{P_2|P_1}(\nu)h \| \leqq \gamma_{P_2|P_1}^{-1} I_{P_2|P_1}(-\nu_I) \| h \|$$

for $h \in \mathfrak{H}_{P_1,\nu}$.

COROLLARY. $J_{P_2|P_1}(\nu)\pi_{P_1,\nu}(x) = \pi_{P_2,\nu}(x)J_{P_2|P_1}(\nu) \quad (x \in G)$.

This is obvious since G acts by right translations.

LEMMA 2. *Suppose P lies between P_1 and P_2 $(P \in \mathscr{P}(A))$. Then*

$$J_{P_2|P_1}(\nu) = J_{P_2|P}(\nu)J_{P|P_1}(\nu)$$

provided $I_{P_2|P_1}(-\nu_I) < \infty$.

It follows from Lemma 2.2 that $I_{P_2|P}(-\nu_I)$ and $I_{P|P_1}(-\nu_I)$ are both finite. Hence $J_{P_2|P}(\nu)$ and $J_{P|P_1}(\nu)$ are defined. Since P lies between P_1, P_2, we have,

$$N_2 \cap \bar{N}_1 = (N \cap \bar{N}_1)(N_2 \cap \bar{N}) .$$

Let dn, dn', dn'' denote the measures $d(N_2 \cap \bar{N}_1)$, $d(N \cap \bar{N}_1)$, $d(N_2 \cap \bar{N})$ respectively. Then $dn = dn'dn''$ $(n = n'n'')$. Since $\gamma_{P_2|P_1} = \gamma_{P_2|P}\gamma_{P|P_1}$, our assertion follows immediately by Fubini's theorem.

---

[3] As usual ν_R and ν_I denote the real and imaginary parts of ν [2(j), § 17].

COROLLARY 1. *Let $\mathfrak{F}_c(P_2 \mid P_1)$ be the open set consisting of all $\nu \in \mathfrak{F}_c$ such that*

$$\langle \nu_I + \rho_1, \alpha \rangle < 0$$

for all $\alpha \in \Sigma(P_2 \mid P_1)$. Then for a fixed $h \in \mathfrak{H}$, the mapping

$$\nu \longmapsto J_{P_2 \mid P_1}(\nu)h$$

from $\mathfrak{F}_c(P_2 \mid P_1)$ to $\mathfrak{H}$ is holomorphic.

This is an easy consequence of Corollary 1 of Lemma 2.5 and what we have seen above. (See also Lemma 3 below.)

Fix a finite subset F of $\mathfrak{E}(K)$. Then since $J_{P_2 \mid P_1}(\nu)$ commutes with π_K, it leaves $\mathfrak{H}_F$ invariant. Let $j_{P_2 \mid P_1}(\nu)$ denote the restriction of $J_{P_2 \mid P_1}(\nu)$ on $\mathfrak{H}_F$. Then $j_{P_2 \mid P_1}$ is a holomorphic function from $\mathfrak{F}_c(P_2 \mid P_1)$ to End $\mathfrak{H}_F$.

COROLLARY 2.

$$\pi_{P_2, \nu}(\alpha) j_{P_2 \mid P_1}(\nu) = j_{P_2 \mid P_1}(\nu)\pi_{P_1, \nu}(\alpha)$$

for $\alpha \in \mathfrak{L}_F$ and $\nu \in \mathfrak{F}_c(P_2 \mid P_1)$.

This is obvious from the corollary of Lemma 1.

We shall see that if σ is an irreducible and square-integrable representation of M, then $j_{P_2 \mid P_1}$ extends to a meromorphic function on $\mathfrak{F}_c$.

Fix $P \in \mathscr{P}(A)$ $(P = MAN)$ and use the notation of [2(j), Thm. 19.1].

LEMMA 3. *Suppose $\nu \in \mathfrak{F}_c$, $P_2 \in \mathscr{P}(A)$, $h \in \mathfrak{H}$ and $I_{P_2 \mid P}(-\nu_I) < \infty$. Then $h' = J_{P_2 \mid P}(\nu)h$ is given by the formula*

$$h'(k) = \gamma_{P_2 \mid P}^{-1} \int_{N_2 \cap \bar{N}} \exp\left\{-((-1)^{1/2}\nu + \rho)(H(\bar{n}))\right\}\sigma\big(\mu(\bar{n})\big)^{-1} h\big(\kappa(\bar{n})^{-1}k\big)d\bar{n} \quad (k \in K),$$

where $d\bar{n} = d(N_2 \cap \bar{N})$.

Fix $k_0 \in K$ and $\bar{n} \in N_2 \cap \bar{N}$. Then $\bar{n} = kman$ where $k = \kappa(\bar{n})$, $m = \mu(\bar{n})$, $a = \exp H(\bar{n})$ and $n \in N$. Hence

$$h(\bar{n}^{-1}k_0) = h(a^{-1}m^{-1}k^{-1}k_0) = \exp\left\{-((-1)^{1/2}\nu + \rho)(\log a)\right\}\sigma(m^{-1})h(k^{-1}k_0)$$

for $h \in \mathfrak{H}_{P,\nu} = \mathfrak{H}$. Therefore

$$\int_{N_2 \cap \bar{N}} h(\bar{n}k_0)d\bar{n} = \int_{N_2 \cap \bar{N}} \exp\left\{-((-1)^{1/2}\nu + \rho)(H(\bar{n}))\right\}\sigma\big(\mu(\bar{n})\big)^{-1} h\big(\kappa(\bar{n})^{-1}k_0\big)d\bar{n}$$

and this implies our assertion.

6. The mapping $T \to \kappa_T$

A function κ from $K \times K$ to End$^0 U$ will be called smooth if:

(1) there exists a finite subset F of $\mathfrak{E}(K_M)$ such that $\kappa(K \times K) \subset$ End U_F,

(2) κ regarded as a function from $K \times K$ to End U_F is $K \times K$-finite.

Let $\mathcal{K}$ be the space of all smooth functions κ from $K \times K$ to $\mathrm{End}^0 U$ such that

$$\kappa(m_2 k_2 : k_1 m_1) = \sigma(m_2)\kappa(k_2 : k_1)\sigma(m_1)$$

for $k_1, k_2 \in K$ and $m_1, m_2 \in K_M$.

Put $\mathfrak{T} = \mathrm{End}^0\,\mathfrak{H}$.

LEMMA 1. *There exists a linear bijection* $T \mapsto \kappa_T$ *of* $\mathfrak{T}$ *onto* $\mathcal{K}$ *such that*

$$(Th)(k_2) = \int_K \kappa_T(k_2 : k_1)h(k_1^{-1})dk_1 \qquad\qquad (k_2 \in K)$$

for all $h \in \mathfrak{H}$.

Given $\kappa \in \mathcal{K}$, define a linear transformation T on $\mathfrak{H}$ by

$$(Th)(k_2) = \int_K \kappa(k_2 : k_1)h(k_1^{-1})dk_1 \qquad\qquad (k_2 \in K)$$

for $h \in \mathfrak{H}$. It follows from the smoothness of κ, that $T \in \mathfrak{T}$ and $T \neq 0$ unless $\kappa = 0$. Conversely given $T \in \mathfrak{T}$, choose a finite subset F of $\mathfrak{E}(K)$ such that $T \in \mathrm{End}\,\mathfrak{H}_F$. Let $h_i\ (1 \leq i \leq r)$ be an orthonormal base for $\mathfrak{H}_F$. For $k_1, k_2 \in K$, consider the linear transformation $\kappa(k_2 : k_1)$ in U given by

$$\kappa(k_2 : k_1)u = \sum_{1 \leq i \leq r} h_i(k_2)\left((T^*h_i)(k_1^{-1}),\, u\right) \qquad\qquad (u \in U)\,.$$

Since h_i and T^*h_i are in $\mathfrak{H}_F$, $\kappa(k_2 : k_1) \in \mathrm{End}^0 U$ and κ is smooth. Then if $h \in \mathfrak{H}$,

$$\int_K \kappa(k_2 : k_1)h(k_1^{-1})dk_1 = \sum_{1 \leq i \leq r} h_i(k_2)(h_i,\, Th) = (Th)(k_2)\,.$$

It is easy to verify that $\kappa \in \mathcal{K}$ and the lemma follows.

We write $\pi(k)$ instead of $\pi_K(k)\ (k \in K)$.

COROLLARY. (1) $\kappa_{T^*}(k_2 : k_1) = \left(\kappa_T(k_1^{-1} : k_2^{-1})\right)^*$.

(2) $\kappa_{T_1 T_2}(k_2 : k_1) = \displaystyle\int_K \kappa_{T_1}(k_2 : k)\kappa_{T_2}(k^{-1} : k_1)dk$.

(3) $\mathrm{tr}\,T = \displaystyle\int_K \mathrm{tr}\,\kappa_T(k : k^{-1})dk$.

(4) *Suppose* $T' = \pi(k_1')T\pi(k_2')\ (k_1', k_2' \in K)$. *Then*

$$\kappa_{T'}(k_2 : k_1) = \kappa_T(k_2 k_1' : k_2' k_1)\,.$$

Here $T, T_1, T_2 \in \mathfrak{T}$ *and* $k_1, k_2 \in K$.

This is a simple exercise.

7. The relation between induced representations and Eisenstein integrals

In [2(i), §26], we have defined a double representation τ of K on $V = C^\infty(K \times K)$. It is unitary with respect to the norm

$$| v |^2 = \int_{K \times K} | v(k_1 \colon k_2) |^2 \, dk_1 dk_2 \qquad\qquad (v \in V) .$$

If F is a finite subset of $\mathfrak{E}(K)$, we denote by $C_F(K \times K) = V_F$ the subspace of all $v \in V$ such that

$$v = \int_K \alpha_F(k)\tau(k)v \, dk = \int_K \alpha_F(k)v\tau(k) \, dk .$$

(See §3 for the definition of α_F.) Then V_F is stable under τ and its dimension is finite. Let τ_F denote the restriction of τ on V_F.

Now suppose σ is an irreducible and square-integrable representation of M on U. Then σ is admissible. Let ω denote the class of σ in $\mathfrak{E}_2(M)$. If τ is a unitary double representation of K on a finite-dimensional Hilbert space V, we put $L(\omega) = \mathcal{C}_\omega(M, \tau_M)$ as usual (see [2(j), § 18]). Then $\dim L(\omega) < \infty$.

Fix a finite subset F of $\mathfrak{E}(K)$ and write (V, τ) for the pair $(C_F(K \times K), \tau_F)$ introduced above. For $T \in \operatorname{End} \mathfrak{H}_F$ define $\psi_T \in L(\omega)$ as follows. If $m \in M$, $\psi_T(m)$ is the element v of $V = C_F(K \times K)$ given by

$$v(k_1 \colon k_2) = \psi_T(k_1 \colon m \colon k_2) - \operatorname{tr} \{ \kappa_T(k_2 \colon k_1)\sigma(m) \} \qquad (k_1, k_2 \in K) .$$

LEMMA 1. *The mapping $T \to \psi_T$ from $\operatorname{End} \mathfrak{H}_F$ to $L(\omega)$ is bijective.*

Let T be an element in $\operatorname{End} \mathfrak{H}_F$ such that $\psi_T = 0$. Fix $k_1, k_2 \in K$. Then

$$\operatorname{tr}\big(\kappa_T(k_2 \colon k_1)\sigma(m)\big) = 0$$

for all $m \in M$. Since σ is irreducible, we conclude from Lemma 3.1 that $\kappa_T(k_2 \colon k_1) = 0$. This proves that $\kappa_T = 0$ and therefore $T = 0$ (Lem. 6.1).

Conversely let ψ be a given element in $L(\omega)$. Then ψ may be regarded as a function on $K \times M \times K$. Let F_0 be the set of all $\delta \in \mathfrak{E}(K_M)$ such that $[\mathfrak{b} \colon \delta] > 0$ for some $\mathfrak{b} \in F$. Then F_0 is a finite set. It follows from the definition of $L(\omega)$ that, for fixed $k_1, k_2 \in K$, the function

$$m \longmapsto \psi(k_1 \colon m \colon k_2)$$

on M is a finite linear combination of K_M-finite matrix coefficients of σ. Since σ is irreducible, we conclude from Lemma 3.1 that there exists a unique element $\kappa(k_2 \colon k_1) \in \operatorname{End}^0 U$ such that

$$\psi(k_1 \colon m \colon k_2) = \operatorname{tr}\big(\kappa(k_2 \colon k_1)\sigma(m)\big)$$

for all $m \in M$. Since $\psi \in L(\omega)$, we conclude easily that $\kappa(k_2 \colon k_1) \in \operatorname{End} U_{F_0}$ and κ lies in $\mathfrak{K}$. Therefore by Lemma 6.1, there exists a unique $T \in \operatorname{End}^0 \mathfrak{H}$ such that $\kappa = \kappa_T$. Applying part (4) of the corollary of Lemma 6.1, we conclude that $T \in \operatorname{End} \mathfrak{H}_F$ and this proves that $\psi = \psi_T$.

Let τ' be the double representation of K on $V = C_F(K \times K)$ defined as follows. If $v \in V$ and $k_1', k_2' \in K$, then $v' = \tau'(k_1')v\tau(k_2')$ is given by

$$v'(k_1: k_2) = v(k_2'k_1: k_2k_1') \qquad (k_1, k_2 \in K).$$

LEMMA 2. *Let $k_1, k_2 \in K$ and $T \in \operatorname{End} \mathfrak{H}_F$. Then*

$$\psi_{\pi(k_1)T\pi(k_2)}(m) = \tau'(k_1)\psi_T(m)\tau'(k_2)$$

for $m \in M$.

This is an immediate consequence of part (4) of the corollary Lemma 6.1.

Fix $P \in \mathscr{P}(A)$ and $\nu \in \mathfrak{F}_c$. Then for $\psi \in L(\omega)$, we consider the Eisenstein integral $E(P: \psi: \nu)$ [2(i), §§ 19, 26].

THEOREM 1. *Let $T \in \operatorname{End} \mathfrak{H}_F$. Then*

$$E(P: \psi_T: \nu: k_1: x: k_2) = \operatorname{tr} T\pi_{P,\nu}(k_1 x k_2)$$

for $k_1, k_2 \in K$ and $x \in G$.

We shall give a proof of this theorem in Section 9. First we need some preparation which will also be useful later on.

8. Some simple properties of $E(P: \psi: \nu)$

Let τ be a unitary double representation of K on a finite-dimensional Hilbert space. We make the following assumptions on V and τ.

(1) V is an associative algebra with an anti-involution $u \mapsto u^*$ ($u \in V$) over $\mathbf{R}$ so that[4]

$$(cu)^* = \bar{c}u^* , \quad (u \cdot v)^* = v^* \cdot u^* \qquad (c \in \mathbf{C}, u, v \in V).$$

(2) τ is compatible with multiplication in V [2(i), p. 154].

(3) There exists a linear mapping $\operatorname{tr}: V \to \mathbf{C}$ such that

$$\operatorname{tr} \tau(k)u = \operatorname{tr} u\tau(k) , \quad \operatorname{tr}(uv) = \operatorname{tr}(vu)$$
$$(u, v) = \operatorname{tr}(u^*v) \qquad (u, v \in V, k \in K).$$

The above assumptions imply the following relation.

(4) $$\big(\tau(k_1)u\tau(k_2)\big)^* = \tau(k_2^{-1})u^*\tau(k_1^{-1}) \qquad (k_1, k_2 \in K, u \in V).$$

For if $v \in V$, we have

$$\begin{aligned}
\big((\tau(k_1)u\tau(k_2))^*, v\big) &= \operatorname{tr}\big(\tau(k_1)u\tau(k_2) \cdot v\big) \\
&= \operatorname{tr}\big(u \cdot \tau(k_2)v\tau(k_1)\big) = \big(u^*, \tau(k_2)v\tau(k_1)\big) \\
&= \big(\tau(k_2^{-1})u^*\tau(k_1^{-1}), v\big) .
\end{aligned}$$

We regard $\mathcal{C}(G, \tau)$ as an algebra under convolution. For $\alpha \in C(G, \tau)$, define $\tilde{\alpha}(x) = \alpha(x^{-1})^*$ ($x \in G$). Then $f \mapsto \tilde{f}$ is an anti-involution (over $\mathbf{R}$) in $\mathcal{C}(G, \tau)$ and a similar statement holds for $\mathcal{C}(M, \tau_M)$. If $P \in \mathscr{P}(A)$ ($P = MAN$) and $\nu \in \mathfrak{F}$, put

---

[4] Here the bar denotes the complex conjugate.

$$f_\nu^{(P)}(m) = \int_{A \times N} f(man) \exp\left\{-((-1)^{1/2}\nu - \rho)(\log a)\right\} da\, dn \qquad (m \in M)$$

for $f \in \mathcal{C}(G, \tau)$. (Here $\rho = \rho_P$.)

LEMMA 1. $f \mapsto f_\nu^{(P)}$ is a homomorphism of $\mathcal{C}(G, \tau)$ into $\mathcal{C}(M, \tau_M)$ and $(\tilde{f})_\nu^{(P)} = (f_\nu^{(P)})^\sim$.

This is verified by a simple computation.

Put $L = {}^0\mathcal{C}(M, \tau_M)$. Then $\dim L < \infty$ [2(i), Thm. 27.3] and L is a two-sided ideal in $\mathcal{C}(M, \tau_M)$ which is stable under the anti-involution $\sim$. Moreover it has the structure of a Hilbert space under the norm

$$\|\psi\|^2 = \int_M |\psi(m)|^2 dm \qquad (\psi \in L).$$

We note that

$$(\phi, \psi) = \operatorname{tr}(\tilde{\phi} * \psi)(1)$$

for $\phi, \psi \in L$.

LEMMA 2. Let $\psi \in L$, $P \in \mathscr{P}(A)$, $\nu \in \mathfrak{F}$ and $f \in \mathcal{C}(G, \tau)$. Then:
(1) $f * E(P\colon \psi\colon \nu) = E(P\colon f_\nu^{(P)} * \psi\colon \nu)$.
(2) $E(P\colon \psi\colon \nu) * f = E(P\colon \psi * f_\nu^{(P)}\colon \nu)$.
(3) $E(P\colon \tilde{\psi}\colon \nu) = E(P\colon \psi\colon \nu)^\sim$.

Fix $g \in \mathcal{C}(G, \tau)$. Then if $h = \tilde{f} * g$, we have (see [2(j), § 18])

$$\begin{aligned}
(f * E(P\colon \psi\colon \nu), g) &= (E(P\colon \psi\colon \nu), h) \\
&= (\psi, h_\nu^{(P)}) = (\psi, (f_\nu^{(P)})^\sim * g_\nu^{(P)}) \\
&= (f_\nu^{(P)} * \psi, g_\nu^{(P)}) = (E(P\colon f_\nu^{(P)} * \psi\colon \nu), g)
\end{aligned}$$

from Lemma 1. This holds in particular for all $g \in C_c^\infty(G, \tau)$ and so (1) follows. The proof of (2) is similar. So we come to (3). Let $g \in C_c^\infty(G, \tau)$. Then

$$(E(P\colon \psi\colon \nu)^\sim, g) = \operatorname{conj}(E(P\colon \psi\colon \nu), \tilde{g}) = \operatorname{conj}(\psi, (\tilde{g})_\nu^{(P)}).$$

But

$$(\psi, (\tilde{g})_\nu^{(P)}) = (\psi, (g_\nu^{(P)})^\sim) = \operatorname{conj}(\tilde{\psi}, g_\nu^{(P)}).$$

Hence

$$(E(P\colon \psi\colon \nu)^\sim, g) = (\tilde{\psi}, g_\nu^{(P)}) = (E(P\colon \tilde{\psi}\colon \nu), g).$$

This being true for all $g \in C_c^\infty(G, \tau)$, our assertion follows.

LEMMA 3. Let $f \in \mathcal{C}(G, \tau)$ and let P be any psgp of G. Then

$$(f_P)^\sim = (\tilde{f})_{\bar{P}}.$$

Let $P = MAN$. Then it follows from the definition of f_P [2(i), § 21] that

$$\lim_{a \xrightarrow{\bar{P}} \infty} \{d_{\bar{P}}(ma)f(m^{-1}a^{-1}) - f_P(m^{-1}a^{-1})\} = 0$$

for $m \in MA$. Therefore

$$\lim_{a \underset{\bar{P}}{\to} \infty} \{ d_{\bar{P}}(ma) \tilde{f}(ma) - (f_P)^{\sim}(ma) \} = 0 \ .$$

This proves that $(\tilde{f})_{\bar{P}} = (f_P)^{\sim}$.

COROLLARY. *Let* $P \in \mathscr{P}(A)$, $\nu \in \mathfrak{F}'$ *and* $\psi \in L$. *Then*

$$c_{\bar{P}|P}(1 \colon \nu) \tilde{\psi} = \left(c_{P|P}(1 \colon \nu) \psi \right)^{\sim}$$

in the notation of Theorm 18.1 of [2(j)].

This follows by applying the above lemma to $f = E(P \colon \nu \colon \psi)$ and making use of part (3) of Lemma 2.

9. Proof of Theorem 7.1

Now take $V = C_F(K \times K)$ and $\tau = \tau_F$ as before in Section 7. Define

$$\operatorname{tr} v = \int_K v(k \colon k^{-1}) dk \ ,$$

$$v^*(k_1 \colon k_2) = \operatorname{conj} v(k_2^{-1} \colon k_1^{-1}) \qquad\qquad (k_1, \, k_2 \in K)$$

for $v \in V$. Moreover define the product $v = v_1 \cdot v_2$ of $v_1, \, v_2 \in V$ by

$$v(k_1 \colon k_2) = \int_K v_1(k_1 \colon k) v_2(k^{-1} \colon k_2) dk \ .$$

Then all the conditions of Section 8 are fulfilled. It is obvious that $L(\omega)$ is a two-sided ideal in L which is stable under $\sim$.

Let $d(\omega)$ denote the formal degree of ω.

LEMMA 1. *Let* $S, \, T \in \operatorname{End} \mathfrak{H}_F$. *Then*

(1) $\psi_S * \psi_T = d(\omega)^{-1} \psi_{TS}$,

(2) $(\psi_T)^{\sim} = \psi_{T^*}$,

(3) $\operatorname{tr} \psi_T(1) = \operatorname{tr} T$,

(4) $(\psi_S, \, \psi_T) = d(\omega)^{-1} \operatorname{tr} S^* T$.

Let $B, \, C \in \operatorname{End}^0 U$. Then it follows from the Schur orthogonality relations for ω that

$$\int_M \operatorname{tr} \sigma(m^{-1}) B \cdot \operatorname{tr} \sigma(mm_0) C dm = d(\omega)^{-1} \operatorname{tr} \left(B \sigma(m_0) C \right)$$

for $m_0 \in M$. Hence if $\phi = \psi_S * \psi_T$,

$$\begin{aligned}
\phi(k_1 \colon m_0 \colon k_2) &= \int_{K \times M} \psi_S(k_1 \colon m \colon k) \psi_T(k^{-1} \colon m^{-1} m_0 \colon k_2) dk dm \\
&= \int_K dk \int_M \operatorname{tr} \sigma(m) \kappa_S(k \colon k_1) \cdot \operatorname{tr} \sigma(m^{-1} m_0) \kappa_T(k_2 \colon k^{-1}) dm \\
&= d(\omega)^{-1} \int_K \operatorname{tr} \{ \kappa_S(k \colon k_1) \sigma(m_0) \kappa_T(k_2 \colon k^{-1}) \} dk \\
&= d(\omega)^{-1} \operatorname{tr} \kappa_{TS}(k_2 \colon k_1) \sigma(m_0) = d(\omega)^{-1} \psi_{TS}(k_1 \colon m_0 \colon k_2)
\end{aligned}$$

from the corollary of Lemma 6.1. This proves (1). (2) and (3) also follow from the same corollary. (4) is a consequence of the fact that

$$(\phi, \psi) = \mathrm{tr}(\tilde{\phi} * \psi)(1) \qquad\qquad (\phi, \psi \in L) .$$

Define $\mathcal{L}$ and $\mathcal{L}_F$ as in Section 3 and put $\check{\alpha}(x) = \alpha(x^{-1})$ $(x \in G)$ for $\alpha \in \mathcal{L}$. For $\alpha \in (\mathcal{L}_F)^{\vee}$ define $\boldsymbol{\alpha} \in C_c^{\infty}(G, \tau)$ by [2(i), § 26]:

$$\boldsymbol{\alpha}(k_1: x: k_2) = \alpha(k_1 x k_2) \qquad\qquad (k_1, k_2 \in K, x \in G) .$$

LEMMA 2. *Let* $\alpha \in (\mathcal{L}_F)^{\vee}$, $P \in \mathcal{P}(A)$, $\nu \in \mathfrak{F}$ *and* $T \in \mathrm{End}\ \mathfrak{H}_F$. *Then*

$$\boldsymbol{\alpha}_{\nu}^{(P)} * \psi_T = \psi_{T\pi_{P,\nu}(\check{\alpha})} ,$$
$$\psi_T * \boldsymbol{\alpha}_{\nu}^{(P)} = \psi_{\pi_{P,\nu}(\check{\alpha})T} .$$

We write $\pi = \pi_{P,\nu}$ and observe that $\pi(\check{\alpha}) \in \mathrm{End}\ \mathfrak{H}_F$. Then we conclude from Lemma 4.1 that

$$\kappa_{\pi(\check{\alpha})}(k_2: k_1) = \int_M \boldsymbol{\alpha}_{\nu}^{(P)}(k_1: m: k_2)\sigma(m^{-1})dm .$$

Therefore (cor. of Lem. 6.1)

$$\kappa_{T\pi(\check{\alpha})}(k_2: k_1) = \int_K \kappa_T(k_2: k^{-1})\kappa_{\pi(\check{\alpha})}(k: k_1)dk$$
$$= \int \boldsymbol{\alpha}_{\nu}^{(P)}(k_1: m: k)\kappa_T(k_2: k^{-1})\sigma(m^{-1})dkdm .$$

Hence if $m_0 \in M$,

$$\psi_{T\pi(\check{\alpha})}(k_1: m_0: k_2) = \mathrm{tr}\{\kappa_{T\pi(\check{\alpha})}(k_2: k_1)\sigma(m_0)\}$$
$$= \int \boldsymbol{\alpha}_{\nu}^{(P)}(k_1: m: k)\mathrm{tr}\{\kappa_T(k_2: k^{-1})\sigma(m^{-1}m_0)\}dkdm$$
$$= \int \boldsymbol{\alpha}_{\nu}^{(P)}(k_1: m: k)\psi_T(k^{-1}: m^{-1}m_0: k_2)dkdm .$$

This shows that

$$\psi_{T\pi(\check{\alpha})} = \boldsymbol{\alpha}_{\nu}^{(P)} * \psi_T .$$

The proof for the second statement is similar.

Now we come to the proof of Theorem 7.1. Put $E(T) = E(P: \psi_T: \nu)$ and $\pi = \pi_{P,\nu}$. Then

$$E(T: 1) = \int_K \tau(k)\psi_T(1)\tau(k^{-1})dk$$

and therefore

$$E(P: \psi_T: \nu: 1: 1: 1) = E(T: 1: 1: 1) = \mathrm{tr}\ \psi_T(1) = \mathrm{tr}\ T$$

from Lemma 1.

Now $\alpha \mapsto \boldsymbol{\alpha}$ is a bijective mapping of $(\mathcal{L}_F)^{\vee}$ onto $C_c^{\infty}(G, \tau)$ [2(i), § 26]. Fix

$\alpha \in (\mathfrak{L}_F)^{\vee}$. Then

$$\boldsymbol{\alpha} * E(T) = E(P: \boldsymbol{\alpha}_{\nu}^{(P)} * \psi_T: \nu) = E\big(T\pi(\check{\alpha})\big)$$

from Lemmas 8.2 and 2. Put $\phi = \boldsymbol{\alpha} * E(T)$. Then it follows from the above result that

$$\phi(1:1:1) = \operatorname{tr} T\pi(\check{\alpha}) .$$

But

$$\phi(1) = \int_G \boldsymbol{\alpha}(x^{-1}) \cdot E(T:x) dx$$

and therefore

$$\phi(1:1:1) = \int \alpha(x^{-1}k^{-1}) E(T:k:x:1) dk dx$$
$$= \int \alpha(x^{-1}k^{-1}) E(T:1:kx:1) dk dx$$
$$= \int \alpha(x^{-1}) E(T:1:x:1) dx .$$

Since

$$\operatorname{tr} T\pi(\check{\alpha}) = \int \alpha(x^{-1}) \operatorname{tr} T\pi(x) \cdot dx$$

and α is arbitrary in $(\mathfrak{L}_F)^{\vee}$, we conclude that

$$E(T:1:x:1) = \operatorname{tr} T\pi(x)$$

and from this the assertion of Theorem 7.1 is obvious.

10. An application of Theorem 7.1

We keep to the notation of Section 9.

LEMMA 1. *Fix* $P \in \mathscr{P}(A)$ *and suppose* ν *is an element in* $\mathfrak{F}'$ *such that* $c_{P|P}(1:\nu)$ *is injective on* $L(\omega)$. *Then* $\pi_{P,\nu}(\mathfrak{L}_F) = \operatorname{End} \mathfrak{H}_F$.

We note that, in view of Lemma 19.3 of [2(j)], ν can actually be chosen so as to satisfy our condition.

Put $\pi = \pi_{P,\nu}$ and $f_T(x) = \operatorname{tr} T\pi(x)$ $(x \in G)$ for $T \in \operatorname{End} \mathfrak{H}_F$. By Lemma 3.1, it is enough to verify that the mapping $T \mapsto f_T$ is injective on $\operatorname{End} \mathfrak{H}_F$. So fix T such that $f_T = 0$ and put

$$\phi = E(P: \psi_T: \nu) .$$

Then $\phi = 0$ by Theorem 7.1 and therefore $\phi_P = 0$. This implies that $c_{P|P}(1:\nu)\psi_T = 0$ and therefore $\psi_T = 0$. But then we conclude from Lemma 7.1 that $T = 0$.

We shall use this lemma to prove later that $\pi_{P,\nu}$ is irreducible for $\nu \in \mathfrak{F}'$

(see Lemma 13.3).

11. Some properties of the j-functions

For $P, Q \in \mathscr{P}(A)$ and $\nu \in \mathfrak{F}_c(Q \mid P)$, put $j_{Q\mid P}(\nu) = E_F J_{Q\mid P}(\nu)E_F$ as in Section 5. We note that $\mathfrak{F}_c(\bar{P} \mid P) \subset \mathfrak{F}_c(\bar{P})$ in the notation of Theorem 19.1 of [2(j)]. Put

$$c(P) = \gamma_{\bar{P}\mid P}\gamma(P)^{-1} \, .$$

It follows from Lemma 2.6 that $c(P)$ is independent of $P \in \mathscr{P}(A)$. Hence we may denote it by $c(A)$ or $c(G/A)$.

LEMMA 1. *Fix $P \in \mathscr{P}(A)$ and $\nu \in \mathfrak{F}_c(\bar{P} \mid P)$. Then*

$$c_{P\mid P}(1\colon \nu)\psi_T = c(A)\psi_{j_{\bar{P}\mid P(\nu)}T}$$

for $T \in \operatorname{End} \mathfrak{H}_F$.

Put $j = j_{\bar{P}\mid P}(\nu)$. Then it follows from Lemma 5.3 that

$$\kappa_{jT}(k_2\colon k_1) = \gamma_{\bar{P}\mid P}^{-1} \int_{\bar{N}} \exp\left\{-\big((-1)^{1/2}\nu + \rho\big)(H(\bar{n}))\right\}\psi_T\big(k_1\colon m\mu(\bar{n})^{-1}\colon \kappa(\bar{n})^{-1}k_2\big)d\bar{n} \, .$$

Our assertion now follows from Theorem 19.1 of [2(j)].

The above result shows that $j_{\bar{P}\mid P}(\nu)$ extends to a meromorphic function on $\mathfrak{F}_c(\bar{P}) \cup \mathfrak{F}_c(\delta)$. Moreover it is holomorphic at every point $\nu \in \mathfrak{F}$. Therefore we conclude from Corollary 1 of Lemma 4.1 and Corollary 2 of Lemma 5.2 that

$$j_{\bar{P}\mid P}(\nu)\pi_{P,\nu}(\alpha) = \pi_{\bar{P},\nu}(\alpha)j_{\bar{P}\mid P}(\nu) \qquad\qquad (\alpha \in \mathfrak{L}_F)$$

for $\nu \in \mathfrak{F}'$.

Let $\mathfrak{F}''$ be the set of all $\nu \in \mathfrak{F}'$ where $j_{\bar{P}\mid P}(\nu)$ is bijective. Since $\det c_{P\mid P}(1\colon \nu)$ is not identically zero [2(j), Lem. 19.3], we conclude from Lemma 1 that $\mathfrak{F}''$ is a dense subset of $\mathfrak{F}$.

LEMMA 2. *Fix $\nu \in \mathfrak{F}''$. Then there exists a scalar $c > 0$ such that*

$$\big(j_{\bar{P}\mid P}(\nu)\big)^*j_{\bar{P}\mid P}(\nu) = c \, .$$

Put $j = j_{\bar{P}\mid P}(\nu)$, $\pi_P = \pi_{P,\nu}$ and $\pi_{\bar{P}} = \pi_{\bar{P},\nu}$. Then

$$\pi_{\bar{P}}(\alpha)j = j\pi_P(\alpha) \qquad\qquad (\alpha \in \mathfrak{L}_F)$$

and therefore, taking adjoints, we get

$$j^*\pi_{\bar{P}}(\tilde{\alpha}) = \pi_P(\tilde{\alpha})j^* \, .$$

Here $\tilde{\alpha}(x) = \operatorname{conj} \alpha(x^{-1})$ $(x \in G)$ and we make use of the fact that π_P and $\pi_{\bar{P}}$ are unitary. This shows that j^*j commutes with $\pi_P(\alpha)$ $(\alpha \in \mathfrak{L}_F)$. But $\pi_P(\mathfrak{L}_F) = \operatorname{End} \mathfrak{H}_F$ by Lemma 10.1. Therefore j^*j is a scalar. It is obvious that $c > 0$.

LEMMA 3. *Let $\nu \in \mathfrak{F}'$. Then*

$$j_{P|\bar{P}}(\nu) = \big(j_{\bar{P}|P}(\nu)\big)^*$$

and

$$c_{\bar{P}|P}(1:\nu)\psi_T = c(A)\psi_{Tj_{P|\bar{P}}(\nu)}$$

for $T \in \operatorname{End} \mathfrak{H}_F$.

Put $j = j_{\bar{P}|P}(\nu)$. Then we conclude from the corollary of Lemma 8.3, Lemma 9.1 and Lemma 1 that

$$c_{\bar{P}|P}(1:\nu)\psi_T = \big(c_{P|P}(1:\nu)\psi_{T^*}\big)^{\sim} = c(A)(\psi_{jT^*})^{\sim} = c(A)\psi_{Tj^*}\,.$$

Hence it remains only to verify that $j_{P|\bar{P}}(\nu) = j^*$.

Replacing P by $\bar{P}$ in the above result, we get

$$c_{P|\bar{P}}(1:\nu)\psi_T = c(A)\psi_{T(j_{P|\bar{P}}(\nu))^*}\,.$$

On the other hand we have the following simple result.

LEMMA 4. *Put $S = E_F$. Then*

$$E(P:\psi_S:\nu) = E(\bar{P}:\psi_S:\nu)$$

for all $\nu \in \mathfrak{F}$.

Assuming this for the moment, we continue with the proof of Lemma 3. Fix $\nu \in \mathfrak{F}'$ as before. Then

$$c_{P|P}(1:\nu)\psi_S = c_{P|\bar{P}}(1:\nu)\psi_S$$

from Lemma 4. But we have seen above that

$$c_{P|P}(1:\nu)\psi_S = c(A)\psi_{j_{\bar{P}|P}(\nu)}\,,$$
$$c_{P|\bar{P}}(1:\nu)\psi_S = c(A)\psi_{(j_{P|\bar{P}}(\nu))^*}\,.$$

In view of Lemma 7.1 this proves that $j_{\bar{P}|P}(\nu) = \big(j_{P|\bar{P}}(\nu)\big)^*$.

Now we come to the proof of Lemma 4 and use the notation of the proof of Lemma 2. It would be enough to consider the case when $\nu \in \mathfrak{F}''$. Then $j = j_{\bar{P}|P}(\nu)$ is bijective and

$$\pi_{\bar{P}}(\alpha)j = j\pi_P(\alpha) \qquad\qquad (\alpha \in \mathfrak{L}_F)\,.$$

Therefore

$$\operatorname{tr}\pi_{\bar{P}}(\alpha) = \operatorname{tr}\pi_P(\alpha) \qquad\qquad (\alpha \in \mathfrak{L}_F)$$

or

$$\operatorname{tr}E_F\pi_{\bar{P},\nu}(x) = \operatorname{tr}E_F\pi_{P,\nu}(x) \qquad\qquad (x \in G)\,.$$

The required result now follows from Theorem 7.1.

LEMMA 5. *Fix $P \in \mathscr{P}(A)$ and $\nu \in \mathfrak{F}'$. Then $j_{\bar{P}|P}(\nu)$, $j_{P|\bar{P}}(\nu)$ commute and* *their product is a scalar operator.*

This is obvious from Lemmas 2 and 3.

12. The μ-function in a special case

We assume in this section that dim $A = \mathrm{prk}\, G + 1$. Then $\mathscr{P}(A) = \{P, \bar{P}\}$.

LEMMA 1. *Suppose $\mathfrak{H}_F \neq \{0\}$ and fix $\nu \in \mathfrak{F}'$. Then there exists a number $\mu_\omega(\nu) > 0$ such that*

$$\mu_\omega(\nu) j_{P|\bar{P}}(\nu) j_{\bar{P}|P}(\nu) = \mu_\omega(\nu) j_{\bar{P}|P}(\nu) j_{P|\bar{P}}(\nu) = 1$$

on $\mathfrak{H}_F$. Moreover $\mu_\omega(\nu)$ is independent of F.

Put $j = j_{\bar{P}|P}(\nu)$. We claim that $j \neq 0$. For otherwise suppose $j = 0$. Fix $T \in \mathrm{End}\, \mathfrak{H}_F$ and put

$$\phi = E(P \colon \psi_T \colon \nu)\,.$$

Since $\mathscr{P}(A) = \{P, \bar{P}\}$ and $j = j^* = 0$, it follows from Lemmas 1 and 3 that

$$\phi_P = \phi_{\bar{P}} = 0\,.$$

On the other hand by [2(j), Lem. 18.3], $\phi_{P'} \sim 0$ for every psgp P' of G which is not associated to P. Therefore $\phi = 0$ [2(i), Lem. 25.2] and we conclude from Theorem 7.1 that

$$\mathrm{tr}\, T\pi_{P,\nu}(x) = 0$$

for all $x \in G$. Substituting $x = 1$ and $T = E_F$, we get

$$0 = \mathrm{tr}\, E_F = \dim \mathfrak{H}_F$$

which contradicts our hypothesis.

Since $j \neq 0$, we conclude from Lemmas 11.3 and 11.5 that

$$j^* j = c$$

where c is a positive scalar. We take $\mu_\omega(\nu) = c^{-1}$.

Now we write j_F and c_F instead of j and c to emphasize their dependence on F. Then

$$j_F^* j_F = c_F\,.$$

Suppose F_1, F_2 are two finite subsets of $\mathscr{E}(K)$ such that $\mathfrak{H}_{F_i} \neq \{0\}$ $(i = 1, 2)$. Put $F = F_1 \cup F_2$. Then it is obvious that

$$j_{F_i} = E_{F_i} j_F$$

and so we conclude immediately that $c_F = c_{F_i}$ $(i = 1, 2)$. This shows that $c_{F_1} = c_{F_2}$ and therefore $\mu_\omega(\nu)$ is independent of F.

COROLLARY.

$$\| c_{P|P}(1 \colon \nu)\psi \|^2 = \| c_{\bar{P}|P}(1 \colon \nu)\psi \|^2 = \mu_\omega(\nu)^{-1} c(A)^2 \| \psi \|^2$$

for $\nu \in \mathfrak{F}'$ and $\psi \in L(\omega)$.

This follows immediately from Lemmas 11.1, 11.3 and 1.

It is obvious from what we have seen in Section 11 that μ_ω extends to a meromorphic function on $\mathfrak{F}_c(\delta)$ which is holomorphic on $\mathfrak{F}'$.

LEMMA 2. *μ_ω is holomorphic on $\mathfrak{F}$.*

Let α be the unique simple root of (P, A). Then Lemma 5.3 shows that $j_{\bar{P}|P}(\nu)$ depends only on $\langle \nu, \alpha \rangle$ and its value remains unchanged if we replace (G, P) by $({}^0 G, {}^0 G \cap P)$. Hence we may assume that prk $G = 0$. Then dim $A = 1$ and therefore $\mathfrak{F}_c$ is the complex plane. We have therefore only to verify that μ_ω does not have a pole at $\nu = 0$.

So let us assume that $\mu = \mu_\omega$ has a pole at zero. Fix $T \in \text{End } \mathfrak{H}_F$ and put

$$\phi = E(P: \psi_T: 0) \,.$$

We know [2(j), § 10] that

$$\phi_P = \lim_{\nu \to 0} E_P(P: \psi_T: \nu) \,,$$
$$\phi_{\bar{P}} = \lim_{\nu \to 0} E_{\bar{P}}(P: \psi_T: \nu) \qquad\qquad (\nu \in \mathfrak{F}') \,,$$

pointwise on MA. Moreover

$$\| c_{P|P}(1: \nu)\psi_T \|^2 = \| c_{\bar{P}|P}(1: \nu)\psi_T \|^2 \longrightarrow 0$$

from the corollary of Lemma 1. This shows that $\phi_P = \phi_{\bar{P}} = 0$ and we conclude, as in the proof of Lemma 1, that $\phi = 0$. But then

$$\text{tr } T\pi_{P,0}(x) = 0 \qquad\qquad (x \in G)$$

from Theorem 7.1. Substituting $x = 1$ and $T = E_F$, we get a contradiction.

13. The μ-function in the general case and irreducibility of representations

We return to the general case of Section 11. Let α be a reduced root of $(\mathfrak{g}, \mathfrak{a})$. Define M_α, $\mathfrak{a}_\alpha$ as in Lemma 2.3. For $\nu \in \mathfrak{F}$, let ${}^*\nu$ denote the restriction of ν on $\mathfrak{a}_\alpha$. If we replace (G, A) by (M_α, A_α) then the conditions of Section 12 are fulfilled. Put $\mu_\alpha(\omega: \nu) = \mu_\omega({}^*\nu)$ where the function μ_ω is defined as in Section 12 for the pair (M_α, M) in place of (G, M). Clearly $\mu_\alpha(\omega: \nu) = \mu_{-\alpha}(\omega: \nu)$.

Suppose P_1, P_2 are two adjacent elements in $\mathscr{P}(A)$ and α is the unique root in $\Sigma(P_2 | P_1)$. We use the notation of Lemma 2.3 and put ${}^*P = P_1 \cap M_\alpha$.

LEMMA 1. *$j_{P_2|P_1}(\nu)$ and $j_{P_1|P_2}(\nu)$ commute and*

$$\mu_\alpha(\omega: \nu) j_{P_1|P_2}(\nu) j_{P_2|P_1}(\nu) = 1 \,.$$

Moreover $j_{P_2|P_1}(\nu) = (j_{P_1|P_2}(\nu))^*$ for $\nu \in \mathfrak{F}'$.

$*P = M \cdot *A \cdot *N$ where $*A = A_\alpha$, $*N = N_\alpha$. Put $*K = M_\alpha \cap K$ and let $*\mathfrak{H}$ be the representation space of

$$\mathrm{Ind}_{K_M}^{*K} (\sigma \mid K_M) .$$

For $h \in \mathfrak{H}_F$ and $k \in K$, define $*h_k \in *\mathfrak{H}$ by

$$*h_k(*k) = h(*k \cdot k) \qquad\qquad (*k \in *K) .$$

We can choose a finite subset $*F$ of $\mathcal{E}(*K)$ such that $*h_k \in *\mathfrak{H}_{*F}$ for all $h \in \mathfrak{H}_F$ and $k \in K$. Since $P_2 \cap M_\alpha = *\bar{P}$, it follows (see Lemma 5.3 and the proof of Lemma 2.4) that

$$*\big(j_{P_2 \mid P_1}(\nu)h\big)_k = j_{*\bar{P} \mid *P}(*\nu)(*h_k)$$

for $\nu \in \mathfrak{F}_c(P_2 \mid P_1)$. But then by analytic continuation this holds also for $\nu \in \mathfrak{F}'$. Similarly

$$*\big(j_{P_1 \mid P_2}(\nu)h\big)_k = j_{*P \mid *\bar{P}}(*\nu)(*h_k) \qquad\qquad (\nu \in \mathfrak{F}') .$$

On the other hand

$$\mu_\alpha(\omega: \nu)j_{*P \mid *P}(*\nu)j_{*P \mid *P}(*\nu) = 1$$

on $*\mathfrak{H}_{*F}$ from Lemma 12.1. Therefore

$$\mu_\alpha(\omega: \nu)*\big(j_{P_2 \mid P_1}(\nu)j_{P_1 \mid P_2}(\nu)h\big)_k = *h_k \qquad\qquad (\nu \in \mathfrak{F}')$$

for all $h \in \mathfrak{H}_F$ and $k \in K$. This proves that

$$\mu_\alpha(\omega: \nu)j_{P_2 \mid P_1}(\nu)j_{P_1 \mid P_2}(\nu) = 1$$

on $\mathfrak{H}_F$. We can obviously interchange P_1 and P_2 in this relation.

Now fix $\nu \in \mathfrak{F}'$. We know from Lemma 11.3 that

$$j_{*\bar{P} \mid *P}(*\nu) = \big(j_{*P \mid *\bar{P}}(*\nu)\big)^* .$$

Since

$$(h_1, h_2) = \int_K (*h_{1k}, *h_{2k})dk \qquad\qquad (h_1, h_2 \in \mathfrak{H}_F) ,$$

it follows from what we have seen above that

$$\big(j_{P_2 \mid P_1}(\nu)h_1, h_2\big) = \big(h_1, j_{P_1 \mid P_2}(\nu)h_2\big)$$

and this completes the proof of the lemma.

Given $\delta > 0$ and $P_1, P_2 \in \mathscr{P}(A)$, let $\mathfrak{F}_c(P_2 \mid P_1, \delta)$ denote the open set consisting of all $\nu \in \mathfrak{F}_c$ such that

$$\langle \nu_I, \alpha \rangle < \delta$$

for all $\alpha \in \Sigma(P_2 \mid P_1)$. Note that $\mathfrak{F}_c(P_2 \mid P_1, \delta)$ is a convex set and it contains $\mathfrak{F}$ as well as the set $\mathfrak{F}_c(P_2 \mid P_1)$ of Corollary 1 of Lemma 5.2. Moreover $\mathfrak{F}_c(P_2 \mid P_1, \delta) = \mathfrak{F}_c$ if $P_1 = P_2$.

THEOREM 1. *Fix* $P_1, P_2 \in \mathscr{P}(A)$. *Then if* δ *is sufficiently small,* $j_{P_2|P_1}$ *extends to a meromorphic function on* $\mathfrak{F}_c(P_2 \mid P_1, \delta)$.

We have seen in Section 11 that this is true if $P_2 = \bar{P}_1$. So in particular our assertion holds if dim $A = 1$. Now first assume that $d(P_2, P_1) = 1$ and keep to the notation of the proof of Lemma 1. Then if $h_1, h_2 \in \mathfrak{H}_F$ and $\nu \in \mathfrak{F}_c(P_2 \mid P_1)$,

$$\left(h_2, j_{P_2|P_1}(\nu)h_1 \right) = \int_K \left({}^*h_{2k}, j_{*\bar{P}|*P}({}^*\nu){}^*h_{1k} \right) dk \ .$$

Since dim* $A = 1$, our theorem holds for $j_{*\bar{P}|*P}$. Since

$$k \longmapsto {}^*h_{ik} \qquad\qquad (i = 1, 2)$$

are continuous mappings of K into $^*\mathfrak{H}_{*F}$, it follows that our statement is true also for $j_{P_2|P_1}$.

So now let us use induction on $d(P_2, P_1)$ and suppose that $d(P_2, P_1) \geqq 2$. Then we can choose $P \in \mathscr{P}(A)$ such that $d(P_2, P) = 1$ and $d(P, P_1) = d(P_2, P_1) - 1$. Then by induction hypothesis the theorem holds for $j_{P_2|P}$ and $j_{P|P_1}$ and we know from Lemma 5.2 that

$$j_{P_2|P_1}(\nu) = j_{P_2|P}(\nu)j_{P|P_1}(\nu)$$

for $\nu \in \mathfrak{F}_c(P_2 \mid P_1)$. From this it is obvious that our theorem is true also for $j_{P_2|P_1}$.

Fix $\delta > 0$ such that $j_{P_2|P_1}$ extends to a meromorphic function on $\mathfrak{F}_c(P_2 \mid P_1, \delta)$ for all $P_1, P_2 \in \mathscr{P}(A)$.

COROLLARY. *Let* P_1, P_2, P *be elements in* $\mathscr{P}(A)$ *such that* P *lies between* P_1 *and* P_2. *Then*

$$j_{P_2|P_1}(\nu) = j_{P_2|P}(\nu)j_{P|P_1}(\nu)$$

for $\nu \in \mathfrak{F}_c(P_2 \mid P_1, \delta)$.

This is obvious from what we have said above.

Put

$$\mu_{P_2|P_1}(\omega: \nu) = \prod_{\alpha \in \Sigma(P_2|P_1)} \mu_\alpha(\omega: \nu)$$

for $P_1, P_2 \in \mathscr{P}(A)$. Clearly $\mu_{P_2|P_1} = \mu_{P_1|P_2}$.

LEMMA 2. *Let* $P_1, P_2 \in \mathscr{P}(A)$ *and* $\nu \in \mathfrak{F}'$. *Then*

$$\mu_{P_1|P_2}(\omega: \nu)j_{P_1|P_2}(\nu)j_{P_2|P_1}(\nu) = 1 \ , \quad j_{P_2|P_1}(\nu) = \left(j_{P_1|P_2}(\nu) \right)^* \ .$$

This follows immediately from Lemma 1 and the corollary of Theorem 1.

Note that $\mu_{\bar{P}|P}(\omega: \nu)$ is independent of $P \in \mathscr{P}(A)$. Hence we denote it by $\mu(\omega: \nu)$. It is clear from Lemma 12.1 that $\mu(\omega: \nu) > 0$ for $\nu \in \mathfrak{F}'$.

We recall that the representation $\pi_{P,\nu}$ $(P \in \mathscr{P}(A), \nu \in \mathfrak{F})$ is unitary (§4).

LEMMA 3. $\pi_{P,\nu}$ *is irreducible for* $P \in \mathcal{P}(A)$ *and* $\nu \in \mathfrak{F}'$. *Moreover its class is independent of* P.

Fix a finite subset F of $\mathfrak{E}(K)$. In view of Lemma 3.1 and Theorem 7.1, it would be enough to verify that the mapping

$$T \longmapsto E(P\colon \psi_T\colon \nu)$$

is injective on End $\mathfrak{H}_F$. So suppose $E(P\colon \psi_T\colon \nu) = 0$. Then

$$j_{\bar{P}|P}(\nu)T = 0$$

from Lemma 11.1 and therefore

$$T = \mu(\omega\colon \nu)j_{P|\bar{P}}(\nu)j_{\bar{P}|P}(\nu)T = 0$$

from Lemma 2. This proves that $\pi_{P,\nu}$ is irreducible.

Now let $P_1, P_2 \in \mathcal{P}(A)$. Then

$$j_{P_2|P_1}(\nu)\pi_{P_1,\nu}(\alpha) = \pi_{P_2,\nu}(\alpha)j_{P_2|P_1}(\nu)$$

for $\alpha \in \mathfrak{L}_F$. (This follows immediately from Corollary 2 of Lemma 5.2 by analytic continuation.) Since $j_{P_2|P_1}(\nu)$ is bijective from Lemma 2, this shows that the two representations of $\mathfrak{L}_F$ on $\mathfrak{H}_F$ (defined by $\pi_i = \pi_{P_i,\nu}$, $i = 1, 2$) are equivalent. Since π_1, π_2 are irreducible and unitary and we can choose F so that $\mathfrak{H}_F \neq \{0\}$, this implies that $\pi_1 \simeq \pi_2$.

COROLLARY. *Fix* $T \in$ End $\mathfrak{H}_F$. *Then*

$$E(P_1\colon \psi_{Tj_{P_2|P_1}(\nu)}\colon \nu) = E(P_2\colon \psi_{j_{P_2|P_1}(\nu)T}\colon \nu)$$

for $P_1, P_2 \in \mathcal{P}(A)$ *and* $\nu \in \mathfrak{F}'$.

For

$$\operatorname{tr}\{Tj_{P_2|P_1}(\nu)\pi_{P_1,\nu}(\alpha)\} = \operatorname{tr}\{T\pi_{P_2,\nu}(\alpha)j_{P_2|P_1}(\nu)\}$$
$$= \operatorname{tr}\{j_{P_2|P_1}(\nu)T\pi_{P_2,\nu}(\alpha)\}$$

for $\alpha \in \mathfrak{L}_F$. The required result now follows from Theorem 7.1.

For $P \in \mathcal{P}(A)$ and $\nu \in \mathfrak{F}'$, let $C(P, \omega, \nu)$ denote the class of the irreducible unitary representation

$$\pi_{P,\omega,\nu} = \operatorname{Ind}_P^G \delta_P^{1/2}\sigma_\nu$$

of G. Define $\mathfrak{w} = \mathfrak{w}(G/A)$ as usual (see [2(j), Thm. 18.1]).

THEOREM 2. *Let* $P_1, P_2 \in \mathcal{P}(A)$, $\omega \in \mathfrak{E}_2(M)$, $\nu \in \mathfrak{F}'$ *and* $s \in \mathfrak{w}$. *Then*

$$C(P_1, \omega, \nu) = C(P_2, s\omega, s\nu) .$$

It is obvious that

$$C(P_1, \omega, \nu) = C(P_1^s, s\omega, s\nu) .$$

Hence it is enough to consider the case $s = 1$ which is covered by Lemma 3.

LEMMA 4. *Under the conditions of Theorem 2, $c_{P_2|P_1}(s:\nu)$ defines a bijection of $L(\omega)$ onto $L(s\omega)$. Moreover*

$$\mu(\omega:\nu)\,\|\,c_{P_2|P_1}(s:\nu)\psi\,\|^2 = c(A)^2\,\|\,\psi\,\|^2$$

for $P \in \mathscr{P}(A)$ and $\psi \in L(\omega)$.

It is obvious that s defines a unitary mapping of $L(\omega)$ onto $L(s\omega)$. Hence, in view of [2(j), Lem. 18.1], it would be sufficient to consider the case $s = 1$.

Fix $T \in \mathrm{End}\,\mathfrak{H}_F$. Then by the corollary of Lemma 3,

$$c_{P_2|P_1}(1:\nu)\psi_{T_1} = c_{P_2|P_2}(1:\nu)\psi_{T_2}$$

where $T_1 = T j_{P_2|P_1}(\nu)$ and $T_2 = j_{P_2|P_1}(\nu)T$. Put $T_3 = j_{\bar{P}_2|P_2}(\nu)T_2$. Then

$$c_{P_2|P_2}(1:\nu)\psi_{T_2} = c(A)\psi_{T_3}$$

from Lemma 11.1. Put

$$\|\,S\,\|^2 = \mathrm{tr}\,SS^* = \mathrm{tr}\,S^*S$$

for $S \in \mathrm{End}\,\mathfrak{H}_F$. Then

$$\|\,T_1\,\|^2 = \mu_{P_2|P_1}(\omega:\nu)^{-1}\,\|\,T\,\|^2$$

from Lemma 2. Similarly since $T_3 = j_{\bar{P}_2|P_2}(\nu)j_{P_2|P_1}(\nu)T$, we have

$$\begin{aligned}
\|\,T_3\,\|^2 &= \mathrm{tr}\,T_3^* T_3 \\
&= \mu_{P_2|P_1}(\omega:\nu)^{-1}\mu(\omega:\nu)^{-1}\,\|\,T\,\|^2 \\
&= \mu(\omega:\nu)^{-1}\,\|\,T_1\,\|^2\,.
\end{aligned}$$

Since $j_{P_2|P_1}(\nu)$ is bijective, this proves that

$$\|\,c_{P_2|P_1}(1:\nu)\psi\,\|^2 = \mu(\omega:\nu)^{-1}c(A)^2\,\|\,\psi\,\|^2$$

for $\psi \in L(\omega)$.

Part II. Maass-Selberg relations and the functional equations

14. The Maass-Selberg relations

Let τ be a unitary double representation of K on a finite-dimensional Hilbert space V. Fix a special vector subgroup A of G and define M, $\mathfrak{F}$, $\mathfrak{w}$ as above. Let $\mathcal{A}(G,\tau)$ and $^0\mathcal{C}(M,\tau_M)$ have their usual meanings [2(i)]. Then $^0\mathcal{C}(M,\tau_M)$ has a natural norm [2(j), § 18].

THEOREM 1. *Fix $\nu \in \mathfrak{F}'$ and let f be an element in $\mathcal{A}(G,\tau)$ such that:*

(1) *$f_{P'} \sim 0$ for every psgp $P' = M'A'N'$ of G such that A' is not conjugate to A under K.*

(2) *For every $P \in \mathscr{P}(A)$, we can choose elements $\phi_{P,s} \in {}^0\mathcal{C}(M,\tau_M)$ ($s \in \mathfrak{w}$) such that*

$$f_P(ma) = \sum_{s \in \mathfrak{w}} \phi_{P,s}(m) \exp\{(-1)^{1/2} s\nu(\log a)\} \qquad (m \in M, a \in A) .$$

Then

$$\|\phi_{P_1,s_1}\| = \|\phi_{P_2,s_2}\|$$

for $P_i \in \mathscr{P}(A)$ and $s_i \in \mathfrak{w}$ $(i = 1, 2)$.

By the Maass-Selberg relations we mean the equalities

$$\|\phi_{P_1,s_1}\| = \|\phi_{P_2,s_2}\|$$

of the above theorem.

COROLLARY. *Suppose $\phi_{P,s} = 0$ for some pair (P, s). Then $f = 0$.*

For then it follows from the Maass-Selberg relations that $f_P = 0$ for every $P \in \mathscr{P}(A)$. Hence we conclude from Lemmas 21.1 and 25.2 of [2(i)] that $f = 0$.

The above corollary shows that the functions $\phi_{P,s}$ are uniquely determined by f. Hence we may denote them by $f_{P,s}$.

15. Some preparatory remarks

We need some preparation before taking up the proof of Theorem 14.1.

A unitary double K-module (or, more briefly, a module) is a pair (V, τ) consisting of a finite-dimensional complex Hilbert space V and a unitary double representation τ of K on V. Two such modules (V, τ) and (V', τ') are said to be isomorphic if there exists a K-isomorphism $v \mapsto v'$ of V onto V' and a number $c > 0$ such that $|v'| = c|v|$ for all $v \in V$. We shall usually denote the module by τ instead of (V, τ). The meaning of the direct sum $\tau \oplus \tau'$ of two modules τ, τ' is clear.

Let P be a statement about modules. We say P is additive if the following two conditions hold.

(1) If τ and τ' are isomorphic, then $P(\tau)$ implies $P(\tau')$.

(2) Suppose $\tau = \tau' \oplus \tau''$. Then $P(\tau)$ holds if and only if both $P(\tau')$ and $P(\tau'')$ hold.

LEMMA 1. *Suppose P is additive. Then in order that P be true for all τ, it is sufficient that $P(\tau)$ be true for all irreducible τ.*

This is obvious.

For any finite subset F of $\mathfrak{E}(K)$, define the module (V_F, τ_F) as in Section 7 so that $V_F = C_F(K \times K)$.

LEMMA 2. *Let τ be an irreducible module. Then τ is isomorphic to a submodule of τ_F for some F.*

Let V be the representation space of τ. Fix a linear function $\lambda \neq 0$ on

V. For any $v \in V$, consider the function κ_v on $K \times K$ given by

$$\kappa_v(k_1: k_2) = \langle \lambda, \tau(k_1)v\tau(k_2) \rangle \qquad (k_1, k_2 \in K).$$

Then if F is suitably chosen, $\kappa_v \in V_F = C_F(K \times K)$ and

$$\kappa: v \longmapsto \kappa_v$$

is a K-homomorphism of V into V_F. Hence ker κ is a submodule of V. Since τ is irreducible and $\lambda \neq 0$, we conclude that κ is injective.

Put

$$(u, v)_0 = (\kappa_u, \kappa_v) \qquad (u, v \in V).$$

This is a positive-definite K-invariant scalar product in V. Since τ is irreducible, this must be a multiple of the old scalar product. Hence the lemma.

COROLLARY. *Let P be an additive statement on modules. Then in order to prove it for all modules, it is enough to verify $P(\tau_F)$ for $F = F_1, F_2, \cdots$ where F_i is an increasing sequence of finite subsets of $\mathcal{E}(K)$ such that*

$$\mathcal{E}(K) = \bigcup_{i \geq 1} F_i.$$

This is obvious from Lemma 2.

16. Proof of Theorem 14.1

Fix $\nu \in \mathfrak{F}'$ and let $\mathcal{A}(\nu)$ denote the set of all $f \in \mathcal{A}(G, \tau)$ which satisfy the hypothesis of Theorem 14.1.

LEMMA 1. *Put $L = {}^0\mathcal{C}(M, \tau_M)$ and fix $P \in \mathcal{P}(A)$. Then*

$$\psi \longmapsto E(P: \psi: \nu)$$

is a bijection of L onto $\mathcal{A}(\nu)$.

Note that the statements of this lemma and Theorem 14.1 are both additive in τ. Hence it is enough to verify them for $\tau = \tau_F$ where F is a finite subset of $\mathcal{E}(K)$ such that[5] $F = F^*$.

So fix F and put $\tau = \tau_F$. We claim that Lemma 1 is equivalent to Theorem 14.1 in this case. We know [2(j), §18] that dim $L < \infty$ and

$$L = \sum_{\omega \in \mathcal{E}_2(M)} L(\omega)$$

where the sum is orthogonal. Fix $\psi \in L$. Then

$$\psi = \sum_\omega \psi_\omega$$

where ψ_ω is the component of ψ in $L(\omega)$ and it follows from Lemma 13.4 that

---

[5] $\mathfrak{b}^*$ is the class contragredient to $\mathfrak{b} \in \mathcal{E}(K)$ and F^* is the image of F under the mapping $\mathfrak{b} \mapsto \mathfrak{b}^*$.

$$\| c_{Q|P}(s:\nu)\psi \|^2 = \sum_\omega \| c_{Q|P}(s:\nu)\psi_\omega \|^2 = c(A)^2 \sum_\omega \mu(\omega:\nu)^{-1} \| \psi_\omega \|^2$$

for $P, Q \in \mathcal{P}(A)$ and $s \in \mathfrak{w}$. This shows that $\| c_{Q|P}(s:\nu)\psi \|$ is independent of P, Q and s. Moreover $c_{Q|P}(s:\nu)$ is bijective on L.

It is obvious that $E(P:\psi:\nu) \in \mathcal{C}(\nu)$. Moreover since $c_{P|P}(1:\nu)$ is bijective on L, it is clear that the mapping of Lemma 1 is injective. So the main assertion of our lemma is that it is surjective.

Now suppose Theorem 14.1 holds. Fix $f \in \mathcal{C}(\nu)$ and put

$$\psi = c_{P|P}(1:\nu)^{-1}f_{P,1}, \qquad g = f - E(P:\psi:\nu).$$

Then $g \in \mathcal{C}(\nu)$ and $g_{P,1} = 0$. Hence by the corollary of Theorem 14.1, $g = 0$ and this proves the surjectivity in Lemma 1.

Conversely suppose Lemma 1 holds. Then $f = E(P:\psi:\nu)$ for some $\psi \in L$. Since $\nu \in \mathfrak{F}'$, it is clear that

$$\phi_{Q,s} = c_{Q|P}(s:\nu)\psi$$

for $Q \in \mathcal{P}(A)$ and $s \in \mathfrak{w}$. Therefore

$$\| \phi_{Q,s} \| = \| c_{Q|P}(s:\nu)\psi \|$$

is independent of (Q, s). This proves the equivalence of Lemma 1 and Theorem 14.1.

LEMMA 2. $\dim \mathcal{C}(\nu) < \infty$.

Fix $f \in \mathcal{C}(\nu)$ and $P \in \mathcal{C}(A)$. Since $\nu \in \mathfrak{F}'$, $s\nu$ $(s \in \mathfrak{w})$ are all distinct and therefore the functions $\phi_{P,s}$ are uniquely determined by f. We may therefore denote them by $f_{P,s}$. Since $f_{P^s} = s(f_P)$, it is clear that

$$f_{P,s} = s(f_{Q,1})$$

where $Q = P^{s^{-1}}$. This means that if $f_{P,1} = 0$ for all $P \in \mathcal{P}(A)$, then $f = 0$. Let $p = [\mathcal{P}(A)]$. Then we conclude that

$$\dim \mathcal{C}(\nu) \leq (\dim L)^p < \infty .$$

For $\alpha \in \mathcal{L}_F$ define $\boldsymbol{\alpha} \in C_c^\infty(G, \tau)$ as in Section 9. (Recall that $F = F^*$.) If $f \in \mathcal{C}(\nu)$, it follows from [2(i), Thm. 21.2] that $\boldsymbol{\alpha} * f \in \mathcal{C}(\nu)$ and

$$(\boldsymbol{\alpha} * f)_{P,s} = \alpha_{s\nu}^{(P)} * f_{P,s}$$

in the notation of Section 9. Since $\alpha \mapsto \boldsymbol{\alpha}$ is a homomorphism of $\mathcal{L}_F$ into $C_c^\infty(G, \tau)$, we may regard $\mathcal{C}(\nu)$ as a left $\mathcal{L}_F$-module.

LEMMA 3. *Fix $P \in \mathcal{P}(A)$ and suppose f is an element in $\mathcal{C}(\nu)$ such that $f_{P,1} = 0$. Then $f = 0$.*

Let $\mathcal{B}$ be the set of all $f \in \mathcal{C}(\nu)$ such that $f_{P,1} = 0$. We have to show that $\mathcal{B} = \{0\}$. Clearly $\mathcal{B}$ is an $\mathcal{L}_F$-submodule of $\mathcal{C}(\nu)$. If $\mathcal{B} \neq \{0\}$, we can, in view of Lemma 2, choose a simple submodule $\mathfrak{B}_0 \neq \{0\}$ of $\mathfrak{B}$. Fix $f_0 \neq 0$ in $\mathfrak{B}_0$ and let E_ω denote the orthogonal projection of L on $L(\omega)$ for $\omega \in \mathfrak{S}_2(M)$. Then (see the proof of Lemma 2) we can choose $Q \in \mathscr{P}(A)$ and $\omega \in \mathfrak{S}_2(M)$ such that $E_\omega(f_0)_{Q,1} \neq 0$. Consider the mapping

$$f \longmapsto E_\omega f_{Q,1}$$

of $\mathfrak{B}_0$ into $L(\omega)$. If $\alpha \in \mathfrak{L}_F$, then

$$E_\omega(\alpha * f)_{Q,1} = E_\omega(\alpha_\nu^{(Q)} * f_{Q,1}) = \alpha_\nu^{(Q)} * E_\omega f_{Q,1}$$

for $f \in \mathfrak{B}_0$. Hence we conclude that the kernel of this mapping is an $\mathfrak{L}_F$-submodule of $\mathfrak{B}_0$. But $\mathfrak{B}_0$ is simple and therefore this mapping must be injective.

Note that an element $f \in \mathfrak{a}(\nu)$ is a function on $K \times G \times K$. Since

$$f(k_1: x: k_2) = f(1: k_1 x k_2: 1) \qquad\qquad (k_1,\ k_2 \in K,\ x \in G)\ ,$$

we may also regard it as a function on G by setting $f(x) = f(1: x: 1)$. Under this identification $\alpha * f$ corresponds to $\alpha * f$ $(\alpha \in \mathfrak{L}_F)$.

Given $f \in \mathfrak{B}_0$, there exists a unique element $S_f \in \operatorname{End} \mathfrak{H}_F$ such that

$$(\psi_T,\ E_\omega f_{Q,1}) = d(\omega)^{-1} \operatorname{tr} T^* S_f$$

for all $T \in \operatorname{End} \mathfrak{H}_F$. (Here the notation is the same as in § 9.) Then if $\alpha \in \mathfrak{L}_F$,

$$
\begin{aligned}
\operatorname{tr}(T^* S_{\alpha \cdot f}) &= d(\omega)(\psi_T,\ \alpha_\nu^{(Q)} * E_\omega f_{Q,1}) \\
&= d(\omega)((\tilde{\alpha})_\nu^{(Q)} * \psi_T,\ E_\omega f_{Q,1}) \\
&= d(\omega)(\psi_{T\pi(\tilde{\alpha})},\ E_\omega f_{Q,1}) \\
&= \operatorname{tr}(\pi(\check{\alpha})T^* \cdot S_f) = \operatorname{tr}(T^* S_f \pi(\check{\alpha}))
\end{aligned}
$$

from Lemma 9.2. Here $\bar{\alpha} = \operatorname{conj} \alpha$ and $\pi = \pi_{Q,\omega,\nu}$. This shows that

$$S_{\alpha \cdot f} = S_f \pi(\check{\alpha})\ .$$

Put

$$\alpha \cdot T = T\pi(\check{\alpha}) \qquad\qquad (\alpha \in \mathfrak{L}_F,\ T \in \operatorname{End} \mathfrak{H}_F)\ .$$

This makes $\operatorname{End} \mathfrak{H}_F$ into a left $\mathfrak{L}_F$-module and then the mapping

$$f \longmapsto S_f$$

defines an $\mathfrak{L}_F$-homomorphism of $\mathfrak{B}_0$ into $\operatorname{End} \mathfrak{H}_F$. Since $\mathfrak{B}_0$ is simple and $S_{f_0} \neq 0$, this mapping is injective.

Let $f_1, \cdots, f_r$ be a base for $\mathfrak{B}_0$. Put $S_i = S_{f_i}$ and choose $T_j \in \operatorname{End} \mathfrak{H}_F$ such that $\operatorname{tr} S_i T_j = \delta_{ij}$ $(1 \leq i,\ j \leq r)$. Put

$$\phi_{ij}(x) = \operatorname{tr}(S_i \pi(x) T_j) \qquad\qquad (x \in G)\ .$$

Then if $\alpha \in \mathfrak{L}_F$,

$$S_{\alpha * f_i} = S_i \pi(\check{\alpha}) = \sum_j \mathrm{tr}\,(S_i \pi(\check{\alpha}) T_j) S_j$$
$$= \sum_j \int_G \alpha(x^{-1})\phi_{ij}(x)dx \cdot S_j \ .$$

The mapping $f \mapsto S_f$ being injective, this implies that

$$\alpha * f_i = \sum_j \int \alpha(x^{-1})\phi_{ij}(x)dx \cdot f_j \ .$$

Therefore

$$(\alpha * f_i)(1) = \sum_j (\alpha * \phi_{ij})(1) \cdot f_j(1)$$

for all $\alpha \in \mathfrak{L}_F$. Put $\mathfrak{L} = C_c^\infty(G)$ and define the mapping $\beta \mapsto \beta_F$ of $\mathfrak{L}$ into $\mathfrak{L}_F$ as in Section 3. It is obvious that

$$f_i = \alpha_F *_K f_i *_K \alpha_F \ , \quad \phi_{ij} = \alpha_F *_K \phi_{ij} *_K \alpha_F \ .$$

Hence we conclude that

$$(\beta * f_i)(1) = \sum_j (\beta * \phi_{ij})(1) \cdot f_j(1)$$

for $\beta \in \mathfrak{L}$. This implies that

$$f_i(x) = \sum_j \phi_{ij}(x) f_j(1) \ .$$

Now put

$$\psi_{ij} = \psi_{T_j S_i} \ , \quad \psi_i = \sum_j \psi_{ij} f_j(1) \qquad\qquad (1 \leqq i,\, j \leqq r) \ .$$

Then $\psi_{ij} \in L(\omega)$ and

$$E(Q: \psi_{ij}: \nu) = \phi_{ij}$$

from Theorem 7.1. So the above relation becomes

$$f_i = E(Q: \psi_i: \nu) \qquad\qquad (1 \leqq i \leqq r) \ .$$

This shows that every $f \in \mathfrak{B}_0$ is of the form

$$f = E(Q: \psi: \nu)$$

with $\psi \in L(\omega)$. But then

$$f_{P,1} = c_{P|Q}(1: \nu)\psi = 0 \ .$$

This implies that $\psi = 0$ and hence $f = 0$. This proves that $\mathfrak{B}_0 = \{0\}$ and so we get a contradiction. This concludes the proof of Lemma 3.

Now we shall prove Lemma 1 for $\tau = \tau_F$. We have only to verify the surjectivity. Fix $f \in \mathcal{Q}(\nu)$ and put

$$\psi = c_{P|P}(1: \nu)^{-1} f_{P,1} \ , \quad g = f - E(P: \psi: \nu) \ .$$

It is clear that $g \in \mathcal{Q}(\nu)$ and $g_{P,1} = 0$. Hence $g = 0$ from Lemma 3 and therefore

$$f = E(P: \psi: \nu) \ .$$

This completes the proof of Lemma 1 and hence also of Theorem 14.1.

17. The functional equations for $E(P\colon \psi\colon \nu)$

We return to the notation of Section 14. Put $L = {}^{\circ}\mathcal{C}(M, \tau_M)$. Then L is the orthogonal sum of $L(\omega)$ for $\omega \in \mathcal{E}_2(M)$.

LEMMA 1. *Fix* $P_1, P_2 \in \mathscr{P}(A)$, $s \in \mathfrak{w}$, $\nu \in \mathfrak{F}'$ *and* $\omega \in \mathcal{E}_2(M)$. *Then* $c_{P_2|P_1}(s\colon \nu)$ *defines a bijection of* $L(\omega)$ *onto* $L(s\omega)$. *Moreover*

$$\mu(\omega\colon \nu)c_{P_2|P_1}(s\colon \nu)^* c_{P_2|P_1}(s\colon \nu) = c(A)^2$$

on $L(\omega)$ *for* $P \in \mathscr{P}(A)$.

It is easy to check that this statement is additive in τ. Hence it follows from Lemma 13.4.

COROLLARY. $\mu(s\omega\colon s\nu) = \mu(\omega\colon \nu)$ *for* $s \in \mathfrak{w}$.

Given $\omega \in \mathcal{E}_2(M)$, we can choose τ so that $L(\omega) \neq \{0\}$. Fix $P \in \mathscr{P}(A)$ and put $Q = P^*$. Then if $\nu \in \mathfrak{F}'$ and $\psi \in L(\omega)$, we have

$$\mu(s\omega\colon s\nu) \, \| \, c_{Q|Q}(1\colon s\nu)s\psi \, \|^2 = c(A)^2 \, \| \, s\psi \, \|^2 = c(A)^2 \, \| \, \psi \, \|^2$$

from the above lemma. On the other hand [2(j), Lem. 18.1]

$$s^{-1}c_{Q|Q}(1\colon s\nu)s = c_{P|P}(1\colon \nu) \, .$$

Hence

$$\| \, c_{Q|Q}(1\colon s\nu)s\psi \, \|^2 = \| \, c_{P|P}(1\colon \nu)\psi \, \|^2 = \mu(\omega\colon \nu)^{-1}c(A)^2 \, \| \, \psi \, \|^2$$

and the desired result follows.

Put

$$^{\circ}c_{Q|P}(s\colon \nu) = c_{Q|Q}(1\colon s\nu)^{-1}c_{Q|P}(s\colon \nu)$$

for $P, Q \in \mathscr{P}(A)$ and $\nu \in \mathfrak{F}_c(\delta)$.

LEMMA 2. $E\big(Q\colon {}^{\circ}c_{Q|P}(s\colon \nu)\psi\colon s\nu\big) = E(P\colon \psi\colon \nu)$ *for* $\psi \in L$ *and* $\nu \in \mathfrak{F}_c(\delta)$.

It is enough to verify this for $\psi \in L(\omega)$ and $\nu \in \mathfrak{F}'$. Put

$$f = E(P\colon \psi\colon \nu) - E\big(Q\colon {}^{\circ}c_{Q|P}(s\colon \nu)\psi\colon s\nu\big) \, .$$

Then it is obvious that $f \in \mathcal{C}(\nu)$ and

$$f_{Q,\cdot} = c_{Q|P}(s\colon \nu)\psi - c_{Q|Q}(1\colon s\nu)^{\circ}c_{Q|P}(s\colon \nu)\psi$$
$$= 0 \, .$$

Hence we conclude from Theorem 14.1 that $f = 0$.

COROLLARY. *Let* $P, P_1, P_2 \in \mathscr{P}(A)$ *and* $s, t \in \mathfrak{w}$. *Then*

$$^{\circ}c_{P_2|P_1}(st\colon \nu) = {}^{\circ}c_{P_2|P}(s\colon t\nu)^{\circ}c_{P|P_1}(t\colon \nu)$$

for $\nu \in \mathfrak{F}_c(\delta)$.

Fix $\psi \in L$ and $\nu \in \mathfrak{F}'$. Then

$$E(P_1: \psi: \nu) = E\big(P: {}^0c_{P|P_1}(t: \nu)\psi: t\nu\big)$$

from Lemma 2. Taking the constant terms of both sides along P_2, we conclude that

$$c_{P_2|P_1}(st: \nu)\psi = c_{P_2|P}(s: t\nu){}^0c_{P|P_1}(t: \nu)\psi$$

and this implies our assertion.

LEMMA 3. ${}^0c_{Q|P}(s: \nu)$ *maps* $L(\omega)$ *into* $L(s\omega)$. *Moreover if* $\nu \in \mathfrak{F}'$, ${}^0c_{Q|P}(s: \nu)$ *is unitary.*

The first statement is obvious from Lemma 1. Now fix $\nu \in \mathfrak{F}'$. If ω_1, ω_2 are two distinct elements in $\mathscr{E}_2(M)$, then $L(\omega_1)$ is orthogonal to $L(\omega_2)$. Hence it is sufficient to verify that

$$\| {}^0c_{Q|P}(s: \nu)\psi \|^2 = \| \psi \|^2$$

for $\psi \in L(\omega)$. Put

$$\phi = {}^0c_{Q|P}(s: \nu)\psi \ .$$

Then $\phi \in L(s\omega)$ and

$$c_{Q|Q}(1: s\nu)\phi = c_{Q|P}(s: \nu)\psi \ .$$

But

$$\| c_{Q|Q}(1: s\nu)\phi \|^2 = c(A)^2\mu(s\omega: s\nu)^{-1} \| \phi \|^2 \ ,$$

$$\| c_{Q|P}(s: \nu)\psi \|^2 = c(A)^2\mu(\omega: \nu)^{-1} \| \psi \|^2$$

from Lemma 1. Since $\mu(s\omega: s\nu) = \mu(\omega: \nu)$ (cor. of Lemma 1), we conclude that $\| \phi \| = \| \psi \|$.

LEMMA 4. *Let* P_1, $P_2 \in \mathscr{P}(A)$ *and suppose* $P' = M'A'N'$ *is a psgp of* G *such that* $P' \supset P_1 \cup P_2$. *Put* ${}^*P_i = M' \cap P_i \, (i = 1, 2)$ *and* ${}^*A = M' \cap A$. *Let* ${}^*\mathfrak{w}$ *be the subgroup of all elements in* $\mathfrak{w}$ *which leave* $\mathfrak{a}'$ *pointwise fixed. Then if* $\nu \in \mathfrak{F}'$,

$$
{}^0c_{P_2|P_1}(s: \nu) = {}^0c_{{}^*P_2|{}^*P_1}(s: {}^*\nu)
$$

for $s \in {}^*\mathfrak{w}$. *Here* ${}^*\nu$ *is the restriction of* ν *on* ${}^*\mathfrak{a}$.

The proof depends on an auxiliary result on Eisenstein integrals. Fix $P \in \mathscr{P}(A)$ and let $P' = M'A'N'$ be a psgp of G containing P. Define ${}^*P = P \cap M'$. Then ${}^*P = M \cdot {}^*A \cdot {}^*N$ where ${}^*A = A \cap M'$.

LEMMA 5.

$$E\big(P': E({}^*P: \psi: {}^*\nu): \nu'\big) = E(P: \psi: \nu) \qquad \text{for } \psi \in L \text{ and } \nu \in \mathfrak{F}_c \ .$$

Here ${}^*\nu$ *and* ν' *are the restrictions of* ν *on* ${}^*\mathfrak{a}$ *and* $\mathfrak{a}'$ *respectively.*

It is enough to verify this for $\nu \in \mathfrak{F}$. Fix $f \in C_c^\infty(G, \tau)$. Then

$$\big(E(P': E(^*P: \psi: {}^*\nu): \nu'), f\big) = \big(E(^*P: \psi: {}^*\nu), f_{\nu'}^{(P')}\big) = \big(\psi, (f_{\nu'}^{(P')})_{*\nu}^{(*P)}\big) \, .$$

This follows immediately from the definition of the Eisenstein integral [2(i), §19]. But

$$(f_{\nu'}^{(P')})_{*\nu}^{(*P)} = f_\nu^{(P)}$$

and

$$(\psi, f_\nu^{(P)}) = \big(E(P: \psi: \nu), f\big)$$

and so our assertion is obvious.

Now we come to the proof of Lemma 4. Since s leaves $\mathfrak{a}'$ fixed, it may be regarded as an element of $\mathfrak{w}(M'/{}^*A)$. Then $s \cdot {}^*\nu = {}^*(s\nu)$ and we conclude from Lemma 2 that

$$E(^*P_1: \psi: {}^*\nu) = E\big(^*P_2: {}^0c_{*P_2|*P_1}(s: {}^*\nu)\psi: {}^*(s\nu)\big) \, .$$

Hence

$$\begin{aligned} E(P_1: \psi: \nu) &= E\big(P': E(^*P_1: \psi: {}^*\nu): \nu'\big) \\ &= E\big(P': E\big(^*P_2: {}^0c_{*P_2|*P_1}(s: {}^*\nu)\psi: {}^*(s\nu)\big): \nu'\big) \\ &= E\big(P_2: {}^0c_{*P_2|*P_1}(s: {}^*\nu)\psi: s\nu\big) \end{aligned}$$

from Lemma 5. On the other hand

$$E(P_1: \psi: \nu) = E\big(P_2: {}^0c_{P_2|P_1}(s: \nu)\psi: s\nu\big)$$

and so the desired result follows from Lemma 16.1.

LEMMA 6. *Fix* $P, Q \in \mathscr{P}(A)$ *and* $s \in \mathfrak{w}$. *Then* ${}^0c_{Q|P}(s: \nu)$ *is everywhere holomorphic and unitary on* $\mathfrak{F}$.

In view of [2(j), Lem. 18.1], it is enough to consider the case $s = 1$. Then by the corollary of Lemma 2 we have the product formula

$$^0c_{Q|P}(1: \nu) = {}^0c_{Q|P_1}(1: \nu){}^0c_{P_1|P}(1: \nu) \qquad\qquad \big(P_1 \in \mathscr{P}(A)\big) \, .$$

Hence it is enough to consider the case $d(Q, P) = 1$. (Note that ${}^0c_{P|P}(1: \nu) = 1$.) Let α be the unique root in $\Sigma(Q \mid P)$. Put $^*P = M_\alpha \cap P$ in the notation of Lemma 2.3. Then

$$^0c_{Q|P}(1: \nu) = {}^0c_{*\bar{P}|*P}(1: {}^*\nu)$$

by Lemma 4 where $^*\nu$ is the restriction of ν on $\mathfrak{a}_\alpha = \mathbf{R}H_\alpha$. (We take $P' = P_{\{\alpha\}}$ in Lemma 4.) Now ${}^0c_{*\bar{P}|*P}(1: {}^*\nu)$ is a meromorphic function of the complex variable $\nu_\alpha = \langle \alpha, \nu \rangle$ defined for all sufficiently small values of $|\operatorname{Im}\nu_\alpha|$. Since it is unitary for ν_α real and $\neq 0$, it is obvious that it cannot have a pole at $\nu_\alpha = 0$. This proves the lemma.

18. Relations between the c- and j-functions

We return to the notation of Section 13.

LEMMA 1. *Fix $T \in \mathrm{End}\ \mathfrak{H}_F$, $P, Q \in \mathscr{P}(A)$ and $\nu \in \mathfrak{F}'$. Then*

$$^{0}c_{Q|P}(1:\nu)\psi_T = \psi_{T'}$$

where

$$T' = j_{P|Q}(\nu)^{-1}Tj_{P|Q}(\nu) \ .$$

This follows from the corollary of Lemma 13.3 if we take into account Lemmas 13.2, 16.1 and 17.2.

COROLLARY.

$$c_{Q|P}(1:\nu)\psi_T = \psi_{T_1}$$

where

$$T_1 = c(A)j_{\bar{Q}|P}(\nu)Tj_{P|Q}(\nu) \ .$$

Since $c_{Q|P}(1:\nu) = c_{Q|Q}(1:\nu)^{0}c_{Q|P}(1:\nu)$, we conclude from Lemma 11.1 that

$$c_{Q|P}(1:\nu)\psi_T = \psi_{T_1}$$

where

$$T_1 = c(A)j_{\bar{Q}|Q}(\nu)j_{P|Q}(\nu)^{-1}Tj_{P|Q}(\nu) \ .$$

But P lies between Q and $\bar{Q}$. Hence (corollary of Theorem 13.1)

$$j_{\bar{Q}|Q}(\nu) = j_{\bar{Q}|P}(\nu)j_{P|Q}(\nu) \ .$$

Therefore

$$T_1 = c(A)j_{\bar{Q}|P}(\nu)Tj_{P|Q}(\nu) \ .$$

19. Rationality of $^{0}c_{Q|P}$

Let τ and L be as in Section 17.

LEMMA 1. *Fix $P \in \mathscr{P}(A)$ and ν, ν' in $\mathfrak{F}'$. Then the four linear transformations*

$$c_{P|P}(1:\nu)\ , \quad c_{\bar{P}|\bar{P}}(1:\nu)\ , \quad c_{\bar{P}|P}(1:\nu')\ , \quad c_{P|\bar{P}}(1:\nu')$$

commute with each other. Moreover

$$c_{P|P}(1:\nu)^* = c_{\bar{P}|\bar{P}}(1:\nu)\ , \quad c_{\bar{P}|P}(1:\nu')^* = c_{P|\bar{P}}(1:\nu')^* \ .$$

It is easy to see that the above statement is additive in τ. Moreover by Lemma 17.1 all these linear transformations map $L(\omega)$ into itself. Therefore our assertion follows from Lemma 13.2 and the corollary of Lemma 18.1.

The following result is of some interest although we shall not make use of it in this paper.

LEMMA 2. *Fix* $P, Q \in \mathscr{P}(A)$ *and* $s \in \mathfrak{w}$. *Then*

$$\nu \longmapsto {}^0c_{Q|P}(s:\nu)$$

defines a rational mapping of $\mathfrak{F}_c$ *into* End L.

Again, in view of [2(j), Lem. 18.1], it is enough to consider the case $s = 1$. Moreover, as we have seen above, it would be sufficient to verify the following lemma in the notation of Section 18.

LEMMA 3. *Fix* $T \in$ End $\mathfrak{H}_F$. *Then*

$$\nu \longmapsto j_{P|Q}(\nu)^{-1} T j_{P|Q}(\nu)$$

defines a rational mapping of $\mathfrak{F}_c$ *into* End $\mathfrak{H}_F$.

Define $\mathfrak{G}$ and $\mathfrak{K}$ as usual (see [2(i), §11]) and let $\tilde{\mathfrak{G}} = K \otimes \mathfrak{G}$ denote the space of all functions $f: K \to \mathfrak{G}$ such that $f(k) = 0$ for all $k \in K$ except a finite number. Then $\tilde{\mathfrak{G}}$ is a complex vector space. For $k \in K$ and $g \in \mathfrak{G}$, let $k \otimes g$ denote the function $f \in \tilde{\mathfrak{G}}$ given by $f(k) = g$, $f(k') = 0$ $(k' \neq k)$. We turn $\tilde{\mathfrak{G}}$ into an associative algebra by introducing a multiplication as follows:

$$(k_1 \otimes g_1) \cdot (k_2 \otimes g_2) = k_1 k_2 \otimes (g_1^{k_2^{-1}} \cdot g_2) \ .$$

Put

$$\mathfrak{H}^0 = \bigcup_{F_0} \mathfrak{H}_{F_0}$$

when F_0 runs over all finite subsets of $\mathscr{E}(K)$. Then if $\pi = \pi_{P,\omega,\nu}$, we associate to it a representation π^0 of $\tilde{\mathfrak{G}}$ on $\mathfrak{H}^0$ as follows. If $h \in \mathfrak{H}^0$,

$$\pi^0(k \otimes g)h = \pi(k)\pi(g)h \qquad\qquad (k \in K, g \in \mathfrak{G}) \ .$$

(Recall that there is a natural action of $\mathfrak{G}$ on $\mathfrak{H}^0$ under π.) We shall usually write π instead of π^0.

Put $\tilde{\mathfrak{K}} = K \otimes \mathfrak{K} \subset \tilde{\mathfrak{G}}$. Then $\tilde{\mathfrak{K}}$ is a subalgebra of $\tilde{\mathfrak{G}}$. Let $\tilde{\mathfrak{K}}_F$ denote the set of all $\kappa \in \tilde{\mathfrak{K}}$ such that $\pi(\kappa)h = 0$ for every $h \in \mathfrak{H}_F$. Then $\tilde{\mathfrak{K}}_F$ is a two-sided ideal in $\tilde{\mathfrak{K}}$ (which is independent of P and ν). Let $\tilde{\mathfrak{G}}_F$ be the set of all $f \in \tilde{\mathfrak{G}}$ such that

$$\tilde{\mathfrak{K}}_F \cdot f \subset \tilde{\mathfrak{G}} \cdot \tilde{\mathfrak{K}}_F \ .$$

Then $\tilde{\mathfrak{G}}_F$ is a subalgebra of $\tilde{\mathfrak{G}}$ and $\mathfrak{H}_F$ is stable under $\pi(\tilde{\mathfrak{G}}_F)$. Let $\pi_F = \pi_{F,P,\nu}$ denote the corresponding representation of $\tilde{\mathfrak{G}}_F$ on $\mathfrak{H}_F$. If π is irreducible as a representation of G on $\mathfrak{H}$, then it is easy to see that π_F is an irreducible representation of $\tilde{\mathfrak{G}}_F$ on $\mathfrak{H}_F$.

Now fix $\nu_0 \in \mathfrak{F}'$. Then by Lemma 13.3, π_{F,P,ν_0} is irreducible. Let T_i $(1 \leq i \leq r)$ be a base for End $\mathfrak{H}_F$. Then we can choose $g_i \in \tilde{\mathfrak{G}}_F$ such that

$$\pi_{F,P,\nu_0}(g_i) = T_i \qquad\qquad (1 \leq i \leq r) \, .$$

On the other hand, it is easy to deduce from Corollary 2 of Lemma 4.1 that for fixed $g \in \widetilde{\mathfrak{G}}_F$ and $Q \in \mathscr{P}(A)$,

$$\nu \longmapsto \pi_{F,Q,\nu}(g)$$

is a polynomial mapping from $\mathfrak{F}_c$ to End $\mathfrak{H}_F$. Hence

$$\pi_{F,P,\nu}(g_i) = \sum_j p_{ij}(\nu) T_j$$

where p_{ij} are polynomial functions on $\mathfrak{F}_c$. Put

$$p(\nu) = \det \big(p_{ij}(\nu)\big)_{1 \leq i, j \leq r} \, .$$

Then p is also a polynomial function on $\mathfrak{F}_c$. Moreover $p_{ij}(\nu_0) = \delta_{ij}$ and hence $p(\nu_0) = 1$. This shows that $p \neq 0$. We can choose polynomial functions p'_{ij} on $\mathfrak{F}_c$ such that if

$$g_i(\nu) = \sum_j p'_{ij}(\nu) g_j \, ,$$

then

$$\pi_{F,P,\nu}\big(g_i(\nu)\big) = p(\nu) T_i \qquad\qquad (1 \leq i \leq r, \, \nu \in \mathfrak{F}_0) \, .$$

Now let $\nu \in \mathfrak{F}'$. Then

$$\pi_{F,P,\nu}\big(g_i(\nu)\big) j_{P|Q}(\nu) = j_{P|Q}(\nu) \pi_{F,Q,\nu}\big(g_i(\nu)\big) \, .$$

This follows immediately from the corollary of Lemma 5.1 by analytic continuation. Hence

$$j_{P|Q}(\nu)^{-1} T_i j_{P|Q}(\nu) = p(\nu)^{-1} \pi_{F,Q,\nu}\big(g_i(\nu)\big)$$

provided $p(\nu) \neq 0$. Since the right hand side is rational in ν, the lemma is proved.

Part III. Explicit determination of the Plancherel measure

20. Evaluation of $(\phi_\alpha)_\nu^{(P)}$

Let $d\nu$ denote the Euclidean measure on $\mathfrak{F} = \mathfrak{a}^*$ which is dual to the Haar measure $da = dA$ on A. Define $\gamma(G/A) = \gamma(P)$ (Lemma 2.6) and $c(G/A) = c(P) = c(A)$ (§ 11) for $P \in \mathscr{P}(A)$. Fix $P_1, P_2 \in \mathscr{P}(A)$ and $\psi \in L(\omega)$.

THEOREM 1. *Put*

$$\phi_\alpha = \int_{\mathfrak{F}} \mu(\omega : \nu) \alpha(\nu) E(P_1 : \psi : \nu) d\nu$$

for $\alpha \in C_c^\infty(\mathfrak{F})$. *Then* $\phi_\alpha \in \mathcal{C}(G, \tau)$ *and*

$$(\phi_\alpha)_\nu^{(P_2)} = \gamma c^2 \sum_{s \in \mathfrak{w}} \alpha(s^{-1}\nu)^0 c_{P_2|P_1}(s : s^{-1}\nu)\psi$$

for $\nu \in \mathfrak{F}$. *Here* $\gamma = \gamma(G/A)$ *and* $c = c(G/A)$.

Let Ω be an open and relatively compact subset of $\mathfrak{F}$. Fix $\beta \in C_c^\infty(\mathfrak{F})$

such that $\beta = 1$ on Ω and put

$$\phi(\nu: x) = \beta(\nu)\mu(\omega: \nu)E(P_1: \psi: \nu: x) \qquad\qquad (\nu \in \mathfrak{F},\ x \in G) .$$

Then if $P \in \mathscr{P}(A)$, $\nu \in \mathfrak{F}'$ and $s \in \mathfrak{w}$, we have

$$\| \mu(\omega: \nu)c_{P|P_1}(s: \nu)\psi \|^2 = c^2\mu(\omega: \nu)\| \psi \|^2$$

from Lemma 17.1. Since μ_ω is holomorphic on $\mathfrak{F}$ (Lem. 12.2), we conclude from [2(j), Thms. 11.1 and 13.1] that $\phi_\alpha \in \mathcal{C}(G, \tau)$ for $\alpha \in C_c^\infty(\Omega)$. The same argument shows that $\alpha \mapsto \phi_\alpha$ is a continuous mapping of $C_c^\infty(\mathfrak{F})$ into $\mathcal{C}(G, \tau)$.

Now first assume that $\alpha \in C_c^\infty(\mathfrak{F}')$. Then

$$\phi_\alpha^{(P_2)}(ma) = \gamma \int_{\mathfrak{F}} \exp\{(-1)^{1/2}\nu(\log a)\}$$
$$\times \sum_{s \in \mathfrak{w}} \alpha(s^{-1}\nu)\mu(\omega: s^{-1}\nu)\cdot c_{P_2|\bar{P}_2}(1: \nu)c_{\bar{P}_2|P_1}(s: s^{-1}\nu)\psi d\nu$$

from [2(j), Thm. 19.2] for $m \in M$, $a \in A$. But

$$c_{P_2|\bar{P}_2}(1: \nu)c_{\bar{P}_2|P_1}(s: s^{-1}\nu) = c_{P_2|\bar{P}_2}(1: \nu)c_{\bar{P}_2|\bar{P}_2}(1: \nu)^0 c_{\bar{P}_2|P_1}(s: s^{-1}\nu)$$
$$= c_{\bar{P}_2|\bar{P}_2}(1: \nu)c_{P_2|\bar{P}_2}(1: \nu)^0 c_{\bar{P}_2|P_1}(s: s^{-1}\nu)$$

from Lemma 19.1. Moreover

$$c_{P_2|\bar{P}_2}(1: \nu)^0 c_{\bar{P}_2|P_1}(s: s^{-1}\nu) = c_{P_2|P_2}(1: \nu)^0 c_{P_2|\bar{P}_2}(1: \nu)^0 c_{\bar{P}_2|P_1}(s: s^{-1}\nu)$$
$$= c_{P_2|P_2}(1: \nu)^0 c_{P_2|P_1}(s: s^{-1}\nu)$$

from the corollary of Lemma 17.2. Since $^0c_{P_2|P_1}(s: s^{-1}\nu)\psi \in L(s\omega)$ (Lem. 17.3), we conclude from Lemmas 17.1 and 19.1 that

$$c_{P_2|\bar{P}_2}(1: \nu)c_{\bar{P}_2|P_1}(s: s^{-1}\nu)\psi = c^2\mu(s\omega: \nu)^{-1}\,^0c_{P_2|P_1}(s: s^{-1}\nu)\psi .$$

But $\mu(s\omega: \nu) = \mu(\omega: s^{-1}\nu)$ (cor. of Lem. 17.1). Therefore

$$\phi_\alpha^{(P_2)}(ma) = \gamma c^2 \int_{\mathfrak{F}} \exp\{(-1)^{1/2}\nu(\log a)\} \sum_{s \in \mathfrak{w}} \alpha(s^{-1}\nu)^0 c_{P_2|P_1}(s: s^{-1}\nu)\psi d\nu$$

and so the statement of the theorem is true in this case.

Now let $\alpha \in C_c^\infty(\mathfrak{F})$. Put $f = \phi_\alpha$. Then we have seen above that $f \in \mathcal{C}(G, \tau)$. Moreover if P' is a psgp of G with prk $P' > \dim A$, then $f^{(P')} = 0$ (cor. of Thm. 13.2 of [2(j)]). Fix $\nu \in \mathfrak{F}$ and put $g = f_\nu^{(P_2)}$. We claim that $g \in {}^0\mathcal{C}(M, \tau_M) = L$. Let $^*P \neq M$ be a psgp of M ($^*P = {}^*M \cdot {}^*A \cdot {}^*N$) and P' the corresponding psgp of G contained in P_2. Then $P' = M'A'N'$ where

$$M' = {}^*M, \quad A' = {}^*A\cdot A, \quad N' = {}^*N\cdot N_2 .$$

Hence if $^*\nu \in ({}^*\mathfrak{a})^*$ and $\nu' = \nu + {}^*\nu$, it is clear that

$$g_{^*\nu}^{(^*P)} = f_{\nu'}^{(P')} = 0 .$$

This shows that $g^{(^*P)} = 0$ and therefore $g \in L$.

Fix $\nu \in \mathfrak{F}$ and consider the mapping

$$T_\nu\colon \alpha \longmapsto (\phi_\alpha)_\nu^{(P_2)}$$

of $C_c^\infty(\mathfrak{F})$ into L. We have seen above that $\alpha \mapsto \phi_\alpha$ is a continuous mapping of $C_c^\infty(\mathfrak{F})$ into $\mathcal{C}(G, \tau)$. Hence [2(i), Lem. 16.1] T_ν may be regarded as a distribution on $\mathfrak{F}$ with values in L. (Recall that $\dim L < \infty$.)

Fix $\nu_0 \in \mathfrak{F}$, $\omega_0 \in \mathcal{E}_2(M)$ and put

$$T_{\omega_0, \nu_0}(\alpha) = E_{\omega_0}(\phi_\alpha)_{\nu_0}^{(P_2)}$$

where E_{ω_0} is the orthogonal projection of L on $L(\omega_0)$. Write $T = T_{\omega_0, \nu_0}$. Then T is a distribution on $\mathfrak{F}$ with values in $L(\omega_0)$.

Extend $\mathfrak{a}$ to a θ-stable Cartan subalgebra $\mathfrak{h}$ of $\mathfrak{g}$ such that $\mathfrak{h}_R = \mathfrak{a}$ and $\mathfrak{h}_I \subset \mathfrak{m}$ (see [2(i), §8] for notation). Let $\lambda = \lambda_\omega$ and $\lambda_0 = \lambda_{\omega_0}$ be the elements in $(-1)^{1/2}\mathfrak{h}_I^*$ which correspond to the infinitesimal characters of ω and ω_0 respectively. For $z \in \mathfrak{Z}$, define the polynomial functions $p(z)$ and $p_0(z)$ on $\mathfrak{F}$ as follows:

$$p(z)\colon \nu \longmapsto \gamma(z\colon \lambda + (-1)^{1/2}\nu)\,, \quad p_0(z)\colon \nu \longmapsto \gamma(z\colon \lambda_0 + (-1)^{1/2}\nu) \qquad (\nu \in \mathfrak{F})$$

where $\gamma = \gamma_{\mathfrak{g}/\mathfrak{h}}$ (see [2(i), §11] for notation).

LEMMA 1. $T(p(z)\alpha) = p_0(z\colon \nu_0)T(\alpha)$ *for* $\alpha \in C_c^\infty(\mathfrak{F})$ *and* $z \in \mathfrak{Z}$.

Note that

$$zE(P\colon \psi\colon \nu) = p(z\colon \nu)E(P\colon \psi\colon \nu)$$

from [2(i), Lem. 19.1]. Hence

$$z\phi_\alpha = \phi_{p(z)\alpha}$$

and therefore

$$T(p(z)\alpha) = E_{\omega_0}(z\phi_\alpha)_{\nu_0}^{(P_2)}\,.$$

But

$$(z\phi_\alpha)^{(P_2)} = \mu(z)(\phi_\alpha)^{(P_2)}$$

where $\mu = \mu_{P_2}$ in the notation of [2(i), Lem. 16.1]. Hence

$$T(p(z)\alpha) = p_0(z\colon \nu_0)T(\alpha)\,.$$

COROLLARY 1. *If* $\nu \in \operatorname{Supp} T$, *then* $\lambda + (-1)^{1/2}\nu = s(\lambda_0 + (-1)^{1/2}\nu_0)$ *for some* $s \in W(\mathfrak{g}/\mathfrak{h})$.

This is obvious from the above result.

COROLLARY 2. *Suppose* $\nu_0 \in \mathfrak{F}'$. *Then* $\operatorname{Supp} T \subset \{s\nu_0\}_{s \in \mathfrak{w}}$.

Put $W = W(\mathfrak{g}/\mathfrak{h})$ and suppose $\nu \in \operatorname{Supp} T$. Then by Corollary 1, $\nu = s'\nu_0$ for some $s' \in W$. Put $H_0 = H_{\nu_0}$, $H = H_\nu$. Then H_0 and H are two points in $\mathfrak{a}$ which are conjugate under W and therefore also under K [2(i), Lem. 5.1].

Let $\mathfrak{z}_0$ and $\mathfrak{z}$ respectively be the centralizers of H_0 and H in $\mathfrak{g}$. Then $\dim \mathfrak{z} = \dim \mathfrak{z}_0$ and $\mathfrak{z} \supset \mathfrak{m} + \mathfrak{a} = \mathfrak{z}_0$ since $\nu_0 \in \mathfrak{F}'$. This shows that $\mathfrak{z} = \mathfrak{z}_0 = \mathfrak{m} + \mathfrak{a}$. Fix $k \in K$ such that $H = H_0^k$. Then k normalizes $\mathfrak{z}$. Since $\mathfrak{a}$ is the intersection of the center of $\mathfrak{z}$ with $\mathfrak{p}$, k also normalizes $\mathfrak{a}$. Hence k defines an element $s \in \mathfrak{w}$ and $sH_0 = H$. This proves that $\nu = s\nu_0$.

Now fix $\nu \in \mathfrak{F}'$ and consider the distribution

$$T_\nu : \alpha \longmapsto (\phi_\alpha)_\nu^{(P_2)}$$

on $\mathfrak{F}$ with values in L. By Corollary 2 above, $\operatorname{Supp} T_\nu \subset \{s\nu\}_{s \in \mathfrak{w}}$. Fix $\beta \in C_c^\infty(\mathfrak{F}')$ such that $\beta = 1$ on some neighborhood of the set $\{s\nu\}_{s \in \mathfrak{w}}$. Then $T_\nu(\alpha) = T_\nu(\beta\alpha)$ and therefore

$$(\phi_\alpha)_\nu^{(P_2)} = \gamma c^2 \sum_{s \in \mathfrak{w}} \alpha(s^{-1}\nu)^0 c_{P_2|P_1}(s: s^{-1}\nu)\psi \qquad \left(\alpha \in C_c^\infty(\mathfrak{F})\right),$$

from the result already proved above. But for a fixed α, both sides are continuous functions of ν (with values in L). For the left side this follows from [2(i), Lem. 16.1] and for the right side from Lemma 17.6. Therefore since the equality holds for $\nu \in \mathfrak{F}'$, it must hold for all $\nu \in \mathfrak{F}$. This proves Theorem 1.

We state the following simple lemma for later use.

LEMMA 2. *Let A' be a special vector subgroup of G such that* $\dim A' \geq \dim A$. *Then if $P' \in \mathscr{P}(A')$ and $\alpha \in C_c^\infty(\mathfrak{F})$,*

$$\phi_\alpha^{(P')} = 0$$

unless A' is conjugate to A under K.

This follows immediately from the corollary of [2(j), Thm. 13.2].

21. The characters $\Theta_{\omega,\nu}$

Fix $P \in \mathscr{P}(A)$, $\omega \in \mathscr{E}_2(M)$, $\nu \in \mathfrak{F}_c$ and put $\pi = \pi_{P,\omega,\nu}$. We intend to determine the character Θ_π of π.

LEMMA 1. *There exists an integer $N \geq 1$ such that*

$$\dim \mathfrak{H}_\mathfrak{b} \leq Nd(\mathfrak{b})^2$$

for all $\mathfrak{b} \in \mathscr{E}(K)$.

We have seen in Section 4 that

$$\dim \mathfrak{H}_\mathfrak{b} = d(\mathfrak{b}) \sum_{\delta \in \mathscr{E}(K_M)} [\mathfrak{b}: \delta][\omega: \delta] \, .$$

Since ω is the class of an irreducible unitary representation of M, we can choose an integer $N \geq 1$ such that

$$[\omega: \delta] \leq Nd(\delta)$$

for all $\delta \in \mathscr{E}(K_M)$. Then

$$\dim \mathfrak{H}_{\mathfrak{b}} \leq Nd(\mathfrak{b}) \sum_{\delta \in \mathcal{E}(K_M)} d(\delta)[\mathfrak{b}: \delta] \leq Nd(\mathfrak{b})^2 \,.$$

It follows from Lemma 1 (see [2(a)]) that for $f \in C_c^\infty(G)$, the operator $\pi(f)$ is of the trace class and the mapping

$$\Theta_\pi: f \longrightarrow \operatorname{tr} \pi(f) \qquad\qquad (f \in C_c^\infty(G))\,,$$

is a central distribution on G.

LEMMA 2. *For $f \in C_c^\infty(G)$, put*

$$f^{\natural}(x) = \int_K f(kxk^{-1})dk \qquad\qquad (x \in G)\,.$$

Then

$$\Theta_\pi(f) = \theta_\omega\big((f^{\natural})^{(P)}_{-\nu}\big)$$

where θ_ω denotes the character of the class ω.

We recall that

$$\alpha^{(P)}_\nu(m) = \int_{A \times N} \alpha(man) \exp\left\{-((-1)^{1/2}\nu - \rho)(\log a)\right\}dadn \qquad (m \in M)$$

for $\alpha \in C_c^\infty(G)$ and $\nu \in \mathfrak{F}_c$.

For a finite subset F of $\mathcal{E}(K)$, define f_F as in Section 3. Then it is clear that

$$f = \lim_F f_F$$

in $C_c^\infty(G)$, where F runs through an increasing sequence of finite subsets F_i of $\mathcal{E}(K)$ such that

$$\bigcup_i F_i = \mathcal{E}(K)\,.$$

Now fix F and put $\alpha = f_F$. Then $T = \pi(\alpha)$ lies in End $\mathfrak{H}_F$ and

$$\Theta_\pi(\alpha) = \operatorname{tr} T = \int_K \operatorname{tr} \kappa_T(k: k^{-1})dk$$

from the corollary of Lemma 6.1. But

$$\kappa_T(k_2: k_1) = \kappa_\alpha(k_2: k_1)$$
$$= \int_{M \times A \times N} \alpha(k_2^{-1}man\,k_1^{-1})\sigma(m) \exp\left\{((-1)^{1/2}\nu + \rho)(\log a)\right\}dmdadn$$

from Lemma 4.1. Hence

$$\Theta_\pi(\alpha) = \theta_\omega\big((\alpha^{\natural})^{(P)}_{-\nu}\big)$$

and the required result follows by taking limits with respect to F.

Define $\mathfrak{h}$ and λ_ω as in the proof of Lemma 20.1 and put $\gamma = \gamma_{\mathfrak{g}/\mathfrak{h}}$.

COROLLARY. $z\Theta_\pi = \gamma\big(z: \lambda_\omega + (-1)^{1/2}\nu\big)\Theta_\pi$ *for* $z \in \mathfrak{Z}$.

This is an immediate consequence of the above result and [2(i), Lem. 16.1].

We can now conclude from [2(i), Thm. 11.1] that Θ_π is a function which we denote by $\Theta_{P,\omega,\nu}$.

LEMMA 3. $\Theta_{P,\omega,\nu}$ is independent of $P \in \mathcal{P}(A)$.

Fix $f \in C_c^\infty(G)$. Then it is obvious from Lemma 2 that

$$\nu \longmapsto \Theta_{P,\omega,\nu}(f)$$

is a holomorphic function on $\mathfrak{F}_c$. Now fix $P_1,\, P_2 \in \mathcal{P}(A)$ and put

$$F(\nu) = \Theta_{P_1,\omega,\nu}(f) - \Theta_{P_2,\omega,\nu}(f) \ .$$

Then F is a holomorphic function on $\mathfrak{F}_c$ which is zero on $\mathfrak{F}'$ by Theorem 13.2. Hence $F = 0$.

In view of the above lemma, we can write $\Theta_{\omega,\nu}$ instead of $\Theta_{P,\omega,\nu}$.

LEMMA 4. Let $s \in \mathfrak{w}$ and $\nu \in \mathfrak{F}_c$. Then

$$\Theta_{s\omega,s\nu} = \Theta_{\omega,\nu} \ .$$

This is deduced in the same way from Theorem 13.2.

Now suppose $\nu \in \mathfrak{F}$. Then it follows from Lemma 2 and [2(i), Lem. 16.1] that $\Theta_{\omega,\nu}$ is a tempered distribution on G. Define

$$(\Theta_{\omega,\nu},\, f) = \int_G \operatorname{conj} \Theta_{\omega,\nu}(x) \cdot f(x) dx \qquad\qquad (f \in \mathcal{C}(G,\, V)) \ .$$

In view of [2(i), Cor. 2, p. 139], this is justified. Similarly put

$$(\theta_\omega,\, g) = \int_M \operatorname{conj} \theta_\omega(m) \cdot g(m) dm \qquad\qquad (g \in \mathcal{C}(M,\, V)) \ .$$

LEMMA 5. Let $f \in \mathcal{C}(G,\, \tau)$, $\nu \in \mathfrak{F}$ and $P \in \mathcal{P}(A)$. Then

$$(\Theta_{\omega,\nu},\, f) = F_0(\theta_\omega,\, f_\nu^{(P)}) \ .$$

Here F_0 is the operator on V given by

$$F_0 v = \int_K \tau(k) v \tau(k^{-1}) dk \qquad\qquad (v \in V) \ .$$

This is obvious from Lemma 2.

22. Computation of $(\Theta_{\omega,\nu},\, \phi_\alpha)$

Fix $\omega_0 \in \mathcal{E}_2(M)$, $\psi \in L(\omega_0)$, $P \in \mathcal{P}(A)$ and put

$$\phi_\alpha = \int \mu(\omega_0\colon \nu)\alpha(\nu)E(P\colon \psi\colon \nu)d\nu$$

for $\alpha \in C_c^\infty(\mathfrak{F})$. Let $\mathfrak{w}(\omega_0)$ be the subgroup consisting of all $s \in \mathfrak{w}$ such that $s\omega_0 = \omega_0$.

THEOREM 1. *Fix $\alpha \in C_c^\infty(\mathfrak{F})$ and put*

$$\hat{f}(\omega:\nu) = (\Theta_{\omega,\nu}, \phi_\alpha)$$

for $\omega \in \mathcal{E}_2(M)$ and $\nu \in \mathfrak{F}$. Then

(1) $\hat{f}(s\omega: s\nu) = \hat{f}(\omega:\nu)$ $(s \in \mathfrak{w})$.

(2) $\hat{f}(\omega:\nu) = 0$ *unless* $\omega = s\omega_0$ *for some* $s \in \mathfrak{w}$.

(3) $\hat{f}(\omega_0:\nu) = \gamma c^2 d(\omega_0)^{-1} F_0 \psi(1) \sum_{s \in \mathfrak{w}(\omega_0)} \alpha(s\nu)$.

Here $d(\omega_0)$ denotes the formal degree of ω_0 and γ and c have the same meaning as in Theorem 20.1.

Put $f = \phi_\alpha$. Then

$$\hat{f}(\omega:\nu) = (\Theta_{\omega,\nu}, f) = F_0(\theta_\omega, f_\nu^{(P)})$$
$$= \gamma c^2 \sum_{s \in \mathfrak{w}} \alpha(s^{-1}\nu) F_0\big(\theta_\omega, {}^0c_{P|P}(s: s^{-1}\nu)\psi\big)$$

from Theorem 20.1. Since ${}^0c_{P|P}(s: s^{-1}\nu)\psi \in L(s\omega_0)$, the second statement is obvious and the first follows from Lemma 21.4. Moreover it is clear that

$$\hat{f}(\omega_0:\nu) = \gamma c^2 \sum_{s \in \mathfrak{w}_0} \alpha(s^{-1}\nu) F_0\big(\theta_{\omega_0}, {}^0c_{P|P}(s: s^{-1}\nu)\psi\big)$$

where $\mathfrak{w}_0 = \mathfrak{w}(\omega_0)$.

On the other hand, by [2(f), Lem. 81],

$$d(\omega_0)(\theta_{\omega_0}, \psi') = \psi'(1)$$

for $\psi' \in L(\omega_0)$. Hence

$$F_0\big(\theta_{\omega_0}, {}^0c_{P|P}(s: s^{-1}\nu)\psi\big) = d(\omega_0)^{-1} F_0 {}^0c_{P|P}(s: s^{-1}\nu)\psi(1)$$

for $s \in \mathfrak{w}_0$. Fix $s \in \mathfrak{w}_0$ and put

$$\psi' = {}^0c_{P|P}(s: s^{-1}\nu)\psi .$$

Then it follows from the definition of the Eisenstein integral that

$$E(P: \psi': \nu: 1) = F_0\psi'(1) .$$

On the other hand we conclude from its functional equation (Lem. 17.2) that

$$E(P: \psi': \nu: 1) = E(P: \psi: s^{-1}\nu: 1) = F_0\psi(1) .$$

Therefore

$$\hat{f}(\omega_0:\nu) = \gamma c^2 d(\omega_0)^{-1} F_0\psi(1) \sum_{s \in \mathfrak{w}_0} \alpha(s\nu)$$

and this proves the theorem.

23. The characters of the discrete series

We assume in this section that rank $G = $ rank K and recall some facts about the discrete series of G. As far as possible we shall keep to the notation of [2(i), § 27].

Let B be a Cartan subgroup of G contained in K and B^* the set of all irreducible characters of B. Fix an order in $(-1)^{1/2}\mathfrak{b}^*$ and for $b^* \in B^*$ define $\lambda(b^*)$ and $\varpi(b^*)$ as in [2(i), §27]. Let $B^{*\prime}$ be the subset of those $b^* \in B^*$ where $\varpi(b^*) \neq 0$. The group $W(G/B)$ operates on $B^{*\prime}$.

THEOREM 1. *There is a unique bijective mapping from $\mathcal{E}_2(G)$ to $W(G/B)\backslash B^{*\prime}$ with the following property. Suppose $\omega \in \mathcal{E}_2(G)$ and $b^* \in B^{*\prime}$ correspond under this mapping. Then*

$$'\Delta(b)\Theta_\omega(b) = (-1)^q \operatorname{sign} \varpi(b^*) \cdot \sum_{s \in W(G/B)} \varepsilon(s)\langle sb^*, b\rangle$$

for $b \in B'$. Here $q = \frac{1}{2} \dim G/K$ and Θ_ω is the character of ω.

This follows immediately from [2(i), Thm. 27.1 and Cor. 1, p. 181].

Put

$$'\Phi_\omega(b) = '\Delta(b)\Theta_\omega(b) \qquad\qquad (b \in B') .$$

Then $'\Phi_\omega$ extends to an analytic function on B.

LEMMA 1. *Let $f \in {}^0\mathcal{C}(G)$. Then*

$$'F_f(b) = \sum_{\omega \in \mathcal{E}_2(G)} (\Theta_\omega, f)'\Phi_\omega(b) \qquad\qquad (b \in B') .$$

This is a restatement of the corollary of [2(i), Lem. 27.4].

Put $\varpi' = e^{-\rho}\varpi \circ e^\rho$ and define the constant c_G as in [2(i), Lem. 27.5]. Then

$$'F_f(1; \varpi') = (-1)^q c_G f(1)$$

for $f \in \mathcal{C}(G)$.

As usual, let $d(\omega)$ $(\omega \in \mathcal{E}_2(G))$ denote the formal degree of ω.

COROLLARY. *Suppose $\omega \in \mathcal{E}_2(G)$ corresponds to $b^* \in B^{*\prime}$. Then*

$$d(\omega) = c_G^{-1} |\varpi(b^*)| [W(G/B)]d(b^*) .$$

Moreover

$$f(1) = \sum_{\omega \in \mathcal{E}_2(G)} d(\omega)(\Theta_\omega, f) \qquad\qquad \text{for } f \in {}^0\mathcal{C}(G) .$$

This is obvious from Corollary 1 of [2(i), Lem. 27.5].

Let V and τ be as in Section 14. Put $L = {}^0\mathcal{C}(G, \tau)$. Then $\dim L < \infty$ and L is the orthogonal sum of $L(\omega)$ $(\omega \in \mathcal{E}_2(G))$. For $f \in \mathcal{C}(G, V)$ and $x \in G$ let $r(x)f$ denote the function $y \mapsto f(yx)$ $(y \in G)$.

LEMMA 2. *Given $f \in \mathcal{C}(G, \tau)$ and $\omega \in \mathcal{E}_2(G)$, let $E_\omega f$ denote the function*

$$x \longmapsto d(\omega)(\Theta_\omega, r(x)f) \qquad\qquad (x \in G) .$$

Then E_ω is a continuous projection of $\mathcal{C}(G, \tau)$ onto $L(\omega)$ and

$$(f_1, E_\omega f_2) = (E_\omega f_1, f_2)$$

for $f_1, f_2 \in \mathcal{C}(G, \tau)$.

Fix $\pi \in \omega$ and let $\mathfrak{H}$ be the representation space of π. We can choose a finite subset F of $\mathcal{E}(K)$ such that $\alpha_F * f = f$ for all $f \in \mathcal{C}(G, \tau)$. Put

$$\Theta_{\omega,F} = \alpha_F * \Theta_\omega \, .$$

Then

$$\Theta_{\omega,F}(x) = \operatorname{tr} E_F \pi(x) \qquad\qquad (x \in G) \, ,$$

in the notation of Section 3. Therefore $\Theta_{\omega,F} \in {}^0\mathcal{C}(G)$ and it is clear that

$$\left(\Theta_\omega, \, r(x)f\right) = \int_G f(y) \operatorname{conj} \Theta_{\omega,F}(yx^{-1}) \cdot dy \, .$$

Taking into account the Schur orthogonality relations for ω, we conclude that E_ω is a continuous projection of $\mathcal{C}(G, \tau)$ onto $L(\omega)$. The last statement of the lemma follows from the above formula and Fubini's theorem.

24. Determination of $\mu(\omega\colon \nu)$ in a special case

Now we return to the general case. Fix a θ-stable Cartan subgroup A of G and use the notation of [2(i), §17]. Then $A = A_I A_R$. Fix $P \in \mathscr{P}(A_R)$ $(P = MA_R N)$, $\omega_0 \in \mathcal{E}_2(M)$, $\psi \in L(\omega_0)$ and put $\mathfrak{F} = \mathfrak{a}_R^*$.

LEMMA 1. *Put*

$$\phi_\alpha = \int_{\mathfrak{F}} \mu(\omega_0\colon \nu)\alpha(\nu)E(P\colon \psi\colon \nu)d\nu$$

for $\alpha \in C_c^\infty(\mathfrak{F})$. Then

$${}'F_{\phi_\alpha}(a_1 a_2) = \gamma c^2 d(\omega_0)^{-1}F_0\psi(1) \sum_{s \in \mathfrak{w}} {}'\Phi_{s\omega_0}(a_1) \int_{\mathfrak{F}} \alpha(\nu) \exp\{(-1)^{1/2}s\nu(\log a_2)\}d\nu$$

for $a_1 \in A_I'$ and $a_2 \in A_R$.

Here γ, c and F_0 have the same meaning as in Theorem 22.1, ${}'\Phi_{s\omega_0}$ is defined as in Section 23 and $A_I' = A'(I) \cap K$.

Fix $a_2 \in A_R$ and put $f = \phi_\alpha$ and

$$g(m) = f^{(P)}(ma_2) \qquad\qquad (m \in M) \, .$$

Then by [2(i), Lem. 8.2],

$${}'F_f(a_1 a_2) = F_0 \cdot {}'F_g(a_1) \qquad\qquad (a_1 \in A_I') \, ,$$

where

$${}'F_g(a_1) = {}'F_g^{M/A_I}(a_1) = {}'\Delta_I(a_1) \int_M g(ma_1 m^{-1})dm \, .$$

On the other hand, it follows from Lemma 20.2 that $g \in {}^0\mathcal{C}(M, \tau_M) = L$. Hence

$${}'F_g(a_1) = \sum_{\omega \in \mathcal{E}_2(M)} (\theta_\omega, g){}'\Phi_\omega(a_1)$$

by Lemma 23.1.

LEMMA 2. *Let* $\omega \in \mathscr{E}_2(M)$. *Then*

$$(\theta_\omega, g) = \int_{\mathfrak{F}} (\theta_\omega, f_\nu^{(P)}) \exp \{(-1)^{1/2}\nu(\log a_2)\}d\nu .$$

It is clear that

$$g(m) = \int_{\mathfrak{F}} f_\nu^{(P)}(m) \exp \{(-1)^{1/2}\nu(\log a_2)\}d\nu .$$

Since Supp α is compact, we conclude from Theorem 20.1 that

$$\int_{\mathfrak{F}} d\nu \int_M |\theta_\omega(m)| |f_\nu^{(P)}(m)| dm < \infty .$$

Hence our assertion follows from Fubini's theorem.

This shows that

$$'F_f(a_1 a_2) = \sum_{\omega \in \mathscr{E}_2(M)} '\Phi_\omega(a_1) \int_{\mathfrak{F}} \hat{f}(\omega : \nu) \exp \{(-1)^{1/2}\nu(\log a_2)\}d\nu$$

$$= \sum_{s \in \mathfrak{w}/\mathfrak{w}_0} '\Phi_{s\omega_0}(a_1) \int_{\mathfrak{F}} \hat{f}(s\omega_0 : \nu) \exp \{(-1)^{1/2}\nu(\log a_2)\}d\nu$$

in the notation of Theorem 22.1. (Here $\mathfrak{w}_0 = \mathfrak{w}(\omega_0)$.) But

$$f(s\omega_0 : \nu) = f(\omega_0 : s^{-1}\nu)$$

and therefore

$$'F_f(a_1 a_2) = [\mathfrak{w}_0]^{-1} \sum_{s \in \mathfrak{w}} '\Phi_{s\omega_0}(a_1) \int_{\mathfrak{F}} \hat{f}(\omega_0 : \nu) \exp \{(-1)^{1/2}s\nu(\log a_2)\}d\nu$$

$$= \gamma c^2 d(\omega_0)^{-1} F_0 \psi(1) \sum_{s \in \mathfrak{w}} '\Phi_{s\omega_0}(a_1) \int_{\mathfrak{F}} \alpha(\nu) \exp \{(-1)^{1/2}s\nu(\log a_2)\}d\nu$$

from Theorem 22.1. This proves Lemma 1.

LEMMA 3. *Let* Γ *be a θ-stable Cartan subgroup of G such that* $\dim \Gamma_R \geqq \dim A_R$. *Then*

$$'F_{\phi_\alpha}^\Gamma = 0$$

for $\alpha \in C_c^\infty(\mathfrak{F})$ *unless* Γ *is conjugate to A in G.*

Suppose Γ is not conjugate to A. Then Γ_R and A_R are not conjugate under K. Hence if $Q \in \mathscr{P}(\Gamma_R)$, we conclude from Lemma 20.2 that $\phi_\alpha^{(Q)} = 0$. This implies our assertion (see [2(i), Lem. 8.2]).

Let Q be the set of all positive roots of $(\mathfrak{g}, \mathfrak{a})$. Then

$$Q = Q_I \cup Q_R \cup Q_c$$

where Q_I, Q_R, Q_c respectively are the subsets of imaginary, real and complex roots in Q. Put

$$\varpi_I = \prod_{\alpha \in Q_I} H_\alpha, \quad \varpi_R = \prod_{\alpha \in Q_R} H_\alpha, \quad \varpi_+ = \prod_{\alpha \in Q_c} H_\alpha.$$

Then $\varpi_{\mathfrak{g}/\mathfrak{a}} = \varpi = \varpi_I \varpi_R \varpi_+.$

LEMMA 4. $[Q_c]$ *is even and*

$$\varpi_+^s = \varpi_+$$

for any element $s \in W(\mathfrak{g}/\mathfrak{a})$ *which maps* $\mathfrak{a}$ *into itself.*

Without loss of generality we may assume that a root α of $(\mathfrak{g}, \mathfrak{a})$ is positive if its restriction on $\mathfrak{a}_R$ is a root of (P, A_R). Then if $\alpha \in Q_c$, it is clear that $\alpha \neq -\theta\alpha$ and $-\theta\alpha \in Q_c$. This shows that $[Q_c]$ is even. Note that if $\Lambda \in (-1)^{1/2}\mathfrak{a}^*$, then

$$\langle -\theta\alpha, \Lambda \rangle = -\langle \alpha, \theta\Lambda \rangle = -\operatorname{conj}\langle \alpha, \Lambda \rangle$$

and therefore

$$(-1)^p \varpi_+(\Lambda) \geq 0$$

where $2p = [Q_c]$. Now fix s as above. Then s permutes the complex roots among themselves and therefore $\varpi_+^s = \eta\varpi_+$ where $\eta = \pm 1$. Choose $\Lambda \in (-1)^{1/2}\mathfrak{a}^*$ such that $\varpi_+(\Lambda) \neq 0$. Then since $s^{-1}\Lambda$ is also in $(-1)^{1/2}\mathfrak{a}^*$, it is clear from the above result that $\eta = 1$.

Let a^* be an element in $A_I^{*\prime}$ which corresponds to ω_0 (under Thm. 23.1 applied to M). Put $\lambda = \lambda(a^*) \in (-1)^{1/2}\mathfrak{a}_I^*$.

THEOREM 1. *Suppose* $\mathfrak{a}$ *is fundamental in* $\mathfrak{g}$. *Then*

$$\mu(\omega_0 : \nu) = \gamma c^2[\mathfrak{w}] \frac{c_M}{c_G} \left| \varpi_+\big(\lambda + (-1)^{1/2}\nu\big) \right|$$

for $\nu \in \mathfrak{F}$.

Here c_G and c_M are positive constants for G and M respectively corresponding to $(-1)^q c$ of [2(i), Thm. 37.1]. Also we recall that $\mathfrak{a}$ is fundamental if $\mathfrak{a}_I$ is a Cartan subalgebra of $\mathfrak{k}$.

In proving this theorem, we may obviously assume that the set Q of positive roots of $(\mathfrak{g}, \mathfrak{a})$ is chosen as above. Put

$$\varpi' = e^{-\rho_I} \varpi \circ e^{\rho_I}$$

where $2\rho_I$ is the sum of all roots in Q_I. Then

$$'F_f(1; \varpi') = (-1)^{q_G} c_G f(1)$$

for $f \in \mathcal{C}(G)$. Here

$$q_G = \tfrac{1}{2}\{\dim G/K - \operatorname{rank} G + \operatorname{rank} K\}.$$

Put $q_M = \tfrac{1}{2} \dim M/K_M$. Since $G = PK$, it is clear that

$$\dim G/K = \dim P/P \cap K = \dim M/K_M + \dim A_R + \dim N .$$

Moreover since $\mathfrak{a}$ is fundamental in $\mathfrak{g}$,

$$\operatorname{rank} G - \operatorname{rank} K = \dim \mathfrak{a} - \dim \mathfrak{a}_I = \dim \mathfrak{a}_R .$$

Also Q_R is empty and $\dim N = [Q_c] = 2p$. Therefore

$$q_G = q_M + p .$$

Now $\mathfrak{w} = W(G/A)/W(M/A_I)$. Put

$$\Phi_\nu(a_1 a_2) = \sum_{s \in \mathfrak{w}} {}'\Phi_{s\omega_0}(a_1) \exp \{(-1)^{1/2} s\nu(\log a_2)\} \qquad (a_1 \in A_I,\ a_2 \in A_R)$$

for $\nu \in \mathfrak{F}$. Then we conclude from the results of Section 23 that

$$\Phi_\nu(1;\ \varpi') = (-1)^{q_M} c_M d(\omega_0) \sum_{s \in W(G/A)/W(M/A_I)} \varpi_+(s\lambda + (-1)^{1/2} s\nu)$$
$$= (-1)^{q_G} c_M [\mathfrak{w}] d(\omega_0) \, | \, \varpi_+(\lambda + (-1)^{1/2}\nu) | .$$

Put $f = \phi_\alpha$ in the notation of Lemma 1. Then

$$(-1)^{q_G} c_G f(1) = {}'F_f(1;\ \varpi') = \gamma c^2 d(\omega_0)^{-1} F_0 \psi(1) \int_{\mathfrak{F}} \alpha(\nu) \Phi_\nu(1;\ \varpi') d\nu$$

$$= (-1)^{q_G} c_M [\mathfrak{w}] \gamma c^2 F_0 \psi(1) \int_{\mathfrak{F}} \alpha(\nu) \, | \, \varpi_+(\lambda + (-1)^{1/2}\nu) | \, d\nu$$

from Lemma 1. On the other hand

$$f(1) = \phi_\alpha(1) = \int_{\mathfrak{F}} \mu(\omega_0:\ \nu) \alpha(\nu) F_0 \psi(1) d\nu .$$

Since $\alpha \in C_c^\infty(\mathfrak{F})$ is arbitrary, we conclude that

$$\mu(\omega_0:\ \nu) F_0 \psi(1) = c_M c_G^{-1} [\mathfrak{w}] \gamma c^2 \, | \, \varpi_+(\lambda + (-1)^{1/2}\nu) | \, F_0 \psi(1)$$

for $\nu \in \mathfrak{F}$. Theorem 1 will be proved if we can choose τ and ψ in such a way that $F_0 \psi(1) \neq 0$.

Take $\tau = \tau_F$ (see §7) where F is a finite subset of $\mathfrak{E}(K)$ such that $\mathfrak{H}_F \neq \{0\}$. (Here $\mathfrak{H}$ is defined as in Section 4 corresponding to $\sigma \in \omega_0$.) Choose $T \in \operatorname{End} \mathfrak{H}_F$ such that $\operatorname{tr} T \neq 0$ (for example $T = E_F$) and put $\psi = \psi_T$. Then $\psi \in L(\omega_0)$. Let $v = F_0 \psi(1)$. Then

$$v(1:1) = \int_K \psi(k:1:k^{-1}) dk = \int_K \operatorname{tr} \kappa_T(k^{-1}:k) dk = \operatorname{tr} T \neq 0$$

from the definition of ψ_T (§7) and the corollary of Lemma 6.1. This shows that $F_0 \psi(1) \neq 0$.

25. Extension of μ, j and c on the complex space

Let us now use the notation of Section 13 and consider the function $\mu_\omega(\nu) = \mu(\omega:\ \nu)$ $(\nu \in \mathfrak{F})$.

THEOREM 1[6]. *μ_ω extends to a meromorphic function on $\mathfrak{F}_c$. Moreover*

---

[6] Cf. the first conjecture of [2(b), §16].

we can choose $\delta > 0$ such that the following two conditions hold.

  (1) *μ_ω is holomorphic on $\mathfrak{F}_c(\delta)$.*

  (2) *There exist numbers c, $r \geqq 0$ such that*

$$| \mu(\omega\colon \nu) | \leqq c(1 + | \nu_R |)^r$$

for all $\nu \in \mathfrak{F}_c(\delta)$.

Fix $P \in \mathscr{P}(A)$. Then (see §13)

$$\mu(\omega\colon \nu) = \prod_{\alpha \in \Sigma(P)} \mu_\alpha(\omega\colon \nu) \ .$$

Hence it is clear that it would be sufficient to verify the above theorem under the assumption that $\operatorname{prk} G = 0$ and $\dim A = 1$. This will be done a little later in Sections 28 and 36. For the moment we assume Theorem 1 and proceed to derive its consequences.

LEMMA 1. *Fix P, $Q \in \mathscr{P}(A)$ and $s \in \mathfrak{w}$. Then $j_{Q|P}(\nu)$ and $c_{Q|P}(s\colon \nu)$ extend to meromorphic functions on $\mathfrak{F}_c$.*

Here the notation is the same as in Section 18.

In view of the results of Section 18, it is enough to prove this for $j_{Q|P}$. Moreover by the corollary of Theorem 13.1, we are reduced to the case when $d(Q, P) = 1$. This in turn reduces to the case when $\dim A = 1$ and $Q = \bar P$ (see the proof of Lemma 13.1). Let α be the unique element in $\Sigma(P)$ in this case. Then $\mathfrak{F}_c$ may be identified with the complex plane under the mapping $\nu \mapsto \langle \nu, \alpha \rangle$. We know that

$$\mu(\omega\colon \nu) j_{P|\bar P}(\nu) j_{\bar P|P}(\nu) = 1$$

and $j_{\bar P|P}$ is holomorphic if $\operatorname{Im} \langle \nu, \alpha \rangle < 0$, while $j_{P|\bar P}$ is holomorphic if $\operatorname{Im} \langle \nu, \alpha \rangle > 0$ (see Lemmas 11.1 and 12.1). Our assertion now follows immediately from Theorem 1.

We now use the notation of Section 22. Fix $\psi \in L(\omega)$, $P \in \mathscr{P}(A)$ and put

$$\phi(\nu\colon x) = \mu(\omega\colon \nu) E(P\colon \psi\colon \nu\colon x) \qquad\qquad (\nu \in \mathfrak{F}, \ x \in G) \ .$$

Extend $\mathfrak{a}$ to a θ-stable Cartan subalgebra $\mathfrak{h}$ of $\mathfrak{g}$ such that $\mathfrak{h}_R = \mathfrak{a}$ and $\mathfrak{h}_I \subset \mathfrak{m}$. Let λ be an element of $(-1)^{1/2}\mathfrak{h}_I^*$ which corresponds to the infinitesimal character of ω.

LEMMA 2. *ϕ is a function on $\mathfrak{F} \times G$ of type $\mathrm{II}'(\lambda)$ in the terminology of [2(j), §9].*

It is easy to deduce from Theorem 1 and [2(j), Lem. 17.1] that ϕ is of type $\mathrm{II}(\lambda)$ (see [2(j), §8]). Hence it remains to verify that ϕ satisfies the condition of [2(j), Thm. 11.1].

Fix $Q \in \mathscr{P}(A)$, $\nu \in \mathfrak{F}'$ and $s \in \mathfrak{w}$. Then

$$f_{Q,s}(\nu) = \mu(\omega:\nu)c_{Q|P}(s:\nu)\psi$$

in the notation of [2(j), Thm. 11.1]. Hence

$$\|f_{Q,s}(\nu)\|^2 = c^2\mu(\omega:\nu)\|\psi\|^2$$

from Lemma 17.1. Since μ_ω is holomorphic on $\mathfrak{F}$, it is clear that $\|f_{Q,s}(\nu)\|$ remains locally bounded on $\mathfrak{F}$ and this proves Lemma 2.

26. Some applications of Theorem 25.1

We keep to the notation of Section 22. Fix $P \in \mathscr{P}(A)$, $\omega \in \mathscr{E}_2(M)$ and $\psi \in L(\omega)$.

THEOREM 1. *Put*

$$\phi_\alpha = \int_{\mathfrak{F}} \mu(\omega:\nu)\alpha(\nu)E(P:\psi:\nu)d\nu$$

for $\alpha \in \mathcal{C}(\mathfrak{F})$. Then $\alpha \mapsto \phi_\alpha$ is a continuous mapping of $\mathcal{C}(\mathfrak{F})$ into $\mathcal{C}(G,\tau)$.

This is an immediate consequence of Lemma 25.2 and the corollary of [2(j), Thm. 13.1].

COROLLARY 1. *Fix $Q \in \mathscr{P}(A)$ and $\nu \in \mathfrak{F}$. Then*

$$(\phi_\alpha)_\nu^{(Q)} = \gamma c^2 \sum_{s \in \mathfrak{w}} \alpha(s^{-1}\nu)^0 c_{Q|P}(s:s^{-1}\nu)\psi$$

for all $\alpha \in \mathcal{C}(\mathfrak{F})$.

Put

$$T(\alpha) = (\phi_\alpha)_\nu^{(Q)} - \gamma c^2 \sum_{s \in \mathfrak{w}} \alpha(s^{-1}\nu)^0 c_{Q|P}(s:s^{-1}\nu)\psi$$

for $\alpha \in \mathcal{C}(\mathfrak{F})$. Then T is a tempered distribution on $\mathfrak{F}$ (with values in L). Since $C_c^\infty(\mathfrak{F})$ is dense in $\mathcal{C}(\mathfrak{F})$, we conclude from Theorem 20.1 that $T = 0$.

COROLLARY 2. *The statements of Theorem 22.1 and Lemma 24.1 remain true for $\alpha \in \mathcal{C}(\mathfrak{F})$.*

This is proved in the same way.

Let $P' = M'A'N'$ be a psgp of G and f an element in $\mathcal{C}(G,\tau)$. Recall that $f^{(P')} \sim 0$ means that

$$\int_{M'} (\phi(m'), f^{(P')}(m'a'))dm' = 0$$

for all $\phi \in {}^0\mathcal{C}(M', \tau_{M'})$ and $a' \in A'$ (see [2(i), §20]). Let $\mathcal{C}_A(G,\tau)$ denote the space of all $f \in \mathcal{C}(G,\tau)$ such that $f^{(P')} \sim 0$ unless A' is conjugate to A under K. It follows from [2(i), Lem. 20.2] that $\mathcal{C}_A(G,\tau)$ is a closed subspace of $\mathcal{C}(G,\tau)$.

LEMMA 1. *The function ϕ_α of Theorem 1 lies in $\mathcal{C}_A(G,\tau)$.*

Put $f = \phi_\alpha$ and assume that A' is not conjugate to A under K. Then by [2(i), Lem. 20.2], it is enough to verify that

$$\big(E(P': \psi': \nu'), f\big) = 0$$

for $\psi' \in L' = {}^0\mathcal{C}(M', \tau_{M'})$ and $\nu' \in (\mathfrak{a}')^*$. Since

$$\big(E(P': \psi': \nu'), f\big) = \big(\psi', f_{\nu'}^{(P')}\big)$$

and the right hand side is a continuous function of ν', it is sufficient to consider the case when $\langle \nu', \alpha' \rangle \neq 0$ for every root α' of (P', A'). Moreover $\dim L' < \infty$ and

$$L' = \sum_{\omega' \in \mathcal{E}_2(M')} L'(\omega') .$$

Hence without loss of generality, we may assume that $\psi' \in L'(\omega')$. Extend $\mathfrak{a}$ and $\mathfrak{a}'$ to θ-stable Cartan subalgebras $\mathfrak{h}$ and $\mathfrak{h}'$ respectively of $\mathfrak{g}$ such that $\mathfrak{h}_R = \mathfrak{a}$, $\mathfrak{h}_I \subset \mathfrak{m}$ and $\mathfrak{h}'_R = \mathfrak{a}'$, $\mathfrak{h}'_I \subset \mathfrak{m}'$. Choose $\lambda \in (-1)^{1/2}\mathfrak{h}_I^*$ and $\lambda' \in (-1)^{1/2}\mathfrak{h}'^*_I$ corresponding to the infinitesimal characters of ω and ω' respectively. Put $\gamma = \gamma_{\mathfrak{g}/\mathfrak{h}}$ and $\gamma' = \gamma_{\mathfrak{g}/\mathfrak{h}'}$. Then

$$zE(P': \psi': \nu') = \gamma'\big(z: \lambda' + (-1)^{1/2}\nu'\big)E(P': \psi': \nu') \qquad (z \in \mathfrak{Z})$$

from [2(i), Lem. 19.1]. Similarly if we denote by $p(z)$ the polynomial function

$$\nu \longmapsto \gamma\big(z: \lambda + (-1)^{1/2}\nu\big)$$

on $\mathfrak{F}$, we have

$$zf = \phi_{p(z)\alpha} .$$

Put

$$T(\alpha) = \big(E(P': \psi': \nu'), \phi_\alpha\big) \qquad \big(\alpha \in \mathcal{C}(\mathfrak{F})\big) .$$

Then T is a tempered distribution on $\mathfrak{F}$ and

$$T\big(p(z)\alpha\big) = \gamma'\big(z: \lambda' + (-1)^{1/2}\nu'\big)T(\alpha) \qquad (z \in \mathfrak{Z}) .$$

Hence if $\nu \in \operatorname{Supp} T$, it follows that $\lambda + (-1)^{1/2}\nu$ is conjugate to $\lambda' + (-1)^{1/2}\nu'$ under G_c. But since this implies that $\mathfrak{a}$ and $\mathfrak{a}'$ are conjugate under K [2(i), Lem. 29.2], we conclude that $T = 0$. This proves the lemma.

27. The Plancherel formula for K-finite functions

Fix V and τ as in Section 14 and put $L = {}^0\mathcal{C}(M, \tau_M)$. For $f \in \mathcal{C}(G, \tau)$, define a function $\hat{f}$ from $\mathcal{E}_2(M) \times \mathfrak{F} \times G$ to V by

$$\hat{f}(\omega: \nu: x) = \big(\Theta_{\omega,\nu}, r(x)f\big) = \big(\Theta_{\omega,\nu}, l(x^{-1})f\big) .$$

Here $\omega \in \mathcal{E}_2(M)$, $\nu \in \mathfrak{F}$, $x \in G$ and $r(x)f$, $l(x)f$ respectively are the functions $y \mapsto f(yx)$, $y \mapsto f(x^{-1}y)$ $(y \in G)$. It is easy to verify that

$$\hat{f}(\omega: \nu: k_1 x k_2) = \tau(k_1)\hat{f}(\omega: \nu: x)\tau(k_2)$$

for $k_1, k_2 \in K$. Hence it follows from the corollary of Lemma 21.1 that $\hat{f}(\omega: \nu)$ is a K-finite eigendistribution of $\mathfrak{Z}$ and therefore it lies in $C^\infty(G, \tau)$.

THEOREM 1. *Fix $P \in \mathscr{P}(A)$ $(P = MAN)$, $\nu \in \mathfrak{F}$ and, for $f \in \mathcal{C}(G, \tau)$, define*

$$g_f(\nu: m) = f_\nu^{(P)}(m) = \int_{A \times N} f(man) \exp\{-((-1)^{1/2}\nu - \rho)(\log a)\}dadn$$

$$(m \in M)$$

and

$$\psi_f(\omega: \nu: m) = d(\omega)(\theta_\omega, r(m)g_f(\nu)) \qquad (\omega \in \mathcal{E}_2(M)) .$$

Then $\psi_f(\omega: \nu) \in L(\omega)$ and

$$d(\omega)\hat{f}(\omega: \nu) = E\big(P: \psi_f(\omega: \nu): \nu\big) .$$

Moreover let $\psi_f(\omega)$ denote the function

$$\nu \longmapsto \psi_f(\omega: \nu)$$

on $\mathfrak{F}$. Then $f \mapsto \psi_f(\omega)$ is a continuous mapping of $\mathcal{C}(G, \tau)$ into $\mathcal{C}(\mathfrak{F}) \otimes L(\omega)$.

We observe that $g_f \in \mathcal{C}(M, \tau_M)$ from [2(i), Lem. 16.1] and therefore $\psi_f(\omega: \nu) \in L(\omega)$ by Lemma 23.2. Fix $\psi \in L(\omega)$ and put

$$h_f(\nu) = (\psi, \psi_f(\omega: \nu)) = \int_M (\psi(m), f_\nu^{(P)}(m))dm$$

$$= \int_A \exp\{-(-1)^{1/2}\nu(\log a)\}da \int_M (\psi(m), f^{(P)}(ma))dm .$$

Then it is clear that $h_f \in \mathcal{C}(\mathfrak{F})$ and $f \mapsto h_f$ is a continuous mapping of $\mathcal{C}(G, \tau)$ into $\mathcal{C}(\mathfrak{F})$. Since $\dim L(\omega) \leq \dim L < \infty$, we conclude that $f \mapsto \psi_f(\omega)$ is a continuous mapping of $\mathcal{C}(G, \tau)$ into $\mathcal{C}(\mathfrak{F}) \otimes L(\omega)$.

Now fix $\nu \in \mathfrak{F}$ and $x \in G$. Then

$$\hat{f}(\omega: \nu: x) = \big(\Theta_{\omega,\nu}, l(x^{-1})f\big)$$

$$= \int_{K \times M \times A \times N} \operatorname{conj} \theta_\omega(m) \cdot f(xkmank^{-1})$$

$$\times \exp\{-((-1)^{1/2}\nu - \rho)(\log a)\}dkdmdadn$$

from Lemma 21.2. Fix $k \in K$ and let $xk = k_0 m_0 a_0 n_0$ $(k_0 \in K, m_0 \in M, a_0 \in A, n_0 \in N)$. Then

$$\int_{A \times N} f(xkmank^{-1}) \exp\{-((-1)^{1/2}\nu - \rho)(\log a)\}dadn$$

$$= \tau(k_0)\left\{\int_{A \times N} f(m_0 m a_0 an) \exp\{-((-1)^{1/2}\nu - \rho)(\log a)\}dadn\right\}\tau(k^{-1})$$

$$= \exp\{((-1)^{1/2}\nu - \rho)(\log a_0)\}\tau(k_0)g_f(\nu: m_0 m)\tau(k^{-1}) .$$

Hence

$$d(\omega) \int_{M \times A \times N} \operatorname{conj} \theta_\omega(m) \cdot f(xkmank^{-1}) \exp\{-((-1)^{1/2}\nu - \rho)(\log a)\} dm\, da\, dn$$
$$= \psi_f(\nu: xk)\tau(k^{-1}) \exp\{((-1)^{1/2}\nu - \rho)(H(xk))\}$$

and therefore

$$d(\omega)\hat{f}(\omega: \nu: x) = E(P: \psi_f(\omega: \nu): \nu: x) \,.$$

This proves the theorem.

The function $\psi_f(\omega)$ (on $\mathfrak{F} \times M$) depends on $P \in \mathscr{P}(A)$. Hence we denote it by $\psi_f(P: \omega)$.

LEMMA 1. *Fix P_1, $P_2 \in \mathscr{P}(A)$, $s \in \mathfrak{w}$ and $\omega \in \mathscr{E}_2(M)$. Then*

$$\psi_f(P_2: s\omega: s\nu) = {}^0 c_{P_2|P_1}(s: \nu)\psi_f(P_1: \omega: \nu)$$

for $f \in \mathcal{C}(G, \tau)$ and $\nu \in \mathfrak{F}$.

It would be enough to verify this for $\nu \in \mathfrak{F}'$. It follows from Lemma 21.4 that $\hat{f}(s\omega: s\nu) = \hat{f}(\omega: \nu)$. Therefore

$$E(P_2: \psi_f(P_2: s\omega: s\nu): s\nu) = E(P_1: \psi_f(P_1: \omega: \nu): \nu)$$
$$= E(P_2: {}^0 c_{P_2|P_1}(s: \nu)\psi_f(P_1: \omega: \nu): s\nu)$$

from Lemma 17.2. Our assertion follows from Lemma 16.1.

Put

$$f_\omega = c^{-2}\gamma^{-1}[\mathfrak{w}]^{-1}d(\omega) \int_{\mathfrak{F}} \mu(\omega: \nu)\hat{f}(\omega: \nu)d\nu$$

for $f \in \mathcal{C}(G, \tau)$ and $\omega \in \mathscr{E}_2(M)$. (Here γ and c have the same meaning as in Theorem 20.1.) Then it follows from Theorem 1 and the results of Section 26 that $f \mapsto f_\omega$ is a continuous linear mapping of $\mathcal{C}(G, \tau)$ into $\mathcal{C}_A(G, \tau)$.

Let Ω be the set of all $\omega \in \mathscr{E}_2(M)$ such that $L(\omega) \neq \{0\}$. Since $\dim L < \infty$, it is clear that Ω is a finite set which is stable under $\mathfrak{w}$. Moreover it follows from Theorem 1 that $f_\omega = 0$ for $f \in \mathcal{C}(G, \tau)$ unless $\omega \in \Omega$. Put

$$f_A = \sum_{\omega \in \mathscr{E}_2(M)} f_\omega \qquad\qquad (f \in \mathcal{C}(G, \tau)) \,.$$

Then $f \mapsto f_A$ is also a continuous linear mapping of $\mathcal{C}(G, \tau)$ into $\mathcal{C}_A(G, \tau)$.

LEMMA 2. *Let $f \in \mathcal{C}(G, \tau)$, $\omega \in \mathscr{E}_2(M)$ and $s \in \mathfrak{w}$. Then*

$$f_{s\omega} = f_\omega \,.$$

This is an immediate consequence of the relations

$$\hat{f}(s\omega: s\nu) = \hat{f}(\omega: \nu) \,, \quad \mu(s\omega: s\nu) = \mu(\omega: \nu) \qquad (\nu \in \mathfrak{F})$$

(see Lem. 21.4 and the corollary of Lem. 17.1).

LEMMA 3. *Fix $f \in \mathcal{C}(G, \tau)$, $P \in \mathscr{P}(A)$, ω, $\omega' \in \mathscr{E}_2(M)$, $\psi \in L(\omega')$ and $\nu \in \mathfrak{F}$. Then*

$$\left(E(P:\psi:\nu),\, f_\omega\right) = \begin{cases} [\mathfrak{w}:\mathfrak{w}(\omega)]^{-1}\!\left(E(P:\psi:\nu,\,)f\right) & \text{if } \omega' = s\omega \ \ \text{for some } s\in\mathfrak{w}\,, \\ 0 & \text{otherwise}\,. \end{cases}$$

Here $\mathfrak{w}(\omega)$ denotes, as before, the stabilizer of ω in $\mathfrak{w}$.

We know from Theorem 1 that

$$f_\omega = c^{-2}\gamma^{-1}[\mathfrak{w}]^{-1}\int_{\mathfrak{F}}\mu(\omega:\nu)E\big(P:\psi_f(P:\omega:\nu):\nu\big)d\nu\,.$$

Therefore

$$\big(E(P:\psi:\nu),\, f_\omega\big) = \big(\psi,\,(f_\omega)_\nu^{(P)}\big)$$
$$= [\mathfrak{w}]^{-1}\sum_{s\in\mathfrak{w}}\big(\psi,\,{}^0c_{P|P}(s:\,s^{-1}\nu)\psi_f(P:\omega:s^{-1}\nu)\big)$$

from Corollary 1 of Theorem 26.1. Hence applying Lemma 1, we get

$$\big(E(P:\psi:\nu),\, f_\omega\big) = [\mathfrak{w}]^{-1}\sum_{s\in\mathfrak{w}}\big(\psi,\,\psi_f(P:s\omega:\nu)\big)\,.$$

If the right hand side is not zero, we can choose $s_0\in\mathfrak{w}$ such that $\omega' = s_0\omega$. Since $f_\omega = f_{s_0\omega}$, there is no loss of generality in assuming that $s_0 = 1$. Then

$$\big(E(P:\psi:\nu),\, f_\omega\big) = [\mathfrak{w}:\mathfrak{w}(\omega)]^{-1}\big(\psi,\,\psi_f(P:\omega:\nu)\big)\,.$$

On the other hand

$$\big(E(P:\psi:\nu),\, f\big) = \big(\psi,\, f_\nu^{(P)}\big) = \big(\psi,\,\psi_f(P:\omega:\nu)\big)$$

from Lemma 23.2 and so this proves Lemma 3.

COROLLARY. $\big(E(P:\psi:\nu),\, f_A\big) = \big(E(P:\psi:\nu),\, f\big)$ *for all* $\psi\in L$.

It is enough to verify this for $\psi\in L(\omega')$ for a given $\omega'\in\mathcal{E}_2(M)$. Then

$$\big(E(P:\psi:\nu),\, f_A\big) = \sum_{\omega\in\mathcal{E}_2(M)}\big(E(P:\psi:\nu),\, f_\omega\big)$$
$$= \sum_{s\in\mathfrak{w}/\mathfrak{w}(\omega')}\big(E(P:\psi:\nu),\, f_{s\omega'}\big)$$
$$= [\mathfrak{w}:\mathfrak{w}(\omega')]\big(E(P:\psi:\nu),\, f_{\omega'}\big)$$
$$= \big(E(P:\psi:\nu),\, f\big)$$

from Lemmas 2 and 3.

Let $\Gamma_1,\cdots,\Gamma_r$ be a complete set of θ-stable Cartan subgroups of G, no two of which are conjugate. Put $A_i = (\Gamma_i)_R$. Then if $i\neq j$, A_i is not conjugate to A_j under K. Let C denote the set $(\Gamma_1,\cdots,\Gamma_r)$ and S the set $(A_1,\cdots,A_r)$.

THEOREM 2. $f = \sum_{A\in S} f_A$ *for all* $f\in\mathcal{C}(G,\tau)$.

Put

$$g = f - \sum_{A\in S} f_A\,.$$

Then $g\in\mathcal{C}(G,\tau)$. In order to prove that $g = 0$, it is enough to verify that $g^{(P)}\sim 0$ for every psgp P of G [2(i), Lem. 20.1]. Fix a psgp $P = MAN$ of G.

We may obviously assume that $^0\mathcal{C}(M, \tau_M) \neq \{0\}$. Then we conclude from [2(i), Thm. 18.1] that A is the split component of a θ-stable Cartan subgroup of G. Hence we can choose $k \in K$ such that $A^k \in S$. Since our problem is not substantially changed if we replace P by P^k, we may assume that $A \in S$. Now

$$g^{(P)} = f^{(P)} - \sum_{A' \in S} (f_{A'})^{(P)}$$

and $f_{A'} \in \mathcal{C}_{A'}(G, \tau)$. Therefore $(f_{A'})^{(P)} \sim 0$ if $A' \neq A$. Hence it would be enough to verify that

$$(f - f_A)^{(P)} \sim 0 \ .$$

But this follows from [2(i), Lem. 20.2] and the corollary of Lemma 3.

We recall that

$$f_A(x) = c(G/A)^{-2}\gamma(G/A)^{-1}[\mathfrak{w}(G/A)]^{-1} \sum_{\omega \in \mathcal{E}_2(M)} d(\omega) \int_{\mathfrak{F}} \mu(\omega: \nu)(\Theta_{\omega,\nu}, \, r(x)f)d\nu$$

for $f \in \mathcal{C}(G, \tau)$. Here $A \in S$ and $P = MAN$ is a psgp in $\mathscr{P}(A)$. Observe that

$$c(G/A) = \gamma(G/A) - [\mathfrak{w}(G/A)] = \mu(\omega: 0) = 1$$

if $A = \{1\}$. Evaluating both sides in Theorem 2 at 1 we get the following result.

THEOREM 3. *Let* $f \in \mathcal{C}(G, \tau)$. *Then*

$$f(1) = \sum_{A \in S} c(G/A)^{-2}\gamma(G/A)^{-1}[\mathfrak{w}(G/A)]^{-1} \sum_{\omega \in \mathcal{E}_2(M)} d(\omega)$$
$$\times \int_{\mathfrak{F}} \mu(\omega: \nu)(\Theta_{\omega,\nu}, \, f)d\nu \ .$$

This is just the Plancherel formula for K-finite functions in $\mathcal{C}(G)$. In fact fix a finite subset F of $\mathcal{E}(K)$ and let $\mathcal{C}_F(G)$ be the space of all $f \in \mathcal{C}(G)$ such that $f = \alpha_F * f * \alpha_F$. Then $\mathcal{C}_F(G)$ may be identified with $\mathcal{C}(G, \tau)$ for $\tau = \tau_F$. [2(i), §26] and therefore the above theorem holds for all $f \in \mathcal{C}_F(G)$. Moreover $\Theta_{\omega,\nu}$, being the character of a unitary representation, is a distribution of positive type. Since

$$\bigcup_F \mathcal{C}_F(G)$$

is dense in $L_2(G)$, it is easy to derive from Theorem 3 the usual Plancherel formula for functions in $L_2(G)$.

We shall see in another paper that the statement of Theorem 2 can be extended to the case when $\dim \tau = \infty$ (see [2(g), Thm. 12]).

28. First reduction in the proof of Theorem 25.1

We now come to the proof of Theorem 25.1. As we saw in Section 25, it is enough to consider the case when prk $G = 0$ and $\dim A = 1$. We shall compute μ_ω in this case explicitly.

Extend $\mathfrak{a}$ to a θ-stable Cartan subalgebra $\mathfrak{h}$ of $\mathfrak{g}$ such that $\mathfrak{a} = \mathfrak{h}_R$ and $\mathfrak{h}_I \subset \mathfrak{m}$. Then

$$\operatorname{rank} G = \dim \mathfrak{h} = 1 + \dim \mathfrak{h}_I .$$

If $\operatorname{rank} G > \operatorname{rank} K$, we conclude that $\mathfrak{h}_I$ is a Cartan subalgebra of $\mathfrak{k}$ and therefore $\mathfrak{h}$ is fundamental in $\mathfrak{g}$. Then by Theorem 24.1, μ_ω is actually a polynomial function on $\mathfrak{F}_c$ and so the statement of Theorem 25.1 is true in this case.

So it remains to consider the case when $\operatorname{rank} G = \operatorname{rank} K$.

29. Remarks on notation

In this section we fix some notation. Let A be a θ-stable Cartan subgroup of G. Then $A = A_I \cdot A_R$. Fix compatible orders in $\mathfrak{a}_R^*$ and $\mathfrak{a}_R^* + (-1)^{1/2}\mathfrak{a}_I^*$ and let Q be the set of all positive roots of $(\mathfrak{g}, \mathfrak{a})$. Define Q_R, Q_I and Q_c as in Section 24. Put

$$'\Delta(a) = \prod_{\alpha \in Q} \left(1 - \xi_\alpha(a^{-1})\right) \qquad (a \in A)$$

and define $'\Delta_I$ and Δ_+ as in [2(i), §17]. If $a \in A$ we write $a = a_1 a_2$ ($a_1 \in A_I$, $a_2 \in A_R$). Put

$$\varepsilon_R(a) = \operatorname{sign} \prod_{\alpha \in Q_R(a_1)} \alpha(\log a_2) \qquad (a \in A' = A \cap G') .$$

Here G' is the set of all regular elements of G and $Q_R(a_1)$ is the set of all $\alpha \in Q_R$ such that $\xi_\alpha(a_1) = 1$. Let

$$\rho = \tfrac{1}{2} \sum_{\alpha \in Q} \alpha , \quad \rho_I = \tfrac{1}{2} \sum_{\beta \in Q_I} \beta .$$

LEMMA 1. $'\Delta_I(a)\Delta_+(a) = \varepsilon_R(a)'\Delta(a) \exp \{\rho(\log a_2)\}$ for $a \in A'$.

This is proved in the same way as [2(d), Lemma 12].

For $f \in \mathcal{C}(G)$, define

$$'F_f(a) = '\Delta_I(a)\Delta_+(a) \int_{G/A_R} f(xax^{-1})dx^* \qquad (a \in A') ,$$

where dx^* is the invariant measure on G/A_R normalized as usual [2(i), p. 116]. Put

$$\varpi = \prod_{\alpha \in Q} H_\alpha , \quad \varpi' = e^{-\rho_I}\varpi \circ e^{\rho_I} .$$

Now suppose G is connected and acceptable [2(c), §18]. Then we define

$$\Delta(a) = \xi_\rho(a)'\Delta(a) \qquad (a \in A)$$

and

$$F_f(a) = \xi_\rho(a_1)'F_f(a) = \varepsilon_R(a)\Delta(a) \int_{G/A_R} f(xax^{-1})dx^* \qquad (a \in A')$$

for $f \in \mathcal{C}(G)$. Then

$$F_f(a; \varpi) = \xi_\rho(a_1)' F_f(a; \varpi') \, .$$

LEMMA 2. $\operatorname{conj} \Delta(a) = (-1)^{r + [Q_R(a_1)]} \Delta(a)$ $(a \in A)$ *where* $r = \frac{1}{2}(\dim \mathfrak{g} - \operatorname{rank} \mathfrak{g})$ *and* $Q_R(a_1)$ *is the set of all roots* $\alpha \in Q_R$ *such that* $\xi_\alpha(a_1) = 1$.

This is a simple consequence of Lemma 1 (cf. [2(e), p. 309]).

30. Some auxiliary lemmas

Let us now assume that rank $G = \operatorname{rank} K$. Fix a Cartan subgroup B of K and let A be a θ-stable Cartan subgroup of G with $\dim A_R = 1$. Let $\mathfrak{z}$ be the centralizer of $\mathfrak{a}_I$ in $\mathfrak{g}$. Then $\mathfrak{z} = \mathfrak{a}_I + \mathfrak{l}$ where $\mathfrak{l} = [\mathfrak{z}, \mathfrak{z}]$ is a three-dimensional simple algebra of noncompact type. Clearly $\mathfrak{z} \cap \mathfrak{k} = \mathfrak{a}_I + \mathfrak{l} \cap \mathfrak{k}$ is a Cartan subalgebra of $\mathfrak{g}$. We say that $\mathfrak{a}$ is related to $\mathfrak{b}$ (or A is related to B) if $\mathfrak{z} \cap \mathfrak{k} = \mathfrak{b}$. It is clear that we can always choose $k \in K^0$ such that A^k is related to B.

A real root of $(\mathfrak{g}, \mathfrak{a})$ may be regarded as a root of $(\mathfrak{l}, \mathfrak{a}_R)$. Hence $(\mathfrak{g}, \mathfrak{a})$ has exactly one positive real root which we denote by α. Put $H' = 2 \, |\alpha|^{-2} \cdot H_\alpha$. This can be completed to a base (H', X', Y') for $\mathfrak{l}$ over $\mathbf{R}$ such that

$$[H', X'] = 2X' \, , \quad [H', Y'] = -2Y' \, , \quad [X', Y'] = H' \, .$$

Moreover we can assume that θ maps (H', X', Y') into $(-H', -Y', -X')$.

Put $y = \exp(-1)^{1/2}(\pi/4)\operatorname{ad}(X' + Y') \in G_c$ and suppose that $\mathfrak{a}$ is related to $\mathfrak{b}$. Then $y^{-1}\mathfrak{a}_c = \mathfrak{b}_c$ and $\beta = \alpha^{y^{-1}}$ is a root of $(\mathfrak{g}, \mathfrak{b})$. Put $H'_\beta = 2\,|\beta|^{-2} \cdot H_\beta$. Then

$$H'_\beta = (-1)^{1/2}(X' - Y') \, , \quad \mathfrak{b} = \mathfrak{a}_I + \mathbf{R}(X' - Y') \, .$$

Put $\mathfrak{b}_1 = \mathfrak{a}_I$ and $\mathfrak{b}_2 = (-1)^{1/2}\mathbf{R}H'_\beta$. Then β may be regarded as a root of $(\mathfrak{l}, \mathfrak{b}_2)$ and therefore

$$\mathfrak{l}_c = \mathbf{C}H_\beta + \mathbf{C}X_\beta + \mathbf{C}X_{-\beta}$$

is the usual notation. Moreover we may assume, without loss of generality that $\mathfrak{a}_R = \mathbf{R}(X_\beta + X_{-\beta})$.

LEMMA 1. *Let* B_β *denote the kernel of* ξ_β *in* B. *Then* $A_I \cap B = B_\beta$.

Let $a \in A_I \cap B$. Then a centralizes $X_\beta + X_{-\beta} \in \mathfrak{a}_R$. Hence $\xi_\beta(a) = 1$. Conversely if $a \in B_\beta$, it is clear that a centralizes $\mathfrak{a}$. This shows that $A_I \cap B = B_\beta$.

COROLLARY. $A_I = B_\beta$ *if* G *is connected.*

Let Ξ be the centralizer of $\mathfrak{b}_1$ in K. Since K is connected and $\mathfrak{b}_1 \subset \mathfrak{k}$, it follows that Ξ is connected. But $\mathfrak{k} \cap \mathfrak{z} = \mathfrak{b}$. Hence $\Xi = B$. Since $A_I = A \cap K \subset \Xi = B$, we conclude that $A_I = B_\beta$.

Put $B_i = \exp \mathfrak{b}_i$ $(i = 1, 2)$. Then B_1, B_2 are closed subgroups of B [2(d), p. 99] and $B_1 B_2 = B^\circ$. Since $\mathfrak{b}_1 \cap \mathfrak{b}_2 = \{0\}$, $A_I \cap B_2$ is a finite group.

LEMMA 2. *Put* $\gamma = \exp\left(\pi(X' - Y')\right)$. *Then* $A_I \cap B_2$ *is a cyclic group with* γ *as generator. Moreover* γ *centralizes* I.

Since B_2 is connected, we may assume for our proof that G is connected. Then by the above corollary, $A_I = B_\beta$. Put

$$b_t = \exp\left(-(-1)^{1/2} t H_\beta'\right) \qquad\qquad (t \in \mathbf{R}) .$$

Then $b_t \in B_\beta$ if and only if $t \in \mathbf{Z}\pi$. Since $\gamma = b_\pi$, this proves our first assertion. The second follows from the fact that $\gamma \in B_\beta$.

Put

$$\rho_\alpha = \frac{2\langle \rho, \alpha \rangle}{|\alpha|^2} \in \mathbf{Z}$$

where ρ is half the sum of all positive roots of $(\mathfrak{g}, \mathfrak{a})$. Let $P = MA_R N$ be a psgp in $\mathscr{P}(A_R)$.

LEMMA 3. γ *centralizes* $\mathfrak{m} + \mathfrak{a}_R$. *Moreover if* G *is connected and acceptable,*

$$\xi_\rho(\gamma) = (-1)^{\rho_\alpha} .$$

We know [2(i), p. 119] that

$$\mathrm{Ad}(\gamma) = \exp\left(\pi(-1)^{1/2} \mathrm{ad}\, H'\right)$$

and therefore γ centralizes $\mathfrak{m} + \mathfrak{a}_R$.

Now suppose G is connected and acceptable. Let $j: G \to G_c$ be an acceptable complexification of G. Then, by the same argument,

$$j(\gamma) = \exp\left\{\pi(-1)^{1/2} H'\right\}$$

and therefore

$$\xi_\rho(\gamma) = \exp\left\{\pi(-1)^{1/2}\rho(H')\right\} = (-1)^{\rho_\alpha} .$$

31. Recapitulation of some earlier results

In this section we recapitulate some results of [2(d)]. So suppose that rank G = rank K and G is connected and acceptable. Fix B as in Section 30.

Let A be a θ-stable Cartan subgroup of G. We denote by $Z(A)$ the subgroup of all $a \in A_I$ with the following property. If $j: G \to G_c$ is any complexification of G [2(c), § 18], then $j(a) \in \exp(-1)^{1/2}\mathfrak{a}_R$. Then $Z(A)$ is a finite subgroup of A_I [2(d), p. 99]. Put $\tilde{A}_R = Z(A) \cdot A_R$.

We now use the notation of Section 29. Put

$$\varpi_A = \varpi , \quad \varpi_0 = \prod_{\alpha \in Q_0} H_\alpha , \quad \Delta_A = \Delta$$

where Q_0 is the complement of Q_R in Q. For $f \in \mathcal{C}(G)$ define

$$F_f^A(a) = \varepsilon_R(a)\Delta_A(a) \int_{G/A_R} f(xax^{-1})dx^*$$

as before and put

$$\varepsilon_A(a) = \varepsilon_R(a)(-1)^{[Q_R(a_1)]} \qquad\qquad (a \in A') \ .$$

It is known that $\varpi_0 F_f^A$ extends (uniquely) to a continuous function on A. Let ϕ_f^A denote the restriction of $\varpi_0 F_f^A$ on $\tilde{A}_R$. Then $f \mapsto \phi_f^A$ is a continuous mapping of $\mathcal{C}(G)$ into $\mathcal{C}(\tilde{A}_R)$ [2(f), § 20].

Put

$$G_A = \bigcup_{x \in G} xA'x^{-1} \ .$$

Then G_A is an open subset of G and

$$\int_{G_A} f(x)dx = [W(G/A)]^{-1} \int_A |\Delta_A(a)|^2\, da \int_{G/A_R} f(xax^{-1})dx^*$$

for $f \in C_c(G)$ [2(f), p. 110].

Define L, Θ_λ and Ψ_λ ($\lambda \in L$) as in [2(d), § 8]. Then (see [2(d), § 15])

$$\varpi(\lambda)\Theta_\lambda(f) = \text{p.v.} \int D^{-1}\Psi_\lambda \nabla_G f\, dx$$

$$= \sum_{\{A\}} [W(G/A)]^{-1} \int_A \Psi_\lambda(a)\varepsilon_A(a)F_f^A(a;\ \varpi_A)da$$

for $\lambda \in L$ and $f \in C_c^\infty(G)$. Here $\{A\}$ denotes a complete set of θ-stable Cartan subgroups A of G no two of which are conjugate. We assume that B is contained in $\{A\}$.

LEMMA 1. *Fix A and $f \in C_c^\infty(G)$. Then*

$$\sum_{\lambda \in L} \int_{A_R} da_2 \left| \int_{A_1} \Psi_\lambda(a)\varepsilon_A(a)F_f^A(a;\ \varpi_A)da_1 \right| < \infty \ .$$

Here da_1 is the normalized Haar measure on the compact group A_I and $da_2 = dA_R$ so that $da = da_1 da_2$. This lemma is an immediate consequence of [2(d), Lem. 15].

Put

$$S_A(f) = \sum_{\lambda \in L} [W(G/A)]^{-1} \int_A \Psi_\lambda(a)\varepsilon_A(a)F_f^A(a;\ \varpi_A)da$$

for $f \in C_c^\infty(G)$.

LEMMA 2. *There exists a tempered distribution T_A on $\tilde{A}_R$ such that*

$$S_A(f) = T_A(\phi_f^A)$$

for all $f \in C_c^\infty(G)$.

This follows from [2(d), § 14] (see also [2(f), § 20]).

COROLLARY 1. S_A is a tempered distribution on G. Moreover if $f \in \mathcal{C}(G)$ and $F_f^A = 0$ then $S_A(f) = 0$.

This is obvious.

COROLLARY 2. $\sum_{\lambda \in L} \varpi(\lambda)\Theta_\lambda(f) = \sum_{\{A\}} S_A(f)$ $(f \in \mathcal{C}(G))$.

For $f \in C_c^\infty(G)$ this is clear from the formula for $\varpi(\lambda)\Theta_\lambda(f)$ given above. But since both sides are tempered distributions [2(e), § 29], this relation must hold for all $f \in \mathcal{C}(G)$.

Fix $f \in C_c^\infty(G)$. Then by [2(d), § 15],

$$S_B(f) = \sum_{\lambda \in L} \int_B \varpi_B F_f^B \cdot \xi_\lambda db = F_f^B(1; \varpi_B) = (-1)^q c_G f(1)$$

where q and c_G have the same meaning as in Section 23. Therefore if $\{A\}_+$ denotes the complement of B in $\{A\}$, we get the following result.

LEMMA 3. $(-1)^q c_G f(1) = \sum_{\lambda \in L} \varpi(\lambda)\Theta_\lambda(f) - \sum_{\{A\}_+} S_A(f)$ for all $f \in \mathcal{C}(G)$.

32. Statement of Theorem 32.1

We now drop all restrictions on G except that rank G = rank K. Fix a Cartan subgroup B of K and let A be a θ-stable Cartan subgroup of G with dim $A_R = 1$. Let α be the unique positive root of $(\mathfrak{g}, \mathfrak{a})$. Define (H', X', Y') and put

$$\gamma = \exp \pi(X' - Y'), \quad \rho_\alpha = \frac{2\langle \rho, \alpha \rangle}{|\alpha|^2}$$

as in Section 30. Let C denote the cyclic group generated by γ and F the order of γ.

For $f \in \mathcal{C}(G)$, put

$$I_A(f) = \frac{1}{F} \sum_r \sum_{c \in C} \int_0^\infty \Phi_r(t)\chi_r(c^{-1})' F_f^A(ch_t; \varpi_A')dt.$$

Here $h_t = \exp tH'$ $(t \in \mathbf{R})$ and r runs over all integers such that

$$0 \leq r < 2F, \quad r \equiv \rho_\alpha F \bmod 2.$$

χ_r is the character of C given by

$$\chi_r(\gamma) = (-1)^{\rho_\alpha} \exp\left((-1)^{1/2}\frac{r}{F}\pi\right)$$

and

$$\Phi_r(t) = \frac{\exp\left\{\left(1 - \frac{r}{F}\right)t\right\} + \exp\left\{-\left(1 - \frac{r}{F}\right)t\right\}}{\exp t - \exp(-t)}.$$

It is easy to see that for a fixed $c \in C$,

$$t \longmapsto {}'F_f^A(ch_t: \varpi_A')$$

is a Schwartz function on $\mathbf{R}$ which changes sign under the reflexion $t \mapsto -t$. Therefore I_A is a tempered distribution on G.

We recall that if H is a topological group, then H^0 denotes the connected component of 1 in H. Let $P = MA_R N$ be a psgp of G in $\mathscr{P}(A_R)$. We intend to prove the following theorem.

THEOREM 1. *Let f be a function in $\mathcal{C}(G)$ satisfying the following two conditions*:

(1) $(\Theta_\omega, f) = 0$ *for all* $\omega \in \mathscr{E}_2(G)$.

(2) *If Γ is a θ-stable Cartan subgroup of G with* $\dim \Gamma_R \geq 1$, *then* ${}'F_f^\Gamma = 0$ *unless Γ is conjugate to A in G.*

Then

$$(-1)^q c_G f(1) = [G: G^0 \cdot A] \frac{[W(G^0/B^0)]}{[W(M/A_I)]} \frac{2}{|\alpha|} I_A(f) .$$

The proof proceeds in three steps. First we show that

$$S_A = -\frac{[W(G/B)]}{[W(M/A_I)]} \frac{2}{|\alpha|} I_A$$

provided G is connected and acceptable. In the second step we drop acceptability and in the third step connectedness.

33. First step in the proof

So let us assume that G is connected and acceptable and keep to the notation of Sections 30, 31 and 32.

LEMMA 1. *Define A and B as in Section 32. Then*

$$S_A = -\frac{[W(G/B)]}{[W(M/A_I)]} \frac{2}{|\alpha|} I_A .$$

The proof requires some preparation. Since A will be kept fixed, we drop it from the notation whenever it is convenient to do so. Also we may assume that A is related to B.

LEMMA 2. *Fix $f \in C_c^\infty(G)$ and $\lambda \in L$. Then*

$$[W(G/A)]^{-1} \int_A \Psi_\lambda(a) \varepsilon_A(a) F_f(a; \varpi) da$$

$$= -[W(M/A_I)]^{-1} \|H'\| \int_0^\infty dt \int_{A_I} \Psi_\lambda(a_1 h_t) F_f(a_1 h_t; \varpi) da_1 .$$

Since α is a real root of $(\mathfrak{g}, \mathfrak{a})$, the corresponding Weyl reflexion s_α lies in $W(G/A)$. Hence

$$\mathfrak{w}(G/A_R) = W(G/A)/W(M/A_I)$$

is of order 2. Also α is the only positive real root and $\xi_\alpha(\gamma) = 1$ by Lemma 30.2. Therefore

$$\varepsilon_A(a_1 h_t) = -\operatorname{sign} t \ .$$

Moreover $dh_t = \| H' \| \, dt$ and the functions $\Psi_\lambda(a)$ and $\varepsilon_A(a)F_f(a; \varpi)$ are invariant under s_α. Our assertion is an immediate consequence of these facts. Now

$$\Psi_\lambda(a) = \sum_{s \in W(G/B)} \xi_{s\lambda}(a_1) \exp\{-|(s\lambda)^y(\log a_2)|\}$$

where y has the same meaning as in Section 30. (This is proved in the same way as [2(d), Lem. 52].) Hence

$$\Psi_\lambda(a_1 h_t) = \sum_{s \in W(G/B)} \xi_{s\lambda}(a_1) \exp\{-t\,|\,s\lambda(H'_\beta)\,|\}$$

for $t \geq 0$. Since L is stable under $W(G/B)$, we conclude from [2(d), Lem. 15] that

$$\sum_{\lambda \in L} \int_0^\infty dt \int_{A_I} \Psi_\lambda(a_1 h_t) F_t(a_1 h_t; \varpi) da_1$$

$$= [W(G/B)] \sum_{\lambda \in L} \int_0^\infty \exp\{-|\lambda(H'_\beta)|\,t\} dt \int_{A_I} \xi_\lambda(a_1) F_f(a_1 h_t; \varpi) da_1 \ .$$

On the other hand $\| H' \| = 2\,|\,\alpha\,|^{-1}$. Therefore we get the following result.

LEMMA 3. *Let* $f \in C_c^\infty(G)$. *Then*

$$S_A(f) = -\frac{[W(G/B)]}{[W(M/A_I)]} \frac{2}{|\alpha|} J_A(f)$$

where

$$J_A(f) = \sum_{\lambda \in L} \int_0^\infty \exp\{-|\lambda(H'_\beta)|\,t\} dt \int_{A_I} \xi_\lambda(a_1) F_f(a_1 h_t; \varpi) da_1 \ .$$

We shall now transform the above expression for $J_A(f)$. Let L_1 be the lattice consisting of all $\lambda \in L$ such that $\langle \lambda, \beta \rangle = 0$. Fix $\mu \in L$, $t \in \mathbf{R}$ and put

$$g(a_1) = [C]^{-1} \sum_{c \in C} \xi_\mu(ca_1) F_f(ca_1 h_t; \varpi) \qquad\qquad (a_1 \in A_I)$$

where $C = A_I \cap B_2$ is the cyclic group generated by γ. Then g is a continuous function on A_I which may also be regarded as a function on A_I/C. On the other hand L_1 may be identified with the character group of $B/B_2 = A_I/C$. Since the Fourier series of g converges absolutely [2(d), p. 100], we conclude that

$$g(1) = \sum_{\lambda \in L_1} \int_{A_I} \xi_\lambda(a_1) g(a_1) da_1 \ .$$

Thus we have obtained the following result.

LEMMA 4. *Fix $\mu \in L$ and $f \in C_c^\infty(G)$. Then*

$$\sum_{\lambda \in L_1} \int_{A_I} \xi_{\mu+\lambda}(a_1) F_f(a_1 h_t; \varpi) da_1 = [C]^{-1} \sum_{c \in C} \xi_\mu(c) F_f(ch_t; \varpi)$$

for $t \in \mathbf{R}$.

Let L_0 be the additive subgroup of L generated by L_1 and β. Then

$$\bigcap_{\lambda \in L_0} \ker \xi_\lambda = B_2 \cap B_\beta = B_2 \cap A_I = C .$$

Hence L/L_0 may be regarded as the character group of C. Therefore both C and L/L_0 are cyclic groups of order F.

Put

$$b_t = \exp\{-(-1)^{1/2} t H'_\beta\} \qquad (t \in \mathbf{R}) .$$

We have seen in Section 30 that $b_t \in B_\beta = A_I$ if and only if $t \in \mathbf{Z}\pi$. Since $b_\pi = \gamma$, it follows that $b_t = 1$ if and only if $t \in \mathbf{Z}\pi F$. Hence there exists a character χ of B_2 such that

$$\chi(b_t) = \exp\{-(-1)^{1/2} 2t/F\} .$$

Since χ extends to a character of B, we can choose $\mu \in L$ such that $\mu(H'_\beta) = 2/F$. Clearly $r\mu$ $(0 \leq r < F)$ are incongruent $\bmod L_0$. Since $[L/L_0] = F$, every element in L can be written uniquely in the form

$$r\mu + \lambda + k\beta \quad (0 \leq r < F, \ k \in \mathbf{Z}, \ \lambda \in L_1) .$$

Hence we conclude from Lemma 4 and [2(d), Lem. 15] that

$$J_A(f) = F^{-1} \sum_{0 \leq r < F} \sum_{k \in \mathbf{Z}} \int_0^\infty \exp\left(-2\left|\frac{r}{F} + k\right| t\right)$$
$$\times \sum_{c \in C} \xi_{r\mu+k\beta}(c) F_f(ch_t; \varpi) dt .$$

But

$$\xi_{r\mu+k\beta}(\gamma) = \xi_{r\mu}(\gamma) = \exp\{-(-1)^{1/2} 2r\pi/F\} ,$$
$$F_f(ch_t; \varpi) = \xi_\rho(c)' F_f(ch_t; \varpi') \qquad (c \in C) ,$$

and $\xi_\rho(\gamma) = (-1)^{\rho\alpha}$ from Lemma 30.3. Moreover since $\gamma^F = 1$,

$$\rho_\alpha F \equiv 0 \bmod 2 .$$

Define the character χ_{2r} of C by

$$\chi_{2r}(\gamma) = (-1)^{\rho\alpha} \exp\{(-1)^{1/2} 2r\pi/F\} \qquad (r \in \mathbf{Z}) .$$

Then we get

$$J_A(f) = F^{-1} \sum_{0 \leq r < F} \sum_{k \in \mathbf{Z}} \int_0^\infty \exp\left(-2\left|\frac{r}{F} + k\right| t\right)$$
$$\times \sum_{c \in C} \chi_{2r}(c^{-1})' F_f(ch_t; \varpi') dt .$$

But

$$\sum_{k \in \mathbf{Z}} \exp\left(-2\left|\frac{r}{F} + k\right| t\right) = \sum_{k \geq 0} \exp\left\{-2\left(\frac{r}{F} + k\right)t\right\}$$

$$+ \sum_{k \geq 1} \exp\left\{-2\left(k - \frac{r}{F}\right)t\right\}$$

$$= \Phi_{2r}(t)$$

for $t > 0$ and $0 \leq r < F$. On the other hand for a fixed $c \in C$,

$$t \longmapsto F'_f(ch_t; \varpi') \qquad\qquad (t \in \mathbf{R})$$

is an odd function of t lying in $C_c^\infty(\mathbf{R})$ [2(d), Cor. 3, p. 103]. Hence

$$\int_0^\infty \Phi_{2r}(t) \, | \, F'_f(ch_t; \varpi') \, | \, dt < \infty$$

and therefore we can interchange the summation over k with integration over t. This shows that

$$J_A(f) = F^{-1} \sum_{0 \leq r < F} \int_0^\infty \Phi_{2r}(t) \sum_{c \in C} \chi_{2r}(c^{-1})' F_f(ch_t; \varpi') dt$$

$$= I_A(f) \, .$$

Since both S_A and I_A are tempered distributions on G, Lemma 1 now follows from Lemma 3.

Now suppose Z is a finite subgroup lying in the center of G and $G_0 = G/Z$. Put $C_1 = C \cap Z$ and let γ_0 denote the image of γ in G_0. Similarly let C_0 denote the image of C in G_0 and F_0 the order of γ_0. Then F_0 is the least positive integer such that $\gamma^{F_0} \in C_1$ and it is clear that γ^{F_0} is a generator of C_1.

Now suppose $f \in \mathcal{C}(G)$ is actually a function on G/Z. Then $'F_f(ch_t; \varpi')$ depends only on $c \bmod C_1$. Fix r ($0 \leq r < F$). Then

$$\sum_{c \in C_1} \chi_{2r}(c) = 0$$

unless $\chi_{2r} = 1$ on C_1, i.e., unless

$$r_0 \equiv \rho_\alpha F_0 \bmod 2$$

where $r_0 = 2rF_0/F$. Note that $0 \leq r_0 < 2F_0$ and

$$\chi_{2r}(\gamma) = (-1)^{\rho_\alpha} \exp\left\{(-1)^{1/2} r_0 \pi/F_0\right\}$$

if this condition holds. Hence

$$F^{-1}\sum_{0 \leq r < F} \Phi_{2r}(t) \sum_{c \in C} \chi_{2r}(c^{-1})' F_f(ch_t; \varpi')$$
$$= F_0^{-1} \sum_s \Phi_s^0(t) \sum_{c_0 \in C_0} \chi_s^0(c_0^{-1})' F_f(c_0 h_t; \varpi') \, .$$

Here s runs over all integers such that

$$0 \leq s < F_0 \, , \qquad s \equiv \rho_\alpha F_0 \bmod 2$$

and χ_s^0 is the character of C_0 given by

$$\chi_s^0(\gamma_0) = (-1)^{\rho_\alpha} \exp\left\{(-1)^{1/2} s\pi/F_0\right\} .$$

Moreover Φ_s^0 is obtained by replacing (r, F) by (s, F_0) in the definition of Φ_r, and $'F_f$ is to be regarded as a function on A/Z on the right hand side.

34. Second step

Let us now drop the condition that G be acceptable. Fix $f \in \mathcal{C}(G)$. Then the discussion at the end of Section 33 shows that the value of $I_A(f)$ remains unchanged if we go over to an acceptable, finite covering group $\widetilde{G}$ of G and regard f as a function on $\widetilde{G}$.

Now for any θ-stable Cartan subgroup A of G with $\dim A_R = 1$, *define*

$$S_A(f) = -\frac{[W(G/B)]}{[W(M/A_I)]}\frac{2}{|\alpha|} I_A(f) \qquad\qquad (f \in \mathcal{C}(G)),$$

in the notation of Section 32. Let $\{A\}_1$ denote a complete set of such Cartan subgroups of G, no two of them being conjugate.

THEOREM 1. *Let f be an element in $\mathcal{C}(G)$ such that:*
(1) $\Theta_\omega(f) = 0$ *for all* $\omega \in \mathcal{E}_2(G)$.
(2) $'F_f^\Gamma = 0$ *for every θ-stable Cartan subgroup Γ of G with* $\dim \Gamma_R > 1$.
Then

$$(-1)^q c_G f(1) = -\sum_{\{A\}_1} S_A(f) .$$

Let $\widetilde{G}$ be an acceptable finite covering group of G. Then if we regard f as a function on $\widetilde{G}$, it continues to satisfy all the conditions of the hypothesis. Moreover $c_{\widetilde{G}} = c_G$. Hence it is enough to consider the case when G is acceptable. But then our assertion follows immediately from Lemmas 31.3, 33.1 and [2(f), Thm. 16].

35. Third step

We now come to the proof of Theorem 32.1. Let $\mathfrak{a}_1 = \mathfrak{a}, \mathfrak{a}_2, \cdots, \mathfrak{a}_p$ be a complete set of θ-stable Cartan subalgebras of $\mathfrak{g}$ such that:
(1) they are all conjugate under G,
(2) no two of them are conjugate under G^0,
(3) $\mathfrak{a}_i$ is related to $\mathfrak{b}$ (see § 30) for $1 \leq i \leq p$.
Let $P = MA_R N$ be a psgp of G in $\mathcal{P}(A_R)$. Choose $y_i \in K$ such that $\mathrm{Ad}(y_i)\mathfrak{a} = \mathfrak{a}_i$ $(1 \leq i \leq p, y_1 = 1)$. Let A_i denote the Cartan subgroup of G corresponding to $\mathfrak{a}_i$. Put

$$M_i = y_i M y_i^{-1} , \quad M_{i0} = M_i \cap G^0 , \quad A_{i0} = A_i \cap G^0 .$$

Then A_{i0} is a Cartan subgroup of G^0 and

$$A_{i0} = y_i A_0 y_i^{-1} , \quad M_{i0} = y_i M_0 y_i^{-1}$$

where $A_0 = A \cap G^0$, $M_0 = M \cap G^0$. Let α_i be the unique positive real root of $(\mathfrak{g}, \mathfrak{a}_i)$. Then

$$[W(M_{i0}/A_{i0I})] = [W(M_0/A_{0I})] , \quad |\alpha_i| = |\alpha| \qquad (1 \leq i \leq p) .$$

For $f \in \mathcal{C}(G)$, put

$$f^\natural(x) = \int_K f(kxk^{-1})dk \qquad (x \in G)$$

and let f_0 denote the restriction of $f^\natural$ on G^0. Then it is obvious that

$$I_A(f) = I_{A_i}(f) = I_{A_{i0}}(f_0) \qquad (1 \leq i \leq p) .$$

Now fix f as in Theorem 32.1. Then by [2(i), Lem. 27.6], f_0 satisfies the hypothesis of Theorem 34.1. Since $f_0(1) = f(1)$ and $c_{G^0} = c_G$, we conclude that

$$(-1)^q c_G f(1) = -\sum_{1 \leq i \leq p} S_{A_{i0}}(f_0) .$$

But

$$S_{A_{i0}}(f_0) = -\frac{[W(G^0/B^0)]}{[W(M_{i0}/A_{i0I})]} \frac{2}{|\alpha_i|} I_{A_{i0}}(f_0)$$

$$= -\frac{[W(G^0/B^0)]}{[W(M_0/A_{0I})]} \frac{2}{|\alpha|} I_A(f) .$$

Hence

$$(-1)^q c_G f(1) = p\frac{[W(G^0/B^0)]}{[W(M_0/A_{0I})]} \frac{2}{|\alpha|} I_A(f) .$$

Let $\tilde{A}$ be the normalizer of $\mathfrak{a}$ in G. Then it is obvious that

$$p = [G: \tilde{A}G^0] = \frac{[G: G^0]}{[\tilde{A}: \tilde{A} \cap G^0]} .$$

But

$$[\tilde{A}: \tilde{A} \cap G^0] = \frac{[\tilde{A}: A_0]}{[\tilde{A} \cap G^0: A_0]} = \frac{[W(G/A)]}{[W(G^0/A_0)]}[A: A_0] .$$

Hence

$$p = [G: G^0 \cdot A]\frac{[W(G^0/A_0)]}{[W(G/A)]} .$$

Since

$$[\mathfrak{w}(G/A)] = [\mathfrak{w}(G^0/A_0)] = 2 ,$$

we conclude that

$$(-1)^q c_G f(1) = [G: G^0 \cdot A]\frac{[W(G^0/B^0)]}{[W(M/A_I)]} \frac{2}{|\alpha|} I_A(f) .$$

This completes the proof of Theorem 32.1.

36. The formula for $\mu(\omega : \nu)$

We keep to the notation of Section 32. Fix $\beta \in C_c^\infty(\mathfrak{F})$ and put

$$f = \phi_\beta = \int_\mathfrak{F} \mu(\omega_0 : \nu)\beta(\nu)E(P : \psi : \nu)d\nu$$

in the notation of Lemma 24.1.

LEMMA 1. *f satisfies the conditions of Theorem 32.1.*

Fix $\omega \in \mathcal{E}_2(G)$ and put

$$T(\beta) = (\Theta_\omega, \phi_\beta) \qquad\qquad (\beta \in C_c^\infty(\mathfrak{F})) \ .$$

Then T is a distribution on $\mathfrak{F}$ (with values in V). It is easy to verify that Supp T is empty (see the proof of Lemma 26.1) and therefore $T = 0$. Our assertion now follows from Lemma 24.3.

Let Q be the set of all positive roots of $(\mathfrak{g}, \mathfrak{a})$. Then $Q_R = \{\alpha\}$ and $\varpi = \alpha\varpi_I\varpi_+$ in the notation of Section 24. For $a^* \in A_I^*$ define the distribution θ_{a^*} on M corresponding to [2(i), p. 177]. Since $L = {}^0\mathcal{C}(M, \tau_M)$ has finite dimension, it is clear that there exists a finite subset F_* of $A_I^{*\prime}$ such that

$$(\theta_{a^*}, \psi_0) = 0$$

for $a^* \in A_I^*$ and $\psi_0 \in L$ unless $a^* \in F_*$. Put

$$\hat{f}(a^* : \nu) = F_0(\theta_{a^*}, f_\nu^{(P)}) \qquad\qquad (a^* \in A_I^*, \nu \in \mathfrak{F})$$

where F_0 has the same meaning as in Lemma 21.5. It follows from Lemma 20.2 that $f_\nu^{(P)} \in L$. Hence $\hat{f}(a^* : \nu) = 0$ unless $a^* \in F_*$.

LEMMA 2.

$${}'F_f(a) = \sum_{a^* \in A_I^*} \langle a^*, a_1\rangle \int_\mathfrak{F} \hat{f}(a^* : \nu) \exp\{(-1)^{1/2}\nu(\log a_2)\}d\nu \qquad for \ a \in A \ .$$

Here the argument is the same as in the proof of Lemma 24.1. Fix $a_2 \in A_R$ and put

$$g(m) = f^{(P)}(ma_2) \qquad\qquad (m \in M) \ .$$

Then $g \in L$ and therefore [2(i), Lem. 18.2]

$${}'F_f(a_1 a_2) = {}'F_g^{M/A_I}(a_1) = \sum_{a^* \in A_I^*} F_0(\theta_{a^*}, g)\langle a^*, a_1\rangle \ .$$

But

$$(\theta_{a^*}, g) = \int (\theta_{a^*}, f_\nu^{(P)}) \exp\{(-1)^{1/2}\nu(\log a_2)\}d\nu$$

and so our assertion follows.

For $a^* \in A_I^*$, define $\lambda(a^*) \in (-1)^{1/2}a_I^*$ as in [2(i), §27] and put

$$\varpi(a^*, \nu) = \varpi\big(\lambda + (-1)^{1/2}\nu\big), \quad \varpi_I(a^*) = \varpi_I(\lambda), \quad \varpi_+(a^*, \nu) = \varpi_+\big(\lambda + (-1)^{1/2}\nu\big)$$

$$(\nu \in \mathfrak{F})$$

with $\lambda = \lambda(a^*)$.

COROLLARY.

$$'F_f(a; \varpi') = \sum_{a^* \in A_I^*} \langle a^*, a_1 \rangle \int_{\mathfrak{F}} \varpi(a^*, \nu) \hat{f}(a^*: \nu) \exp\{(-1)^{1/2}\nu(\log a_2)\} d\nu .$$

This is obvious.

Let C^* be the character group of C (see § 32). Fix $a^* \in A_I^*$ and $\chi \in C^*$. Since σ_{a^*} is an irreducible representation of A_I with character a^*, we denote by $[a^*: \chi]$ the multiplicity of χ in the restriction of σ_{a^*} to C. It is clear that

$$[C]^{-1} \sum_{c \in C} \langle a^*, c \rangle \chi(c^{-1}) = [a^*: \chi] .$$

We recall that F is the order of C. Let S denote the set of all integers r such that

$$0 \leq r < 2F , \qquad r \equiv \rho_\alpha F \bmod 2 .$$

Then we have defined in Section 32 a bijective mapping $r \mapsto \chi_r$ of S onto C^*. Put $\Phi_{\chi_r} = \Phi_r$. Then

$$I_A(f) = \sum_{\chi \in C^*} [C]^{-1} \sum_{c \in C} \int_0^\infty \Phi_\chi(t) \chi(c^{-1})' F_f(ch_t; \varpi') dt .$$

Put

$$\nu_\alpha = \frac{2\langle \nu, \alpha \rangle}{|\alpha|^2} = \nu(H')$$

and recall that $\hat{f}(a^*: \nu) = 0$ unless $a^* \in F_*$. Moreover for a fixed $c \in C$, $'F_f(ch_t; \varpi')$ is an odd function of t. Therefore we conclude from the corollary of Lemma 2 that

$$I_A(f) = \sum_{a^* \in A_I^*} \sum_{\chi \in C^*} [a^*: \chi] \int_0^\infty dt \int_{\mathfrak{F}} (-1)^{1/2} \varpi(a^*: \nu) \hat{f}(a^*: \nu) \Phi_\chi(t) \sin \nu_\alpha t \cdot d\nu .$$

LEMMA 3.

$$\int_0^\infty dt \int_{\mathfrak{F}} (-1)^{1/2} \varpi(a^*: \nu) \hat{f}(a^*: \nu) \Phi_\chi(t) \sin \nu_\alpha t \cdot d\nu$$

$$= (-1)^{p+1} \frac{|\alpha|^2}{4} \varpi_I(a^*) \int_{\mathfrak{F}} \mu_0(\chi: \nu) |\varpi_+(a^*, \nu)| \hat{f}(a^*: \nu) d\nu ,$$

where

$$p = \tfrac{1}{2}[Q_c] , \qquad \nu_\alpha = \frac{2\langle \nu, \alpha \rangle}{|\alpha|^2}$$

and

$$\mu_0(\chi: \nu) = \pi \nu_\alpha \sinh \pi \nu_\alpha \left\{ \cosh \pi \nu_\alpha - \frac{(-1)^{\rho_\alpha}}{2} (\chi(\gamma) + \chi(\gamma^{-1})) \right\}^{-1} .$$

Since

$$(-1)^{1/2}\varpi(a^*,\nu) = -\frac{|\alpha|^2}{2}\nu_\alpha\varpi_I(a^*)\varpi_+(a^*,\nu)$$

and

$$\varpi_+(a^*,\nu) = (-1)^p\,|\,\varpi_+(a^*,\nu)\,|\,,$$

this follows immediately from the corollary of Lemma 42.1.

Fix a real number $t \geq -1$ and put

$$u(z) = \frac{z \sinh z}{\cosh z + t} \qquad\qquad (z \in \mathbf{C})\,.$$

Then u is a meromorphic function on the complex plane which is holomorphic
on the real axis. For $a^* \in A_I^*$, define

$$\mu_0(a^*:\nu) = d(a^*)^{-1}\mathrm{tr}\left\{\frac{\pi\nu_\alpha \sinh \pi\nu_\alpha}{\cosh \pi\nu_\alpha - \dfrac{(-1)^{\rho_\alpha}}{2}\{\sigma_{a^*}(\gamma) + \sigma_{a^*}(\gamma^{-1})\}}\right\}$$

where σ_{a^*} is, as before, an irreducible representation of A_I with character
a^*. Then $\mu_0(a^*)$ is a meromorphic function on $\mathfrak{F}_c$ which is holomorphic on $\mathfrak{F}$.

COROLLARY.

$$I_A(f) = (-1)^{p+1}\frac{|\alpha|^2}{4}\sum_{a^* \in A_I^*} d(a^*)\varpi_I(a^*)\int_{\mathfrak{F}} \mu_0(a^*:\nu)\,|\,\varpi_+(a^*,\nu)\,|\,f(a^*:\nu)d\nu\,.$$

This follows immediately from what we have seen above if we observe
that

$$d(a^*)\mu_0(a^*:\nu) = \sum_{\chi \in C^*} [a^*:\chi]\mu_0(\chi:\nu)\,.$$

Define q_G and q_M as before (see the proof of Theorem 24.1).

LEMMA 3. $p + 1 = q_G - q_M$.

For $P = MA_RN$ and $\dim N = [Q_R] + [Q_c] = 1 + 2p$. Moreover

$$\dim G/K = \dim P/K \cap P = \dim M/K_M + \dim A_R + \dim N\,.$$

Therefore

$$2q_G = 2q_M + 2 + 2p$$

and this implies our assertion.

Fix $a^* \in A_I^{*\prime}$ and let ω be the element in $\mathfrak{E}_2(M)$ which corresponds to a^*
under Theorem 23.1. Put

$$\mu_0(\omega:\nu) = [W(M/A_I)]^{-1}\sum_{s \in W(M/A_I)} \mu_0(sa^*:\nu)\,,$$
$$\varpi_+(\omega:\nu) = \varpi_+(a^*,\nu)\,.$$

(We recall that, by Lemma 24.4, ϖ_+ is invariant under $W(M/A_I)$.) Now

$$d(\omega) = c_M^{-1} \, | \, \varpi_I(a^*) \, | \, [\, W(M/A_I)] d(a^*)$$

from the corollary of Lemma 23.1. Hence

$$d(a^*)\varpi_I(a^*)\hat{f}(a^* \colon \nu) = (-1)^{q_M} c_M [\, W(M/A_I)]^{-1} d(\omega)\hat{f}(\omega \colon \nu)$$

in the notation of Theorem 22.1. Therefore we conclude from Lemma 3 and the corollary of Lemma 2 that

$$I_A(f) = (-1)^{q_G} c_M \frac{|\alpha|^2}{4} \sum_{\omega \in \mathcal{E}_2(M)} d(\omega) \int_{\mathfrak{J}} \mu_0(\omega \colon \nu) \, | \, \varpi_+(\omega \colon \nu) \, | \, \hat{f}(\omega \colon \nu) d\nu \, .$$

Let s_α be the Weyl reflexion corresponding to the root α of $(\mathfrak{g}, \mathfrak{a})$. Then $s_\alpha = \mathrm{Ad}(k)$ on $\mathfrak{a}$ where

$$k = \exp\left(\frac{\pi}{2}(X' - Y')\right)$$

in the notation of Section 30. Since $\gamma = k^2$, it is easy to verify that

$$\langle s_\alpha a^*, \, c \rangle = \langle a^*, \, c \rangle \qquad\qquad (a^* \in A_I^*, \, c \in C)$$

and therefore

$$\mu_0(s_\alpha a^* \colon \nu) = \mu_0(a^* \colon \nu) = \mu_0(a^* \colon -\nu) \, .$$

This implies that

$$\mu_0(\omega \colon \nu) = \mu_0(s\omega \colon \nu) = \mu_0(\omega \colon s\nu) \qquad\qquad (\omega \in \mathcal{E}_2(M), \, s \in \mathfrak{w}) \, .$$

Hence we conclude from Theorem 22.1 that

$$I_A(f) = (-1)^{q_G} c_M \frac{|\alpha|^2}{4} [\mathfrak{w} \colon \mathfrak{w}(\omega_0)] d(\omega_0) \int_{\mathfrak{J}} \mu_0(\omega_0 \colon \nu) \, | \, \varpi_+(\omega_0 \colon \nu) \, | \, \hat{f}(\omega_0 \colon \nu) d\nu$$

$$= (-1)^{q_G} c_M \frac{|\alpha|^2}{2} \gamma(G/A_R) F_0 \psi(1) \int_{\mathfrak{J}} \mu_0(\omega_0 \colon \nu) \, | \, \varpi_+(\omega_0 \colon \nu) \, | \, \beta(\nu) d\nu \, .$$

(We observe that $c(G/A_R) = 1$ since $\dim A_R = 1$.) On the other hand

$$f(1) = F_0 \psi(1) \int_{\mathfrak{J}} \mu(\omega_0 \colon \nu)\beta(\nu) d\nu \, .$$

Therefore since β is an arbitrary element in $C_c^\infty(\mathfrak{F})$, we conclude from Theorem 32.1 (see also the end of §24) that

$$\mu(\omega_0 \colon \nu) = \frac{c_M}{c_G} \frac{[G \colon G^0 \cdot A][\, W(G^0/B^0)]}{[\, W(M/A_I)]} \, | \, \alpha \, | \, \gamma(G/A_R) \cdot \mu_0(\omega_0 \colon \nu) \, | \, \varpi_+(\omega_0 \colon \nu) \, |$$

$$(\nu \in \mathfrak{F}) \, .$$

The assertion of Theorem 25.1 can now be verified immediately from the definition of $\mu_0(\omega_0)$ if we take into account Lemma 4 below.

For $\varepsilon > 0$ let D_ε denote the region consisting of all $z \in C$ with $| \, \mathrm{Im} \, z \, | < \varepsilon$. Let U denote the set of all meromorphic functions u on the complex plane satisfying the following conditions. There exist numbers $\varepsilon > 0$, $s \geq 0$ (depending on u) such that:

(1) u is holomorphic on D_ε,

(2) $\sup_{z \in D_\varepsilon} |u(z)| (1 + |z|)^{-s} < \infty$.

For a real number $t \geq -1$, put

$$u_t(z) = \frac{z \sinh z}{\cosh z + t}.$$

LEMMA 4. *U is a ring which contains all polynomials as well as the functions u_t ($t \geq -1$).*

This is obvious.

Part IV. Commuting algebras of induced representations

37. A lemma on surjectivity

We return to the notation of Section 14. Fix a psgp $P = MAN$ in $\mathscr{P}(A)$, $\omega \in \mathscr{E}_2(M)$ and $\nu_0 \in \mathfrak{F} = \mathfrak{a}^*$. Let $\mathfrak{w}_0 = \mathfrak{w}(\omega, \nu_0)$ be the subgroup of all $s \in \mathfrak{w}$ such that $s\omega = \omega$ and $s\nu_0 = \nu_0$. Put $L = {}^0\mathcal{C}(M, \tau_M)$ and define $L(\omega)$ as usual [2(j), § 18]. Then $L(\omega)$ is stable under ${}^0c_{P|P}(s: \nu_0)$ for $s \in \mathfrak{w}_0$ (Lemmas 17.3 and 17.6). Let $\chi(s)$ denote the restriction of ${}^0c_{P|P}(s: \nu_0)$ on $L(\omega)$. Then we conclude from the corollary of Lemma 17.2 that χ is a representation of $\mathfrak{w}_0$ on $L(\omega)$.

Let E_ω denote the projection of $\mathcal{C}(M, \tau_M)$ on $L(\omega)$ given by (Lemma 23.2),

$$E_\omega g: m \longmapsto d(\omega)(\theta_\omega, r(m)g) \qquad (m \in M)$$

for $g \in \mathcal{C}(M, \tau_M)$.

LEMMA 1. *Let $L(\omega, \nu_0)$ denote the space of all $\psi \in L(\omega)$ such that*

$${}^0c_{P|P}(s: \nu_0)\psi = \psi$$

for all $s \in \mathfrak{w}(\omega, \nu_0)$. Then $f \mapsto E_\omega f_{\nu_0}^{(P)}$ is a surjective mapping of $\mathcal{C}(G, \tau)$ onto $L(\omega, \nu_0)$.

Fix $\psi \in L(\omega)$ and $f \in \mathcal{C}(G, \tau)$. Then if $s \in \mathfrak{w}_0$,

$$\left(\psi, {}^0c_{P|P}(s: \nu_0)E_\omega f_{\nu_0}^{(P)}\right) = \left({}^0c_{P|P}(s^{-1}: \nu_0)\psi, f_{\nu_0}^{(P)}\right)$$

from the results of Section 17. But the right hand side

$$= \left(E\big(P: {}^0c_{P|P}(s^{-1}: \nu_0)\psi: \nu_0\big), f\right)$$
$$= \left(E(P: \psi: \nu_0), f\right) \text{ from Lemma 17.2 },$$
$$= (\psi, f_{\nu_0}^{(P)}) = (\psi, E_\omega f_{\nu_0}^{(P)}).$$

This being true for all $\psi \in L(\omega)$, we conclude that $E_\omega f_{\nu_0}^{(P)} \in L(\omega, \nu_0)$.

Conversely fix $\psi \in L(\omega, \nu_0)$ and choose an open neighborhood U of ν_0 in $\mathfrak{F}$ such that $U \cap sU = \varnothing$ for $s \in \mathfrak{w}(\omega)$ unless $s \in \mathfrak{w}_0$. Put

$$f = \int_{\mathfrak{F}} \mu(\omega: \nu)\alpha(\nu)E(P: \psi: \nu)d\nu$$

where $\alpha \in C_c^\infty(U)$. Then $f \in \mathcal{C}(G, \tau)$ and

$$E_\omega f_\nu^{(P)} = \gamma c^2 \sum_{s \in \mathfrak{w}(\omega)} \alpha(s^{-1}\nu)^0 c_{P|P}(s: s^{-1}\nu)\psi \qquad (\nu \in \mathfrak{F})$$

from Theorem 20.1. Therefore

$$E_\omega f_{\nu_0}^{(P)} = \gamma c^2 [\mathfrak{w}_0]\alpha(\nu_0)\psi \;,$$

and if we choose α so that $\gamma c^2[\mathfrak{w}_0]\alpha(\nu_0) = 1$, we get $E_\omega f_{\nu_0}^{(P)} = \psi$. This proves the lemma.

COROLLARY. *The mapping*

$$f \longrightarrow E_\omega f_{\nu_0}^{(P)}$$

from $C_c^\infty(G, \tau)$ to $L(\omega, \nu_0)$ is surjective.

Since $C_c^\infty(G, \tau)$ is dense in $\mathcal{C}(G, \tau)$ and $\dim L(\omega, \nu_0) \leqq \dim L < \infty$, this is an immediate consequence of the above lemma.

Fix a finite subset F of $\mathfrak{E}(K)$ and take $\tau = \tau_F$. Let us now use the notation of Lemma 9.2.

LEMMA 2. *Let π_F denote the representation of $\mathfrak{L}_F$ on $\mathfrak{H}_F$ defined by $\pi = \pi_{P,\omega,\nu_0}$. Then under the bijection*

$$T \longmapsto \psi_T$$

of End $\mathfrak{H}_F$ *onto $L(\omega)$, $\pi_F(\mathfrak{L}_F)$ corresponds to $L(\omega, \nu_0)$.*

Let $\alpha \in (\mathfrak{L}_F)^\vee$ and $T \in$ End $\mathfrak{H}_F$. Then

$$(E_\omega \alpha_{\nu_0}^{(P)}) * \psi_T = \alpha_{\nu_0}^{(P)} * \psi_T = \psi_{T\pi_F(\check{\alpha})}$$

from Lemma 9.2. Put $T = I$ where I is the identity mapping of $\mathfrak{H}_F$. Then we conclude from Lemma 9.1 that

$$E_\omega \alpha_{\nu_0}^{(P)} = d(\omega)\psi_{\pi_F(\check{\alpha})} \;.$$

Our assertion now follows from the corollary of Lemma 1.

38. The centralizer of $\pi_F(\mathfrak{L}_F)$

We keep to the notation of Lemma 37.2. Fix $s \in \mathfrak{w}(\omega)$ and choose a representative y in K for s. Recall that $\sigma \in \omega$ and U is the representation space of σ. Since $s\omega = \omega$, we can choose a unitary transformation L_y in U such that

$$\sigma(y^{-1}my) = L_y^{-1}\sigma(m)L_y \qquad (m \in M) \;.$$

We define a unitary operator S on $\mathfrak{H}$ as follows. If $h \in \mathfrak{H}$, then

$$(Sh)(k) = L_y h(y^{-1}k) \qquad (k \in K) \;.$$

It is clear that S commutes with $\pi(k)$ $(k \in K)$ and therefore maps $\mathfrak{H}_F$ into itself.

LEMMA 1. *Let* $T \in \mathrm{End}\ \mathfrak{H}_F$. *Then*

$$s\psi_T = \psi_{STS^{-1}}.$$

Put $T' = STS^{-1}$. We shall compute its kernel $\kappa_{T'}$ (Lemma 6.1). Let $h \in \mathfrak{H}$. Then

$$\begin{aligned}
(T'h)(k_2) &= L_y(TS^{-1}h)(y^{-1}k_2)\\
&= \int L_y\kappa_T(y^{-1}k_2: k_1)(S^{-1}h)(k_1^{-1})dk_1\\
&= \int L_y\kappa_T(y^{-1}k_2: k_1)L_y^{-1}h(yk_1^{-1})dk_1\\
&= \int L_y\kappa_T(y^{-1}k_2: k_1y)L_y^{-1}h(k_1^{-1})dk_1.
\end{aligned}$$

This implies that

$$\kappa_{T'}(k_2: k_1) = L_y\kappa_T(y^{-1}k_2: k_1y)L_y^{-1} \qquad (k_1, k_2 \in K)$$

and therefore

$$\begin{aligned}
\psi_{T'}(k_1: m: k_2) &= \mathrm{tr}\{L_y\kappa_T(y^{-1}k_2: k_1y)L_y^{-1}\sigma(m)\}\\
&= \mathrm{tr}\{\kappa_T(y^{-1}k_2: k_1y)\sigma(y^{-1}my)\}\\
&= \psi_T(k_1y: y^{-1}my: y^{-1}k_2) \qquad (m \in M).
\end{aligned}$$

Hence

$$\psi_{T'}(m) = \tau(y)\psi_T(y^{-1}my)\tau(y^{-1})$$

and this proves that $\psi_{T'} = s\psi_T$.

Recall that $L(\omega)$ is an algebra under convolution (§ 9).

COROLLARY. *Fix* $s \in \mathfrak{w}(\omega)$, $\nu \in \mathfrak{F}$ *and* $P_1, P_2 \in \mathscr{P}(A)$. *Then* ${}^0c_{P_2|P_1}(s: \nu)$ *defines an automorphism of the algebra* $L(\omega)$.

It follows from [2(j), Lem. 18.1] that

$${}^0c_{P_2|P_1}(s: \nu) = s^0c_{P_3|P_1}(1: \nu)$$

where $P_3 = P_2^{s^{-1}}$. Hence it would be enough to consider the case $s = 1$. If $\nu \in \mathfrak{F}'$, our assertion follows from Lemmas 9.1 and 18.1. The rest is obvious from Lemma 17.6.

We return to the notation of Section 37.

LEMMA 2. *Fix* $\nu_0 \in \mathfrak{F}$. *Then for every* $s \in \mathfrak{w}(\omega, \nu_0)$, *we can choose* $\gamma_s \in \mathrm{GL}(\mathfrak{H}_F)$ *such that:*

(1) ${}^0c_{P|P}(s: \nu_0)\psi_T = \psi_{\gamma_s T\gamma_s^{-1}}$ *for all* $T \in \mathrm{End}\ \mathfrak{H}_F$,

(2) $\gamma_s^*\gamma_s = 1$.

Here γ_s^* denotes the adjoint of γ_s.

Fix $s \in \mathfrak{w}(\omega, \nu_0) = \mathfrak{w}_0$. Then ${}^0 c_{P|P}(s : \nu_0)$ defines an automorphism of $L(\omega)$ which, by Lemma 9.1, corresponds to an automorphism of the simple algebra $\operatorname{End} \mathfrak{H}_F$. Our assertion now follows immediately if we observe that ${}^0 c_{P|P}(s : \nu_0)$ is unitary.

COROLLARY. *The mapping $s \mapsto \gamma_s$ defines a projective representation of* $\mathfrak{w}(\omega, \nu_0)$.

This is obvious.

Let Γ be the subspace of $\operatorname{End} \mathfrak{H}_F$ spanned by γ_s $(s \in \mathfrak{w}_0)$. Then Γ is a subalgebra of $\operatorname{End} \mathfrak{H}_F$ and $\Gamma^* = \Gamma$.

THEOREM 1. Γ *is exactly the centralizer of* $\pi_F(\mathfrak{L}_F)$ *in* $\operatorname{End} \mathfrak{H}_F$.

Let $\mathfrak{C}$ be the centralizer of Γ in $\operatorname{End} \mathfrak{H}_F$. Fix $T \in \operatorname{End} \mathfrak{H}_F$. Then $T \in \mathfrak{C}$ if and only if $\gamma_s T \gamma_s^{-1} = T$ for all $s \in \mathfrak{w}_0$. But in view of Lemmas 2 and 37.2, this is equivalent to the condition that $T \in \pi_F(\mathfrak{L}_F)$. Hence $\mathfrak{C} = \pi_F(\mathfrak{L}_F)$. But since $\Gamma = \Gamma^*$, we know that Γ is the centralizer of $\mathfrak{C}$ in $\operatorname{End} \mathfrak{H}_F$. This proves the theorem.

Let $\chi(s)$ $(s \in \mathfrak{w}_0)$ denote the restriction of ${}^0 c_{P|P}(s : \nu_0)$ on $L(\omega)$. We have seen in Section 37 that χ is a representation of $\mathfrak{w}_0$. Let $\mathfrak{w}_0(F)$ denote the kernel of χ. It is obvious that $\dim \Gamma \leq [\mathfrak{w}_0 : \mathfrak{w}_0(F)]$.

We recall that $\pi = \pi_{P,\omega,\nu_0}$ is a unitary representation of G on $\mathfrak{H}$. Let $\mathfrak{T}$ denote the algebra of all bounded operators on $\mathfrak{H}$ which commute with $\pi(x)$ for all $x \in G$. $\mathfrak{T}$ is called the commuting algebra of π and $\dim \mathfrak{T}$ the self intertwining number of π.

LEMMA 3. $\dim \mathfrak{T} \leq \max_F [\mathfrak{w}_0 : \mathfrak{w}_0(F)]$. *Moreover π is irreducible if and only if* $\mathfrak{w}_0 = \mathfrak{w}_0(F)$ *for all* F.

For any F, let $\mathfrak{T}_F$ denote the centralizer of $\pi_F(\mathfrak{L}_F)$ in $\operatorname{End} \mathfrak{H}_F$. If $T \in \mathfrak{T}$, then T commutes with $\pi(k)$ $(k \in K)$ and therefore $\mathfrak{H}_F$ is stable under T. Let T_F denote the restriction of T on $\mathfrak{H}_F$. Then $T_F \in \mathfrak{T}_F$. Let $\mathfrak{T}(F)$ denote the subspace of all $T \in \mathfrak{T}$ such that $T_F = 0$. Then

$$\dim \mathfrak{T}/\mathfrak{T}(F) \leq \dim \mathfrak{T}_F \leq [\mathfrak{w}_0 : \mathfrak{w}_0(F)]$$

from Theorem 1. Note that $\mathfrak{T}(F_2) \subset \mathfrak{T}(F_1)$ if $F_1 \subset F_2$. Moreover $\bigcup_F \mathfrak{H}_F$ is dense in $\mathfrak{H}$ and therefore

$$\bigcap_F \mathfrak{T}(F) = \{0\} \, .$$

This implies that

$$\dim \mathfrak{T} = \max_F \dim \mathfrak{T}/\mathfrak{T}(F) \leq \max_F [\mathfrak{w}_0 : \mathfrak{w}_0(F)] \, .$$

Now suppose π is irreducible. Then π_F is irreducible and therefore, by

Theorem 1, Γ contains only scalar operators. Hence we conclude from Lemma 2 that $\mathfrak{w}_0(F) = \mathfrak{w}_0$.

LEMMA 4. *Fix* $s \in \mathfrak{w}$ *such that* $s\nu_0 = \nu_0$. *Then*

$$^0c_{P_2|P_1}(s: \nu_0) = {}^0c_{P_2|P}(1: \nu_0)^0c_{P|P}(s: \nu_0)^0c_{P|P_1}(1: \nu_0)$$

for $P_1, P_2 \in \mathscr{P}(A)$.

This follows from the corollary of Lemma 17.2.

COROLLARY. $\mathfrak{w}_0(F)$ *is independent of* $P \in \mathscr{P}(A)$.

This is clear from Lemma 4 if we observe that

$$^0c_{Q|P}(1: \nu_0)^0c_{P|Q}(1: \nu_0) = 1$$

for $Q \in \mathscr{P}(A)$.

39. Closer study of a special case

We return to the notation of the beginning of Section 37.

LEMMA 1. *Suppose* $\dim A = 1$ *and* $\mu(\omega: \nu_0) = 0$. *Then* $\nu_0 = 0$, $[\mathfrak{w}] = 2$, $s\omega = \omega$ *and* $^0c_{P|P}(s: 0) = 1$ *on* $L(\omega)$ *for* $s \in \mathfrak{w}$.

Since $\mu(\omega: \nu) > 0$ for $\nu \in \mathfrak{F}'$, it is clear that prk $G = 0$ and $\nu_0 = 0$. Let α be the unique simple root of (P, A). Put $H = 2|\alpha|^{-2} \cdot H_\alpha$ and identify $\mathfrak{F}$ with $\mathbf{R}$ under the mapping $\nu \mapsto \nu(H)$. Define $a_t = \exp tH$ $(t \in \mathbf{R})$ and regard t as a coordinate on A. Fix $\psi \neq 0$ in $L(\omega)$. Since $\mu(\omega: 0) = 0$, we conclude from Lemma 17.1 that $c_{P|P}(1: \nu)\psi$ has a pole at $\nu = 0$. Let $k \geq 1$ be the order of this pole. Then we have the power-series expansion

$$\nu^k c_{P|P}(1: \nu)\psi = \psi_0 + \nu\psi_1 + \cdots$$

around $\nu = 0$. Here $\psi_0, \psi_1, \cdots$ are in $L(\omega)$ and $\psi_0 \neq 0$. Put $f(\nu) = E(P: \psi: \nu)$. If $[\mathfrak{w}] = 1$, we conclude that

$$\nu^k f_P(\nu) = \nu^k c_{P|P}(1: \nu)\psi \exp\{(-1)^{1/2}\nu t\} \qquad (\nu \neq 0) \,.$$

Making ν tend to zero, we deduce from [2(j), Lem. 10.7] that $\psi_0 = 0$. This contradiction proves that $[\mathfrak{w}] = 2$. Let $\mathfrak{w} = (1, s)$. Then we conclude again from Lemma 17.1 that

$$\nu^k c_{P|P}(s: \nu)\psi = \psi_0' + \nu\psi_1' + \cdots$$

where $\psi_0', \psi_1', \cdots$ are in $L(s\omega)$. Hence

$$\begin{aligned}
\nu^k f_P(\nu) &= \nu^k c_{P|P}(1: \nu)\psi \exp\{(-1)^{1/2}\nu t\} + \nu^k c_{P|P}(s: \nu)\psi \exp\{-(-1)^{1/2}\nu t\} \\
&\equiv (\psi_0 + \psi_0') + \{(\psi_1 + \psi_1') + (-1)^{1/2}(\psi_0 - \psi_0')t\}\nu \bmod \nu^2 \,.
\end{aligned}$$

Since $k \geq 1$, we conclude from [2(j), Lem. 10.7] that $\psi_0 + \psi_0' = 0$ and this implies that $s\omega = \omega$. We claim that $k = 1$. For otherwise we would have

$$\psi_1 + \psi_1' + (-1)^{1/2}(\psi_0 - \psi_0')t = 0 .$$

This being true for all t, we could conclude that $\psi_0 - \psi_0' = 0$. This would imply that $\psi_0 = \psi_0' = 0$, giving a contradiction. Hence $k = 1$ and

$$\nu f_P(\nu) \equiv 2(-1)^{1/2}\nu t\psi_0 + \nu(\psi_1 + \psi_1') \bmod \nu^2 .$$

Therefore

$$f_P(0) = 2(-1)^{1/2}t\psi_0 + \psi_1 + \psi_1' .$$

Since $\psi_0 \neq 0$, this shows that $f_P(0) \neq 0$. This proves that the mapping $\psi \mapsto E(P\colon \psi\colon 0)$ is injective on $L(\omega)$.

Let C denote the restriction of $^0c_{P|P}(s\colon 0)$ on $L(\omega)$. Since $s^2 = 1$, it is clear that $C^2 = 1$. We claim that $C = 1$. For otherwise we could choose $\psi \neq 0$ in $L(\omega)$ such that $C\psi = -\psi$. Then

$$E(P\colon \psi\colon 0) = E(P\colon C\psi\colon 0) = -E(P\colon \psi\colon 0)$$

from Lemma 17.2. Hence $E(P\colon \psi\colon 0) = 0$, contradicting the result obtained above. This completes the proof of the lemma.

COROLLARY. *Under the above conditions $\mu(\omega\colon \nu)$ has a zero of order 2 at $\nu = 0$.*

This is an immediate consequence of Lemma 17.1 and the fact that $k = 1$.

40. A bound for the intertwining number

Let us now return to the general case of Section 37. Let Σ denote the set of all reduced roots of $(\mathfrak{g}, \mathfrak{a})$ and Σ_∞ the subset of those $\alpha \in \Sigma$ for which $\mu_\alpha(\omega\colon \nu_0) = 0$.

LEMMA 1. *Σ_∞ is stable under $\mathfrak{w}_0 = \mathfrak{w}(\omega, \nu_0)$.*

Let $\alpha \in \Sigma$ and $\nu \in \mathfrak{F}$. Then (see §13) $\mu_\alpha(\omega\colon \nu)$ depends only on $\langle \nu, \alpha \rangle$ and $\mu_\alpha(\omega\colon \nu) > 0$ unless $\langle \nu, \alpha \rangle = 0$. Now fix $\alpha \in \Sigma_\infty$ and $s \in \mathfrak{w}_0$. Then we have to verify that $s\alpha \in \Sigma_\infty$. Choose $P \in \mathscr{P}(A)$ and let $\alpha_1, \alpha_2, \cdots, \alpha_r$ be all the distinct elements in $\Sigma(P)$. We may assume that $\alpha = \alpha_1$. Let j be the unique index such that $s\alpha = \pm\alpha_j$. We have to show that $\alpha_j \in \Sigma_\infty$.

Fix $\nu \in \mathfrak{F}$ such that

$$\langle \alpha, \nu \rangle = 0 , \quad \langle \alpha_i, \nu \rangle \neq 0 \qquad (2 \leq i \leq r) .$$

Then

$$\mu_\alpha(\omega\colon \nu) = \mu_\alpha(\omega\colon 0) = \mu_\alpha(\omega\colon \nu_0) = 0 .$$

Hence

$$\mu(\omega\colon \nu) = \prod_{1 \leq i \leq r} \mu_{\alpha_i}(\omega\colon \nu) = 0 .$$

But $s\omega = \omega$ and so we conclude from the corollary of Lemma 17.1 that

$$0 = \mu(\omega: \nu) = \mu(s\omega: s\nu) = \mu(\omega: s\nu) \,.$$

This implies that

$$\mu_{\alpha_i}(\omega: s\nu) = 0$$

for some i. But it is clear that

$$\langle s\nu, \alpha_i \rangle \neq 0$$

unless $i = j$. Hence

$$0 = \mu_{\alpha_j}(\omega: s\nu) = \mu_{\alpha_j}(\omega: 0) = \mu_{\alpha_j}(\omega: \nu_0) \,.$$

This proves that $\alpha_j \in \Sigma_\infty$.

For any $\alpha \in \Sigma$, let s_α denote the reflexion (in $\mathfrak{a}$) in the hyperplane $\alpha = 0$ so that

$$s_\alpha H = H - 2\alpha(H)\,|\alpha|^{-2} \cdot H_\alpha \qquad\qquad (H \in \mathfrak{a})\,.$$

LEMMA 2. *Suppose $\alpha \in \Sigma_\infty$. Then $s_\alpha \in \mathfrak{w}_0$.*

Define M_α, A_α as in Lemma 2.3 and apply Lemma 39.1 to (M_α, A_α) in place of (G, A). Since $\mu_\alpha(\omega: \nu_0) = 0$, we conclude that $\langle \nu_0, \alpha \rangle = 0$, $s_\alpha \in \mathfrak{w}$ and $s_\alpha \omega = \omega$.

Let $\mathfrak{w}_\infty$ be the subgroup of $\mathfrak{w}_0$ generated by s_α for all $\alpha \in \Sigma_\infty$. Then it follows from Lemma 1 that $\mathfrak{w}_\infty$ is a normal subgroup of $\mathfrak{w}_0$. Define $\mathfrak{w}_0(F)$ as in Section 38.

LEMMA 3. $\mathfrak{w}_\infty \subset \mathfrak{w}_0(F)$.

Fix $a \in \Sigma_\infty$ and choose $P \in \mathcal{P}(A)$ such that α is a simple root of (P, A). In view of the corollary of Lemma 38.4, it is sufficient to verify that $^0c_{P|P}(s_\alpha: \nu_0) = 1$ on $L(\omega)$. Fix $\psi \in L(\omega)$ and put $^*P = M_\alpha \cap P$. Since $\langle \nu_0, \alpha \rangle = 0$, we conclude from Lemmas 17.4 and 39.1 that

$$^0c_{P|P}(s_\alpha: \nu_0)\psi = {}^0c_{*P|*P}(s_\alpha: 0)\psi = \psi \,.$$

This proves the lemma.

COROLLARY. *Fix $P \in \mathcal{P}(A)$. Then the self intertwining number of π_{P,ω,ν_0} cannot exceed $[\mathfrak{w}_0: \mathfrak{w}_\infty]$.*

This is obvious from Lemma 38.3. It follows from [3] and [4] that when G is linear this intertwining numbers is equal to $[\mathfrak{w}_0: \mathfrak{w}_\infty]$.

41. Irreducibility of the fundamental series

As an application of the above theory, we shall prove a result which can be regarded as a generalization of a theorem of Wallach [10, Thm. 4.1] and Želobenko [11] on complex groups.

Let B be a θ-stable Cartan subgroup of G which is fundamental. Fix

$P \in \mathscr{P}(B_R)$. Then $P = MB_R N$.

THEOREM 1. *Fix $\omega \in \mathscr{E}_2(M)$ and $\nu \in \mathfrak{F} = \mathfrak{b}_R^*$. Then $\pi_{P,\omega,\nu}$ is irreducible.*

Put $A = B_R$. It follows from Lemma 38.3 and the corollary of Lemma 38.4 that the statement of the theorem is independent of $P \in \mathscr{P}(A)$. Hence we may assume that $\langle \nu, \alpha \rangle \geq 0$ for all $\alpha \in \Sigma(P)$. Let Φ denote the set of simple roots α of (P, A) such that $\langle \nu, \alpha \rangle = 0$. Put $(P', A') = (P, A)_\Phi$. Then $P' = M'A'N'$. Let

$$(*P, *A) = (M' \cap P, M' \cap A) , \quad *\mathfrak{w} = \mathfrak{w}(M'/*A) .$$

Then $*\mathfrak{w}$ may be regarded as the subgroup consisting of those elements of $\mathfrak{w}$ which leave $\mathfrak{a}'$ fixed. Let $\mathfrak{w}(\nu)$ be the set of all $s \in \mathfrak{w}$ such that $s\nu = \nu$. It is easy to verify that $\mathfrak{w}(\nu) = *\mathfrak{w}$ (see the proof of [2(j), Lem. 22.3]). Hence

$$\mathfrak{w}_0 = \mathfrak{w}(\omega, \nu) \subset *\mathfrak{w}$$

and therefore

$$^0c_{P|P}(s : \nu) = {}^0c_{*P|*P}(s : 0) \qquad\qquad (s \in \mathfrak{w}_0)$$

from Lemma 17.4. So it would be sufficient to verify that $^0c_{*P|*P}(s : 0) = 1$ on $L(\omega)$ for $s \in \mathfrak{w}_0$. Now $M' \cap B$ is clearly a fundamental Cartan subgroup of M' and it remains to show that the representation π' of M' corresponding to $(*P, \omega, 0)$ (obtained by inducing from $*P$ to M') is irreducible. Thus we are reduced to the case $\nu = 0$.

Fix an element $b^* \in B_I^{*'}$ which corresponds to ω under Theorem 23.1. Put $\lambda = \lambda(b^*)$ so that $\lambda \in (-1)^{1/2}\mathfrak{b}_I^*$. Let $\mathfrak{z}_\lambda$ denote the centralizer of H_λ in $\mathfrak{g}$. Since $b^* \in B_I^{*'}$, it is clear that $(\mathfrak{z}_\lambda, \mathfrak{b})$ has no imaginary roots. On the other hand $\mathfrak{b}$ is fundamental in $\mathfrak{g}$ and therefore $(\mathfrak{g}, \mathfrak{b})$ has no real roots. This shows that every root of $(\mathfrak{z}_\lambda, \mathfrak{b})$ is complex.

Put $W = W(\mathfrak{g}/\mathfrak{b})$ and $W(\lambda) = W(\mathfrak{z}_\lambda/\mathfrak{b})$. Then $W(\lambda)$ is the stabilizer of λ in W. Let W_0 be the subgroup of all elements in W which commute with the Cartan involution θ. Then

$$\mathfrak{w} = W_0/W_1$$

where $W_1 = W(\mathfrak{m}_1/\mathfrak{b})$ $(\mathfrak{m}_1 = \mathfrak{m} + \mathfrak{a})$. Note that $W(\lambda) \cap W_1 = \{1\}$ since $(\mathfrak{z}_\lambda, \mathfrak{b})$ has no imaginary root. Put $W_0(\lambda) = W_0 \cap W(\lambda)$ and let $\mathfrak{w}(\lambda)$ be the image of $W_0(\lambda)$ in $\mathfrak{w}$. Since $W_0(\lambda) \cap W_1 = \{1\}$, the projection of W_0 on $\mathfrak{w}$ defines an isomorphism of $W_0(\lambda)$ on $\mathfrak{w}(\lambda)$.

LEMMA 2. $\mathfrak{w}(\omega) \subset \mathfrak{w}(\lambda)$.

Fix $s \in \mathfrak{w}(\omega)$ and choose $\sigma \in W_0$ such that s is the image of σ in $\mathfrak{w} = W_0/W_1$. Since $s\omega = \omega$, it is clear that $\sigma\lambda$ is conjugate to λ under W_1.

Hence we can choose $t \in W_1$ such that $t\sigma \in W_0 \cap W(\lambda) = W_0(\lambda)$. This proves that $s \in \mathfrak{w}(\lambda)$.

LEMMA 3. *Let α be a reduced root of $(\mathfrak{z}_\lambda, \mathfrak{a})$. Then the reflexion s_α lies in $\mathfrak{w}(\lambda)$. Moreover $\mathfrak{w}(\lambda)$ is generated by all such elements s_α.*

Since $(\mathfrak{z}_\lambda, \mathfrak{b})$ has no imaginary roots, $\mathfrak{a} = \mathfrak{b}_R$ is maximal abelian in $\mathfrak{z}_\lambda \cap \mathfrak{p}$. This implies the lemma.

Let α_0 be a reduced root of $(\mathfrak{z}_\lambda, \mathfrak{a})$. Then there exists a root β of $(\mathfrak{z}_\lambda, \mathfrak{b})$ such that $\beta = \alpha_0$ on $\mathfrak{a}$. Let α be the reduced root of $(\mathfrak{g}, \mathfrak{a})$ corresponding to α_0. Put

$$(*P, *A) = (M_\alpha \cap P, M_\alpha \cap A)$$

where M_α is defined as in Lemma 2.3. Since B_I is a Cartan subgroup of K, it is clear that

$$*B = B \cap M_\alpha = *A \cdot B_I$$

is a fundamental Cartan subgroup of M_α. Therefore by Theorem 24.1,

$$\mu_\alpha(\omega: 0) = c_\alpha \left| \prod_\gamma \lambda(H_\gamma) \right|$$

where c_α is a positive number and γ runs over all positive complex roots of $(\mathfrak{m}_\alpha, *\mathfrak{b})$. On the other hand there exists a number $r \geq 1$ such that $\beta = \alpha_0 = r\alpha$ on $\mathfrak{a}$. Therefore β is a root of $(\mathfrak{m}_\alpha, *\mathfrak{b})$. But since β is a root of $(\mathfrak{z}_\lambda, \mathfrak{b})$, it is complex and $\lambda(H_\beta) = 0$. This shows that $\mu_\alpha(\omega: 0) = 0$ and therefore $s_\alpha \in \mathfrak{w}_\infty$. Since $s_\alpha = s_{\alpha_0}$, we conclude from Lemma 3 that

$$\mathfrak{w}(\lambda) \subset \mathfrak{w}_\infty \subset \mathfrak{w}(\omega) \, .$$

Combining this with Lemma 2, we get

$$\mathfrak{w}(\lambda) = \mathfrak{w}_\infty = \mathfrak{w}(\omega)$$

and therefore $\pi_{P,\omega,0}$ is irreducible from the corollary of Lemma 40.3.

42. Appendix

LEMMA 1.

$$\int_{-\infty}^{\infty} \frac{\exp \alpha t + \exp(-\alpha t)}{\exp t - \exp(-t)} \sin \lambda t \, dt = \frac{\pi \sinh \pi \lambda}{\cosh \pi \lambda + \cos \pi \alpha}$$
$$for \ \alpha, \lambda \in \mathbf{R} \quad and \quad |\alpha| < 1 \, .$$

Fix α. Then

$$\left| \frac{\exp \alpha t + \exp(-\alpha t)}{\exp t - \exp(-t)} \right| \leq 4 \exp\{(|\alpha| - 1)|t|\}$$

for $|t| \geq 1$. Hence the integral converges and defines a continuous function of λ. Hence it is enough to consider the case $\lambda > 0$.

Fix $\lambda > 0$ and put

$$f(z) = \frac{\exp\{(\alpha + i\lambda)z\}}{\exp z - \exp(-z)}$$

where z is a complex variable. Then f has simple poles at $z = n\pi i$ $(n \in \mathbf{Z})$ and

$$\operatorname{Res}_{z=n\pi i} f(z) = \frac{(-1)^n}{2} \exp\{(\alpha + i\lambda)n\pi i\} = \frac{1}{2} \exp\{(\alpha - 1)n\pi i - n\pi\lambda\} \ .$$

Now suppose $z = x + iy$ $(x, y \in \mathbf{R})$ and $|x| \geqq 1$, $y \geqq 0$. Then

$$|f(z)| \leqq 2 \exp\{(\alpha x - \lambda y) - |x|\} \leqq 2 \exp\{-(1 - |\alpha|)|x| - \lambda y\} \ .$$

Let N be a large positive integer, T a large positive real number and ε a small positive real number. Consider the rectangle in the complex plane with the four vertices $-T$, T, $T + (N + \frac{1}{2})\pi i$, $-T + (N + \frac{1}{2})\pi i$. Introduce a small semicircular dent of radius ε around $z = 0$ so as to exclude the origin. Integrate $f(z)$ along this contour and let $\varepsilon \to 0$, $T \to \infty$, $N \to \infty$. Then we get

$$\text{p.v.} \int_{-\infty}^{\infty} \frac{\exp\{(\alpha + i\lambda)t\}}{\exp t - \exp(-t)} \, dt = \frac{1}{2}\pi i + \pi i \sum_{n \geqq 1} \exp\{(\alpha - 1)n\pi i - n\pi\lambda\}$$

$$= \frac{\pi i}{2} \frac{\sinh \pi\lambda - i \sin \pi\alpha}{\cosh \pi\lambda + \cos \pi\alpha} \ .$$

Therefore

$$\int_{-\infty}^{\infty} \frac{\exp \alpha t}{\exp t - \exp(-t)} \sin \lambda t \, dt = \frac{\pi}{2} \frac{\sinh \pi\lambda}{\cosh \pi\lambda + \cos \pi\alpha}$$

and this implies our assertion.

As usual let $\mathcal{C}(\mathbf{R})$ denote the Schwartz space of $\mathbf{R}$.

COROLLARY. *Let $f \in \mathcal{C}(\mathbf{R})$. Then*

$$\int_{-\infty}^{\infty} \frac{\exp \alpha t + \exp(-\alpha t)}{\exp t - \exp(-t)} dt \int_{-\infty}^{\infty} \lambda f(\lambda) \sin \lambda t \, d\lambda = \int_{-\infty}^{\infty} \frac{\pi\lambda \sinh \pi\lambda}{\cosh \pi\lambda + \cos \pi\alpha} f(\lambda) d\lambda$$

for $-1 \leqq \alpha \leqq 1$.

Put

$$g(t) = \int_{-\infty}^{\infty} \lambda f(\lambda) \sin \lambda t \, d\lambda \qquad\qquad (t \in \mathbf{R}) \ .$$

Then $g \in \mathcal{C}(\mathbf{R})$ and $g(0) = 0$. Therefore

$$\int_{-\infty}^{\infty} |\coth t| \, |g(t)| \, dt < \infty \ .$$

This shows that the left side in the corollary is a continuous function of α. On the other hand $\cos \pi\alpha \geqq -1$ and

$$\frac{\lambda \sinh \pi\lambda}{\cosh \pi\lambda - 1} = \lambda \coth \frac{\pi\lambda}{2}.$$

Therefore

$$0 \leq \frac{\lambda \sinh \pi\lambda}{\cosh \pi\lambda + \cos \pi\alpha} \leq \lambda \coth \frac{\pi\lambda}{2} \leq c(1 + |\lambda|) \qquad (\lambda \in \mathbf{R})$$

where c is a suitable constant. Therefore the right hand side of the corollary is also a continuous function of α. Hence it is enough to verify the equality when $|\alpha| < 1$.

But if $|\alpha| < 1$, it is clear that the double integral on the left is absolutely convergent. Therefore the required result follows from Lemma 1 by Fubini's theorem.

THE INSTITUTE FOR ADVANCED STUDY, PRINCETON, N.J.

REFERENCES

[1] J. ARTHUR, Intertwining integrals for cuspidal parabolic subgroups, preprint.
[2] HARISH-CHANDRA, (a) Representations of semisimple Lie groups, III, Trans. A.M.S. **76** (1954), 234-253.
    (b) Spherical functions on a semisimple Lie group, II, Amer. J. Math. **80** (1958), 553-613.
    (c) Invariant eigendistributions on a semisimple Lie group, Trans. A.M.S. **119** (1965), 457-508.
    (d) Two theorems on semisimple Lie groups, Ann. of Math. **83** (1966), 74-128.
    (e) Discrete series for semisimple Lie groups, I, Acta Math. **113** (1965), 241-318.
    (f) Discrete series for semisimple Lie groups, II, Acta Math. **116** (1966), 1-111.
    (g) Harmonic analysis on semisimple Lie groups, Bull. A.M.S. **78** (1970), 529-551.
    (h) On the theory of the Eisenstein integral, in *Lecture Notes in Math.*, Vol. 266, pp. 123-149, Springer-Verlag, 1971.
    (i) Harmonic analysis on real reductive groups, I, J. Funct. Anal. **19** (1975), 104-204.
    (j) Harmonic analysis on real reductive groups II, to appear in Inv. Math.
[3] A. KNAPP, Commutativity of intertwining operators II, Bull. A.M.S. **82** (1976), 271-273.
[4] A. KNAPP and E. M. STEIN, (a) Intertwining operators for semi-simple Lie groups, Ann. of Math. **93** (1971), 489-578.
    (b) Singular integrals and the principal series III, IV, Proc. Nat. Acad. Sci. USA **71** (1974), 4622-4624; **72** (1975), 2459-2461.
[5] R. A. KUNZE and E. M. STEIN, Uniformly bounded representations I, II, III, Amer. J. Math. **82** (1960), 1-62; **83** (1961), 723-786; **89** (1967), 385-442.
[6] F. BRUHAT, Sur les représentations induites des groupes de Lie, Bull. Soc. Math. France **84** (1956), 97-205.
[7] S. G. GINDIKIN and F. I. KARPELEVIČ, Plancherel measure of Riemannian symmetric spaces of nonpositive curvature, Sov. Math. **3** (1962), 962-965.
[8] G. SHIFFMAN, Intégrales d'entrelacement et fonctions de Whittaker, Bull. Soc. Math. France **99** (1971), 3-72.
[9] P. C. TROMBI, On Harish-Chandra's theory of the Eisenstein integral for real semisimple Lie groups, preprint.
[10] N. WALLACH, Cyclic vectors and irreducibility for principal series representations, Trans. A.M.S. **158** (1971), 107-112.
[11] D. P. ŽELOBENKO, The analysis of irreducibility in the class of elementary representations of a complex semi-simple Lie group, Math.-USSR-Izvestija **2** (1968), 105-128.

(Received February 6, 1976)

The characters

of reductive p-adic groups

Harish-Chandra

The Institute for Advanced Study

The motivation for the results which I am going to discuss comes from the Lefschetz principle, which says that whatever is true for real groups should also be true for p-adic groups. So let me begin by describing some known results in the real case.

Let G be a compact, connected, real semisimple Lie group and $\mathfrak{g}$ its Lie algebra. Fix a Cartan subalgebra $\mathfrak{h}$ and define Fourier transforms on $\mathfrak{g}$ and $\mathfrak{h}$ as follows:

$$\hat{f}(Y) = \int_{\mathfrak{g}} e^{iB(Y, X)} f(X) \, dX \qquad (Y \in \mathfrak{g}, \quad f \in \mathscr{C}(\mathfrak{g})),$$

$$\hat{g}(H') = \int_{\mathfrak{h}} e^{iB(H', H)} g(H) \, dH \qquad (H' \in \mathfrak{h}, \quad g \in \mathscr{C}(\mathfrak{h})).$$

Here B is the Killing form of $\mathfrak{g}$ and $\mathscr{C}(\mathfrak{g})$ and $\mathscr{C}(\mathfrak{h})$ the corresponding Schwartz spaces. It is convenient to assume that the Euclidean measures dX and dH on $\mathfrak{g}$ and $\mathfrak{h}$ respectively are self-dual. This means that

$$f(0) = \int_{\mathfrak{g}} \hat{f}(X) \, dX, \qquad g(0) = \int_{\mathfrak{h}} \hat{g}(H) \, dH$$

175

for all $f \in \mathscr{C}(\mathfrak{g})$ and $g \in \mathscr{C}(\mathfrak{h})$. Put

$$\phi_f(H) = \pi(H) \int_G f(x.H)dx \qquad (H \in \mathfrak{h},\, f \in \mathscr{C}(\mathfrak{g})).$$

Here $x.H = Ad(x)H$, dx is the normalized Haar measure on G and

$$\pi = \prod_{\alpha > 0} \alpha$$

where α runs over all positive roots of $(\mathfrak{g},\, \mathfrak{h})$ (under some order). Then $f \mapsto \phi_f$ is a continuous mapping of $\mathscr{C}(\mathfrak{g})$ into $\mathscr{C}(\mathfrak{h})$ and the following theorem establishes a relation between this mapping and the Fourier transform [3, §5].

Theorem 1   *Let $r = \frac{1}{2}\dim \mathfrak{g}/\mathfrak{h}$. Then*

$$\phi_f = i^r(\phi_f)^{\wedge} \qquad \text{for } f \in \mathscr{C}(\mathfrak{g}).$$

Corollary

$$\int_G e^{i\,B(x.H,\,H')}\,dx = \frac{c \sum_{s \in W} \varepsilon(s) e^{i\,B(sH,\,H')}}{\pi(H)\pi(H')} \qquad \text{for } H,\,H' \in \mathfrak{h}.$$

Here W is the Weyl group of $(\mathfrak{g},\, \mathfrak{h})$ and c a constant given by

$$c = i^r[W]^{-1}\langle \pi,\,\pi \rangle$$

in the notation of [3, p. 90]. ($[W]$ denotes the order of W.) The striking resemblance between the above corollary and Weyl's character formula is obvious.

Now let us drop the assumption that G is compact. Let A be the Cartan subgroup of G corresponding to $\mathfrak{h}$. Fix an invariant measure dx^* on G/A and let $\mathfrak{h}'$ be the set of all points $H \in \mathfrak{h}$ where $\pi(H) \neq 0$. Put

$$\phi_f(H) = \pi(H) \int_{G/A} f(x.H)\,dx^* \qquad (H \in \mathfrak{h}')$$

for $f \in \mathscr{C}(\mathfrak{g})$. This integral is well defined and ϕ_f is a C^{α} function on $\mathfrak{h}'$. Fix $H_0 \in \mathfrak{h}'$ and consider the mapping

$$T : f \to \phi_f(H_0) \qquad (f \in \mathscr{C}(\mathfrak{g})).$$

Then T is a tempered and G-invariant distribution on $\mathfrak{g}$.

Let $\mathfrak{g}_c$ be the complexification of $\mathfrak{g}$ and $S(\mathfrak{g}_c)$ the symmetric algebra over $\mathfrak{g}_c$. For $X \in \mathfrak{g}$, define a translation-invariant differential operator $\partial(X)$ on $\mathfrak{g}$ by

$$(\partial(X)f)(Y) = \left| \frac{d}{dt} f(Y + tX) \right|_{t=0} \qquad (Y \in \mathfrak{g})$$

for $f \in C^\infty(\mathfrak{g})$. Then ∂ can be extended (uniquely) to an isomorphism of $S(\mathfrak{g}_c)$ onto the algebra of all translation-invariant differential operators on $\mathfrak{g}$. (These are also known as differential operators with constant coefficients.) We may identify $\mathfrak{g}$ with its dual by means of the Killing form. Then $S(\mathfrak{g}_c)$ becomes the algebra of all polynomial functions on $\mathfrak{g}$.

Consider the orbit $H_0{}^G = G.H_0$ and let μ denote the invariant measure on this orbit so that

$$\mu(f) = \int_{G/A} f(x.H_0)\, dx^* \qquad (f \in \mathscr{C}(\mathfrak{g})).$$

Then μ is a tempered measure on $\mathfrak{g}$ and

$$T(f) = \pi(H_0)\mu(\hat{f}) = \pi(H_0)\hat{\mu}(f)$$

where $\hat{\mu}$ is the Fourier transform of μ. This shows that $T = \pi(H_0)\hat{\mu}$.

Let $I(\mathfrak{g}_c)$ be the algebra of all G-invariants in $S(\mathfrak{g}_c)$. It is obvious that

$$p\mu = p(H_0)\mu \qquad (p \in I(\mathfrak{g}_c)).$$

Taking Fourier transforms, we get

$$p(H_0)\hat{\mu} = (p\mu)^\wedge = \partial(p^*)\hat{\mu}$$

where p^* is the polynomial function $X \mapsto p(-iX)$ $(X \in \mathfrak{g})$. Therefore T is an invariant distribution on $\mathfrak{g}$ which satisfies the differential equations

$$\partial(p)T = p(iH_0)T \qquad (p \in I(\mathfrak{g}_c)).$$

But then by a general theorem [4, p. 11], T is a locally summable function on $\mathfrak{g}$ which is analytic on the regular set $\mathfrak{g}'$.

Now assume for a moment that A is compact. Let W denote the Weyl group of the pair $(\mathfrak{g}_c, \mathfrak{h}_c)$ and W_G the subgroup of those elements which have a representative in G. Then on $\mathfrak{h}'$, T is given by a formula of the form [6, §16]

$$T(H) = \pi(H)^{-1} \sum_{s \in W} \varepsilon(s) c_s\, e^{i\, B(sH_0,\, H)} \qquad (H \in \mathfrak{h}').$$

Here c_s are constants such that $c_{ts} = c_s$ for $t \in W_G$. The analogy with the compact case is clear.

But what does this have to do with the theory of characters? Let $\mathfrak{g}_0$ be the set of all $X \in \mathfrak{g}$ such that $|\mathrm{Im}\,\lambda| < \pi = 3.14\ldots$ for every eigenvalue λ of $\mathrm{ad}\,X$. Then $\mathfrak{g}_0$ is an open G-invariant neighborhood of zero in $\mathfrak{g}$. Put $G_0 = \exp \mathfrak{g}_0$. Then G_0 is an open subset of G and the exponential mapping defines an analytic diffeomorphism of $\mathfrak{g}_0$ onto G_0. Put

$$\xi(X) = \left|\det\{(e^{\mathrm{ad}X/2} - e^{-\mathrm{ad}X/2})/\mathrm{ad}\,X\}\right|^{1/2} \qquad (X \in \mathfrak{g}_0).$$

Then under the exponential mapping the Haar measure dx on G_0 is related to the Euclidean measure dX on $\mathfrak{g}_0$ by the formula

$$dx = \xi(X)^2 \, dX.$$

For $\phi \in C^\infty(\mathfrak{g}_0)$, define $f_\phi \in C^\infty(G_0)$ by

$$f_\phi(\exp X) = \xi(X)^{-1}\phi(X) \qquad (X \in \mathfrak{g}_0).$$

Then

$$\int \phi_1 \phi_2 \, dX = \int f_{\phi_1} \cdot f_{\phi_2} \, dx$$

for $\phi_1 \in C^\infty(\mathfrak{g}_0)$ and $\phi_2 \in C_c^\infty(\mathfrak{g}_0)$. Given a distribution T on G_0, define a distribution τ_T on $\mathfrak{g}_0$ by

$$\tau_T(\phi) = T(f_\phi) \qquad (\phi \in C_c^\infty(\mathfrak{g}_0)).$$

Now G operates on G_0 by conjugation and it is obvious that if T is G-invariant, so is τ_T.

Let $\mathfrak{Z}$ be the algebra of all differential operators on G which commute with both left and right translations of G. Suppose Θ is an invariant distribution on G_0 and $z \in \mathfrak{Z}$. Then what is the relation between τ_Θ and $\tau_{z\Theta}$? The answer is quite simple [5, p. 477]. There exists an algebra isomorphism

$$z \mapsto p_z$$

from $\mathfrak{Z}$ to $I(\mathfrak{g}_c)$ such that

$$\tau_{z\Theta} = \partial(p_z)\tau_\Theta.$$

Now suppose Θ is an eigendistribution of $\mathfrak{Z}$. Then

$$z\Theta = \chi(z)\Theta \qquad (z \in \mathfrak{Z})$$

where χ is a homomorphism of $\mathfrak{Z}$ into $\mathbf{C}$. Then if we write $\theta = \tau_\Theta$ and $\chi(p_z) = \chi(z)$, we get

$$\partial(p)\theta = \chi(p)\theta \qquad (p \in I(\mathfrak{g}_c)).$$

This shows that θ is an invariant eigendistribution of $\partial(I(\mathfrak{g}_c))$ on $\mathfrak{g}_0$.

The characters of G are invariant eigendistributions of $\mathfrak{Z}$. Hence, in order to find them, one is led by the above procedure to the construction of invariant eigendistributions of $\partial(I(\mathfrak{g}_c))$ on $\mathfrak{g}_0$. But, as we have seen, the Fourier transform of the invariant measure on an orbit gives a solution to this problem. This suggests that there is an intimate connection between the multiplicative Fourier analysis on the group G and the additive Fourier analysis on its Lie algebra $\mathfrak{g}$. In fact, the characters of the discrete series are actually constructed precisely by the method outlined above. Moreover, quite similar results have been discovered by Kirillov for nilpotent groups [10].

We would now like to apply the same philosophy in the p-adic case. So it is natural to expect that the characters are closely related to the Fourier transforms of the invariant measures on orbits.

Let Ω be a p-adic field. I shall have to assume that Ω has characteristic zero. We use the terminology of [7]. Let $\mathbf{G}$ be a connected reductive Ω-group and G the subgroup of all Ω-rational points in $\mathbf{G}$. Then G, with its usual topology, is a locally compact, totally disconnected, unimodular group. Let $\mathfrak{g}$ be the Lie algebra of G. Then $\mathfrak{g}$ is a vector space over Ω of finite dimension and G operates on $\mathfrak{g}$ by means of the adjoint representation. If ω is any subset of $\mathfrak{g}$, we denote by $J(\omega)$ the space of all G-invariant distributions T on $\mathfrak{g}$ such that Supp $T \subset \mathrm{Cl}\,\omega^G$. (Here ω^G is the orbit of ω under G and Cl denotes closure.) Let L be a lattice (i.e. an open, compact, additive subgroup) in $\mathfrak{g}$. Then $C_c(\mathfrak{g}/L) \subset C_c^\infty(\mathfrak{g})$. For any distribution T on $\mathfrak{g}$, let $j_L T$ denote the restriction of T on $C_c(\mathfrak{g}/L)$. The following remarkable theorem of Howe [8, §4] plays a very important role in harmonic analysis.

Theorem 2 (Howe) *Suppose ω is a compact subset in $\mathfrak{g}$ and L a lattice in $\mathfrak{g}$. Then* $\dim j_L J(\omega) < \infty$.

In many proofs this theorem makes up for the absence of differential operators in the p-adic case.

Fix a nontrivial character χ of the additive group of Ω. Also fix a non-degenerate, symmetric, G-invariant, bilinear form B on $\mathfrak{g}$ (with values in Ω). For $f \in C_c^\infty(\mathfrak{g})$, define

$$\hat{f}(Y) = \int_{\mathfrak{g}} \chi(B(Y, X)) f(X)\, dX \qquad (Y \in \mathfrak{g}).$$

Here dX is a Haar measure on the additive group of $\mathfrak{g}$. Then $f \mapsto \hat{f}$ is a bijective mapping of $C_c^\infty(\mathfrak{g})$ onto itself. If T is a distribution on $\mathfrak{g}$, its Fourier transform $\hat{T}$ is given by $\hat{T}(f) = T(\hat{f})$ $(f \in C_c^\infty(\mathfrak{g}))$.

Let l be the (absolute) rank of G. If t is an indeterminate, let $\eta(X)$ denote the coefficient of t^l in $\det(t\text{-ad }X)$ $(X \in \mathfrak{g})$. Then η is a polynomial function on $\mathfrak{g}$ which is not identically zero. Let $\mathfrak{g}'$ be the set of all $X \in \mathfrak{g}$ such that $\eta(X) \neq 0$.

Theorem 3 *Let ω be a compact subset of $\mathfrak{g}$ and T a distribution in $J(\omega)$. Then $\hat{T}$ is a locally summable function on $\mathfrak{g}$ which is locally constant on $\mathfrak{g}'$. Moreover $|\eta|_\mathfrak{p}^{1/2} T$ is locally bounded on $\mathfrak{g}$.*

The analogy with the real case [4] is clear.

Fix $X_0 \in \mathfrak{g}$ and consider the orbit $\mathfrak{o} = X_0{}^G$. Let G_{X_0} denote the stabilizer of X_0 in G. Then G_{X_0} is unimodular and therefore G/G_{X_0} has an invariant measure dx^*. By a theorem of Deligne and Rao [11], there exists a measure $\mu_\mathfrak{o}$ on $\mathfrak{g}$ given by

$$\mu_\mathfrak{o}(f) = \int_{G/G_{X_0}} f(x.X_0)\, dx^* \qquad (f \in C_c(\mathfrak{g})).$$

Since $\mu_\mathfrak{o} \in J(X_0)$, we conclude from Theorem 3 that $\hat{\mu}_\mathfrak{o}$ is a function.

Let π be an irreducible representation of G on a Hilbert space $\mathfrak{H}$. π is said to be admissible if, for every open subgroup K of G, $\dim \mathfrak{H}_K < \infty$. Here $\mathfrak{H}_K$ is the space of all vectors in $\mathfrak{H}$ which are left fixed by K. Suppose π is admissible. For $f \in C_c{}^\infty(G)$ define

$$\pi(f) = \int_G f(x)\pi(x)\, dx$$

where dx is the Haar measure on G. Then $\pi(f)$ is an operator of finite rank. Put

$$\Theta_\pi(f) = \operatorname{tr} \pi(f) \qquad (f \in C_c{}^\infty(G)).$$

Then Θ_π is an invariant distribution on G which is called the character of π.

Let $D(x)$ denote the coefficient of t^l in $\det(t + 1 - \operatorname{Ad}(x))$ $(x \in G)$ and G' the set of all $x \in G$ where $D(x) \ne 0$.

Theorem 4 *Suppose π is irreducible and admissible. Then Θ_π is a locally summable function on G which is locally constant on G'. Moreover $|D|_\mathfrak{p}^{1/2}\Theta_\pi$ is locally bounded on G.*

It is possible to describe the behavior of Θ_π around singular points. We give here the result for $x = 1$. Recall that $\mathfrak{g} \subset gl(n, \Omega)$ for some $n \ge 1$. Let $\mathcal{N}$ be the set of all elements of $\mathfrak{g}$ which are nilpotent (as matrices). Then $\mathcal{N}$ is the union of a finite number of G-orbits which we call the nilpotent orbits.

Theorem 5 *We can choose an open neighborhood U of zero in $\mathfrak{g}$ and, for each nilpotent orbit $\mathfrak{o}$, a complex constant $c_\mathfrak{o}$ such that*

$$\Theta_\pi(\exp X) = \sum_\mathfrak{o} c_\mathfrak{o}\hat{\mu}_\mathfrak{o}(X) \qquad (X \in U).$$

The constants $c_\mathfrak{o}$ are unique and we denote them by $c_\mathfrak{o}(\pi)$. In particular we consider the constant $c_0(\pi)$ corresponding to the nilpotent orbit $\mathfrak{o} = \{0\}$.

Let Z denote the maximal split torus contained in the center of G. Fix a Haar measure dx^* on G/Z. Let π be an irreducible unitary representation of G on $\mathfrak{H}$. As usual we say that π is square-integrable if

$$\int_{G/Z} |(\phi, \pi(x)\psi)|^2 \, dx^* < \infty$$

for $\phi, \psi \in \mathfrak{H}$. Then there exists a positive number $d(\pi)$, called the formal degree of π, such that

$$\int_{G/Z} |(\phi, \pi(x)\psi)|^2 \, dx^* = d(\pi)^{-1} |\phi|^2 |\psi|^2$$

for all $\phi, \psi \in \mathfrak{H}$. A square-integrable representation is always admissible. It is called supercuspidal if we can choose $\phi, \psi \neq 0$ in $\mathfrak{H}$ such that the function

$$x \mapsto (\phi, \pi(x)\psi)$$

on G has compact support mod Z.

Theorem 6   *There exists a real number $c \neq 0$ such that*

$$c_0(\pi) = cd(\pi)$$

for every supercuspidal representation π.

It is natural to believe that this relation actually holds for all square-integrable π. Let $l_0 = \dim A_0/Z$ where A_0 is a maximal split torus of G. Then if π_0 is the Steinberg representation [7, §15], we expect that

$$c = (-1)^{l_0} d(\pi_0)^{-1}$$

Theorem 7   *We can choose $Q \neq 0$ in $\mathbf{Z}$ such that $Qc_0(\pi) \in \mathbf{Z}$ for every supercuspidal π.*

Corollary   *It is possible to normalize the Haar measure dx^* on G/Z so that $d(\pi)$ is an integer for every supercuspidal π.*

This has been proved for $GL(n, \Omega)$ by Howe [9] for arbitrary characteristic.

Bernshtein has shown (without any assumption on the characteristic) that every irreducible unitary representation of G is admissible [1, 2]. In our case this result is an easy consequence of the above corollary.

References

1. I. N. Bernshtein, All reductive *p*-adic groups are tame, *J. Functional Anal.* **8** (1974), 91–93.
2. P. Cartier, Les représentations des groupes reductifs *p*-adiques et leurs caractères, *Sem. Bourbaki* (1975), No. 471.
3. Harish-Chandra, Differential operators on a semisimple Lie algebra, *Amer. J. Math.* **79** (1957), 87–120.
4. Harish-Chandra, Invariant eigendistributions on a semisimple Lie algebra, *Inst. Hautes Études Sci. Publ. Math.* **27** (1965), 5–54.
5. Harish-Chandra, Invariant eigendistributions on a semisimple Lie group, *Trans. Amer. Math. Soc.* **119** (1965), 457–508.
6. Harish-Chandra, Discrete series for semisimple Lie groups I, *Acta Math.* **113** (1965), 241–318.
7. Harish-Chandra, Harmonic analysis on reductive *p*-adic groups, *in* "Harmonic Analysis on Homogeneous Spaces," pp. 167–192. Amer. Math. Soc., Providence, Rhode Island, 1973.
8. R. Howe, Two conjectures about reductive *p*-adic groups, *in* "Harmonic Analysis on Homogeneous Spaces," pp. 377–380. Amer. Math. Soc., Providence, Rhode Island, 1973.
9. R. Howe, The Fourier transform and germs of characters (case of Gl_n over a *p*-adic field), *Math. Ann.* **208** (1974), 305–322.
10. C. C. Moore, Representations of solvable and nilpotent groups and harmonic analysis on nil and solvmanifolds, *in* "Harmonic Analysis on Homogeneous Spaces." Amer. Math. Soc., Providence, Rhode Island, 1973.
11. R. R. Rao, Orbital integrals in reductive groups, *Ann. of Math.* **96** (1972), 505–510.

AMS (MOS) 1970 subject classification: 22E50

The Plancherel Formula for Reductive p-adic Groups

§1. Introduction

In [4] I have given a derivation of the Plancherel formula for a real reductive group, which presupposes a considerable knowledge of the discrete series and its characters. Since our understanding of the corresponding objects in the $\mathfrak{p}$-adic case is still very rudimentary, it may seem somewhat surprising that this method can be so modified as to yield the Plancherel formula for $\mathfrak{p}$-adic groups. Of course the philosophy of cusp forms remains the driving force in both cases. That this philosophy works, at least up to a point, in the context of supercusp forms, is clear from [2]. However supercusp forms, since they do not exhaust the discrete series, are not quite enough. So we have to develop a similar theory for cusp forms and this requires a systematic use of the weak inequality just as in the real case.

Let $\mathcal{C}(G)$ denote the Schwartz space [2, §14] of G and $\mathcal{C}^w(G)$ the subspace of $\mathcal{C}(G)$ [2, p. 172] consisting of those elements which satisfy the weak inequality. (P, A) being a p-pair in G ($P = MN$), we define a linear mapping $f \mapsto f_P^w$ of $\mathcal{C}^w(G)$ into $\mathcal{C}^w(M)$ and call f_P^w the weak constant term of f along P. Moreover we write $f_P^w \sim 0$ if the following condition holds. Fix $\omega \in \mathcal{E}_2(M)$ and let h be any matrix coefficient of ω. Then

$$(h, \phi f_P^w)_M = \int_M \operatorname{conj} h(m) \cdot \phi(m) f_P^w(m)\, dm = 0$$

for $\phi \in C_c(M/{}^0M)$.

Similarly we define a linear mapping $g \mapsto g^{(P)}$ of $\mathcal{C}(G)$ into $\mathcal{C}(M)$ by

$$g^{(P)}(m) = \delta_P(m)^{1/2} \int_N g(mn)\, dn \qquad (m \in M)$$

and write $g^{(P)} \sim 0$ if

$$(h, g^{(P)})_M = 0$$

for all matrix coefficients h of the discrete series of M. Then the following two lemmas play a crucial role in our theory.

Lemma 1. *Let f be an element in $\mathcal{C}^w(G)$ such that $f_P^w \sim 0$ for every p-pair (P, A) in G. Then $f = 0$.*

This is proved by a fairly easy induction on $\dim G$.

Lemma 2. *Suppose g is an element in $\mathcal{C}(G)$ and $g^{(P)} \sim 0$ for every p-pair (P, A) in G. Then $g = 0$.*

The proof, which again proceeds by induction, is more difficult in this case.

Let $^0\mathcal{C}(G)$ be the subspace of all $g \in \mathcal{C}(G)$ such that $g^{(P)} = 0$ for every p-pair $(P, A) \neq (G, Z)$. Fix a character χ of Z and let $^0\mathcal{C}(G, \chi)$ denote the space of all functions g on G such that:

1) $g(xz) = g(x)\chi(z)$ $(x \in G, z \in Z)$,
2) $\phi g \in {}^0\mathcal{C}(G)$ for all $\phi \in C_c(G/{}^0G)$.

Let K_0 be an open compact subgroup of G.

Lemma 3. $\dim {}^0\mathcal{C}(G//K_0, \chi) < \infty$.

This is the analogue of Theorem 27.3 of [3]. Put

$$\|g\|^2 = \int_{G/Z} |g(x)|^2 \, dx^* \qquad (g \in {}^0\mathcal{C}(G, \chi))$$

and let $^0L_2(G/K_0, \chi)$ denote the completion of $^0\mathcal{C}(G/K_0, \chi)$ with respect to this norm and $^0\lambda$ the representation of G on $^0L_2(G/K_0, \chi)$ defined by left translations.

Lemma 4. *For any $\alpha \in C_c^\infty(G)$, the operator $^0\lambda(\alpha)$ is of the Hilbert–Schmidt class.*

The proof of Lemma 2 is based on Lemma 3 and that of Lemma 3 on Lemma 4. Therefore Lemma 4, which is quite similar to a result of Gelfand and Piatetsky–Shapiro [1, p. 9, Thm. 2] in the theory of automorphic forms, emerges as the dominating fact here.

Let S be a complete set of special tori [2, p. 176] in G, no two of which are conjugate in G. For $A \in S$, let $\mathcal{C}_A(G)$ denote the space of all $g \in \mathcal{C}(G)$ with the following property. If (P', A') is a p-pair in G, then $g^{(P')} \sim 0$ unless A' is conjugate to A in G.

Lemma 5. $\mathcal{C}(G) = \sum_{A \in S} \mathcal{C}_A(G)$ *where the sum is orthogonal.*

It turns out that $\mathcal{C}_Z(G) = {}^0\mathcal{C}(G)$ and $\mathcal{C}_A(G)$ consists of wave-packets built from Eisenstein Integrals corresponding to $P \in \mathscr{P}(A)$ and $\omega \in \mathcal{E}_2(M)$ $(P = MN)$. Let S' denote the set of all $A \neq Z$ in S. In order to carry out the reduction from Lemma 2 to Lemma 3, one has to find a formula for the projection $f \mapsto f_A$ of $\mathcal{C}(G)$ onto $\mathcal{C}_A(G)$ $(A \in S')$ and show that

$$^0f = f - \sum_{A \in S'} f_A$$

lies in $^0\mathcal{C}(G)$. It turns out (see §13) that

$$f_A(x) = c^{-2} \gamma^{-1}[w]^{-1} \int_{\mathcal{E}_2(M)} d(\omega) \mu(\omega)(\Theta_\omega, r(x)f) \, d\omega \qquad (x \in G).$$

Moreover $^0f = f_Z$ and the above formula is also true for $A = Z$. Therefore the Plancherel formula is an immediate consequence of the relation

$$f = \sum_{A \in S} f_A.$$

There is one important difference here from the real case which should be mentioned. In the $\mathfrak{p}$-adic case every connected component of $\mathcal{E}_2(M)$ is compact. This is not true in the real case and therefore, in order to develop an adequate theory of wave-packets, we were obliged to investigate the behavior of μ at infinity [4, Thm. 25.1] and this investigation had to rely on a good knowledge of the characters of the discrete series. Because of compactness, this problem does not arise in the $\mathfrak{p}$-adic case. On the other hand, although we obtain a product formula for μ (Theorem 5), we do not get an explicit expression for μ_α.

Our results should be compared with those of Macdonald [5], who has obtained an explicit Plancherel formula for functions in $C_c(G//K)$.

§2. The Canonical Measure on $\mathcal{E}_2(G)$

Let Ω be a $\mathfrak{p}$-adic field and G the group of all Ω-rational points of a connected reductive Ω-group $\mathbf{G}$. As far as possible we shall keep to the notation and terminology of [2]. Let $\mathcal{E}_c(G)$ denote the set of all equivalence classes of irreducible admissible representations of G and $\mathcal{E}(G)$ the subset of the unitary classes. If π is an admissible representation put $\mathcal{C}(\omega) = \mathcal{C}(\pi)$ [2, p. 170] where ω is the class of π.

Let Z be the maximal split torus lying in the center of G and $\mathfrak{z}$ the real Lie algebra of Z. Then in [2, §7] we have defined a homomorphism H_G of G into $\mathfrak{z}$. Let ^0G denote the kernel of H_G and $\mathcal{E}_{2c}(G)$ the set of all $\omega \in \mathcal{E}_c(G)$ such that

$$\int_{^0G} |f(x)|^2 \, dx < \infty$$

for every $f \in \mathcal{C}(\omega)$. Put $\mathcal{E}_2(G) = \mathcal{E}(G) \cap \mathcal{E}_{2c}(G)$.

Let $\mathfrak{z}^*$ be the space dual to $\mathfrak{z}$. For $\nu \in \mathfrak{z}_c^*$ let χ_ν denote the one-dimensional representation of G given by

$$\chi_\nu(x) = q^{i\langle \nu, H_G(x) \rangle} \qquad (x \in G).$$

If $\omega \in \mathcal{E}_{2c}(G)$, let ω_ν denote the class of the representation $\pi \otimes \chi_\nu$ $(\pi \in \omega)$. The mapping $(\nu, \omega) \mapsto \omega_\nu$ defines an action of the additive group $\mathfrak{z}_c^*$ on $\mathcal{E}_{2c}(G)$. Fix $\omega \in \mathcal{E}_{2c}(G)$ and let $\mathfrak{o}_c$ denote the orbit of ω under this action and L^* the stabilizer of ω in $\mathfrak{z}_c^*$. Then it is easy to verify that L^* is a lattice in $\mathfrak{z}^*$ and therefore $\mathfrak{o}_c$ may be identified with the complex Lie group $\mathfrak{z}_c^*/L^*$. This defines the structure of a complex manifold on $\mathfrak{o}_c$. We now introduce a complex structure on $\mathcal{E}_{2c}(G)$ by demanding that each $\mathfrak{z}_c^*$-orbit be an open submanifold of $\mathcal{E}_{2c}(G)$. If $\omega \in \mathcal{E}_2(G)$, then $\mathfrak{o} = \mathfrak{o}_c \cap \mathcal{E}_2(G)$ is the orbit of ω under $\mathfrak{z}^*$. Hence $\mathfrak{o} \simeq \mathfrak{z}^*/L^*$ and it is a real submanifold of $\mathfrak{o}_c$. Note that $H_G(Z)$ is a lattice in $\mathfrak{z}$. Let $H_G(Z)^*$ denote the dual lattice in $\mathfrak{z}^*$. Then $H_G(Z)^*$ consists of all points $\lambda \in \mathfrak{z}^*$ such that $\chi_\lambda(z) = 1$ for all $z \in Z$. Let $d\nu$ denote the Euclidean measure on $\mathfrak{z}^*$ normalized so as to make the total measure of

$\mathfrak{z}^*/H_G(Z)^*$ equal to 1. Then[1] $d\nu$ defines a measure on $\mathfrak{z}^*/L^*$ and therefore also on $\mathfrak{o}$. Since $\mathcal{E}_2(G)$ is the disjoint union of the open orbits $\mathfrak{o}$, we get in this way a measure on $\mathcal{E}_2(G)$ which we denote by $d\omega$. We shall call it the canonical measure on $\mathcal{E}_2(G)$.

Put $^0Z = {}^0G \cap Z$. Then $\mathfrak{z}^*/H_G(Z)^*$ may be identified with the character group of the discrete group $Z/^0Z$. Hence the canonical measure corresponds to the normalized Haar measure on this compact character group.

§3. The Weak Constant Term

Fix a p-pair (P, A) in G $(P = MN)$, $\omega \in \mathcal{E}_2(M)$ and define

$$\pi = \mathrm{Ind}_P^G \delta_P^{1/2} \sigma.$$

Here $\sigma \in \omega$ and it has been extended to a representation of P by means of the isomorphism $P/N \simeq M$. Then π is unitary and its class is independent of the choice of P in $\mathcal{P}(A)$. Let $C_M^G(\omega)$ denote the class of π and Θ_ω the character of $C_M^G(\omega)$. Then Θ_ω is a tempered and invariant distribution on G. Moreover if ω is unramified in G, π is irreducible and therefore $C_M^G(\omega) \in \mathcal{E}(G)$ [2, Thm. 28].

Fix A_0 and K as in [2, §10, p. 178] and define Ξ and σ as usual [2, §14]. We shall say that a complex-valued function f on G satisfies the weak inequality if there exist numbers $c, r \geqslant 0$ such that

$$|f(x)| \leqslant c\Xi(x)(1 + \sigma(x))^r$$

for all $x \in G$. (It is not difficult to show that this definition is actually independent of the choice of A_0 and K.) Let $\mathcal{C}^w(G)$ denote the space of all functions in $\mathcal{C}(G)$ which satisfy the weak inequality.

Let (P, A) be a p-pair in G $(P = MN)$. Put

$$\beta(a) = \inf_\alpha \langle \alpha, H_M(a) \rangle \qquad (a \in A)$$

where α runs over all roots of (P, A). We write $a \underset{P}{\to} \infty$ if $\beta(a) \to +\infty$.

Theorem 1. *Fix $f \in \mathcal{C}^w(G)$. Then there exists a unique element $\phi \in \mathcal{C}^w(M)$ such that*

$$\lim_{\substack{a \to \infty \\ P}} |\delta_P(ma)^{1/2} f(ma) - \phi(ma)| = 0$$

for all $m \in M$.

We call ϕ the weak constant term of f along P and denote it by f_P^w.

---

[1] The assertion made on p. 177 of [2] that $L^* = H(M)^*$ is a mistake. This was pointed out to me by A. Silberger.

Let ${}^{0}\mathcal{C}^{w}(G)$ denote the space of all $f \in \mathcal{C}^{w}(G)$ with the following property. If (P, A) is any p-pair in G, then $f_{P}^{w} = 0$ unless $P = G$. Put

$$(\alpha, \beta) = (\alpha, \beta)_{G} = \int_{G} \mathrm{conj}\,\alpha(x)\cdot\beta(x)\,dx \qquad (\alpha, \beta \in C(G)),$$

provided the integral is convergent. Given $f \in \mathcal{C}^{w}(G)$, we write $f \sim 0$ if

$$(g, \phi f) = 0$$

for all $\phi \in C_{c}(G/{}^{0}G)$, $g \in \mathcal{C}(\omega)$ and $\omega \in \mathcal{E}_{2}(G)$. (It follows from the results of [2, §14] that the corresponding integral is convergent.)

Theorem 2. *Let f be an element in $\mathcal{C}^{w}(G)$. Suppose $f_{P}^{w} \sim 0$ for every p-pair (P, A) in G. Then $f = 0$.*

This corresponds to Lemma 9 of [2].

§4. Maass–Selberg Relations

Let τ be a smooth and unitary double representation of K on a finite-dimensional Hilbert space V. If $B(G)$ is a subspace of $C(G)$, we define [2, §10]

$$B(G, \tau) = (B(G) \otimes V) \cap C(G, \tau).$$

Let (P, A) be a semistandard p-pair in G $(P = MN)$ and f an element in $\mathcal{C}^{w}(G, \tau)$. Then $f_{P}^{w} \in \mathcal{C}^{w}(M, \tau_{M})$ and $f_{P}^{w} \sim 0$ if

$$(g, \phi f_{P}^{w})_{M} = \int_{M} (g(m), \phi(m) f_{P}^{w}(m))\,dm = 0$$

for all $\phi \in C_{c}(M/{}^{0}M)$, $g \in \mathcal{C}(\omega, \tau_{M})$ and $\omega \in \mathcal{E}_{2}(M)$.

Fix $\omega \in \mathcal{E}_{2}(M)$. Then $\mathcal{C}(\omega, \tau_{M})$ has a natural pre-hilbertian structure given by

$$\|g\|^{2} = (g, g)_{M/A} = \int_{M/A} |g(m)|^{2}\,dm^{*} \qquad (g \in \mathcal{C}(\omega, \tau_{M}))$$

where dm^{*} is the Haar measure on M/A. Put $\mathfrak{w} = \mathfrak{w}(G/A)$ and assume that ω is unramified in G. Then the sum

$$\sum_{s \in \mathfrak{w}} \mathcal{C}(s\omega, \tau_{M})$$

is direct. Put $C(\omega) = C_{M}^{G}(\omega)$. Then it is easy to verify that $\mathcal{C}(C(\omega)) \subset \mathcal{C}^{w}(G)$.

Theorem 3. *Fix $\omega \in \mathcal{E}_{2}(M)$ and suppose that it is unramified in G. Then an element $f \in \mathcal{C}^{w}(G, \tau)$ lies in $\mathcal{C}(C(\omega), \tau)$ if and only if it satisfies the following two conditions:*

(1) *(P', A') being a semistandard p-pair in G, $f_{P'}^{w} \sim 0$ unless A' is conjugate to A in G.*

(2) *If $P \in \mathcal{P}(A)$, then*
$$f_{P}^{w} \in \sum_{s \in \mathfrak{w}} \mathcal{C}(s\omega, \tau_{M}).$$

Now suppose $f \in \mathcal{C}(C(\omega), \tau)$ and let $f_{P,s}$ denote the component of f_P^w in $\mathcal{C}(s\omega, \tau_M)$. Then

$$\|f_{P_1, s_1}\| = \|f_{P_2, s_2}\|$$

for $P_1, P_2 \in \mathcal{P}(A)$ and $s_1, s_2 \in \mathfrak{w}$.

§5. Normalization of Measures

We now introduce the following convention for normalizing invariant measures. Let H be a closed unimodular subgroup of G, dh a Haar measure on H and dx^* an invariant measure on G/H. We shall say that dh and dx^* are standard if

$$\int_{H \cap K} dh = 1, \qquad \int_{K/H \cap K} dx^* = 1$$

Let dx denote a Haar measure on G. Then if dx, dh, dx^* are all standard, we have $dx = dx^* dh$. From now on all such measures are meant to be standard unless explicitly stated otherwise.

Fix a standard torus A and put

$$\gamma(P) = \int_{\overline{N}} q^{-2\langle \rho, H(\overline{n})\rangle} \, d\overline{n} \qquad (P \in \mathcal{P}(A))$$

in the notation of [2, Thm. 23]. It can be shown that $\gamma(P)$ is actually independent of $P \in \mathcal{P}(A)$. Hence we denote it by $\gamma(G/A)$.

Every root α of A in G can be regarded as an element of $\mathfrak{a}^*$. We call α reduced if $r\alpha$ is not a root for $0 < r < 1$ ($r \in R$). Suppose α is a reduced root. Then α is an element of $X(A)$ [2, §7]. Let A_α be the maximal torus contained in the kernel of α. Then A_α is a standard torus. Let M_α be the centralizer of A_α in G. Replacing (G, K, A) by $(M_\alpha, K \cap M_\alpha, A)$, we define $\gamma_\alpha = \gamma(M_\alpha/A)$. It is clear that $\gamma_\alpha = \gamma_{-\alpha}$.

Fix $P \in \mathcal{P}(A)$ and let $\Sigma(P)$ denote the set of all reduced roots of (P, A). Put

$$c(G/A) = \gamma(G/A)^{-1} \prod_{\alpha \in \Sigma(P)} \gamma_\alpha.$$

We observe that $c(G/A) = 1$ if $\dim A/Z = 1$.

§6. The Function μ

We return to the notation of §4. For $P_1, P_2 \in \mathcal{P}(A)$, define $\tau_{P_1|P_2}$ as in [2, §11] and put

$$L(\omega, P) = \mathcal{C}(\omega, \tau_{P|P}), \qquad \mathcal{L}(\omega, P) = \mathcal{C}(\omega, \tau_{P|\overline{P}})$$

for $\omega \in \mathcal{E}_2(M)$ and $P \in \mathcal{P}(A)$. Given $\psi \in L(\omega, P)$, we define the element $E(P : \psi) \in \mathcal{C}(G, \tau)$ exactly as before [2, §11]. Then $E(P : \psi)$ lies in $\mathcal{C}^w(G, \tau)$.

Theorem 4. *Fix $P_1, P_2 \in \mathcal{P}(A)$, $\omega \in \mathcal{E}_2(M)$ and suppose that ω is unramified in G.*

Then there exist unique linear mappings $c_{P_2|P_1}(s:\omega)$ of $L(\omega, P_1)$ into $\mathcal{L}(s\omega, P_2)$ ($s \in \mathfrak{w}$) such that

$$\left(E(P_1:\psi)\right)^{w}_{P_2} = \sum_{s \in \mathfrak{w}} c_{P_2|P_1}(s:\omega)\psi$$

for all $\psi \in L(\omega, P_1)$. Moreover $c_{P_2|P_1}(s:\omega)$ is bijective and there exists a number $\mu(\omega) > 0$ such that

$$\mu(\omega)\|c_{P_2|P_1}(s:\omega)\psi\|^2 = c(G/A)^2\|\psi\|^2$$

for all $P_1, P_2 \in \mathcal{P}(A)$, $s \in \mathfrak{w}$ and $\psi \in L(\omega, P_1)$. Finally $\mu(\omega) = \mu(s\omega)$ ($s \in \mathfrak{w}$).

It can be shown that $c_{P_2|P_1}(s)$ and μ extend (uniquely) to meromorphic functions on $\mathcal{E}_{2c}(M)$, μ is everywhere holomorphic on $\mathcal{E}_2(M)$ and $\mu(\omega) \geq 0$ for $\omega \in \mathcal{E}_2(M)$.

Fix a reduced root α of A in G and let μ_α denote the function on $\mathcal{E}_{2c}(M)$ corresponding to μ when the pair (G, A) is replaced by (M_α, A) in the notation of §5. Fix $P \in \mathcal{P}(A)$.

Theorem 5.

$$\mu = \prod_{\alpha} \mu_\alpha$$

where α runs over all reduced roots of (P, A).

§7.   Wave-Packets

Fix $P \in \mathcal{P}(A)$ and let $\mathfrak{o}$ be an $\mathfrak{a}^*$-orbit in $\mathcal{E}_2(M)$. Let ψ be a function which assigns to every $\omega \in \mathfrak{o}$ an element $\psi(\omega) \in L(\omega, P)$. Fix a point $\omega_0 \in \mathfrak{o}$ and define a function ψ_0 from $\mathfrak{a}^*$ to $L(\omega_0, P)$ as follows.

$$\psi_0(\nu) = \psi(\omega_{0\nu})\chi_\nu^{-1} \qquad (\nu \in \mathfrak{a}^*)$$

where

$$\chi_\nu(m) = q^{i\langle \nu, H_M(m)\rangle} \qquad (m \in M).$$

We say that ψ is of class C^∞ if ψ_0 is of class C^∞. Let $C^\infty(\mathfrak{o}:P)$ denote the space of all such ψ. Let L^* be the stabilizer of ω_0 in $\mathfrak{a}^*$. Then under the mapping $\psi \mapsto \psi_0$ we may identify $C^\infty(\mathfrak{o}:P)$ with

$$C^\infty(\mathfrak{a}^*/L^*)\otimes L(\omega_0, P).$$

This enables us to define a topology in $C^\infty(\mathfrak{o}:P)$. The definition of $C^\infty(\mathfrak{o}:P)$ and its topology are clearly independent of the choice of ω_0. Note that $\mathfrak{o}$ is compact and we have the canonical measure (§2) $d\omega$ on $\mathcal{E}_2(M)$.

Define the Schwartz space $\mathcal{C}(G)$ as usual [2, §14].

Theorem 6. *Put*

$$f_\psi = \int_{\mathfrak{o}} \mu(\omega)E(P:\psi(\omega))\,d\omega$$

for $\psi \in C^\infty(\mathfrak{o}:P)$. Then $\psi \mapsto f_\psi$ is a continuous mapping of $C^\infty(\mathfrak{o}:P)$ into $\mathcal{C}(G, \tau)$.

§8.   The Mapping E_ω

Let T be a tempered distribution on G. Define

$$(T, f) = \operatorname{conj} T(\bar{f}) \qquad (f \in \mathcal{C}(G))$$

where $\bar{f} = \operatorname{conj} f$. Moreover if

$$f = \sum_{1 \leqslant i \leqslant r} f_i \otimes v_i \in \mathcal{C}(G) \otimes V \qquad (f_i \in \mathcal{C}(G),\ v_i \in V)$$

we set

$$(T, f) = \sum_i (T, f_i) v_i \in V.$$

Let r denote the right-regular representation of G, $d(\omega)$ the formal degree and Θ_ω the character of $\omega \in \mathcal{E}_2(G)$ [2, §3].

Theorem 7. *Fix $\omega \in \mathcal{E}_2(G)$ and $f \in \mathcal{C}(G)$. Then there exists a unique element $f_\omega \in \mathcal{C}(\omega)$ such that*

$$(g, f_\omega)_{G/Z} = (g, f)_G$$

for all $g \in \mathcal{C}(\omega)$. f_ω is given by the formula

$$f_\omega(x) = d(\omega)(\Theta_\omega, r(x)f) \qquad (x \in G).$$

Let E_ω denote the mapping $f \mapsto f_\omega$. Then $E_\omega \otimes 1$ defines a linear mapping of $\mathcal{C}(G, \tau)$ into $\mathcal{C}(\omega, \tau)$, which we shall again denote by E_ω.

§9.   The Function $\hat{f}(P : \omega)$

We return to the notation of §7. For $f \in \mathcal{C}(G, \tau)$, define

$$f^{(P)}(m) = \delta_P(m)^{1/2} \int_N f(mn)\, dn \qquad (m \in M)$$

as usual [2, §14]. Then $f^{(P)} \in \mathcal{C}(M, \tau_{P|P})$. Fix $\omega \in \mathcal{E}_2(M)$ and put

$$\hat{f}(P : \omega) = E_\omega f^{(P)}.$$

Then $\hat{f}(P : \omega) \in L(\omega, P)$.

Lemma 6. *Fix an $\mathfrak{a}^*$-orbit $\mathfrak{o}$ in $\mathcal{E}_2(M)$. Then if $f \in \mathcal{C}(G, \tau)$, the function*

$$\omega \mapsto \hat{f}(P : \omega) \qquad (\omega \in \mathfrak{o})$$

lies in $C^\infty(\mathfrak{o} : P)$. Moreover

$$E(P : \hat{f}(P : \omega) : x) = d(\omega)(\Theta_\omega, r(x)f) \qquad (x \in G)$$

for $\omega \in \mathcal{E}_2(M)$.

We recall that Θ_ω is the character of $C_M^G(\omega)$ (§3).
Put

$$f_\mathfrak{o}(x) = c^{-2}\gamma^{-1}[\mathfrak{w}]^{-1} \int_\mathfrak{o} \mu(\omega)\, d(\omega)(\Theta_\omega, r(x)f)\, d\omega \qquad (x \in G)$$

for $f \in \mathcal{C}(G, \tau)$, where (§5)

$$c = c(G/A), \qquad \gamma = \gamma(G/A), \qquad \mathfrak{w} = \mathfrak{w}(G/A).$$

Then it follows from Theorem 6 and Lemma 6 that $f_{\mathfrak{o}} \in \mathcal{C}(G, \tau)$.

Lemma 7. $f \mapsto f_{\mathfrak{o}}$ *is a continuous endomorphism of* $\mathcal{C}(G, \tau)$. *Moreover*

$$\hat{f}_{\mathfrak{o}}(P:\omega) = [\mathfrak{w}]^{-1} \sum_{s \in \mathfrak{w}} \xi_{\mathfrak{o}}(s\omega)\hat{f}(P:\omega)$$

for $P \in \mathscr{P}(A)$, $\omega \in \mathscr{E}_2(M)$ *and* $f \in \mathcal{C}(G, \tau)$. *Here* $\xi_{\mathfrak{o}}$ *is the characteristic function of* $\mathfrak{o}$.

§10. The Condition $f^{(P)} \sim 0$

Let $f \in \mathcal{C}(G)$. Then we write $f \sim 0$ if $(\phi, f)_G = 0$ for all $\phi \in \mathcal{A}(\omega)$ and $\omega \in \mathscr{E}_2(G)$. Note that $\mathcal{A}(G) \cap \mathcal{C}(G) = \{0\}$ unless $Z = \{1\}$. Hence this definition agrees with the earlier one (§3) when $f \in \mathcal{A}^w(G) \cap \mathcal{C}(G)$. If $f \in \mathcal{C}(G, \tau)$, the condition $f \sim 0$ means that

$$(\phi, f)_G = 0$$

for all $\phi \in \mathcal{A}(\omega, \tau)$ and $\omega \in \mathscr{E}_2(G)$.

A being a special torus, define $\mathcal{C}_A(G)$ as in §1. It is easy to verify that $\mathcal{C}_A(G)$ is a closed two-sided ideal in $\mathcal{C}(G)$ (regarded as an algebra under convolution).

Fix a semistandard p-pair (P, A) in G ($P = MN$).

Lemma 8. *Let f be an element in* $\mathcal{C}(G, \tau)$. *Then the following conditions on f are equivalent.*

(1) $f^{(P)} \sim 0$.
(2) $(E(P:\psi), f)_G = 0$ *for all* $\psi \in L(\omega, P)$ *and* $\omega \in \mathscr{E}_2(M)$.
(3) $(f_\psi, f)_G = 0$ *in the notation of Theorem 6, for all* $\mathfrak{a}^*$-*orbits* $\mathfrak{o}$ *in* $\mathscr{E}_2(M)$ *and* $\psi \in C^\infty(\mathfrak{o}:P)$.

Applying this lemma one can obtain the following result.

Lemma 9. *Fix (P, A) and $\mathfrak{o}$ as in Theorem 6. Then for any* $\psi \in C^\infty(\mathfrak{o}:P)$, *the function* f_ψ *lies in* $\mathcal{C}_A(G, \tau)$.

§11. Statements of Theorems 8 and 10

Let $\mathcal{C}^w(G)$ denote the space of all $f \in C(G)$ satisfying the following conditions:

(1) There exists an open compact subgroup K_0 in G such that $f \in C(G//K_0)$.
(2) f satisfies the weak inequality (§3).
(3) $\phi f \in \mathcal{C}(G)$ for all $\phi \in C_c(G/{}^0G)$.

Let (P, A) be a p-pair in G $(P = MN)$. For $f \in \mathcal{C}^w(G)$, define

$$f^{(P)}(m) = \delta_P(m)^{1/2} \int_N f(mn)\, dn \qquad (m \in M)$$

as before. Since $N \subset {}^0G$, it is easy to verify that $f^{(P)} \in \mathcal{C}^w(M)$. Let ${}^0\mathcal{C}^w(G)$ denote the space of all $f \in \mathcal{C}^w(G)$ such that $f^{(P)} = 0$ for every p-pair $(P, A) \neq (G, Z)$. We have defined the space ${}^0\mathcal{C}^w(G)$ in §3.

Theorem 8. ${}^0\mathcal{C}^w(G) \subset {}^0\mathcal{C}^w(G)$ *and an element of* ${}^0\mathcal{C}^w(G)$ *lies in* ${}^0\mathcal{C}^w(G)$ *if and only if it is Z-finite.*

The first statement is an easy consequence of Jacquet's theorem [2, Thm. 4]. The only difficult point is to show that a Z-finite function in ${}^0\mathcal{C}^w(G)$ is Hecke finite [2, p. 172]. We shall sketch a proof of this in §14.

Note that $\mathcal{C}(G) \subset \mathcal{C}^w(G)$ and if $f \in \mathcal{C}^w(G)$ and $\phi \in C_c(G/{}^0G)$ then $\phi f \in \mathcal{C}(G)$. We write $f \underset{*}{\sim} 0$ if $\phi f \sim 0$ for every $\phi \in C_c(G/{}^0G)$. This means that

$$(g, \phi f)_G = 0$$

for all $\phi \in C_c(G/{}^0G)$, $g \in \mathcal{C}(\omega)$ and $\omega \in \mathcal{E}_2(G)$.

Lemma 10. *Let f be an element in $\mathcal{C}^w(G)$ such that $f^{(P)} \underset{*}{\sim} 0$ for every p-pair (P, A) in G. Then $f = 0$.*

Lemmas 2 and 10 are easy consequences of Theorem 8.

If K_0 is an open compact subgroup of G and $B(G)$ a subspace of $C(G)$, define

$$B(G/K_0) = B(G) \cap C(G/K_0), \qquad B(G//K_0) = B(G) \cap C(G//K_0).$$

Put ${}^0\mathcal{C}(G) = \mathcal{C}(G) \cap {}^0\mathcal{C}^w(G)$.

Theorem 9. *Fix an open compact subgroup K_0 of G. Then there exists a compact subset Ω_0 of $\mathcal{E}_2(G)$ with the following property. Let ω be an element in $\mathcal{E}_2(G)$. Then $E_\omega f = 0$ for all $f \in {}^0\mathcal{C}(G//K_0)$ unless $\omega \in \Omega_0$.*

Let $\mathcal{E}_2(G : K_0)$ denote the set of all $\omega \in \mathcal{E}_2(G)$ such that $[\omega : 1_{K_0}] \geq 1$. Here 1_{K_0} denotes the class of the trivial representation of K_0 and $[\omega : 1_{K_0}]$ the multiplicity of 1_{K_0} in ω.

Theorem 10. $\mathcal{E}_2(G : K)$ *is a compact subset of* $\mathcal{E}_2(G)$.

It is easy to see that Theorems 9 and 10 are equivalent.

Theorem 11. *Fix $\theta \in {}^0\mathcal{C}^w(G)$. Then*

$$\int_{G/Z} dy^* \left| \int_G f(x)\theta(yxy^{-1})\, dx \right| < \infty$$

for $f \in \mathcal{C}(G)$. Put

$$\Theta(f) = \int_{G/Z} dy^* \int_G f(x)\theta(yxy^{-1})\, dx \qquad (f \in \mathcal{C}(G)).$$

Then Θ is a tempered distribution on G.

Theorems 8 and 9 are the two main steps in the proof of the Plancherel formula. Their proof depends on Theorem 11.

§12. The Mapping $f \mapsto {}^0 f$

Assuming Theorem 11, we shall now sketch the proof of Theorems 8 and 9. We proceed by induction on $\dim G$. Fix a semistandard p-pair $(P, A) \neq (G, Z)$ in G $(P = MN)$. Then Theorem 9 holds for M and therefore if $f \in \mathcal{C}(G, \tau)$, the function

$$\omega \mapsto \hat{f}(P : \omega) \qquad (\omega \in \mathcal{E}_2(M))$$

has its support in a compact set which is independent of f. Put

$$f_A = c^{-2}\gamma^{-1}[\mathfrak{w}]^{-1} \int_{\mathcal{E}_2(M)} \mu(\omega) E(P : \hat{f}(P : \omega))\, d\omega$$

in the notation of §9. Then we conclude from Lemmas 6, 7, and 9 that $f \mapsto f_A$ is a continuous linear mapping of $\mathcal{C}(G, \tau)$ into $\mathcal{C}_A(G, \tau)$ which is independent of the choice of $P \in \mathcal{P}(A)$. Moreover it follows from Lemma 7 that

$$(f_A)\hat{\ }(P : \omega) = \hat{f}(P : \omega) \qquad (\omega \in \mathcal{E}_2(M),\ P \in \mathcal{P}(A)).$$

Let S' be a complete set of standard tori $A \neq Z$ in G, no two of which are conjugate under K. Put

$$^0 f = f - \sum_{A \in S'} f_A \qquad (f \in \mathcal{C}(G, \tau)).$$

Lemma 11. *$f \mapsto {}^0 f$ is a continuous projection of $\mathcal{C}(G, \tau)$ onto ${}^0\mathcal{C}(G, \tau)$.*

Fix $A \in S'$ and $P \in \mathcal{P}(A)$ $(P = MN)$. Then it is easy to verify that $({}^0 f)^{(P)} \sim 0$. But by induction hypothesis Theorem 8 and therefore also Lemma 2 are applicable to M. Hence we conclude without difficulty that $({}^0 f)^{(P)} = 0$. This proves that $^0 f \in {}^0\mathcal{C}(G, \tau)$. If $f \in {}^0\mathcal{C}(G, \tau)$, it is obvious that $\hat{f}(P : \omega) = 0$ and therefore $f_A = 0$. Hence $^0 f = f$ in this case.

Lemma 12. *Let $f \in \mathcal{C}(G, \tau)$ and $g \in {}^0\mathcal{C}^w(G, \tau)$. Then*

$$(f_A, g) = 0 \qquad (A \in S').$$

First suppose $g \in {}^{0}\mathcal{C}(G, \tau)$. Fix $P \in \mathcal{P}(A)$ and $\omega \in \mathcal{E}_2(M)$. Then

$$\left(E(P: \hat{f}(P: \omega)), g \right) = \left(\hat{f}(P: \omega), g^{(P)} \right)_M = 0$$

and our assertion is an easy consequence of this fact. The rest follows by an approximation argument.

Lemma 13. *Let A_1, A_2 be two distinct elements in S'. Then*

$$\left(f_{A_1}, f_{A_2} \right) = 0 \qquad (f \in \mathcal{C}(G, \tau)).$$

This is an immediate consequence of Lemmas 8 and 9.

§13. The Projections E_A and ${}^{0}E$

The results of §12 may now be formulated in a slightly different form as follows. For $f \in \mathcal{C}(G)$ and $A \in S'$, define

$$f_A(x) = c^{-2} \gamma^{-1} [\mathfrak{w}]^{-1} \int_{\mathcal{E}_2(M)} \mu(\omega) \, d(\omega) (\Theta_\omega, r(x)f) \, d\omega \qquad (x \in G)$$

in the notation of §9. Then $f \mapsto f_A$ is a continuous projection of $\mathcal{C}(G)$ into $\mathcal{C}_A(G)$ which we denote by E_A. Since Θ_ω is an invariant distribution, it is clear that E_A commutes with the translations of G. Put

$$ {}^{0}E = I - \sum_{A \in S'} E_A$$

where I is the identity mapping of $\mathcal{C}(G)$. Then ${}^{0}E$ is a continuous projection of $\mathcal{C}(G)$ onto ${}^{0}\mathcal{C}(G)$.

Lemma 14. ${}^{0}\mathcal{C}(G)$ *is orthogonal to* $E_A \mathcal{C}(G)$ *$(A \in S')$.*

This follows from Lemma 12.

Corollary 1. *Let $f \in \mathcal{C}(G)$ and $g \in {}^{0}\mathcal{C}(G)$. Then $g * f_A = f_A * g = 0$.*

Fix $h \in \mathcal{C}(G)$. Then

$$(h, g * f_A) = (\tilde{g} * h, f_A)$$

where $\tilde{g}(x) = \operatorname{conj} g(x^{-1})$. It is clear that $\tilde{g} \in {}^{0}\mathcal{C}(G)$ and since ${}^{0}E$ commutes with the translations of G, we conclude that $\tilde{g} * h \in {}^{0}\mathcal{C}(G)$ and therefore

$$(\tilde{g} * h, f_A) = 0.$$

This implies that $g * f_A = 0$. The proof for $f_A * g$ is similar.

Corollary 2. *Let $f \in \mathcal{C}(G)$ and $g \in {}^{0}\mathcal{C}^w(G)$. Then*

$$f * g = {}^{0}f * g.$$

If $g \in {}^0\mathcal{C}(G)$, this is obvious from Corollary 1. The rest follows by an approximation argument.

Corollary 3. *Let $f, g \in \mathcal{C}(G)$. Then*

$$ {}^0(f * g) = {}^0f * {}^0g, \quad {}^0(\tilde{f}) = \left({}^0f\right)^{\tilde{}} . $$

Since ${}^0E$ commutes with the translations of G,

$$ {}^0E(f * g) = f * {}^0Eg = {}^0f * {}^0g $$

from Corollary 2. Since $\Theta_\omega (\omega \in \mathcal{E}_2(M))$ is the character of a unitary representation, it is obvious that

$$ (\tilde{f})_A = (f_A)^{\tilde{}} \quad (A \in S'). $$

The second statement now follows immediately.

§14.   Proof of Lemma 4

Let $\hat{Z}$ be the character group of Z. Fix $\chi \in \hat{Z}$ and an open compact subgroup K_0 of G and let ${}^0\mathcal{C}(G/K_0, \chi)$ denote the subspace of all $f \in {}^0\mathcal{C}^w(G/K_0)$ such that

$$ f(xz) = f(x)\chi(z) \quad (x \in G, \ z \in Z). $$

Define ${}^0L_2(G/K_0, \chi)$ and ${}^0\lambda$ as in §1. We shall now give a proof of Lemma 4.
   Fix $\alpha \in C_c^\infty(G)$ and $f \in {}^0\mathcal{C}(G/K_0, \chi)$. Then

$$ \alpha * f = {}^0\alpha * f $$

from Cor. 2 of Lemma 14. Put

$$ {}^0\alpha_\chi(x) = \int_Z {}^0\alpha(xz^{-1})\chi(z)\, dz \quad (x \in G) $$

and

$$ \kappa_\alpha(x : y) = \int_{K_0} {}^0\alpha_\chi(xky^{-1})\, d_0k \quad (x, y \in G). $$

Here d_0k is the normalized Haar measure on K_0 so that

$$ \int_{K_0} d_0k = 1. $$

If $g = {}^0\lambda(\alpha)f$, we have

$$ g(x) = \int_{G/Z} \kappa_\alpha(x : y)f(y)\, dy^* \quad (x \in G). $$

Since the characteristic function of K_0 lies in $C_c^\infty(G)$ and ${}^0\alpha_\chi \in {}^0\mathcal{C}^w(G)$, we conclude from Theorem 11 that

$$ \int_{G/Z} |\kappa_\alpha(x : x)|\, dx^* < \infty. $$

Put $\beta = \tilde{\alpha} * \alpha$. Then ${}^0\beta = ({}^0\alpha)^{\tilde{}} * {}^0\alpha$ (Cor. 3 of Lemma 14) and therefore

$$\kappa_\beta(x:y) = \int_{G/Z} \mathrm{conj}\, \kappa_\alpha(u:x) \cdot \kappa_\alpha(u:y)\, du^* \qquad (x, y \in G).$$

Hence

$$\int |\kappa_\alpha(u:x)|^2\, du^*\, dx^* = \int \kappa_\beta(x:x)\, dx^* < \infty.$$

This proves that ${}^0\lambda(\alpha)$ is of the Hilbert–Schmidt class.

Let $\mathcal{E}_2(G: K_0: \chi)$ denote the set of all $\omega \in \mathcal{E}_2(G: K_0)$ such that χ is the central exponent of ω [2, p. 175]. It follows from Lemma 4 that $\dim {}^0\mathcal{C}(G//K_0, \chi) < \infty$ and $\mathcal{E}_2(G: K_0: \chi)$ is a finite set. Theorems 8 and 10 are easy consequences of these facts.

§15.   Identification of ${}^0f$ with f_Z

We return to the notation of §12 and consider the p-pair (G, Z). Then

$$\hat{f}(G: \omega) = E_\omega f \qquad (\omega \in \mathcal{E}_2(G))$$

for $f \in \mathcal{C}(G, \tau)$. Put

$$f_Z(x) = \int_{\mathcal{E}_2(G)} d(\omega)(\Theta_\omega, r(x)f)\, d\omega \qquad (x \in G).$$

Then by Theorems 6 and 10, $f \mapsto f_Z$ is a continuous mapping of $\mathcal{C}(G, \tau)$ into itself.

Lemma 15. $f_Z = {}^0f$ for all $f \in \mathcal{C}(G, \tau)$.

We know from Lemma 9 that $f_Z \in \mathcal{C}_Z(G, \tau)$. Put $g = f_Z - {}^0f$. Then if $A \in S'$ and $P \in \mathcal{P}(A)$, it is clear that $g^{(P)} \sim 0$. Moreover if $\omega \in \mathcal{E}_2(G)$ and $h \in \mathcal{C}(\omega, \tau)$,

$$(h, f_Z) = (h, f) = (h, {}^0f).$$

This follows from Lemmas 7 and 12. This shows that $g \sim 0$. Hence we conclude from Lemma 2 that $g = 0$. Put

$$S = S' \cup \{Z\}.$$

Corollary. $f = \Sigma_{A \in S} f_A$ for all $f \in \mathcal{C}(G, \tau)$.

This is obvious from the above lemma.

§16.   The Plancherel Formula

We can now state the Plancherel formula as follows. Let S be a complete set of standard tori in G no two of which are conjugate in G. For $A \in S$ and $f \in \mathcal{C}(G)$, define

$$f_A(x) = c^{-2}\gamma^{-1}[\mathfrak{w}]^{-1}\int_{\mathscr{E}_2(M)} \mu(\omega)\,d(\omega)\,(\Theta_\omega, r(x)f)\,d\omega \qquad (x \in G)$$

where M is the centralizer of A in G and

$$c = c(G/A), \qquad \gamma = \gamma(G/A), \qquad \mathfrak{w} = \mathfrak{w}(G/A)$$

as before. Then $f \mapsto f_A$ is a continuous projection of $\mathcal{C}(G)$ onto $\mathcal{C}_A(G)$ and

$$f = \sum_{A \in S} f_A.$$

Finally

$$\mathcal{C}(G) = \sum_{A \in S} \mathcal{C}_A(G)$$

where the sum is orthogonal.

References

1. Harish-Chandra, Automorphic forms on semisimple Lie groups, *Lecture notes in Math.*, vol. **62**, Springer-Verlag, 1968.
2. —, Harmonic Analysis on reductive p-adic groups, in Harmonic Analysis on Homogeneous Spaces, 167–192, Amer. Math. Soc., Providence, 1973.
3. —, Harmonic analysis on real reductive groups I, *J. Functional Analysis* **19** (1975), 104–204.
4. —, Harmonic analysis on real reductive groups. III. The Maass–Selberg relations and the Plancherel formula. *Ann. of Math.* **104**, 117–201.
5. I. G. Macdonald, Spherical functions on a group of p-adic type, Ramanujan Institute Publications No. 2, University of Madras, 1971.

Corrections to
"The Plancherel Formula for Reductive p-adic Groups"

Although the main results of this paper are correct, there are some errors which ought to be rectified.

For any $f \in C(G)$ and $x \in G$, let $l(x)f$ and $r(x)f$ denote the functions $y \mapsto f(x^{-1}y)$ and $y \mapsto f(yx)$ $(y \in G)$ respectively. Let (P, A) be a p-pair in G $(P = MN)$ and $(\bar{P}, A)$ $(\bar{P} = M\bar{N})$ the opposite p-pair. Then if $f \in \mathcal{C}(G)$, it is easy to verify that

$$(l(n)f)_P = (f(\bar{n})f)_P = f_P \qquad (n \in N, \ \bar{n} \in \bar{N})$$

and

$$(l(m)f)_P = \delta_P(m)^{1/2} l(m)f_P, \qquad (r(m)f)_P = \delta_P(m)^{-1/2} r(m)f_P \qquad (m \in M).$$

Here f_P is defined as in [2].

Put

$$\beta(a) = \inf_\alpha \langle \alpha, H_M(a) \rangle \qquad (a \in A)$$

where α runs over all roots of (P, A). Let a be a variable element in A. We write $a \xrightarrow[P]{} \infty$ if 1) $\beta(a) \to +\infty$ and 2) there exists a number $\varepsilon > 0$ such that $\beta(a) \geq \varepsilon\sigma(a)$. (Note that this condition is vacuous when $(P, A) = (G, Z)$.) In the statement of Theorem 1 (p. 8) "$\lim$"$_{a \to \infty}$ should be replaced by "$\lim$"$_{a \xrightarrow[P]{} \infty}$.

Fix $f \in \mathcal{C}^w(G)$. Since $G = PK = \bar{P}K$, it follows from what we have said above that the following conditions are equivalent.

1) $(l(x)r(y)f)_P^w \sim 0$ for all $x, y \in G$.
2) $(l(k)r(k')f)_P^w \sim 0$ for all $k, k' \in K$.

Now every p-pair is conjugate in G to a semistandard p-pair. Hence the following three conditions on f are equivalent.

1) $(l(x)f)_P^w \sim 0$ for every $x \in G$ and every p-pair (P, A).
2) $(r(x)f)_P^w \sim 0$ for every $x \in G$ and every p-pair (P, A).
3) $(l(k)r(k')f)_P^w \sim 0$ for all $k, k' \in K$ and every semistandard p-pair (P, A).

Lemma 1 (and Theorem 2) should be corrected to read as follows.

Lemma 1. *Let f be an element in $\mathcal{C}^w(G)$ such that $(r(x)f)_P^w \sim 0$ for all $x \in G$ and every p-pair (P, A) in G. Then $f = 0$.*

An equivalent form is as follows.

Lemma 1'. *Let f be an element in $\mathcal{Q}^w(G, \tau)$ such that $f_P^w \sim 0$ for every semistandard p-pair (P, A) in G. Then $f = 0$.*

$^0\mathcal{C}(G)$ should be defined as the subspace of all $g \in \mathcal{C}(G)$ such that

$$\int_N g(xn)\, dn = 0 \qquad (x \in G)$$

for every p-pair $(P, A) \neq (G, Z)$ $(P = MN)$. Lemma 2 should be corrected to read as follows:

Lemma 2. *Let g be an element in $\mathcal{C}(G)$ such that $(r(x)g)^{(P)} \sim 0$ for all $x \in G$ and every p-pair (P, A) in G. Then $g = 0$.*

An equivalent form is as follows.

Lemma 2'. *Let g be an element in $\mathcal{C}(G, \tau)$ such that $g^{(P)} \sim 0$ for every semistandard p-pair (P, A) in G. Then $g = 0$.*

Fix a special torus A. Then the correct definition of $\mathcal{C}_A(G)$ is as follows. It consists of all functions $f \in \mathcal{C}(G)$ with the following property. If (P', A') is a p-pair in G and A' is not conjugate to A, then $(r(x)f)^{(P')} \sim 0$ for all $x \in G$. It is easy to see that $\mathcal{C}_A(G, \tau)$ consists of all $f \in \mathcal{C}(G, \tau)$ with the following property. If (P', A') is any semistandard p-pair, then $f^{(P')} \sim 0$ unless A' is conjugate to A.

$^0\mathcal{Q}^w(G)$ (p. 9) should be defined as the space of all $f \in \mathcal{Q}^w(G)$ with the following property. If $(P, A) \neq (G, Z)$ is a p-pair in G, then $(r(x)f)_P^w \sim 0$ for all $x \in G$.

There are some small mistakes in §7 which are easily corrected. For $\psi \in L(\omega, P)$ and $\nu \in \mathfrak{a}_c^*$, define $\psi_\nu = \psi \chi_\nu$. Then $\psi \mapsto \psi_\nu$ is a linear bijection of $L(\omega, P)$ onto $L(\omega_\nu, P)$. In this way the additive groups $\mathfrak{a}^*$ and $\mathfrak{a}_c^*$ operate on the spaces

$$\mathfrak{S} = \bigcup_{\omega \in \mathfrak{o}} L(\omega, P), \qquad \mathfrak{S}_c = \bigcup_{\omega \in \mathfrak{o}_c} L(\omega, P)$$

respectively. Fix $\omega_0 \in \mathfrak{o}$ and let L^* be the stabilizer of ω_0 in $\mathfrak{a}^*$. Then L^* operates on $L(\omega_0, P)$ and we may identify $\mathfrak{S}_c$ with

$$\mathfrak{a}_c^* \times_{L^*} L(\omega_0, P).$$

Then $\mathfrak{S}_c$ becomes a holomorphic vector bundle over $\mathfrak{o}_c \simeq \mathfrak{a}_c^*/L^*$. Similarly

$$\mathfrak{S} \simeq \mathfrak{a}^* \times_{L^*} L(\omega_0, P)$$

is an analytic vector bundle over $\mathfrak{o}$ and $C^\infty(\mathfrak{o}\colon P)$ is just the space of all its C^∞ cross-sections.

The definition of $\mathcal{C}^w(G)$, (§11) needs to be modified. Put

$$\sigma_*(x) = \inf_{z \in Z} \sigma(xz) \qquad (x \in G).$$

Then $\mathcal{C}^w(G)$ should be defined as the space of all $f \in C(G)$ satisfying the following two conditions.

1) $f \in C(G//K_0)$ for some open compact subgroup K_0 of G.
2) We can choose $r_0 \geqslant 0$ such that

$$\sup_G |f(x)| \Xi(x)^{-1}(1 + \sigma_*(x))^r (1 + \sigma(x))^{-r_0} < \infty$$

for every $r \geqslant 0$.

Since $G/{}^0G.Z$ is finite, we can choose a subset Y of G such that $G = YZ$ and Y is the union of a finite number of cosets of ${}^0G$. Then 2) is equivalent to the following condition.

2') We can choose $r_0 \geqslant 0$ and, for any given $r \geqslant 0$, a constant $c_r \geqslant 0$ such that

$$|f(yz)| \leqslant c_r \Xi(y)(1 + \sigma(y))^{-r}(1 + \sigma(z))^{r_0}$$

for all $y \in Y$ and $z \in Z$.

One verifies that if (P, A) is a p-pair in G $(P = MN)$ and $f \in \mathcal{C}^w(G)$, then $f^{(P)} \in \mathcal{C}^w(M)$. ${}^0\mathcal{C}^w(G)$ should be defined as the space of all $f \in \mathcal{C}^w(G)$ such that

$$\int_N f(xn)\, dn = 0 \qquad (x \in G)$$

for every p-pair $(P, A) \neq (G, Z)$ $(P = MN)$.

Lemma 10 should be corrected to read as follows.

Lemma 10. *Let f be a function in $\mathcal{C}^w(G)$ such that $(r(x)f)^{(P)} \underset{*}{\sim} 0$ for all $x \in G$ and every p-pair (P, A) in G. Then $f = 0$.*

Admissible invariant distributions on reductive p-adic groups

by Harish-Chandra

Table of contents

§1. Introduction

Let Ω be a $\mathfrak{p}$-adic field and G the group of all Ω-rational points of a connected reductive Ω-group [1(b)]. Then G, with its usual topology, is a locally compact, totally disconnected, unimodular group. Let dx denote the Haar measure of G. We shall use the terminology of [1(b)] without further comment.

Let π be an admissible and irreducible representation of G and Θ_π the character of π. Let G' be the set of all points $x \in G$ where $D_G(x) \neq 0$ [1(b), §15]. Then G' is an open, dense subset of G whose complement has measure zero and it can be shown that there exists a locally constant function F_π on G' such that

$$\Theta_\pi(f) = \int_G f(x)F_\pi(x)dx \qquad (f \in C_c^\infty(G')) \ .$$

One would like to prove that F_π is locally summable on G and that this relation actually holds for all $f \in C_c^\infty(G)$. The main object of these lectures is to sketch a proof of these facts when Ω has characteristic zero.

We assume from now on that $\mathrm{char}\, \Omega = 0$.

Theorem 1. <u>Let Θ_π denote the character of an admissible and irreducible representation of G. Then Θ_π is a locally summable function on G which is locally constant on G'. Moreover the function</u>

$$|D_G|_{\mathfrak{p}}^{\frac{1}{2}}\Theta_\pi$$

<u>is locally bounded on</u> G.

We shall see presently that it is possible to describe the behavior of Θ_π around singular points somewhat more precisely.

Let $\mathfrak{g}$ be the Lie algebra of G. Then $\mathfrak{g}$ is a vector space over Ω of finite dimension and G operates on $\mathfrak{g}$ by the adjoint representation. If $x \in G$ and $X \in \mathfrak{g}$, put

$$x \cdot X = {}^{x}X = \mathrm{Ad}(x)X \ .$$

(If no misunderstanding is possible, we shall write xX instead of $x \cdot X$.) Let T be a distribution on $\mathfrak{g}$. Then ${}^{x}T$ is defined to be the distribution $f \longmapsto T(f^{x})$ where

$$f^{x}(X) = f(x \cdot X) \qquad\qquad (X \in \mathfrak{g}, \ f \in C_c^{\infty}(\mathfrak{g})) \ .$$

T is said to be G-invariant if ${}^{x}T = T$ for all $x \in G$. Let J denote the space of all G-invariant distributions on $\mathfrak{g}$.

For two subsets $S \subset G$ and $\omega \subset \mathfrak{g}$, put

$$S_\omega = \bigcup_{x \in S} \mathrm{Ad}(x)\omega$$

and let $J(\omega)$ denote the space of all $T \in J$ such that $\mathrm{Supp}\, T \subset Cl^{G}\omega$.

Let R denote the ring of all ($\mathfrak{p}$-adic) integers in Ω. By a lattice in $\mathfrak{g}$ we mean an open, compact R-submodule of $\mathfrak{g}$. If L is a lattice, we may regard $C_c(\mathcal{g}/L)$ as a subspace of $C_c^\infty(\mathcal{g})$. For any distribution T on $\mathcal{g}$, let $T_L = j_L T$ denote the restriction of T on $C_c(\mathcal{g}/L)$. The following remarkable theorem of Howe [2(c), (d)] is crucial for our theory.

Theorem 2 (Howe). <u>Let ω be a compact set and L a lattice in $\mathcal{g}$. Then</u>

$$\dim j_L J(\omega) < \infty \ .$$

Fix a symmetric, nondegenerate, G-invariant bilinear form B on $\mathcal{g}$ with values in Ω. Since G is reductive, this is possible. Also fix a character $\chi \neq 1$ of the additive group of Ω. Let dX denote a Haar measure on the additive group of $\mathcal{g}$ and define

$$\hat{f}(Y) = \int_{\mathcal{g}} \chi(B(Y,\ X))f(X)dX \qquad\qquad (Y \in \mathcal{g})$$

for $f \in \mathcal{D} = C_c^\infty(\mathcal{g})$. Then $f \longmapsto \hat{f}$ is a linear bijection of $\mathcal{D}$ onto itself. For any distribution T on $\mathcal{g}$, define the distribution $\hat{T}$ by

$$\hat{T}(f) = T(\hat{f}) \qquad\qquad (f \in \mathcal{D}) \ .$$

$\hat{T}$ is called the Fourier transform of T.

Let ℓ denote the (absolute) rank of $\mathcal{g}$ and $\eta_{\mathcal{g}}(X)$ the coefficient of t^ℓ in the polynomial

$$\det(t - \mathrm{ad}X) \qquad\qquad (X \in \mathcal{g})$$

where t is an indeterminate. Then $\eta_{\mathfrak{g}} = \eta$ is a polynomial function on $\mathfrak{g}$ and $\eta \neq 0$. Let $\mathfrak{g}'$ be the set of all points $X \in \mathfrak{g}$ where $\eta(X) \neq 0$.

Theorem 3. Let ω be a compact subset of $\mathfrak{g}$ and T an element in $J(\omega)$. Then there exists a locally summable function F on $\mathfrak{g}$ with the following properties:

1) $\widehat{T}(f) = \int_{\mathfrak{g}} F f \, dX$ for all $f \in \mathcal{D}$.

2) F is locally constant on $\mathfrak{g}'$.

3) $|\eta|_{\mathfrak{p}}^{\frac{1}{2}} F$ is locally bounded on $\mathfrak{g}$.

By an orbit (or more precisely a G-orbit) in $\mathfrak{g}$, we mean a set of the form $^G X$ where X is an element of $\mathfrak{g}$. Fix an orbit $\mathcal{O}$ and a point $X_o \in \mathcal{O}$. Let G_{X_o} denote the centralizer of X_o in G. Then G_{X_o} is unimodular and therefore the homogeneous space G/G_{X_o} has an invariant measure dx^* which is unique up to a constant factor. By a theorem of Deligne and Rao [3], the integral

$$\mu_{\mathcal{O}}(f) = \int_{G/G_{X_o}} f(x \cdot X_o) \, dx^* \qquad (f \in C_c(\mathfrak{g}))$$

is well defined. Hence $\mu_{\mathcal{O}}$ is a positive measure on $\mathfrak{g}$ which is uniquely determined by the orbit $\mathcal{O}$ up to a constant factor. It follows from Theorem 3 that $\widehat{\mu}_{\mathcal{O}}$ is a function.

Let $M_n(\Omega)$ denote the space of all $n \times n$ matrices with coefficients in Ω. Then one can speak of semisimple, unipotent or nilpotent elements of

$M_n(\Omega)$. Since both G and $\mathcal{g}$ are subsets of $M_n(\Omega)$ (for a suitable n),   such terms are applicable to their elements also. Let $\mathcal{N}$ be the set of all nilpotent elements in $\mathcal{g}$. Then $\mathcal{N}$ is the union of a finite number of G-orbits which are called the nilpotent orbits.

By a G-domain in $\mathcal{g}$ we mean a G-invariant subset of $\mathcal{g}$ which is both open and closed. Let[1] $\mathcal{O}(0)$ denote the set of all nilpotent G-orbits in $\mathcal{g}$.

Theorem 4.[2] <u>Let ω be a compact subset of</u> $\mathcal{g}$. <u>Then there exists a G-domain</u> D <u>containing zero with the following property. For every</u> $T \in J(\omega)$, <u>we can choose complex numbers</u> $c_{\theta}(T)$ ($\theta \in \mathcal{O}(0)$) <u>such that</u>

$$\widehat{T} = \sum_{\theta} c_{\theta}(T)\widehat{\mu}_{\theta}$$

<u>on</u> D. <u>Moreover if</u> V <u>is any neighborhood of zero in</u> $\mathcal{g}$, <u>the functions</u> $\widehat{\mu}_{\theta}$ ($\theta \in \mathcal{O}(0)$) <u>are linearly independent on</u> $V \cap \mathcal{g}'$.

We can now describe the behavior of a character around singular points by means of Theorem 4. Fix a semisimple point γ in G and let M and $\mathcal{m}$ denote the centralizers of γ in G and $\mathcal{g}$ respectively. Define Θ_{π} as in Theorem 1.

Theorem 5. <u>We can choose unique complex numbers</u> $c_{\xi}(\pi)$ <u>such that</u>

$$\Theta_{\pi}(\gamma \exp Y) = \sum_{\xi} c_{\xi}(\pi)\widehat{\nu}_{\xi}(Y)$$

for all $Y \in \mathfrak{m}$ sufficiently near zero. Here ξ runs over all nilpotent M-orbits in $\mathfrak{m}$, ν_ξ is the M-invariant measure on $\mathfrak{m}$ corresponding to ξ and $\hat{\nu}_\xi$ is the Fourier transform of ν_ξ on $\mathfrak{m}$.

Now consider the special case when $\gamma = 1$. Then $M = G$, $\mathfrak{m} = \mathfrak{g}$ and $\{0\}$ is a nilpotent G-orbit in $\mathfrak{g}$. Let $c_0(\pi)$ denote the coefficient corresponding to this orbit. Assume further that π is supercuspidal (and unitary) and let $d(\pi)$ denote the formal degree of π.

Theorem 6. <u>There exists a real number $c \neq 0$ such that</u>

$$c_0(\pi) = cd(\pi)$$

<u>for every supercuspidal</u> π.

It seems likely that this relation actually holds for all square-integrable π. This would imply that

$$c = (-1)^{\ell_0} d(\pi_0)^{-1}$$

where π_0 is the Steinberg representation and $\ell_0 = \dim A_0/Z$ in the notation of [1(b), §15].

Theorem 7. <u>It is possible to normalize the Haar measure on G/Z in such a way that</u> $d(\pi)$ <u>is an integer for every supercuspidal</u> π.

For $GL(n)$ this has been proved by Howe [2(d)] for arbitrary characteristic.

This paper is divided into three parts. In Part I we concentrate on Fourier transforms on the Lie algebra. The main object here is to derive Theorems 3 and 4 from Howe's theorem (Theorem 2) and our argument depends in an essential way on Theorems 10 and 13. Part II is devoted to an extension of Theorem 2 which is necessary for the application of these results to the group. In Part III we recall the main points of Howe's "Kirillov theory" [2(a), (b)] and introduce the concept of an admissible distribution. This concept, which was suggested by the work of Howe (see [1(b), §16] and [2(a), (b), (d)]) is central to our method.

Some of these results have been announced in a brief note [1(c)].

Part I. Fourier transforms on the Lie algebra

§2. The mapping $f \longrightarrow \phi_{\bar{f}}$

The proof of Theorem 3 is quite long and proceeds in several steps.

Fix a Cartan subalgebra $\mathfrak{h}$ of $\mathfrak{g}$ and put $\mathfrak{h}' = \mathfrak{h} \cap \mathfrak{g}'$. Let A denote the split component of the corresponding Cartan subgroup of G. $\mathfrak{h}$ is called elliptic if A lies in the center of G.

For $f \in \mathfrak{J}$, we define a function ϕ_f on $\mathfrak{h}'$ by[3]

$$\phi_f(H) = |\eta(H)|^{\frac{1}{2}} \int_{G/A} f(xH)dx^* \qquad (H \in \mathfrak{h}')$$

where dx^* is the invariant measure on G/A. Then ϕ_f is locally constant on $\mathfrak{h}'$. Sometimes it would be convenient to write $\phi_f^{\mathfrak{h}}$ or $\phi_f^{G/\mathfrak{h}}$ instead of ϕ_f.

Fix an element $H_o \in \mathfrak{h}'$.

Theorem 8. <u>The distribution</u>

$$f \longmapsto \phi_{\bar{f}}(H_o) \qquad (f \in \mathfrak{J})$$

<u>is a locally summable function on</u> $\mathfrak{g}$ <u>which is locally constant on</u> $\mathfrak{g}'$. <u>Let</u> F <u>denote this function. Then</u> $|\eta|^{\frac{1}{2}}F$ <u>is locally bounded on</u> $\mathfrak{g}$.

Fix an open compact subgroup K of G of Bruhat-Tits [1(a), Thm. 5, p. 16] and $P \in \mathscr{P}(A)$ $(P = MN)$. Put

$$\bar{f}(X) = \int_K f(kX)dk \qquad (X \in \mathfrak{g})$$

$$f_P(Y) = \int_{\mathfrak{n}} \bar{f}(Y+Z)dZ \qquad (Y \in \mathfrak{m})$$

for $f \in \mathcal{G}$. Here $\mathfrak{m}$ and $\mathfrak{n}$ are the Lie algebras of M and N respectively, dk the normalized Haar measure on K and dZ the Haar measure on the additive group of $\mathfrak{n}$. Clearly $f_P \in C_c^\infty(\mathfrak{m})$. Let $(f_P)^\wedge$ denote the Fourier transform of f_P on $\mathfrak{m}$ (defined by means of χ and the restriction of B on $\mathfrak{m}$). Then it is easy to verify that $(\hat{f})_P = (f_P)^\wedge$ and

$$\phi_f = \phi_{f_P}^{M/\mathfrak{h}} \qquad\qquad (f \in \mathcal{G}) \, . \,^{4)}$$

Now in order to prove Theorem 8, we proceed by induction on $\dim G$. If $P \neq G$, the induction hypothesis is applicable to M. Hence the important case is when $\mathfrak{h}$ is elliptic.

By a cusp form on $\mathfrak{g}$, we mean an element $f \in \mathcal{G}$ with the following property. Let (P, A) be a p-pair in G $(P = MN)$. Then if $P \neq G$,

$$\int_{\mathfrak{n}} f(Y+Z)dZ = 0$$

for all $Y \in \mathfrak{m}$. Let $\overset{o}{\mathcal{G}}$ denote the space of all cusp forms.

We assume that $\mathfrak{h}$ is elliptic. Put

$$\mathfrak{g}_{\mathfrak{h}} = {}^G(\mathfrak{h}')$$

and fix $\theta \in C_c^\infty(\mathfrak{g}_{\mathfrak{h}})$.

Lemma 1. $\hat{\theta} \in \overset{o}{\mathcal{G}}$ and

$$\int_{G/Z} dx^* \left| \int_{\mathfrak{g}} f(X)\hat{\theta}(xX)dX \right| < \infty$$

<u>for</u> $f \in \mathcal{D}$.

Here Z is the maximal split torus lying in the center of G and dx^* an invariant measure on G/Z. We omit the proof which is not difficult.

Put

$$\Theta(f) = \int_{G/Z} dx^* \int_{\mathfrak{g}} f(X)\hat{\theta}(xX)dX \qquad\qquad (f \in \mathcal{D}) \ .$$

Then $\Theta \in J(\omega)$ where $\omega = \text{Supp}\,\hat{\theta}$.

Theorem 9. Θ <u>is a locally summable function on</u> $\mathfrak{g}$ <u>which is locally constant</u> <u>on</u> $\mathfrak{g}'$.

This is proved in the same way as Theorem 16 (p. 92) of [1(a)]. The fact that $\hat{\theta}$ is a cusp form plays a crucial role in this proof.

We can now complete the proof of the first part of Theorem 8. As we have seen, it is enough to consider the case when $\mathfrak{h}$ is elliptic. Fix an open compact set U in $\mathfrak{g}$. Then we can choose a lattice L in $\mathfrak{g}$ such that $\hat{f} \in C_c(\mathfrak{g}/L)$ for all $f \in C_c^\infty(U)$. Applying Theorem 2, it is easy to conclude that there exists an open neighborhood ω_o of H_o in $\mathfrak{h}'$ such that $\phi_{\hat{f}}$ is constant on ω_o for every $f \in C_c^\infty(U)$.

On the other hand the mapping $(x, H) \longmapsto xH$ from $G \times \omega_o$ to $\mathfrak{g}$ is everywhere submersive. Hence [1(a), p. 49] we have a linear mapping $a \longmapsto f_a$ from $C_c^\infty(G \times \omega_o)$ to $\mathcal{D}$ such that

$$\int_{G \times \omega_o} a(x:H)F(xH)dxdH = \int_{\mathscr{g}} f_a(X)F(X)dX$$

for all $F \in C(\mathscr{g})$. (Here dH is the Haar measure on the additive group of $\mathscr{h}$.)
Fix $a \in C_c^{\infty}(G \times \omega_o)$ such that

$$\int a(x:H)|\eta(H)|^{-\frac{1}{2}}dxdH = 1$$

and put $\theta = f_a$. Then $\mathrm{Supp}\, \theta \subset {}^G\omega_o \subset \mathscr{g}_{\mathscr{h}}$. Now define Θ as in Theorem 9.
Then it is not difficult to verify the following result.

Lemma 2. $\Theta(f) = \phi_f(H_o)$ <u>for</u> $f \in C_c^{\infty}(U)$.

The first statement of Theorem 8 is now a consequence of Theorem 9.
Moreover the second statement of Theorem 8 would follow from Lemma 2 as
soon as we can verify that $|\eta|^{\frac{1}{2}}\Theta$ is bounded on $\mathscr{g}$.

§3. Proof of Theorem 10

Let $\mathscr{D}_o$ be the space of all $f \in \mathscr{D}$ such that $\phi_f^{\mathscr{h}} = 0$ for every Cartan
subalgebra $\mathscr{h}$ of $\mathscr{g}$. The next step is to prove the following theorem.

Theorem 10. $T(f) = 0$ <u>for</u> $T \in J$ <u>and</u> $f \in \mathscr{D}_o$.

Before sketching the proof, we need some preparation.
For $f \in \mathscr{D}$ and $t \in \Omega^X$ define $f_t(X) = f(t^{-1}X)$ $(X \in \mathscr{g})$. If T is a
distribution on $\mathscr{g}$, define the distribution $\rho(t)T$ by[5)]

$$\langle \rho(t)T, \ f \rangle = \langle T, \ f_t \rangle \qquad\qquad (f \in \mathcal{D}) \ .$$

$\mathcal{O}$ being any G-orbit in $\mathcal{g}$, put

$$d(\mathcal{O}) = \dim \mathcal{g}/\mathcal{Z}_X, \quad r(\mathcal{O}) = \dim \mathcal{Z}_X$$

where $\mathcal{Z}_X$ is the centralizer in $\mathcal{g}$ of a point $X \in \mathcal{O}$. We call $d(\mathcal{O})$ the dimension and $r(\mathcal{O})$ the rank of $\mathcal{O}$.

As before let $\mathcal{O}(0)$ denote the set of all nilpotent G-orbits in $\mathcal{g}$. The following lemma is an easy consequence of the Jacobson-Morosow theorem.

Lemma 3. <u>Fix</u> $\mathcal{O} \in \mathcal{O}(0)$. <u>Then</u>

$$\rho(t^2)\mu_{\mathcal{O}} = |t|^{d(\mathcal{O})} \mu_{\mathcal{O}}$$

<u>for</u> $t \in \Omega^X$.

For any integer d, let $\mathcal{N}_d$ denote the union of all nilpotent orbits $\mathcal{O}$ with $d(\mathcal{O}) \leq d$. Then if $\mathcal{O}$ is a nilpotent orbit of dimension d, $\mathcal{O} \cup \mathcal{N}_{d-1}$ is closed and $\mathcal{O}$ is open in $\mathcal{N}_d$. The next lemma is a simple consequence of these facts.

Lemma 4. <u>The measures</u> $\mu_{\mathcal{O}}$ $(\mathcal{O} \in \mathcal{O}(0))$ <u>form a base for</u> $J(\mathcal{N})$.

We shall now prove Theorem 10 by induction on $\dim G$. Let $\mathcal{Z}$ denote the center of $\mathcal{g}$. Fix a semisimple element γ in $\mathcal{g}$ and let $\mathcal{m}$ and M denote the centralizers of γ in $\mathcal{g}$ and G respectively. Assume that $\gamma \notin \mathfrak{z}$. Then the induction hypothesis is applicable to M and one can prove the following lemma.

Lemma 5. <u>There exists a</u> G-domain V <u>containing</u> γ <u>with the following property. If</u> $f \in \mathcal{O}_o$ <u>and</u> Supp $f \subset V$, <u>then</u> $T(f) = 0$ <u>for all</u> $T \in J$.

Observe that if $\mathcal{O}$ is any G-orbit, then $Cl\,\mathcal{O}$ contains a semisimple element. Moreover the set of all G-domains is stable under complementation, finite unions and finite intersections. Therefore the following corollary is immediate.

Corollary. <u>Let</u> f <u>be an element in</u> $\mathcal{O}_o$ <u>such that</u> Supp f <u>does not meet</u> $\mathfrak{z} + \mathcal{N}$. <u>Then</u> $T(f) = 0$ <u>for all</u> $T \in J$.

On the other hand it is easy to reduce the proof of Theorem 10 to the case when $\mathfrak{z} = \{0\}$. Assuming that $\mathfrak{z} = \{0\}$, one can sharpen the above corollary to the following lemma.

Lemma 6. <u>Suppose</u> f <u>is an element in</u> $\mathcal{O}_o$ <u>such that</u> $\mu_{\mathcal{O}}(f) = 0$ <u>for all</u> $\mathcal{O} \in \mathcal{O}(0)$. <u>Then</u> $T(f) = 0$ <u>for</u> $T \in J$.

Let J_o be the space of all $T \in J$ such that $T(f) = 0$ for $f \in \mathcal{O}_o$.

Lemma 7. $J = J_o + J(\mathcal{N})$.

Put

$$J_1 = J_o + J(\mathcal{N})$$

and let $\mathcal{B}_1$ be the subspace of all $f \in \mathcal{B}_o$ such that $\mu_\sigma(f) = 0$ for $\sigma \in \mathcal{O}(0)$. Then[6]

$$\dim \mathcal{B}_o/\mathcal{B}_1 \leq [\mathcal{O}(0)] < \infty \ .$$

The bilinear form

$$(T, \ f) \longrightarrow T(f)$$

on $J_1 \times \mathcal{B}_o$ may also be regarded as a form on

$$(J_1/J_o) \times (\mathcal{B}_o/\mathcal{B}_1)$$

and as such it is nondegenerate from Lemma 6. Let Λ be the space of all linear functions on $\mathcal{B}_o/\mathcal{B}_1$. Then it is clear that

$$\dim \Lambda = \dim \mathcal{B}_o/\mathcal{B}_1 = \dim J_1/J_o \ .$$

For $T \in J$, let λ_T denote the restriction of T on $\mathcal{B}_o$. Then $\lambda_T = 0$ on $\mathcal{B}_1$ from Lemma 6. Hence $T \longmapsto \lambda_T$ may be regarded as a linear mapping of J into Λ. Clearly J_o is the kernel of this mapping. Hence

$$\dim J/J_o \leq \dim \Lambda = \dim J_1/J_o < \infty$$

and this implies that $J_1 = J$.

Lemma 8. _Suppose_ $f \in \mathcal{B}_o$ _and_ $t \in \Omega^\times$. _Then_ f_t _and_ $\hat{f}$ _are also in_ $\mathcal{B}_o$.

It follows from the first part of Theorem 8 (which has been proved) that $\hat{f} \in \mathcal{B}_o$. The corresponding statement for f_t is obvious.

Now we come to the proof of Theorem 10. In view of Lemmas 4 and 7, it is enough to verify that $\mu_\theta \in J_o$ for every $\theta \in \mathcal{O}(0)$. Suppose this is false. Fix $\theta \in \mathcal{O}(0)$ such that $T = \mu_\theta$ is not in J_o. Then by Lemma 8 we can choose $f \in \mathcal{B}_o$ such that $T(\hat{f}) \neq 0$. But since $\hat{T} \in J$, we conclude from Lemma 7 that

$$\hat{T} = T_o + S$$

where $T_o \in J_o$ and $S \in J(\mathcal{N})$. Hence

$$\hat{T}(g) = S(g) \qquad\qquad (g \in \mathcal{B}_o) \ .$$

But it is easy to verify that

$$(h_t)^\wedge = |t|^n (\hat{h})_{t^{-1}} \qquad\qquad (h \in \mathcal{B},\ t \in \Omega^\times)$$

where $n = \dim \mathcal{g}$. Hence

$$S(f_{t^2}) = \langle T, (f_{t^2})^\wedge \rangle = |t|^{2n} \langle \rho(t^{-2})T, \hat{f} \rangle$$
$$= |t|^{2n-d} T(f) = |t|^{2n-d} S(f) \qquad\qquad (t \in \Omega^\times)$$

from Lemma 3, where $d = d(\theta)$.

On the other hand

$$S = \sum_i c_i \mu_i$$

where $c_i \in \underset{\sim}{C}$, $\mu_i = \mu_{\sigma_i}$ and $\sigma_1, \ldots, \sigma_s$ are all the distinct nilpotent G-orbits. Therefore by Lemma 3,

$$\rho(t^2)S = \sum_i c_i |t|^{d_i} \mu_i$$

where $d_i = d(\sigma_i)$. Therefore we conclude that

$$\sum_i c_i \mu_i(f) |t|^{d_i} = \sum_i c_i \mu_i(f) |t|^{2n-d} \qquad (t \in \Omega^{\times}) \ .$$

Now we may obviously assume that $\underset{\sim}{g} \neq \{0\}$. Then $\ell = \text{rank}\, \underset{\sim}{g} > 0$. Put $r = r(\sigma)$ and $r_i = r(\sigma_i)$. Then $r_i \geq \ell$ for every i and therefore

$$d_i - (2n-d) = -(r_i + r) \leq -2\ell < 0 \ .$$

Since distinct quasi-characters of $\Omega^{\times}$ are linearly independent, we conclude from the above relation that

$$\sum_i c_i \mu_i(f) = 0 \ .$$

But then

$$T(\hat{f}) = \hat{T}(f) = S(f) = 0 \ ,$$

which contradicts the definition of f. This completes the proof of Theorem 10.

§4. Some consequences of Theorem 10

We shall now derive some consequences of Theorem 10.

Lemma 9. _Let_ V _be a_ G-domain in $\mathcal{g}$. _Then the following two conditions on an element_ $f \in \mathcal{B}$ _are equivalent._

(1) $\mu_{\sigma}(f) = 0$ _for every regular_[7] _orbit_ σ _in_ V.

(2) $T(f) = 0$ _for every_ $T \in J(V)$.

Put $f_V = Ff$ where F is the characteristic function of V. Then if (1) holds it is clear that $f_V \in \mathcal{B}_0$. Therefore the lemma follows from Theorem 10.

Corollary. _Let_ ω _be a compact set,_ L _a lattice and_ V _a_ G-domain in $\mathcal{g}$. _Assume that_ $Cl^G \omega \supset V$. _Then we can choose a finite number of regular orbits_ σ_i $(1 \leq i \leq r)$ _in_ V _such that_ $j_L \mu_{\sigma_i}$ _span_ $j_L J(V)$.

This is a consequence of Theorem 2 and Lemma 9.

On the other hand we have the following simple lemma.

Lemma 10. _Let_ ω _be a compact set in_ $\mathcal{g}$. _Then we can choose another compact set_ ω_0 _and a_ G-domain D _in_ $\mathcal{g}$ _such that_

$$G_{\omega_0} \supset D \supset \omega .$$

We can now prove the first two parts of Theorem 3. Define D as in Lemma 10. Then $J(D) \supset J(\omega)$. Fix an open compact subset U of $\mathcal{g}$. Then we can choose a lattice L in $\mathfrak{g}$ such that $\hat{f} \in C_c(\mathfrak{g}/L)$ for all $f \in C_c^\infty(U)$.

Moreover, by the corollary of Lemma 9, we can choose regular orbits $\mathfrak{G}_i$ $(1 \leq i \leq r)$ in D such that $j_L \mu_{\mathfrak{G}_i}^{\cdot}$ span $j_L J(D)$. Now fix $T \in J(\omega)$. Then $\hat{T}$ is a linear combination of $\hat{\mu}_{\mathfrak{G}_i}$ on U. On the other hand we have verified the first part of Theorem 8 and therefore know that $\hat{\mu}_{\mathfrak{G}_i}$ are locally summable functions on $\mathfrak{g}$ which are locally constant on $\mathfrak{g}'$. This implies a similar statement about $\hat{T}$ on U.

<u>Remark</u>. Once Theorem 8 is proved completely, it would follow from the above proof that the function $|\eta|^{\frac{1}{2}} T$ is locally bounded on $\mathfrak{g}$.

§5. Proof of Theorem 4

Let γ be a semisimple element in $\mathfrak{g}$ and $\mathcal{O}(\gamma)$ the set of all G-orbits $\mathfrak{G}$ in $\mathfrak{g}$ such that $\gamma \in C\ell\,\mathfrak{G}$. Then $\mathcal{O}(\gamma)$ is a finite set and the next two lemmas are not difficult to prove.

Lemma 11. <u>Let f be an element in $\mathcal{D}$ such that $\mu_{\mathfrak{G}}(f) = 0$ for every</u> $\mathfrak{G} \in \mathcal{O}(\gamma)$. <u>Then if $\mathfrak{h}$ is a Cartan subalgebra of $\mathfrak{g}$ containing γ, we can choose a neighborhood ω of γ in $\mathfrak{h}$ such that</u>

$$\phi_f^{\mathfrak{h}} = 0$$

<u>on</u> $\omega \cap \mathfrak{h}'$.

Fix a compact neighborhood ω_o of γ in $\mathfrak{g}$.

389

Lemma 12. <u>We can choose</u> G-<u>domains</u> V_i $(i \geq 1)$ <u>containing</u> γ <u>such that</u>

$$G_{\omega_o} \supset V_1 \supset V_2 \supset \ldots$$

<u>and</u>

$$\bigcap_{i \geq 1} V_i = \bigcup_{\sigma \in \mathcal{O}(\gamma)} \sigma.$$

Let L be a lattice in $\mathfrak{g}$.

Corollary. <u>There exists an index</u> i_o <u>such that</u> $j_L J(V_i)$ <u>is spanned by</u> $j_L \mu_\sigma$ $(\sigma \in \mathcal{O}(\gamma))$ <u>for all</u> $i \geq i_o$.

By Theorem 2 we can choose i_o such that $j_L J(V_i) = j_L J(V_{i_o})$ for all $i \geq i_o$. Let λ be a linear function on $j_L J(\omega_o)$ such that $\lambda(\mu_{\sigma, L}) = 0$ for every $\sigma \in \mathcal{O}(\gamma)$. We have to show that $\lambda = 0$ on $j_L J(V_{i_o})$. Again by Theorem 2 there exists an element $f \in C_c(\mathfrak{g}/L)$ such that $\lambda(T_L) = T(f)$ for all $T \in J(\omega_o)$. Put

$$W = \bigcap_{i \geq 1} V_i$$

and suppose $\lambda \neq 0$ on $j_L J(V_{i_o})$. Then by the corollary of Lemma 9, we can choose, for every $i \geq i_o$, a regular orbit σ_i in V_i such that $\mu_{\sigma_i}(f) \neq 0$. Since $V_i \subset G_{\omega_o}$, we can arrange, by selecting a subsequence, that the following condition holds. There exists a Cartan subalgebra $\mathfrak{h}$ of $\mathfrak{g}$ and elements $H_o \in \mathfrak{h}$ and $H_i \in \mathfrak{h}' \cap V_i$ such that

$$\phi_f^{\partial}(H_i) \neq 0$$

and H_i converges to H_o in $\mathfrak{h}$ as $i \longrightarrow \infty$. Therefore we conclude from Lemma 11 that $\mu_{\theta}(f) \neq 0$ for some orbit $\theta \in \mathcal{O}(H_o)$. It is clear that $H_o \in W$ and therefore $\theta \in \mathcal{O}(\gamma)$. This gives a contradiction since

$$\mu_{\theta}(f) = \lambda(\mu_{\sigma, L}) = 0 \ .$$

We shall now follow Howe [2(d)] in our proof of Theorem 4.

Lemma 13. <u>Let</u> ω <u>be a compact subset of</u> $\mathfrak{g}$. <u>Then we can choose a lattice</u> L <u>in</u> $\mathfrak{g}$ <u>such that</u> $j_L J(\omega)$ <u>is contained in the space spanned by</u> $\mu_{\sigma, L}$ $(\theta \in \mathcal{O}(0))$.

Fix a lattice $L_o \supset \omega$. It would be enough to prove the lemma for L_o instead of ω. Put $R' = R \cap \Omega^{\times}$. Then it is obvious from Theorem 2 that we can choose $t_o \in R'$ such that

$$j_{L_o} J(t_o^2 L_o) = \bigcap_{t \in R'} j_{L_o} J(t L_o) \ .$$

By Lemma 10 we can choose a G-domain V in $\mathfrak{g}$ such that $0 \in V \subset {}^G L_o$. Since $t_1 L_o \subset V$ for some $t_1 \in R'$, it is clear that

$$\bigcap_{t \in R'} j_{L_o} J(t L_o) = \bigcap_{t \in R'} j_{L_o} J(t V) \ .$$

Hence we conclude from the corollary of Lemma 12 that $j_{L_o} J(t_o^2 L_o)$ is spanned by μ_{σ, L_o} $(\sigma \in \mathcal{O}(0))$.

Now let $T \in J(L_o)$. Then it is easy to verify that $\rho(t_o^{-2})T \in J(t_o^2 L_o)$ (in the notation of Lemma 3). Hence we can choose complex numbers c_σ such that

$$j_{L_o}(\rho(t_o^{-2})T - \sum_\sigma c_\sigma \mu_\sigma) = 0 \ .$$

Put $L = t_o^2 L_o$. Then the above relation implies that

$$j_L(T - \sum_\sigma c_\sigma \rho(t_o^2)\mu_\sigma) = 0$$

and the desired result now follows from Lemma 3.

On the other hand we have the following simple result.

Lemma 14. <u>For any lattice</u> L <u>in</u> $\mathfrak{g}$, $\mu_{\sigma,L}$ ($\sigma \in \mathcal{O}(0)$) <u>are linearly inde-</u><u>pendent</u>.

Let $\sigma_1, \ldots, \sigma_s$ be all the distinct nilpotent G-orbits. Put $\mu_i = \mu_{\sigma_i}$. It follows from Lemma 4 that we can choose $f_j \in \mathcal{B}$ such that $\mu_i(f_j) = \delta_{ij}$. Fix a lattice L_o in $\mathfrak{g}$ such that $f_j \in C_c(\mathfrak{g}/L_o)$ ($1 \le j \le s$). Choose $t \in \Omega^\times$ such that $L \subset t^2 L_o$ and put $g_j = (f_j)_{d_j t^2}$. Then $g_j \in C_c(\mathfrak{g}/L)$ and we conclude from Lemma 3 that $\mu_i(g_j) = |t|^{d_i} \delta_{ij}$ where $d_i = d(\sigma_i)$. This implies the lemma.

Corollary. <u>Let</u> V <u>be an open neighborhood of zero in</u> $\mathfrak{g}$. <u>Then the functions</u> $\hat{\mu}_\sigma$ ($\sigma \in \mathcal{O}(0)$) <u>are linearly independent on</u> $V \cap \mathfrak{g}'$.

In order to prove Theorem 4, choose L as in Lemma 13 and fix an open compact neighborhood U of zero in $\mathfrak{g}$ such that $\hat{f} \in C_c(\mathfrak{g}/L)$ for every $f \in C_c^\infty(U)$. For a given $T \in J(\omega)$, we can, by Lemma 13, choose complex numbers c_ϑ $(\vartheta \in \mathcal{O}(0))$ such that

$$T_L = \sum_\vartheta c_\vartheta \, \mu_{\vartheta, L} \ .$$

Then

$$\hat{T}(f) = T(\hat{f}) = \sum_\vartheta c_\vartheta \, \hat{\mu}_\vartheta \, (f) \qquad\qquad (f \in C_c^\infty(U)) \ .$$

On the other hand by Lemma 10 there exists a G-domain D such that

$$0 \in D \subset {}^G U \ .$$

This proves Theorem 4.

§6. Application of the induction hypothesis

In order to complete the proof of Theorem 8, it remains to verify the following result.

Theorem 11. <u>Let</u> ϑ <u>be a regular orbit in</u> $\mathfrak{g}$. <u>Then the function</u> $|\eta|^{\frac{1}{2}}\hat{\mu}_\vartheta$ <u>is locally bounded on</u> $\mathfrak{g}$.

Before proving this theorem, we derive some of its consequences.

Corollary 1. <u>Let</u> ω <u>be a compact subset of</u> $\mathfrak{g}$. <u>Then for any</u> $T \in J(\omega)$, <u>the function</u> $|\eta|^{\frac{1}{2}}\hat{T}$ <u>is locally bounded on</u> $\mathfrak{g}$.

This follows from the remark at the end of §4.

Corollary 2. <u>Let $\mathcal{O}$ be any G-orbit in $\mathcal{g}$. Then the function $|\eta|^{\frac{1}{2}}\hat{\mu}_{\mathcal{O}}$ is locally bounded on $\mathcal{g}$.</u>

This is obvious from Corollary 1.

The proof of Theorem 11 proceeds by induction on dim G. Let $\mathcal{g}_e$ denote the set of all $X \in \mathcal{g}'$ such that the centralizer of X in $\mathcal{g}$ is an elliptic Cartan subalgebra. Then $\mathcal{g}_e$ is an open G-invariant subset of $\mathcal{g}'$. A regular G-orbit $\mathcal{O}$ is called elliptic if $\mathcal{O} \subset \mathcal{g}_e$.

Lemma 15. <u>Suppose $\mathcal{O}$ is a regular G-orbit in $\mathcal{g}$ which is not elliptic. Then the function $|\eta|^{\frac{1}{2}}\hat{\mu}_{\mathcal{O}}$ is locally bounded on $\mathcal{g}$.</u>

As we have seen in §2, this follows from the induction hypothesis.

Fix an element $\theta \in \mathcal{D}$ such that

$$\int_{G/Z} dx^* \left| \int_{\mathcal{g}} f(X)\theta(xX)dX \right| < \infty$$

for all $f \in \mathcal{D}$. (Here the notation is the same as in Lemma 1.)

Theorem 12. <u>Define $\Theta \in J$ by</u>

$$\Theta(f) = \int_{G/Z} dx^* \int_{\mathcal{g}} f(X)\theta(xX)dX \qquad\qquad (f \in \mathcal{D}) \ .$$

<u>Then Θ is a locally summable function on $\mathcal{g}$ which is locally constant on $\mathcal{g}'$. Moreover $|\eta|^{\frac{1}{2}}\Theta$ is bounded on $\mathcal{g}$.</u>

It is clear that $\hat{\Theta} \in J(\mathrm{Supp}\,\hat{\theta})$. The first statement therefore follows from the part of Theorem 3 which has been proved. Since $\Theta \in J(\mathrm{Supp}\,\theta)$, it would be enough to verify that $|\eta|^{\frac{1}{2}}\Theta$ is locally bounded on $\mathfrak{g}$.

Lemma 16. Theorems 11 and 12 are equivalent.

Corollary 1 of Theorem 11 implies Theorem 12. The reverse implication follows from Lemma 2.

We shall prove Theorem 12 also by induction on $\dim G$.

Fix a semisimple element γ in $\mathfrak{g}$ $(\gamma \notin \mathfrak{z})$.[8] Then by considering the centralizer of γ in G, it is not difficult to derive the following result from the induction hypothesis.

Lemma 17. $|\eta|^{\frac{1}{2}}\Theta$ remains bounded on some neighborhood of γ in $\mathfrak{g}$.

Corollary. Let $\mathcal{O}$ be a regular G-orbit in $\mathfrak{g}$. Then $|\eta|^{\frac{1}{2}}\hat{\mu}_{\mathcal{O}}$ remains bounded on some neighborhood of γ in $\mathfrak{g}$.

In view of Lemma 15, we may assume that $\mathcal{O}$ is elliptic. Then our assertion follows from Lemmas 17 and 2.

§7. Reformulation of the problem and completion of the proof

We shall now approach our problem from a somewhat different angle.

Fix an open compact subgroup K of G and let dk denote the normalized Haar measure of K. Let $\mathfrak{h}_1$, $\mathfrak{h}_2$ be two Cartan subalgebras of $\mathfrak{g}$ and A_2 the split component of the Cartan subgroup of G corresponding to $\mathfrak{h}_2$. Let $x \longmapsto x^*$ denote the projection of G on $G^* = G/A_2$ and dx^* an invariant measure on G^*.

Lemma 18. Fix compact sets $\omega_i \subset \mathfrak{h}_i'$ $(i = 1, 2)$. Then we can choose a compact set C^* in G^* with the following property. Suppose $H_i \in \omega_i$ and $x \in G$. Then the integral

$$\int_K \chi(B(kH_1, \; xH_2)) dk$$

is zero unless $x^* \in C^*$.

The proof is rather long and so we do not give it here.

Put

$$\Phi(H_1 : H_2) = |\eta(H_1)|^{\frac{1}{2}} |\eta(H_2)|^{\frac{1}{2}} \int_{G/A_2} dx^* \int_K \chi(B(kH_1, \; xH_2)) dk$$

for $H_i \in \mathfrak{h}_i'$. Then it follows from Lemma 18 that Φ is a locally constant function on $\mathfrak{h}_1' \times \mathfrak{h}_2'$.

Theorem 13. Φ is locally bounded on $\mathfrak{h}_1 \times \mathfrak{h}_2$.

Fix $H_2 \in \mathfrak{h}_2'$ and put $\mu_{H_2} = \mu_\theta$ where $\theta = {}^G(H_2)$ and the measure μ_θ is normalized so as to correspond to dx^*. Then it is not difficult to prove the following result.

Lemma 19. $\Phi(H_1 : H_2) = |\eta(H_1)|^{\frac{1}{2}} |\eta(H_2)|^{\frac{1}{2}} \hat{\mu}_{H_2}(H_1)$ $\qquad (H_i \in \mathcal{F}_i', \; i = 1, 2).$

It is clear from this lemma that Theorem 13 implies Theorem 11.

Lemma 20. <u>Put</u> $\omega_i' = \omega_i \cap \mathfrak{g}'$ <u>where</u> ω_i <u>is a compact subset of</u> $\mathcal{F}_i$ $(i = 1, 2)$. <u>Then if</u> $\omega_1 \cap \mathfrak{z} = \emptyset$, Φ <u>remains bounded on</u> $\omega_1' \times \omega_2'$.

Let U be an open compact subset of $\mathfrak{g}$ containing ω_1 and L a lattice in $\mathfrak{g}$ such that $\hat{f} \in C_c(\mathfrak{g}/L)$ for all $f \in C_c^\infty(U)$. Put

$$\phi_g = \phi_g^{\mathfrak{z}_2} \qquad\qquad (g \in \mathfrak{G})$$

and for $H \in \omega_2'$, let σ_H denote the distribution on U given by

$$\sigma_H(f) = \phi_{\hat{f}}(H) = |\eta(H)|^{\frac{1}{2}} \mu_H(\hat{f}) \qquad\qquad (f \in C_c^\infty(U)) \; .$$

By Theorem 2 we can choose $\gamma_i \in \omega_2'$ $(1 \le i \le r)$ such that $\sigma_i = \sigma_{\gamma_i}$ form a base for the space spanned by σ_H $(H \in \omega_2')$. Choose $f_j \in C_c^\infty(U)$ such that $\sigma_i(f_j) = \delta_{ij}$. Then

$$\sigma_H = \sum_j \sigma_H(f_j)\sigma_j \qquad\qquad (H \in \omega_2') \; .$$

Since $\omega_1 \subset U$, we conclude from Lemma 19 that

$$\Phi(H_1 : H_2) = \sum_j \phi_{\hat{f}_j}(H_2)\Phi(H_1 : \gamma_j) \qquad (H_i \in \omega_i', \; i = 1, 2) \; .$$

But we know [1(a), Thm. 13, p. 64] that

$$\sup_{H \in \mathfrak{h}_2'} |\phi_g(H)| < \infty$$

for $g \in \mathfrak{G}$. Therefore there exists a number $c > 0$ such that

$$|\Phi(H_1 : H_2)| \le c \sum_j |\Phi(H_1 : \gamma_j)|$$

for all $H_i \in \omega_i'$ $(i = 1, 2)$. The required result now follows from Lemma 19 and the corollary of Lemma 17.

Let $\mathfrak{g}_1$ be the derived algebra of $\mathfrak{g}$. It is clear that

$$\Phi(H_1 + Z_1 : H_2 + Z_2) = \chi(B(Z_1, Z_2))\Phi(H_1 : H_2)$$

for $H_i \in \mathfrak{h}_i'$ and $Z_i \in \mathfrak{z}$ $(i = 1, 2)$. Fix a norm[9] on $\mathfrak{g}$ and a number $T \ge 1$ and let ω_i denote the set of all points $H \in \mathfrak{h}_i \cap \mathfrak{g}_1$ with $|H| \le T$. Then in order to prove Theorem 13, it is sufficient to verify that Φ remains bounded on $\omega_1' \times \omega_2'$.

Let U_1 denote the set of all $H_1 \in \omega_1$ with $|H_1| \ge 1$. Then U_1 is a compact subset of ω_1 and $U_1 \cap \mathfrak{z} = \emptyset$. Put $U_1' = U_1 \cap \mathfrak{g}'$. Then by Lemma 20 we can choose a number $b > 0$ such that $|\Phi| \le b$ on $U_1' \times \omega_2'$. On the other hand every element in ω_1' is of the form tH_1 where $t \in R'$ and $H_1 \in U_1'$. Moreover $t\omega_2' \subset \omega_2'$. Therefore

$$|\Phi(tH_1 : H_2)| = |\Phi(H_1 : tH_2)| \le b \qquad (H_2 \in \omega_2')$$

and we conclude that $|\Phi| \le b$ on $\omega_1' \times \omega_2'$. This proves Theorem 13 and therefore also Theorems 11 and 12.

Theorems 3 and 8 are now fully established.

§8. Some results on Shalika's germs

Fix a Cartan subalgebra $\mathfrak{h}$ of $\mathfrak{g}$. The following theorem is a slightly sharpened form of a result of Shalika [4].

Theorem 14. There exist functions Γ_{σ} ($\sigma \in \mathcal{O}(0)$) on $\mathfrak{h}'$ with the following two properties:

(1) $\Gamma_{\sigma}(t^2 H) = |t|^{r(\sigma)-\ell} \Gamma_{\sigma}(H)$ $\qquad (t \in \Omega^{\times}, H \in \mathfrak{h}')$.

(2) For every $f \in \mathcal{B}$, we can choose a neighborhood ω of zero in $\mathfrak{h}$ such that

$$\phi_f^{\mathfrak{h}} = \sum_{\sigma} \mu_{\sigma}(f) \Gamma_{\sigma}$$

on $\omega \cap \mathfrak{h}'$.

These functions are unique. They take only real values and they satisfy, in addition, the following conditions.

(i) $\Gamma_{\sigma}(Z+H) = \Gamma_{\sigma}(H)$ $(Z \in \mathfrak{z}, H \in \mathfrak{h}')$.

(ii) Γ_{σ} is locally constant on $\mathfrak{h}'$.

(iii) Γ_{σ} is locally bounded on $\mathfrak{h}$.

We recall that $\mathcal{O}(0)$ is the set of all nilpotent G-orbits, $\mathfrak{z}$ is the center of $\mathfrak{g}$, $\ell = \operatorname{rank} \mathfrak{g}$ and $r(\sigma)$ has been defined in §3. We shall call Γ_{σ} the σ-germ on $\mathfrak{h}$.

Corollary. <u>Let</u> L <u>be a lattice in</u> $\mathfrak{g}$. <u>Then there exists a neighborhood</u> ω <u>of</u> <u>zero in</u> $\mathfrak{h}$ <u>such that</u>

$$\phi_f^{\mathfrak{h}} = \sum_{\sigma} \mu_{\sigma}(f) \Gamma_{\sigma}$$

<u>on</u> $\omega \cap \mathfrak{h}'$ <u>for all</u> $f \in C_c(\mathfrak{g}/L)$.

Fix a compact neighborhood ω_o of zero in $\mathfrak{g}$ and choose a lattice L_o in $\mathfrak{g}$ so that the pair (ω_o, L_o) fulfills the conditions of Lemma 13. Let $\sigma_1, \ldots, \sigma_s$ be all the distinct nilpotent G-orbits. Put $\mu_i = \mu_{\sigma_i}$ and $\Gamma_i = \Gamma_{\sigma_i}$. By Lemma 14 we can choose $f_j \in C_c(\mathfrak{g}/L_o)$ such that $\mu_i(f_j) = \delta_{ij}$. Moreover by Theorem 14 there exists a neighborhood ω_1 of zero in $\omega_o \cap \mathfrak{h}$ such that

$$\phi_{f_i} = \Gamma_i \qquad\qquad (1 \le i \le s)$$

on $\omega_1 \cap \mathfrak{h}'$. Then it follows from Lemma 13 that

$$\phi_f = \sum_i \mu_i(f) \Gamma_i$$

on $\omega_1 \cap \mathfrak{h}'$ for all $f \in C_c(\mathfrak{g}/L_o)$.

Now fix $t \in \Omega^{\times}$ such that $L_o \subset t^2 L$ and put $\omega = t^{-2}\omega_1$. Then the required result follows from Lemma 3 and part (1) of Theorem 14.

Let A denote the split component of the Cartan subgroup of G corresponding to $\mathfrak{h}$ and M and $\mathfrak{m}$ the centralizers of A in G and $\mathfrak{g}$ respectively. Let $\eta_{\mathfrak{m}}$ be the polynomial function on $\mathfrak{m}$ which is the analogue of η on $\mathfrak{g}$. Then $\mathfrak{h}' \subset \mathfrak{h}''$ where $\mathfrak{h}''$ is the set of all points $H \in \mathfrak{h}$ with $\eta_{\mathfrak{m}}(H) \neq 0$.

For any nilpotent M-orbit ξ in $\mathfrak{m}$, let ν_ξ denote the corresponding M-invariant measure on $\mathfrak{m}$ and Γ_ξ the function on $\mathfrak{h}''$ defined by Theorem 14 (applied to $(M, \mathfrak{h})$ instead of $(G, \mathfrak{g})$). Fix $\sigma \in \mathcal{O}(0)$. Then it follows from Theorem 14 and Lemma 4 that we can choose $f \in \mathcal{A}$ and an open neighborhood ω of zero in $\mathfrak{h}$ such that

$$\phi_f = \Gamma_\sigma$$

on $\omega \cap \mathfrak{h}'$. Now fix $P \in \mathcal{P}(A)$. Then we have seen in §2 that

$$\phi_f = \phi_{f_P}^{M/\mathfrak{h}}$$

on $\mathfrak{h}'$. Hence

$$\Gamma_\sigma = \sum_\xi \nu_\xi(f_P)\Gamma_\xi$$

on $\omega \cap \mathfrak{h}'$ if ω is sufficiently small. Therefore the following result is obvious.

Lemma 20. Γ_σ <u>is a linear combination of</u> Γ_ξ <u>where</u> ξ <u>runs over those nilpotent</u> M-<u>orbits in</u> $\mathfrak{m}$ <u>for which</u> $r(\xi) = r(\sigma)$.

Corollary 1. $\Gamma_\sigma = 0$ <u>if</u> $d(\sigma) < \dim G/M$.

Let $n = \dim G$ and $m = \dim M$. Then

$$r(\theta) = n - d(\theta), \quad r(\xi) = m - d(\xi) \ .$$

Therefore the condition $r(\theta) = r(\xi)$ implies that

$$d(\theta) = n - m + d(\xi) \geq n - m \ .$$

Let Γ_0 denote the function Γ_θ for $\theta = \{0\}$.

Corollary 2. $\Gamma_0 = 0$ <u>unless</u> $\mathfrak{h}$ <u>is elliptic.</u>

This is obvious from Corollary 1.

$$\S 9. \quad \text{Proof of Theorem 15}$$

For $\theta \in \mathcal{O}(0)$ put

$$\psi_\theta = |\eta|^{\frac{1}{2}} \mu_\theta$$

and for any Cartan subalgebra $\mathfrak{h}$, let $\Gamma_\theta^{\mathfrak{h}}$ denote the θ-germ on $\mathfrak{h}$ corresponding to Theorem 14. We return to the notation of Theorem 13.

Lemma 21. <u>Fix a compact subset</u> ω_1 <u>of</u> $\mathfrak{h}_1$. <u>Then we can choose a</u> <u>neighborhood</u> ω_2 <u>of zero in</u> $\mathfrak{h}_2$ <u>such that</u>

$$\Phi(H_1 : H_2) = \Sigma \psi_\theta (H_1) \Gamma_\theta^{\mathfrak{h}_2}(H_2)$$

<u>for</u> $H_i \in \omega_i \cap \mathfrak{h}_i'$ $(i = 1, 2)$.

We may assume that ω_1 is open in $\mathfrak{h}_1$. Let U be an open compact subset of $\mathcal{g}$ containing ω_1. Fix a lattice L in $\mathcal{g}$ such that $f \in C_c(\mathcal{g}/L)$ for $f \in C_c^\infty(U)$. Then by the corollary of Theorem 14, we can choose a neighborhood ω_2 of zero in $\mathcal{h}_2$ such that

$$\phi_f = \sum_\vartheta \mu_\vartheta(f)\Gamma_\vartheta \qquad (f \in C_c(\mathcal{g}/L))$$

on $\omega_2' = \omega_2 \cap \mathcal{h}_2'$. (Here $\phi_f = \phi_f^{\mathcal{h}_2}$ and $\Gamma_\vartheta = \Gamma_\vartheta^{\mathcal{h}_2}$.) Hence

$$\phi_f(H_2) = \sum_\vartheta \hat{\mu}_\vartheta(f)\Gamma_\vartheta(H_2)$$

for $f \in C_c^\infty(U)$ and $H_2 \in \omega_2'$. Our assertion now follows from Lemma 19.

Lemma 22. $\quad \psi_\vartheta(t^2 X) = |t|^{r(\vartheta)-\ell} \psi_\vartheta(X) \qquad (t \in \Omega^\times,\ X \in \mathcal{g}')$ <u>for</u> $\vartheta \in \vartheta(0)$.

This is immediate from Lemma 3.

Define $\mathcal{g}_e$ as in §6 and put

$$\Phi(X_1 : X_2) = |\eta(X_1)|^{\frac{1}{2}} |\eta(X_2)|^{\frac{1}{2}} \int_{G/Z} dx^* \int_K \chi(B(kX_1,\ xX_2))dk$$

for $X_1,\ X_2 \in \mathcal{g}_e$. (Here K and dk have the same meaning as in Lemma 18 and Z and dx^* as in Theorem 12.)

Lemma 23. $\quad \Phi \in C^\infty(\mathcal{g}_e \times \mathcal{g}_e)$ <u>and</u>

$$\Phi(X_1 : X_2) = \Phi(X_2 : X_1)$$

<u>for</u> $X_1,\ X_2 \in \mathcal{g}_e$.

This is obvious from Lemma 18.

Corollary. _Let_ $\mathfrak{h}_1$, $\mathfrak{h}_2$ _be two elliptic Cartan subalgebras of_ $\mathfrak{g}$ _and_ d _an_ _integer_ ≥ 0. _Then_

$$\sum_{d(\sigma)=d} \psi_\sigma(H_1)\Gamma_\sigma^{\mathfrak{h}_2}(H_2) = \sum_{d(\sigma)=d} \psi_\sigma(H_2)\Gamma_\sigma^{\mathfrak{h}_1}(H_1)$$

for $H_i \in \mathfrak{h}_i$ $(i = 1, 2)$. _Here_ σ _runs over all nilpotent_ G-_orbits in_ $\mathfrak{g}$ _such_ _that_ $d(\sigma) = d$.

This follows from Lemmas 21, 22 and 23 if we take into account part (1) of Theorem 14.

Lemma 24. _Let_ $\mathfrak{h}_i$ $(1 \leq i \leq r)$ _be a complete set of Cartan subalgebras of_ _no two of which are conjugate under_ G. _Suppose_ c_σ $(\sigma \in \mathcal{O}(0))$ _are complex_ _numbers such that_

$$\sum_\sigma c_\sigma \Gamma_\sigma^{\mathfrak{h}_i} = 0 \qquad (1 \leq i \leq r)$$

Then $c_\sigma = 0$ _for all_ σ.

By Lemma 4 we can choose $f \in \mathcal{A}$ such that $\mu_\sigma(f) = c_\sigma$ for all $\sigma \in \mathcal{O}(0)$. Moreover by Theorem 14 we can choose neighborhoods ω_i of zero in $\mathfrak{h}_i$ such that

$$\phi_f^{\mathfrak{h}_i} = \sum_\sigma \mu_\sigma(f)\Gamma_\sigma^{\mathfrak{h}_i} = 0$$

on $\omega_i' = \omega_i \cap \mathfrak{h}_i'$ $(1 \leq i \leq r)$. It is not difficult to see that there exists a compact neighborhood ω of zero in $\mathcal{g}$ such that ${}^G\omega \cap \mathfrak{h}_i \subset \omega_i$ $(1 \leq i \leq r)$. By Lemma 10 we can choose a G-domain V in $\mathcal{g}$ such that

$$0 \in V \subset {}^G\omega \ .$$

Put $f_0 = Ff$ where F is the characteristic function of V. Then it is clear that

$$\phi_{f_0}^{\mathfrak{h}_i} = 0 \qquad\qquad (1 \leq i \leq r) \ .$$

Hence we conclude from Theorem 10 that $\mu_\theta(f_0) = 0$ $(\theta \in \mathcal{O}(0))$. But since $F = 1$ on $\mathcal{N}$, we get

$$c_\theta = \mu_\theta(f) = \mu_\theta(f_0) = 0 \ .$$

Define μ_0 and $\Gamma_0^{\mathfrak{h}}$ corresponding to the orbit $\mathfrak{r} = \{0\}$.

Theorem 15. <u>There exists a real number</u> $c \neq 0$ <u>such that</u>

$$\Gamma_0^{\mathfrak{h}}(H) = c\,|\eta(H)|^{\frac{1}{2}} \qquad\qquad (H \in \mathfrak{h}')$$

<u>for every elliptic Cartan subalgebra</u> $\mathfrak{h}$ <u>of</u> $\mathcal{g}$.

Clearly

$$\mu_0(f) = f(0) \qquad\qquad (f \in \mathcal{D})$$

and therefore $\mu_0 = 1$ and $\psi_0 = |\eta|^{\frac{1}{2}}$. It follows from Lemma 24 and Cor. 2

of Lemma 20 that there exists an elliptic Cartan subalgebra $\mathfrak{h}_1$ such that $\Gamma_0^{\mathfrak{h}_1} \neq 0$. Let $\mathfrak{h}_2$ be any elliptic Cartan subalgebra of $\mathfrak{g}$. Since $\{0\}$ is the only nilpotent G-orbit of dimension zero, we conclude from the corollary of Lemma 23 that

$$|\eta(H_1)|^{\frac{1}{2}}\Gamma_0^{\mathfrak{h}_2}(H_2) = |\eta(H_2)|^{\frac{1}{2}}\Gamma_0^{\mathfrak{h}_1}(H_1)$$

for $H_i \in \mathfrak{h}_i'$ ($i = 1, 2$). This implies our assertion if we recall that $\Gamma_0^{\mathfrak{h}_1}$ takes only real values.

Part II. An extension of Howe's theorem

§10. The space J(V, t, L)

Fix a norm[9] on $\mathfrak{g}$ and for any non-empty subset ω of $\mathfrak{g}$, put

$$|\omega| = \sup_{X\epsilon\omega}|X| \ .$$

For $t > 0$, let $\mathfrak{g}(t)$ denote the set of all $X \epsilon \mathfrak{g}$ with $|X| \le t$. Moreover let S be the set of all $X \epsilon \mathfrak{g}$ with $|X| = 1$.

Let L be a lattice in $\mathfrak{g}$, V a neighborhood of $\mathcal{N} \cap S$ in S and t a positive real number. For $X \epsilon \mathfrak{g}$ let f_X be the characteristic function of the coset $X+L$ in $\mathfrak{g}$. Then we denote by J(V, t, L) the subspace of all $T \epsilon J$ with the following property. If $X \epsilon \mathfrak{g}$ and $|X| \ge t$, then $T(f_X) = 0$ unless[10] $X \epsilon \Omega V$.

Fix (V, t, L) as above. By C(V, t, L) we mean the following condition on (V, t, L).

C(V, t, L): <u>Suppose</u> $X \epsilon \Omega V$ <u>and</u> $|X| \ge t$. <u>Put</u>

$$\phi_a = \int_G a(x)(f_X)^x dx \qquad\qquad (a \epsilon C_c^\infty(G)) \ ,$$

<u>where</u> f_X <u>is the characteristic function of</u> $X+L$ <u>on</u> $\mathfrak{g}$. <u>Then we can choose</u> a <u>such that:</u>

1) $\int_G a dx \ne 0$,

2) $\phi_a \epsilon C_c(\mathfrak{g}/L)$,

3) Supp $\phi_a \subset \mathcal{g}(t')$ for some t' such that

$$0 < t' < |X| \ .$$

The following theorem provides the justification for this condition.

Theorem 16. Suppose $C(V, t, L)$ holds. Then

$$\dim j_L J(V, t, L) < \infty \ .$$

It is enough to verify the following lemma which imitates Howe's original argument.

Lemma 25. Suppose $C(V, t, L)$ holds and T is an element in $J(V, t, L)$ such that $T(f_X) = 0$ for all $X \in \mathcal{g}(t)$. Then $j_L T = 0$.

For otherwise suppose $j_L T \neq 0$. Then $T(f_X) \neq 0$ for some $X \in \mathcal{g}$. Clearly $|X| > t$. Choose X so that $T(f_X) \neq 0$ and $|X|$ has the least possible value. Then since $T \in J(V, t, L)$ and $|X| > t$, it follows that $X \in \Omega V$. Now choose $a \in C_c^\infty(G)$ as in the statement of the condition $C(V, t, L)$. Then

$$T(\phi_a) = \int a\,dx . T(f_X) \neq 0 \ .$$

On the other hand, Supp $\phi_a \subset \mathcal{g}(t')$ with $0 < t' < |X|$ and $\phi_a \in C_c(\mathcal{g}/L)$. Therefore it follows from the definition of X that $T(\phi_a) = 0$. This contradiction proves the lemma.

It is obvious that

$$J(V,\ t,\ L) \subset J(V,\ t',\ L)$$

for $t' \geq t$. Put

$$J(V,\ \infty,\ L) = \bigcup_{t>0} J(V,\ t,\ L)\ .$$

If $t_o \geq t$ and V_o is a neighborhood of $\mathcal{N} \cap S$ in V, then it is clear that $C(V,\ t,\ L)$ implies $C(V_o,\ t_o,\ L)$.

Put

$$J_o = \bigcup_{\omega} J(\omega)$$

where ω runs over all compact subsets of $\mathcal{g}$ and let

$$J_o(V,\ t,\ L) = J_o \cap J(V,\ t,\ L)\ .$$

Theorem 17. <u>Suppose</u> $C(V,\ t,\ L)$ <u>holds.</u> <u>Choose a lattice</u> Λ <u>in</u> $\mathcal{g}$ <u>such that</u> $|\Lambda| < 1$ <u>and</u>

$$V_o = (\mathcal{N} \cap S) + \Lambda \subset V\ .$$

<u>Then</u>

$$j_L J(V_o,\ t_o,\ L) = j_L J_o(V_o,\ t_o,\ L)$$

<u>for all</u> t_o <u>sufficiently large.</u>

We shall give a proof of this theorem in §11.

Corollary. $j_L J(V_o, \infty, L) \subset j_L J_o$.

§11. Proof of Theorem 17

Let $S(\mathfrak{g})$ be the symmetric algebra over $\mathfrak{g}$ and $I(\mathfrak{g})$ the subalgebra of all G-invariants in $S(\mathfrak{g})$. By means of the bilinear form B, we may identify elements of $S(\mathfrak{g})$ with polynomial functions on $\mathfrak{g}$. It is easy to see that one can choose homogeneous elements $p_i \neq 0$ $(1 \leq i \leq r)$ in $I(\mathfrak{g})$ with the following two properties:

(1) $\mathcal{N}$ is exactly the set of all points $X \in \mathfrak{g}$ where $p_i(X) = 0$ $(1 \leq i \leq r)$.

(2) Let U be any subset of $\mathfrak{g}$ such that $|p_i|$ $(1 \leq i \leq r)$ remain bounded on U. Then we can choose a compact set ω in $\mathfrak{g}$ such that $U \subset {}^G\omega$. Let d_i be the degree of p_i. Put

$$\phi(X) = \max_{1 \leq i \leq r} |p_i(X)|^{1/d_i} \qquad (X \in \mathfrak{g}) \ .$$

Then ϕ is a continuous function from $\mathfrak{g}$ to $\mathbb{R}$ and $\phi(X) \geq 0$. Moreover $\phi(cX) = |c|\phi(X)$ $(c \in \Omega)$ and $\mathcal{N}$ is exactly the set of points where ϕ is zero. Finally ϕ is locally constant on $\mathfrak{g}$ outside $\mathcal{N}$.

Now we come to the proof of Theorem 17. Let W be the complement of V_o in S. Then W is an open and compact set and $W + \Lambda = W$. Moreover ϕ is locally constant on W. Hence we can choose a lattice $\Lambda_o \subset \Lambda$ such that $\phi(X+\lambda) = \phi(X)$ for all $X \in W$ and $\lambda \in \Lambda_o$. Fix $t_o \geq t$ such that $c^{-1}L \subset \Lambda_o$

if $|c| \geq t_o$ ($c \in \Omega^\times$). Then the following lemma is obvious.

Lemma 26. <u>Suppose</u> X <u>is an element in</u> $\mathcal{O}$ <u>such that</u> $|X| \geq t_o$ <u>and</u> $X \notin \Omega V_o$. <u>Then</u> $\phi(X+\lambda) = \phi(X)$ <u>for</u> $\lambda \in L$.

Let q be the number of elements in the residue field of Ω. For $\nu \geq 0$ ($\nu \in \underset{\sim}{Z}$), put

$$\Phi_\nu(X) = \begin{cases} 1 & \text{if } \phi(X) \leq q^\nu , \\ 0 & \text{otherwise} \end{cases} \qquad (X \in \mathcal{O}) .$$

Then Φ_ν is a locally constant function on $\mathcal{O}$.

Lemma 27. <u>Fix</u> $t_1 \geq t_o$, $T \in J(V_o, t_1, L)$ <u>and put</u> $T_\nu = \Phi_\nu T$ ($\nu \geq 0$). <u>Then</u> $T_\nu \in J_o(V_o, t_1, L)$.

It is obvious that $T_\nu \in J_o$. Now suppose $X \in \mathcal{O}$, $|X| \geq t_1$ and $X \notin \Omega V_o$. Then $\Phi_\nu f_X = \Phi_\nu(X) f_X$ by Lemma 26 and therefore

$$T_\nu(f_X) = \Phi_\nu(X) T(f_X) = 0 .$$

This proves that $T_\nu \in J_o(V_o, t_1, L)$.

Corollary. <u>Suppose</u> C(V, t, L) <u>holds</u>. <u>Then</u>

$$j_L J(V_o, t_1, L) = j_L J_o(V_o, t_1, L)$$

<u>for</u> $t_1 \geq t_o$.

Since $C(V, t, L)$ implies $C(V_o, t_1, L)$, we conclude from Theorem 16 that $\dim j_L J(V_o, t_1, L) < \infty$. Now fix $T \in J(V_o, t_1, L)$ and put $T_\nu = \Phi_\nu T$. It is obvious that for any fixed $f \in \mathfrak{O}$, $T_\nu(f) = T(f)$ for all ν sufficiently large. This shows that $j_L T_\nu$ converges to $j_L T$ in the finite-dimensional space $j_L J(V_o, t_1, L)$ as $\nu \longrightarrow \infty$. Clearly this implies the corollary and hence also proves Theorem 17.

$$\S 12. \quad \text{Reduction of the condition } C(V, t, L)$$

Let L be a lattice in $\mathfrak{g}$. For $X \in \mathfrak{g}$ define f_X as before ($\S 10$) and let $C(L)$ denote the following condition on L.

$C(L)$: Given $X_o \in \mathcal{N} \cap S$, we can choose a neighborhood W of X_o in S and a number $t > 0$ with the following properties. Put

$$\phi_a = \int_G a(x)(f_X)^x dx \qquad\qquad (a \in C_c^\infty(G))$$

where X is any element in ΩW with $|X| \geq t$. Then we can choose a such that:

1) $\int a\, dx \neq 0$,

2) $\phi_a \in C_c(\mathfrak{g}/L)$,

3) $\mathrm{Supp}\, \phi_a \subset \mathfrak{g}(t')$ for some t' such that

$$0 < t' < |X| \quad .$$

Lemma 28. <u>Suppose</u> $C(L)$ <u>holds.</u> <u>Then we can choose a neighborhood</u> V <u>of</u> $\mathcal{N} \cap S$ <u>in</u> S <u>and a number</u> $t > 0$ <u>such that</u> $C(V, t, L)$ <u>holds.</u>

This is a simple consequence of the fact that $\mathcal{N} \cap S$ is compact.

Fix a maximal split torus A_o in G and an open compact subgroup K of Bruhat-Tits [1(a), p. 16] corresponding to A_o. Let $X(A_o)$ denote as usual the group of all rational characters of A_o which are defined over Ω. A lattice L in $\mathcal{G}$ will be said to be adapted to (K, A_o) if there exists a base X_i $(1 \leq i \leq n)$ for $\mathcal{G}$ over Ω and elements $\chi_i \in X(A_o)$ such that:

(1) $\mathrm{Ad}(a)X_i = \chi_i(a)X_i$ $(a \in A_o, \ 1 \leq i \leq n)$,

(2) $L = \sum_i R X_i$,

(3) $^K L = L.$

Let L^* be the lattice dual to L. Then L^* consists of all points $\mu \in \mathcal{G}$ such that

$$\chi(B(\mu, \lambda)) = 1$$

for all $\lambda \in L$. It is easy to check that if L is adapted to (K, A_o), the same holds for L^*.

Lemma 29. <u>Let</u> L <u>be a lattice in</u> $\mathcal{G}$ <u>which is adapted to</u> (K, A_o). <u>Then</u> $C(L)$ <u>holds.</u>

This lemma is, in fact, the main step in Howe's proof of his theorem (Theorem 2).

It follows from a result of Tits that there always exists a lattice L which is adapted to (K, A_o). By multiplying it by an element in R' such a lattice can be made arbitrarily small.

We shall say that L is well adapted if it is adapted to (K, A_o) for a suitable choice of (K, A_o). Combining Lemmas 28 and 29 we get the following theorem.

Theorem 18. <u>Let L be a well adapted lattice in $\mathcal{Y}$. Then we can choose a neighborhood V of $\mathcal{N} \cap S$ in S and a number t > 0 such that C(V, t, L) holds.</u>

Part III.   Theory on the group

§13.   Representations of compact groups

Let K be a compact group. We denote by $\mathcal{C}(K)$ the set of all equivalence classes of irreducible representations of K. For $\underline{d} \in \mathcal{C}(K)$, let $\xi_{\underline{d}}$ denote the character and $d(\underline{d})$ the degree of $\underline{d}$. F being a finite subset of $\mathcal{C}(K)$, we put

$$\xi_F = \sum_{\underline{d} \in F} \xi_{\underline{d}}$$

and denote by $\mathcal{A}(F)$ the subspace of $C(K)$ spanned by the left and right translates of ξ_F. Suppose L is a closed subgroup of K and $\underline{d} \in \mathcal{C}(K)$, $\delta \in \mathcal{C}(L)$. Then $[\underline{d} : \delta]$ stands for the multiplicity of δ in the restriction of $\underline{d}$ to L.

Let G be a topological group and K_1, K_2 two compact subgroups of G. F_i being finite subsets of $\mathcal{C}(K_i)$ $(i = 1, 2)$, we say that F_1, F_2 interact if there exist elements $\underline{d}_i \in F_i$ and $\delta \in \mathcal{C}(K_1 \cap K_2)$ such that $[\underline{d}_i : \delta] \geq 1$ $(i = 1, 2)$. Put[11)]

$$[F_1 : F_2] = \int_{K_1 \cap K_2} \operatorname{conj} \xi_{F_1}(k) \cdot \xi_{F_2}(k)\,dk$$

where dk is the normalized Haar measure[12)] of $K_1 \cap K_2$.

Lemma 30. $[F_1 : F_2]$ <u>is an integer</u> ≥ 0. <u>Moreover</u> $[F_1 : F_2] = [F_2 : F_1]$ <u>and</u> $[F_1 : F_2] \neq 0$ <u>if and only if</u> F_1, F_2 <u>interact.</u>

If $x \in G$ and S is any subset of G, we put $^x S = xSx^{-1}$. Let K be a compact subgroup of G. Then $k \longmapsto {}^{x}k$ $(k \in K)$ is an isomorphism of K onto ${}^{x}K$. Hence it defines a bijection

$$\underline{d} \longmapsto {}^{x}\underline{d} \qquad\qquad (\underline{d} \in \mathscr{E}(K))$$

of $\mathscr{E}(K)$ onto $\mathscr{E}({}^{x}K)$.

Suppose F_1, F_2 are as in Lemma 30 and $x \in G$. We say that x intertwines F_2 with F_1 if $[F_1 : {}^{x}F_2] \neq 0$. Let S be a subset of G. We say that S intertwines F_2 with F_1 if $[F_1 : {}^{x}F_2] \neq 0$ for some $x \in S$.

Lemma 31. <u>Suppose</u> $[F_1 : F_2] = 0$. <u>Then</u>

$$\int_{K_1 \cap K_2} \operatorname{conj} f_1(k) \cdot f_2(k)dk = 0 \ .$$

<u>for</u> $f_i \in \mathscr{A}(F_i)$ $(i = 1, 2)$.

Corollary. <u>Fix</u> $x \in G$ <u>and let</u> f <u>be a complex-valued function on</u> $K_1 \times K_2$ <u>such that the function</u>

$$(k_1, k_2) \longmapsto f(k_1 x k_2) \qquad\qquad (k_i \in K_i, i = 1, 2)$$

<u>lies in</u> $\mathscr{A}(F_1) \otimes \mathscr{A}(F_2)$. <u>Then if</u> $f \neq 0$, x <u>intertwines</u> F_2 <u>with</u> F_1.

Let K_0, K_1, K_2 be three compact subgroups of G such that $K_1 \cap K_2$ is open in both K_1 and K_2.

Lemma 32. <u>Fix</u> $\underline{d}_i \in \mathscr{E}(K_i)$ $(i = 0, 1)$ <u>and let</u> F_2 <u>be the set of all</u> $\underline{d}_2 \in \mathscr{E}(K_2)$

such that $[\underline{d}_1 : \underline{d}_2] \neq 0$. Then the condition $[\underline{d}_o : \underline{d}_1] \neq 0$ implies that $[\underline{d}_o : F_2] \neq 0$.

The proofs, which we omit, are not difficult.

Let K be an open compact subgroup of G and dk the normalized Haar measure on K. Assume that G is totally disconnected [1(b), §2].

Lemma 33. Let Θ be a distribution of K. Put

$$\Theta_{\underline{d}} = d(\underline{d})\Theta * \xi_{\underline{d}} \qquad\qquad (\underline{d} \in \mathscr{E}(K))$$

where the star denotes convolution on K with respect to dk. Then

$$\Theta = \sum_{\underline{d} \in \mathscr{E}(K)} \Theta_{\underline{d}} \ .$$

Moreover if Θ is K-invariant,

$$\Theta_{\underline{d}} = \Theta(\operatorname{conj} \xi_{\underline{d}}) \cdot \xi_{\underline{d}} \ .$$

This is an immediate consequence of the Plancherel formula for K.

§14. Admissible distributions

Let G be a totally disconnected topological group [1(b), §§2, 3] which we assume to be unimodular. Let Θ be a distribution on an open set U of G, K_o an open compact subgroup of G and γ an element of U. We say that Θ is (G, K_o)-admissible at γ if:

(1) $\gamma K_o \subset U$,

(2) For any open subgroup K of K_o and $\underline{d} \in \mathscr{C}(K)$,

$$\Theta * \xi_{\underline{d}} = 0$$

on γK_o unless G intertwines 1_{K_o} with $\underline{d}$.

Here the star denotes convolution and 1_{K_o} is the class of the trivial representation of K_o.

Suppose Θ is (G, K_o)-admissible at γ and K_1 is an open subgroup
of K_o. Then it is easy to deduce from Lemma 32 that Θ is (G, K_1)-admissible
at every point in γK_o.

We shall say that Θ is G-admissible at γ if it is (G, K_o)-admissible
at γ for some open compact subgroup K_o of G.

By an admissible distribution on G, we mean a distribution which is
G-admissible at every point.

Let π be an irreducible admissible representation of G on V and
Θ_π its character [1(b), §3]. Let K_o be an open compact subgroup of G and
V_o the space of all elements in V which are left fixed by K_o. Then if K_o
is sufficiently small, $V_o \neq \{0\}$ and it is easy to verify that Θ_π is (G, K_o)-
admissible at every point.

§15. Statement of the main results

Now let G have the same meaning as in Part I. It operates on
itself by conjugation. Our object is to prove the following two theorems.

Theorem 19. <u>Let</u> Θ <u>be a</u> G-invariant distribution on an open, G-invariant subset U of G. <u>Let</u> γ <u>be a semisimple element in</u> U. <u>Then if</u> Θ <u>is</u> G-admissible at γ, <u>it coincides with a locally summable function around</u> γ.

Let M and $\mathfrak{m}$ be the centralizers of γ in G and $\mathfrak{g}$ respectively.

Theorem 20. <u>Under the above conditions we can choose unique complex numbers</u> c_ξ <u>such that</u>

$$\Theta(\gamma \exp Y) = \sum_\xi c_\xi \hat{\nu}_\xi(Y)$$

<u>for</u> Y <u>sufficiently near zero in</u> $\mathfrak{m}$. <u>Here</u> ξ <u>runs over all nilpotent</u> M-orbits <u>in</u> $\mathfrak{m}$, ν_ξ <u>is the</u> M-invariant measure[13] <u>on</u> $\mathfrak{m}$ <u>corresponding to</u> ξ <u>and</u> $\hat{\nu}_\xi$ <u>is the Fourier transform of</u> ν_ξ <u>on</u> $\mathfrak{m}$.

§16. Recapitulation of Howe's theory

We shall now recall some results of Howe [2(a)].

As before let Z denote the maximal split torus contained in the center of G. Also let $D(x) = D_G(x)$ ($x \in G$) denote the coefficient of t^ℓ in the polynomial

$$\det(t - \mathrm{Ad}(x) + 1)$$

where t is an indeterminate and $\ell = \mathrm{rank}\ G$. Then it is not difficult to see that there exists a G-domain $\mathfrak{g}_0$ in $\mathfrak{g}$ satisfying the following conditions.

(1) $t\,\mathfrak{g}_0 \subset \mathfrak{g}_0$ for $t \in R$ and the exponential mapping is defined on $\mathfrak{g}_0$.
Put $G_0 = \exp \mathfrak{g}_0$. Then G_0 is an open, closed and G-invariant subset of G
and $X \longmapsto \exp X$ is an analytic ($\mathfrak{p}$-adic) diffeomorphism of $\mathfrak{g}_0$ on G_0.
Moreover

$$\exp(\mathrm{Ad}(x)X) = x \exp X \cdot x^{-1} \qquad\qquad (x \in G,\ X \in \mathfrak{g}_0)$$

(2) If Q is any compact subset of G then $G_0 \cap (QZ)$ is compact.

(3) $|D(\exp X)|_{\mathfrak{p}} = |\eta(X)|_{\mathfrak{p}}$ for $X \in \mathfrak{g}_0$.

Let p be the rational prime dividing the prime $\mathfrak{p}$ of Ω and e the
ramification index of $\mathfrak{p}$ over p. Put[14]

$$\beta = \begin{cases} \left[\dfrac{3e}{2(p-1)}\right] + 1 & \text{if } p \neq 3\ , \\[2mm] e+1 & \text{if } p = 3\ . \end{cases}$$

Fix an element ϖ in R of order 1. Suppose L is a lattice in $\mathfrak{g}$ such that

    1) $L \subset \mathfrak{g}_0$,

    2) $[L,\ L] \subset \varpi^{\beta} L$.

Then $K = \exp L$ is an open compact subgroup of G.

By a s-lattice[15] we mean a lattice L satisfying the above two con-
ditions. If $p = 2$, we shall further assume that $L \subset 2\,\mathfrak{g}_0$ and $[L,\ L] \subset 2^3 L$.
Then in every case $K = \exp L$ and $K^{1/2} = \exp \tfrac{1}{2}L$ are open compact subgroups
of G and K is normal in $K^{1/2}$. (Note that $L = \tfrac{1}{2}L$ and $K = K^{1/2}$ unless
$p = 2$.) Hence $K^{1/2}$ operates on K by conjugacy and therefore also on $\mathfrak{E}(K)$.

Let $\mathscr{C}^{1/2}(K)$ denote the set of all $K^{1/2}$-orbits in $\mathscr{C}(K)$. Since $[K^{1/2} : K] < \infty$, every such orbit is a finite subset of $\mathscr{C}(K)$.

Fix $\underline{d} \in \mathscr{C}^{1/2}(K)$ and let $\underline{d}_i$ ($1 \le i \le N$) be all the distinct elements of $\mathscr{C}(K)$ lying on the orbit $\underline{d}$. It is clear that $d(\underline{d}_i)$ is independent of i. Put $d(\underline{d}) = d(\underline{d}_i)$ and

$$\xi_{\underline{d}} = \sum_i \xi_{\underline{d}_i} \ .$$

We call $d(\underline{d})$ the degree and $\xi_{\underline{d}}$ the character of $\underline{d}$.

Fix a s-lattice L in $\mathfrak{g}$ and let L^* denote the dual lattice (see §12). Define $K = \exp L$ and $K^{1/2} = \exp \tfrac{1}{2}L$ as above. Then L^* is stable under $\mathrm{Ad}(K^{1/2})$ and therefore $K^{1/2}$ operates on $\mathfrak{g}/L^*$.

Theorem 21 (Howe). <u>There is a bijection</u> $\underline{d} \longmapsto \mathcal{O}_{\underline{d}}$ <u>from</u> $\mathscr{C}^{1/2}(K)$ <u>to the set of all</u> $K^{1/2}$-<u>orbits in</u> $\mathfrak{g}/L^*$ <u>such that</u>

$$d(\underline{d})\xi_{\underline{d}}(\exp \lambda) = \sum_{X \in \mathcal{O}_{\underline{d}/L^*}} \chi(B(X, \lambda)) \qquad (\lambda \in L)$$

<u>and</u>

$$d(\underline{d}) = [K : K_X]^{1/2}$$

<u>Here</u> X <u>is any element in</u> $\mathcal{O}_{\underline{d}}$ <u>and</u> K_X <u>is the subgroup of all elements</u> $k \in K$ <u>such that</u>

$$\mathrm{Ad}(k)X \in X + L^*$$

It is convenient to regard a subset of $\mathcal{G}/L^{*}$ as a subset of $\mathcal{G}$ (by identifying it with its pre-image). This holds in particular for $\sigma_{\underline{d}}$.

Suppose L_1, L_2 are two s-lattices in $\mathcal{G}$. Put $K_i = \exp L_i$ $(i = 1, 2)$.

Corollary (Howe). <u>Fix</u> $\underline{d}_i \in \mathcal{E}^{1/2}(K_i)$ <u>and let</u> σ_i <u>denote the</u> $K_i^{1/2}$-<u>orbit in</u> $\mathcal{G}/L_i^{*}$ <u>corresponding to</u> $\underline{d}_i$. <u>Then an element</u> $x \in G$ <u>intertwines</u> $\underline{d}_2$ <u>with</u> $\underline{d}_1$ <u>if and only if</u>

$$\sigma_1 \cap {}^{x}(\sigma_2) \neq \emptyset \ .$$

§17. Application to admissible invariant distributions

We return to the notation of §15. Put $\mathcal{q} = (\mathrm{Ad}(\gamma)-1)\mathcal{G}$. Since γ is semisimple, $\mathcal{G} = \mathcal{m} + \mathcal{q}$ where the sum is direct. Moreover $\mathcal{m}$ and $\mathcal{q}$ are orthogonal under the bilinear form B and the restriction of B on $\mathcal{m}$ is nondegenerate. Hence $\mathcal{m}$ is reductive.

Put

$$D_{G/M}(m) = \det(\mathrm{Ad}(m)-1)_{\mathcal{G}/\mathcal{m}} \qquad (m \in M)$$

and let M' be the set of all $m \in M$ where $D_{G/M}(\gamma m) \neq 0$. Put $U_M = M' \cap (\gamma^{-1}U)$. Then U_M is an open and M-invariant neighborhood of 1 in M. The mapping

$$(x, m) \longmapsto x\gamma mx^{-1}$$

of $G \times U_M$ into U is everywehre submersive and $U_o = {}^G(\gamma U_M)$ is an open

G-invariant neighborhood of γ in U. Hence there exists a linear mapping

$a \longmapsto f_a$ of $C_c^\infty(G \times U_M)$ onto $C_c^\infty(U_o)$ such that

$$\int_{G \times U_M} a(x : m)F(x\gamma mx^{-1})dxdm = \int_G f_a(x)F(x)dx$$

for all $F \in C(G)$. (dx and dm are the Haar measures on G and M respectively.)

Let Θ be any G-invariant distribution on U. Then [1(a), p. 56] there exists a

unique M-invariant distribution θ on U_M such that

$$\Theta(f_a) = \theta(\beta_a) \qquad\qquad (a \in C_c^\infty(G \times U_M))$$

where

$$\beta_a(m) = \int_G a(x : m)dx \qquad\qquad (m \in M) \ .$$

Let K_o be an open compact subgroup of G such that the distribution

Θ of Theorem 19 is (G, K_o)-admissible at γ. Fix an open compact subgroup

M_o of M such that $M_o \subset U_M \cap K_o$ and let K be any open subgroup of K_o.

Fix $\beta \in C_c^\infty(M_o)$ and define $a \in C_c^\infty(G \times M_o)$ by

$$a(x : m) = a_K(x)\beta(m) \qquad\qquad (x \in G, m \in M) \ .$$

Here a_K is the characteristic function of K divided by the total measure of K.

Then $\beta_a = \beta$ and therefore

$$\theta(\beta) = \Theta(f_a) = \sum_{\underline{d} \in \mathscr{E}(K)} \int f_a(x)\Theta_{\underline{d}}(x)dx$$

from the Plancherel formula for K. Here

$$\Theta_{\underline{d}} = d(\underline{d})\Theta * \xi_{\underline{d}} \qquad\qquad (\underline{d} \in \mathscr{E}(K))$$

and the convolution on K is taken with respect to the normalized Haar measure on K. Moreover we have to observe that

$$\text{Supp } f_a \subset {}^K(\gamma M_o) \subset K_o \gamma K_o \subset U \ .$$

It is obvious that the function $\Theta_{\underline{d}}$ is K-invariant. Hence it follows from the definition of a that

$$\int f_a(x)\Theta_{\underline{d}}(x)dx = \int \beta(m)\Theta_{\underline{d}}(\gamma m)dm \ .$$

Therefore

$$\theta(\beta) = \sum_{\underline{d} \in \mathscr{E}(K)} \int \beta(m)\Theta_{\underline{d}}(\gamma m)dm \qquad\qquad (\beta \in C_c^\infty(M_o)) \ .$$

Normalize dm so that M_o has total measure 1. Then the following result is obvious from Lemma 33.

Lemma 34. $\theta = \sum_{\delta \in \mathscr{E}(M_o)} \theta(\text{conj } \xi_\delta)\cdot \xi_\delta$ __on__ M_o __and__

$$\theta(\text{conj } \xi_\delta) = \sum_{\underline{d} \in \mathscr{E}(K)} \int_{M_o} \text{conj } \xi_\delta(m) \cdot \Theta_{\underline{d}}(\gamma m)dm \qquad (\delta \in \mathscr{E}(M_o)) \ .$$

For any finite subset F of $\mathscr{E}(K)$, put

$$\Theta_F = \sum_{\underline{d} \in F} \Theta_{\underline{d}} \ .$$

Lemma 35. <u>Let</u> F <u>be a finite subset of</u> $\mathcal{E}(K)$. <u>Fix</u> $\delta \in \mathcal{E}(M_o)$ <u>and suppose</u> $[F : \delta] = 0$. <u>Then</u>

$$\int_{M_o} \mathrm{conj}\ \xi_\delta(m) \cdot \Theta_F(\gamma m)dm = 0 \ .$$

Fix $m_o \in M_o$. It would be enough to verify that

$$\int_{M_o \cap K} \mathrm{conj}\ \xi_\delta(m_o m) \cdot \Theta_F(\gamma m_o m)dm = 0 \ .$$

Put $f(k) = \Theta_F(\gamma m_o k)$ $(k \in K)$ and $g(m) = \xi_\delta(m_o m)$ $(m \in M_o)$. Then $f \in \mathcal{A}(F)$ and $g \in \mathcal{A}(\delta)$ and the required result follows from Lemma 31.

Let F_i $(i \geq 1)$ be finite subsets of $\mathcal{E}(K)$ such that $\mathcal{E}(K)$ is the disjoint union of F_i. Then

$$\theta(\beta) = \sum_i \int \beta(m)\Theta_{F_i}(\gamma m)dm \qquad\qquad (\beta \in C_c^\infty(M_o)) \ .$$

Lemma 36. <u>Fix</u> $\delta \in \mathcal{E}(M_o)$. <u>Then</u> $\theta(\mathrm{conj}\ \xi_\delta) = 0$ <u>unless we can choose</u> i <u>such that:</u>

1) δ <u>interacts with</u> F_i,

2) $\Theta_{F_i}(\gamma m) \neq 0$ <u>for some</u> $m \in M_o$.

This is obvious from Lemmas 34 and 35.

§18. First step of the reduction from G to M

Choose an M-domain $\mathfrak{m}_o$ in $\mathfrak{m} \cap \mathfrak{g}_o$ satisfying conditions similar to those described in §16 for $\mathcal{g}_o$ and such that $\exp \mathcal{m}_o \subset U_M$. We shall use $\mathcal{m}_o$ to define s-lattices in $\mathcal{m}$.

Now fix s-lattices L and Λ in $\mathcal{g}$ and $\mathcal{m}$ respectively. Put $K = \exp L$, $M_o = \exp \Lambda$ and let $p_{\mathcal{m}}$ and $p_{\mathcal{q}}$ denote the projections of $\mathcal{g}$ on $\mathcal{m}$ and $\mathcal{q}$ respectively corresponding to the direct sum $\mathcal{g} = \mathcal{m} + \mathcal{q}$.

Lemma 37. <u>Suppose</u> $\underline{d} \epsilon \mathcal{E}^{1/2}(K)$ <u>and</u> $\delta \epsilon \mathcal{E}^{1/2}(M_o)$. <u>Then</u> $\underline{d}$ <u>and</u> δ <u>interact if and only if</u>

$$(p_{\mathcal{m}} \, \mathcal{O}_{\underline{d}}) \cap \mathcal{O}_\delta \neq \emptyset \ .$$

This is an easy consequence of Theorem 21.

As we have already observed in §1, G and $\mathcal{g}$ may be regarded as subsets of $M_n(\Omega)$. Put

$$|x| = \max_{i,j} |x_{ij}| \qquad\qquad (x \epsilon M_n(\Omega))$$

where x_{ij} are the coefficients of the matrix. This defines, in particular, a norm on $\mathcal{g}$.

Put $K_o = \exp L_o$ where L_o is a s-lattice in $\mathcal{g}$. By choosing L_o sufficiently small, we may assume that:

1) $K_o^{1/2} \subset GL(n, R)$.

2) $|(Ad(\exp \lambda)-1)X| \leq |\lambda| \, |X|$ for $\lambda \epsilon \frac{1}{2}L_o$, $X \epsilon \mathcal{g}$.

3) Θ is (G, K_o)-admissible at γ.

Let $\underset{\sim}{M}$ be the centralizer of γ in $\underset{\sim}{G}$ and $\underset{\sim}{M}^o$ the connected component of 1 in $\underset{\sim}{M}$. Put $M^o = \underset{\sim}{M}^o \cap G$. Fix a s-lattice Λ in $\mathfrak{m}$ which is well adapted (see §12) with respect to M^o. Put $M_o = \exp \Lambda$. By choosing Λ-sufficiently small, we may assume that:

1) $\Lambda \subset L_o$, $M_o \subset U_M$.

2) There is a number $c > 0$ such that

$$|(\mathrm{Ad}(\gamma m)-1)Y| \geq c|Y|$$

for $m \in M_o$ and $Y \in \mathcal{y}$.

3) $|\tfrac{1}{2}\Lambda| < 1$ and

$$|(\mathrm{Ad}(m)-1)Z| \leq |\tfrac{1}{2}\Lambda||Z|$$

for $m \in M_o^{1/2}$ and $Z \in \mathfrak{m}$.

Put $L_\nu = \varpi^\nu L_o$, $K_\nu = \exp L_\nu$ ($\nu \in Z$, $\nu \geq 0$) and let Φ_ν denote the set of all $\underline{d} \in \mathcal{E}^{1/2}(K_\nu)$ such that:

1) $\mathcal{v}_{\underline{d}} \cap \mathcal{N} \neq \emptyset$,

2) $^{\gamma m}(\mathcal{v}_{\underline{d}}) \cap \mathcal{v}_{\underline{d}} \neq \emptyset$ for some $m \in M_o$.

As before let S be the set of all $X \in \mathcal{g}$ with $|X| = 1$.

Lemma 38. Fix a neighborhood V of $\mathcal{N} \cap S \cap \mathfrak{m}$ in $S \cap \mathfrak{m}$. Then we can choose $\nu_o \geq 0$ with the following property. Suppose $\nu \geq \nu_o$, $\underline{d} \in \Phi_\nu$ and $|X| \geq q^{2\nu}$ for some $X \in \mathcal{v}_{\underline{d}}$. Then

$$P_{\mathfrak{m}}\, \mathcal{v}_{\underline{d}} \subset \Omega V \ .$$

The proof of this lemma is rather long and technical. Therefore we do not give it here.

§19. Second step

Put $\sigma_0 = L_0^*$. Then by Theorem 21, σ_0 corresponds to 1_{K_0} regarded as an element of $\mathcal{E}^{1/2}(K_0)$. Fix $\nu \geq 0$ and $\underline{d} \in \mathcal{E}^{1/2}(K_\nu)$. Then we conclude from the corollary of Theorem 21 that $\Theta_{\underline{d}} = 0$ on γK_0 unless

$$\sigma_{\underline{d}} \cap {}^G(\sigma_0) \neq \emptyset \ .$$

For any endomorphism T of $\mathfrak{g}$ let $|T|$ denote the bound of T so that

$$|T| = \sup_{X \in \mathfrak{g}(1)} |TX| \ .$$

Put $q^r = \max(|p_{\mathfrak{m}}|, |p_{\mathfrak{q}}|)$ and $\mathfrak{m}(t) = \mathfrak{m} \cap \mathfrak{g}(t)$ $(t > 0)$. Then $r \geq 0$. Since Λ is well adapted (with respect to M^o) the same holds for its dual Λ^* in $\mathfrak{m}$ (see §12). Therefore by Theorem 18 we can choose a neighborhood V of $\mathcal{N} \cap S \cap \mathfrak{m}$ in $S \cap \mathfrak{m}$ and a number $t_o > 0$ such that $C(V, t_o, \Lambda^*)$ holds. Moreover we may assume that

$$V = \mathcal{N} \cap S \cap \mathfrak{m} + \mathfrak{m}(\varepsilon)$$

for some ε $(0 < \varepsilon < 1)$.

Put

$$\theta(\delta) = \theta(\mathrm{conj}\ \xi_{\delta_0}) \qquad\qquad (\delta \in \mathcal{E}^{1/2}(M_0))$$

where δ_o is any element of $\mathcal{E}(M_o)$ lying in the orbit δ. Since θ is M-invariant, this definition is legitimate. Let F_ν $(\nu \geq 0)$ denote the set of all $\delta \in \mathcal{E}^{1/2}(M_o)$ such that $\sigma_\delta \subset \mathcal{M}(q^{2\nu+r})$. Then F_ν is a finite set.

Lemma 39. __Fix__ $\nu > 0$ __sufficiently large.__ __Then if__ $\delta \in \mathcal{E}^{1/2}(M_o)$, $\delta \notin F_\nu$ __and__ $\theta(\delta) \neq 0$, __we can conclude that__

$$\sigma_\delta \subset \Omega V$$

__and__ $|Z| > q^{2\nu} \geq t_o$ __for every__ $Z \in \sigma_\delta$.

The proof, which we omit, depends on Lemmas 36, 37 and 38.

§20. Completion of the proof

For $f \in C_c^\infty(\mathcal{M})$, define $f_o \in C_c^\infty(U_M)$ as follows.

$$f_o(\exp Z) = f(Z) \qquad\qquad (Z \in \mathcal{M}_o)$$

and $f_o = 0$ outside $\exp \mathcal{M}_o$. Put

$$\theta_o(f) = \theta(f_o) \qquad\qquad (f \in C_c^\infty(\mathcal{M})) \ .$$

Then θ_o is an M-invariant distribution on $\mathcal{M}$ and $\mathrm{Supp}\ \theta_o \subset \mathcal{M}_o$.

Lemma 40. __Let__ f_Z __denote the characteristic function of__ $Z + \Lambda^*$ __on__ $\mathcal{M}$ $(Z \in \mathcal{M})$. __Then if__ $\delta \in \mathcal{E}^{1/2}(M_o)$ __and__ $Z \in \sigma_\delta$,

$$\theta(\delta) = v(\Lambda^*)^{-1} d(\delta)\hat{\theta}_o(f_{-Z})$$

<u>where</u>

$$v(\Lambda^*) = \int_{\Lambda^*} du \ .$$

Here $\hat{\theta}_o$ is the Fourier transform of θ_o and du the Haar measure on the additive group of $\mathcal{m}$. This lemma is a simple consequence of Howe's formula for the characters (Theorem 21).

Fix ν as in Lemma 39.

Lemma 41. <u>Suppose</u> Z <u>is an element in</u> $\mathcal{m}$ <u>such that</u> $|Z| > q^{2\nu + r}$ <u>and</u> $Z \notin \Omega V$. <u>Then</u>

$$\hat{\theta}_o(f_Z) = 0 \ .$$

There is a unique $\delta \in \mathcal{L}^{1/2}(M_o)$ such that $-Z \in \mathcal{v}_\delta$. Since $|Z| > q^{2\nu + r}$, it is clear that $\delta \notin F_\nu$ and our assertion follows from Lemmas 39 and 40.

Lemma 41 shows that $\hat{\theta}_o \in J(V, \infty, \Lambda^*)$ (in the notation of §10 applied to $(M^o, \mathcal{m})$ in place of $(G, \mathcal{g})$). Hence we conclude from the corollary of Theorem 17 that there exists an element $\tau \in J_o$ such that $\theta_o = \hat{\tau}$ on Λ. Theorems 19 and 20 now follow immediately from Theorems 3 and 4.

Let $\mathcal{v}$ be a G-orbit in G. Then the closure of $\mathcal{v}$ contains a semi-simple element γ and Theorem 1 is a consequence of this fact.

§21. Formal degree of a supercuspidal representation

Fix a Haar measure dx^* on G/Z where Z is the maximal split torus

lying in the center of G. For any square-integrable representation π of G,
let $d(\pi)$ denote the formal degree of π [1(b), §3] defined by means of dx^*.

A Cartan subgroup Γ of G is called elliptic if Γ/Z is compact. Define
D as in §16 and let G' be the set of all $x \in G$ where $D(x) \neq 0$. Fix an elliptic
Cartan subgroup Γ of G and put $\Gamma' = \Gamma \cap G'$. If f is an element in $C^\infty(G)$
such that Supp f is compact mod Z, we define

$$F_f(\gamma) = |D(\gamma)|^{1/2} \int_{G/Z} f(x\gamma x^{-1})dx^* \qquad (\gamma \in \Gamma') \ .$$

Then F_f is a locally constant function on Γ'.

Let π be an irreducible, unitary, supercuspidal representation of G
[1(b), §6]. Define $\mathcal{A}(\pi)$ as usual [1(b), §3] and let Θ_π denote the character of
π. We know from Theorem 1 that Θ_π is a function.

Lemma 42. $F_f(\gamma) = d(\pi)^{-1} f(1) |D(\gamma)|^{1/2} \Theta_\pi(\gamma)$ $(\gamma \in \Gamma')$ __for__ $f \in \mathcal{A}(\pi)$.

This follows from the Schur orthogonality relations [1(b), §3] by standard
arguments.

Define $\mathcal{G}_o$ and G_o as in §16 and let dX denote a Haar measure on
the additive group of $\mathcal{G}$. Then it follows from condition (3) on $\mathcal{G}_o$ (§16) that
the Haar measure dx on G can be normalized so as to correspond to dX
under the exponential mapping. Let θ_π denote the distribution on $\mathcal{G}_o$ given
by

$$\theta_\pi(a_o) = \Theta_\pi(a) \qquad (a \in C_c^\infty(G_o))$$

where $a_0 \in C_c^\infty(\mathcal{g}_0)$ is defined by $a_0(X) = a(\exp X)$ $(X \in \mathcal{g}_0)$. Then it is obvious that

$$\theta_\pi(a_0) = \int a_0(X)\theta_\pi(X)dX$$

where

$$\theta_\pi(X) = \Theta_\pi(\exp X) \qquad\qquad (X \in \mathcal{g}_0) \ .$$

Fix $f \in \mathcal{A}(\pi)$ and define the function g on $\mathcal{g}$ by

$$g(X) = \begin{cases} f(\exp X) & \text{if } X \in \mathcal{g}_0 \ , \\ 0 & \text{otherwise} \ . \end{cases}$$

Then it follows from condition (2) on $\mathcal{g}_0$ (§16) that $g \in C_c^\infty(\mathcal{g}_0)$. Let $\mathcal{h}$ be the Lie algebra of Γ. It is clear that

$$F_f(\exp H) = \phi_g^{\mathcal{h}}(H) \qquad\qquad (H \in \mathcal{h}' \cap \mathcal{g}_0) \ .$$

Hence we obtain the following result from Lemma 42 and Theorem 14.

Lemma 43. <u>There exists a neighborhood</u> ω <u>of zero in</u> $\mathcal{h} \cap \mathcal{g}_0$ <u>such that</u>

$$|\eta(H)|^{1/2}\theta_\pi(H)g(0) = d(\pi) \sum_{\vartheta \in \mathcal{O}(0)} \mu_\vartheta(g)\Gamma_\vartheta^{\mathcal{h}}(H)$$

<u>for all</u> $H \in \omega \cap \mathcal{h}'$.

On the other hand we know from Theorem 5 that

$$\theta_\pi = \sum_{\sigma} c_\sigma(\pi)\hat{\mu}_\sigma$$

around zero. Let 0 denote the orbit $\sigma = \{0\}$. We may assume that
$\mu_0(h) = h(0)$ $(h \in C_c^\infty(\mathfrak{g}))$. Hence $\hat{\mu}_0 = 1$ and Lemma 43 can be expressed in the
notation of §9 as follows.

$$\{|\eta(H)|^{1/2}c_0(\pi) + \sum_{\sigma \neq 0} c_\sigma(\pi)\psi_\sigma(H)\}g(0)$$

$$= d(\pi)\{g(0)\Gamma_0^{\flat}(H) + \sum_{\sigma \neq 0} \mu_\sigma(g)\Gamma_\sigma^{\flat}(H)\} \qquad (H \in \omega \cap \mathfrak{y}') \ .$$

Since 0 is the only nilpotent G-orbit of dimension zero, we conclude from
Lemma 22 and part (1) of Theorem 14 that

$$c_0(\pi) = cd(\pi)$$

where c is defined as in Theorem 15. This proves Theorem 6.

Put $K = \exp L$ where L is a s-lattice in $\mathfrak{g}$ and let $m_\pi(L)$ denote
the multiplicity of the trivial representation of K in π.

Lemma 44. $m_\pi(L) = v(L)^{-1}\theta_\pi(a_L)$ where a_L is the characteristic function
of L and

$$v(L) = \int_L dX$$

Let dk denote the normalized Haar measure on K and a_K the
characteristic function of K. It is obvious that

$$\int_K dx = \int_L dX = v(L)$$

and therefore $dk = v(L)^{-1}dx$. Now $m_\pi(L) = \operatorname{tr} E$ where

$$E = \int_K \pi(k)dk = v(L)^{-1}\int a_K(x)\pi(x)dx \ .$$

Hence $m_\pi(L) = v(L)^{-1}\Theta_\pi(a_K) = v(L)^{-1}\theta_\pi(a_L)$.

Now assume that L is so small that

$$\theta_\pi = \sum_\sigma c_\sigma(\pi)\hat{\mu}_\sigma$$

on L. Then

$$\theta_\pi(a_L) = \sum_\sigma c_\sigma(\pi)\mu_\sigma((a_L)^\wedge) \ .$$

But

$$(a_L)^\wedge = v(L)a_{L^*}$$

where a_{L^*} is the characteristic function of the dual lattice L^*. Now replace L by $t^2 L$ ($t \in R'$) and observe that $(t^2 L)^* = t^{-2}L^*$ and

$$\mu_\sigma(a_{t^{-2}L^*}) = |t|^{-d(\sigma)}\mu_\sigma(a_{L^*})$$

from Lemma 3. This gives the following result.

Lemma 45. <u>If</u> L <u>is sufficiently small,</u>

$$m_\pi(\varpi^{2j}L) = \sum_\sigma c_\sigma(\pi)q^{jd(\sigma)}\mu_\sigma(a_{L^*})$$

<u>for all</u> $j \geq 0$.

Let $d_o = 0 < d_1 < \ldots < d_r$ be all the integers d such that $d = d(\vartheta)$ for some nilpotent G-orbit ϑ . Put

$$c(d) = \sum_{d(\vartheta)=d} c_{\vartheta}(\pi)\mu_{\vartheta}(a_{L^*}) \ .$$

Then we conclude from Lemma 45 that

$$\sum_{0 \leq i \leq r} c(d_i)q^{jd_i} = m_\pi(\varpi^{2j}L) \in \mathbb{Z} \qquad\qquad (0 \leq j \leq r) \ .$$

Put $x_i = q^{d_i}$ $(0 \leq i \leq r)$. Then

$$\sum_{0 \leq i \leq r} c(d_i)x_i^j \in \mathbb{Z} \qquad\qquad (0 \leq j \leq r)$$

and

$$\det(x_i^j)_{0 \leq i, j \leq r} = \prod_{0 \leq i < j \leq r} (x_j - x_i) = \Omega \neq 0 \ .$$

Hence $\Omega c(d_i) \in \mathbb{Z}$ $(0 \leq i \leq r)$. But

$$c(0) = c_0(\pi)\mu_0(a_{L^*}) = c_0(\pi) = cd(\pi)$$

from Theorem 6 and therefore we conclude that $\Omega cd(\pi) \in \mathbb{Z}$. This proves Theorem 7.

The Institute for Advanced Study
    Princeton, N. J.

References

1. Harish-Chandra, (a) Harmonic analysis on reductive p-adic groups,
   Lecture Notes in Math. vol. 162, Springer-Verlag, 1970.

   (b) Harmonic analysis on reductive p-adic groups, in Harmonic Analysis
   on Homogeneous Spaces, 167-192, Amer. Math. Soc., Providence, 1973.

   (c) The characters of reductive p-adic groups, to appear in Contributions
   to Algebra, Academic Press, New York.

2. R. E. Howe, (a) Kirillov theory for compact p-adic groups, 1970 (preprint).

   (b) Some qualitative results on the representation theory of GL(n) over
   a p-adic field, 1972 (preprint).

   (c) Two conjectures about reductive p-adic groups, in Harmonic Analysis
   on Homogeneous Spaces, 377-380, Amer. Math. Soc., Providence, 1973.

   (d) The Fourier transform and germs of characters (case of Gl_n over
   a p-adic field), Math. Annalen 208 (1974), 305-322.

3. R. R. Rao, Orbital integrals in reductive groups, Ann. of Math. 96 (1972),
   505-510.

4. J. A. Shalika, A theorem on semi-simple p-adic groups, Ann. of Math. 95
   (1972), 226-242.

Footnotes

1) An orbit $\mathcal{O}$ is nilpotent if and only if $Cl\,\mathcal{O}$ contains zero. This justifies our notation (see also §5).

2) Cf. [2(d), prop. 3].

3) We drop the subscript $\mathfrak{p}$ and write $|t|$ instead of $|t|_{\mathfrak{p}}$ $(t \in \Omega)$ when no confusion is possible.

4) We have tacitly assumed here that the various measures are suitably normalized.

5) If T is a distribution on $\mathfrak{g}$, it is sometimes convenient to write $\langle T, f \rangle$ for $T(f)$ $(f \in \mathcal{D})$.

6) $[S]$ denotes the number of elements in a set S.

7) An orbit is called regular if it is contained in $\mathfrak{g}'$.

8) Recall that $\mathfrak{z}$ is the center of $\mathfrak{g}$.

9) We assume that $|cX| = |c|_{\mathfrak{p}} |X|$ $(c \in \Omega, X \in \mathfrak{g})$. Moreover if $X \neq 0$ we can choose $c \in \Omega^{\times}$ such that $|cX| = 1$.

10) ΩV is the set of all elements in $\mathfrak{g}$ of the form cv $(c \in \Omega, v \in V)$.

11) conj c denotes the complex conjugate of $c \in \underset{\sim}{C}$.

12) Note that this definition of $[F_1 : F_2]$ is consistent with the definition of $[\underline{d} : \delta]$ given above.

13) The centralizer $\underset{\sim}{M}$ of γ in $\underset{\sim}{G}$ is, in general, not connected. But since the connected component of 1 is a subgroup of finite index, this does not present a serious problem and we can apply the results of Part I. In particular $\hat{\nu}_\xi$ are functions.

14) As usual $[r]$ denotes the greatest integer $n \leq r$ $(r \in \underset{\sim}{R})$.

15) This term is meant to signify that the lattice is small in a technical sense.

Reprinted from
*Proceedings of the 1977 annual seminar of
the Canadian Mathematical Congress*
Edited by W. Rossmann. Queen's Papers in
Pure and Applied Math. No. **48**, 1978, 281–347

A submersion principle and its applications

By

HARISH-CHANDRA

1. Introduction

Let G be a real reductive group and π an irreducible admissible representation of G. Let Θ denote the character of π. We recall that Θ is a distribution on G defined by

$$\Theta(f) = \operatorname{tr} \pi(f) \quad (f \in C_e^\infty(G)).$$

It is well known that Θ is a locally summable function on G which is analytic on the set G' of regular elements. Fix $\gamma_0 \in G'$ and let Γ be the Cartan subgroup of G containing γ_0. Put

$$\Gamma' = G' \cap \Gamma \text{ and } G_\Gamma = \bigcup_{x \in G} x \Gamma' x^{-1}.$$

The mapping $(x, \gamma) \mapsto x\gamma x^{-1}$ of $G \times \Gamma'$ into G is everywhere submersive. Since Θ is invariant under all inner automorphisms of G, one proves easily that Θ defines a distribution θ on Γ'. Let $\mathscr{Z}$ be the algebra of all differential operators on G which commute with both left and right translations. Then Θ satisfies the differential equations

$$z\Theta = \chi(z)\Theta \quad (z \in \mathscr{Z}),$$

where χ is the infinitesimal character of π. We can transcribe these equations in terms of θ. In this way we obtain a system of differential equations for θ on Γ'. It turns out that this system is elliptic and therefore θ is an analytic function on Γ'. But this implies that Θ coincides with an analytic function on G_Γ.

We would like to prove a similar result in the p-adic case. Let Ω be a p-adic field and G the group of all Ω-rational points of a connected reductive Ω-group $\mathbf{G}$ [3]. Then G, with its usual topology, is a locally compact, totally disconnected, unimodular group. Let dx denote its Haar measure. We shall use the terminology of [3] without further comment.

Let π be an admissible irreducible representation of G. Then for every $f \in C_e^\infty(G)$, the operator

$$\pi(f) = \int_G f(x)\pi(x)\,dx$$

95

has finite rank. Put

$$\Theta(f) = \operatorname{tr} \pi(f).$$

Then Θ is a distribution on G. Let G' be the set of all points $x \in G$ where $D_G(x) \neq 0$ ([3], § 15). Then G' is an open, dense subset of G whose complement has measure zero. We intend to show that Θ coincides with a locally constant function F on G'. This means that

$$\Theta(f) = \int f(x) F(x) \, dx,$$

for all $f \in C_\bullet^\infty(G')$.

In case char $\Omega = 0$, this fact was first proved by Howe by making use of his Kirillov theory for p-adic groups. Moreover when char $\Omega = 0$, it is known that Θ is a locally summable function on G [4, 5]. But these methods, which make extensive use of Lie algebras and the exponential mapping, do not seem to work in characteristic p. We shall therefore construct a proof on a totally different principle.

2. The submersion principle

Let P be a parabolic subgroup of G and $x \mapsto x^*$ the projection of G on $G^* = G/P$.

Theorem 1 (the submersion principle). *Fix $\gamma \in G'$. Then the mapping*

$$x \mapsto (x\gamma x^{-1})^*$$

from G to G^* is everywhere submersive.

When char $\Omega = 0$, the proof is very easy. Moreover Borel assures me that this principle is actually true over an arbitrary field.

Let us verify it when char $\Omega = 0$. Put

$$\phi_\gamma : x \mapsto (x\gamma x^{-1})^*.$$

Then $\quad \phi_\gamma(xy) = \phi_{y\gamma y^{-1}}(x) \qquad (x, y \in G).$

Hence it is enough to prove that ϕ_γ is submersive at $x = 1$.

Fix a split component A of P and let $P = MN$ be the corresponding Levi decomposition of P. Let $(\bar{P}, A)$ denote the p-pair opposite to (P, A). Then $\bar{P} = M\bar{N}$. We denote the Lie algebras of $G, P, M, N, \bar{N}$ by $\mathfrak{g}, \mathfrak{p}, \mathfrak{m}, \mathfrak{n}, \bar{\mathfrak{n}}$ respectively. Then

$$\mathfrak{p} = \mathfrak{m} + \mathfrak{n}, \quad \mathfrak{g} = \bar{\mathfrak{n}} + \mathfrak{m} + \mathfrak{n},$$

and we have to verify that

$$(\operatorname{Ad}(\gamma^{-1}) - 1)\,\mathfrak{g} + \mathfrak{p} = \mathfrak{g}.$$

Fix a symmetric, nondegenerate, G-invariant bilinear form B on $\mathfrak{g}$ with values in Ω. Since G is reductive, this is possible. Let Γ be the Cartan subgroup of G determined by γ and $\mathfrak{c}$ the Lie algebra of Γ. Then $\mathfrak{c} = \ker(\operatorname{Ad}(\gamma) - 1)$. For any linear subspace $\mathfrak{q}$ of $\mathfrak{g}$, let $\mathfrak{q}^\perp$ denote the space of all $Y \in \mathfrak{g}$ such that $B(X, Y) = 0$ for all $X \in \mathfrak{q}$. Then

$$\mathfrak{p}^\perp = \mathfrak{n}, \quad ((\mathrm{Ad}\,(\gamma^{-1}) - 1)\,\mathfrak{g})^\perp = \ker\,(\mathrm{Ad}\,(\gamma) - 1) = \mathfrak{t}.$$

Therefore $((\mathrm{Ad}\,(\gamma^{-1}) - 1)\,\mathfrak{g} + \mathfrak{p})^\perp \subset \mathfrak{t} \cap \mathfrak{n} = \{0\}$,
and this implies the desired result.

3. The function $f_{a,\,y}$

Fix γ and P as in theorem 1. Then the mapping*

$$(x, p) \mapsto {}^x\gamma \cdot p \qquad (x \in G, p \in P)$$

of $G \times P$ into G is everywhere submersive. Hence ([2], p. 49) there exists a unique linear mapping

$$a \mapsto f_{a,\,\gamma} \qquad \big(a \in C_o^\infty\,(G \times P)\big)$$

from $C_o^\infty\,(G \times P)$ to $C_o^\infty\,(G)$ such that

$$\int_{G \times P} a\,(x : p)\,F\,({}^x\gamma \cdot p)\,dx\,d_l p = \int_G f_{a,\,\gamma}\,(x)\,F\,(x)\,dx,$$

for all $F \in C_o^\infty\,(G)$. Here $d_l p$ is the left Haar measure on P.

Lemma 1. Fix $a \in C_o^\infty\,(G \times P)$. Then the mapping

$$y \mapsto f_{a,\,y} \quad (y \in G')$$

from G' to $C_o^\infty\,(G)$ is locally constant.

The mapping

$$(y, x, p) \mapsto (y, {}^x y \cdot p)$$

from $G' \times G \times P$ to $G' \times G$ is submersive. Hence there exists a unique linear mapping

$$\beta \mapsto \phi_\beta \quad (\beta \in C_o^\infty\,(G' \times G \times P))$$

from $C_o^\infty\,(G' \times G \times P)$ to $C_o^\infty\,(G' \times G)$ such that

$$\int \beta\,(y : x : p)\,\Phi\,(y : {}^x y \cdot p)\,dy\,dx\,d_l p = \int \phi_\beta\,(y : x)\,\Phi\,(y : x)\,dy\,dx$$

for all $\Phi \in C_o^\infty\,(G' \times G)$. Now let

$$\Phi\,(y : x) = \lambda\,(y)\,F\,(x),$$

where $\lambda \in C_o^\infty\,(G')$ and $F \in C_o^\infty\,(G)$. Then

$$\int_G \lambda\,(y)\,dy \int_{G \times P} \beta\,(y : x : p)\,F\,({}^x y \cdot p)\,dx\,d_l p = \int_G \lambda\,(y)\,dy \int_G \phi_\beta\,(y : x)\,F\,(x)\,dx.$$

---

* We write ${}^x y = x\gamma x^{-1}$ for $x, y \in G$.

This being true for all λ, we conclude that

$$\int_{G \times P} \beta(y : x : p) F(^{\bullet}y \cdot p) \, dx d_1 p = \int \phi_\beta(y : x) F(x) \, dx$$

for all $y \in G'$ and $F \in C_o^\infty(G)$.

Now fix $y_0 \in G'$ and put

$$\beta(y : x : p) = \mu(y) \, a(x : p)$$

where $a \in C_o^\infty(G \times P)$, $\mu \in C_o^\infty(G')$ and $\mu(y_0) = 1$. Let G_0 be a neighbourhood of y_0 in G' such that $\mu = 1$ on G_0. Then $\beta(y : x : p) = a(x : p)$ for $y \in G_0$ and therefore

$$\int_{G \times P} \beta(y : x : p) \ F(^{\bullet}y \cdot p) \, dx d_1 p = \int a(x : p) F(^{\bullet}y \cdot p) \, dx d_1 p$$

$$= \int f_{a, \, y}(x) F(x) \, dx.$$

Hence $\quad \int f_{a, \, y}(x) \ F(x) \, dx = \int \phi_\beta(y : x) F(x) \, dx$

for $y \in G_0$ and $F \in C_o^\infty(G)$. This shows that

$$f_{a, \, y}(x) = \phi_\beta(y : x) \quad (y \in G_0, x \in G).$$

Since $\phi_\beta \in C_o^\infty(G' \times G)$, our assertion is now obvious.

4. The operator $T_{\bullet}$

Fix a minimal p-pair (P, A) in G $(P = MN)$ and let K be an open compact subgroup of G of Bruhat-Tits ([2], p. 16) corresponding to A. Let π be an admissible representation of G on V. We recall that $\mathrm{End}^0 V$ is defined to be the subspace of all $T \in \mathrm{End} \, V$ such that the mappings

$$x \mapsto \pi(x) T, \ x \mapsto T\pi(x),$$

of G into $\mathrm{End} \, V$ are both smooth.

Let dk denote the normalised Haar measure on K.

Theorem 2. Let π be an admissible representation of G on V such that V is a finite G-module under π. For $x \in G'$, define

$$T_{\bullet} = \int_K \pi(kxk^{-1}) \, dk.$$

Then $T_{\bullet} \in \mathrm{End} \, {}^0V$ and $x \mapsto T_{\bullet}$ is a smooth mapping from G' to $\mathrm{End} \, {}^0V$.

Let K_0 be an open and normal subgroup of K and V_0 the subspace of all vectors in V, which are left fixed by K_0. Then $\dim V_0 < \infty$. By choosing K_0 sufficiently small, we may assume that V is spanned by elements of the form $\pi(x) v \, (x \in G, v \in V_0)$.

Let M^+ denote the set of all $m \in M$ such that $\langle a, H_M(m) \rangle \geqslant 0$ for every root a of (P, A) (see [3], § 7) for the definition of H_M). Then $G = KM^+ \cdot K$. Put $A^+ = A \cap M^+$. Since M/A is compact, we conclude that $M^+ \subset CA^+$ where C is a compact subset of M. Hence it is clear that we can choose an open compact subgroup P_0 of P such that $m^{-1}P_0 m \subset K_0$ for all $m \in M^+$.

Let a denote the characteristic function of $K \times P_0$ and put $f_y = f_{a, y}$ ($y \in G'$) in the notation of lemma 1. Then $y \mapsto f_y$ is a smooth mapping from G' to $C_c^\infty(G)$ and

$$\int_{K \times P_0} F(^k y \cdot p)\, dkdp = \int_G f_y(x)\, F(x)\, dx,$$

for all locally summable functions F on G. (Here dp is the normalised Haar measure on P_0.) From this we deduce immediately that

$$\int_{K \times P_0} \pi(^k y \cdot p)\, dkdp = \int_G f_y(x)\, \pi(x)\, dx = \pi(f_y).$$

Let V_y be the smallest K-invariant subspace of V containing $\pi(f_y) V$. Since $f_y \in C_c^\infty(G)$, it is clear that $\dim V_y < \infty$.

Since K_0 is normal in K, V_0 is stable under $\pi(K)$. Therefore since $G = KM^+K$, V is spanned by $\pi(KM^+) V_0$. Fix $k \in K$, $m \in M^+$ and $v \in V_0$. Then

$$T_y \pi(km)\, v = \pi(k)\, T_y \pi(m)\, v \qquad (y \in G').$$

But $\quad \pi(f_y)\, \pi(m)\, v = T_y \int_{P_0} \pi(pm)\, vdp = T_y \pi(m) \int_{P_0} \pi(m^{-1}pm)\, vdp = T_y \pi(m)\, v$

since $m^{-1} P_0 m \subset K_0$. Therefore

$$T_y \pi(km)\, v = \pi(k)\, \pi(f_y)\, \pi(m)\, v \in V_y.$$

This shows that $T_y V \subset V_y$. Since T_y commutes with $\pi(k)$ ($k \in k$), $\dim V_y < \infty$ and π is admissible, it is now clear that $T_y \in \mathrm{End}^0 V$.

Fix k, m, v as above. Since the mapping $y \mapsto f_y$ is smooth, it follows from the result obtained above that the mapping

$$y \mapsto T_y \pi(km)\, v = \pi(k)\, \pi(f_y)\, \pi(m)\, v\ (y \in G'),$$

from G' to V is also smooth. The second statement of the theorem is now obvious if we recall that V is spanned by $\pi(KM^+) V_0$.

Corollary. Let Θ denote the character of π. Then Θ coincides on G' with the locally constant function

$$x \mapsto \mathrm{tr}\, T_x \qquad (x \in G').$$

Put $f^0(x) = \int_K f(kxk^{-1})\, dk \qquad (x \in G),$

for $f \in C_c^\infty(G)$. Then

$$\Theta(f) = \Theta(f^0) = \mathrm{tr}\, \pi(f^0).$$

But $\qquad \pi(f^0) = \int\limits_{G} f(x)\, T_\bullet\, dx$

for $\qquad f \in C_\bullet^\infty(G')$. Hence

$$\Theta(f) = \int\limits_{G} f(x)\,\mathrm{tr}\, T_\bullet\, dx \qquad (f \in C_\bullet^\infty(G')).$$

5. Some applications

Let Γ be a Cartan subgroup of G. For $\gamma \in \Gamma' = \Gamma \cap G'$ and $f \in C_\bullet^\infty(G)$, define

$$F_f(\gamma) = |D(\gamma)|_\flat^{\frac{1}{2}} \int\limits_{G/A_\Gamma} f(x\gamma x^{-1})\, dx*$$

where $D = D_G$, A_Γ is the split component of Γ and $dx*$ is an invariant measure on G/A_Γ.

Theorem 3. *Let K_0 be an open compact subgroup of G. Given $\gamma_0 \in \Gamma'$, we can choose a neighbourhood ω of γ_0 in Γ' such that F_f is constant on ω for every $f \in C_\bullet(G/K_0)$.*

This result had been proved by Howe some years ago in the case char $\Omega = 0$.

Without loss of generality, we may assume that K_0 is a normal subgroup of K. Let us now use the notation of § 4. Then

$$\int\limits_{K \times P_0} F\,({}^k\gamma \cdot p)\, dk\, dp = \int\limits_{G} f_\gamma(x)\, F(x)\, dx,$$

for $\gamma \in \Gamma'$ and $F \in C_\bullet^\infty(G)$. Fix $f \in C_\bullet(G/K_0)$ and put

$$g(x) = \int\limits_{K} f(kxk^{-1})\, dk \qquad (x \in G).$$

Since K_0 is normal in K, $g \in C_\bullet(G//K_0)$. Fix $m \in M^+$ and let $F(x) = g(m^{-1} xm)$ $(x \in G)$. Then

$$\int\limits_{K \times P_0} g(m^{-1} \cdot {}^k\gamma \cdot pm)\, dk\, dp = \int f_\gamma(x)\, g(m^{-1} xm)\, dx.$$

But $pm = m \cdot m^{-1} pm \in mK_0$. Hence

$$\int\limits_{K} g(m^{-1} \cdot {}^k\gamma \cdot m)\, dk = \int f_\gamma(x)\, g(m^{-1} xm)\, dx \qquad (\gamma \in \Gamma').$$

Fix a neighbourhood Γ_0 of γ_0 in Γ' such that $f_\gamma = f_{\gamma_0}$ for $\gamma \in \Gamma_0$. This is possible by lemma 1. Then

$$\int\limits_{K} g(m^{-1} \cdot {}^k\gamma \cdot m)\, dk = \int\limits_{K} g(m^{-1} \cdot {}^k\gamma_0 \cdot m)\, dk \qquad (\gamma \in \Gamma_0).$$

By standard arguments, the proof of theorem 3 is reduced to the case when Γ is elliptic. Then $A_\Gamma = Z$ where Z is the maximal split torus lying in the centre of G. We know from the work of Bruhat and Tits that

$$K \backslash G/KZ \simeq M^+/^0 MZ.$$

Therefore $F_f(\gamma) = |D(\gamma)|_{\mathfrak{p}}^{\frac{1}{2}} \int\limits_{G/Z} f(x\gamma x^{-1}) \, dx^*$

$$= |D(\gamma)|_{\mathfrak{p}}^{\frac{1}{2}} \sum_{m \in M^+/^0MZ} \mu(m) \int\limits_K g(m^{-1} \cdot {}^k\gamma \cdot m) \, dk,$$

where $\mu(m)$ is the Haar measure of KmK. Let ω be a neighbourhood of γ_0 in Γ_0 such that $|D(\gamma)|_{\mathfrak{p}}$ is constant for $\gamma \in \omega$. Then

$$F_f(\gamma) = F_f(\gamma_0) \qquad (\gamma \in \omega).$$

Since ω is independent of $f \in C_c(G/K_0)$, the theorem is proved.

Theorem 3 makes it possible to prove Lemma 13 of ([3], § 16) without any restriction on char Ω.

I believe it is important, from the point of view of harmonic analysis, to obtain an analogue of theorem 5 of ([1], p. 32). We give below a result which may be regarded as a step in this direction.

Take $K_0 = K$ and define f_γ ($\gamma \in \Gamma'$) as in § 4. Then $f_\gamma \geq 0$. Put

$$\beta(\gamma) = \sup_{\mathfrak{g}} f_\gamma(x) \qquad (\gamma \in \Gamma')$$

and define Ξ as in ([3], § 14).

Theorem 4. Let ω be a compact subset of Γ. Then we can choose a positive number c such that

$$\int\limits_K \Xi(m^{-1} \cdot {}^k\gamma \cdot m) \, dk \leq c\beta(\gamma) \, \Xi(m)^2,$$

for all $m \in M^+$ and $\gamma \in \omega' = \omega \cap \Gamma''$.

It is obvious that

$$\text{Supp } f_\gamma \subset \bigcup_{k \in K} ({}^k\gamma \cdot P_0). \qquad (\gamma \in \Gamma').$$

Therefore $\text{Supp } f_\gamma \subset {}^K\omega \cdot P_0 = C$ (say)

for $\gamma \in \omega'$. Since C is compact, we can choose a finite number of elements y_i ($1 \leq i \leq r$) in G such that

$$\text{Supp } f_\gamma \subset \bigcup_{1 \leq i \leq r} y_i K$$

for all $\gamma \in \omega'$.

Now fix $\gamma \in \omega'$, $m \in M^+$ and put

$$F(x) = \Xi(m^{-1} xm) \qquad (x \in G)$$

in the relation

$$\int\limits_{K \times P_0} F({}^k\gamma \cdot p) \, dk \, dp = \int\limits_G f_\gamma(x) \, F(x) \, dx.$$

Observe that

$$F({}^k\gamma \cdot p) = \Xi(m^{-1} \cdot {}^k\gamma \cdot pm) = \Xi(m^{-1} \cdot {}^k\gamma \cdot m) \qquad (p \in P_0)$$

since $m^{-1} pm \in K$. Therefore

$$\int_K \Xi (m^{-1} \cdot {}^k\gamma \cdot m)\, dk \leq \beta (\gamma) \sum_i \int_K \Xi (m^{-1} y_i\, km)\, dk$$

$$= \beta (\gamma) \sum_i \Xi (m^{-1} y_i)\, \Xi (m)$$

from the identity

$$\int_K \Xi (xky)\, dk = \Xi (x)\, \Xi (y) \qquad (x, y \in G).$$

We can choose a number $c_1 \geq 1$ such that

$$\Xi (xy_i) \leq c_1\, \Xi (x) \qquad (1 \leq i \leq r)$$

for all $x \in G$. Put $c = rc_1$ and observe that $\Xi (x^{-1}) = \Xi (x)$. Then we get

$$\int_K \Xi (m^{-1} \cdot {}^k\gamma \cdot m) \leq c\beta (\gamma)\, \Xi (m)^2$$

and this proves our assertion.

One would like to verify that

$$\sup_{\gamma \in \hat{\omega}'} |\, D (\gamma)\, |_{\mathfrak{p}}^{\frac{1}{2}}\, \beta (\gamma) < \infty.$$

I believe this to be true but do not have a proof.

References

[1] Harish-Chandra 1966 Discrete series for semisimple Lie groups, II. *Acta Math.* **116** 1–111.

[2] Harish-Chandra 1970 *Harmonic analysis on reductive p-adic groups, Lecture Notes in Math.* (Berlin and New York: Springer-Verlag) Vol. 162.

[3] Harish-Chandra 1973 Harmonic analysis on reductive *p*-adic groups, in *Harmonic analysis on homogeneous spaces* (Providence: Am. Math. Soc.), pp. 167–192.

[4] Harish-Chandra 1977 The characters of reductive *p*-adic groups, in *Contributions to algebra* (New York: Academic Press), pp. 175–182.

[5] Harish-Chandra 1978 Admissible invariant distributions on reductive *p*-adic groups, Lie theories and their applications, *Queen's papers in pure and applied mathematics,* No. 48 (1978), Queen's University, Kingston, Ontario, pp. 281-347.

Institute for Advanced Study,
Princeton, New Jersey 08540, USA

Reprinted from
Papers dedicated to the memory of V. K. Patodi,
Indian Academy of Sciences, Bangalore,
and the Tata Institute of Fundamental Research, Bombay
1980, pp. 95–102

STUDIES IN APPLIED MATHEMATICS
ADVANCES IN MATHEMATICS SUPPLEMENTARY STUDIES, VOL. 8

Supertempered Distributions on Real Reductive Groups

HARISH-CHANDRA

The Institute for Advanced Study
Princeton, New Jersey

DEDICATED TO IRVING SEGAL

1. INTRODUCTION

It is well known that the concept of a cusp form plays a very central role in harmonic analysis on reductive groups. This is so whether one works over a global field, a local field, or a finite field. When G is a real reductive group, for example, the cusp forms lead directly to the discrete series. In general there exist quite unmistakable links between cusp forms, on the one hand, and the discrete spectrum and elliptic orbits, on the other.

We should like to extend this philosophy of cusp forms to distributions and to explore these connections a bit further. Although we deal here only with real groups, a significant portion of this theory holds good also in the p-adic case. Nevertheless, some crucial obstacles have yet to be overcome there. Personally, I regard the real case as a trial run that might yield some useful hints about the much less understood p-adic case, and this is how my interest in supertempered distributions first arose.

Roughly speaking, a cusp form f is a function whose constant term f_P is zero for every parabolic subgroup (psgp) $P \neq G$. One would like to imitate this definition for a distribution. Note, however, that, in order to define f_P, we have to impose some conditions on f, namely that f be $\mathfrak{z}$-finite and should satisfy the weak inequality. Similarly, if $\mathfrak{I}(G)$ is the space of all invariant, $\mathfrak{z}$-finite and tempered distributions on G and $P = MN$ a parabolic subgroup of G, it is possible to define the constant term $\Theta_P \in \mathfrak{I}(M)$ for Θ in $\mathfrak{I}(G)$. We say that Θ is supertempered if $\Theta_P = 0$ for all $P \neq G$. As one would expect, the characters of the discrete series are supertempered. But there are others besides.

139

On the other hand, no nonzero supertempered distributions exist unless G has an elliptic Cartan subgroup.

We shall see that supertempered distributions resemble the discrete series characters in some ways (see Section 4). They are associated in a natural fashion to elliptic representations and are parameterized by the characters of a compact Cartan subgroup (Section 6). Moreover they appear discretely in the formula for the Fourier transform of the invariant measure on an elliptic orbit (Section 8). These distributions were first introduced some years ago in a somewhat different context [3, Section 8] and have been studied more recently by Herb and Sally [10].

2. The Constant Term of a Distribution

Let G be a real reductive group [5, Section 3]. We shall use the terminology and notation of [5] without comment. Recall that $\mathscr{A}(G)$ is the space of all C^∞ functions f on G satisfying the following two conditions [5, p. 175].

(1) f is $\mathfrak{Z}$-finite.
(2)   For every pair $g_1, g_2 \in \mathfrak{G}$, the function

$$x \longmapsto f(g_1; x; g_2) \qquad (x \in G)$$

satisfies the weak inequality.

Let (P, A) be a p-pair in G $(P = MN)$. Then given $f \in \mathscr{A}(G)$, there exists a unique element $\phi \in \mathscr{A}(M)$ with the following property. For any $m \in M$,

$$\left| \delta_P(ma)^{1/2} f(ma) - \phi(ma) \right| \to 0$$

as $a \underset{P}{\to} \infty$ [5, Section 21]. (Here δ_P is the module of P.) ϕ is called the the constant term of f along P and we denote it by f_P.

For $y \in G$ and $f \in C_c^\infty(G)$, define

$$l(y)f : x \to f(y^{-1}x), \quad r(y)f : x \to f(xy)$$
$$f^y : x \to f(^y x) \qquad (x \in G),$$

where $^y x = yxy^{-1}$. If T is a distribution on G, we write

$$\langle T, f \rangle = T(f), \qquad (T, f) = \operatorname{conj}\langle T, \operatorname{conj} f \rangle.$$

Let $\mathscr{C}(G)$ denote the Schwartz space of G and $\mathfrak{I}(G)$ the space of all invariant, $\mathfrak{Z}$-finite, tempered distributions on G [5, Section 12]. Fix $\Theta \in \mathfrak{I}(G)$ and put

$$E_f(x) = (\Theta, r(x)f) \qquad (x \in G)$$

for $f \in \mathscr{C}(G)$. Then $E_f \in \mathscr{A}(G)$. Fix a p-pair (P, A) in G $(P = MN)$ and consider the mapping

$$T : f \mapsto E_{f,\bar{P}}(1) \qquad (f \in \mathscr{C}(G)).$$

Here $\bar{P} = \theta(P)$ and $E_{f,\bar{P}} = (E_f)_{\bar{P}}$. Then T is a tempered distribution on G, and it is easy to verify that

$$\langle T, r(n)f \rangle = \langle T, l(\bar{n})f \rangle = \langle T, f^m \rangle = \langle T, f \rangle$$

for $n \in N$, $\bar{n} \in \bar{N} = \theta(N)$, and $m \in M$. Observe that $\bar{N}MN$ is an open subset of G. For $f \in C_c^\infty(\bar{N}MN)$ define

$$h_f(m) = \delta_P(m)^{1/2} \int_{\bar{N} \times N} f(\bar{n}mn)d\bar{n}\,dn \qquad (m \in M).$$

(We normalize all Haar measures in accordance with [5, Section 7].) Extend δ_P to a function on G by the rule

$$\delta_P(kp) = \delta_P(p) \qquad (k \in K, p \in P)$$

and put

$$\gamma(P) = \int_{\bar{N}} \delta_P(\bar{n})^{-1}\,d\bar{n} < \infty.$$

Let dx denote the standard Haar measure on G. Then [6, p. 45]

$$dx = \gamma(P)^{-1}\delta_P(m)\,d\bar{n}\,dm\,dn$$

if $x = \bar{n}mn$. By considering the restriction of T on $\bar{N}MN$ it is possible to prove the following theorem.

THEOREM 1.   *There exists an element $\theta \in \mathfrak{I}(M)$ such that* ,

$$\gamma(P)E_{f,\bar{P}}(1) = (\theta, h_f)$$

for all $f \in C_c^\infty(\bar{N}MN)$.

We call θ the constant term of Θ along P and denote it by Θ_P. Observe that Θ and Θ_P are functions on G and M, respectively [5, Section 11]. We shall now describe some properties of Θ_P.

Let $l = \text{rank } G$. If t is an indeterminate, let $D_G(x)$ denote the coefficient of t^l in

$$\det(t - 1 + Ad(x)) \qquad (x \in G).$$

Then D_G is an analytic function on G. Similarly define D_M on M. Let $\mathscr{A}(A)$ denote the space of all functions $\phi \in C^\infty(A)$ such that

(1) ϕ is $\mathfrak{A}$-finite.
(2)   There exist constants $c, r \geq 0$ such that

$$|\phi(a)| \leq c(1 + \sigma(a))^r \qquad (a \in A).$$

THEOREM 2.   *Let M' be the set of all points $m \in M$ where $D_M(m) \neq 0$. For any $m \in M'$, there exists a unique element $\phi_m \in \mathscr{A}(A)$ such that*

$$\lim_{a \underset{P}{\to} \infty} |\delta_P(ma)^{1/2}\Theta(ma) - \phi_m(a)| = 0.$$

Θ_P *is given by the relation*

$$\Theta_P(m) = \phi_m(1) \qquad (m \in M').$$

COROLLARY.   *Let ω be a compact subset of M'. Then*

$$\lim_{a \underset{P}{\to} \infty} \sup_{m \in \omega} |\delta_P(ma)^{1/2}\Theta(ma) - \Theta_P(ma)| = 0.$$

Let $(P', A') > (P, A)$ $(P' = M'N', P = MN)$ be two p-pairs in G. Put $*P = M' \cap P$. Then

$$(\Theta_{P'})_{*P} = \Theta_P.$$

We call this the transitivity of the constant term (cf. [5, p. 153]).

THEOREM 3.   *Let P_1, P_2 be two psgps of G with the same split component A. Then $\Theta_{P_1} = \Theta_{P_2}$.*

This implies that Θ_P is invariant under $\mathfrak{w}(G/A)$.

3.   THE MAPPING I_M^G

By a special pair in G we mean a pair of the form (A, M), where A is a special vector subgroup of G [5, p. 114] and M the centralizer of A in G. For $\Theta \in \mathfrak{J}(G)$, put

$$j_M \Theta = \Theta_P$$

for $P \in \mathscr{P}(A)$. Then j_M is a linear mapping from $\mathfrak{I}(G)$ to $\mathfrak{I}(M)$. We shall now define a linear mapping in the opposite direction.

Fix $P \in \mathscr{P}(A)$ and for $f \in \mathscr{C}(G)$, put [5, Section 16]

$$g_f = (f^{\natural})^{(P)},$$

where

$$f^{\natural} = \int_K f^k \, dk.$$

Then $f \mapsto g_f$ is a continuous mapping from $\mathscr{C}(G)$ to $\mathscr{C}(M)$. If $\tau \in \mathfrak{I}(M)$, define a distribution T on G by

$$\langle T, f \rangle = \langle \tau, g_f \rangle \qquad (f \in \mathscr{C}(G)).$$

It is not difficult to show that $T \in \mathfrak{I}(G)$, and it is independent of the choice of P. We denote it by $I_M^G \tau$. Then I_M^G is a linear mapping from $\mathfrak{I}(M)$ to $\mathfrak{I}(G)$.

4. SUPERTEMPERED DISTRIBUTIONS

By a supertempered distribution on G we mean an element Θ of $\mathfrak{I}(G)$ such that $\Theta_P = 0$ for all psgps $P \neq G$. Let G' be the set of all regular elements of G. We know [5, Section 12] that an invariant $\mathfrak{Z}$-finite distribution Θ is tempered if and only if

$$\sup_{x \in G'} |D_G(x)|^{1/2} |\Theta(x)| (1 + \sigma(x))^{-r} < \infty$$

for some $r \geq 0$. The following theorem justifies our terminology.

THEOREM 4.   *Assume* $\operatorname{prk} G = 0$. *Then an element* $\Theta \in \mathfrak{I}(G)$ *is supertempered if and only if*

$$\sup_{\gamma \in \Gamma'} |D_G(\gamma)|^{1/2} |\Theta(\gamma)| (1 + \sigma(\gamma))^r < \infty$$

for every Cartan subgroup Γ *of* G *and all* $r \geq 0$. (*Here* $\Gamma' = \Gamma \cap G'$.)

Let Z be the split component of G. A Cartan subgroup Γ is called elliptic if Γ/Z is compact. An element of G is said to be elliptic if it is contained in some elliptic Cartan subgroup. Let G_e denote the set of all elliptic elements in G'. Then G_e is open in G. (Note that $G_e = \varnothing$ unless $\operatorname{rank} G = \operatorname{rank} K + \operatorname{prk} G$.) If T is a distribution on G, we denote by T_e the restriction of T on G_e. ($T_e = 0$ by convention when $G_e = \varnothing$.)

Moreover if T_1, T_2 are two distributions on G we write $T_1 \underset{e}{\sim} T_2$ if $T_1 = T_2$ on G_e.

THEOREM 5.   *Let Θ be a supertempered element in $\Im(G)$. Then if $\Theta_e = 0$ we can conclude that $\Theta = 0$.*

It is clear from Theorem 4 that the characters of the discrete series are supertempered. Conversely we conclude from the above two theorems that supertempered distributions behave formally like the characters of the discrete series.

5.   FROBENIUS RECIPROCITY

Let π be an irreducible, unitary and tempered representation of G on a Hilbert space V. Let V_∞ be the subspace of all C^∞ vectors in V and (P, A) a p-pair in G ($P = MN$). Fix u, $v \in V_\infty$ and consider the function

$$f : x \mapsto (u, \pi(x)v) \qquad (x \in G).$$

Then $f \in \mathscr{A}(G)$. Put

$$(u, v)_P = f_P(1).$$

This defines a sesquilinear from on $V_\infty \times V_\infty$. Let $V_\infty(\bar{P})$ be the subspace of all $v \in V_\infty$ such that $(u, v)_P = 0$ for all $u \in V_\infty$. Then $\pi(\bar{n})v - v$ and $\pi(X)v$ lie in $V_\infty(\bar{P})$ for all $\bar{n} \in \bar{N}$, $X \in \bar{\mathfrak{n}}$ and $v \in V_\infty$. Put

$$V_P = V_\infty / V_\infty(\bar{P}), \qquad V_{\bar{P}} = V_\infty / V_\infty(P).$$

Then our form actually defines a nondegenerate sesquilinear form on $V_{\bar{P}} \times V_{\bar{P}}$.

Let p denote the projection of V_∞ on V_P. Define a representation π_P of M on V_P (in the algebraic sense) as follows:

$$\pi_P(m) \cdot pv = \delta_P(m)^{1/2} p \cdot \pi(m)v \qquad (m \in M, v \in V_\infty).$$

Similarly define $\pi_P(Y)$ ($Y \in \mathfrak{m}$) by

$$\pi_P(Y) \cdot pv = p \cdot (\pi(Y) + \langle \rho_P, Y \rangle)v,$$

where

$$\langle \rho_P, Y \rangle = \tfrac{1}{2} \operatorname{tr}(ad Y)_{\mathfrak{n}}.$$

Let V^0 be the subspace of all K-finite vectors in V. Then $V^0 \subset V_\infty$. Put $V_P^0 = pV^0$. It is easy to see that V_P^0 is stable under $\pi_P(\mathfrak{m})$ and

$\pi_P(K_M)$ $(K_M = K \cap M)$. Actually it turns out that V_P^0 is a finitely generated admissible $(\mathfrak{m}, K_M)$-module [13, Section 5.8]. Hence there is associated to it a global character. Let Θ_π denote the character of π.

THEOREM 6.   *The global character associated to the $(\mathfrak{m}, K_M)$-module V_P^0 is $(\Theta_\pi)_P$.*

This is similar to a result of Hecht [8, Theorem (1.3)].

Let σ be a tempered, irreducible, unitary representation of M on U and U^0 the subspace of all K_M-finite vectors in U. Put [7, p. 129]

$$\pi_\sigma = \mathrm{Ind}_P^G(\delta_P^{1/2}\sigma).$$

We denote the representation space of π_σ by $\mathfrak{H}$. Let $\mathfrak{H}^0$ denote the subspace of all K-finite vectors in $\mathfrak{H}$. Then V^0 and $\mathfrak{H}^0$ are admissible $(\mathfrak{g}, K)$-modules while V_P^0 and U^0 are admissible $(\mathfrak{m}, K_M)$-modules.

THEOREM 7   (Frobenius reciprocity).   *There exists a linear bijection*

$$\mu \mapsto T_\mu$$

from $\mathrm{Hom}_{(\mathfrak{m}, K_M)}(V_P^0, U^0)$ *to* $\mathrm{Hom}_{(\mathfrak{g}, K)}(V^0, \mathfrak{H}^0)$ *such that*

$$(T_\mu v)(1) = \mu p v \qquad (v \in V^0).$$

Here p is the projection of V^0 on V_P^0.

Let (A, M) be a special pair (Section 3) and $\mathscr{E}_2(M)$ the set (possibly empty) of all equivalence classes of irreducible, unitary representations of M that are square-integrable mod A. If $\omega \in \mathscr{E}_2(M)$ we shall denote by θ_ω the character of ω. By a triple we mean a triple of the form (A, M, ω) where (A, M) is a special pair and $\omega \in \mathscr{E}_2(M)$.

Let (A, M, ω) be a triple. Fix $P \in \mathscr{P}(A)$ and put

$$\pi_\omega = \mathrm{Ind}_P^G \delta_P^{1/2}\omega.$$

Then the class of π_ω is independent of the choice of P and $I_M^G \theta_\omega$ is the character of π_ω.

THEOREM 8.   *Let π be an irreducible, unitary, tempered representation of G. Then $[\pi_\omega:\pi] \geq 1$ if and only if $[(\Theta_\pi)_P:\theta_\omega] \geq 1$. Now suppose*

$$n = [\pi_\omega:\pi] \geq 1.$$

Then

$$(\Theta_\pi)_P = n[\mathfrak{w}_\infty] \sum_{s \in \mathfrak{w}/\mathfrak{w}(\omega)} \theta_{s\omega}.$$

Finally, n is independent of π (provided $[\pi_\omega:\pi] \geq 1$) and $n = 1$ if an elliptic Cartan subgroup of M is abelian.

Here $[\pi_\omega:\pi]$ denotes the multiplicity of π in π_ω. Similarly $[(\Theta_\pi)_P:\theta_\omega]$ denotes the coefficient of θ_ω when $(\Theta_\pi)_P$ is written as a linear combination of irreducible characters of M. $\mathfrak{w} = \mathfrak{w}(G/A)$ and $\mathfrak{w}(\omega)$ is the subgroup of all $s \in \mathfrak{w}$ such that $s\omega = \omega$. Finally $\mathfrak{w}_\infty$ is the subgroup of $\mathfrak{w}(\omega)$ defined as follows. Let Σ_∞ be the set of all reduced roots α of (P, A) such that $\mu_\alpha(\omega) = 0$ [7, Section 40]. Then the Weyl reflection s_α lies in $\mathfrak{w}(\omega)$ and $\mathfrak{w}_\infty$ is the subgroup of $\mathfrak{w}(\omega)$ generated by s_α for all $\alpha \in \Sigma_\infty$.

The proof of this theorem depends on a result of Zuckerman (14, Theorem 5.7]. The statement about $n = 1$ may be regarded as a generalization of a theorem of Knapp [12].

6.   THE BIJECTION FROM $K\backslash\Omega_e$ TO $W(G/B)\backslash B^{*\prime}$

Now suppose rank $G = $ rank K and fix a Cartan subgroup B of G such that $B \subset K$. Choose a system of positive roots for $(\mathfrak{g}, \mathfrak{b})$ and let B^* denote the set of all irreducible characters of B. We use the notation of [5, Section 27].

THEOREM 9.   *For every $b^* \in B^*$ there exists a unique supertempered element $\Theta_{b^*} \in \mathfrak{I}(G)$ such that*

$$'\Delta(b)\Theta_{b^*}(b) = \sum_{s \in W(G/B)} \varepsilon(s)\langle sb^*, b\rangle \qquad (b \in B \cap G').$$

Let $W_{b^*}(G/B)$ be the subgroup of all $s \in W(G/B)$ such that $sb^* = b^*$.

COROLLARY. $\Theta_{sb^*} = \varepsilon(s)\Theta_{b^*}$ *for $s \in W(G/B)$. Moreover $\Theta_{b^*} = 0$ if and only if $\varepsilon(s) = -1$ for some $s \in W_{b^*}(G/B)$.*

Let $\mathfrak{I}_{st}(G)$ denote the space of all supertempered elements in $\mathfrak{I}(G)$.

THEOREM 10. $\Theta_{b^*}(b^* \in B^*)$ *span* $\mathfrak{I}_{st}(G)$.

Let Ω be the set of all triples (A, M, ω) (see Section 5). Then K operates on Ω as follows. If $y \in K$, then

$$y \cdot (A, M, \omega) = ({}^{y}A, {}^{y}M, y \cdot \omega),$$

where $({}^{y}A, {}^{y}M) = (yAy^{-1}, yMy^{-1})$ and $y \cdot \omega$ is the image of ω under the bijection from $\mathscr{E}_2(M)$ to $\mathscr{E}_2({}^{y}M)$ corresponding to the isomorphism $m \mapsto {}^{y}m$ $(m \in M)$ of M with ${}^{y}M$. To abbreviate we write ω instead of (A, M, ω). Then the class of π_ω (see Section 5) depends only on the K-orbit of ω. We say that ω is elliptic if $(\Theta_\pi)_e \neq 0$ for some irreducible component π of π_ω.

For $\omega \in \Omega$, let $\mathfrak{I}(G:\omega)$ denote the subspace of $\mathfrak{I}(G)$ spanned by Θ_π for all irreducible components π of π_ω. Put

$$\mathfrak{I}_{st}(G:\omega) = \mathfrak{I}_{st}(G) \cap \mathfrak{I}(G:\omega)$$

and let Ω_e denote the set of all elliptic elements in Ω.

THEOREM 11. *Fix $\omega \in \Omega$. Then $\dim \mathfrak{I}_{st}(G:\omega)$ is 1 or 0 according as $\omega \in \Omega_e$ or not. Moreover, if ω is elliptic we can choose $b^* \in B^*$ such that*

$$\mathfrak{I}_{st}(G:\omega) = \mathbf{C}\Theta_{b^*}.$$

Now suppose $\omega \in \Omega_e$ and fix b^* as in the above theorem.

THEOREM 12. *If π is any irreducible component of π_ω then*

$$\Theta_\pi \underset{e}{\sim} \varepsilon(\pi)\Theta_{b^*}$$

where $\varepsilon(\pi) = \pm 1$.

Let $B^{*\prime}$ denote the set of all $b^* \in B^*$ such that $\Theta_{b^*} \neq 0$. Then the above correspondence actually defines a bijection from $K\backslash\Omega_e$ to $W(G/B)\backslash B^{*\prime}$. This may be regarded as an extension of [7, Theorem 1, p. 164].

Fix $(A, M, \omega) \in \Omega$, $P \in \mathscr{P}(A)$ and define π_ω as in Section 5. Let $\mathfrak{C}$ denote the commuting algebra of π_ω. For every $s \in \mathfrak{w}(\omega)$, it is possible to define in a standard way an element $\zeta_s \in \mathfrak{C}$ such that $s \mapsto \zeta_s$ is a projective representation of $\mathfrak{w}(\omega)$ [7, Section 38]. By construction ζ_s is uniquely determined up to a unimodular scalar factor.

Since rank $G =$ rank K, there exists a unique element $s_\theta \in \mathfrak{w}$ such that

$$s_\theta H = \theta H = -H \qquad (H \in \mathfrak{a}).$$

Let $\mathfrak{w}_0(\omega)$ be the subgroup of all $s \in \mathfrak{w}(\omega)$ such that ζ_s lies in the center of $\mathfrak{C}$.

THEOREM 13. *In order that* $(A, M, \omega) \in \Omega_e$, *it is necessary and sufficient that*

(1) $\mu(\omega) > 0$, *and*
(2) $s_\theta \in \mathfrak{w}_0(\omega)$.

Here μ is the Plancherel density [7, Section 25]. Note that (2) implies that A lies in the kernel of ω.

Now suppose $\omega \in \Omega_e$ and let $b^* \in B^{*\prime}$ correspond to ω (under Theorem 11). Since $s_\theta^2 = 1$, we can normalize $\zeta_\theta = \zeta_{s_\theta}$ so that $\zeta_\theta^2 = 1$. This determines it uniquely up to sign. Then

$$\mathrm{tr}(\zeta_\theta \pi_\omega(\check{f})) = \pm n[\mathfrak{w}_0(\omega)](\Theta_{b^*}, f) \qquad (f \in C_c^\infty(G)),$$

where $\check{f}(x) = f(x^{-1})$ $(x \in G)$ and $n = [\pi_\omega : \pi]$ for any irreducible component π of π_ω (see Theorem 8). Moreover $[\mathfrak{w}_0(\omega)]$ is the number of inequivalent irreducible components of π_ω and

$$n^2[\mathfrak{w}_0(\omega)] = \dim \mathfrak{C} = [\mathfrak{w}(\omega)].$$

The work of Herb and Sally [11] is closely related to the results of this section.

7.   ORTHOGONALITY AND COMPLETENESS

We keep to the notation of Section 6. A representation π of G will be called elliptic if it is irreducible, unitary and tempered and $(\Theta_\pi)_e \neq 0$. Put

$$\Phi_\pi(b) = {}'\Delta(b)\Theta_\pi(b) \qquad (b \in B \cap G').$$

Then Φ_π extends to an analytic function on B. For any class function ϕ on B and $s \in W(G/B)$, define

$$\phi^s(b) = \phi({}^y b)\xi_{s^{-1}\rho - \rho}(b) \qquad (b \in B)$$

in the notation of [5, Section 27] where y is a representative for s in G. Then

$$\Phi_\pi^s = \varepsilon(s)\Phi_\pi \qquad (s \in W(G/B)).$$

Let $L_2^{\natural}(B)$ denote the space of all class functions ϕ in $L_2(B)$ such that $\phi^s = \varepsilon(s)\phi$ $(s \in W(G/B))$.

THEOREM 14. *Let π_1, π_2 be two elliptic representations of G. Then either*

$$\Phi_{\pi_1} = \pm \Phi_{\pi_2}$$

or Φ_{π_1} is orthogonal to Φ_{π_2} in $L_2(B)$. Moreover the space spanned by Φ_π for all elliptic π is dense in $L_2^{\natural}(B)$.

This is obvious from Theorem 12.

8. FOURIER TRANSFORM OF ORBITAL INTEGRALS

Let $\mathrm{Car}_\theta(G)$ denote the set of all θ-stable Cartan subgroups of G. Fix $A \in \mathrm{Car}_\theta(G)$. Then $A = A_I \cdot A_R$ [5, Section 8]. Let A^*, A_I^* and A_R^* denote the sets of irreducible unitary characters of A, A_I, and A_R respectively. Then $A^* = A_I^* \times A_R^*$. Put $\mathfrak{F} = \mathfrak{a}_R^*$ and, for $v \in \mathfrak{F}$, let $\eta(v)$ denote the element in A_R^* given by

$$\langle \eta(v), \exp H \rangle = e^{(-1)^{1/2}\langle v, H \rangle} \qquad (H \in \mathfrak{a}_R).$$

We identify $\mathfrak{F}$ with A_R^* by means of η and give A_I^* the discrete topology. Then $A^* = A_I^* \times \mathfrak{F}$ becomes a real manifold.

Put $S_c(\mathfrak{a}_R^*) = S(\mathfrak{F}_c)$ where $S(\mathfrak{F}_c)$ is the symmetric algebra over $\mathfrak{F}_c$. For every $p \in \mathfrak{F}_c$, we get a translation-invariant differential operator $\partial(p)$ on $\mathfrak{F}$ [2, Section 3] and hence also on A_R^*. In this way $\partial(\mathfrak{a}_R^*)$ may be regarded as the Lie algebra of A_R^*. Then $\mathfrak{S}(\mathfrak{a}_R^*) = \partial(S_c(\mathfrak{a}_R^*))$ is the algebra of all translation-invariant differential operators on A_R^*.

Let M be the centralizer of A_R in G and define ^0M as in [5, Section 2]. We say that A is weakly polarized (or w-polarized) if we have chosen a system of positive roots for $(^0M, A_I)$. Assume that A is w-polarized and $a^* \in A^*$. Then $a^* = (a_1^*, v)$ where $a_1^* \in A_I^*$ and $v \in \mathfrak{a}_R^*$. Put

$$\lambda(a^*) = \lambda(a_1^*) + (-1)^{1/2}v,$$

where $\lambda(a_1^*)$ is defined as in [5, p. 176]. Note that $\mathfrak{a}_c$ and $\mathfrak{a}_c^*$ have natural Hilbert-space structures and

$$|\lambda(a^*)|^2 = |\lambda(a_1^*)|^2 + |v|^2.$$

We regard elements of $\mathfrak{S}(\mathfrak{a}_R^*)$ as differential operators on $A^* = A_I^* \times A_R^*$.

Let $\mathscr{C}(A^*)$ denote the space of all $f \in C^\infty(A^*)$ such that

$$\mathbf{s}_{r,u}(f) = \sup_{a^*}(1 + |\lambda(a^*)|)^r |f(a^*; u)| < \infty$$

for all $u \in \mathfrak{S}(\mathfrak{a}_R^*)$ and $r \geq 0$. The seminorms $\mathbf{s}_{r,u}(r \geq 0, u \in \mathfrak{S}(\mathfrak{a}_R^*))$ define the structure of a locally convex space on $\mathscr{C}(A^*)$.

Define $W(G/A)$ as usual [5, p. 201] and let π_I denote the product of all positive imaginary roots of $(\mathfrak{g}, \mathfrak{a})$. (Recall that A is w-polarized.) Then

$$s\pi_I = \varepsilon_I(s)\pi_I \qquad (s \in W(G/A)),$$

where $\varepsilon_I(s) = \pm 1$. Now $W(G/A)$ operates on A^*. In fact if $a^* = (a_1^*, v)$ $(a_1^* \in A_I^*, v \in \mathfrak{a}_R^*)$, then

$$sa^* = (sa_1^*, sv) \qquad (s \in W(G/A)),$$

where sa_1^* is defined as in [5, p. 177]. Let $\mathscr{C}^\natural(A^*)$ be the subspace of all $f \in \mathscr{C}(A^*)$ such that

$$f(sa^*) = \varepsilon_I(s)f(a^*) \qquad (s \in W(G/A), a^* \in A^*).$$

Then $\mathscr{C}^\natural(A^*)$ is closed in $\mathscr{C}(A^*)$.

Fix $a^* = (a_1^*, v)$ in A^* and put

$$\theta_{a^*}(ma) = \theta_{a_1^*}(m)\exp[(-1)^{1/2}\langle v, \log a \rangle] \qquad (m \in {}^0M, a \in A_R),$$

where $\theta_{a_1^*}$ is defined as in Theorem 9 applied to $({}^0M, A_I)$ in place of (G, B). Then $\theta_{a^*} \in \mathfrak{I}(M)$. Put $\Theta_{a^*} = I_M^G \theta_{a^*}$. Then

$$\Theta_{sa^*} = \varepsilon_I(s)\Theta_{a^*} \qquad (s \in W(G/A)).$$

Now define

$$\hat{f}_A(a^*) = (\Theta_{a^*}, f) \qquad (a^* \in A^*)$$

for $f \in \mathscr{C}(G)$. Then

$$f \mapsto \hat{f}_A$$

is a continuous mapping from $\mathscr{C}(G)$ to $\mathscr{C}^\natural(A^*)$.

Let $d_R a$ denote the Haar measure on A_R (normalized as in [5, Section 7]) and $d_R a^*$ the dual measure on its character group A_R^*. Since $A^* = A_I^* \times A_R^*$, we can define a measure da^* on A^* by

$$\int_{A^*} \phi(a^*)\,da^* = \sum_{a_1^* \in A_I^*} \int_{A_R^*} \phi(a_1^* : a_2^*)\,d_R a_2^* \qquad (\phi \in C_c(A^*)).$$

We call da^* the standard measure on A^*. For $f \in \mathscr{C}(G)$, define $'F_f^A$ as in [5, Section 17].

If A and B are two elements in $\mathrm{Car}_\theta(G)$, we write $A \succ B$ (or $B \prec A$) if $\mathfrak{w}(\mathfrak{a}_R | \mathfrak{b}_R) \neq \varnothing$ [5, p. 112]. Let Γ be a complete set of θ-stable Cartan subgroups of G, no two of which are conjugate under K. Fix a w-polarization for every $A \in \Gamma$ and put $A' = A \cap G'$.

THEOREM 15.   *Fix $B \in \Gamma$. Then there exist unique continuous functions $\Phi_{B|A}$ on $B' \times A^*$ $(A \in \Gamma)$ satisfying the following three conditions:*

(1)   *We can choose numbers $c, r \geq 0$ such that*

$$\left|\Phi_{B|A}(b:a^*)\right| \leq c(1 + |\lambda(a^*)|)^r,$$

(2) $\Phi_{B|A}(b:sa^*) = \varepsilon_I(s)\Phi_{B|A}(b:a^*)$ $(s \in W(G/A))$,

(3) $'F_f^B(b) = \displaystyle\sum_{A \in \Gamma} \int_{A^*} \Phi_{B|A}(b:a^*)\hat{f}_A(a^*)\,da^*$ $(f \in \mathscr{C}(G))$.

Here $b \in B'$, $a^* \in A^*$ and da^* is the standard measure on A^*.

Let $_0B$ be the set of all $b \in B$ such that $\xi_\beta(b) \neq 1$ for every singular imaginary root β of $(\mathfrak{g}, \mathfrak{b})$ [5, Section 8]. Then it turns out that $\Phi_{B|A}$ is a C^∞ function on $_0B \times A^*$ and the following additional conditions hold.

(4) $\Phi_{B|A} = 0$ unless $B \prec A$.
(5)   Given $u \in \mathfrak{S}(\mathfrak{b}_c)$ [5, p. 131], $v \in \mathfrak{S}(\mathfrak{a}_R^*)$, we can choose numbers $c, r, s \geq 0$ such that

$$\left|\Phi_{B|A}(b; u : a^*; v)\right| \leq c(1 + \sigma(b))^s(1 + |\lambda(a^*)|)^r$$

for all $b \in {_0B}$ and $a^* \in A^*$.

The proof of these results proceeds by induction on $\dim G$. The important case is when $B \subset K$. I understand from J. G. Arthur that the uniqueness part of Theorem 15 is a consequence of his general results on Fourier transforms [1, pp. 387–388]. For "stabilized invariant integrals" Herb [9, 10] has obtained a formula similar to (3) under the assumption that G is a group of classical type.

There exists a homomorphism $\gamma': \mathfrak{Z} \to \mathfrak{S}(\mathfrak{b}_c)$ [5, p. 147] such that

$$'F_{zf}^B(b) = {}'F_f^B(b; \gamma'(z))   (b \in B', z \in \mathfrak{Z})$$

for all $f \in \mathscr{C}(G)$. Fix $a^* \in A^*$ and observe that Θ_{a^*} is an eigendistribution of $\mathfrak{Z}$. Let χ_{a^*} denote the corresponding homomorphism of $\mathfrak{Z}$ into $\mathbf{C}$. Then we can conclude that

$$\Phi_{B|A}(b; \gamma'(z) : a^*) = \Phi_{B|A}(b:a^*)\chi_{a^*}(z)   (b \in {_0B}, z \in \mathfrak{Z}).$$

These differential equations play a central role in the determination of the functions $\Phi_{B|A}$ and the idea of using them for this purpose goes back to Langlands [4, pp. 545–546].

It is easy to see that

$$(6) \quad \Phi_{B|B}(b:b^*)=[W(G/B)]^{-1} \sum_{s \in W(G/B)} \varepsilon_I(s)\langle sb^*,b\rangle \quad (b^* \in B^*, b \in B).$$

Fix $A \in \Gamma$, $a^* \in A_I^*$, $b \in {}_0B$ and $u \in \mathfrak{S}(\mathfrak{b}_c)$. Then

(7)   The function

$$v \mapsto \Phi_{B|A}(b;u:a^*:v) \qquad (v \in \mathfrak{a}_R^*)$$

extends to a meromorphic function on $(\mathfrak{a}_R^*)_c$.

Let α be a real root of $(\mathfrak{g},\mathfrak{a})$ and c a real number. Then we denote by $\pi(\alpha:c)$ the hyperplane in $(\mathfrak{a}_R^*)_c$ given by the equation

$$\langle \alpha, v \rangle = (-1)^{1/2}c.$$

It is possible to show that all poles of the above meromorphic functions are simple and they lie along hyperplanes of the form $\pi(\alpha:c)$ $(c \neq 0)$. Moreover one can give a procedure for computing these functions explicitly.

Now assume that $B \subset K$. Then the contribution of B to the right-hand side of (3) in Theorem 15 is

$$\sum_{b^* \in B^*} \Phi_{B|B}(b:b^*)(\Theta_{b^*},f).$$

This shows that each supertempered distribution Θ_{b^*} $(b^* \in B^{*\prime})$ makes a discrete contribution to the Fourier transform of the orbital integral ${}'F_f^B$.

REFERENCES

1. J. G. ARTHUR, Harmonic analysis of invariant distributions, *in* "Lie Theories and Their Applications," Queen's papers in pure and applied mathematics, No. 48 (1978), pp. 384–393, Queen's University, Kingston.
2. HARISH-CHANDRA, Invariant differential operators and distributions on a semisimple Lie algebra, *Amer. J. Math.* **86** (1964), 534–564.
3. HARISH-CHANDRA, Two theorems on semisimple Lie groups, *Ann. of Math.* **83** (1966), 74–128.
4. HARISH-CHANDRA, Harmonic analysis on semisimple Lie groups, *Bull. Amer. Math. Soc.* **78** (1970), 529–551.
5. HARISH-CHANDRA, Harmonic analysis on real reductive groups I, *J. Funct. Anal.* **19** (1975), 104–204.

6. HARISH-CHANDRA, Harmonic analysis on real reductive groups II, *Invent. Math* **36** (1976), 1–55.
7. HARISH-CHANDRA, Harmonic analysis on real reductive groups III, *Ann. of Math.* **104** (1976), 117–201.
8. H. HECHT, On characters and asymptotics of representations of a real reductive Lie group, *Math. Ann.* **242** (1979), 103–126.
9. R. A. HERB, Fourier inversion of a stabilized invariant integral, preprint (1978).
10. R. A. HERB, Fourier inversion and the Plancherel theorem for semisimple real Lie groups, preprint (1979).
11. R. A. HERB AND P. J. SALLY, JR., Singular invariant eigendistributions as characters in the Fourier transform of invariant distributions, *J. Funct. Anal.* **33** (1979), 195–210.
12. A. W. KNAPP, Commutativity of intertwining operators II, *Bull. Amer. Math. Soc.* **82** (1976), 271–273.
13. N. R. WALLACH, Representations of semi-simple Lie groups and Lie algebras, *in* "Lie Theories and Their Applications," Queen's papers in pure and applied mathematics, No. 48 (1978), pp. 154–246, Queen's University, Kingston.
14. G. ZUCKERMAN, Tensor products of finite and infinite dimensional representations of semisimple Lie groups, *Ann. of Math.* **106** (1977) 295–308.